Environmental Science and Engineering

Series Editors

Ulrich Förstner, Buchholz, Germany

Wim H. Rulkens, Department of Environmental Technology, Wageningen, The Netherlands

The ultimate goal of this series is to contribute to the protection of our environment, which calls for both profound research and the ongoing development of solutions and measurements by experts in the field. Accordingly, the series promotes not only a deeper understanding of environmental processes and the evaluation of management strategies, but also design and technology aimed at improving environmental quality. Books focusing on the former are published in the subseries Environmental Science, those focusing on the latter in the subseries Environmental Engineering.

Sijing Wang · Runqiu Huang · Rafig Azzam ·
Vassilis P. Marinos
Editors

Engineering Geology for a Habitable Earth: IAEG XIV Congress 2023 Proceedings, Chengdu, China

Volume 1: Engineering Geomechanics of Rock and Soil Masses

Springer

Editors
Sijing Wang
Department of Hydraulic Engineering
Tsinghua University
Beijing, China

Runqiu Huang
Chengdu University of Technology
Chengdu, Sichuan, China

Rafig Azzam
Department of Engineering Geology
and Hydrogeology
RWTH Aachen University
Aachen, Germany

Vassilis P. Marinos
School of Civil Engineering
National Technical University of Athens
Zografou, Greece

ISSN 1863-5520 ISSN 1863-5539 (electronic)
Environmental Science and Engineering
ISBN 978-981-99-9056-6 ISBN 978-981-99-9057-3 (eBook)
https://doi.org/10.1007/978-981-99-9057-3

This Springer imprint is published by the registered company Springer Nature Singapore Pte Ltd.
The registered company address is: 152 Beach Road, #21-01/04 Gateway East, Singapore 189721, Singapore

Academic Committee

Academic Committee Chairs

Sijing Wang
Runqiu Huang

Academic Committee Members

Anthony Bowden
Bin Shi
Bo An-Jang
Carlos Delgado
Charles W. W. Ng
Chungsik Yoo
Defang Kong
Dingcheng Huang
Faquan Wu
Fawu Wang
Helen Reeves
Hengxing Lan
Hsein Juang
Huiming Tang
Janusz Wasowski
Jean Hutchinson
Jean-Alain Fleurisson
Jian Yang
Jianbing Peng
Jianmin Zhang
Jianxin Hua

Jinxiu Yan
John Ludden
Julien Cohen-Waeber
Kyoji Sassa
Lansheng Wang
Manchao He
Martin Culshaw
Moshood Niyi Tijani
Nicola Casagli
Niek Rengers
Peng Cui
Qing Wang
Rafig Azzam
Ranjan Kumar Dahal
Resat Ulusay
Ricardo Oliveira
Roger Frank
Scott F. Burns
Shitian Wang
Shutao Yang
Vassilis P. Marinos
Victor Manuel Hernandez Madrigal
Wei Wu
Wei Zhang
Xiao Li
Xiating Feng
Yong-Seok Seo
Yueping Yin
Yusheng Gao
Zelian Chen
Zuyu Chen

Organizing Committee Chairs

Qiang Xu
Qiangbing Huang
Shengwen Qi
Xiangjun Pei
Xiaoqing Chen
Yuyong Jiao
Xuanmei Fan

Organizing Committee Members

Chaojun Ouyang
Chaosheng Tang
Jianjun Zhao
Liangqing Wang
Ning Liang
Wen Zhang
Yuhuan Song

Preface

The XIV Congress of the International Association for Engineering Geology and the Environment (XIV IAEG Congress 2023) was successfully held in Chengdu, China, from September 21 to 27, 2023. Focusing on the main theme "Engineering Geology for a Habitable Earth", researchers and practitioners worldwide from academia, industry, and government have joined us in this prestigious event. Based on the topics discussed at the congress, the proceedings are organized into six volumes as follows:

- Volume 1: Engineering Geomechanics of Rock and Soil Masses
- Volume 2: Geohazard Mechanisms, Risk Assessment and Control, Monitoring and Early Warning
- Volume 3: Active Tectonics, Geomorphology, Climate and Geoenvironmental Engineering Geology
- Volume 4: Technological Innovation and Applied for Engineering Geology
- Volume 5: Megacity Development and Preservation of Cultural Heritage Engineering Geology
- Volume 6: Marine and Deep Earth Engineering Geology

Meanwhile, on behalf of the organizing committee, we would also like to express our deepest appreciation to the technical program committee members, reviewers, session chairs, and volunteers for their strong support for congress.

Last but not least, our gratitude also goes to the editors and press for their great support to the congress.

IAEG XIV Congress 2023 Organizing Committee

Beijing, China — Sijing Wang
Chengdu, China — Runqiu Huang
Aachen, Germany — Rafig Azzam
Zografou, Greece — Vassilis P. Marinos
September 2023

Contents

Chapter 1
New Applications of Field Permeability Tests in Single Observation Well or Borehole

Lu Zhang and Robert P. Chapuis

Abstract A single observation well or borehole installed in an aquifer is commonly used for monitoring water level and collecting water samples for chemical and microbiological analyses. Field permeability (Lefranc) tests are conducted within the same well or borehole to determine the local hydraulic conductivity K, an essential parameter in engineering geology and environmental projects. The tests were also applied to analyze the anisotropy coefficient of the aquifer and evaluate the groutability of the soil ground. The quality of the well or borehole significantly impacts the reliability of these tests and the accuracy of derived K values. The paper first reviews the two categories of field permeability tests, namely Type Constant-Head and Type Variable-Head tests. It subsequently presents several novel applications of these test to establish a systematic and practical approach for assessing the quality of the well or borehole. The interpretation and comparison of steady-state and transient-state test data of Type Constant-Head tests are utilized to regulate the installation quality of the well. The imperfect condition of the well screen/filter zone is evaluated by analyzing the field permeability tests of opposite flows. The numerical modeling of aquifer-single well systems is conducted using seep/w to provide perfect well condition and quantify the impact of well screen blocking on the apparent K value. The results indicate that even a slight decrease in K value can correspond to a significant levels of screen blockage, which can introduce great errors in contaminant transport predictions displayed by MODFLOW, resulting in inaccurate risk assessment and environment management. Therefore, it is imperative to implement quality control for observation wells and boreholes.

Keywords Field permeability test · Observation well · Hydraulic conductivity · Well quality assessment

L. Zhang (✉)
Jilin University, Changchun, China
e-mail: lu_zhang@jlu.edu.cn

R. P. Chapuis
Polytechnique Montreal, Montreal, Canada
e-mail: robert.chapuis@polymtl.ca

S. Wang et al. (eds.), *Engineering Geology for a Habitable Earth: IAEG XIV Congress 2023 Proceedings, Chengdu, China*, Environmental Science and Engineering,
https://doi.org/10.1007/978-981-99-9057-3_1

1.1 Introduction

The field hydraulic conductivity K of the aquifer is necessary for such engineering and environmental problems as dam seepage, waste site infiltration, and groundwater contamination (Chapuis 1990). A pumping test is frequently used to estimate the large-scale field K value to avoid soil disturbance (Todd and Mays 2005). However, a pumping test requires a pumping well and several observation wells (or piezometers) at different distances, which is not always justified or even possible due to the limitation of testing area or cost (Kruseman and Ridder 1994). In this case, the more efficient field permeability test can be applied in a single observation well or borehole, which is a reliable method to obtain the mid-scale local K value for any aquifer or the large-scale field K value for homogeneous aquifers because the scale effect is negligible (Chapuis et al. 2005; Dallaire 2004).

The French engineer Lefranc (1936, 1937), for the first time, proposed the field permeability test performed in a single casing borehole and its interpretation method to calculate the local K value of aquifer or aquitard. The test consists of Type Constant-Head and Type Variable-Head tests, also named Lefranc test. Type Constant-Head test generates a water flow with a constant water head through the filter zone (cavity or lantern) of a given shape at the bottom of the borehole. It is interpreted by Lefranc's method, which connects the K value with flow rate, constant head difference, and shape factor of the filter zone. Type Variable-Head test consists of generating water with a variable hydraulic head flowing through the lantern and observing the variations of the water level. The corrected Hvorslev's and relative velocity plot methods were proved to interpret the Type Variable-Head test in the most reliable way (Chapuis 2015).

The constant water head difference in the well casing of a Type Constant-Head test was initially achieved by continuously adding or withdrawing water until the flow rate becomes nearly constant (Lefranc 1936, 1937; Cassan 2005; Monnet 2015). Later, the second and easier way to keep a constant head difference during the test is to start with a constant flow rate, which gives the test another name: a constant-rate test (Cassan 2005; ISO 22282-2 2014). The two ways to perform the test are presented in Fig. 1a and b. Type Constant-Head test consists of injection and discharge tests corresponding to the opposite flows in well casing during the test (Zhang et al. 2019). The implementation of the Type Constant-Head test is more efficient and requires less equipment compared to the pumping test, which can be done as the geotechnical investigation progresses. It is, however, not so well known as it is not included in the US standards and its original version is in French.

Unlike the continuous water flow of a constant rate during a Type Constant-Head test, Type Variable-Head test, also known as slug test, generates a sudden change of water volume in the well by quickly adding or withdrawing water, and the subsequent falling or rising of water level in well casing corresponds to a falling-head (slug-injection, slug-out or baildown) or rising-head (slug-withdrawal or slug-in) test (Fig. 1.2), respectively. Sometimes, field practitioners believe that they performed

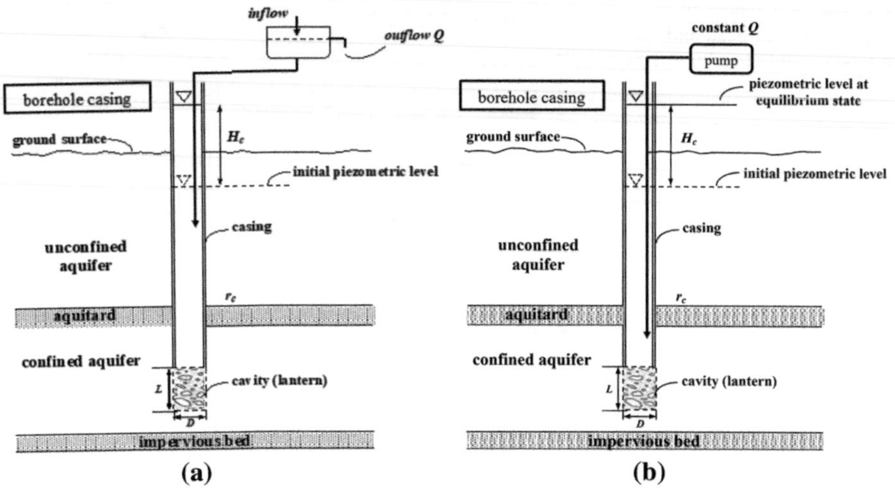

Fig. 1.1 Two ways to conduct a Type Constant-Head test **a** to maintain a constant water head until reaching a steady-state flow, or **b** to inject water at a constant rate until reaching a constant head (constant-flow test)

arising-head test after withdrawing water constantly rather than suddenly with a long-time groundwater sampling (Type Constant-Head test), and interpret the results with corrected Hvorslev's and relative velocity plot method. This manipulation was proved acceptable in theory (Zhang and Chapuis 2019), that is, the Type Variable-Head test can be generated by either a sudden or a continuous water level change and interpreted by its interpretation methods. Type Variable-Head test was not recommended in aquifers with high permeability previously, as the test is too rapid to manually record the water level variation and elapsed time. At present, it is achievable with the pressure transducer taking measurements every 0.5 s.

Besides the single borehole, the field permeability test can also be performed in a single driven permeameter (Chapuis and Chenaf, 2010), observation well (monitoring well), piezometer (e.g., Chapuis 1998), or double- packer permeameter (Mathias and Butler 2007). For the test in a driven casing, water flows through either the end of the drill casing (CAN/BNQ 2501-130 2014) or the filter zone (gravel pack) below the driving shoe (CAN/BNQ 2501-135 2014). For the test in a well pipe, water seeps through the well screen with an artificial or natural filter zone around the screen (Zhang et al. 2019). The manipulation and interpretation of the field permeability test have been included in several textbooks (Cassan 1980, 2005; Monnet 2015) and the French, International, and Canadian standards (AFNOR NF P94-132 2000; CAN/BNQ 2501-135 2014; ISO 22282-2 2014). Except for obtaining the K value, the field permeability test was also used to assess the groutability of the soil ground (Cambefort 1987; Bell 2004) and the anisotropy coefficient of the aquifer (Cassan 2000; Lafhaj and Shahrour 2004).

Being the flow channel between the aquifer and the outside, the observation well or borehole must have a good quality, otherwise, it significantly impacts the reliability of

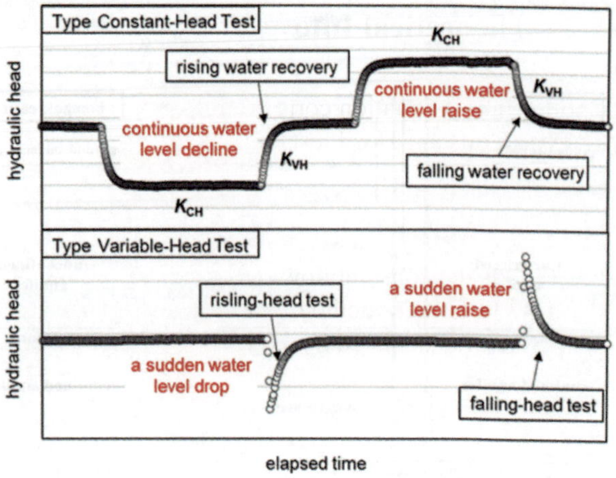

Fig. 1.2 Schematics of **a** type constant-head and **b** type variable-head test data

the field permeability tests and the accuracy of derived K values. After the installation of a well, the well is developed to remove sediments and fine particles left from the borehole drilling or washed off from the aquifer. The adjacent natural soil of the well might be remolded during the drilling and washing and mix with the filter material around the well screen. For the observation well, the well development is much slighter than the pumping wells, resulting in a higher potential of clogging (Baptiste and Chapuis 2015). Nielsen and Schalla (2005) reported that over 65% of observation wells installed in North America since the late 1970s were poorly installed or developed. The poor well installation and clogged well screens lead to erroneous K values.

In this paper, new applications of field permeability tests in a single observation well or borehole are presented, to provide a practical approach to assess the quality of the installation of well and the condition of well screen/ filter zone. The way to evaluate the installation of well screen proposed by the authors is summarized herein to compose the first steps of the approach. The observation well in reality is rarely in perfect condition due to negative problems that occurred in installation, development, and testing. Therefore, all types of field permeability tests in a perfect observation well are numerically modeled by the code seep/w. Then, the results of field permeability tests conducted in real observation wells in sand aquifers are presented. The results of field permeability tests with opposite flows that appeared through the well screen were compared and analyzed for perfect and imperfect observation wells. These compose the following steps of the approach. The practical approach in steps provides practitioners systematic guidance for the assessment of well quality. In addition, the effect of blocking of the well screen on the apparent K values of the aquifer is quantified for Type Constant-Head and Type Variable-Head tests in numerical ways. The influence of not-so-big decreased K values upon contaminant transport is also studied by the code MODFLOW, to display the great impact of small discrepancy of K values.

1.2 Theoretical Interpretation

The Lefranc's equation corresponds to the steady state, i.e., the equilibrium portion of Type Constant-Head test curve in Fig. 1.2, which gives

$$Q = cK_{CH}H_c \tag{1.1}$$

where Q is the constant flow rate, KCH represents the hydraulic conductivity calculated by Lefranc's method, Hc is the constant water head difference, and c is the shape factor, calculated by

$$c = \frac{2\pi L}{\ln(2L/D)} \quad \text{if } L/D > 4 \tag{1.2}$$

$$c = 2\pi D\sqrt{\frac{L}{D} + \frac{1}{4}} \quad \text{if } 1 \leq L/D \leq 8 \tag{1.3}$$

The governing equation of the transient flow right after starting a Type Constant-Head test was given by Cassan (2005)

$$H = \frac{Q}{cK} + \left(H_i - \frac{Q}{cK}\right) \cdot e^{\left[-\frac{cK}{S_{inj}}(t-t_i)\right]} \tag{1.4}$$

where H_i is the water head difference at time t_i, $S_{inj} = \pi d^2/4$ is the internal cross-section area of the observation well or borehole, d is the inner diameter. It corresponds to the first phase of a Type Constant-Head test.

When the test reaches equilibrium, $H_i = Hc = Q/(cK)$, the second term of Eq. 1.4 becomes 0 and Eq. 1.4 is equivalent to Eq. 1.1, which represents the second phase of the Type Constant-Head test.

The third phase or water recovery starts when the pump is stopped, when $Q = 0$, Eq. 1.4 is simplified to:

$$H = H_{ie}^{\left[-\frac{cK}{S_{inj}}(t-t_i)\right]} \tag{1.5}$$

Rewriting it as

$$\ln\left(\frac{H}{H_i}\right) = -\frac{cK_{VH1}}{S_{inj}}(t - t_i) \tag{1.6}$$

which governs the recovery flow of the Type Constant-Head test. It is the same as Hvorslev's equation for the Type Variable-Head test and gives a KVH1 value. Eq. 1.6 indicates that Hvorslev's semi-log plot is a straight line in theory, however, usually shows either upward or downward curves due to piezometric errors.

The relative velocity v is written as

Table 1.1 Comparison of type variable-head test and type constant-head test

	Type variable-head test	Type constant-head test
ΔH generation	A sudden of change	A constant flow rate
	–	Equation 1.4
Equilibrium	No steady state	A steady state
	–	Equation 1.1
	–	$K = K_{CH}$
Recovery	A rising or falling water head	
	Equations 1.6 and 1.8	
	$K = K_{VH1}$	$K = K_{VH2}$

$$v = \frac{dH}{dt} = -\frac{cK}{S_{inj}}H \tag{1.7}$$

in which v and head difference H between tj and $tj+1$ are calculated by the two consecutive measurements:

$$\frac{H_{j+1} + H_j}{2} \text{ and } v = \frac{H_{j+1} + H_j}{t_{j+1} - t_j}$$

Equation 1.7 changes to Eq. 1.8 of the equation of relative velocity plot by adding a correction for the piezometric errors of H_0, the difference between the apparent H value and the true value H_r.

$$H = -\frac{S_{inj}}{cK_{VH2}}\frac{dH}{dt} + H_0 \tag{1.8}$$

From Eq. 1.8, the relative velocity plot is theoretically straight with an intercept H_0 value, and yields a K_{VH2} value After correction, H_r replace the H value in Eq. 1.6, and the curved Hvorslev's plot becomes straight.

Type Variable-Head test and the recovery of Type Constant-Head test provide lots of $H - t$ data, which made the data selection of relative velocity plot method user-dependent. Chiasson (2005) proposed automatic ways to correct the Hvorslev's semi-log graph, using data Solver in Excel to find the best-fit line of ln(H – $H0$) versus t, which is user-independent and saves time (Table 1.1).

1.3 Well Screen Installation Assessment

The recovery phase of a Type Constant-Head test can be regarded as a Type Variable-Head test and interpreted by the methods of corrected Hvorslev's plot and relative graph plot. The detailed theoretical derivation has been given by the authors in previous work (Zhang and Chapuis 2019). In theory and from the results of observation wells in good condition, first, the K_{CH} value calculated by Lefranc's equation is equal to the K_{VH1} valuc determined from corrected Hvorslev's semi-log plot, and equal to and the K_{VH2} value derived from the relative graph plot. Second, the relative velocity and corrected Hvoslev's semi-log plots should be straight. Third, the CH point (H_c, Q/S_{inj}) locates on the relative velocity plots.

In reality, however, the observation wells are sometimes poorly designed and installed, for example, thc screen as partially immersed in water before groundwater sampling and testing, which causes a change in shape factor, or the flow rate is set too high, creating parasitic head losses and a large seepage face in the screen. The results of such wells have been discussed by the authors (Zhang and Chapuis 2019), which can yield quite different K_{CH}, K_{VH1}, and K_{VH2} values, curved or scattered relative velocity plots and mislocated CH points. Therefore, these three abnormalities are defined as the criteria to evaluate the well design and installation, which are normally related to the installation of well screen.

Therefore, the manipulations are concluded to compose the first steps of the approach of well-screen quality assessment in the paper. First, a Type Constant-Head test must be conducted if it is a groundwater sampling case, that is, a constant flow rate Q and a constant head difference at equilibrium must be recorded, calculating a K_{CH} value. Then the recovery data versus time after the pump stops must be documented and interpreted to give K_{VH1} and K_{VH2} values. At last, using the three criteria to check if the well screen is poorly installed or not. If the screen is poorly installed, it yields wrong K values which should be abandoned or corrected if possible. If the screen is perfectly installed, the next step is to check the blocking/erosion of the screen/filter zone.

1.4 Well Screen Quality Assessment

1.4.1 Field Permeability Tests in Perfect Observation Well

In reality, the observation well has more or less imperfect conditions. Therefore, the numerical model built through seep/w is used herein to simulate the Type constant-head test and type variable-head test with raised or declined water head difference, respectively.

The aquifer-single well model is axisymmetric (x, z), where the x-axis directs the radius of the aquifer and the z-axis represents the vertical elevation. The aquifer is located from the ground surface of $z = 0$ m to the impervious layer of $z = 4$ m with an initial water level of $z = 3$ m which was indicated by applying a far boundary of hydraulic head $H = 3$ m for all field permeability tests. The aquifer radius is 3 m in the *x-direction*. A single observation well is installed in the aquifer from $z = 0$ m to $z = 4.2$ m and has an inner and outer radius of 2.6 and 3 cm, which corresponds to a schedule 40 2-inch well pipe. The screen is located from the bottom of the aquifer with a length of 1 m. No filter pack is surrounding the well screen.

Type Variable-Head test with two opposite flows was first implemented in the aquifer-single well model under transient state. The falling-head test was initiated after raising a water level of 1 m in the pipe in 5 s, achieved by adding a boundary function of *H*on well screen and *t* and then removing it. The boundary function of the rising- head test, on the contrary, is a monotonic decreasing function of $H - t$, to generate a subsequent rising water level in the well pipe. The variation of water level with time were recorded and interpreted by corrected Hvorslev's and relative velocity methods to calculate the *K* values of falling-head and rising-head tests (Figs. 1.3 and 1.4).

Type Constant-Head test with two opposite flows was then conducted at steady-state conditions in the same aquifer-single well model. The injection and discharge tests were started by applying constant *H* of 4 m and 2 m, respectively, on well screen.

The falling-head test gives a *K* value of 1.63×10^{-4} m/s, equivalent to that yielded from the rising-head test by the corrected Hvorslev's method. After the selection of data points, the relative velocity plots gave close values of 1.83×10^{-4} and 1.81×10^{-4} m/s for falling-head and rising-head tests. The difference in *K* values between the two interpretation methods is due to user-dependent data selection and the time step setting of numerical models. For each interpretation method, the falling-head and rising-head tests yield the same *K* values. Injection and discharge tests also give the same *K* value of 9.89×10^{-5} m/s. The identical *K* values of each type of

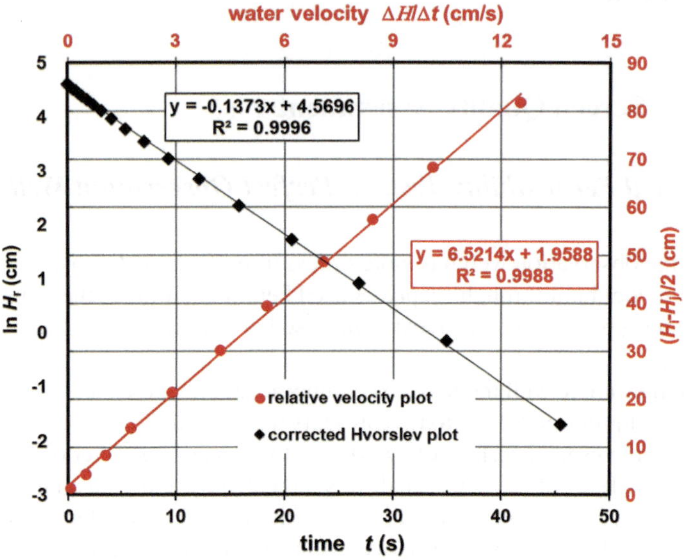

Fig. 1.3 Interpretation by corrected Hvorslev's and relative velocity methods for rising-head test in the model

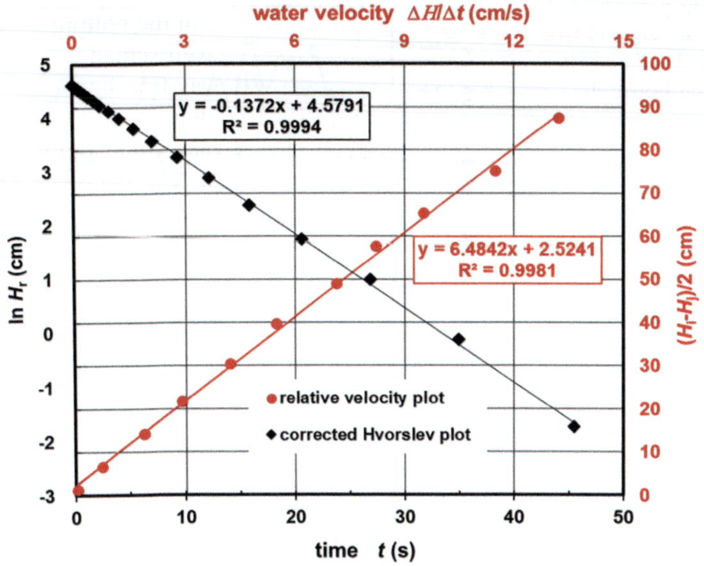

Fig. 1.4 Interpretation by corrected Hvorslev's and relative velocity methods for falling-head test in the model

field permeability test of opposite flows are in accordance with the theory because the manipulations and physical mechanisms of falling-head and rising-head tests are identical, so are the injection and discharge tests. For the mid-scale field permeability test, the volume strain of the aquifer during test is null, and thus the only difference is the opposite water flow directions. Another phenomenon is that in the modelled perfect aquifer-single well system, Type Variable-Head test yielded greater K values compared to Type Constant-Head test. The observation well is considered as a very high permeable material in the model, and the screen boundary was applied on the central axis, thus a radius of 3 cm high permeable material may contribute to the value the apparent K. Type Variable-Head testis more rapid and has a smaller influence radius than that of Type Constant-Head test, and thus easier to be affected to yield a larger K value.

1.4.2 Field Permeability Tests in Real Observation Wells

A total number of 116 field permeability tests were conducted in 29 single observation wells installed in two sand aquifers, giving K values ranging from 7.45×10^{-6} to 9.02×10^{-5} m/s. Figure 1.5 plots the percentage decrease in K values of Type Variable-Head Tests compared to that of Type Constant-Head Tests. Each set contains 29 test data, which makes 58 groups of comparative groups.

Fig. 1.5 The percentage decrease in *K* values of type variable-head compared to type constant-head tests

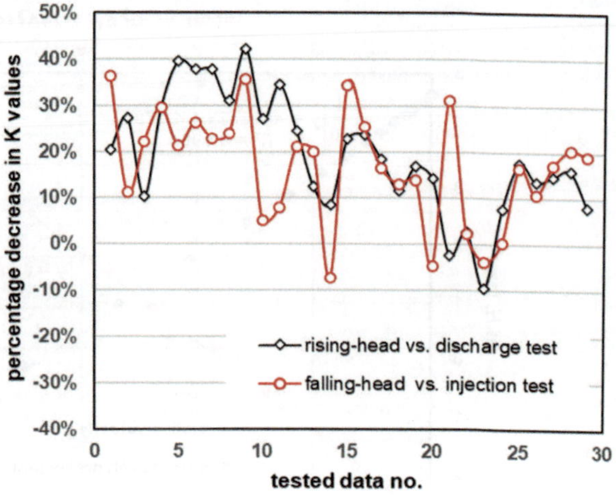

All Type Variable-Head Tests yield greater *K* values than Type Constant-Head Tests except four opposite results that Type Variable-Head test *K* values are 2, 4, 7, and 10% smaller than Type Constant-Head test *K* values. This discrepancy might be due to error of manual manipulation, gas bubble generation or slight change in water filter zone during the test. In detail, the *K* values of 27 rising-head tests are 3–42% larger than that of the discharge tests, and the *K* values of 27 falling-head tests are 3–42% larger than that of the injection tests. It is consistent with the results of field permeability tests performed in the perfect single observation well in the numerical aquifer-single well model.

In reality, when a well screen is surrounded by an artificial filter zone (gravel pack), the apparent local *K* value given by the field permeability test is influenced by the permeabilities of both gravel pack and aquifer. If only a well screen exists, a created (natural) filter zone around the screen is gradually generated due to the washing of small particles. The artificial and natural filter zone has larger porosity compared to the aquifer, which results in a greater permeability. Compared to Type Variable-Head Test, Type Constant-Head Test has longer test duration and larger influence radius, and hence is less affected by the highly pervious filter zone. As a result, the *K* values of rising-head and falling-head tests are generally larger than that of the discharge and injection tests.

In Fig. 1.6, the black plot shows the percentage decreases in *K* values of rising-head tests compared to that of the falling-head tests. 25 out of 29 comparative groups indicate that the rising-head test *K* values are 2.87–22.63% larger than the falling-head test *K* values. The red plot shows that *K* values of discharge tests are 1.56–31.02% larger than that of the injection test. The rising-head and discharge tests have water flowing from the aquifer to the screen, which may carry fine particles from filter zone or natural aquifer, leading to an erosion and a higher apparent *K* value. The water of falling-head and injection tests flow from screen to the aquifer and may cause screen clogging by displacing the fine sediments existed in the well.

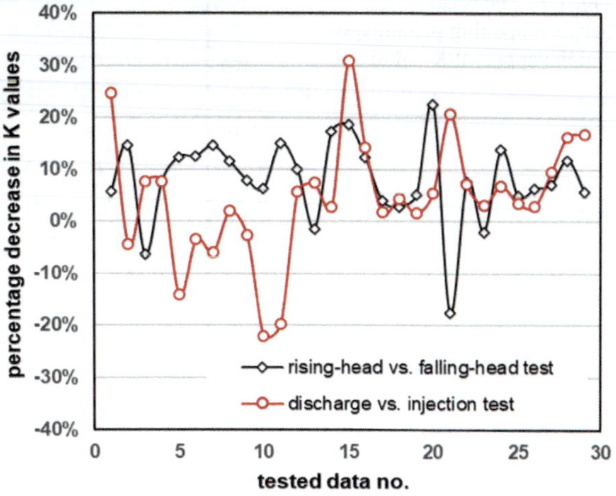

Fig. 1.6 The percentage decreases in K values of rising-head and discharge tests respectively compared to falling-head and injection tests

However, over half of the first 14 test data shows opposite results. The first 14 test data are acquired from Type Constant-head test conducted in smaller observation wells installed in a smaller-scaled aquifer. The flow rate might be relatively high to carry larger particles, which cause the outside blocking of the screen. Therefore, the condition of the well screen/filter zone can be assessed by comparing the field permeability test of two-directional flows, which contributes to the following steps of the approach of well screen quality assessment.

The largest percentage decrease of K values of the field permeability test results is 31.2% (Fig. 1.6), calculated from $K = 1.86 \times 10^{-5}$ and $K = 1.28 \times 10^{-5}$ m/s of a discharge test and an injection test (two opposite flows), respectively. The two values are in the same order of magnitude, the difference of which seems to be insignificant. Due to the not-so-big difference, researchers and engineers are prone to underestimate the importance of imperfect conditions of water filter zone, either well screen blocking or the adjacent soil erosion in the derived K values. It is difficult to quantify imperfect condition of water filter zone, the established numerical aquifer-well model is therefore employed to the study the effect of well screen blocking on the hydraulic conductivity.

1.5 Effect of Imperfect Well Screen Condition

1.5.1 Field Permeability Tests in Blocked Screens

The well screen blockings of 10, 20, 30, 40, 50, 60, 70, 80, and 90% were simulated with the aquifer-single well model by decreasing the elevation of boundary condition applied on the screen. Figure 1.7 presents the apparent hydraulic conductivities

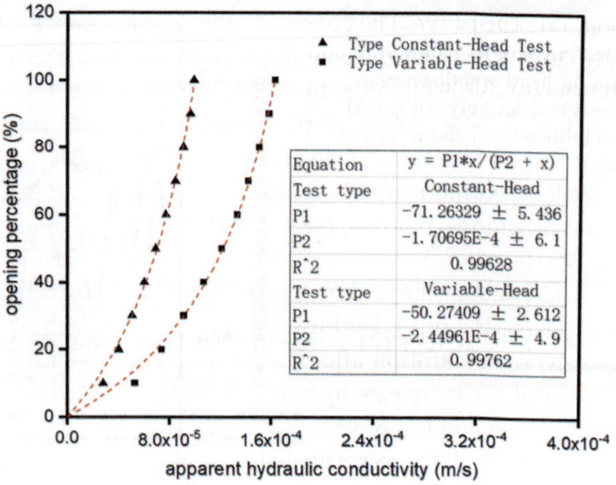

Fig. 1.7 The relationship between opening percentage and the apparent *K* values

derived from Type Constant-Head and Type Variable-Head tests in single observation well under zero to full blocking conditions corresponding to 100–0% opening area. In the case of 10–100% screen opening, Type Variable-Head tests yield larger values of 5.27×10^{-5}–1.63×10^{-4} m/s compared to that of Type Constant-Head test of 2.80×10^{-5}–9.89×10^{-5} m/s. The two plots both present a hyperbolic equation of $y = (p_1x)/(p_2 + x)$ between the opening percentage and the corresponding apparent hydraulic conductivity, of which the R^2 values are 0.99628 and 0.99762 for Type Constant-Head and Type Variable-Head tests, respectively. The values of P1 and P2 are displayed in Fig. 1.7. It is indicated from the hyperbolic equation that when $x = 0$, i.e., the well screen is fully blocked, the apparent hydraulic conductivity of the y-axis becomes zero because no water flows through the screen to the aquifer. If the well screen is in perfect condition ($x = 100\%$), they value equals the local *K* of the aquifer, as discussed before.

Taking the Type Constant-Head test in this model with varied blocking percentage for example, no blocking condition yield a *K* value of 9.89×10^{-5} m/s, and 50% blocking percentage corresponds to an apparent *K* value of 6.89×10^{-5} m/s. A decrease of 11% in *K* can correspond to a 50% blocking in this case, which is unexpected. The small difference is normally overlooked; however, the not-so-substantial difference may cause errors in certain aspects of engineering or environment project design and management.

1.5.2 Influence of No-so-big Differences on Contaminant Transport

The contamination source of Cr was released simultaneously from 22 well points located evenly in an approximate rectangle shape on the target zone, simulated by the code MODFLOW. The concentration of released Cr ranges from 0.02 to 362.42 mg/L. Figure 1.8 shows the contaminant plume after 1 year when the source was released continuously through the first 24 h in the aquifer with (a) $K = 9.89 \times 10^{-5}$ m/s (no blocking), (b) $K = 6.89 \times 10^{-5}$ m/s (50% blocking), (c) $K = 1.24 \times 10^{-5}$ m/s (99% blocking). Figure 1.9 shows the contaminant plume after 1 year when the source was released continuously through the 5 days in the aquifer with (a) $K = 9.89 \times 10^{-5}$ m/s (no blocking), (b) $K = 6.89 \times 10^{-5}$ m/s (50% blocking), (c) $K = 1.24 \times 10^{-5}$ m/s (99% blocking).

The concentration shown on the fringe of the contaminated groundwater plume is 0.02 and 0.2 mg/L for Figs. 1.8 and 1.9, respectively. The upper region reflects the highest concentration after releasing the contaminant source. The greatest concentration of Cr increases from 0.26, 0.32 to 0.42 mg/L for the case of 1 day release (Fig. 1.8) when the K values decreased from 9.89×10^{-5}, 6.89×10^{-5}, to 1.24×10^{-5} m/s, and the vertical distances of the three plumes are 20, 19.5 and 16 m, respectively. For the case of 5-day release, the concentration of the upper region is much higher, showing a maximum value of 2.4, 2.6 to 2.8 mg/L (Fig. 1.9) when the

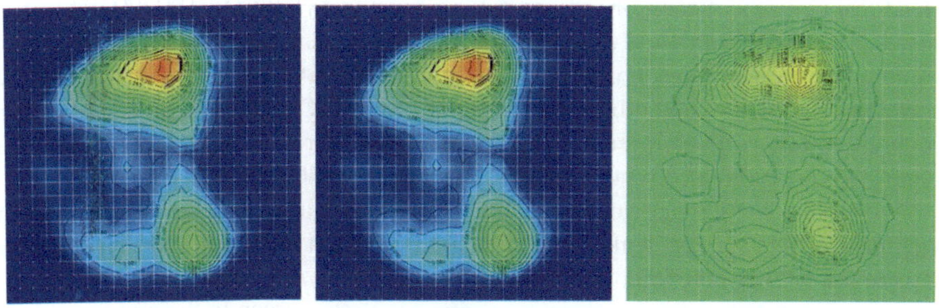

(a) (b) (c)

Fig. 1.8 Cr(VI)-contaminated groundwater plume after 1 year with 1 day source release in aquifers with **a** $K = 9.89 \times 10^{-5}$ m/s, **b** $K = 6.89 \times 10^{-5}$ m/s, **c** $K = 1.24 \times 10^{-5}$ m/s

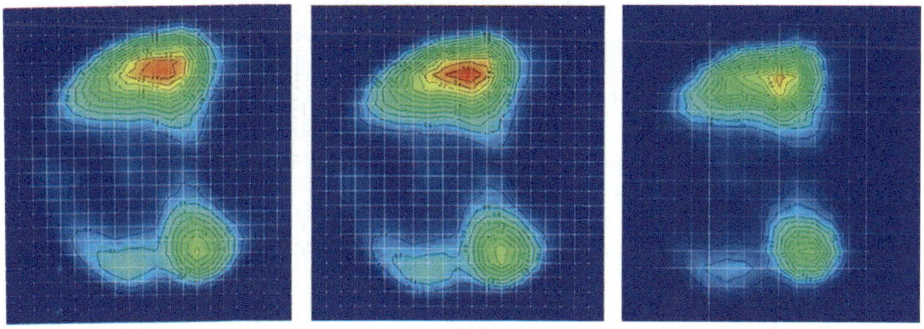

(a) (b) (c)

Fig. 1.9 Cr(VI)-contaminated groundwater plume after 1 year with 5 days source release in aquifers with **a** $K = 9.89 \times 10^{-5}$ m/s, **b** $K = 6.89 \times 10^{-5}$ m/s, **c** $K = 1.24 \times 10^{-5}$ m/s

K values decreased from 9.89×10^{-5}, 6.89×10^{-5}, to 1.24×10^{-5} m/s, and the vertical distances of the three plumes are 20, 19.5 and 16 m, respectively. As a result, the slightly decreased K value can cause an obvious increase of Cr concentration and a decrease in plume scale, which indicates an erroneous prediction in contaminant transport. It may lead to an invalid design of contaminant treatment or environment management projects. Therefore, it is recommended for the practitioners to pay more attention to the importance of well quality assessment as poorly installed well and clogged well screen may yield inaccurate K values.

1.6 Conclusion

The paper first reviewed four types of field permeability tests in a single observation or borehole. From the perspective of the testing principle, it is divided into Type Constant-Head and Type Variable-Head tests. With varied flow direction, the Type Constant-Head test includes discharge test and injection test, and the Type Variable-Head test consists of rising-head and falling-head test. Type Constant-Head test is solved by Lefranc's method to obtain the local K value, which equals the in-situ K value when the aquifer is homogeneous. Type Variable-Head test is interpreted by corrected Hvorslev's method. The K values of the Type Constant-Head test are more accurate to reflect the permeability of the tested aquifer as it has a larger influence radius to be less affected by the high permeable artificial or natural filter zone.

A systematic and practical approach to assess the quality of a single observation well or borehole was presented by conducting different types of field permeability tests in the well or borehole. It is concluded in steps:

(1) conducting a Type Constant-Head test to acquire a complete test curve (refer to Fig. 1.2) with both steady and transient states;
(2) using Lefranc's method to interpret the steady-state curve and obtain a K_{CH};
(3) applying corrected Hvorslev's method and relative velocity plot method to interpret the transient-state (recovery) data and obtain K_{VH1} and K_{VH2} values and a relative velocity plot;
(4) checking if $K_{CH} = K_{VH1} = K_{VH2}$, if the velocity graphs are linear, and if the CH point falls on linear velocity graph;
(5) if it is yes, the design and installation of the observation well screen or the borehole filter zone is good, and viceversa;

For the observation well and bore of good design and installation, the condition of well screen and filter during the test is still necessary to be assessed because the blocking in well screen and erosion in filter zone yield inaccurate K values, which may lead to great errors in such as contamination analysis and engineer design. Therefore, the next steps are,

(6) conducting field permeability test with opposite flows in the same observation well;

(7) comparing and analyzing the yielded K values of different flow directions;
(8) if they are the same, the condition of well screen or borehole during a test is good, viceversa;
(9) when the rising-head and discharge tests yield larger *K* values, it indicates that either the water from the aquifer to pipe carries fine particles and tiny gas bubbles from the filter zone to facilitate the flow, or the water from the pipe to the aquifer displaces the fine sediments in the well pipe and creates tiny bubbles to block parts of the screen or the voids of the filter pack;
(10) in some cases, the injection and falling-head tests give larger *K* values, which indicates that the particles carried by the discharge and rising-head tests are larger than the screen slots, which block the screen from outside.

Through these steps, the condition of observation well screen and filter zone before testing and during testing can be assessed. The extent of blocking and erosion is difficult to quantify in reality except using some visual equipment like groundwater cameras, establishing transparent aquifer model, etc.

A numerical aquifer-single well model was then made to giveaway to quantify the effect of well screen blocking. For both Type Constant-Head and Type Variable-Head tests, the opening percentage (equals 100%-blocking percentage) and the corresponding apparent hydraulic conductivity were found to follow a hyperbolic equation of $y = (p_1 x)/(p_2 + x)$. For the model, the apparent K values decreased 3–30% when the screen is blocked 10–90%, which indicates that a small decrease in *K* value may correspond to a severe screen blocking.

Unfortunately, the 30% decrease may not attract too much attention of field practitioners as it corresponds to the percentage decrease from K values of 9.89×10^{-5} to 2.80×10^{-5} m/s, which are still in the same order of magnitude. Therefore, the development of Cr contamination plume based on this insignificant *K* decrease was presented. It is resulted that the decreased *K* value leads to smaller scale of the plume and larger concentration of Cr. It seems to be small but big enough to generate great error in contamination simulation. Therefore, to assess the quality observation well is very important. The recommended stepped is easily to apply on the assessment.

References

AFNOR NF P94-132 (2000) Sols: reconnaissance et essais—essai d'eau Lefranc. AFNOR (Replaced by ISO 22282-2), Paris

Baptiste N, Chapuis RP (2015) What maximum permeability can be measured with a monitoring well? Eng Geol 184:111–118

Bell FG (2004) Engineering geology and construction. CRC Press, New York

Cambefort H (1987) Grouts and grouting. Ground Eng 32:1–32

CAN/BNQ 2501-130 (2014) Soils—determination of permeability at the end of a casing. National Standard of Canada, Ottawa

CAN/BNQ 2501-135 (2014) Soils—determination of permeability by the Lefranc method. National Standard of Canada, Ottawa

Cassan M (1980) Les essais d'eau dans la reconnaissance des sols. Eyrolles, Paris

Cassan M (2000) Application des essais Lefranc à l'évaluation du coefficient d'anisotropie hydraulique des sols aquifères. Rev Fr Géotech 12(90):25–43

Cassan M (2005) Les essais de perméabilitésur site dans la reconnaissance des sols. Presses des Ponts, Paris

Chapuis RP (1990) Sand–bentonite liners: predicting permeability from laboratory tests. Can Geotech J 27(1):47–57

Chapuis RP (1998) Overdamped slug test in monitoring wells: review of interpretation methods with mathematical, physical, and numerical analysis of storativity influence. Can Geotech J 35(5):697–719

Chapuis RP (2015) Overdamped slug tests in aquifers: the three diagnostic graphs for a user-independent interpretation. Geotech Test J 38(4):474–489

Chapuis RP, Chenaf D (2010) Driven field permeameters: reinventing the wheel? Geotech News 28(1):37

Chapuis RP, Dallaire V, Marcotte D, Chouteau M, Acevedo N, Gagnon F (2005) Evaluating the hydraulic conductivity at three different scales within anunconfined sand aquifer at Lachenaie, Quebec. Can Geotech J 42(4):1212–1220

Chiasson P (2005) Methods of interpretation of borehole falling-head tests performed in compacted clay liners. Can Geotech J 42(1):79–90

Dallaire V (2004) Étude des effets d'échelledansun aquifèrealluvionnaire (Master thesis)

ISO 22282-2 (2014) Geotechnical investigation and testing: geohydraulic testing—part 2: water permeability tests in a borehole using open systems. ISO, Geneva

Kruseman GP, de Ridder NA (1994) Analysis and evaluation of pumping test data. International Institute for Land Reclamation and Improvement, Wageningen

Lafhaj Z, Shahrour I (2004) Détermination de l'anisotropie de perméabilité des sols par l'essai Lefranc. Rev Fr Géotech 21(109):99–103

Lefranc E (1936) Procédéde mesure dela perméabilitédessolsdans lesnappes aquifèreset application au calcul du débit des puits. Le Génie Civil 109(15):306–308

Lefranc E (1937) La théorie des poches absorbantes et son application àla détermination du coefficient de perméabilité en place et au calcul du débit desnappes d'eau. Le Génie Civil 111(20):409–413

Mathias SA, Butler AP (2007) Shape factors for constant -head double -packer permeameters. Water Resour Res 43(6):11542

Monnet J (2015) In situtests in geotechnical engineering. Wiley, Amsterdam

Nielsen DM, Schalla R (2005) Design and installation of ground-water monitoring wells. In: Practical handbook of environmental site characterization and ground-water monitoring. CRC Press, pp 649–816

Todd DK, Mays LW (2005) Groundwater hydrogeology. Willey, New York

Zhang L, Chapuis PR (2019) Recovery test after a constant-head test in a monitoring well: interpretation methods and new findings. Eng Geol 259:105150

Zhang L, Chapuis PR, Marefat V (2019) Field permeability tests with inward and outward flow in confined aquifer. Geotech Test J 42(6):234

Chapter 2
Variability Analysis of Bearing Capacity of Piles with Slowly Varying Q–s Curves

Gaofeng He, Sixin Wang, Jinhui Li, and Xiaoheng Zhang

Abstract Ultimate bearing capacity of foundation pile is of great importance for pile design and optimization, but it may not be feasible due to economic and safety considerations, especially the destructive test. A multi-parametric inversion method is proposed based on a hybrid algorithm that combines SA (simulated annealing algorithm) with Powell's method, which utilizes the non-failed static load–settlement (Q–s) curves of 36 pipe piles from a specific field to invert the 16 surrounding soil parameters from a global optimization perspective. The results demonstrate that the maximum displacement value of the pile head and actual load-holding duration have negative correlation with predicted ultimate bearing capacity of piles with the same type at the same site. Their values range from 1.3 to 2.1 times of the ultimate bearing capacity of the pile body material based on the soil parameters inversion, which contributes to optimizing pile foundation design and furnishing data-driven support to AI algorithms for addressing intricate nonlinear problems.

Keywords Load-movement curve · Multi-parameters inversion · Bearing capacity prediction

G. He · S. Wang
Shenzhen Institute of Building Research Co., Ltd, Shenzhen 518049, China
e-mail: happyhgf336@sjtu.edu.cn

G. He · J. Li (✉)
Department of Civil and Environmental Engineering, Harbin Institute of Technology (Shenzhen), Shenzhen 518055, China
e-mail: jinhui.li@hit.edu.cn

X. Zhang
Shenzhen Jianyan Testing Co., Ltd, Shenzhen 518031, China

S. Wang et al. (eds.), *Engineering Geology for a Habitable Earth: IAEG XIV Congress 2023 Proceedings, Chengdu, China*, Environmental Science and Engineering,
https://doi.org/10.1007/978-981-99-9057-3_2

2.1 Introduction

Artificial Intelligence (AI) algorithms, with their powerful nonlinear optimization capabilities, have found significant applications in the analysis of the ultimate bearing capacity of foundation piles (Teh et al. 1997; Momeni et al., 2014; Moayedi et al. 2018, 2019; Armaghani et al. 2022). However, their success relies heavily on the quality of samples and input data. The existing static load tests, which have been used extensively, typically push the pile head to a maximum test load twice the design characteristic value due to economic and safety considerations, rather than the real ultimate bearing capacity and corresponding displacement (Zhang et al. 2011; Nguyen et al. 2013). This results in a Q–s curve that mostly presents slowly-varying type, which hinders the determination of the ultimate bearing capacity of piles directly. Even though design requirements can be met, the required specific parameters for optimizing the pile design and their operation and maintenance management for the long-term consideration need more consideration. It is also difficult to distinguish the bearing capacity difference between piles with non-failed static load tests, or analyze specific reasons for piles with failed static load tests. Furthermore, when using AI algorithms in predicting the ultimate bearing capacity of the foundation pile, the input mechanical parameters mostly include three parts (1) shear strength of the soil, like cohesion c, internal friction angle ϕ, and hammering number N of the standard penetration test, (2) pile characteristics, such as length L, pile type and cross-sectional area, and (3) the construction parameters, for instance, hammer weight and drop height (Momeni et al. 2014; Benali et al. 2021). And the Q–s curve data are regarded as output, rather than the pile bearing capacity as a result of lacking instrumented and destructive field study (Zhang et al. 2011; Fellenius 2021). Although predicted values can achieve good agreement with experimental data by using more sophisticated algorithms, their lack of physical basis and high-quality data of destructive static load test limits their further practical interpretation and application (Shahin 2016).

Due to the limitations and difficulties of conducting static load tests and obtaining accurate ultimate bearing capacity values, numerical analysis and theoretical calculations are widely used (Zhang et al. 2010; Zhou et al. 2019). Prior to carrying out numerical simulations based on static load test curves, it is crucial to analyze the soil characteristics surrounding the pile using inversion methods to obtain its accurate parameters (Bradshaw 2016; Fellenius 2023). However, the homogeneous assumption used in the inverse analysis process does not accurately reflect the actual non-uniform nature of the surrounding soil (Zhang et al., 2010; Zhang and Zhang, 2012; Fellenius 2013). This results in an average value that fails to capture the nuances of the actual conditions. And the use of Q–s curve from a single pile to derive the average value of the inverse parameters is inadequate in considering the relationships between different piles within the same site. Because local overfitting may arise to make it difficult to obtain complete, globally applicable inverse parameters. Unfortunately, inhomogeneous geotechnical layers containing multiple parameters cannot be effectively analyzed based on just one single pile.

This paper aims to address these deficiencies through a global optimization approach to predicting the ultimate bearing capacity of a single pile using measured Q–s curves. The study utilizes static load data from 36 pipe piles at a project site with the same pile diameter and soil strata but varying thickness in each layer. Firstly, with actual four layers of geotechnical bodies, a total of 16 parameters involving pile shaft and tip resistance are inverted employing a multi-parameter optimization algorithm. Secondly, after validation with the test data, the optimized inversion parameters are incorporated into the theoretical solution of the load transfer method to predict the ultimate bearing capacity of the piles and analyze their differences.

2.2 Methodology

2.2.1 *Data*

Our study aims to investigate the behavior of PHC-AB500(125) prestressed high-strength concrete pipe piles under static load by analyzing all the 36 Q–s curves collected from a construction site in a typical granite region of Shenzhen, China. The pile lengths ranged from 11.62 to 28.12 m, and there were 10 loading levels ranging from 0 to 4400 kN, with the first load being twice the graded load of 440 kN, followed by 5 unloading levels. All the Q–s curves are derived from the non-failed static load test with a slowly varying tendency, shown in Fig. 2.1. The site soil consisted of four layers, including the Quaternary Holocene artificial fill layer (Q4ml, above the design elevation of pile head), the Quaternary alluvial layer silty clay (Q4al + pl), the Quaternary slope layer (Q4dl) gravelly clay and residual layer (Q4el) gravelly clay. The bedrock was composed of Yanshan-age granite (γ53), where the fully weathered granite was considered the pile tip bearing layer. The depths of the multilayer soils can be obtained from the adjacent boring holes.

2.2.2 *Simulation Procedure*

The load transfer method is a well-established approach to analyze the load bearing mechanism of a pile, which consists of modeling the pile as a sequence of pile segments (elastic units), and simulating the interaction between each pile segment and the soil using a nonlinear spring. It is crucial to select suitable transfer function parameters and determine the calculation parameters based on soil properties by using the 36 Q–s data. In our study, t–z model of bilinear model and q–z model of strain-softening response are employed for the soil-pile interaction at the pile shaft and tip, respectively, demonstrated in Fig. 2.2, recommended by Zhang and Zhang (2012). The essential parameters of the transfer function include the present displacement Sb and ultimate displacement Sbu at the pile tip, the compressive rigidity of the

Fig. 2.1 Static load test data of 36 pipe piles

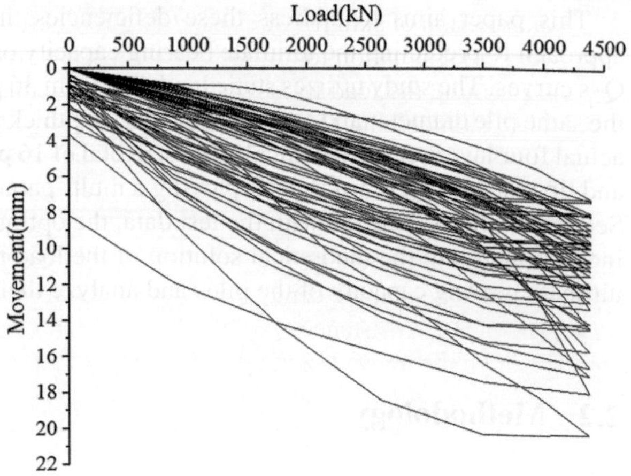

pile-tip soil in the first and second stages of the load–displacement curve k1 and k2, respectively, as well as ultimate displacement Ssu, corresponding ultimate load Qsu, and ratio of the residual unit skin friction to the limiting unit shaft resistance βs for pile shaft soils. Deriving the transfer parameters that match the actual soil layer can be challenging in actual engineering, particularly when there are multiple soil layers.

The load transfer method is a commonly used approach for investigating pile tip and shaft resistance. The following steps are required to use the method:

(1) Obtain the characteristic parameters of the pile, including its length (L), diameter (D), circumference (U), cross-sectional area (A), and modulus of elasticity (E).
(2) The pile is divided into N units from top to bottom, with the length of each unit (ΔL) depending on the required calculation accuracy. N must ensure uniform

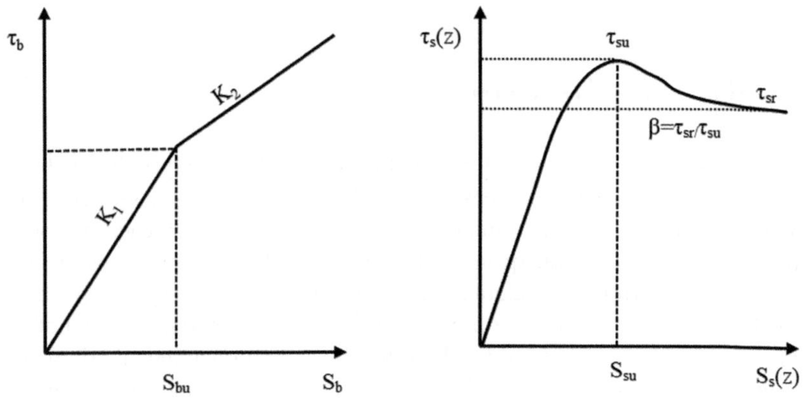

Fig. 2.2 t/q–z models used in the back calculation

distribution of skin friction along the pile shaft despite being inconsistent with the actual situation to minimize calculation error.

(3) Determine the load transfer function for each layer of geotechnical medium. This includes the Qs–S function (load-movement of the pile shaft) and the Qb–Sb function (pile-tip resistance), where Sb is the pile-tip displacement and Sz is the pile displacement at section z.

(4) Assume the pile tip displacement of the nth unit is Sbn, equivalent to the rigid body displacement of pile section n at this point, and the initial value of the displacement at the center point of pile section n is assumed to be Scn = Sbn. The pile tip resistance Qbn and skin friction Qsn can be obtained respectively through the above t–z models.

(5) Calculate the pile head pressure Qtn of pile segment n. According to the equilibrium condition, Qtn = Qbn + ΔL·U·Qsn. The average axial force of pile section n is Qpn = (Qtn + Qbn)/2. The compressive deformation of pile section n is ΔSbn = Qpn − ΔL/(A·E) (under the condition of no plastic damage of pile body), and the displacement of pile head of pile section n is Stn = Sbn + ΔSbn, while the displacement at the center is Scn = (Qtn + 3Qbn) − ΔL/(8A·E)

(6) Compare the initial displacement value Scn at the center of pile segment n (see step 4) with the corrected displacement Scn′ and we have Scn′ = Scn + Sbn.

(7) If the difference (Scn′–Scn) is less than 10–6 m, then use Scn′ as the new value for the displacement of the center of pile segment n. If not, repeat steps 4–6 until the difference is less than 10–6 m.

(8) Iterate through steps 4–7 to obtain the friction resistance and displacement along the pile shaft and at the pile tip for pile segment n.

(9) Use Qbn and Stn as the pile tip axial force and movement of pile section n − 1, respectively.

(10) Repeat steps 4–9 to calculate the pile head displacement and load for the remaining pile sections until the pile head displacement (St1) and pile head load (Pt1) of the first unit at the pile head are obtained.

In the process of load transfer of the pile, a hybrid algorithm for parametric inversion was used in our study by combining Powell method and Simulated Annealing (SA) algorithm.

2.3 Case Study

2.3.1 Inversion Parameters and Range of Values

Setting reasonable bounds for each computational parameter is critical to ensuring that the inverse analysis yields meaningful results that have practical applications.

Table 2.1 Inversion parameters and their values

Inversion parameters	Numbers	Ranges of values	Units
Pile tip displacement S_b	1	0.01–0.30 (first load level) 0.01–10.00 (other levels)	mm
Ultimate pile tip displacement S_{bu}	1	5–10	mm
Compressive rigidity of the pile-tip soil in the first stage k_1	1	1.2e6–2.5e6	kPa/m
Compressive rigidity of the pile-tip soil in the second stage k_2	1	0.3e6–1.5e6	kPa/m
Ultimate shaft displacement $S_{su1\sim4}$	4	3–15 or 0.5–2.5%D (pile diameter D)	mm
Ultimate unit shaft resistance $\tau_{su1\sim4}$	4	silty clay (20–35); gravelly clay (Q_4^{dl}) (30–70); gravelly clay (Q_4^{el}) (45–100); fully weathered granite (70–130)	kPa
Ratio of the residual unit skin friction to the limiting unit shaft resistance $\beta_{s1\sim4}$	4	0.77–0.97	\

In this project site, 16 parameters were inverted across four layers of soil based on the t/q − z model. The range of these parameters was established by considering both reference materials and field test results (Zhang and Zhang 2012), seen in Table 2.1.

2.3.2 *Multi-parameters Inversion*

Inversion procedure. The inversion of pile-soil interaction model parameters using static load test data is a complex process, especially for larger loads. When using static load data from only one pile, a considerable number of parameters must be inverted, leading to reduced accuracy and overfitting or insufficient generalization. To address these issues, we propose a strategy for the multi-parameter inversion of the multi-layered soil surrounding a pile. The strategy is further explained using two steps:

In the first step, the multi-parameter inversion is performed on the static load data of multiple piles for the first stage load of 880 kN. The load transfer method uses the pile tip displacement in the compressive static load test as the initial condition, thereby affecting the distribution of the Q–s curve. However, due to the variation in the actual values of pile tip displacement of different piles and with different load classes, there cannot be enough constraints for achieving accurate inversion of all pile tip displacements. The challenge to inverse problems, however, is that the solutions are non-unique, and there can be many different combinations of parameters that can produce the same result (Bradshaw 2016). In order to address this, we assume that the corresponding pile tip displacement of the 36 piles with a bearing layer of fully weathered granite loaded at 880 kN are all small and equal, ranging from 0.01 to 0.06

mm. This assumption is based on some instrumentation data listed in (Zhang and Zhang 2012; Fellenius 2023) and the fact that load is transferred to mainly pile shaft rather than pile tip at the first stage of load, with the corresponding measured pile head displacement ranging from 0.8 to 2.2 mm. This study considers the characteristics of the load transfer and justifies the assumption, allowing for the appropriate parameter inversion value to be obtained using the static load data of each pile.

In the second step, as the load level increases, the soil on the pile side gradually enters the limit state, while the soil at the pile tip takes more responsibility for supporting the increasing load on the pile head. The soil parameter values along the pile shaft already obtained in the first step can be fixed and served as new constraints for calculating the soil parameters at the pile tips (Sb, Sbu, K1, K2). By doing so, the lack of accuracy of multi-parameter inversion is fulfilled using data among individual piles, and also the possible overfitting probability can be reduced.

Finally, the inversion results for a total of 16 parameters of pile shaft/tip resistance parameters of the four geotechnical layers at this site can be obtained, shown in Table 2.2.

Validation. Verification of the inversion parameters can be achieved by comparing the difference between the calculated and measured values of pile head displacement for load classes other than 880 and 4400 kN, as shown in Fig. 2.3. The predicted values match well with the actual static load data, indicating that the overall situation and reliability of the inversion results can be accurately reflected and verified.

Small difference of the pile head displacement values can be seen under the same loading conditions. Moreover, the change trends of the Q–s curves are consistent, with the straight line section and the slope change section being well reflected. There are fewer cases with large differences but evenly distribution in the results. For example, piles 11#, 17#, and 31# have larger predicted displacements, while piles 3#, 5#, and 21# have smaller ones. These results may be due to several factors. Firstly, the static load test results of all the same type of piles at the site can take into account the influence of soil parameters around the pile and pile length comprehensively. The inverse parameters are thus the global optimal solutions, which also satisfy the

Table 2.2 Multi-parameter inversion results

Inversion parameters	Units	Actual four layers (mean/median)			
		Fully weathered granite	Gravelly clay (Q_4^{el})	Gravelly clay (Q_4^{dl})	Silty clay
S_b (maximum load)	mm	5.53	/	/	/
S_{bu}	mm	6.85	/	/	/
k_1	MPa/mm	1.83	/	/	/
k_2	MPa/mm	0.73	/	/	/
S_{su}	mm	4.01	8.48	7.16	6.33
τ_{su}	kPa	84.01	64.00	51.52	24.96
β_s	/	0.80	0.90	0.92	0.79

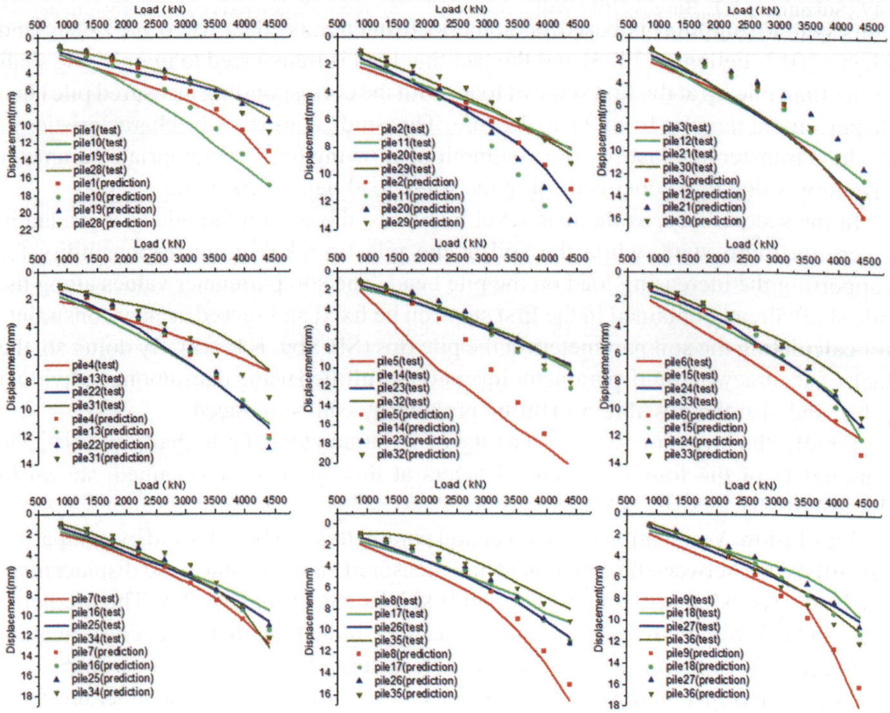

Fig. 2.3 Displacement prediction based on inversion analysis

situation of most piles, rather than the local optimal solutions of single pile. Secondly, the discrete nature of the same soil layer in the site cannot be completely considered. Thirdly, the static load test data of the pipe pile is related to many other different parameters, such as the supporting effect of the geotechnical body around the pile on the foundation pile (reflected in different pile-soil interaction models, resting time, groundwater level), and construction conditions (e.g. different pile residual stress distribution caused by construction). While the calculated values only take into account the differences in the main factor of the thickness of each soil layer.

2.3.3 Forward Prediction

The model is capable of predicting the long-term settlement of pile foundations, and the calculated results align with the measured results shown in Fig. 2.4. After eliminating slightly deviated values, the predicted values are incorporated into the load transfer equation to calculate the pile head displacement under greater loads, as depicted in Fig. 2.5. This paper employs a standard of 40mm pile head displacement in predicting the ultimate bearing capacity of foundation piles. The predicted Q–s curve

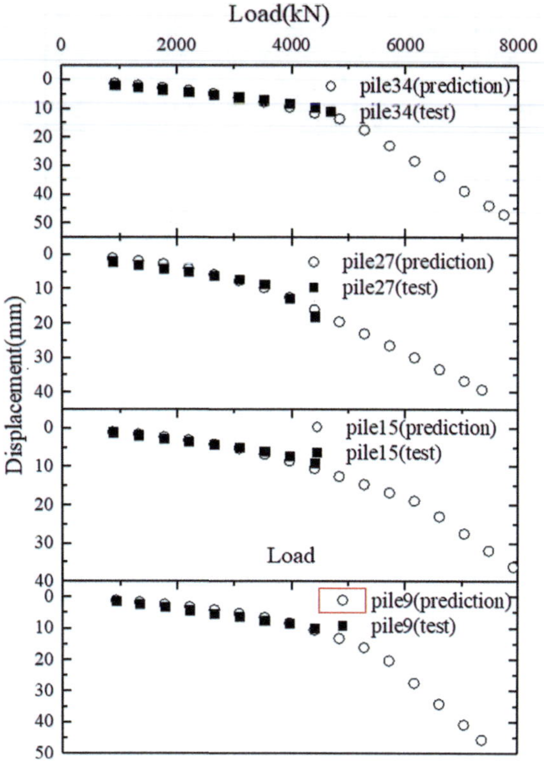

Fig. 2.4 Some forward prediction results

for a maximum load of 8800 kN is presented without any instances of displacement values at a certain load level exceeding five times that of the previous level. However, some of the maximum displacement values have exceeded 40 mm. Hence, it is possible to infer the ultimate bearing capacity of piles at the site.

Notably, the strain softening model of the surrounding soil and pile interaction mainly accounts for the changes in skin friction under a small pile head load. The soil at the pile tip plays a greater role with the increasing load on the pile head. If the pile-tip soil displacement Sb surpass the ultimate displacement Sbu when k1 and k2 are similar, it is indicative of not evident folded section of the Q–s curve. Such piles exhibit a less pronounced pile-tip softening effect, as characterized by the nearly linear Q–s curves recorded for pile 27# in forward modelling. In Fig. 2.5, we present a statistical analysis of its relationship with the total load-holding duration, maximum displacement at the pile head, pile length, and resting period.

The results show that the predicted ultimate bearing capacity (corresponding to a 40mm pile head displacement) ranges from 5874 to 9047 kN with an average of 7598 kN, which is less than twice the designed ultimate bearing capacity of 4400 kN. This value is found to have a negative correlation with the total load-holding duration and the maximum displacement value on the pile head in the static load test, but not with the pile length and resting period. The longer load-holding duration in the static

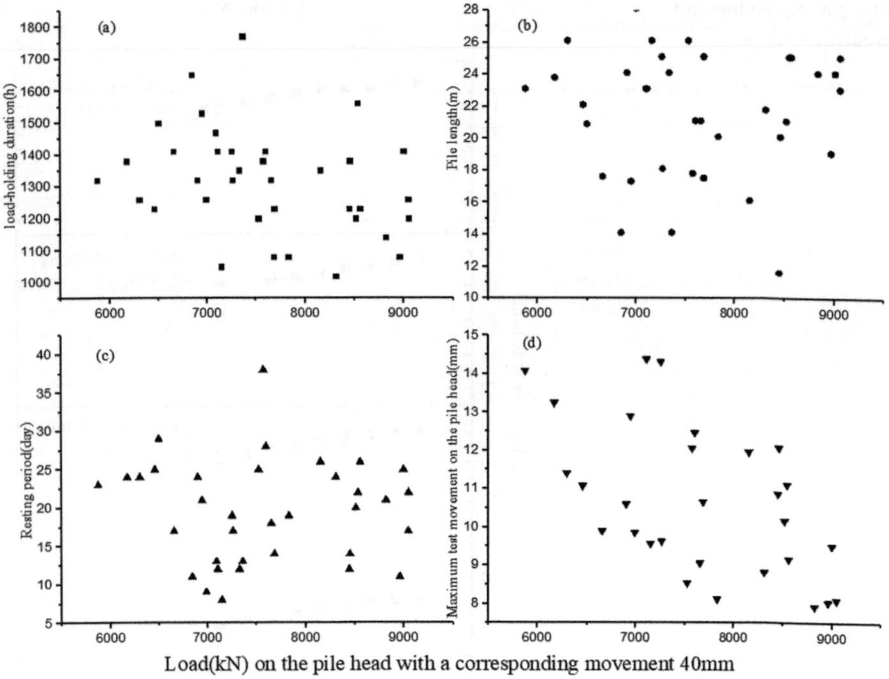

Fig. 2.5 Relationships between predicted ultimate bearing capacity and other factors

load test implies that the force of soil around the pile is weaker or develops more gradually during the early stage of pile-soil contact. While the pile-soil interaction is intensified and attains a stable state later, thereby resulting in a similar negative correlation with the maximum displacement on the pile head.

The pile length is an essential factor to study the stratification characteristics of the soil around the pile under similar conditions at the same site. However, the resting period does not significantly affect the predicted ultimate bearing capacity of the foundation pile as the pile strength and resting period are not closely interrelated for prefabricated piles. Or the geotechnical parameters around the foundation pile at the site during the static load test vary minimally over time.

2.4 Discussion

The selection of appropriate inversion parameters is crucial for accurate analysis of soil parameters surrounding the pile and identification of factors influencing ultimate bearing capacity.

Firstly, multi-layer soils inversion with diverse parameters can reflect the actual site conditions and account for interpreting differences in static load test data. To secure sufficient inversion parameters and constraints, 16 inversion parameters were considered, with 3 friction resistance parameters for each of the 4 soil layers, along with 4 parameters of one pile tip soil. By using the static load data from the entire site, better constraint shortcomings can be achieved from the global optimization perspective, though accurate fitting of each static load data may not be accomplished. Nonetheless, results of this inversion method showed that most of the calculated and measured Q–s curves agreed closely, with even errors between the remaining data. Although there are no specific pile internal force or displacement test results for comparison, the use of pile head displacements at other levels of loading (excluding the first and last levels) as validation is sufficient for most cases. This outcome can serve as the basis of next prediction step.

Secondly, the accuracy of the pile-soil interaction model is vital for reliable predictions of pile head displacement and friction parameters. The double-folded t–z model, based on existing engineering data and experience, has been found to yield consistent results. By optimizing the distribution of pile tip/shaft parameters within a specified range of values, a clearer range of values can be achieved through parameter inversion. Such optimization also helps detect statistical patterns of dispersion of geomechanical parameters located around the pile. While the curved pile-soil interaction model is potentially more accurate, the bifold model is preferred in this study due to simplicity and the avoidance of excessive increase of inversion parameters. The bifold model has been verified to meet the requirements of accurate predictions (Chen et al. 2022).

Thirdly, the influence of construction factors on bearing capacity of foundation piles is multifaceted. (1) The construction process and resting period impact the soils surrounding the pile, particularly in the case of extruded piles. There are deviations in the residual stress of each foundation pile construction (Dias 2018; Chen et al. 2022). (2) The non-uniformity of site geotechnical parameters can significantly affect the behavior of foundation piles. It can arise due to the differing modes of pile tip/side soil resistance. The actual distribution of the surrounding soils is typically selected based on survey holes near the foundation piles, which may lead to some degree of inaccuracy. (3) While the load magnitude and specific stability criteria for each level of load is consistent, the load-holding duration for each level of load may vary. (4) In some cases, multiple solutions may be present. For instance, the ultimate friction resistance range of each soil layer may overlap, the inversion results with abnormal ranking should be eliminated first. (5) To ensure accurate results, verification of data is crucial. This includes verifying the predicted ultimate bearing capacity of the foundation pile through the use of failed static load test data from the same site or internal force and pile tip displacement test results. In the absence of such verification, the predicted result can serve as a preliminary indicator for discerning the differences in ultimate bearing capacity between different foundation piles.

2.5 Conclusion

This paper proposes a solution to obtain the variability between the ultimate bearing capacity of each foundation pile with the same pile type at the same site.

(1) The proposed global optimization approach of multi-parameter inversion is well used in the typical granite area. And the bifold t–z model for fully weathered granite in Shenzhen can be used and the average first and second stage pile-tip stiffness k1, k2 can be taken as 1.8 and 0.7 MPa/mm respectively.
(2) 36 static load test data sets of the PHC-AB500(125) pile type at the same site were used for predicting ultimate bearing capacity of the foundation pile, which ranges from 1.3 to 2.1 times the ultimate bearing capacity of the pile body material.
(3) The variation in ultimate bearing capacity among piles of the same type and at the same site shows a negative correlation with two parameters: the maximum displacement value at the pile head and the actual load-holding duration observed in static load tests. Due to the limited quantity of available data, a precise analysis of the factors impacting the ultimate bearing capacity of piles of the same type at the same site is untenable. Nevertheless, utilizing the method of parameter inversion and forward prediction of ultimate bearing capacity of foundation pile can furnish additional data-driven support to AI algorithms for addressing intricate nonlinear problems.

Acknowledgements This work was supported in part by the National Key Research and Development Program of China under Grant 2019YFB2102703. The authors would like to thank Shenzhen Jianyan Testing Co., Ltd for acquiring 36 static loading test data and the reference data.

References

Armaghani DJ, Harandizadeh H, Momeni E et al (2022) An optimized system of GMDH-ANFIS predictive model by ICA for estimating pile bearing capacity. Artif Intell Rev 55:2313–2350

Benali A, Hachama M, Bounif A et al (2021) A TLBO-optimized artificial neural network for modeling axial capacity of pile foundations. Eng Comput 37(2):675–684

Bradshaw AS (2016) Pile load transfer from static load test inversion. DFI J J Deep Found Instit 10(1):34–40

Chen H, Li L, Li J, Sun D (2022) A rigorous elastoplastic load-transfer model for axially loaded pile installed in saturated modified Cam-clay soils. Acta Geotech 17:635–651

Dias T, Bezuijen A (2018) Load-transfer method for piles under axial loading and unloading. J Geotech Geoenviron Eng 144(1):04017096.1-04017096.9

Fellenius BH (2013) Simplified non-linear approach for single pile settlement analysis, discussion. Can Geotech J 50(6):685–687

Fellenius BH (2021) Results of an instrumented static loading test. Application to design and compilation of an international survey. J Deep Found Instit 15(1):71–87

Fellenius BH, Harris DE, Anderson DG (2004) Static loading test on a 45 m long pipe pile in Sandpoint, Idaho. Can Geotech J 41:613–628

Fellenius BH (2023) Basics of foundation design (electronic edition). www.Fellenius.net

Lee IM, Lee JH (1996) Prediction of pile bearing capacity using artificial neural networks. Comput Geotech 18(3):189–200

Moayedi H, Armaghani D et al (2018) Optimizing an ANN model with ICA for estimating bearing capacity of driven pile in cohesionless soil. Eng Comput 34:347–356

Moayedi H, Mosallanezhad M, Rashid ASA et al (2019) A systematic review and meta-analysis of artificial neural network application in geotechnical engineering: theory and applications. Neural Comput Appl 32:495–518

Momeni E, Nazir R, Armaghani DJ et al (2014) Prediction of pile bearing capacity using a hybrid genetic algorithm-based ANN. Measurement 57:122–131

Nguyen GD, Einav I, Zhang C (2013) The end-bearing capacity of piles penetrating into crushable soils. Geotechnique 63(5):341–354

Shahin MA (2016) State-of-the-art review of some artificial intelligence applications in pile foundations. Geosci Front 7:33–44

Teh CI, Wong KS, Goh A et al (1997) Prediction of pile capacity using neural networks. J Comput Civil Eng 11(2):129–138

Zhang QQ, Zhang ZM (2012) A simplified nonlinear approach for single pile settlement analysis. Can Geotech J 49(11):1256–1266

Zhang QQ, Zhang ZM, He JY (2010) A simplified approach for settlement analysis of single pile and pile groups considering interaction between identical piles in multilayered soils. Comput Geotech 37(7–8):969–976

Zhang Z-M, Zhang Q-Q, Yu F (2011) A destructive field study on the behavior of piles under tension and compression. J Zhejiang Univ Sci A Appl Phys Eng 12(4):291–300

Zhang QQ, Li SC, Li LP, Chen YJ (2013) Simplified method for settlement prediction of pile groups considering skin friction softing and end resistance hardening. Chin J Rock Mech Eng 32(03):615–624

Zhou WJ, Wang LZ, Guo Z, Liu JW, Rui SJ (2019) A novel t-z model to predict the pile responses under axial cyclic loadings. Comput Geotech 112:120–134

Chapter 3
Sand-Well Immersion Test on Deep-Seated Collapsible Loess in Shenhe Loess Plateau

Xiangyang Hu, Dengli Pan, Cheng Zhao, Bifu Sun, Hui Wu, Shengjun Shao, and Zhibin Hou

Abstract To study the collapsibility of deep-seated collapsible loess strata in the Shenhe plateau, a 64-day sand-well immersion test was carried out. This test assessed the water transmission law of deep-seated collapsible loess, self-weight collapsible deformation of strata at different depths, and the earth surface settlement deformation. The test results indicate that the maximum collapsible deformation of shallow marked points in the site is 6.2 mm. Accordingly, the site is judged to be a non-self-weight collapsible site, which was consistent with the evaluation results on large-scale trial pit immersion tests in adjacent sites, verifying that the results of the sand-well immersion test have high reference value for collapsibility evaluation. The suggested value for correction coefficient $\beta_0\prime$ of the calculated value for self-weight collapse deformation of loess strata in Shenhe plateau is 0.16, and the research results can provide a reference for engineering design and construction in the loess area of Shenhe plateau.

Keywords Sand-well immersion test · Deep-seated loess · Collapsibility evaluation · Shenhe loess plateau

X. Hu · D. Pan (✉) · C. Zhao · H. Wu · Z. Hou
Powerchina Northwest Engineering Corporation Limited, Xi'an, China
e-mail: pandengli@nwh.cn

X. Hu
e-mail: 0309@nwh.cn

B. Sun · S. Shao
Institute of Geotechnical Engineering, Xi'an University of Technology, Xi'an, China
e-mail: sjshao@xaut.edu.cn

S. Wang et al. (eds.), *Engineering Geology for a Habitable Earth: IAEG XIV Congress 2023 Proceedings, Chengdu, China*, Environmental Science and Engineering,
https://doi.org/10.1007/978-981-99-9057-3_3

3.1 Introduction

Collapsible loess is the Quaternary sediment. Soil strength decreases after soaking or saturation by water due to the combined action of soil mass's self-weight stress or additional external stress. The structure is rapidly damaged, and significant settlement occurs (Qian 1985; Xu et al. 2007; Feng et al. 2015). Accurate understanding and evaluation of loess collapsibility are critical in engineering loess areas because they are the basis for designing foundations on loess soils.

There are mainly three methods of studying loess collapsibility: laboratory compression test, field immersion test on trial pits, and sand-well immersion test. The layerwise summation method is often adopted in the laboratory compression test. The values of the self-weight collapsibility and collapsibility of the site are calculated and corrected from the measured self-weight collapsibility and collapsibility coefficients. The site's collapsibility type and the foundation's collapsibility grade can be determined based on the self-weight collapsibility and collapsibility. This method is widely applied, guiding numerous engineering construction projects. Through extensive in-situ trial pit immersion tests, the measured value for the self-weight collapsible deformation of loess at a site and the self-weighted collapsibility depth can be directly obtained (Ma et al. 2014). The method of sand-well immersion test (Shao et al. 2014) is a new method proposed by Shao et al. Multiple sand-well immersion tests have been carried out in Yang Village in the Shaoling plateau, Yuedengge Village in the Guanzhong area, Fanjiawo Village in Lanzhou (Shao et al. 2016; Wang et al. 2018; Li et al. 2019). In combination with the evaluation results from three methods, i.e., laboratory compression test, trial pit immersion test, and sand-well immersion test, the results of the sand-well tests on these three sites were consistent with those of the in-situ trial pit immersion test, which verifies the feasibility of sand-well immersion test in evaluating collapsibility. In some cases, the collapsibility evaluation results obtained through the first two methods may be inconsistent. For example, the collapsibility type of the test site in Yang Village, Chang'an District of Xi'an in Shaanxi, China (Xi'an Subway Limited Liability Company 2013) is inconsistent according to the results of the laboratory compression test and the field trial pit immersion test. Shao et al. (2015) analyzed the differences between the two methods. The variations were mainly caused by the differences in stress, immersion, loess collapsibility, stratum structure distribution, loess structure, and other factors.

Considering that the evaluation results from traditional laboratory tests are inaccurate and the field trial pit immersion tests require a high cost, long cycle, and are challenging, an in-situ sand-well immersion test was carried out at the site of Changning base in Shenhe plateau. This test evaluated loess collapsibility in key construction projects in Xi'an, Shaanxi Province of China—the Changning base construction project by Northwest Hydropower and New Energy Technology Industry Center. Surface and subsurface settlement, water transmission, and the development process of cracks around the sand well in the immersion process were monitored. The results from the sand-well immersion test, the trial pit immersion test of adjacent sites, and the corresponding laboratory test were compared and analyzed to comprehensively

Table 3.1 Stratum lithology of the test site

Stratum	Depth/m	Lithology
① Q_4^{ml}	0–1	Fill
② Q_3^{eol}	1–9.5	Loess
③ Q_3^{el}	9.5–12.6	Paleosol
④ Q_2^{eol}	12.6–22.8	Loess
⑤ Q_2^{el}	22.8–25.0	Paleosol

determine collapsible types of the site. The test has practical guiding significance and application value for the construction of Changning base and even the future investigation and design of loess foundations in the Shenhe plateau area. Furthermore, it verifies the feasibility and practicability of the sand-well immersion test method for collapsibility evaluation.

3.2 Test Overview

3.2.1 *Site Conditions*

The site for this sand-well immersion test is located in the Chang'an District of Xi'an, Shaanxi, China, which is adjacent to Chengnan Avenue in the east, Changxiang Street in the west, Shenhe Fifth Road in the north, and Shenhe Fourth Road in the south. The ground elevation of the site is about 473 m, the overall terrain is relatively smooth, and the geomorphic unit belongs to the first-level loess plateau (Shenhe plateau). The geotechnical investigation indicated that the groundwater depth is about 30–33 m, and the stratum lithology within a depth of 25 m in the test site is shown in Table 3.1.

3.2.2 *Test Scheme*

Figure 3.1 shows the layout of settlement and moisture monitoring points of sand-well immersion tests.

A total of 14 marked points for settlement observation were laid out in the sand well: seven deep marked points, six shallow marked points, and one central marked point. The deep marked points were numbered as S1–S7, of which the depths were 3, 7, 10, 13, 15, 21, and 28 m, respectively. They were evenly distributed on a circle with a radius of 1.5 m, centered at the center of the sand well. These were used to measure the collapsible deformation of soil layers with different depths. The shallow marked points were numbered as Q1–Q6, with burial depths of 0.5 m. They were used to measure the surface settlement in the radial direction of the sand well. There was

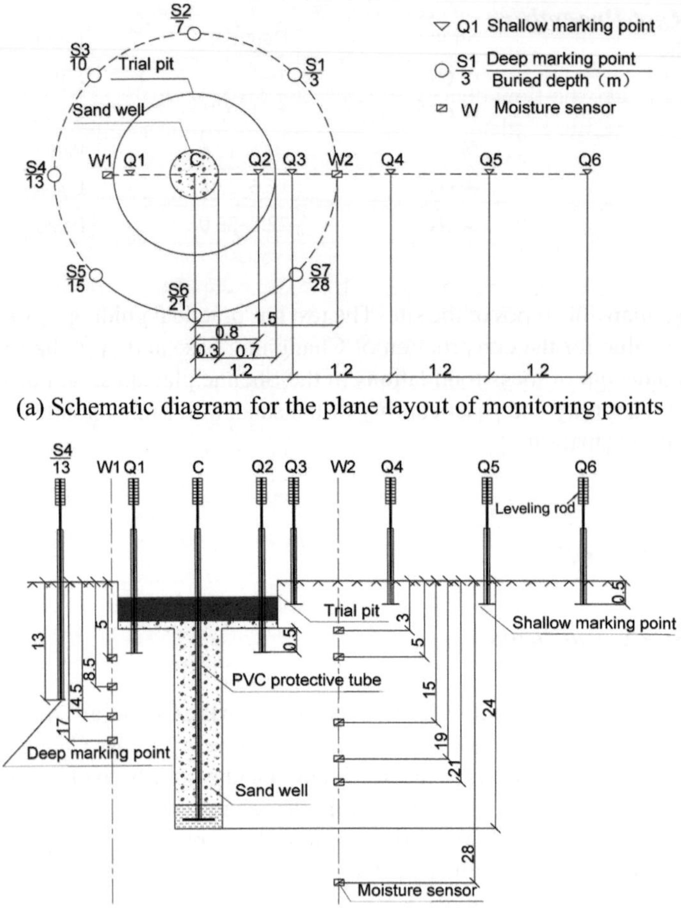

(a) Schematic diagram for the plane layout of monitoring points

(b) Schematic diagram for the section layout of monitoring points

Fig. 3.1 Layout diagram for monitoring points of sand-well immersion test (unit: m)

one central marked point, numbered C, which was used to monitor the deformation law of the strata below the well bottom.

Two groups of soil moisture sensors were laid out in the sand well, 10 in total. They are denoted as W1 and W2, as shown in Fig. 3.1. The sensors in the W1 group were located at the edge of the trial pit, which was 1 m away from the center of the sand well, and the burial depths of the moisture sensors were 5, 8.5, 14.5, and 17 m, respectively. W2 group contained six moisture sensors distributed in the vertical direction 1.5 m away from the well center. The moisture sensors in the W2 group and the deep marked points were distributed on the same circumference to better reflect the relationship between the change of moisture content and collapsible deformation. Their burial depths were 3, 5, 15, 19, 21, and 28 m, respectively.

3.2.3 Test Procedure

Figure 3.2 shows the flow chart for the whole process of the sand-well immersion test, and the specific implementation steps are shown in the references (Shao et al. 2016; Wang et al. 2018). An electronic level was used to observe the leveling rod to read the settlement data in this test. This method has high efficiency, fast reading, and high precision, with no error caused by artificial reading.

In the early stage of the test, water supplied by a pipe plays a dominant role, supplemented by a water truck. Due to the greater depth of the sand well, this intermittent mode cannot ensure a stable water head in the trial pit. In the middle and later stages, the water supply is implemented by a water hose, which can maintain the constant water level in the trial pit of the sand well. Liu (1985) showed through the intermittent immersion test that under a given pressure, for a particular specimen, its total collapsible deformation was a specific value, which was equal to the collapsible deformation for full immersion under this pressure but had nothing to do with the number of times of immersion. Therefore, even if it is in the mode of intermittent water supply, as long as the test soil layer is guaranteed to be saturated, and the collapsibility of the soil layer can be fully developed.

The observation lasted 64 days from May 16 to July 18, 2021, including three days for initial value observation before water injecting, 51 days for injecting water in the trial pit, and ten days for observation after the water supply was cut off. Water supply started on May 19, 2021, and was cut off on July 8, 2021. The total volume of water injection was 8027 m^3. All settlement monitoring points and moisture sensors were observed at a fixed time every day by fixed personnel and instruments except for when the weather did not allow for observations, as heavy rain and strong wind.

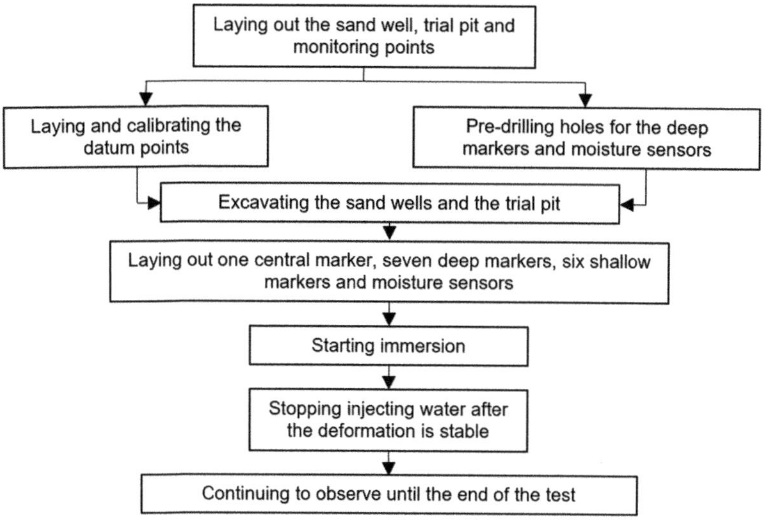

Fig. 3.2 Flowchart for the sand-well immersion testing process

The well was then measured once every 2 days in the later stage of the test. The observed data were counted and calculated, and the variation curves were drawn.

3.3 Analysis of Test Results

3.3.1 Analysis of Variation of Moisture Content in Strata

The data collected by the moisture sensors used in the sand well immersion test were the volumetric moisture content of the soil, and the changing curves for moisture content by each moisture sensor were drawn. Figure 3.3 shows the variation of moisture sensors in group W1 with time at 1 m away from the center of the sand well. In terms of the time from the initial monitoring to the abrupt change of water content, the change of moisture content was first monitored by moisture sensors at 17 and 14.5 m on the second day of immersion. Then, the moisture sensor at 5 m changed on the sixth day, while the sensor at a burial depth of 8 m was always in the process of slowly increasing in the early stage of immersion. The abrupt change in the reading on the moisture sensors indicated that the forward line for immersion reached the corresponding position on the moisture sensor. Thus, the deep-seated moisture content sensor detected the moisture changing earlier than the upper sensor, indicating that the forward line for immersion first reached the deep soil layer and then extended upward from the deep layer to the shallow layer. This was because the core of the sand well immersion test was to directly lead the water into the well bottom through the gap between the gravels in the well. The stratum below the well bottom was the first to be saturated by water. With the increase in the water level, the burial depth of the stratum affected by moisture decreased continuously until the moisture content in the shallow soil layer changed.

Figure 3.3 shows that the moisture content at 8.5, 14.5, and 17 m had a second abrupt change on the 41st day. This was because when water was supplied through the water hose, the water head in the trial pit changed from an intermittent water head in the early stage to a stable water head with a height of 80 cm, as shown in Fig. 3.2. This indicates that the water head pressure had a more significant influence on water diffusion, and a larger water head pressure could make the soil saturated more completely.

In Fig. 3.3, the change law of soil moisture content at 17 m is significantly different from that at the other three locations. First, it was sensitive to immersion and water supply cut-off: the moisture content on the first day of immersion and the first day of water supply cut-off changed abruptly. The variation difference reached over 50%; second, the moisture content during the water immersion period was maintained above 78%, far higher than the other three locations. The main reason for this case was that when the moisture sensor was laid, the plain backfilling soil was not dense, and large pores developed in the soil; free water could quickly fill the pores during water immersion and rapidly dissipate when the water supply was stopped.

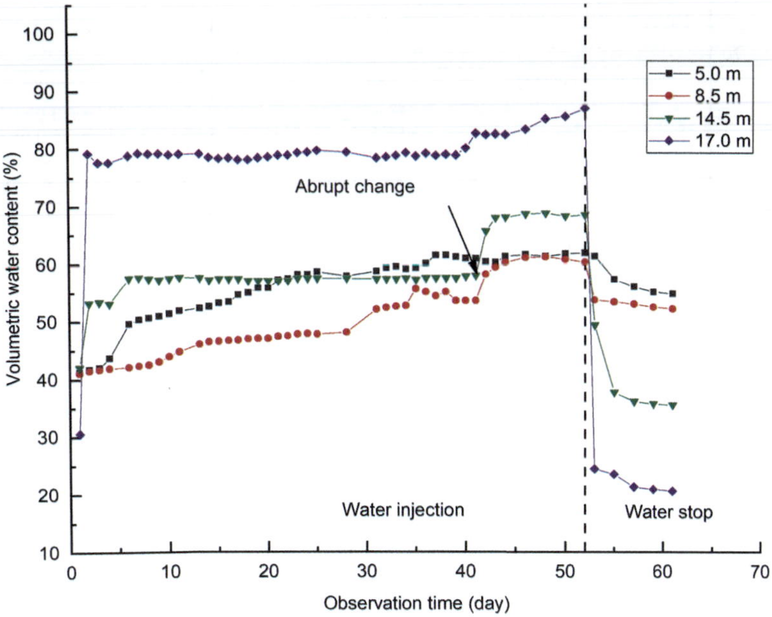

Fig. 3.3 Variation curves for moisture sensors in the W1 group

The moisture sensors in the W2 group were laid in the vertical direction 1.5 m away from the center of the sand well. Figure 3.4 shows the change of the volumetric moisture content in the soil layer at this location with time. As shown in Fig. 3.4, the change law on moisture content in the W2 group was similar to that in the W1 group. The deep-seated moisture content sensors detected the change of moisture content earlier. On the second day after immersion started, moisture content sensors buried at 28, 21, 19, and 15 m detected moisture changes, and afterward, the volumetric moisture content in soil increased rapidly to a saturation state. The soil layer at 28 m reached a state close to saturation first on the second day of immersion, and those at 21, 19, and 15 m reached the saturation state on the third, fourth and fifth day, respectively. After reaching the saturation state, the soil layers at 19 and 21 m kept fluctuating near the saturation line. This was because the water in the early stage of the test was transported by water pipe and water truck, and the intermittent water supply failed to continuously maintain the water head in the trial pit. After 41 days, under the action of a continuously stable high water head, the moisture content in the stratum was stable until the water supply was cut off. Compared with the four deep-seated moisture monitoring sensors, the one at a depth of 5 m did not show a rapid rising stage after monitoring moisture variation but developed slowly, indicating that it was near the saturation line.

The moisture sensor at 3 m was an exception. It rose slightly in the early stage of immersion but declined continuously in the middle and late stages of immersion. This was because the burial depth of the moisture sensor was relatively shallow and

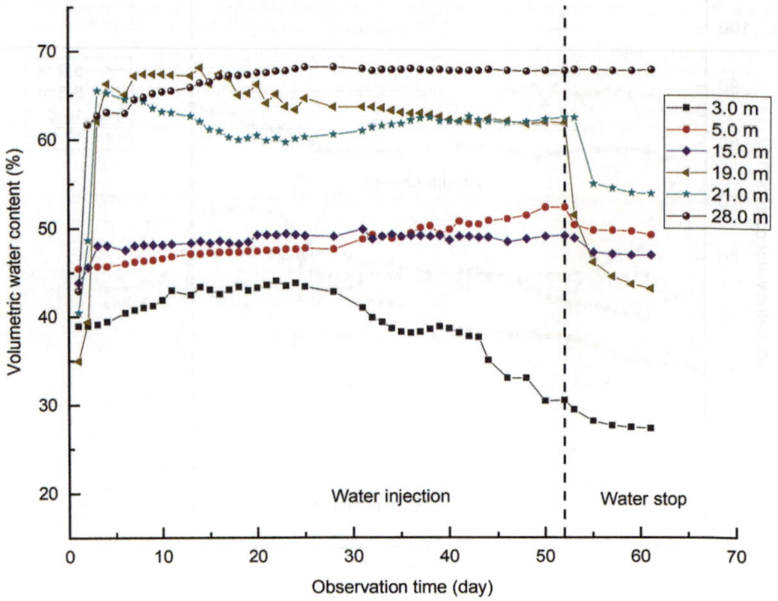

Fig. 3.4 Variation curves for moisture sensors in the W2 group

located above the saturation line, with the moisture evaporation capacity and supply capacity achieving equilibrium. Therefore, the reading in the early stage remained unchanged. The shallow soil mass showed a larger void ratio, the soil was loose, and voids were connected to the atmosphere. In addition, the later stage was within the hot summer season with higher moisture evaporation, so the reading in the later stage kept decreasing.

The above analysis shows that the readings on moisture sensors and variation laws are related to multiple factors, such as burial depth, density of backfill soil, water head pressure, distance from the sand well center, and others. However, the change in the readings on moisture sensors in groups W1 and W2 showed similar variations: when the forward line of immersion reached the burial location of moisture sensors, the measured value on the moisture sensors would show a sudden increase, and then show smaller fluctuations near the saturation line after reaching the peak; after the water supply was stopped, the moisture sensors would decline at different degrees and then maintained at a relatively stable value.

3.3.2 Analysis of Collapsible Deformation

Analysis of Settlement at the Central Marked Point. According to the monitoring data, the variation curves were drawn for cumulative collapsibility, daily collapsibility, and daily water consumption at marked point C of settlement in the center of

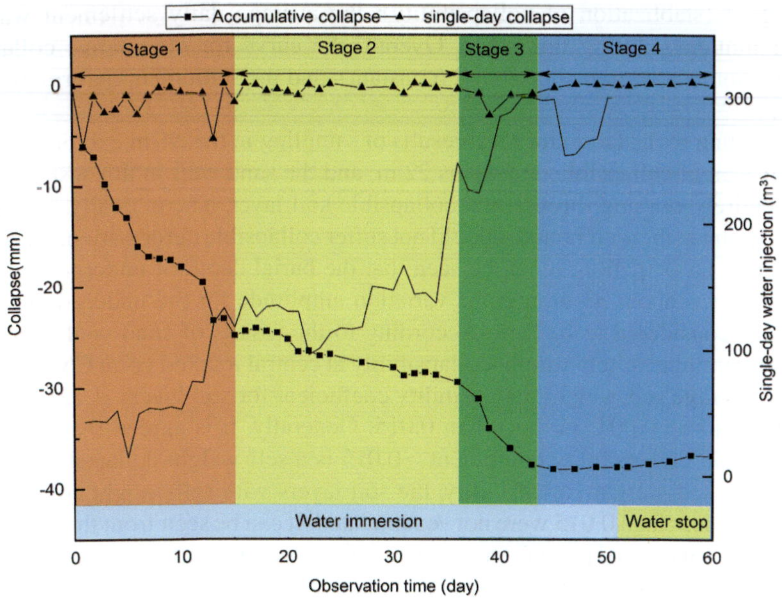

Fig. 3.5 Curves for collapsibility at the central marked point of sand well and daily water consumption

sand well with time. As shown in Fig. 3.5, the cumulative self-weight collapsibility of the soil layer below the bottom of the sand well was 38 mm.

The settlement process was divided into four stages according to the settlement rate of the central marked point:

Stage 1 (severe collapsibility): at the beginning of immersion, water directly flowed into the bottom of the sand well through pores between the gravels, and the marked point for settlement changed rapidly. With the continuous rise in water level in the sand well, the large pores in soil mass below the well bottom were constantly filled with water and then suffered severe collapses. The maximum collapsibility in a single day was 6.1 mm. Although the daily water injection volume was not significant, the curve at this stage was relatively steep, the collapse rate was fast, and the daily settlement volume was large.

Stage 2 (slow collapsibility): after severe collapsible deformation, the mesopores and small pores in soil were constantly filled with water with an increase in daily water consumption. The soil mass was further saturated, the collapsible deformation developed slowly, and the daily settlement volume was smaller compared to Stage 1.

Stage 3 (collapsibility with pressure): the daily water consumption continued to increase. The water level in the trial pit continued to rise, and the effect of water head pressure strengthened. The depth range of water in the well diffusing to the strata below the well bottom increased, and the collapsible settlement continued to develop. Overall, the cumulative collapsibility increased.

Stage 4 (stablization of collapsibility): the average daily settlement was less than 1 mm/days during this stage. Overall, the curve for cumulative collapsible deformation tended to be horizontal (constant), and the settlement deformation was stable.

According to the laboratory test results of sampling in the 24-m exploratory well, the self-weight collapsible depth was 22 m, and the sand well in this test was 24 m deep, entirely running through the collapsible soil layer. Theoretically, the central marked point at the well bottom should not suffer collapsible deformation. According to the field investigation, it can be seen that the burial depth of underground water at the site was about 33 m, and the variation amplitude for the underground water level was considered to be 2 m. According to the results of sand-well immersion tests, the cumulative self-weight collapsibility at central marked point C was 38 mm, and the average self-weight collapsibility coefficient for soil layers at 22–31 m was calculated to be 0.004, far less than 0.015. Generally, it is agreed that loess with a self-weight collapsibility coefficient $\geq$0.015 is a self-weight collapsible loess. In calculating self-weight collapsibility, the soil layers with self-weight collapsibility coefficients less than 0.015 were not accumulated. It can be seen from the collapsible deformation of marked point C that even though the average self-weight collapsibility coefficient of the strata was less than 0.015, under the saturated self-weight pressure, the measured collapsibility volume of the strata still cannot be ignored.

Analysis of Settlement at Shallow Marked Points. A self-weight collapsible loess site will suffer settlement under the influence of immersion. The settlement amount on the top surface of each soil layer reflects the total self-weight settlement of the collapsible soil layer in the lower part of the layer. Therefore, the surface settlement is the largest, and it has the most significant influence on the building structures.

During the test, no cracks caused by collapsibility were seen around the trial pit. Figure 3.6 is the photo of the test on the 45th day after immersion. According to the measured settlement data, the curves for cumulative deformation of shallow marked points for monitoring ground settlement with time were drawn, as shown in Fig. 3.7 (Q3 marked point was not successfully completed, so it was not taken into account). Figure 3.8 shows the curve for the single-day deformation of each shallow marked point. Figure 3.9 is accumulative subsidence curve for ground deformation after one and 60 days of immersion.

It can be seen from Fig. 3.7 that the changes of shallow marked points of Q2, Q4 and Q6 were very small in the immersion stage, and the stage of water supply cut-off, and the final settlement was less than 1 mm by the time when observations were stopped. Compared with the settlement variations for shallow marked points Q2, Q4 and Q6, those for shallow marked points Q1 and Q5 were larger, being 3.5 and 6.2 mm, respectively. Figure 1a shows marked points Q1 and Q2 were very close to each other, both located in the trial pit and at the same distance from the center of the sand well. However, there was a tremendous difference between the two in collapsibility after immersion, reflecting the existence of uneven settlement in the loess site. Marked point Q5 was 3.6 m away from the center of the sand well, with the maximum collapsibility, indicating that the radial influence range of the sand-well

Fig. 3.6 The photo of the site for sand-well immersion test

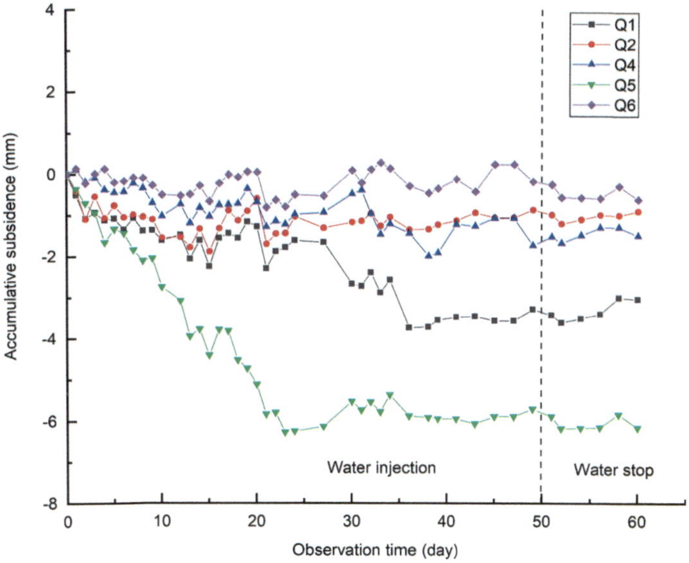

Fig. 3.7 Variation curve for settlement at shallow marked points with time

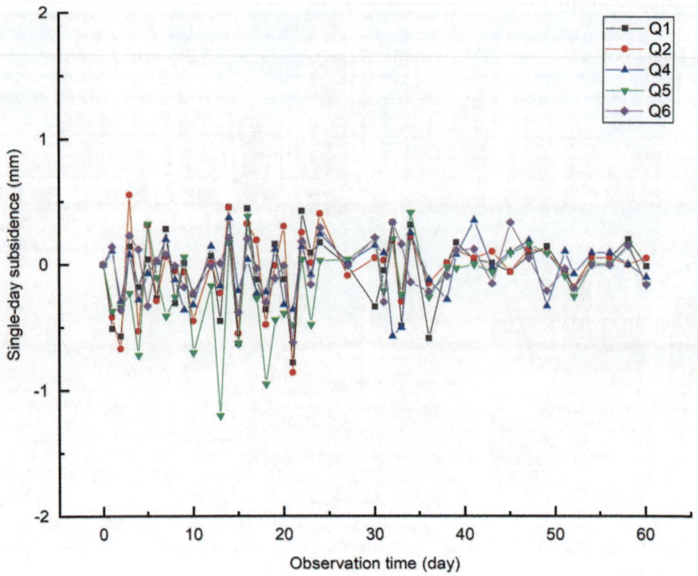

Fig. 3.8 Curve for single-day deformation of shallow marked points with time

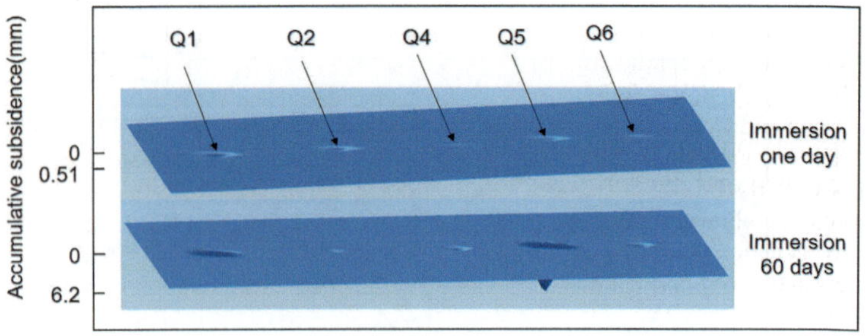

Fig. 3.9 Accumulative surface subsidence curve of different immersion days

immersion (from the center of the sand well) was at least 3.6 m. The deformation of each marked point in a single day was mainly concentrated in the range of 1 mm, with a significantly smaller variation amplitude.

Analysis of Settlement at Deep Marked Points. Deep settlement reflects the collapsibility of soil layers at various depths below the ground. It is necessary to evaluate the depth of self-weight collapsibility and embedment depth of foundation and select foundation design parameters. In this test, a total of seven deep marked points were laid out to measure the self-weight collapsibility at different depths, which lasted for 60 days and were observed 43 times in total. Figure 3.10 shows the variation curve for collapsible deformation of deep marked points with time in the immersion process.

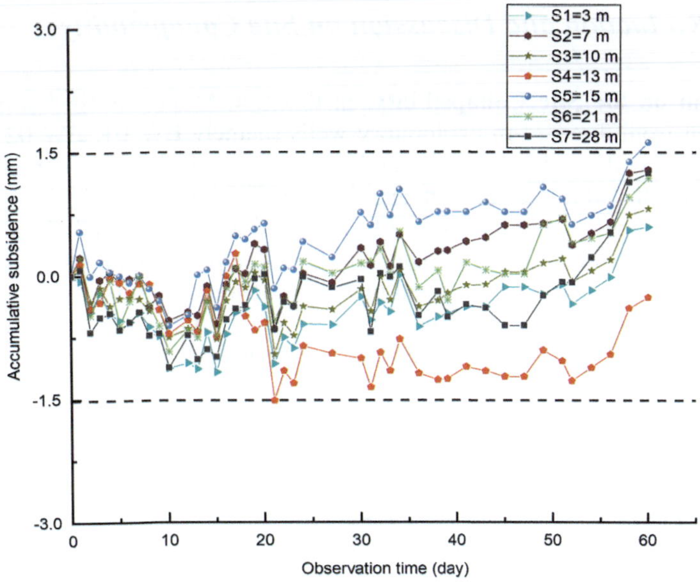

Fig. 3.10 Deformation curves for deep marked points

The cumulative collapsibility of all deep marked points was very small throughout the entire test process. They generally increased, except for S4, and the deformation fluctuated between −1.5 and 1.5 mm. The burial depth of S4 was 13 m. According to the previous data on exploration well, 12.7–22.8 m consists of about 10 m of Middle Pleistocene loess, and the soil of this layer has greater collapsibility than that of any other layers. The deep marked point of S4 showed slight settlement deformation.

According to the preliminary investigation data on the exploratory well, between depths 9.7–12.7 m and 22.8–24.8 m in the test site consisted of paleosol layers. They showed high strength and weak expansivity, and they can be explained by rock bridge theory. The loess layers have better permeability than paleosol layers, so during water infiltration, the loess layer first reached a saturation state and deformed. In contrast, the paleosol layer had not deformed yet and had greater mechanical strength and weaker water penetration than the loess layer, providing support for the upper soil layers. In addition, the paleosol layer had weaker expansibility, and it could expand slightly when its immersion reached a saturation state. Therefore, the characteristics of deep marked points rising instead of falling may be related to the paleosol with a greater thickness.

3.3.3 Evaluation and Discussion on Site Collapsibility

Evaluation on the Site Collapsibility. In this test, 56 undisturbed soil samples were taken from three 24-m exploratory wells (namely EW 01, EW 02, EW 03) for laboratory collapsibility tests. It can be seen from the variation curves between self-weight collapsibility coefficient and depth in Fig. 3.11 that only the self-weight collapsibility coefficient of Q2 loess of layer ③ was greater than 0.015, and the self-weight collapsible coefficients of other soil layers were less than 0.015 (except for individual data points). This indicated that only the middle-Pleistocene Q2 loess of layer ③ in the strata of this section was self-weight collapsible loess, which was mainly distributed in the depth range of 12.6–22.8 m.

Laboratory tests showed that the maximum burial depth of the collapsible soil layer was 22 m. The soil layer with the self-weight collapsibility coefficient $\delta_{zs} > 0.015$ was the loess of the ③ layer, and the maximum self-weight collapsibility coefficient reached 0.045. According to the current loess standard, the calculated value for self-weight collapsibility was counted from the ground, 176–255 mm, according to the regional correction coefficient $\beta_0 = 0.9$ recommended by the current loess standard. Therefore, the site was judged as a self-weight collapsible site.

In the sand well immersion test, five shallow marked points were laid to measure the self-weight collapsibility of the site. It could be seen above that the variation of each shallow marked point was minimal in the immersion stage and the stage of

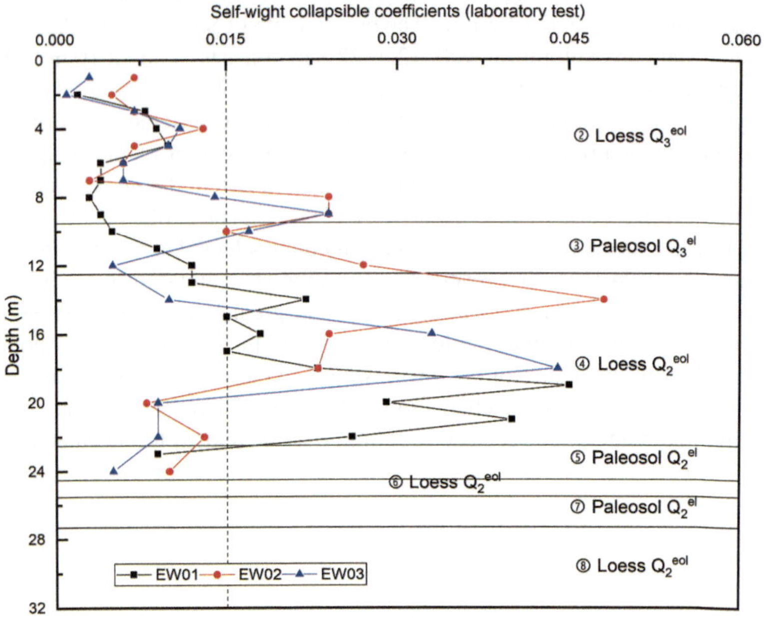

Fig. 3.11 Variation of stratum structure of the site and self-weight collapsibility coefficients with depth

stopping water injection. When observation stopped, the maximum settlement value was 6.2 mm, so the site was judged to be a non-self-weight collapsible site.

Evaluation on Collapsibility of the Adjacent Sites. By collecting the data on adjacent engineering sites, it was found that the field trial pit immersion tests were conducted both in the new campus of Xi'an University of Finance and Economics, only 2–3 km away from the test site, and the south extension section of Xi'an Metro Line 2 (China Jikan Institute of Engineering Investigations and Design 2004; Xu et al. 2021). The geomorphic units of these two sites both belong to the Shenhe plateau, with basically similar stratum structures.

According to the results of field sampling from exploratory wells and laboratory soil test, the average corrected calculation values for self-weight collapsibility of the two sites were 216 and 196 mm (β0 = 0.9), respectively, indicating that the two sites were self-weight collapsible loess sites.

The diameter of the trial pit for the field immersion test in Xi'an University of Finance and Economics was 30 m. The immersion time lasted 73 days, the deformation observation time lasted 88 days, and the cumulative water immersion was about 27,000 m^3. During the immersion saturation state period with a length of 73 days, no collapsible settlement phenomenon occurred. The trial pit and surrounding strata were lifted by several millimeters during the immersion process, with the maximum uplift of 9 mm. The test site in the south extension section of Xi'an Metro Line 2 was a circular trial pit with a diameter of 30 m and filled with water for 35 days. Observations lasted for 60 days after the water supply was cut off. The test results indicated that the maximum self-weight collapsibility of the soil layers in the trial pit was 32 mm. Therefore, the collapsibility type of these two sites could be determined as non-self-wight collapsible loess sites according to the field trial pit test results.

Comparison of Evaluation Methods. The laboratory soil test and trial pit immersion test have a certain application reference value for evaluating loess collapsibility in practical engineering. According to the laboratory test and sand-well immersion test results of this site and the evaluation results of the adjacent sites, the collapsibility evaluation methods were compared to verify the rationality and practicality of the method of sand-well immersion test. The comparison results are shown in Table 3.2. Referring to the current loess standard (GB 50025-2018 2018), the ratio $\beta_0\prime$ of the measured self-weight collapsibility and the calculated value for self-weight collapsibility (not multiplied by β_0) was calculated according to Formula (3.1), with the calculation results shown in Table 3.2.

$$\beta_0\prime = \frac{\Delta_{zs}\prime}{\sum_{i=1}^{n} \delta_{zsi} h_i} = \frac{\Delta_{zs}\prime \beta_0}{\Delta_{zs}} \tag{3.1}$$

where, $\beta_0\prime$ denotes the recommended value for correction coefficient in this region; $\Delta_{zs}\prime$ denotes the measured value for self-weight collapsibility; Δ_{ZS} denotes the calculated value for self-weight collapsibility. The correction coefficient β_0 for the Guanzhong area is taken as 0.9; δ_{zsi} denotes the self-weight collapsibility coefficient for the i-th layer of soil, and h_i denotes the thickness of the i-th layer of soil.

Table 3.2 Comparison of different evaluation methods

Test site	Changing base		The new campus of Xi'an university of finance and economics		The south extension section of Xi'an Metro Line 2	
Evaluation method	Laboratory test	Sand-well immersion test	Laboratory test	Trial pit immersion test	Laboratory test	Trial pit immersion test
Self-weight collapse Δ_{ZS}/mm	213	6	216	9*	196	31
Collapsibility type	Self-weight	Non-self-weight	Self-weight	Non-self-weight	Self-weight	Non-self-weight
Correction coefficient $\beta_0\prime$	/	0.03	/	0	/	0.16

Note * in the table denotes uplift

According to the test results of the three sites and their comparison results, the following conclusions can be drawn:

(1) The calculation results on self-weight collapsibility through laboratory tests in the three engineering sites were similar, all greater than 70 mm. The sites were all determined to be self-weight collapsible sites. The judgment given based on the sand-well immersion test in the Changning base was consistent with the field trial pit immersion test in the other two sites. The three sites were non-self-weight collapsible sites and verify that the sand-well test can reference collapsibility evaluation.
(2) The correction coefficient $\beta_0\prime$ calculated from the tests was far smaller than the correction coefficient of 0.9 recommended in the standard for the Guanzhong area. It indicated that the distribution of loess strata in the Guanzhong area was uneven. It is crucial to scientifically evaluate the impact of loess collapsibility on the safety and cost of the engineering. Considering the engineering safety, the correction coefficient $\beta_0\prime$ was taken as 0.16, the maximum among the values for the three sites.

3.4 Conclusion

In this paper, based on the Changning base project, sand-well immersion test was carried out in representative site, and the following conclusions were mainly drawn:

(1) In the sand-well immersion test, water was directly guided into the strata buried at the well bottom through the voids between the gravels in the well. This unique way of water immersion led to the case that deep-seated moisture sensors monitored moisture change earlier than shallow-seated sensors, showing that the forward line for immersion arrived in the deep soil layer first. With the water level continuously rising, the forward line decreased in depth.
(2) The soil layer within the influence range of immersion suffered collapsible deformation under the condition of sand-well immersion, and the deformation at the central marked point for settlement was the largest. The boundary constraint effect on the saturated gravels in the sand well was smaller than that on the soil columns in the saturated area around the sand well. Due to the influence of rock bridging of a paleosol and other factors on each deep and shallow marked point for settlement, no apparent collapsible settlement occurred, and the test site was determined to be a non-self-weight collapsible loess site.
(3) The rationality and practicability of the sand-well immersion test method were verified by comparing and analyzing the laboratory test results, sand-well immersion test, and trial pit immersion test on three adjacent sites. The results showed that the sand-well immersion test had a significant reference value for the evaluation of collapsibility.
(4) The correction coefficients obtained from the field tests in the three sites were smaller than the recommended value of 0.9 in the specification. Considering the uneven distribution of loess collapsibility and engineering safety in the Shenhe

plateau, it was advised that the correction coefficient $\beta_{0'}$ should be 0.16 in the Shenhe plateau area.

Acknowledgements This research was supported by the Changning base construction project of Northwest Hydropower and New Energy Technology Industry Center. We also would like to express our sincere gratitude to the two anonymous reviewers for their valuable comments, which have greatly improved this paper.

References

China Jikan Institute of Engineering Investigations and Design (2004) Report on soaking test of collapsible loess site of the first phase of the new campus of Xi'an Institute of Finance and Economics. China Jikan Institute of Engineering Investigations and Design, Xi'an

Feng SJ, Du FL, Shi ZM et al (2015) Field study on the reinforcement of collapsible loess using dynamic compaction. Eng Geol 185:105–115. https://doi.org/10.1016/j.enggeo.2014.12.006

GB 50025-2018 (2018) Code for building construction in collapsible loess regions

Li J, Shao SJ, She FT et al (2019) Application research of sand well immersion test in collapsibility evaluation of loess tunnel foundations. Chin J Rock Mech Eng 38:1937–1944

Liu MZ (1985) Collapsible loess intermittent water immersion test. Chin J Geotech Eng 98:47–54

Ma Y, Wang JD, Peng SJ et al (2014) Immersion tests on characteristics of deformation of self-weight collapsible loess under overburden pressure. Chin J Geotech Eng 36:537–546

Qian HJ (1985) Collapsible loess foundation. China Architecture and Architecture Press, Beijing

Shao SJ, Li J, Li GL et al (2015) Evaluation method for self-weight collapsible deformation of large thickness loess foundation. Chin J Geotech Eng 37:965–978

Shao SJ, Li J, Shao J et al (2016) In-situ sand well immersion tests on self-weight collapsible loess site with large depth. Chin J Geotech Eng 38:1549–1558

Shao SJ, Li J, Li GL et al (2014) Test method for sand soil immersion for collapsible deformation of loess

Wang YX, Shao SJ, Han CL, et al. (2018). Application of sand drain immersion tests on collapsible loess. Chin J Geotech Eng

Xi'an Subway Limited Liability Company (2013) Research results report on Xi'an subway engineering through the collapsible loess engineering characteristics and countermeasures. Xi'an Subway Limited Liability Company, Xi'an

Xu ZJ, Lin ZG, Zhang MS (2007) Loess in China and loess landslides. Chin J Rock Mech Eng 26:1297–1312. https://doi.org/10.3321/j.issn:1000-6915.2007.07.001

Xu Y, Wang J, Gu T et al (2021) Study on the loess immersion test of metro line 2 in Xi'an, Shaanxi Province, China. Adv Civil Eng 2021:1–14. https://doi.org/10.1155/2021/6654391

Chapter 4
Determining the Geotechnical Parameters of Rock Mass via Non-invasive Geophysical Methods

Muhammad Hasan and Yanjun Shang

Abstract Rock mass reliability classification is carried out via geotechnical parameters, such as rock mass integrity index (Kv), volume joint number of rock mass (JV), rock core index (RCI), average joint spacing (dP), modulus of elasticity (E) and rock quality designation (RQD) etc. Such geomechanical parameters, however, are traditionally obtained from the expensive drilling tests. Therefore, it is never an easy task to acquire engineering parameters in the frequent borehole tests. Furthermore, such tests cannot provide volumetric evaluation of subsurface geological units. So, an attempt to minimize lots of drilling tests and to correctly estimate geotechnical parameters is essential for appropriate designs of engineering infrastructures. We propose an empirical-based geophysical approach of non-invasive geophysical method, namely controlled-source audio-frequency magneto telluric (CSAMT). Useful equations are obtained by the integration of limited boreholes and geophysical data, which provides determination of geomechanical parameters for all geophysical measurements over the complete site where even no well is available. This approach offers a thorough insight into the subsurface geology by 2D and 3D mapping of engineering parameters, eradicates the subsurface ambiguities due to limited wells, and fills a gap between the perfect geotechnical models and the inadequate drilling tests.

Keywords Rock mass integrity index (Kv) · Geotechnical parameters · Geophysical methods · Rock mass quality

---

M. Hasan (✉)
State Key Laboratory of Mountain Hazards and Engineering Resilience, Institute of Mountain Hazards and Environment, Chinese Academy of Sciences, Chengdu 610299, China
e-mail: hasan.mjiinnww@gmail.com

China-Pakistan Joint Research Center on Earth Sciences, CAS-HEC, Islamabad, Pakistan

M. Hasan · Y. Shang
Key Laboratory of Shale Gas and Geoengineering (KLSGG), Institute of Geology and Geophysics, Chinese Academy of Sciences, Beijing 100029, China
e-mail: jun94@mail.iggcas.ac.cn

University of Chinese Academy of Sciences, Beijing 100049, China

S. Wang et al. (eds.), *Engineering Geology for a Habitable Earth: IAEG XIV Congress 2023 Proceedings, Chengdu, China*, Environmental Science and Engineering,
https://doi.org/10.1007/978-981-99-9057-3_4

4.1 Introduction

Accurate assessments of rock strength is highly challenging in engineering geological studies. The characteristics of a rock mass mainly depend on its internal structure (Bieniawski 1973; Han et al. 2020). The solutions of many underground engineering problems rely on rock mass quality evaluation which depends on rapid measurements, fine description and visual interpretation of rock mass structure (Barton et al. 1974; Masahiro et al. 1999; Zolfaghari et al. 2015). The detailed valuation of foundation rocks is crucial while undertaking major engineering projects (Hoek and Brown 1997; Wang et al. 2020a; Ji et al. 2019). The safe and cost-effective engineered structures rely on proper classification of subsurface rock units (Shin et al. 1999; Lu and Liu 2009; Li et al. 2017). Rock mass strength evaluation is a top priority during the development of an engineering infrastructure (kagi et al. 2001; Vásárhelyi and Ván 2006; Lowson and Bieniawski 2013). Determining the bearing capacity of foundation rock to prevent failure is, therefore, integral in the infrastructure design (Yuji et al. 2006; Rahmati et al. 2014; Pytel and Pałac-Walko 2015; Rehman et al. 2018).

At present, the rock mass strength is evaluated via many comprehensive rock mass indices, such as the rock mass integrity coefficient (Kv), the joint number volume of rock mass (JV), the rock mass quality designation (RQD), the modulus of elasticity (E), and average joint spacing (dP) (Palmstrom 2005; Salaamah et al. 2018; Wang et al. 2020b). Such engineering parameters are determined via the traditional geological/drilling techniques (Deere 1964). The conventional geological methods to evaluate the rock mass strength are complicated and cumbersome. In such methods, first the location of observation points is located; afterwards the measurements are performed one by one in a site with dire working environment. The traditional methods suffer inefficiency due to high labor intensity, weighty equipment, low data coverage, high risk, and steep topographic constraints. The traditional approaches provide point-scale measurements only vertically and hence may cause uncertainties in the evaluations of geological models. The site planners need the alternative methods which are more economical and can reliably assess the rock strength by decreasing lots of costly boreholes.

In recent decades, the surface geophysical methods, namely controlled source audio frequency magnetotelluric (CSAMT), gravity, electromagnetic (TDEM), magnetic, seismic and electrical resistivity tomography (ERT) have been commonly used in geotechnical studies (Wang et al. 2015; Loperte et al. 2016; Zhamaletdinov 2020; Zhang et al. 2021). Geophysical methods are economical and non-invasive, and are capable to ease not only the risk but also the redundant expenses by suggesting suitable places for the necessary drilling tests. Compared to the conventional boreholes techniques, geophysical investigation provides detailed imaging of the subsurface geology units via volumetric measurements. At present, CSAMT has become the most suitable and economical method in engineering geological investigations to acquire detailed information of the subsurface at large depths.

Several researchers, in the past, have used empirical relationships for geotechnical and geophysical properties (Singh and Rao 2005; Sharma and Singh 2008; Nourani et al. 2017; Olayanju et al. 2017; Salaamah et al. 2018). However, none of the past investigations has estimated geotechnical parameters (Kv etc.) from the geophysical methods (such as CSAMT). In this novel approach, we propose that the limited borehole tests are conducted only at the essential locations (Fig. 1a) to acquire geomechanical parameters. Then, more reliable geophysical surveys (CSAMT) are carried out to estimate geotechnical parameters for complete coverage of large area (Fig. 1a). This is the only contribution, which uses CSAMT to estimate rock mechanical parameters. Our innovative work provides new findings to evaluate rock mass quality and reduces a gap between limited data and perfect geotechnical models for smooth construction of large-scale engineering infrastructures.

4.2 Methods

4.2.1 CSAMT Survey

CSAMT is a hybrid electrical resistivity technique. In CSAMT, electrical resistivity is acquired via the electric/magnetic field resulting either from earth's natural fields or from unnatural electromagnetic waves (Wang et al. 2015). This method computes subsurface resistivity via magnetometers and electric dipoles (Loperte et al. 2016). The apparent/field resistivity is acquired when the magnetic and electric fields are concurrently observed at the target of the ground material (Zhamaletdinov 2020; Zhang et al. 2021). When the local power-down structure is fixed, the lower the electromagnetic wave frequency, the greater the penetration depth, which can reflect the deep geoelectric characteristics (Loperte et al. 2016; Zhamaletdinov 2020).

The CSAMT survey was performed along eleven profiles in JUNO site, Guangdong, China (Fig. 1a). The pole distance of 20 m was used for tensor measurements. About 8–12 times, the frequency bands were used in the survey. Resistivity data was acquired using distance spacing of 50 m. The magnetic and electric fields were observed via the magnetic probe and non-polarized electrode, respectively. Frequency ranges between 10 Hz and 92 kHz, and the depth (granite area, resistivity is 1000 Ωm, and frequency is calculated at 100 Hz) is about 1500 m. The azimuth deviation remains less than 3° for the magnetic probe and electrode compass reading. The geophysical measurements were acquired using four channels. The obtained resistivity in the field is apparent resistivity of materials but not true resistivity of the structures. Consequently, inversion procedure of apparent resistivity was performed to obtain true resistivity along each CSAMT profile.

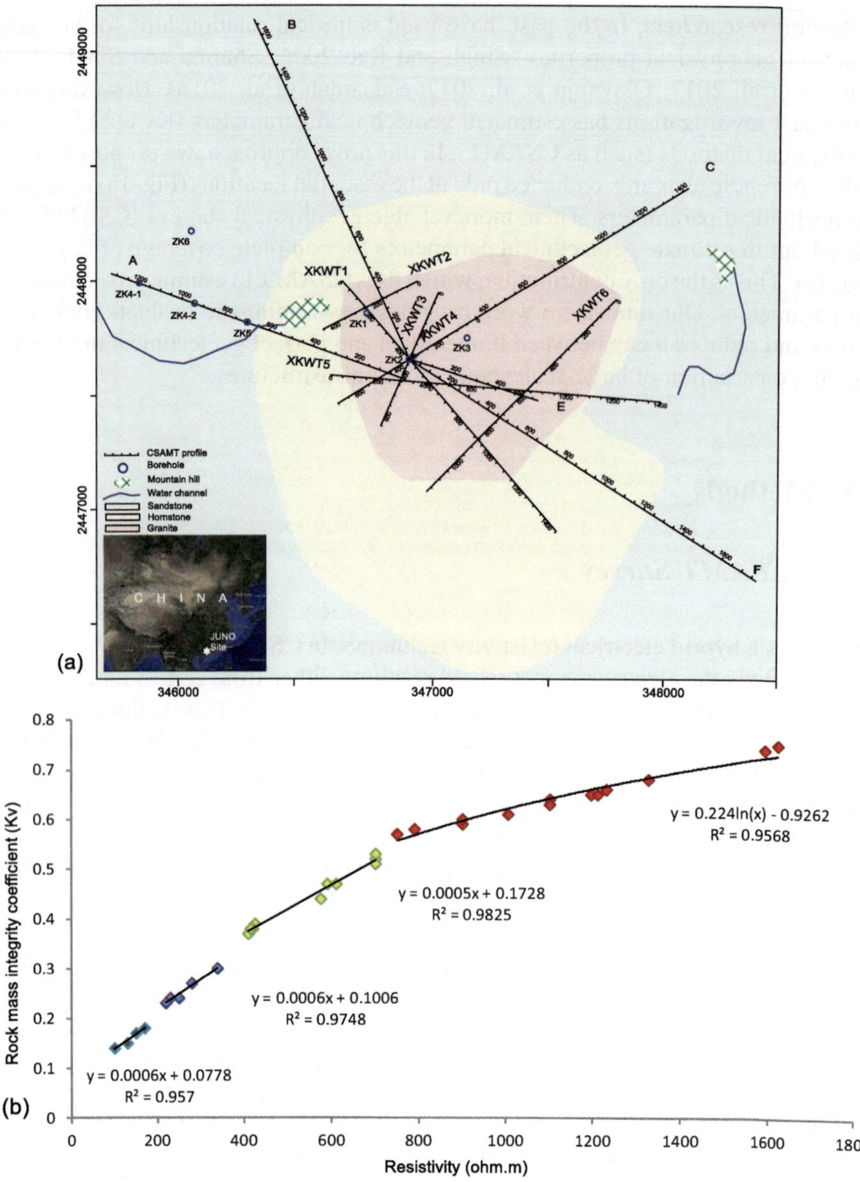

Fig. 4.1 The** **a** Location of JUNO site in South China; A, B, C, E, F, XKWTI, XKWT2, XKWT3, XKWT4, XKWT5, XKWT6 are names of geophysical profiles, **b** Empirical correlation between resistivity obtained from CSAMT and Kv obtained from boreholes

4.2.2 *Rock Mass Integrity Index (Kv)*

Since acoustic wave velocity is a comprehensive reflection of geomechanical properties such as rock mass structure and weathering degree, therefore the velocity of engineering rock mass is very useful to assess the rock integrity (Cai et al. 2020). The rock mass integrity index (Kv), among all geotechnical indices, is more efficient and reliable to appraise the load-bearing potency of foundation rocks (Cai et al. 2020). Kv is square of the ratio of Vp to Vpr, given by:

$$Kv = \left(\frac{Vpm}{Vpr}\right)^2 (0 \leq Kv \leq 1) \tag{4.1}$$

where, Vp is the velocity of rock unit, and Vpr is the velocity of integral/fresh rock. Kv reflects the integrity and fragmentation of the rock unit relative to the rock mass. Obviously, when there are joint cracks or fractures in the rock mass, the wave propagation path will change due to the difference of the wave impedance, so that the rock unit velocity is significantly reduced. When the obtained Kv is close to 1, it reveals the complete integrity of rock mass.

Firstly, Eq. 4.1 was used to determine Kv for the limited borehole tests. Afterwards, we correlated the measured Kv with the electrical resistivity obtained from CSAMT (Fig. 1b). The following empirical equations were obtained by the integration between drilling-based Kv and CSAMT-based resistivity:

$$y = 0.0006(x) + 0.0778 \tag{4.2}$$

$$y = 0.0006(x) + 0.1006 \tag{4.3}$$

$$y = 0.0005(x) + 0.1728 \tag{4.4}$$

$$y = 0.224\ln(x) - 0.9262 \tag{4.5}$$

where y shows kv, and x indicates resistivity in Ωm. Equation 4.2 was used to evaluate sandstone using Kv < 0.20 and resistivity <200 Ωm. We delineated hornstone using Eq. 4.3 for Kv 0.20–0.35 and resistivity 200–400 Ωm. Equation 4.4 was used to estimate Kv 0.35–0.53 using resistivity 400–700 Ωm for the weathered granite. Similarly, we used Eq. 4.5 to evaluate fresh granite via Kv ranging from 0.53 to1.00 and electrical resistivity >700 Ωm.

Table 4.1 Integration between geophysical and borehole data

Resistivity (Ωm)	Kv	Subsurface geological layer
Less than 200	0.00–0.20	Sandstone
200–400	0.20–0.35	Hornstone
400–700	0.35–0.53	Weathered granite
Greater than 700	0.53–1.00	Fresh granite

4.3 Results and Discussions

4.3.1 Geophysics-Borehole Integration

The geology knowledge of the investigated site was obtained by seven boreholes, namely W1–W7 (Fig. 1a). The borehole information constricted the subsurface geology for 4 different layers, for instance fresh/integral or unweathered granite, sandstone, hornstone, and fractured or weathered granite. However, the drilling data delineate the subsurface geological layers at shallow depth for seven point locations only, and so produce suspicions in the evaluation of rock units both laterally and vertically over large area. In most of cases, the budget of project site does not allow to drill so many boreholes to obtain accurate geological information of the entire area. To solve this problem, first we performed eleven CSAMT profiles to cover the entire area and afterwards we correlated the geophysical data with the available limited drilling data. The geophysics-borehole correlation enabled us to delineate the subsurface geological layers along the CSAMT profiles through definite ranges of subsurface resistivity, i.e., sandstone with resistivity less than 200 Ωm, hornstone with resistivity 200–400 Ωm, weathered granite for resistivity 400–700 Ωm and fresh granite/unweathered granite via resistivity greater than 700 Ωm (Table 4.1).

4.3.2 Rock Mass Quality Evaluation

There are numerous indices for the rock strength assessment such as rock mass integrity index (Kv), modulus of elasticity (E) and rock core index (RCI) etc. Kv is the most consistent engineering index. Traditionally, such indices are acquired via the boreholes. However, in this contribution, we obtained Kv by CSAMT via integration between resistivity and borehole data. Eleven 2D CSAMT models (Fig. 2a) were constructed to attain Kv for the complete investigated terrain. The subsurface resistivity remains between 0 and 5000 Ωm in the project site. High resistivity suggests rock mass of good quality, whereas lo resistivity shows rock mass of low quality. Higher resistive rocks (with resistivity 5000 Ωm) were mainly delineated along profile C, E and F, while lower resistive rocks (with resistivity 3000 Ωm) are mostly revealed along rest of the profiles. The rock mass integrity coefficient measured by seven drilling tests evaluated the subsurface for various geological units with specific value range of Kv, such as sandstone via Kv 0–0.20, hornstone for Kv from 0.20 to 0.35, weathered granite having Kv between 0.35 and 0.53 and fresh granite via Kv ranging from 0.53 to 1 (Table 4.1).

The estimated Kv values generated the 2D Kv model for each CSAMT profile (Fig. 2b). Kv varies between 0.05 and 1 in the entire study area (Fig. 2b). Profile

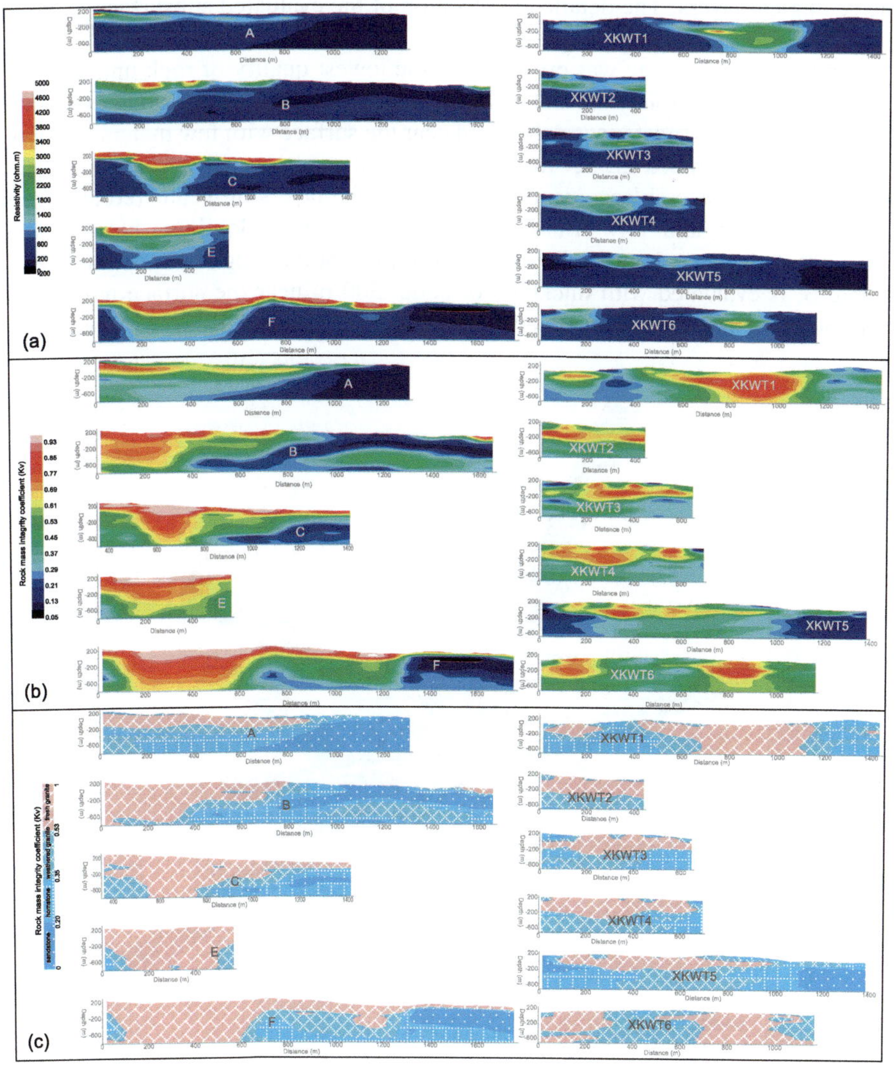

Fig. 4.2 The** **a** 2D maps of CSAMT for different surveyed lines, **b** 2D maps of obtained Kv along CSAMT profiles, and **c** 2D Kv maps evaluated for sandstone, hornstone, weathered granite and integral granite

A reveals the lowest Kv, while profile F suggests the highest Kv. The obtained Kv models (Fig. 2b) were interpreted for the distinct subsurface geological layers for depth of about 800 m (Fig. 2c). For 3D view, all 2D Kv models along eleven surveyed lines were incorporated (Fig. 4.3) and then evaluated the subsurface rock mass quality via distinct layers (Fig. 4.4). For 1D mapping of Kv, the average Kv of all profiles was obtained and plotted via interpretation of distinct subsurface layers of various rocks mass rating (Fig. 4.5). The evaluated 2D Kv models offer thorough insights of the geological layers. The detailed description of the evaluated subsurface layers is given in the following sections.

Sandstone. Sandstone was evaluated as the lowest quality of rock unit with Kv varying between 0.00 and 0.20 and resistivity below 200 Ωm. From boreholes data of Kv, a thin layer of sandstone is delineated near the surface with few meters of depth. However, the estimated 2D Kv models reveal that sandstone is not only evaluated near surface but also at large depths, even under the layers of weathered and fresh granite (Figs. 2c, and 4.4). Along profile A, sandstone is dominant between 800 and 1300 m distance from surface to the bottom of 800 m depth. Sandstone, along profile B, is evaluated with thickness of about 300 m near the surface at distance from 805 to 1620 m. For surveyed line C, about 200 m thick layer of sandstone

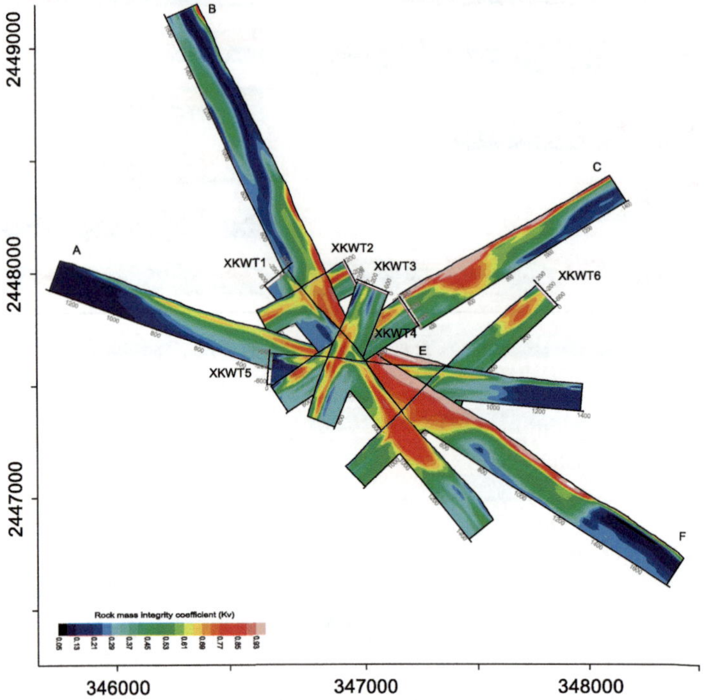

Fig. 4.3 Integration of all 2D Kv models (3D view). Kv increases from black blue to red white on color scale

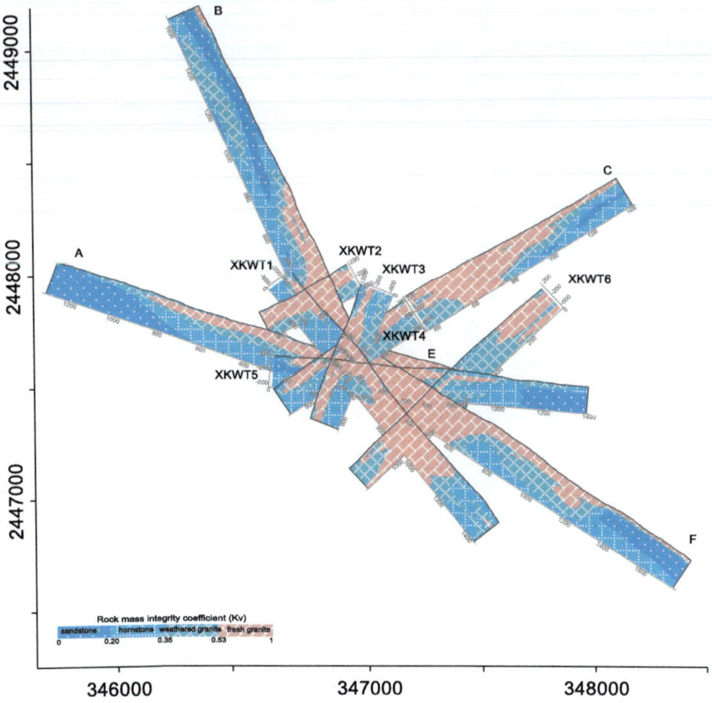

Fig. 4.4 Integration of all 2D Kv maps (3D view) interpreted for sandstone with Kv less than 0.20, hornstone with Kv between 0.20 and 0.35, weathered granite with Kv 0.35–0.53, and fresh granite with Kv from 0.53 to 1

is delineated at depth of about 400 m from the surface for distance between 1180 and 1390 m. For surveyed line F, 500 m thick sandstone was evaluated just below the ground at distance between 1295 and 1695 m. Similarly, next to surveyed line XKWT5, sandstone was mainly delineated at 1100–1400 m distance up to 800 m depth. However, rest of the Kv profiles did not expose any sandstone layer. The average imaging of Kv suggests sandstone domination in southern regions of the investigated terrain mainly along profile A (Fig. 4.5).

Hornstone. Hornstone was appraised via Kv varying between 0.20 and 0.35 and subsurface resistivity between 200 and 400 Ωm. The measured Kv from the limited data does not identify hornstone at large depth as in the estimate Kv models (Figs. 2c, and 4.4). Along profile A, hornstone is evaluated with 400 m thickness at the bottom under granite between a distance of 195 and 795 m. For surveyed line B, under sandstone and granite, hornstone is dominant at bottom with 400 m thickness especially at distance from 390 to 995 m. Hornstone, for surveyed line C, is evaluated at bottom under 200 m thick layer of sandstone/granite between distance ranging from 795 to 1390 m. For surveyed line F, hornstone of about 400 m thickness is delineated at bottom under sandstone and granite for distance from 645 to 1690 m. For surveyed line XKWT1, hornstone is dominant from surface to

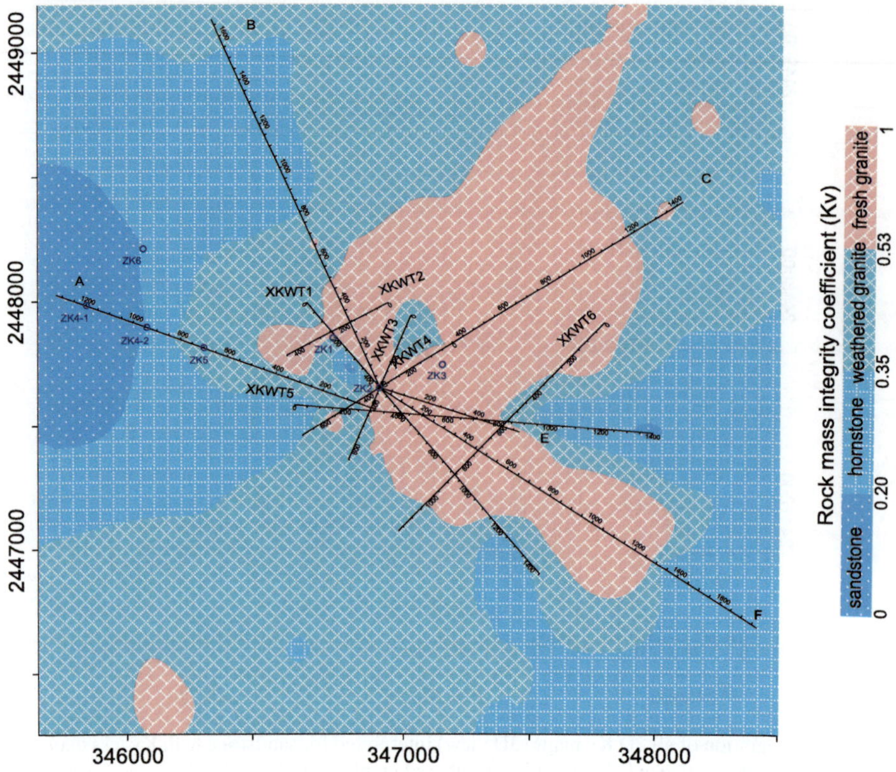

Fig. 4.5 Mapping of average Kv interpreted for sandstone with Kv less than 0.20, hornstone with Kv between 0.20 and 0.35, weathered granite with Kv 0.35–0.53, and fresh granite with Kv from 0.53 to 1

the bottom especially for distance of 0–500 and 1100–1420 m. For surveyed line XKWT3, hornstone is mainly evaluated at 400–625 m distance under granite with 400 m thickness. Along profile XKWT4, hornstone is exposed at 400–700 m distance under 400 m thick granite layer. Similarly along profile XKWT5, hornstone is mainly delineated for distance of 0–395 and 895–1090 m under 350 m thick granite layer. The results of average Kv imaging in Fig. 4.5 suggest that hornstone is prevailing in NW and SE regions especially along profiles A, B, F, XKWT1 and XKWT5. The evaluated hornstone is suggested as the poor quality of rock mass after sandstone.

Granite. The weathered granite was evaluated by Kv between 0.35 and 0.53 and resistivity range of 400–700 Ωm, whereas the fresh granite was delineated via Kv 0.53–1.00 and resistivity greater than 700 Ωm. The 2D/3D imaging of Kv obtained by CSAMT provides a thorough insight of the weathered and fresh granite over the entire study area (Figs. 2c, and 4.4). Fresh/unweathered granite suggests rock unit of good strength, while weathered or fractured rock (granite) provides rock unit of low strength followed by hornstone and sandstone. For surveyed line A, granite rock is evaluated for distance between 0 and 890 m, such as integral granite of

300 m thickness near the surface and weathered granite of 200–600 m thickness below fresh granite. Along profile B, weathered granite is evaluated at 100–800 m distance with 200 m thickness below fresh granite and at 1000–1550 m distance with 400 m thickness, while fresh granite of about 300–1000 m thickness is evaluated near the surface at 0–800 m distance. Fresh granite along profile C is exposed with depth ranging from 200 to 1000 m, while weathered granite along this profile is evaluated below fresh granite at 350–550 m distance with 500 m thickness and at 750–1200 m distance with about 300 m thickness. Profile E almost entirely exposed by granite with two 500 m thick zones of weathered granite at both sides of the profile below fresh granite. Fresh granite along profile F is dominant at 100–600 m distance, whereas weathered granite along the same profile is revealed at 600–1250 m with 200–500 m thickness below fresh granite. Fresh granite along profile XKWT1 is dominant at 600–1100 m distance, while weathered granite is evaluated mainly at 400–600 and 1100–1200 m distance. The fresh granite along profiles XKWT2-5 is dominant near surface up to about 400 m, whereas thick layer of weathered granite along these profiles is revealed below fresh granite. Similarly, fresh granite along profile XKWT6 is mainly exposed at 0–300 and 700–1000 m distance, while weathered granite along this profile is dominant at 300–650 m distance. The results of average Kv mapping (Fig. 4.5) suggest that integral granite rock is mainly dominant in NE and central regions especially along profile C, E and F.

4.4 Conclusions

Our contribution presents a novel research for the estimation of geotechnical parameters. The load-bearing capacity of rock units is appraised via geomechanical indices, namely rock mass quality coefficient (Kv). Such parameters are conventionally by drilling tests. Alternatively, we estimate the rock mechanical parameters using CSAMT method through practical incorporation of subsurface resistivity with borehole measurements. The estimated rock mechanical parameters, compared with the limited drilling tests, provide comprehensive and precise evaluation of rock unit strength for large areas. The geological strengths of various rock units were assessed via wide ranges of resistivity and Kv values, i.e., sandstone with Kv 0.00–0.20 and resistivity below 200 Ωm, hornstone via Kv range of 0.20–0.35 and resistivity 200–400 Ωm, weathered granite using Kv range of 0.35–0.53 and subsurface resistivity 400–700 Ωm and fresh or unweathered granite for Kv range of 0.53–1.00 and subsurface resistivity above 700 Ωm. The estimated Kv provides rock mass quality evaluation for 800 m depth from the mean sea level. Our interpretations suggest that sandstone is prevailing in western part of the study area mainly along profile A, whereas fresh or integral granite rock is dominant in NE and central regions especially along profile C, E and F. Fresh granite is evaluated as the most integral rock unit suitable for infrastructure development, while sandstone, fractured granite and hornstone show relatively low integrity of rock unit and thus are unsuitable for infrastructure design in the study area. The obtained empirical equations can be used in the terrains having same subsurface conditions. This research offers comprehensive appraisal of rock mass reliability via engineering parameters, and therefore reduces lots of expensive boreholes.

Acknowledgements This research was financially supported by Research Fund for International Young Scientists (RFIS-I) of National Natural Science Foundation of China (Grant No. 42350410442). We are grateful to the State Key Laboratory of Mountain Hazards and Engineering Resilience, Institute of Mountain Hazards and Environment, Chinese Academy of Sciences, and the Key Laboratory of Shale Gas and Geoengineering, Institute of Geology and Geophysics, Chinese Academy of Sciences, for providing the research facilities for this work.

References

Barton N, Lien R, Lunde J (1974) Engineering classification of rock masses for the design of tunnel support. Rock Mech 6:189–236

Bieniawski Z (1973) Engineering classification of jointed rock masses. Civil Eng South Afr 15:1154

Cai J, Jibin L, Ting X, Yongyue S (2020) Study on fast test and calculation method of integrity coefficient of surrounding rock during tunnel construction. IOP Conf Ser Earth Environ Sci 474:042014

Deere DU (1964) Technical description of rock cores for engineering purpose. Rock Mech Eng Geol 1:17–22

Han W, Li G, Sun Z, Luan H, Liu C, Wu X (2020) Numerical investigation of a foundation pit supported by a composite soil nailing structure. Symmetry 12:252

Hoek E, Brown ET (1997) Practical estimates of rock mass strength. Int J Rock Mech Min Sci 34:1165–1186

Ji F, Shi Y, Li R, Zhou C, Zhang N, Gao J (2019) Modified Q-index for prediction of rock mass quality around a tunnel excavated with a tunnel boring machine (TBM). Bull Eng Geol Environ 78:3755–3766

Kagi W, Sano A, Shinji M, Nishi T, Nakagawa K (2001) A new rock mass classification method at tunnel face for tunnel support system. Doboku Gakkai Ronbunshu 2001:121–134

Li SC, Wu J, Xu ZH, Li LP (2017) Unascertained measure model of water and mud inrush risk evaluation in karst tunnels and its engineering application. KSCE J Civ Eng 21:1170–1182

Loperte A, Soldovieri F, Palombo A, Santini F, Llapenna V (2016) An integrated geophysical approach for water infiltration detection and characterization at Monte Cotugno rock-fill Dam (southern Italy). Eng Geol 211:162–170

Lowson A, Bieniawski Z (2013) Critical assessment of RMR based tunnel design practices: a practical engineer's approach. In: Proceedings of the SME, rapid excavation and tunnelling conference, Washington, DC, pp 23–26

Lu J, Liu X (2009) Construction techniques for water and sand gushing section in Xiushan Tunnel on Yuxi-Mengzi railway. Tunnel Constr 3:524

Masahiro N, Koji M, Hiroshi Y, Takuro N, Kazuo N, Koji N (1999) A new proposal of evaluation system for tunnel face based on the analysis of the observation records. J Jpn Soc Civil Eng 623:131–141

Nourani MH, Moghadder MT, Safari M (2017) Classification and assessment of rock mass parameters in Choghart iron mine using P-wave velocity. J Rock Mech Geotech Eng 9:318–328

Olayanju GM, Mogaji KA, Lim HS, Ojo TS (2017) Foundation integrity assessment using integrated geophysical and geotechnical techniques: case study in crystalline basement complex, southwestern Nigeria. J Geophy Eng 14(3):32–64

Palmstrom A (2005) Measurements of and correlations between block size and rock quality designation (RQD). Tunn Undergr Space Technol 20:362–377
Pytel W, Pałac-Walko B (2015) Geomechanical safety assessment for transversely isotropic rock mass subjected to deep mining operations. Can Geotech J 52:1477–1489
Rahmati A, Faramarzi L, Sanei M (2014) Development of a new method for RMR and Q classification method to optimize support system in tunneling. Front Struct Civil Eng 8:448–455
Rehman H, Naji AM, Kim J-J, Yoo H-K (2018) Empirical evaluation of rock mass rating and tunneling quality index system for tunnel support design. Appl Sci 8:782
Salaamah AF, Fathani TF, Wilopo W (2018) Correlation of P-wave velocity with rock quality designation (RQD) in volcanic rocks. J Appl Geol 3(2):62–72
Sharma PK, Singh TN (2008) A correlation between P-wave velocity, impact strength index, slake durability index and uniaxial compressive strength. Bull Eng Geol Environ 67(1):17–22
Shin HS, Han KC, Sunwoo C, Choi SO, Choi YK (1999) Collapse of a tunnel in weak rock and the optimal design of the support system. Paper presented at the 9th ISRM Congress, Paris
Singh M, Rao KS (2005) Empirical methods to estimate the strength of jointed rock masses. Eng Geol 77:127–137
Vásárhelyi B, Ván P (2006) Influence of water content on the strength of rock. Eng Geol 84(1):70–74
Wang R, Yin C, Wang M, Di Q (2015) Laterally constrained inversion for CSAMT data interpretation. J Appl Geophys 121:63–70
Wang C, Jiang Y, Liu R, Wang C, Zhang Z, Sugimoto S (2020a) Experimental study of the nonlinear flow characteristics of fluid in 3D rough-walled fractures during shear process. Rock Mech Rock Eng 53:2581–2604
Wang X, Yuan W, Yan Y, Zhang X (2020b) Scale effect of mechanical properties of jointed rock mass: a numerical study based on particle flow code. Geotech Eng 21:259–268
Yuji W, Tatsuo K, Masaki K, Kenichi H (2006) Solution with modified perceptron to tunnel cutting face evaluation problems. Geoinformatics 17:61–70
Zhamaletdinov AA (2020) A method for quantifying static shift distortions using a magnetic field of controlled source (CSAMT). Seism Instr 56:555–563
Zhang M, Farquharson CG, Li C (2021) Improved controlled source audio-frequency magnetotelluric method apparent resistivity pseudo-sections based on the frequency and frequency–spatial gradients of electromagnetic fields. Geophys Prospect 69:474–490
Zolfaghari A, Sohrabi Bidar A, Maleki Javan MR, Haftani M, Mehinrad A (2015) Evaluation of rock mass improvement due to cement grouting by Q-system at Bakhtiary dam site. Int J Rock Mech Min Sci 74:38–44

Chapter 5
Ways to Improve the Accuracy of Determining the Preconsolidation Pressure

Vladimir Matveev and Violetta Shanina

Abstract The main factors, both objective and subjective, which lead to a decrease in the accuracy of determining the preconsolidation pressure are considered. Factors are manifested at various stages of research, both in the preparation of the test and in the interpretation of the test. At the preparation stage, groups of factors are identified due to the quality of samples, equipment, and the choice of the test scheme. At the stage of interpretation, groups of factors are identified due to the choice of the method of graphical constructions (Casagrande, Becker, Pacheco Silva, Wang and Frost), the operator, computational methods, and decision-making. Ways to improve accuracy by choosing more gentle testing schemes and more precise methods of graphical constructions are proposed. Methods of minimizing the influence of subjective factors are proposed, primarily through the use of technical means and automation of routine activities of the operator performing test processing and interpretation of results.

Keywords Preconsolidation pressure · Test scheme · Subjective factors · Graphical constructions · Ranking distance

5.1 Introduction

Various modern nonlinear soil models use overconsolidation parameters, such as overconsolidation ratio OCR and pre-overburden pressure *POP*. Inaccuracies in determining these parameters affect the accuracy of determining the initial stress–strain state. The accuracy of determining the *POP* and *OCR* parameters depends on the accuracy of determining the preconsolidation pressure σ_c and the effective

V. Matveev (✉) · V. Shanina
Lomonosov Moscow State University, Moscow, Russia
e-mail: vladimir@matveev.email

V. Matveev
Geotechnical Graduate Lab LLC, Moscow, Russia

S. Wang et al. (eds.), *Engineering Geology for a Habitable Earth: IAEG XIV Congress 2023 Proceedings, Chengdu, China*, Environmental Science and Engineering,
https://doi.org/10.1007/978-981-99-9057-3_5

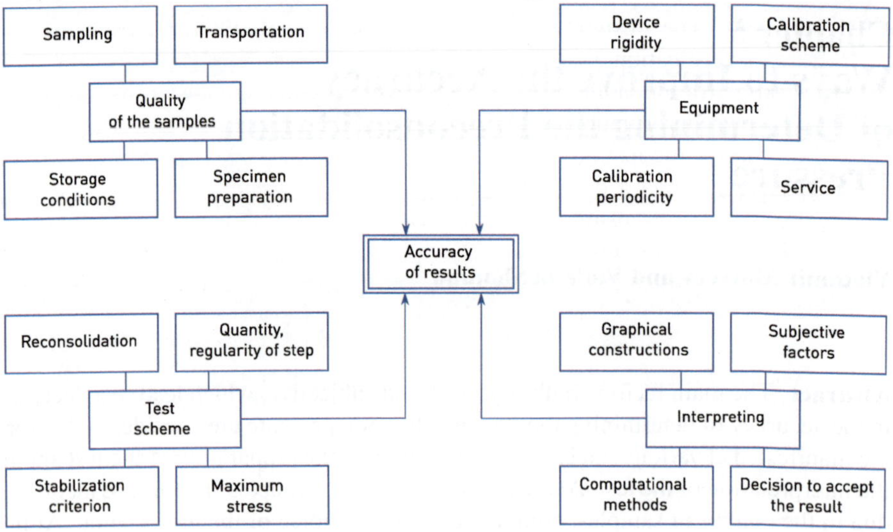

Fig. 5.1 Classification of factors affecting the accuracy of determining the preconsolidation pressure

overburden pressure σ_0. Consideration of ways to improve the accuracy of determining effective overburden pressure is beyond the scope of this study. It should be noted that isotropic preconsolidation pressure is used in the models, since it is convenient to consider separately the spherical stress tensor and the stress deviator. The preconsolidation pressure determines the shape and size of the yield surface.

The proposed methods for improving the accuracy of σ_c are separately considered for the stage of preparation and conduct of tests and for the stage of processing and interpretation of the results obtained. The classification of factors affecting the accuracy of determining the preconsolidation pressure is presented in the form of a diagram (Fig. 5.1).

5.2 Ways to Improve Accuracy in Preparation for Tests

The quality of the samples significantly affects the reliability of the test results. Samples should be taken without disturbing their structure, carefully stored and transported in accordance with current standards. In the case of samples of unsatisfactory preservation, the inflection on the compression curve from the over consolidation to the normally consolidation state of the soil becomes unexpressed, which makes it difficult to interpret the results. Various criteria are proposed for quality control of samples, for example, Lunne et al. (1997).

Equipment. The design of the instruments, the calibration scheme and the frequency of verification also affect the test results. It is important to use a separate calibration for unloading in order to take into account the hysteresis loop of the odometer's own deformations in methods using the unloading branch. An important condition for the test is the impossibility of lateral expansion, which may not be observed in standard odometers at high values. It is not easy to establish the correction, since the transmitted lateral stress is an internal characteristic of the soil due to the lateral stress coefficient K_0. The simplest solution is to use more rigid odometers (rings with a larger wall thickness). It is also possible to implement with flexible walls, but with mandatory feedback to compensate for lateral expansion. It is important that the radial deformation does not exceed 0.005% of the sample diameter (Lade Poul 2016). The load system of devices for determining the preconsolidation pressure should be able to create a vertical load significantly exceeding the expected preconsolidation pressure.

Selection of the test scheme. The factors of the test scheme include the presence of a reconsolidation procedure, the count of steps, their values and the regularity of their increments, the stabilization criterion and the maximum specified stress.

The reconsolidation procedure is an important stage of soil testing, which will allow the sample to be brought into a state close to its natural state and also partially compensate for the violation of its structure during sampling and sample preparation.

For all graphical methods, the accuracy of the constructions directly depends on the number of experimental points, and the accuracy of determining the preconsolidation pressure directly depends on the density of points in the area of the expected preconsolidation pressure. The load increment ratio $LIR = \Delta\sigma/\sigma$ should be small so as not to destroy the structure of the soil (Crawford 1964) and simplify graphical constructions. It is also possible to use a kinematic loading mode with a given deformation rate (CRS method) (Melnikov and Popkov 2016).

The load increment ratio *LIR* steps must be constant during the test. If, for example, the steps increase by a fixed amount, then this leads to an increase in the frequency of points in the area of high stresses, and, as a consequence, to a distortion of the compression curve in this area. As a result, a reverse bend of the compression curve is observed, which does not allow to reliably determine the normal consolidation line (Crawford 1964).

The stabilization criterion for saturated soils is the completion of primary consolidation, which is determined either by the graphical methods of Taylor or Casagrande, or by the dispersion of excess pore pressure in the sample. If the test scheme adopts the criterion of conditional stabilization in time and (or) deformation, this may lead to additional deformations due to volumetric creep, and the compression curve will be distorted (Umar and Sadrekarimi 2016).

The maximum stress should obviously be greater than the expected preconsolidation pressure σ_c, and by so much as to guarantee access to the normal consolidation line NCL. However, it is worth taking into account that the linear law of compressibility has a range of applicability beyond which the compression curve can become concave.

5.3 Ways to Improve Accuracy in Interpreting Tests

The most common and frequently used methods for determining the preconsolidation pressure are methods based on the interpretation of the compression curve itself or the compression curve transformed by differentiation (for example, the Janbu method (Janbu 1969)) or integration (energy methods), as well as various axis logarithms. Since the transition from elastic to plastic behavior is not always unambiguously identified, many researchers have proposed various geometric constructions (Kramarenko et al. 2014). The equations of an ideal adiabatic process are often used as a theoretical justification (Butterfield 2011; Wang and Frost 2004).

In addition to interpreting the compression curve, a number of researchers have proposed using the dependencies of other characteristics directly or indirectly related to the transition through the flow surface. Methods based on consolidation parameters, pore pressure (under kinematic loading), lateral stress coefficient (K_0-consolidation in triaxial compression devices (Bishop and Henkel 1962; Lade Poul 2016) and odometers with the ability to measure lateral stress (Trufanov et al. 2014), as well as on the nature of deviatory loading along the trajectory of standard triaxial compression (Kell 2021) are proposed. As a rule, such methods have a theoretical justification, and the determination of the preconsolidation pressure is reduced to the search for the extremum of dependence, which greatly simplifies interpretation. The disadvantages should include the need for special devices and sensors to record these parameters during the test.

Choosing a method for graphical constructions. Different methods of graphical constructions have different accuracy (Boone 2010; Butterfield 1979; Grozic et al. 2003; Kirstein et al. 2017; Matveev and Shanina 2021; Paniagua et al. 2016; Umar and Sadrekarimi 2016). According to studies Matveev and Shanina (2021), the most reliable results are obtained by graphical constructions proposed by Boone (2010), Burland (1990), Becker et al. (1987), Wang and Frost (2004), as well as a group of bilogarithmic methods (Butterfield 1979; Oikawa 1987; Onitsuka et al. 1995). The advantages of the above methods are: relative simplicity of constructions, the use of unloading branches (Boone, Wang—Frost), the use of natural relative deformations (bilogarithmic, Wang—Frost).

Ranking Distance (*RD*) was used to assess the accuracy and precision of the methods (Fig. 5.2):

$$\mathrm{RD} = \sqrt{\left[1 - \mu\left(\frac{\sigma_{pp}}{\sigma_{pi}}\right)\right]^2 + \left[\sigma\left(\frac{\sigma_{pp}}{\sigma_{pi}}\right)\right]^2}$$

Simple constructions are easy to interpret, and also make it easy to automate them. At the same time, the Pacheco Silva method (1970), despite its apparent simplicity, has a systematic deviation from the set value of the preconsolidation pressure (Matveev and Shanina 2021).

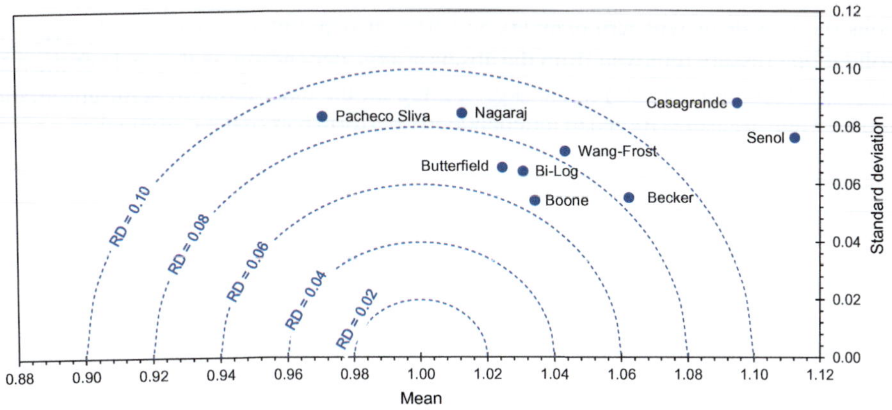

Fig. 5.2 Diagram of ranking distance method for graphical construction on studies

The unloading branch corresponds to the decompression of the soil and reflects its elastic behavior. The quality of the samples practically does not affect the nature of unloading, which is an undoubted advantage of its use.

The advantage of using natural (logarithmic, true) relative deformations ε is their applicability in the field of not only small deformations, unlike engineering relative deformations ε. To calculate the natural relative deformations under compression, the formula $\overline{\varepsilon} = -\ln(1-\varepsilon)$ is used (Lade Poul 2016). A feature of the tests for determining the preconsolidation pressure is loading to fairly high values of vertical stress (according to GOST R 58326-2018 2018 at least 8 MPa, many times higher than the maximum loads of conventional compression tests), which leads to significant deformations, and, therefore, it is impossible to neglect the assumptions of engineering relative deformation. It is easy to show that the value $\ln(1+e)$, used in bilogarithmic methods, is equivalent to the natural relative deformation: $\ln(1+e) = -\overline{\varepsilon}+\ln(1+e_0)$.

Subjective factors presuppose any mental activity of the operator regarding the conduct of graphical constructions (determination of points and segments, drawing lines, tangents and secants, their intersections), preliminary evaluation of the result and the establishment of the need to repeat the construction (Paniagua et al. 2016).

To reduce the influence of subjective factors and improve the efficiency of results processing, various automation tools are used, both in the form of simple user spreadsheets and as part of software complexes.

The correctness of the understanding of the methodology. The use of methods involves a detailed study of the construction procedure and requires appropriate competencies from the operator. Automation tools significantly simplify the operator's work and to some extent provide protection from incorrect actions.

The resolution of vision during manual processing. It is believed that the average person is able to distinguish objects with the naked eye up to one angular minute. At the same time, it is necessary to take into account the low accuracy of graphic constructions made with a ruler and pencil. This also includes the value of the preconsolidation pressure removed from the abscissa axis, depending on the scale of the test schedule and the frequency of strokes applied on the axis. Even in semi-automatic mode, when the user is forced to independently perform part of the graphical constructions, the ability to scale individual sections of the constructions improves their quality.

Complexity of calculations and operator inattention. As a rule, graphical methods do not require complex calculations, which is their advantage. Errors can occur when interpolating the preconsolidation pressure taken from the graph between the strokes on the axis, especially on a logarithmic scale. Even the simplest automation tools, such as spreadsheets, allow you to completely eliminate errors caused by this factor.

Uncertainty in constructions arises when the choice of points and the methods of drawing the tangent are ambiguous, and formal conditions are satisfied by a number of constructions that differ in the result. The solution is to adopt clear, non-contradictory instructions and procedures for the implementation of these actions.

In the case of automation of actions, the operator's responsibility for the subjective choice during interpretation is shifted to the developer implementing the algorithm. At the same time, the developer should be guided by reasonable decision-making criteria (Matveev and Shanina 2021).

Computational methods that increase accuracy. The Casagrande method (1936) is unstable to changing the proportions of the axes' scales, since the constructions use a bisector that does not retain its properties during such transformations. The bisector is a way of averaging the angle of inclination of the horizontal straight line and tangent to the point with the smallest radius of curvature (Matveev and Shanina 2021). The use of a bisector in the space of quantities of different dimensions is doubtful, since it is necessary to establish a proportionality coefficient between them (Clementino 2005). As an improvement of this construction, it is proposed to use the median (Matveev and Shanina 2021), in which case the intersection point will not depend on the selected axis scale. At the same time, the result is comparable because the tangent of half an angle and half the tangent of an entire angle for graphical constructions are approximately equal in the angle range used.

For some constructions, it is required to draw segments for conditionally linear sections, for example, for the primary compression line. Instead of drawing a secant connecting the two extreme points of the segment, it is proposed to use the least squares method, which allows you to smooth out random deviations of the extreme points and takes into account the points inside the segment.

Graphical constructions, especially in the methods of Casagrande and Pacheco Silva, require a smooth continuous line. Interpolation, both linear and cubic, does not provide significant advantages. One of the solutions is the approximation of the compression curve in a given area by an analytical function. The analytical method of curve assignments makes it possible to uniquely identify points of curvature and determine tangents. In energy methods, the transition from deformation to deformation energy by direct integration, rather than by the trapezoid method, also increases the accuracy of the constructions (Matveev and Shanina 2021). Other effective methods of computational mathematics are also used, for example, differentiation by regularization (Chartrand 2011), integration by Simpson, Gauss and others.

The decision to accept the result. For a preliminary assessment of the result and the establishment of the need to repeat the construction, experience and some training of the operator is certainly required. Test schedules may have a complicated form, processing in automatic mode of which will give an incorrect result. The operator needs experience to recognize and correct an incorrect interpretation. Most likely, the future belongs to artificial intelligence that can help the responsible person make a decision.

5.4 Conclusions

To increase the level of reliability of determining the parameters of overconsolidation, the following measures are proposed: high-quality sampling and proper storage of samples; timely maintenance, regular calibration, periodic verification of measuring instruments and test equipment; use a test scheme that provides for reconsolidation and gentle loading.

Automation of constructions and the use of a combination of the most accurate methods will minimize the influence of subjective factors, thus providing a statistically controlled, predictable result in conditions of repeatability.

References

Becker DE, Crooks JHA, Been K, Jefferies MG (1987) Work as a criterion for determining in situ and yield stresses in clays. Can Geotech J 4:549–564

Bishop AW, Henkel DJ (1962) The measurement of soil properties in the triaxial test

Boone SJ (2010) A critical reappraisal of "preconsolidation pressure" interpretations using the oedometer test. Can Geotech J 3:281–296

Burland JB (1990) On the compressibility and shear strength of natural clays. Géotechnique 40(3):329–378

Butterfield R (1979) A natural compression law for soils (an advance on e − log p'). Géotechnique 24(4):469–480

Butterfield RL (2011) An improved model of soil response to load, unload and re-load cycles in an oedometer. Soils Found 51:253–263

Casagrande A (1936) The determination of the preconsolidation load and its practical significance. In: Proceedings of the 1st international conference on soil mechanics and foundation engineering. Harvard Printing Office, Cambridge, pp 60–64

Clementino RV (2005) Discussion of "an oedometer test study on the preconsolidation stress of glaciomarine clays." Can Geotech J 42(3):972–974. https://doi.org/10.1139/t05-010

Crawford CB (1964) Interpretation of the consolidation test. J Soil Mech Found Div ASCE 90(5):87–102

Gost R 58326-2018 (2018) Soils. Laboratory method for determining the over consolidation characteristics. Standartinform, Moscow

Grozic JL, Lunne T, Pande S (2003) An oedometer test study on the preconsolidation stress of glaciomarine clays. Can Geotech J 40(5):857–872. https://doi.org/10.1139/t03-043

Janbu N (1969) The resistance concept applied to deformation of soils. In: Balkema AA (ed) Proceedings of the 7th international conference on soil mechanics and foundation engineering, Mexico City, 25–29 August 1969. Rotterdam, pp 191–196

Kell C (2021) SIGMA/W material model series: hardening soil model. https://www.seequent.com/sigma-w-material-model-series-hardening-soil-model/. Accessed 03 June 2022

Kirstein AL, Salloum J, Tareq S (2017) Evaluation of different methods for identification of preconsolidation pressure for Champlain Sea Clay Using CRS Tests

Kramarenko VV, Nikitenkov AN, Molokov VY (2014) About the preconsolidation pressure of clay soils in Tomsk region. Izvestiya TGU 5:1–16

Lade Poul V (2016) Triaxial testing of soil. Wiley, Amsterdam

Lunne TA, Berre T, Strandvik S (1997) Sample disturbance effects in soft low plastic norwegian clay

Matveev VV, Shanina VV (2021) The influence of subjective factors on the accuracy of graphical constructions to interpret the preconsolidation pressure in IL-tests. In: Proceedings of 3rd European regional conference of IAEG/Athens/Greece/6–10 October 2021, vol 2—National Group of Greece of IAEG Athens, pp 10–17

Melnikov RV, Popkov AA (2016) Using Becker method to determine preconsolidation pressure at CRS oedometer test. Geotechnika 6:54–59

Oikawa H (1987) Compression curve of soft soils. J Jpn Geotech Soc Soils Found 3:99–104

Onitsuka K, Hong Z, Hara Y, Yoshitake S (1995) Interpretation of oedometer test data for natural clays. Soils Found 35(3):61–70. https://doi.org/10.3208/sandf.35.61

Pacheco Silva F (1970) A new graphical construction for determination of the preconsolidation stress of a soil sample. In: Proceedings of the 4th Brazilian conference on soil mechanics and foundation engineering. Rio de Janeiro, pp 225–232

Paniagua P, L'Heureux J-S, Yang S, Lunne T (2016). Study on the practices for preconsolidation stress evaluation from oedometer tests. In: Proceedings of the 17th nordic geotechnical meeting. Reykjavik

Chartrand R (2011) Numerical differentiation of noisy, nonsmooth data. ISRN Appl Math 2011:1–11. https://doi.org/10.5402/2011/164564

Trufanov AN, Shulyat'ev OA, Rostovtsev AV, Gabsalyamov GU (2014) Methods for determination of soil overconsolidation parameters and their practical use in the conditions of St. Petersburg. Eng Surv 11:32–39

Umar M, Sadrekarimi A (2016) Accuracy of determining pre-consolidation pressure from laboratory tests. Can Geotech J 3:441–450. https://doi.org/10.1139/cgj-2016-0203

Wang L, Frost JD (2004) Dissipated strain energy method for determining preconsolidation pressure. Can Geotech J 4:760–768

Chapter 6
Thermal and Shear-Rate Effects in Landslides: From the Classics to the Future

Gianvito Scaringi and Marco Loche

Abstract The frictional resistance of geomaterials has been shown to depend on the rate of shearing. Both weakening and strengthening have been observed, and various mechanisms that take the mineralogy and stress level into account have been proposed. Evidence also exists that temperature plays an important role in defining the frictional resistance as well as its dependence on the rate of shearing. Temperature and rate-dependent mechanisms can indeed control landslide runouts, causing runaway sliding or prolonged slow creep displacements. We reviewed the literature on the matter and noted that, while studies on shear-rate effects are abundant, systematic classifications useful in predicting landslide fates in a variety of lithologies and environments are lacking. As for the thermal effects, these are well studied with respect to large and fast-runout landslides, where frictional heating plays a major role. Conversely, little is known about changes in slope stability (prior to failure, remobilization, or reactivation) in relation to changes in ground temperature caused by varying boundary conditions, such as changes in groundwater temperature or heat transfer from the surface under seasonal or long-term climatic changes. Results of our preliminary experiments, targeting clay-rich materials, demonstrate an important effect of temperature on the residual shear strength, which is coupled with changes in shear-rate response. Catchment-scale statistical analyses also reveal that surface temperature can be correlated with landslide activity in space and time. We conclude by suggesting that landslide modelling approaches in the future should account for thermal and shear-rate effects. In physically-based modelling, this may be achieved via coupled thermo-hydro-mechanical formulations, in which the constitutive model includes a time-dependent (e.g., thermo-viscous) component.

Keywords Rate effect · Temperature · Landslide · Clay · Velocity weakening · Thermo-mechanical coupling

G. Scaringi (✉) · M. Loche
Institute of Hydrogeology, Engineering Geology and Applied Geophysics, Faculty of Science, Charles University, Prague, Czech Republic
e-mail: gianvito.scaringi@natur.cuni.cz

S. Wang et al. (eds.), *Engineering Geology for a Habitable Earth: IAEG XIV Congress 2023 Proceedings, Chengdu, China*, Environmental Science and Engineering,
https://doi.org/10.1007/978-981-99-9057-3_6

6.1 Shear-Rate Effects: Learning from the Classics

Landslide runout predictions via shear velocity-dependent formulations are not commonly found in the literature, and there have been few attempts to integrate soil/rock mechanics frameworks with approaches commonly used in fault mechanics, which often involve state-and-rate models (Handwerger et al. 2016; Scaringi et al. 2018b). Typically, large catastrophic landslides display shear rate-dependent responses, where their runaway behavior is attributed to velocity-weakening mechanisms (Goren et al. 2010; Lucas et al. 2014; Alonso et al. 2016). Under lower normal stresses and shear displacement rates, which are especially common in soil landslides (thickness up to 20 m, displacement rate up to 10^{-3} m/s) (e.g., Larsen et al. 2010), shear rate-dependent behaviors are more elusive; in fact, the literature reports contradictory results (Saito et al. 2006; Li et al. 2013; Scaringi and Di Maio 2016; Scaringi et al. 2018a; Duque et al. 2023) which could be explained, at least in part, with the difficulty of identifying clear signals from typically "noisy" test outputs.

Tika et al. (1996) first reviewed these behaviors in a classic paper, focusing on the residual shear strength of both granular and clayey soils. The authors pointed out that shear-rate effects are typically negligible in granular soils (*neutral rate effect*), while both shear-rate strengthening (*positive rate effect*) and weakening (*negative rate effect*) are observed in clay-bearing materials. Consistently with a discussion by Lupini et al. (1981), the authors connected the observed effects to "modes" of shearing: a turbulent shear mode should produce neutral or negative shear rate effects depending on the normal stress; a "transitional" shear mode should mainly yield negative effects; a laminar shear mode (common in saturated clays) may produce either negative or positive effects, possibly depending on the nature of the minor soil components.

Scaringi et al. (2018a, b) enriched Tika et al. (1996)'s work by including results of relaxation tests and rotary shear experiments. They highlighted that shear-rate dependent behaviors may be more complex than previously observed. Focusing on illitic-smectitic materials undergoing ring-shear tests, for instance, the authors revealed a non-monotonic behavior consisting of a weakening followed by a strengthening as the shear displacement rate increased. Shear strength oscillations were reported by Hu et al. (2017) under relatively fast shearing ($0.02–2 \cdot 10^{-3}$ m/s), which were attributed to the migration of fine particles away from the shear zone and to changes in hydraulic conductivity. Yet, when submitted to large shear rates and normal stresses ($>10^{-2}$ m/s, >0.5–1 MPa), all experiments indicated important weakening due to a variety of processes, including acoustic fluidization, frictional heating, and thermal pressurization (Wang et al. 2017; Scaringi et al. 2018b). However, owing to changes in temperature and effective stress (through changes in pore fluid pressure), and alterations in soil properties (by crushing, recrystallization), analyses and interpretations in these conditions are only possible in terms of total stresses, power density, or thermodynamic parameters (e.g., Alonso et al. 2016; Acosta et al. 2018; Deng et al. 2020b).

In some cases, shear-rate effects are small (less than 10% of variation in strength under an increase in shear displacement rate by an order of magnitude) but some trends can still be identified (Hu et al. 2018; Scaringi et al. 2018a). In some other cases, scattering of data is large and conclusions cannot be drawn (Scaringi and Di Maio 2016). Uncertainties can be reduced by an appropriate choice of testing device, acquisition system, and normal stress (Roháč et al. 2020; Duque et al. 2023). Chemical composition and temperature of the water bath also should be controlled because they can influence both the available resistance and its sensitivity to the rate of shearing significantly (Di Maio et al. 2015; Scaringi and Di Maio 2016). Considering all of the above, it can be suggested that a statistical approach to determine the significance of strength values and trends could be used; however, this approach is not common in the literature (Tiwari and Latha 2019; Roháč et al. 2020) and, as a result, the extent to which published empirical correlations or models are applicable remains unconstrained. Similarly, it cannot be judged whether comparisons among studies are significant or not.

6.2 Thermal Effects: A Jump into the Future

In the past decades, the role of temperature in geotechnics has become more prominent, as witnessed by the establishing of the new subfield of "energy geotechnics". Heat transfers in the ground can influence the static response of foundations; soil themselves are investigated for their role as heat sinks or sources for heat pumps aimed at the heating/cooling of buildings or underground spaces (Vieira et al. 2017; Amatya et al. 2012). Changes in temperature in the ground can produce changes in water pressures (positive or negative—suctions) and can alter the water retention capacity. This can result in a stress–strain field that sums to—and perhaps interferes with—the hydro-mechanically induced one (Rotta Loria and Laloui 2016). There is a flourishing literature focusing on the performance of geological repositories of spent nuclear fuel in which thermal effects in the ground (host rock) and engineered barrier layers (bentonite, bentonite–sand mixtures) are taken into account (e.g., Pusch 1992; Gens et al. 2009). For instance, the performance of bentonite barriers is lower at high temperatures, since such conditions cause a decrease in the bentonite swelling potential, reducing the self-sealing ability of discontinuities which is critical in limiting the migration of radionuclides (Villar and Lloret 2004).

In engineering geology, thermo-hydro-mechanical coupling is also known to play a role in the mechanics of basal shear zones of very deep and fast landslides (e.g., Veveakis et al. 2007), where conditions approach those of seismic faults. In the latter, processes controlled by temperature are well known and studied (e.g., Di Toro et al. 2011; Han et al. 2007). In general, however, thermal gradients—and, hence, heat flows—occur everywhere on slopes, controlled by changes in thermal boundary conditions or exothermal/endothermal processes within the domain (Scaringi and Loche 2022).

At the scale of the representative elementary volume in the laboratory, it can be shown than considering thermo-hydro-mechanical processes helps to achieve a more accurate depiction of the observed phenomena (e.g., Hueckel and Baldi 1990; Romero et al. 2005; Sun et al. 2020, 2022). The magnitude of these processes is particularly large in clay-bearing materials, so that even small variations in temperature can cause measurable changes in the hydro-mechanical response (Loche and Scaringi 2022a). Variations in the order of a few to tens of degrees are common in the shallow subsurface under seasonal and long-term climatic forcing; therefore, it seems reasonable to hypothesize that changes in temperature may affect the stability condition of slopes as well as the fate of existing landslide bodies (Scaringi and Loche 2022).

However, cases of temperature-activated landslides are not at all common in the literature, with few notable examples (Shibasaki et al. 2016; Loche et al. 2022b); yet, it must be noted that very few researchers have put an effort into seeking for evidence of such cases thus far. This may be reasonable even in the present situation of rapid climatic changes because most researchers have been focusing on the major role of climatic extremes in terms of enhanced hydro-mechanical forcing (Gariano and Guzzetti 2016; Scaringi and Loche 2022). As a result, slope stability and landslide research is yet to incorporate considerations and knowledge from studies on thermo-hydro-mechanical coupling in geomaterials, with the consequence that the direct role played by temperature changes and heat fluxes is being neglected. This task is, nevertheless, complex because while samples in the laboratory are typically homogeneous, the situation of real slopes is much more complex owing to heterogeneities, layering, and sets of discontinuities on the one hand, and boundary conditions that vary over various periods with various amplitudes on the other. Again, certain specific processes are recognized and taken into account: think of the role of heating–cooling cycles coupled with periodic wetting and drying conditions, which are typical in weathered rock layers and can be made even more complicated by cyclic freezing and thawing in (seasonally) cold climates. Note that the presence of discontinuities can serve as preferential pathways for the transmission of pore water pressures to large depths, but also of contaminants and heat where air or water fluxes are significant (Melis et al. 2020; Loche et al. 2021; Hu et al. 2018).

The limited availability of advanced modeling options that incorporate at least selected thermo-hydro-mechanical effects into slope stability analysis is another factor, together with the complexity of slope conditions, that limits the available evidence, or search for evidence, of thermally-activated processes in slopes (with the exception, of course, of issues related to freezing and thawing of water). Scaringi and Loche (2022) identified various links between features of landslides and the possible significance of thermal processes (Table 6.1). These can serve as a basis for prioritizing directions of model development and future research.

At an even larger scale (catchment-wide and regional analysis), physically-based models used in slope stability are very simplified owing to difficulties in obtaining input data (boundary conditions and calibrated parameters) with sufficient spatial and temporal resolution (Jeandet et al. 2019; Dreyfus et al. 2013; Salvatici et al. 2018). In fact, the goal of such modeling efforts becomes to only grasp first-order controls

Table 6.1 Landslide features and the possible influence of temperature on soil characteristics and behaviors (mod. from Scaringi and Loche 2022)

Landslide features	Possible influence of temperature	
Depth of the shear zone	Shallow versus deep	Sensitivity to short-period vs. long-period oscillations of temperature; importance of cyclic response of hydraulic parameters vs. importance of monotonic response of strength parameters
Thickness of the shear zone	Thin versus thick	Simple vs. complex response as soil parameters are affected by temperature oscillations of a single vs. various periods
Sliding mechanism	Planar versus rotational	Simple vs. complex response as soil parameters are affected by temperature oscillations of a single vs. various periods; moreover, stresses or flows between parts of the landslide body can occur
Material	Clay-rich versus clay-poor	Enhanced vs. reduced sensitivity to temperature, linked with the physico-chemical response of the microstructure (bound water) vs. changes in bulk water density and viscosity; persisting response in terms of pore water pressure excess vs. quick dissipation through highly conductive macro-porosity
Stage of movement	Initiation, triggering versus runout, reactivation	Importance of thermal effects on peak and critical state strength vs. residual shear strength; major vs. negligible role of thermal sensitivity of volumetric and hydraulic parameters
Velocity	Slow-moving (creeping) versus fast	Role of changing boundary conditions (e.g., climate forcing) vs. role of frictional heating
State of consolidation	Normally versus over-consolidated	Thermal sensitivity of critical and residual shear strength vs. peak strength, dilatancy and softening after peak (and enhanced heterogeneity in vertical direction due to loss of overconsolidation in the shear zone

of landslide patterns to inform risk reduction strategies or, in a broader perspective, predictions of landslide evolution. The same goal is pursued through the more widely available and fast-developing geostatistical and machine-learning approaches. These can successfully characterize spatio-temporal probabilities of landslide occurrence in wide regions based on lithological, hydrological, geometric, and other factors (Fan et al. 2019b). The use of spatially-distributed geotechnical parameters beyond simple physical characterizations is rare (Reichenbach et al. 2018); the use of thermal data per se or as a proxy for geotechnical parameters has basically been unexplored to date. In principle, this should not be a demanding task considering that thermal data are widely available from satellites and are already used for evaluations concerning land use and vegetation coverage. These data can be correlated, at a more detailed scale, with measurements of physical and geomechanical properties, as suggested by current studies (Loche et al. 2021, 2022a; Racek et al. 2023).

6.3 Promising Evidence

A recent work by Loche et al. (2022b) discussed the use of geostatistical approaches as a valid aid in identifying "hidden" processes responsible for changes in landslide patterns and trends. In particular, the authors explored the epicentral area of the 2008 Wenchuan earthquake in Sichuan, China through a detailed, publicly available multitemporal landslide inventory (Fan et al. 2019a). For the first time, the authors included the land surface temperature, a widely available remote sensing product, as one of the covariates in their numerical study, which was based on a generalized additive model in order to account for complex nonlinear relationships among the covariates (e.g., Goetz et al. 2015; Steger et al. 2016). The model was capable of displaying the fading legacy effect of the shaking over the years (Fig. 6.1) and, at the same time, the emerging of an increasingly important role of the surface temperature, which became one of the significant controls of the remobilization of coseismic landslides in the studied region. The model, in particular, highlighted that higher temperatures corresponded to more abundant remobilizations, suggesting an underlying thermal weakening mechanism of the slope materials. It is also worth noting that the patterns of the variable effects of the land surface temperature were uncorrelated with those of the normalized difference vegetation index, suggesting that higher surface temperatures did not necessarily correspond to unvegetated areas (possibly less stable owing to lack of root reinforcement). The authors discussed that explaining physical processes on the basis of geostatistical modeling results remains challenging. However, recent progresses towards explainable data-driven modeling may produce important advances in the future (Dahal and Lombardo 2023).

Moving down to the scale of an individual slope, Loche and Scaringi (2022b) conducted a conceptual slope stability analysis using finite element code and including thermo-mechanical effects via sequential coupling. As surface temperature changes in response to the alternation of seasons and, in the long term, to climate change, the temperature at each location within the ground changes with increasing delay and reduced amplitude of oscillations as the depth increases. Because the ground material is sensitive to temperature, its shear strength will exhibit a change in time consistent with that of temperature, with the sign of the change depending on whether the material weakens or strengthens. The authors hypothesized a case of a clayey soil that weakens by up to 1.5% per °C of warming, backed by experimental evidence (see the following paragraph). Over the domain of a hypothetical slope, the authors then conducted a stability analysis, and evaluated how the factor of safety changes with time. They observed, for curvilinear slip surfaces reaching depths of 6–10 m, changes in factor of safety by ~ 0.14 during the seasons, to which a change by ~ 0.08 could be added due to climate change in a scenario with + 5 °C of warming by the end of the century (Fig. 6.2). The study demonstrated that, for slopes with still a reasonable safety margin (e.g., FS = 1.1–1.2), failure may occur in the future by a solely thermal mechanism of weakening. This warming-induced failure would not be captured by the "classic" hydro-mechanical approaches that only look at changes in pore water pressures under hydrological forcing.

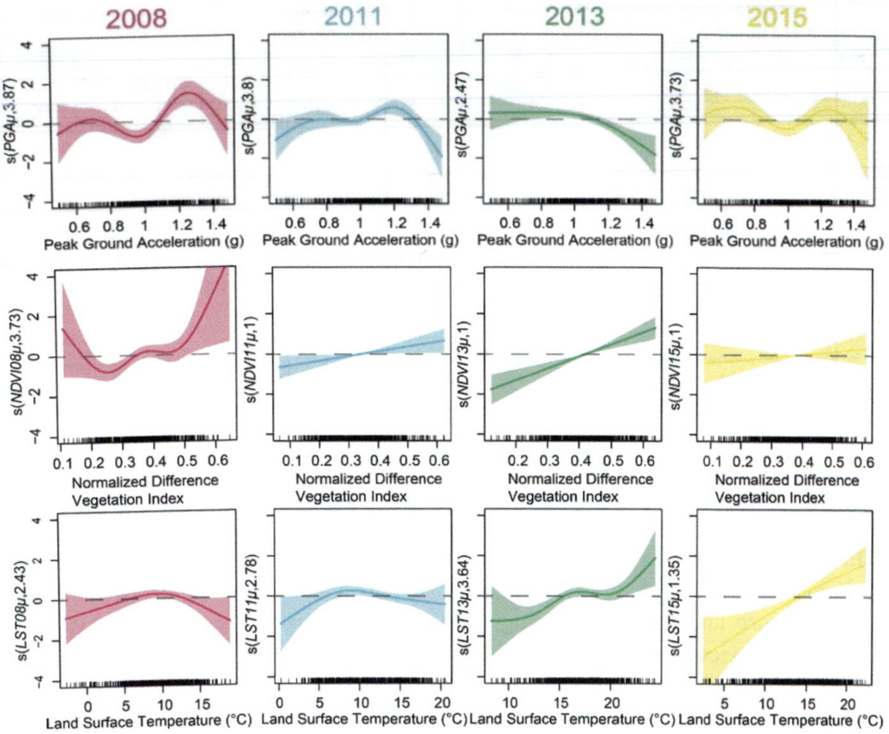

Fig. 6.1 Variable effects of selected covariates in the model of landslide susceptibility in the epicentral area of the 2008 Wenchuan Earthquake in Sichuan, China (mod. from Loche et al. 2022b): PGA—peak ground acceleration, NDVI—normalized difference vegetation index, LST—land surface temperature. The covariates are taken as average values over slope units

Fig. 6.2 Factor of safety (FS = 0.7–1.5) evaluated for potential slip surfaces reaching the depth of at least 6 m in the modelling by Loche and Scaringi (2022b): the baseline "no temperature accounted" was obtained by neglecting soil thermal sensitivity; other lines represent current hypothetical conditions according to the seasons ("first" spring, etc.), and conditions by the end of the century after a warming of 5 °C without changes in seasonal amplitude ("last" spring, etc.)

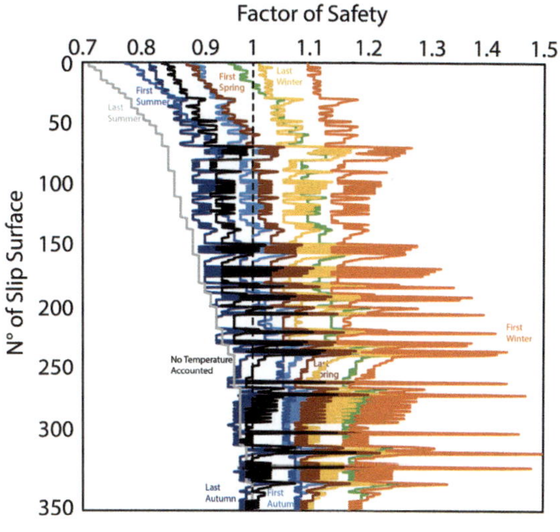

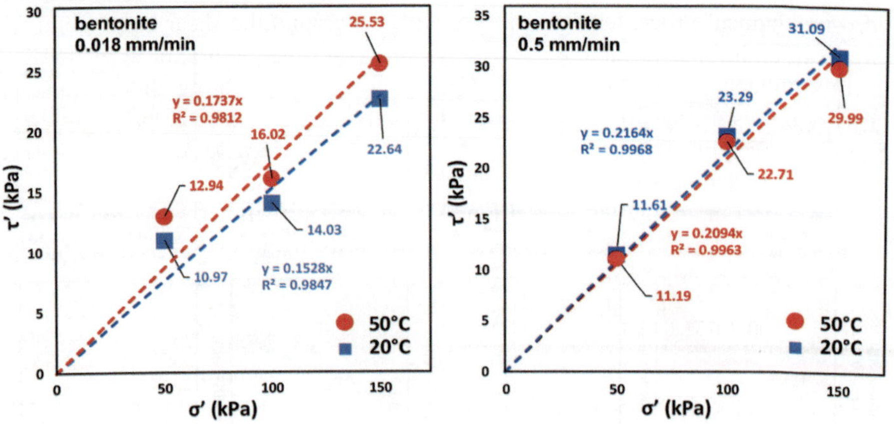

Fig. 6.3 Temperature effects observed by Loche and Scaringi (2022a) during ring-shear testing of a Ca–Mg bentonite under different temperatures, rates of shearing, and normal stresses

In the laboratory, Loche and Scaringi (2023) obtained experimental evidence of changes in the residual shear strength of pure clays under combined changes of temperature and rate of shearing. The authors utilized a commonly available commercial device—a Bromhead-type ring-shear machine—which was modified to achieve control over the temperature of the water bath, which was changed within the range of 20–50 °C while the samples were being sheared at a defined rate. In addition to observing different responses depending on the mineralogy of the tested samples (e.g., Ca–Mg bentonite versus kaolin), the authors noted a change in the response to heating and cooling according to the rate of shearing (Fig. 6.3). For instance, upon slow shearing (0.018 mm/min), the tested bentonite exhibited significant strengthening under normal stresses typical of landslides (50–150 kPa) in likely a drained condition. However, no strengthening was seen under moderately fast shearing (0.5 mm/min); instead, limited weakening was observed. Upon even faster shearing (1.8, 44.5 mm/min), results became inconsistent, displaying important weakening first, and then again strengthening. Under this testing condition, however, arising of turbulent shearing with disruption of the shear zone and possible porosity change-induced pore water pressure excess cannot be excluded.

6.4 Conclusion

In this contribution, we reviewed "classic" knowledge and recent advances in the evaluation of shear rate and thermal effects relevant to landslide behavior. Despite the elegant and effective classification proposed by Tika et al. (1996), changes in the available shear strength upon changes in shear displacement rate underly complex hydro-mechanical processes occurring within and around the shear zone. Temperature is an additional variable that further complicates the picture, in ways that are not fully understood and/or accounted for. Upon very fast shearing and under large normal stresses, that is, under large energy levels (Deng et al. 2020b, 2023), frictional heating typically produces dramatic weakening via a variety of processes. Conversely, under more common slope failure conditions, the main driver of temperature changes—in a range certainly different from that of the mineral-melting temperatures in giant landslides (e.g., Deng et al. 2020a)—is the ground-atmosphere interface. This interface is subject to a variety of processes that include water and air flows carrying heat through advection and convection, on top of heat transfers via irradiation and conduction in the solid and fluid phases. At depths of typical landslide shear zones, temperature may still vary seasonally (but is insensitive to daily fluctuations or short-lived hot or cold spells) by several degrees and, importantly, is affected by long-term changes such as those driven by global warming. The possibility of landslides induced or reactivated by changes in temperature is seldom considered by engineering geologists and, consequently, appropriate modeling tools have not been developed. However, we argue that the recent evidence of temperature sensitivity of the shear strength of landslide materials, combined with that of a sensitivity to temperature of shear rate effects, should prompt more extensive studies in order to either incorporate these effects into common modeling practices or excluding them if judged insignificant, but only after thorough investigation.

Acknowledgements The authors acknowledge support by the Czech Science Foundation (GAČR grant No. GJ 20-28853Y) and by the Czech Ministry of Education, Youth and Sport (MŠMT grant No. LL2316).

References

Acosta M, Passelègue FX, Schubnel A, Violay M (2018) Dynamic weakening during earthquakes controlled by fluid thermodynamics. Nat Commun 9:3074. https://doi.org/10.1038/s41467-018-05603-9

Alonso EE, Zervos A, Pinyol NM (2016) Thermo-poro-mechanical analysis of landslides: from creeping behaviour to catastrophic failure. Géotechnique 66:202–219. https://doi.org/10.1680/jgeot.15.LM.006

Amatya BL, Soga K, Bourne-Webb PJ, Amis T, Laloui L (2012) Thermo-mechanical behaviour of energy piles. Géotechnique 62:503–519. https://doi.org/10.1680/geot.10.P.116

Dahal A, Lombardo L (2023) Explainable artificial intelligence in geoscience: a glimpse into the future of landslide susceptibility modeling. Comput Geosci 176:105364. https://doi.org/10.1016/j.cageo.2023.105364

Deng Y, He S, Scaringi G, Lei X (2020a) Mineralogical analysis of selective melting in partially-coherent rockslides: bridging solid and molten friction. J Geophys Res Solid Earth 12:e2020JB019453. https://doi.org/10.1029/2020JB019453

Deng Y, Yan S, Scaringi G, Liu W, He S (2020b) An empirical power density-based friction law and its implications for coherent landslide mobility. Geophys Res Lett 47:e2020GL087581. https://doi.org/10.1029/2020GL087581

Deng Y, Fan X, Scaringi G, Wang D, He S (2023) Effect of particle crushing and thermally induced pressurization on rockslide mobility. Landslides 20:1535–1546. https://doi.org/10.1007/s10346-023-02053-3

Di Maio C, Scaringi G, Vassallo R (2015) Residual strength and creep behaviour on the slip surface of specimens of a landslide in marine origin clay shales: influence of pore fluid composition. Landslides 12:657–667. https://doi.org/10.1007/s10346-014-0511-z

Di Toro G, Han R, Hirose T, De Paola N, Nielsen S, Mizoguchi K, Ferri F, Cocco M, Shimamoto T (2011) Fault lubrication during earthquakes. Nature 471:494–498. https://doi.org/10.1038/nature09838

Dreyfus D, Rathje EM, Jibson RW (2013) The influence of different simplified sliding-block models and input parameters on regional predictions of seismic landslides triggered by the Northridge earthquake. Eng Geol 163:41–54

Duque J, Loche M, Scaringi G (2023) Rate-dependency of residual shear strength of soils: implications for landslide evolution. Géotech Lett 13:1–8. https://doi.org/10.1680/jgele.23.00004

Fan X, Scaringi G, Domènech G, Yang F, Guo X, Dai L, He C, Xu Q, Huang R (2019a) Two multi-temporal datasets that track the enhanced landsliding after the 2008 Wenchuan earthquake. Earth Syst Sci Data 11:35–55. https://doi.org/10.5194/essd-11-35-2019

Fan X, Scaringi G, Korup O, West AJ, van Westen CJ, Tanyas H, Hovius N, Hales TC, Jibson RW, Allstadt KE, Zhang L, Evans SG, Xu C, Li G, Pei X, Xu Q, Huang R (2019b) Earthquake-induced chains of geologic hazards: patterns, mechanisms, and impacts. Rev Geophys 57:421–503. https://doi.org/10.1029/2018RG000626

Gariano SL, Guzzetti F (2016) Landslides in a changing climate. Earth Sci Rev 162:227–252. https://doi.org/10.1016/j.earscirev.2016.08.011

Gens A, Sánchez M, Guimaraes LDN, Alonso EE, Lloret A, Olivella S, Villar MV, Huertas F (2009) A full-scale in situ heating test for high-level nuclear waste disposal: observations, analysis and interpretation. Géotechnique 59:377

Goetz JN, Brenning A, Petschko H, Leopold P (2015) Evaluating machine learning and statistical prediction techniques for landslide susceptibility modeling. Comput Geosci 81:1–11. https://doi.org/10.1016/j.cageo.2015.04.007

Goren L, Aharonov E, Anders MH (2010) The long runout of the heart mountain landslide: heating, pressurization, and carbonate decomposition. J Geophys Res 115:7113. https://doi.org/10.1029/2009JB007113

Han R, Shimamoto T, Hirose T, Ree J-H, Ando J-I (2007) Ultralow friction of carbonate faults caused by thermal decomposition. Science 316:878–881. https://doi.org/10.1126/science.1139763

Handwerger AL, Rempel AW, Skarbek RM, Roering JJ, Hilley GE (2016) Rate-weakening friction characterizes both slow sliding and catastrophic failure of landslides. Proc Natl Acad Sci 113:10281–10286. https://doi.org/10.1073/pnas.1607009113

Hu W, Xu Q, Wang G, Scaringi G, Mcsaveney M, Hicher P-Y (2017) Shear resistance variations in experimentally sheared mudstone granules: a possible shear-thinning and thixotropic mechanism. Geophys Res Lett 44:11040–11050. https://doi.org/10.1002/2017GL075261

Hu W, Scaringi G, Xu Q, Van Asch TWJ, Huang R, Han W (2018) Suction and rate-dependent behaviour of a shear-zone soil from a landslide in a gently-inclined mudstone-sandstone sequence in the Sichuan basin, China. Eng Geol 237:1–11. https://doi.org/10.1016/j.enggeo.2018.02.005

Hueckel T, Baldi G (1990) Thermoplasticity of saturated clays: experimental constitutive study. J Geotech Eng 116:1778–1796. https://doi.org/10.1061/(ASCE)0733-9410(1990)116:12(1778)

Jeandet L, Steer P, Lague D, Davy P (2019) Coulomb mechanics and relief constraints explain landslide size distribution. Geophys Res Lett 46:4258–4266. https://doi.org/10.1029/2019GL082351

Larsen IJ, Montgomery DR, Korup O (2010) Landslide erosion controlled by hillslope material. Nat Geosci 3:247–251. https://doi.org/10.1038/ngeo776

Li Y, Chan LS, Yeung AT, Xiang X (2013) Effects of test conditions on shear behaviour of composite soil. Proc Inst Civ Eng Geotech Eng 166:310–320

Loche M, Scaringi G, Blahůt J, Melis MT, Funedda A, Da Pelo S, Erbì I, Deiana G, Meloni MA, Cocco F (2021) An infrared thermography approach to evaluate the strength of a rock cliff. Remote Sens 13:1265. https://doi.org/10.3390/rs13071265

Loche M, Scaringi G, Blahůt J, Hartvich F (2022a) Investigating the potential of infrared thermography to inform on physical and mechanical properties of soils for geotechnical engineering. Remote Sens 14:4067. https://doi.org/10.3390/rs14164067

Loche M, Scaringi G, Yunus AP, Catani F, Tanyaş H, Frodella W, Fan X, Lombardo L (2022b) Surface temperature controls the pattern of post-earthquake landslide activity. Sci Rep 12:988. https://doi.org/10.1038/s41598-022-04992-8

Loche M, Scaringi G (2022b) Temperature continuously controls the stability of clay slopes. https://doi.org/10.48550/arXiv.2211.09046

Loche M, Scaringi G (2023) Temperature and shear-rate effects in two pure clays: Possible implications for clay landslides. Results in Engineering 20:101647. https://doi.org/10.1016/j.rineng.2023.101647

Lucas A, Mangeney A, Ampuero JP (2014) Frictional velocity-weakening in landslides on Earth and on other planetary bodies. Nat Commun 5:4417. https://doi.org/10.1038/ncomms4417

Lupini JF, Skinner AE, Vaughan PR (1981) The drained residual strength of cohesive soils. Geotechnique 31:181–213

Melis MT, Da Pelo S, Erbì I, Loche M, Deiana G, Demurtas V, Meloni MA, Dessì F, Funedda A, Scaioni M, Scaringi G (2020) Thermal remote sensing from UAVs: a review on methods in coastal cliffs prone to landslides. Remote Sens 12:1971. https://doi.org/10.3390/rs12121971

Pusch R (1992) Use of bentonite for isolation of radioactive waste products. Clay Miner 27:353–361

Racek O, Balek J, Loche M, Vích D, Blahůt J (2023) Rock surface strain in situ monitoring affected by temperature changes at the Požáry Field Lab (Czechia). Sensors 23:2237. https://doi.org/10.3390/s23042237

Reichenbach P, Rossi M, Malamud BD, Mihir M, Guzzetti F (2018) A review of statistically-based landslide susceptibility models. Earth Sci Rev 180:60–91. https://doi.org/10.1016/j.earscirev.2018.03.001

Roháč J, Scaringi G, Boháč J, Kycl P, Najser J (2020) Revisiting strength concepts and correlations with soil index properties: insights from the Dobkovičky landslide in Czech Republic. Landslides 17:597–614. https://doi.org/10.1007/s10346-019-01306-4

Romero E, Villar MV, Lloret A (2005) Thermo-hydro-mechanical behaviour of two heavily overconsolidated clays. Eng Geol 81:255–268

Rotta Loria AF, Laloui L (2016) Thermally induced group effects among energy piles. Géotechnique 67:374–393. https://doi.org/10.1680/jgeot.16.P.039

Saito R, Fukuoka H, Sassa K (2006) Experimental study on the rate effect on the shear strength. Disaster mitigation of debris flows, slope failures and landslides. Universal Academic Press, Tokyo, pp 421–427

Salvatici T et al (2018) Regional physically based landslide early warning modelling: soil parameterisation and validation of the results. Nat Hazards Earth Syst Sci Discuss 12:1–23. https://doi.org/10.5194/nhess-2017-425

Scaringi G, Di Maio C (2016) Influence of displacement rate on residual shear strength of clays. Proced Earth Planet Sci 16:137–145. https://doi.org/10.1016/j.proeps.2016.10.015

Scaringi G, Loche M (2022) A thermo-hydro-mechanical approach to soil slope stability under climate change. Geomorphology 401:108108. https://doi.org/10.1016/j.geomorph.2022.108108

Scaringi G, Hu W, Xu Q (2018a) Discussion on: "experimental study of residual strength and the index of shear strength characteristics of clay soil." Eng Geol 242:218–221. https://doi.org/10.1016/j.enggeo.2018.06.021

Scaringi G, Hu W, Xu Q, Huang R (2018b) Shear-rate-dependent behavior of clayey bimaterial interfaces at landslide stress levels. Geophys Res Lett 45:766–777. https://doi.org/10.1002/2017GL076214

Shibasaki T, Matsuura S, Okamoto T (2016) Experimental evidence for shallow, slow-moving landslides activated by a decrease in ground temperature: landslides affected by ground temperature. Geophys Res Lett 43:6975–6984. https://doi.org/10.1002/2016GL069604

Steger S, Brenning A, Bell R, Petschko H, Glade T (2016) Exploring discrepancies between quantitative validation results and the geomorphic plausibility of statistical landslide susceptibility maps. Geomorphology 262:8–23. https://doi.org/10.1016/j.geomorph.2016.03.015

Sun H, Mašín D, Najser J, Scaringi G (2020) Water retention of a bentonite for deep geological radioactive waste repositories: high-temperature experiments and thermodynamic modeling. Eng Geol 269:105549. https://doi.org/10.1016/j.enggeo.2020.105549

Sun H, Scaringi G, Mašín D, Najser J (2022) An experimental investigation on the swelling behavior of compacted B75 bentonite. Eng Geol 296:106452. https://doi.org/10.1016/j.enggeo.2021.106452

Tika TE, Vaughan PR, Lemos LJLJ (1996) Fast shearing of pre-existing shear zones in soil. Géotechnique 46:197–233. https://doi.org/10.1680/geot.1996.46.2.197

Tiwari G, Latha GM (2019) Reliability analysis of jointed rock slope considering uncertainty in peak and residual strength parameters. Bull Eng Geol Environ 78:913–930. https://doi.org/10.1007/s10064-017-1141-1

Veveakis E, Vardoulakis I, Toro GD (2007) Thermoporomechanics of creeping landslides: the 1963 Vaiont slide, northern Italy. J Geophys Res Earth Surf 112:702. https://doi.org/10.1029/2006JF000702

Vieira A, Alberdi-Pagola M, Christodoulides P, Javed S, Loveridge F, Nguyen F, Cecinato F, Maranha J, Florides G, Prodan I, Van Lysebetten G, Ramalho E, Salciarini D, Georgiev A, Rosin-Paumier S, Popov R, Lenart S, Erbs Poulsen S, Radioti G (2017) Characterisation of ground thermal and thermo-mechanical behaviour for shallow geothermal energy applications. Energies 10:2044. https://doi.org/10.3390/en10122044

Villar MV, Lloret A (2004) Influence of temperature on the hydro-mechanical behaviour of a compacted bentonite. Appl Clay Sci 26:337–350

Wang YF, Dong JJ, Cheng QG (2017) Velocity-dependent frictional weakening of large rock avalanche basal facies: implications for rock avalanche hypermobility? J Geophys Res Solid Earth 122:1648–1676. https://doi.org/10.1002/2016JB013624

Chapter 7
The Classification Methods of Anisotropic Rock Mass and Applications

Songfeng Guo, Shengwen Qi, Charalampos Saroglou, Xin Yu, and Bowen Zheng

Abstract Rock mass classification methods are useful tools for engineering geologists to evaluate the quality of jointed rock mass in nature, and have been widely used in the engineering construction. Nevertheless some special rock mass such as anisotropic rock mass or bimrocks. In this study, we reviewed the existing classification tools and introduced a recently developed classification method for anisotropic rock mass. Both the anisotropic degree and rock mass quality can be determined with this method. And then the we evaluate the rock mass engineering slopes along a railway using this method for layered rock mass along with a combined method for non-layered rock mass. The reliability and accessibility of the method may expect an extensive application in future.

Keywords Anisotropic rock mass · Rock quality · Classification · Rock mass stability · Rock mechanical behavior

S. Guo (✉) · S. Qi · C. Saroglou · X. Yu · B. Zheng
Key Laboratory of Shale Gas and Geoengineering, Institute of Geology and Geophysics, Chinese Academy of Sciences, Beijing 100029, China
e-mail: guosongfeng@mail.iggcas.ac.cn

Innovation Academy of Earth Science, Chinese Academy of Sciences, Beijing 100029, China

University of Chinese Academy of Sciences, Beijing 100049, China

S. Qi
e-mail: qishengwen@mail.iggcas.ac.cn

C. Saroglou
e-mail: saroglou@central.ntua.gr

X. Yu
e-mail: yuxin@mail.iggcas.ac.cn

B. Zheng
e-mail: zhengbowen@mail.iggcas.ac.cn

C. Saroglou
School of Civil Engineering, National Technical University of Athens, Athens, Greece

S. Wang et al. (eds.), *Engineering Geology for a Habitable Earth: IAEG XIV Congress 2023 Proceedings, Chengdu, China*, Environmental Science and Engineering,
https://doi.org/10.1007/978-981-99-9057-3_7

7.1 Introduction

The rock mass classification is an effective and common used tool to assess the quality and then stability of rock mass engineering. Engineering rock mass quality evaluation mainly includes three categories: the surrounding rock quality classification of underground engineering, the rock mass quality evaluation of slope engineering and the rock mass quality evaluation of dam foundation engineering. The research has gone through nearly a century of development history, among which the research on rock mass quality evaluation of underground engineering was carried out earlier and more perfect than other projects.

Before the 1970s, the international evaluation of rock mass quality mostly focused on single-index qualitative or quantitative classification, with low reliability. Deere's (1964) RQD classification was a representative example. After the 1970s, the classification of rock mass quality has developed from qualitative to quantitative, from single factor to multi-factor. The representative method is Barton's Q system (Barton et al. 1974; Barton 2013; Bar and Barton 2017), Bieninawski's (1973) RMR classification and Hoek's GSI surrounding rock classification (Hoek and Brown 1997; Chen 2007) etc.

The research on rock mass quality evaluation in China began in the 1960s and 1970s. In 1979, Gu and Huang, former Institute of Geology, Chinese Academy of Sciences, put forward the rock mass quality classification method "Z system" based on a large number of rock mass engineering practices (Gu 1979; Gu and Huang 1979a, b). Subsequently, the state and various industries gradually proposed their own classification standards for rock mass quality, including the "Code for Geological Survey of Water Conservancy and Hydropower Engineering", the HC classification of surrounding rocks in underground chambers (The National Standards Compilation Group of People's Republic of China 2009), the BQ classification of the national standard for engineering rock mass classification (Wu and Liu 2012), the CSMR classification of the China Water Conservancy and Hydropower Slope Engineering Registration Group (Chen 2004). Saroglou et al. (2019) revised the RMR classification method and proposed an ARMR classification method suitable for anisotropic rock masses. Sonmez et al. (2004) established the relationship between elastic modulus and rock strength in slope engineering based on GSI.

Among them, the international BQ method is a general rock mass quality classification method, which has been promoted and implemented as a national standard for rock mass classification, and it is stipulated that engineering rock mass classification should be carried out in two steps by combining qualitative and quantitative methods. The basic rock mass quality BQ shall be determined first, and then determine the rock mass grade in combination with the characteristics of the specific project. The National Standard of the People's Republic of China "Engineering Rock Mass Grading Standards" promulgated in 2014 (Wu and Liu 2012) Changjiang Academy of Sciences, Changjiang Water Resources Commission, 2014 is the highest basic standard of the country, which is proposed on the basis of summarizing various rock mass classification methods and a large number of engineering practices at home and abroad. It is applicable to the rock mass classification of various industries and types of rock engineering, and is the basic basis for formulating rock mass classification standards for various industries. In this standard, the BQ classification method is established, which highlights the commonness of engineering rock mass, namely, the basic quality of rock mass, and is classified by a combination of qualitative and quantitative methods, which is highly scientific and practical. In the study of slope stability, it is more suitable for the evaluation and analysis of rock mass quality of non-layered slopes.

However, due to the directional structure of the layered slope rock mass, the strength, deformation, and even permeability of the rock mass show differences in different directions, which has the anisotropy of the rock mass. There are a large number of sedimentary rocks and parametamorphic rocks distributed in and around the Qinghai-Tibet Plateau. However, most of the existing rock mass quality classification methods are formulated based on the assumption of isotropic conditions, which are not suitable for anisotropic rock mass engineering, which will easily lead to deviations in rock mass quality evaluation, improper support measures, huge economic losses and even threats to life safety. Therefore, in recent years anisotropy system A-BQ based on the national standard BQ and ARMR based on RMR were proposed to solve the classification problem of anisotropic rock mass, including the degree of anisotropy and rock mass quality (Saroglou et al. 2019; Guo et al. 2020a, b). Several common engineering rock mass quality evaluation methods are shown in Table 7.1.

7.2 Rock Mass Quality Evaluation Method for Non-layered Rock Mass

The international BQ-RMR-GSI method is based on the geological strength index (GSI) and uses the RMR method and the international BQ method to classify rock mass quality. The Geological Strength Index (GSI) rock mass classification system is a method for directly classifying jointed rock masses to determine their mechanical parameters.

Based on the investigation of rock mass joint surface characteristics and rock mass structure, combined with indoor rock mechanics tests, the Hoek–Brown strength criterion is used to classify rock masses and finally estimate rock mass mechanical parameters (Hoek and Brown 1997, 2019; Marinos and Hoek 2000; Hoek et al. 2005). As shown in Fig. 7.1, E. Hoek et al. divided the rock mass structure into complete or massive structure, mosaic structure, fragmented structure/disturbance/crack, granular structure, layered/shear zone, and classified it into very good, good, average, poor and very poor etc. according to the surface condition of the structural plane. The two main quantitative indexes of joint spacing and joint surface roughness coefficient (JRC) value of rock mass structure are used to quantify GSI, and the fine evaluation of GSI rock mass indicators is realized.

Table 7.1 Quality classification method of engineering rock mass

Method	Parameters	Related formula	Related charts	Founder	Country	Fields applications
Q system	Rock quality index (RQD)	$Q = RQD/Jn \cdot Jr/Ja \cdot Jw/SRF$	Classification and grade of surrounding rock of Q system	Barton et al. (1974) (last modified 2002)	Norway	Tunnels, mines, foundations
	Number of joint groups Jn	Where, RQD/Jn represents the integrity of rock mass; Jr/Ja, indicating the filling characteristics and weathering alteration degree of cracks; Jw/SRF represents the reduction of water and geostress on rock mass quality	RQD value table of rock quality index			
	Joint roughness Jr		Value table of joint group number Jn			
	Joint alteration value Ja		Joint roughness Jr value table			
	Joint moisture reduction coefficient Jw		Value table of joint alteration coefficient Ja			
	Stress reduction factor (SRF)		Value table of joint water reduction coefficient Jw			
			Stress reduction factor SRF value table			
RMR system	R1 uniaxial compressive strength of rock	$RMR = R1 + R2 + R3 + R4 + R5 + R6$	Rock mass quality classification by RMR classification method	Bieniawski (1973) (last modified 1989-USA)	South Africa	Tunnels, mines, slopes, foundations

(continued)

Table 7.1 (continued)

Method	Parameters	Related formula	Related charts	Founder	Country	Fields applications
	R2 rock quality index (RQD)	RMR is the sum of the scoring values of various indicators (see the table RMR classification parameters and scoring values for details), and at the same time consider the influence of the occurrence of the main structural plane (R6) to modify it and correct the scoring value. For structural plane characters (R4), the specific description list of structural plane character classification (R4) can be used for detailed scoring	RMR classification parameters and scoring values			
	R3 structural plane spacing		Detailed description list of structural plane character classification (R4)			
	R4 structural surface condition		Main structural planes impact correction table			
	R5 groundwater condition		Influence of structural plane direction on the project			
	R6 structural plane orientation					
GSI system	Joint spacing of rock mass structure	The GSI rock mass classification system is used to determine the GSI value of the rock mass, and then combined with the Hoek–Brown strength criterion to calculate the rock mass parameters	Quantitative description of geological strength index of rock mass by GIS	Hoek and Brown (2019)		

(continued)

Table 7.1 (continued)

Method	Parameters	Related formula	Related charts	Founder	Country	Fields applications
	JRC value of structural plane	$\sigma_1 = \sigma_3 + \sigma_c(m_b \frac{\sigma_3}{\sigma_c} + s)^{\alpha}$	Standard roughness profile and its JRC value			
		σ_1, σ_3 are the maximum principal stress and the minimum principal stress respectively; σ_C is the uniaxial compressive strength of rock block; mb, s, α Is the Hoek Brown constant of rock mass				
		$m_b = m_i \exp\left(\frac{GSI - 100}{28 - 14D}\right)$ $S = \exp\left(\frac{GSI - 100}{28 - 3D}\right)$ $\alpha = \frac{1}{2} + \frac{1}{6}\left(e^{-\frac{GSI}{15}} + e^{-\frac{20}{3}}\right)$				
		Mi is the Hoek Brown constant of complete rock, D is the rock mass disturbance coefficient, generally 0 for no disturbance, 0.5 for partial disturbance and 1 for complete disturbance				
		$E_m = \left(1 - \frac{D}{2}\right)\sqrt{\frac{\sigma_{ci}}{100}} \times 10^{\frac{GSI-10}{40}} (\sigma_{ci} < 100\ \text{Mpa})$ $E_m = \left(1 - \frac{D}{2}\right)\sqrt{\frac{\sigma_{ci}}{100}} \times 10^{\frac{GSI-10}{40}} (\sigma_{ci} > 100\ \text{Mpa})$				
Z system	Integrity of rock mass	$Z = K_v \times f \times S$	Z system rock mass quality classification method	Gu and Huang (1979a, b)	China	Tunnels, mines, foundations
	Shear strength of structural plane	K_v is the integrity coefficient, $K_v = V_m^2/V_r^2$, V_r is the longitudinal wave velocity of elastic wave propagation in rock, and V_m is the longitudinal wave velocity of elastic wave in rock mass composed of similar rocks; F is the friction coefficient of structural surface, f = tan φ, φ Is the friction angle of the structural plane; S is the strength coefficient of rock block, $S = R_c/100$, and R_c is the saturated uniaxial compressive strength of rock				
	Strength of rock block					

(continued)

Table 7.1 (continued)

Method	Parameters	Related formula	Related charts	Founder	Country	Fields applications
National standard BQ system	Type and extensibility of main structural planes	$[BQ] = BQ - 100(K_4 + \lambda K_5)$	Main structural plane types and extensibility correction coefficient of slope engineering λ	Wu and Liu (2012)	China	General classification method of rock mass quality
	Development degree of groundwater in the slope	$K_5 = F_1 \times F_2 \times F_3$	Groundwater impact correction coefficient for slope engineering			
	Relationship between occurrence of structural plane and slope surface	Where: λ-Slope engineering main structural surface type and extensibility correction coefficient K4-Slope engineering groundwater impact correction coefficient K5-Correction factor for the influence of the occurrence of main structural planes in slope engineering; F1-Coefficient reflecting the influence of the relationship between the inclination of the main structural plane and the inclination of the slope F2-Coefficient to reflect the influence of inclination angle of main structural plane F3-Coefficient reflecting the relationship between the slope angle and the main structural plane angle				

(continued)

Table 7.1 (continued)

Method	Parameters	Related formula	Related charts	Founder	Country	Fields applications
CSMR system	Uniaxial compressive strength	F_1 and F_3 reflect the difference between the inclination and dip angle between the slope surface and the control structural plane; F_2 reflects the dip angle of the structural plane; F_4 is the coefficient reflecting the blasting excavation method; a is the height correction factor; λ Is the condition coefficient of the structural plane; H is the slope height; Hr is the critical height of 80 m; Y = H/Hr	Adjustment value of discontinuity occurrence Adjustment coefficient of structural plane	Chen (2004)	China	Tunnels, mines, foundations
	RQD					
	Joint condition					
	Joint spacing					
	Groundwater					
A-BQ system	Rock strength anisotropy	$R_{c\max} = 22.82 I_{s(50)\perp}^{0.75}$ $R_{c\min} = 22.82 I_{s(50)\parallel}^{0.75}$	Determination Method of Rock Mass Integrity Index $K_{v\text{-}an}$ of Anisotropic Structure	Guo et al. (2020a, b)	China	Tunnels, slopes (anisotropic rock quality evaluation)
		$I_{s(50)\perp}^{0.75}$ and $I_{s(50)\parallel}^{0.75}$ is (50) under the load perpendicular to and parallel to the inherent anisotropic plane of the rock mass (such as bedding, gneissic, etc.). When it is difficult to obtain the experimental data of Rc, it can be determined according to the uniaxial compressive strength under dry conditions: $R_c = 0.759\sigma_c - dry$	Method for Determination of Rock Mass Integrity Index $K_{v\text{-}an}$ Contributed by Non-Anisotropic Structural Planes			
	Structural anisotropy	$K_{v-an} = \frac{V_{m-an}^2}{V_r^2}$ $K_{v-ra} = \frac{V_{m-ra}^2}{V_r^2}$	Classification according to the degree of anisotropy of I_{BQan}			

(continued)

Table 7.1 (continued)

Method	Parameters	Related formula	Related charts	Founder	Country	Fields applications
		K_{v-an} denotes the integrity index derived from the anisotropic structure, and K_{v-ra} denotes the integrity index due to discontinuities except the anisotropic structure. V_r represents the P-wave velocity of the intact rock, while $V_{m\text{-}an}$ and $V_{m\text{-}ra}$ represent the P-wave velocity of the rock mass. Rock mass integrity can be divided into five categories according to K_v				
		$ABQ_{max} = 100 + 3R_{c\,max} + 250K_{v-ra}$ $ABQ_{min} = 100 + 3R_{c\,min} + 250K_{v-an}$ $I_{BQan} = \frac{ABQ_{max}}{ABQ}$				
		The degree of anisotropy of rock mass is divided into five grades according to I_{BQan}				

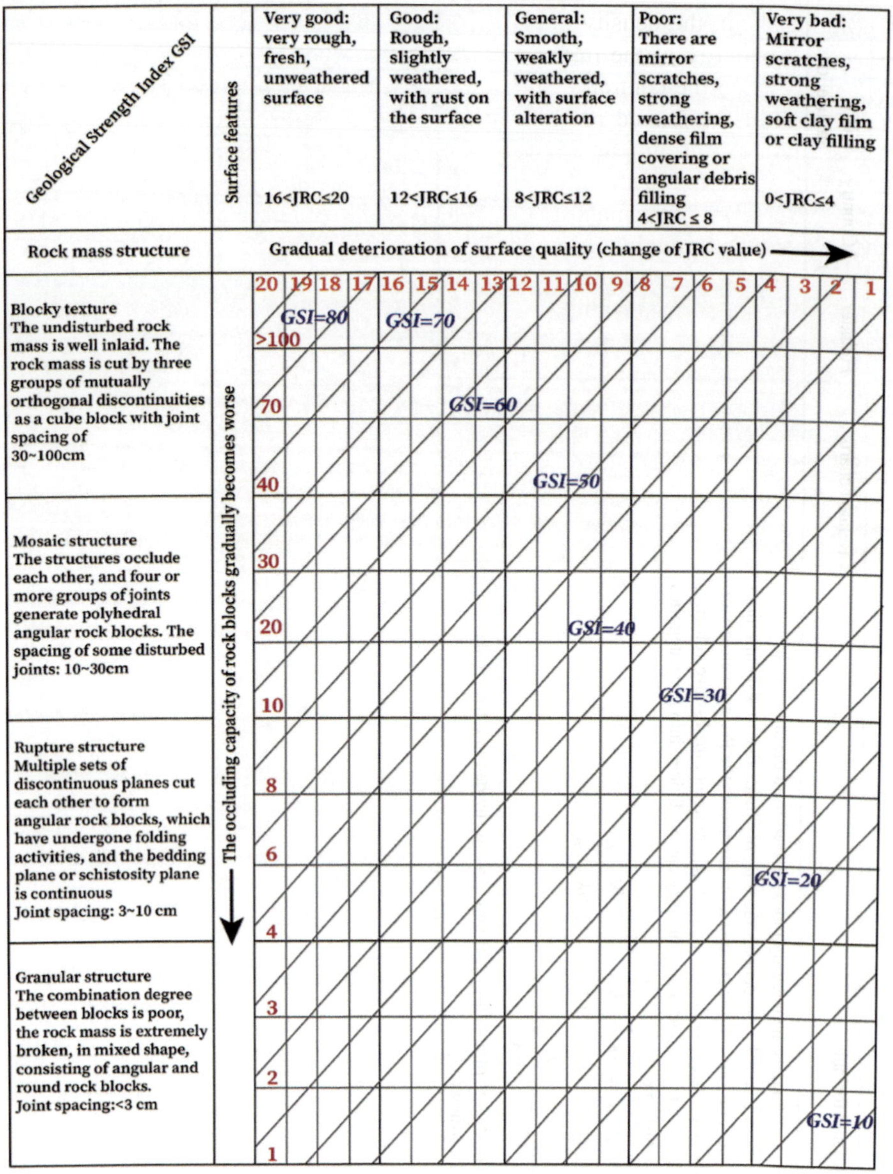

Fig. 7.1 Quantitative description of rock mass geological strength index GSI (Hu et al. 2012)

Among them, the density of joints and fissures in the rock mass structure can reflect the integrity of the rock mass. The denser the joints, the more broken the rock mass is, on the contrary, the more complete rock mass it is. The quantitative description of the jointed rock mass structure can be achieved by measuring the size of the joint spacing, and more accurate GSI values can be obtained. JRC is a quantitative description index of the roughness and undulation characteristics of joint surface, and is one of the main controlling factors affecting the mechanical properties of rock mass discontinuity. Barton and Choubey (1977) proposed the concept of JRC for the first time, divided the JRC value into 10 grades according to the roughness of the structural surface within 0–20, and established the contour curve of a typical rough undulating surface as the empirical evaluation standard of JRC, as shown in Fig. 7.2 (Barton and Choubey 1977).

Based on the investigation of the rock mass structure, joint spacing and JRC value at the site, the GSI of rock mass at the entrance and exit slope of the tunnel can be taken according to Fig. 7.1. RMR is a commonly used rock mass classification method in the world, and its value has a certain relationship with BQ, etc., and Hoek et al. made the value of GSI universally applicable by linking GSI with RMR. The basic relationship between RMR and GSI is as follows (Dongxu 2016):

$$\mathrm{RMR} = \mathrm{GSI} + 5 \tag{7.1}$$

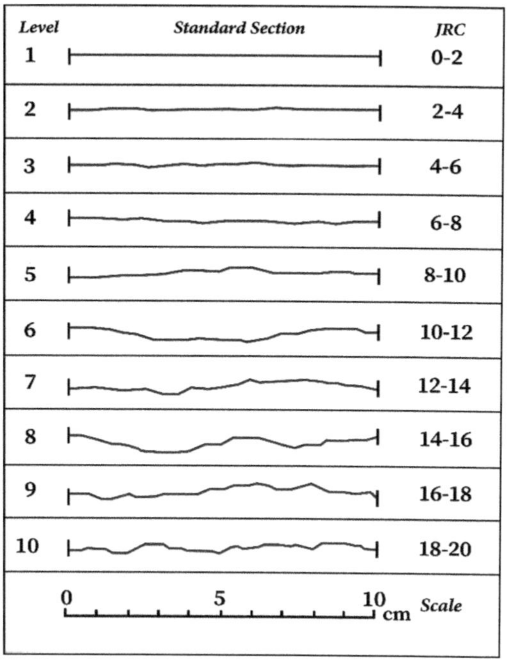

Fig. 7.2 Standard roughness profiles and their JRC values (Barton and Choubey 1977)

Statistical analysis of a large number of measured data shows that there is a linear relationship between the BQ value and the RMR value (Cai et al. 2001), and the BQ value of the rock mass can be obtained by the following fitting relationship:

$$BQ = (RMR - 21.375)/0.089 \tag{7.2}$$

The BQ value of the basic quality index of rock mass is based on 103 typical projects as a sampling population, using multiple stepwise regression and discriminant analysis methods, and considering the influence of groundwater, the occurrence of main structural planes and natural stress, establishing a modified expression of BQ, and then grading the rock mass quality according to the BQ value (Table 7.2).

Table 7.2 Rock mass quality classification by BQ method (Wu and Liu 2012)

Rock mass quality grade	Qualitative characteristics of rock mass quality	Rock mass basic quality index BQ
I	Hard rock, complete rock mass	> 550
II	Hard rock, relatively complete rock mass; Relatively hard rock, complete rock mass	550–451
III	Hard rock, relatively broken rock mass; Harder rock or soft and hard interbedded, the rock mass is relatively complete; Softer rock, complete rock mass	450–351
IV	Hard rock, broken rock mass; Hard rock, rock mass is relatively broken or broken; Soft rock or hard rock interbedding, dominated by soft rock, the rock mass is complete or relatively broken; Soft rock, complete or intact rock mass	350–251
V	Soft rock, broken rock mass; Soft rock, the rock mass is relatively broken or broken; All extremely soft rocks and all extremely broken rocks	< 250

7.3 Classification Method of Anisotropic Rock Mass

The A-BQ method incorporates two basic sets of rating factors, including strength anisotropy and structural anisotropy. The degree of anisotropy of the rock mass is characterized by the ratio of the maximum quality score to the minimum quality score, and is adjusted by the surrounding rock stress. The rock mass quality score depends on the key factors for the appearance of the anisotropic structure and the correction factors for the stress state and groundwater conditions; The quality of the rock mass is characterized by the quality score, which is divided into five grades.

The saturated uniaxial compressive strength R_c is used as the strength parameter in the BQ method, and the maximum and minimum saturated uniaxial compressive strengths (R_{cmax} and R_{cmin}) are used in the A-BQ method. R_{cmax} can be determined by a point load test perpendicular to the anisotropic plane, and R_{cmin} can be determined by a point load test parallel to the anisotropic plane, $I_{s(50)}$ is the point load strength of a rock specimen with a diameter of 50 mm (Franklin 1985; Singh et al. 2012). Based on a large number of rock mechanics tests, the China Academy of Railway Construction Sciences proposed the equation (GB50218-2014), and established the relationship between R_{cmax} and R_{cmin} and the anisotropic point load strength index (Wu and Liu 2012) as follows:

$$R_{c\max} = 22.82 I_{s(50)\perp}^{0.75} \tag{7.3}$$

$$R_{c\min} = 22.82 I_{s(50)\parallel}^{0.75} \tag{7.4}$$

where R_c can be determined by performing uniaxial compression tests on saturated intact rock samples in different loading directions of the anisotropy plane, $I_{s(50)\perp}^{0.75}$ and $I_{s(50)\parallel}^{0.75}$ are the axial and radial compression tests perpendicular to and parallel to the anisotropy plane, respectively acquired $I_{s(50)}$ (Saroglou and Tsiambaos 2007). When it is difficult to determine the experimental value of Rc, it can be determined from the uniaxial compressive strength under dry conditions $\sigma_{c-\text{dry}}$ (Vasarhelyi 2003):

$$R_c = 0.759 \sigma_{c-\text{dry}} \tag{7.5}$$

The degree of structural plane development in the rock mass is determined by the rock mass integrity index K_v. Rock mass integrity can be evaluated by various parameters, including joint spacing, RQD, volumetric joint number (J_v), joint group number (J_n), joint length (J_L), and block volume (V_b). In A-BQ, the maximum and minimum values of K_v in different directions are determined by the following formulas:

$$K_{v-an} = \frac{V_{m-an}^2}{V_r^2} \tag{7.6}$$

$$K_{v-ra} = \frac{V_{m-ra}^2}{V_r^2} \tag{7.7}$$

$K_{v\text{-}an}$ indicates the integrity index originating from the anisotropic structure, and $K_{v\text{-}ra}$ indicates the integrity index due to discontinuities other than the anisotropic structure. V_r represents the compressional wave velocity of the intact rock, as determined by ultrasonic testing of the core. $V_{m\text{-}an}$ and $V_{m\text{-}ra}$ denote the compressional wave velocity of the rock mass, determined by in situ wave velocity measurements perpendicular and parallel to the anisotropic structure, respectively.

Rock mass integrity can be classified into five categories according to K_v (Table 7.3), which can also be roughly determined using the spacing of anisotropic structures if it is difficult to measure in-situ velocity (Saroglou et al. 2019), while also considering RQD (RQD_{an}) derived from anisotropic structure (Deere 1964). For other structural planes except anisotropic structures, modified RQD (RQD_{ra}) and volume joint number ($J_{v\text{-}ra}$) excluding anisotropic structures can be used to determine (Table 7.4).

In order to exclude the influence of high-strength rock mass with densely developed structural planes, if the rock mass strength is high, the following conditions should be followed (Wu and Liu 2012):

When

$$\text{When } R_{c\max} > 90K_{v-ra} + 30,\quad R_{c\max} = 90K_{v-ra} + 30 \tag{7.8}$$

$$\text{When } R_{c\min} > 90K_{v-an} + 30,\quad R_{c\min} = 90K_{v-an} + 30 \tag{7.9}$$

Table 7.3 Determination method of rock mass integrity index Kv-an of anisotropic structure [modified from GB 50218-2014; (Saroglou et al. 2019)]

Anisotropic structure plane spacing/m	QUR	Kv-an	Describe
>1	90–100	0.75–1	Whole
0.4–1	75–90	0.55–0.75	Relatively complete
0.2–0.4	50–75	0.35–0.55	Relatively broken
0.2–0.04	25–50	0.15–0.35	Broken
<0.04	0–25	0–0.15	Extremely broken

Table 7.4 Determination method of rock mass integrity index Kv-ra contributed by non-anisotropic structural plane [modified according to (50218GB/T 2014)]

$J_{v\text{-}ra}$/m^{-3}	RQD_{ra}	$K_{v\text{-}ra}$
<3	90–100	0.75–1
3–10	75–90	0.55–0.75
10–20	50–75	0.35–0.55
20–35	25–50	0.15–0.35
>35	0–25	0–0.15

In order to reduce the impact of low-strength rock mass with undeveloped structural planes, if the rock mass strength is low, the following conditions should be followed:

$$\text{When } K_{v-ra} > 0.04R_{c\,\max} + 0.4, \quad K_{v-ra} = 0.04R_{c\,\max} + 0.4 \tag{7.10}$$

$$\text{When } K_{v-an} > 0.04R_{c\,\min} + 0.4, \quad K_{v-an} = 0.04R_{c\,\min} + 0.4 \tag{7.11}$$

Referring to the rock mass classification equation (GB50218-2014) in the BQ method, the maximum and minimum ABQ values of anisotropic rock mass in different directions can be determined by the following formulas:

$$ABQ_{\max} = 100 + 3R_{c\,\max} + 250K_{v-ra} \tag{7.12}$$

$$ABQ_{\min} = 100 + 3R_{c\,\min} + 250K_{v-an} \tag{7.13}$$

The ratio of the maximum and minimum values of ABQ is used to express the degree index of anisotropy IBQ_{an}:

$$I_{BQan} = \frac{ABQ_{\max}}{ABQ_{\min}} \tag{7.14}$$

Referring to the rock anisotropy classification method (Saroglou and Tsiambaos 2007), the degree of anisotropy of rock mass is divided into five grades (Table 7.5), and the threshold range is obtained from a large number of rock tests.

Studies have shown that confining pressure has a significant impact on the degree of anisotropy (Ramamurthy and Arora 1994; Vutukuri and Hossaini 1995; Nasseri et al. 2003; Guo et al. 2013; Zhou et al. 2014; Guo et al. 2020a, b), and the degree of anisotropy decreases with increasing confining pressure. Saroglou suggested that the classification of the degree of anisotropy should be adjusted according to the confining pressure (Table 7.6) (Saroglou et al. 2019).

In slope engineering, it is necessary to use the aspect of the anisotropic structure (α_2), the dip angle of the anisotropic structure (γ_2) and the dip angle of the slope (γ_s) to determine the quality ABQ_s of the anisotropic rock mass in the slope engineering.

Table 7.5 Classification according to the degree of anisotropy of IBQ_{an}

I_{BQan}	Grading	Descriptions
≤1.05	A	Isotropic rock mass
1.1–1.5	B	Weakly anisotropic rock mass
1.5–2.0	C	Moderately anisotropic rock mass
2.0–2.5	D	Strongly anisotropic rock mass
>2.5	E	Extremely anisotropic rock mass

Table 7.6 Confining pressure correction (Saroglou et al. 2019)

Confining pressure	σ_3/σ_c	Correction
Low	<0.15	No correction
Medium	0.15–0.4	Anisotropy level reduced by one
High	>0.4	Anisotropy level reduced by two

Table 7.7 Classification of Anisotropic Rock Mass Engineering Quality (50218GB/T 2014)

ABQ_s	Rock mass quality grade	Description
>550	I	Excellent
550–451	II	Good
450–351	III	Medium
350–251	IV	Difference
≤250	V	Extremely bad

$$\text{When } \alpha_2 \geq 30°,\ \gamma_2 \leq 20°,\ \gamma_2 > \gamma_s, \text{ABQ}_s = \text{ABQ}_{max};$$
$$\text{when } \alpha_2 \leq 10°,\ \gamma_2 \geq 35°,\ \gamma_2 < \gamma_s, \text{ABQ}_s = \text{ABQ}_{min};$$

under other conditions,

$$ABQ_s = 1/2(ABQ_{max} + ABQ_{min}) \tag{7.15}$$

According to the revised ABQs (GB50218-2014), the anisotropic rock mass quality is divided into five grades, as shown in Table 7.7.

7.4 Application of the Classification Methods on the Engineering Slopes

The engineering slopes along one railway tunnels can be divided into 10 layered slopes and 88 non-layer slopes (Tan 2022). There are 10 non-layered and, and the international BQ-RMR-GSI method is used to evaluate the rock mass quality of the non-layered slope, and the A-BQ method is used to comprehensively consider the anisotropy characteristics of the rock to evaluate the Slope for rock mass quality evaluation.

7.5 Non-layered Slopes

The geotectonic unit at the entrance slope of Yigong Tunnel and Layue Tunnel are both Boshula Ridge magmatic arc, which are close to the Jiali-Zayu Fault. In recent years, the fault zone is still active, and both belong to high, steep, complex and non-layered rock slopes (see Fig. 7.3).

The slope at the entrance of Yigong Tunnel mainly contains biotite monzonite, which belongs to the ultra-high steep slope with fracture structure, has 5 groups of structural planes, and has a potential wedge-shaped failure trend. The entrance slope of Layue Tunnel mainly contains gneiss and schist, which belongs to the block structure ultra-high near vertical slope, with 5 groups of structural planes, and has a potential complex wedge failure trend. The GSI value range determined according to the field investigation results of the structural characteristics of the slope rock mass is shown in Table 7.8.

The GSI, RMR, BQ values and the rock mass quality classification of two typical high-steep complex non-layered slope rock masses can be determined based on Table 7.8 (Table 7.9).

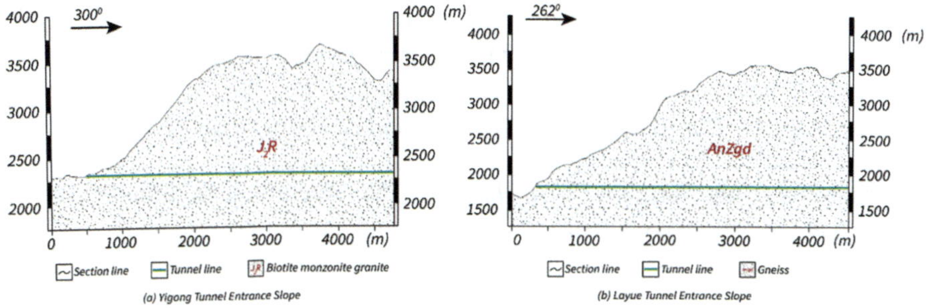

Fig. 7.3 Engineering geological section of typical high steep complex non-layered rock slope

Table 7.8 Structural characteristics of typical high steep complex non-layered rock mass slopes

Slope name	Lithology	Rock mass structure	Slope type	Number of structural plane groups	Deformation failure trend	GSI value range
Yigong tunnel entrance	Biotite monzonite granite	Fragmentation structure	Super steep slope	5	Wedge failure	40–50
Layue tunnel entrance	Gneiss and Schist	Blocky texture	Extra high suberect	5	Complex wedge failure	50–60

Table 7.9 Rock mass quality classification results of typical high steep complex non-layered slopes

Slope name	GSI value	RMR value	BQ value	Rock mass quality grade
Yigong tunnel entrance	45	50	322	Grade IV
Layue tunnel entrance	55	60	434	Grade III

7.6 Stratified Slopes

According to the geological map data (Figs. 7.4 and 7.5), the stratum at the entrance slope of tunnel 2# is mainly the Upper Triassic Qugasi Formation (T_3q) and the Middle Triassic Yidun Group ($T_{1-2}Y$). The lithology of Qugasi Formation (T_3q) is light crystalline limestone, crustaceous limestone interbedded with sandy slate, basic volcanic rock, and the bottom contains conglomerate. The occurrence of the stratum is 64° ∠ 55°; the lithology of Yidun Group ($T_{1-2}Y$) is phyllite, metamorphic sand plate, mixed with limestone, medium acid-basic volcanic rock, and the rock occurrence is 69° ∠ 42°.

The rock mass quality evaluation parameters and classification results of the slope are shown in Table 7.10.

According to the laboratory test results, the maximum and minimum values of σ_{c-dry} the tunnel slope rock are 97.66 and 30 MPa, respectively, and R_{cmax} and R_{cmin} are calculated by formula (7.5) to be 74.13 and 22.77 MPa, respectively.

According to the field joint spacing survey results, $K_{v\text{-}ra}$ and $K_{v\text{-}an}$ are 0.75 and 0.55 respectively. The ABQ_{max} and ABQ_{min} calculated by formula (7.12) and (7.13) are 509.89 and 305.81 respectively.

According to the formula (7.14), I_{BQan} is 1.67. According to Table 7.10, it is known that the rock mass belongs to Class medium anisotropic rock mass.

According to α_2, γ_2 and γ_s, the rock mass of the slope belongs to other conditions, and $ABQ_s = 407.85$ is calculated by formula (7.15).

According to Table 7.9, the rock mass quality of the slope is Class III and described as "medium".

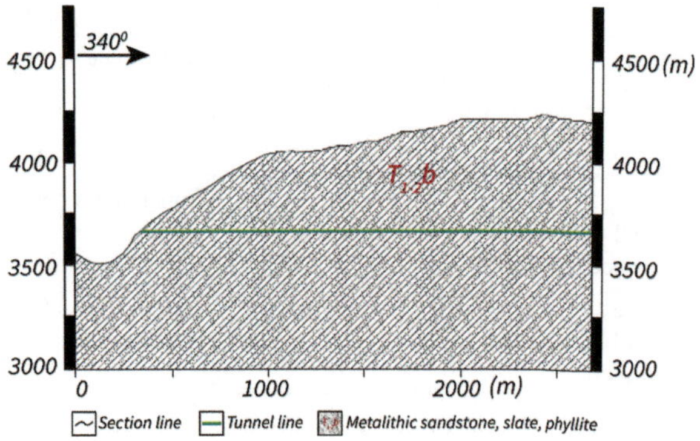

Fig. 7.4 Engineering geological profile

Fig. 7.5 Tunnel 2 # entrance slope (Inclination angle 301°)

Table 7.10 Quality evaluation parameters of rock mass A-BQ at the slope of tunnel 2# entrance

Evaluation parameters	Numerical value
R_{cmax}	74.13 MPa
R_{cmin}	22.77 MPa
$K_{v\text{-}ra}$	0.75
$K_{v\text{-}an}$	0.55
ABQ_{max}	509.89
ABQ_{min}	305.81
I_{BQan}	1.67
Confining pressure correction	No correction required
Anisotropy degree	C
Rock quality grade	Class III rock mass

Based on the above analysis process, this paper evaluates and classifies the rock mass quality of the tunnel entrance and exit slopes along the railway. The location distribution of the rock mass quality classification results of the tunnel entrance and exit slopes is shown in Fig. 7.5. Among them, there are 88 non-stratified slopes, 8 of which are grade I, 37 grade II, 24 grade III, and 19 as grade IV; 10 layered slopes are all grade III (Table 7.11).

Table 7.11 Rock mass classification results of tunnel entrance and exit slopes in the study area

Rock mass quality grade	Slope name	Number of slopes	Proportion/%
I	Zhugangshan Tunnel Entrance, Honglashan Tunnel Exit, Mangkangshan Tunnel Exit, Yigong Tunnel Exit, Tongmai Tunnel Entrance, Tongmai Tunnel Exit, Chalong Tunnel Entrance, Lulang Tunnel Entrance	8	8.16
II	Exit of Tianquan Tunnel, Exit of Wangjialin Tunnel, Entrance of Xinfang Tunnel, Exit of Yakou Tunnel, Entrance of Duimenshan Tunnel, Exit of Duimenshan Tunnel, Exit of Erlangshan Tunnel, Entrance of Guodashan Tunnel, Exit of Guodashan Tunnel, Entrance of Anliangba Tunnel, Gaoer Temple Tunnel Entrance, Exit of Gaoer Temple Tunnel, Entrance of Pamling Tunnel, Exit of Pamling Tunnel, Entrance of Wangjia Tunnel, Exit of Baizi Village Tunnel, Exit of Ma'anshan Tunnel, Entrance 1 of Yingjinshan Tunnel, Exit 1 of Yingjinshan Tunnel, Yingjinshan Tunnel 2 Entrance, Yingjinshan Tunnel Exit 2, Kazilashan Tunnel 1 Entrance, Deda Tunnel Exit, Tunnel 3 Exit, Tunnel 6 Exit, Tunnel 7 Exit, Tunnel 8 Entrance, Tunnel 8 Exit, Tunnel 9 Entrance, Tunnel 9 Exit, Gong Entrance of Jue Tunnel, Entrance of Gayi Tunnel, Exit of Gayi Tunnel, Entrance of Hongla Mountain Tunnel, Entrance of Mangkang Mountain Tunnel, Entrance of Changdu Tunnel, Exit of Changdu Tunnel	37	37.76
III	Exit of Liziping Tunnel, Entrance of Erlangshan Tunnel, Entrance of Baolingshan Tunnel, Exit of Baolingshan Tunnel, Entrance of Kangding Tunnel, Exit of Kangding Tunnel, Entrance of Zheduoshan Tunnel, Exit of Zheduoshan Tunnel, Exit of Dongeruo No. 1 Tunnel, Tunnel 1 Entrance, Exit of Tunnel 1, Entrance of Tunnel 2, Exit of Tunnel 2, Entrance of Tunnel 3, Exit of Gonggyak Tunnel, Entrance of Intraek Tunnel, Exit of Intraek Tunnel, Entrance of Tunnel 10, Exit of Tajou Tunnel, Exit of Golasan Tunnel, Kang Ok Tunnel Entrance, Kangyu Tunnel Exit, Chada Tunnel Entrance, Chada Tunnel Exit, Bashula Ridge Tunnel Entrance, Bashula Ridge Tunnel Exit, Duoji Tunnel Entrance, Duoji Tunnel Exit, Domug Tunnel Entrance, Domug Tunnel Exit, Yigong Tunnel Entrance, Chalong Tunnel Exit, Layue Tunnel Entrance, Layue Tunnel Exit	34	34.69
IV	Mengdingshan tunnel entrance, Mengdingshan tunnel exit, Baitashan tunnel entrance, Baitashan tunnel exit, Xinmiaozi tunnel entrance, Xinmiaozi tunnel exit, Jianfengshan tunnel entrance, Jianfengshan tunnel exit, Zhaiziding tunnel entrance, Zhaiziding tunnel exit, Daping tunnel entrance, Daping tunnel exit, Tianquan tunnel entrance, Wangjialin tunnel entrance, Dongerluo 2 tunnel entrance, Dongerluo 2 tunnel exit, Bangda tunnel entrance, Xiali tunnel entrance, Xiali tunnel exit	19	19.39

7.7 Conclusion

In this study, the research status of rock quality evaluation method are thoroughly reviewed, and a recently developed ABQ method for anisotropic rock mass is introduced. All the required parameters and the way to reach are depicted in detail, and both the anisotropic degree and rock mass quality can be determined by legible classification. This method are then applied to evaluate the anisotropic rock mass engineering slopes along a railway in southwest China, together with a combined method for nonlayered rock mass. The applications confirm the reliability and accessibility of the method and possible extensive application in future.

References

50218GB/T (2014) Engineering rock mass classification standard.Ministry of Housing and Urban-Rural Development of the People's Republic of China; General Administration of Quality Supervision, Inspection and Quarantine of the People's Republic of China

50487GB (2009) Code for engineering geological investigation of water. The National Standards Compilation Group of People's Republic of China

Bar N, Barton N (2017) The Q-slope method for rock slope engineering. Rock Mech Rock Eng 50(12):3307–3322

Barton N (2013) Shear strength criteria for rock, rock joints, rockfill and rock masses: problems and some solutions. J Rock Mech Geotech Eng 5(4):249–261

Barton N, Choubey V (1977) The shear strength of rock joints in theory and practice. Rock Mech 10(1):1–54

Barton N, Lien R, Lunde J (1974) Engineering classification of rock masses for the design of tunnel support. Rock Mech 6(4):189–236

Bieniawski ZT (1973) (1973) Engineering classification of jointed rock masses. Civil Eng Siviele Ingenieurswese 12:335–343

Cai B, Yu Y, Wu XM (2001) The relationship between the engineering rock mass classification standard and the Q classification method and RMR classification method and the estimation of deformation parameters. J Rock Mech Eng 12(S1):1677–1679

Chen JZ (2007) Study on rock mass quality classification of Shuangjiangkou Hydropower Station project

Deere DU (1964) Technical description of rock cores for engineering purpose. Rock Mech Eng Geol 1(1):17–22

Dongxu Z (2016) A review of quantitative methods for geological strength indexes. Min Resour Geol 30(6):1024–1029

Franklin JA (1985) Suggested method for determining point load strength. In: International journal of rock mechanics and mining sciences and geomechanics abstracts. Elsevier, pp 51–60

Gu DZ (1979) Fundamentals of rock-mass engineering geomechanics. China Science Press, Beijing, pp 75–87

Gu DZ, Huang DC (1979a) Classification of rock mass structure and determination of its mass coefficient. Hydrogeol Eng Geol 21(2):8–13

Gu DZ, Huang DC (1979b) Classification of rock mass structure and determination of its quality coefficient. Hydrogeol Eng Geol 2:10–15

Guo SF, Qi SW, Huang XL (2013) Anisotropy of rock mass strength and stress conditions for its transformation. J Rock Mech Eng 32(S2):3222–3227

Guo SF, Qi SW, Saroglou C (2020a) A-BQ, a classification system for anisotropic rock mass based on China National Standard. J Central South Univ 27(10):3090–3102
Guo SF et al (2020b) Numerical study on the progressive failure of heterogeneous geomaterials under varied confining stresses. Eng Geol 269:105556
Hoek E, Brown ET (1997) Practical estimates of rock mass strength. Int J Rock Mech Min Sci 34(8):165–186
Hoek E, Brown ET (2019) The Hoek-Brown failure criterion and GSI–2018 edition. J Rock Mech Geotech Eng 11(3):445–463
Hoek E, Marinos PG, Marinos VP (2005) Characterisation and engineering properties of tectonically undisturbed but lithologically varied sedimentary rock masses. Int J Rock Mech Min Sci 42(2):277–285
Hu JH et al (2012) Morphology characteristics of joint surface in rock mass. J Central South Univ 19(10):2918–2925
Marinos P, Hoek E (2000) GSI: a geologically friendly tool for rock mass strength estimation. In: ISRM international symposium. OnePetro
Nasseri MHB, Rao KS, Ramamurthy T (2003) Anisotropic strength and deformational behavior of Himalayan schists. Int J Rock Mech Min Sci 40(1):3–23
Ramamurthy T, Arora VK (1994) Strength predictions for jointed rocks in confined and unconfined states. In: International journal of rock mechanics and mining sciences and geomechanics abstracts. Elsevier, pp 9–22.
Saroglou C et al (2019) ARMR, a new classification system for the rating of anisotropic rock masses. Bull Eng Geol Environ 78(5):3611–3626
Saroglou H, Tsiambaos G (2007) Classification of anisotropic rocks. In: Proceedings of the 11th ISRM Congress. OnePetro
Singh TN, Kainthola A, Venkatesh A (2012) Correlation between point load index and uniaxial compressive strength for different rock types. Rock Mech Rock Eng 45(2):259–264
Sonmez H, Gokceoglu C, Ulusay R (2004) Indirect determination of the modulus of deformation of rock masses based on the GSI system. Int J Rock Mech Min Sci 41(5):849–857
Tan B (2022) Study on mechanical parameters of fractured rock mass of Sichuan-Tibet Railway project. J Yangtze River Acad Sci 39(6):76–81
Vasarhelyi B (2003) Some observations regarding the strength and deformability of sandstones in dry and saturated conditions. Bull Eng Geol Environ 62(3):245–249
Vutukuri VS, Hossaini SMF (1995) Modification to the original Hoek-Brown strength criterion. In: Rock foundation, pp 133–137
Wu AQ, Liu FZ (2012) Application and progress of the national standard classification standard for engineering rock mass. J Rock Mech Eng 31(8):1513–1523
Zhou Y et al (2014) Macro and meso analysis of jointed rock mass triaxial compression test by using equivalent rock mass (ERM) technique. J Central South Univ 21(3):1125–1135

Chapter 8
Geotechnical Properties of Burned Soils: A Case Study of a Pine-Forest Fire in Southwestern Türkiye

Tümay Kadakci Koca

Abstract The geotechnical properties of soils such as consistency limits and shear strength are closely related to the erodibility. It is well known that erosion risk is promoted after forest fires. However, the changes in the geotechnical properties of topsoil after forest fires have not been extensively studied in the literature. In addition, the existing studies show contradictory results. Since forest fires exert randomly distributed burning temperatures depending on the slope, aspect, vegetation type, and height, different burn severity degrees can be observed in one fire perimeter. This results in variations in soil parameters due to the fire. In this study, the effect of a pine forest fire on the geotechnical properties of sandy soils was addressed. A total of 69 unburned and burned soil samples were collected from different depths of the soil profile. It was found that the samples taken below 4 cm depth have not shown statistically significant differences due to burning. However, the samples collected from the top organic layer (0–4 cm) showed signs of disaggregation and accordingly, consistency limits increased while shear strength decreased. These results contributed to the existing knowledge that cohesionless shallow soils, typically regosols, respond to natural burning distinctly.

Keywords Regosols · Forest fire · Consistency limits · Shear strength · Burn severity

8.1 Introduction

Wildfires are more frequently destroying the forests in mountainous areas due to climate change. As a result of burning under high temperatures, the physical, chemical, mineralogical, hydrological, and mechanical properties of soils change. Hence, the risk of erosion and potential of mass movement increases (Riestenberg and

T. Kadakci Koca (✉)
Engineering Faculty, Geological Engineering Department, Muğla Sıtkı Koçman University, Muğla, Türkiye
e-mail: tumaykoca@gmail.com

S. Wang et al. (eds.), *Engineering Geology for a Habitable Earth: IAEG XIV Congress 2023 Proceedings, Chengdu, China*, Environmental Science and Engineering,
https://doi.org/10.1007/978-981-99-9057-3_8

Sovonick-Dunford 1983; Wu and Sidle 1995; Inbar et al. 1997; DeBano 2000; González- Pérez et al. 2004; Shakesby and Doerr 2006; Fox et al. 2007; Cerdà and Doerr 2008; Zavala et al. 2010; Dlapa et al. 2015; Thomaz 2021). Shear strength is a crucial parameter to evaluate the erodibility of soils (Léonard and Richard 2004). The relations between the shear strength and landslides after wildfires have also been investigated (Parise and Canon 2012; Moody et al. 2013). These studies reported that a decrease in shear strength accompanies increased sediment transportation. Nyman et al. (2013) investigated the shear strength of soils from different depths 6 months and one year after a forest fire. They determined that shear strength increases in time after the fire and it always stabilizes below the depth of 6 cm. Additionally, the shear strength of soil would decrease due to the loss of vegetation roots which contributes to the overall shear resistance (Gabet 2003; Sidle and Bogaard 2016). Consistency limits (liquid limit and plastic limit) are also key parameters to evaluate the shear strength of soil (Deng et al. 2017). However, the geotechnical properties such as consistency limits and shear strength of burned sandy soils have received little attention. Abu-Zreig et al. (2001) investigated CH, CL, and MH-type soils that burned in the laboratory between the temperatures of 0 and 400 °C. They found that liquid limit (LL) and plastic limit (PL) did not significantly change below 100 °C while LL and PL decreased following the temperatures 200 °C and 100 °C, respectively. Similarly, Tan et al. (2004) determined that the LL and PL of CH-type soils rapidly decrease between 100 °C and 300 °C whereas the decrement slows down between 300 °C and 1000 °C. The interesting finding is that soil behaves as non-plastic if burned over 400 °C. This suggests that burning above 400 °C can result in the fusion of clay particles and leads to aggregation-induced non-plasticity. The above studies investigated the clayey soils which are likely to behave distinctly from sandy soils when burned. Stoof (2011) measured the in-situ shear strength of burned sandy soils developed from schist and quartzite instantly after the fire and 1 year after the fire by the vane shear test device. She found that shear strength increases immediately after the fire and decreases below its unburned value after 1 year. However, she determined that the shear strength is inversely proportional to the natural moisture content of the soil. Therefore, it is impossible to draw any conclusion about the effect of fire on the shear strength if the tests are performed under natural moisture content. Secci et al. (2014) studied soils developed over granitic bedrock. They investigated the changes in shear strength parameters of soils collected from 30 cm and 1.3 m deep after the fire. They revealed that the cohesion decreased, and internal friction angle increased. However, the sampling depths are too deep to reflect the effect of fire since the heat produced during the fire is likely to be transferred commonly to a depth of no more than 5 cm (Bentley and Fenner 1958; DeBano et al. 1979; Zavala et al. 2014). Therefore, the changes they reported can be strongly related to the physical properties of the soil (e.g., moisture content, particle size). Vacchiano et al. (2014) studied soil samples having %50 and more sand and low clay content. They concluded that the LL of topsoil only changes for moderate to high burn severity (>220 °C). In other words, LL tends to increase in moderate severity and decrease in high severity. However, these changes were not found to be statistically significant according to the post hoc tests. Tadza and Noraini (2016) examined the burned and unburned samples gathered from 20 cm depth and reported that LL decreases and PL increases after natural fire burning. These changes were attributed to the infiltration of ash into the pores of soils and the increase in the organic carbon content. However, theoretically, the increase in organic carbon should have resulted in an increment in LL. Haake (2020) investigated the shear strength of gravelly sandy soils after the fire. She found no statistically significant change in the shear strength of soils collected from 5 to 8 cm in different burn severity locations (high, moderate, low severities, and unburned). However, the highest shear strength was declared for the soils collected from high burn severity locations. More recently, Kadakci Koca (2023) studied the consistency limits of gravelly sandy soils developed from limestone bedrock. She found that particularly LL but also PL increased after the forest fire. Above mentioned studies reported contradictory results for consistency limits and shear strength of burned soils which suggests that different types of soils respond to burning distinctly. In addition, fire intensity is an influencing factor for the level of change. This study, therefore, addresses this research gap by investigating the effect of naturally occurring moderate-high intensity (<350 °C) forest fires on the geotechnical properties of sandy soils developed over different types of bedrock.

8.2 Study Area

The southwestern region of Türkiye is one of the potential hotspots for climate change in the world. The lowest and highest elevations in the study area are 680.2 m and 1153.2 m, respectively. The slopes are mainly south facing, and the maximum slope angle is 30.15°. The forest fire occurred in the black pine (Pinus nigra L.) and red pine (Pinus brutia Ten.) forest in August 2021. The hot and dry winds with a speed of 15.84 km/h provided favorable conditions for the fire to spread and increase in intensity. The temperature and relative humidity in the vicinity (Kavaklıdere village, 37°26′49″ N and 28°21′48 ″E) on the day of the initiation of the fire were 35.4 °C and 15%, respectively (TSMS 2022). The exposed rock units are schist-quartzite, phyllite, marble, limestone, and conglomerate-sandstone-mudstone intercalations from oldest to youngest (Göktaş1998; Gürer et al. 2013; Gül 2015) (Fig. 8.1). The area is referred to as having moderate chemical weathering, according to the classification proposed by Peltier (1950) based on the annual temperature and precipitation. The soils developed from these bedrocks are mainly regosols according to (FAO 2015). These soils are known to be shallow, and poor in organic content and clay content which results in low aggregate stability.

Fig. 8.1 The parameters used for qualitative description of soil burn severity recommended by Parsons et al. (2010) **a** ground cover: amount and condition, **b** ash color and depth, **c** soil structure, **d** roots

Table 8.1 The distribution of samples among depth and bedrock

Depth	Bedrock[a]			
	L	S	P	C
0–4 cm (organic layer, O)	7	6	12	8
4–10 cm (A horizon, A)	8	9	12	7
Total sample number	15	15	24	15

[a] Limestone-L; Schist-S; Phyllite-P; conglomerate, sandstone, mudstone intercalations-C)

8.3 Methods

8.3.1 Soil Sampling

During a forest fire, temperatures can reach a maximum at the surface (1100 °C) and decrease progressively with depth through organic topsoil (850 °C) and mineral soil (150 °C) (DeBano et al. 1979; Cerdà and Robichaud 2009). The depth of influence may vary due to the heat produced during the fire and the thermal conductivity of the soil. For this reason, soil sampling in this study was performed for the top organic layer of 0–4 cm which consists of ash, litter, and mineral soil, and 4–10 cm which involves mineral soil, less amount of ash, no litter. This sampling methodology enabled defining the influence depth of the fire. Seven months after the fire between March and May 2022, 55 soil samples were collected from 29 locations throughout the fire perimeter following this methodology. On the other hand, 14 unburned samples were gathered from eight locations adjacent to the burned areas with similar lithology, type of soil, relief, aspect, and vegetation cover (Table 8.1). The burn severity levels in the field were qualitatively defined according to the field guide recommended by Parsons et al. (2010) (Fig. 8.1). Subsequently, unburned locations were symbolized by U while low, moderate, and high burn severity levels were symbolized by L, M, and H, respectively.

8.3.2 Laboratory Tests

Initially, sieve analyses were performed following ASTM D422-63 (2017) standard to obtain the grain size distribution and mean grain size (D50) of the soil samples (Fig. 8.2a). Before sieve analyses, the collected soil samples were dried at 65 °C in the oven for 24 h, contrary to the recommended temperature in the standard. The 65 °C drying temperature was selected to maintain the organic content, not to alter the calcite and clay minerals, principally for unburned samples. In conjunction with the Atterberg limit tests, the soil samples were classified according to the USCS classification (ASTM D2487-17e1 2017). To determine the LL of the sandy soils Casagrande test device and AASHTO T 89 (2022) Method B was followed (Fig. 8.2c). Method

B was chosen because it allows the determination of the LL of a greater number of soil samples in relatively less time than the standard method. Afterward, the PL test was performed according to ASTM D4318-17e1 (2018) standard (Fig. 8.2b).

Since the soil specimens are sandy soils with very low fines content, cohesion will be negligible. Hence, the shear strength equation relying on the Mohr–Coulomb failure criterion will be as follows (Eq. 8.1).

$$\tau(\text{kPa}) = \sigma \tan \phi \tag{8.1}$$

where ϕ is the internal friction angle, σ is the normal stress. ASTM D3080-04 (2012) standard was followed and disturbed soil samples with particle sizes smaller than 3 mm were utilized. The primary objective was to examine whether the burn severity level affected variations in shear strength, rather than determining the shear strength of the undisturbed sample. To eliminate the effect of moisture content, the samples were dried at 65 °C in an oven before testing. Afterward, the soil sample was put into the shear mold and compacted using a square hammer with three strikes before

Fig. 8.2 **a** Sieve shaker, **b** plastic limit test sample, **c** Casagrande liquid limit test device, **d** shear box test device

the test. Horizontal displacement and shear load measurements were recorded at specific time intervals by employing a loading rate of 1 mm/min (Fig. 8.2d). Three samples derived from the same specimen which have similar volumes and weights were subjected to three different normal stress levels (2.73 kPa, 5.45 kPa, 13.63 kPa). Lower values for normal stress were chosen to simulate surface stress conditions, considering that erosion predominantly occurs on the surface. Consequently, shear stress-normal stress graphs were produced, failure envelopes were generated, and shear strength parameters were determined for each sample group. In the statistical analysis, the internal friction angle (ϕ) and shear strength (τ) value under a 2.73 kPa load were considered.

8.3.3 *Data Analyses*

In data analyses, outliers were omitted. One-way ANOVA Scheffe test was used to test the effect on each soil parameter of two factors separately: burn severity level and sampling depth. The Scheffe test is commonly preferred if the sample sizes of groups are not the same. Normality and homogeneity of variance assumptions were tested by the Shapiro–Wilk test and the Levene test, respectively. If the distribution is normal but the homogeneity of variance is not satisfied, ANOVA Tamhane's T2 was used. Kruskal–Wallis and Mann–Whitney U non-parametric tests were also employed on the non-normally distributed datasets. Consequently, the hypotheses that there are significant differences in soil parameters depending on the burn severity and sampling depth were tested. The p-value less than 95% confidence level (0.05) indicated that the hypothesis can be accepted or vice versa.

8.4 Results and Discussion

The samples from the organic layer (O) involve an average sand and fines percentage of 75.67% and 15.65%, respectively. On the other hand, A horizon (A) constitutes an average sand and fines percentage of 74.74% and 11.05%, respectively. The LL and PL of the O and A samples have also shown statistically significant differences as well as the shear strength ($p < 0.05$). Subsequently, the organic layer and A horizon were typically classified as SM and SW-SM according to USCS.

ANOVA tests for the hypothesis that soil sampling depth affects the soil parameters independent from the burn severity degree (LL, PL, τ, ϕ) have been accepted with a 95% confidence level. Since the main objective was to find out if the burn severity degree influences the soil parameters separately among O and A samples, post hoc test results were evaluated to control this hypothesis. It is interesting to note that none of the soil parameters showed a variance owing to the burn severity degree among the A samples ($p > 0.05$). This result indicated that the heat produced by the fire did not affect the soils at 4–10 cm depth. This finding should also be supported by further chemical and mineralogical examinations of soil samples collected from the A horizon as stated by Kadakci Koca et al. (2024).

One-way ANOVA tests showed that there is a significant difference in sand percentage and mean grain size (D50) among the O samples considering the burn severity levels ($p < 0.05$). Apart from the lithological units, burning decreased the sand size particles in all locations and hence the mean grain sizes (Fig. 8.3) Similar results have been reported by several researchers (Albalasmeh et al. 2013; Rhoades et al. 2004; Vacchiano et al. 2014). Rhoades et al. (2004) explained this process by the disintegration of sand particles into finer grain sizes with heating and burning.

The consistency limits also have been determined to be changing in relation to the burning degree (Fig. 8.3). Some samples shift from low plasticity to moderate plasticity or moderate plasticity to high plasticity with the increasing burn severity. However, a slight decrease in high severity when compared to moderate severity is pronounced (Fig. 8.4). This result can be attributed to the textural changes due to fire-induced disintegration. One-way ANOVA tests also supported that LL and PL of O samples have considerable variations due to burning degree ($p < 0.05$). At a close distance to the study area, Kadakci Koca (2023) found that consistency limits increase in burned samples independent from the burn severity degree. This increment is slightly higher than that of the findings of this study. A possible explanation for that difference is that the variable clay fraction and organic matter content among the two soils types.

One-way ANOVA tests revealed that burning caused differences in both τ and ϕ values of O samples ($p < 0.05$). There is a very limited number of studies that investigated the shear strength of burned soils. Furthermore, these studies reported contradictory results due to the effect of moisture content variations. Only, Secci et al. 2014, argued that ϕ significantly increased due to burning nevertheless the reasons have not been discussed. Besides, the samples were collected from a considerable depth at which the fire effect should have been limited. It is only possible that ϕ can increase in the case of aggregation of soil particles under high temperatures. In that study, τ values under 2.73 kPa (0.028 kg/cm^2) normal stress shows significant variations (Fig. 8.5a). It can be declared that the τ decreases with increasing burning severity. On the other hand, even the ϕ value also tends to decrease with an increasing

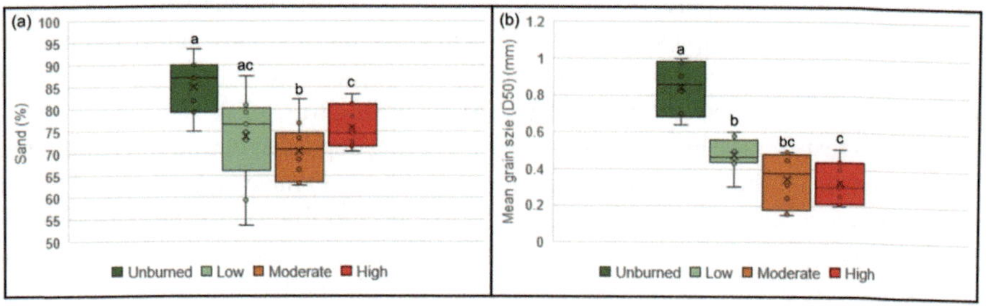

Fig. 8.3 Boxplot of **a** Sand (%), **b** mean grain size (mm) of O samples and results of post hoc test ($p = 0.05$). Boxes with no letters in common are significantly different from each other

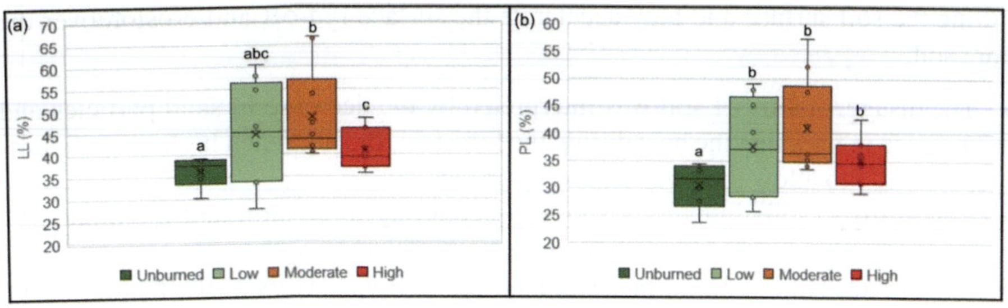

Fig. 8.4 Boxplot of **a** LL (%), **b** PL (%) of O samples and results of post hoc test ($p = 0.05$). Boxes with no letters in common are significantly different from each other

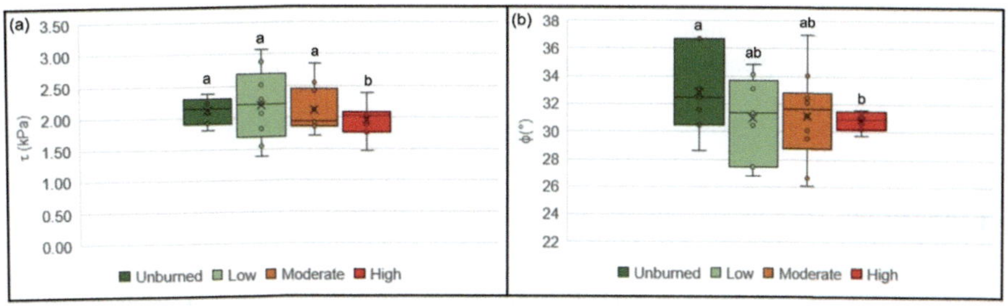

Fig. 8.5 Boxplot of **a** τ (kPa), **b** ϕ (°) of O samples and results of post hoc test ($p = 0.05$). Boxes with no letters in common are significantly different from each other

burning degree, this change is only prominent between unburned and high burn severity samples (Fig. 8.5b). Indeed, the trend of change in τ and ϕ well accords with the changes in particle size distribution. Another perspective has been pointed out by Tadza and Noraini (2016) that plasticity decreases, and water retention increases after burning. For this reason, shear strength will show a decreasing trend. However, the cohesive soils may lose their cohesion and consequently, shear strength may decrease by obeying the Mohr–Coulomb Criterion. When compared to the results of this study, plasticity increased by approximately 20%, ϕ decreased by 4% and therefore the shear strength decreased by 9%.

8.5 Conclusions

This study investigated the effect of a forest fire on the geotechnical properties of sandy soils, typically regosols, developed from different bedrock. These parameters are closely related to the erodibility of the soils which allows the evaluation of a fire-induced erosion risk. The below-given results imply that the changed soil geotechnical properties due to burning are not possessing a high-level of erosion risk whereas the loss of vegetation will most probably increase the intensity of raindrops hitting the soil surface and may accelerate the surface run-off and erosion over the bare soil.

- The disaggregation of soil was interpreted by the decrease in sand particles and increase in fines that the heat during the forest fire did not reach high temperatures (>475 °C) to allow the fusion of soil particles.
- Regosol-type soils inherently show susceptibility to erosion and the moderate-intensity forest fire led to a slight decrease in shear strength.
- Particle size reduction was promoted during the fire, and this resulted in enhanced plasticity and decreased shear strength. Different factors such as chemical and mineralogical content can be investigated to undoubtedly reveal if the particle size reduction is the main reason for shear strength decrement.

Acknowledgements This study is a part of project number 121Y561 which was funded by The Scientific and Technological Research Council of Türkiye (TÜBİTAK).

References

AASHTO T89 (2022) Standard method of test for determining the liquid limit of soils method B. American Association of State Highway and Transportation Officials, USA

Abu-Zreig M (2001) Factors affecting sediment trapping in vegetated filter strips: simulation study using VFSMO. Hydrol Process 15(8):1477–1488. https://doi.org/10.1002/hyp.220

Albalasmeh AA, Berli M, Shafer DS, Ghezzehei TA (2013) Degradation of moist soil aggregates by rapid temperature rise under low intensity fire. Plant Soil 362(1–2):335–344. https://doi.org/10.1007/s11104-012-1408-z

ASTM D2487-17e1 (2017) Standard practice for classification of soils for engineering purposes (Unified Soil Classification System). ASTM International, West Conshohocken, PA

ASTM D3080-04 (2012) Standard Test Method for Direct Shear Test of Soils Under Consolidated Drained Conditions. ASTM International, West Conshohocken, PA

ASTM D422-63 (2017) Standard Test Method for Particle-Size Analysis of Soils. ASTM International, West Conshohocken, PA

ASTM D4318–17e1 (2018) Standard test methods for liquid limit, plastic limit and plasticity index of soils. ASTM International, West Conshohocken, PA.

Bentley JR, Fenner RL (1958) Soil temperatures during burning related to postfire seedbeds on woodland range. J for 56:737–740. https://doi.org/10.1093/jof/56.10.737

Cerdà A, Doerr SH (2008) The effect of ash and needle cover on surface runoff and erosion in the immediate post-fire period. CATENA 74:256–263. https://doi.org/10.1016/j.catena.2008.03.010

Cerdà A, Robichaud P (2009) Fire effects on soils and restoration strategies. CRC Press, Boca Raton, FL

DeBano LF (2000) The role of fire and soil heating on water repellency in wildland environments: a review. J Hydrol 231–232(9):195–206. https://doi.org/10.1016/S0022-1694(00)00194-3

DeBano LF, Rice RM, Conrad CE (1979) Soil heating in chaparral fires: effects on soil properties, plant nutrients, erosion, and runoff. Pacific Southwest Forest and Range Experiment Station, Berkeley, CA

Deng Y, Cai C, Xia D, Ding S, Chen J, Wang T (2017) Soil Atterberg limits of different weathering profiles of the collapsing gullies in the hilly granitic region of southern China. Solid Earth 8(2):499–513. https://doi.org/10.5194/se-8-499-2017

Dlapa P, Bodi MB, Mataix-Solera J, Cerdà A, Doerr SH (2015) Organic matter and wettability characteristics of wildfire ash from Mediterranean conifer forests. CATENA 135:369–376. https://doi.org/10.1016/j.catena.2014.06.018

FAO (2015) World Reference Base for Soil Resources International soil classification system for naming soils and creating legends for soil maps. http://www.fao.org/3/a-i3794e.pdf. Accessed 19 Mar 2023

Fox DM, Darboux F, Carrega P (2007) Effects of fire-induced water repellency on aggregate stability, splash erosion, and saturated hydraulic conductivity for different size fractions. Hydrol Process 21:2377–2384. https://doi.org/10.1002/hyp.6758

Gabet EJ (2003) Post-fire thin debris flows: sediment transport and numerical modelling. Earth Surf Process Landf 28:1341–1348. https://doi.org/10.1002/esp.590

Göktaş F (1998) Stratigraphy and sedimentology of Neogene sedimentation around Mugla (SW Anatolia). Mineral and Research Institute, Ankara, Turkey

González-Pérez JA, González-Vila FJ, Almendros G, Knicker H (2004) The effect of fire on soil organic matter: a review. Environ Internat 30:855–870. https://doi.org/10.1016/j.envint.2004.02.003

Gül M (2015) Lithological properties and environmental importance of the quaternary colluviums (Muğla, SW Turkey). Environ Earth Sci 74:4089–4108. https://doi.org/10.1007/s12665-015-4506-4

Gürer ÖF, Sangu E, Özburan M, Gürbüz A, Sarıca-Filoreau N (2013) Complex basin evolution in the Gökova Gulf region: implications on the Late Cenozoic tectonics of SW Turkey. Int J Earth Sci 102(8):2199–2221. https://doi.org/10.1007/s00531-013-0909-1

Haake S (2020) Burn severity and its impact on soil properties: a study of the 2016 Erskine fire in the southern Sierra Nevada. Dissertation, California State University, Bakersfield

Inbar M, Wittenberg L, Tamir M (1997) Soil erosion and forestry management after wildfire in a Mediterranean woodland, Mt. Carmel, Israel. Int J Wildland Fire 7:285–294. https://doi.org/10.1071/WF9970285

Kadakci Koca T (2023) Studying the effects of forest fire on consistency limits of sandy soils: a case study, Kozağaç, Muğla. J Geol Eng 46(2):81–97. https://doi.org/10.24232/jmd.1221946

Kadakci Koca T, Küçükuysal C, Gül M, Esetlili T (2024) A comprehensive approach to soil burn severity mapping for erosion susceptibility assessment. CATENA, Under Review

Léonard J, Richard G (2004) Estimation of runoff critical shear stress for soil erosion from soil shear strength. CATENA 57(3):233–249. https://doi.org/10.1016/j.catena.2003.11.007

Moody J, Shakesby R, Robichaud P, Cannon S, Martin D (2013) Current research issues related to post-wildfire runoff and erosion processes. Earth Sci Rev 122:10–37. https://doi.org/10.1016/j.earscirev.2013.03.004

Nyman P, Sheridan GJ, Moody JA, Smith HG, Noske PJ, Lane PN (2013) Sediment availability on burned hillslopes. J Geophys Res Earth Surf 118:2451–2467. https://doi.org/10.1002/jgrf.20152

Parise M, Cannon SH (2012) Wildfire impacts on the processes that generate debris flows in burned watersheds. Nat Hazards 6(1):217–227. https://doi.org/10.1007/s11069-011-9769-9

Parsons A, Robichaud PR, Lewis SA, Napper C, Clark JT (2010) Field guide for mapping post-fire soil burn severity. U.S. Department of Agriculture, Forest Service, Rocky Mountain Research Station, Fort Collins, CO

Peltier LC (1950) The geographic cycle in periglacial regions as it is related to climatic geomorphology. Ann Assoc Am Geogr 40:214–236. https://doi.org/10.2307/2561059

Rhoades CC, Meier AJ, Rebertus AJ (2004) Soil properties in fire-consumed log burnout openings in a Missouri oak savanna. For Ecol Manag 192:277–284. https://doi.org/10.1016/j.foreco.2004.01.012

Riestenberg MM, Sovonick-Dunford S (1983) The role of woody vegetation on stabilizing slopes in the Cincinnati area. Geol Soc Am Bull 94:504–518. https://doi.org/10.1130/0016-7606(1983)94%3c506:TROWVI%3e2.0.CO;2

Secci R, Calcina SV, Ranieri G, Uras G (2014) Analysis of the stability variation of a slope crossed by forest fire. Int J Civ Eng 3(1):41–50

Shakesby RA, Doerr SH (2006) Wildfire as a hydrological and geomorphological agent. Earth Sci Rev 74:269–307. https://doi.org/10.1016/j.earscirev.2005.10.006

Sidle RC, Bogaard TA (2016) Dynamic earth system and ecological controls of rainfall-initiated landslides. Earth Sci Rev 159:275–291. https://doi.org/10.1016/j.earscirev.2016.05.013

Stoof CR (2011) Fire effects on soil and hydrology. Dissertation, Wageningen University, Wageningen, Netherlands

Tadza M, Noraini MYA (2016) Effect of wildfire on soil-water characteristics of natural slope containing temperature sensitive silica-alumina polymorph minerals. Aust J Basic Appl Sci 10(17):157–163

Tan O, Yilmaz L, Zaimoglu AS (2004) Variation of some engineering properties of clays with heat treatment. Mat Lett 58(7):1176–1179. https://doi.org/10.1016/j.matlet.2003.08.030

Thomaz EL (2021) Effects of fire on the aggregate stability of clayey soils: a meta-analysis. Earth Sci Rev 221:103802. https://doi.org/10.1016/j.earscirev.2021.103802

Turkish State Meteorological Service (TSMS) (2022) Statistical precipitation and temperature records. https://www.mgm.gov.tr/veridegerlendirme/ilve-ilceler-istatistik.aspx?m=MUGLA. Accessed 19 Mar 2023

Vacchiano G, Stanchi S, Marinari G, Ascoli D, Zanini E, Motta R (2014) Fire severity, residuals and soil legacies affect regeneration of Scots pine in the Southern Alps. Sci Total Environ 472:778–788. https://doi.org/10.1016/j.scitotenv.2013.11.101

Wu TH, Sidle RC (1995) A distributed slope stability model for steep forested basins. Water Resour Res 31:2097–2110. https://doi.org/10.1029/95WR01136

Zavala LM, Granged AJ, Jordán A, Bárcenas-Moreno G (2010) Effect of burning temperature on water repellency and aggregate stability in forest soils under laboratory conditions. Geoderma 158(3–4):366–374. https://doi.org/10.1016/j.geoderma.2010.06.004

Zavala LM, de Celis R, Jordán A (2014) How wildfires affect soil properties. A brief review. Cuad Investig Geogr 40:311–331. https://doi.org/10.18172/cig.2522

Chapter 9
Engineering Classification for the Determination of the Uniaxial Compressive Strength of Carbonate Fault Rocks

Vasileios Kallimogiannis and Charalampos Saroglou

Abstract The carbonate fault rocks exhibit severe transformations when subjected to fault zone deformation. The parent materials are transformed into fault rocks that comprise fragments and secondary (usually weak) matrix. In the present study, specific types of carbonate fault rocks are investigated, and their mechanical properties are quantified using a novel methodology. These types include the parent/host rock, welded fault breccias, unwelded fault breccias and matrix material. Carbonate fault rocks from western Lefkada island, Greece, were studied in the laboratory to establish a relationship between their Uniaxial Compressive Strength and their textural features, i.e., the geometrical aspects of their fragments and matrix (size, shape, distribution). The results indicate that the Uniaxial Compressive Strength of the welded/unwelded breccias is a function of the corresponding strength of their components (fragments and matrix) and the matrix ratio (the percentage of matrix in a given volume). Due to the weak nature of the matrix, the Uniaxial Compressive Strength of the matrix cannot be readily measured. Therefore, suggestions for the determination of its Uniaxial Compressive Strength based on field assessment are proposed.

Keywords Carbonate fault rocks · Welded breccias · Unwelded breccias · Matrix · Uniaxial compressive strength · Heterogeneity

---

V. Kallimogiannis (✉) · C. Saroglou
School of Civil Engineering, Department of Geotechnical Engineering, National Technical University of Athens, 15780 Athens, Greece
e-mail: vkallim@central.ntua.gr

C. Saroglou
e-mail: saroglou@central.ntua.gr

C. Saroglou
School of Earth Science & Environment, University of Leeds, Leeds, UK

S. Wang et al. (eds.), *Engineering Geology for a Habitable Earth: IAEG XIV Congress 2023 Proceedings, Chengdu, China*, Environmental Science and Engineering,
https://doi.org/10.1007/978-981-99-9057-3_9

9.1 Introduction

The structure and mechanical properties of carbonate rocks are prone to severe alternation when subjected to tectonic disturbance. In the vicinity of a fault zone, parent rocks are transformed into fault rocks which have been sheared at low or high crustal levels (Sibson 1977). A typical fault comprises three major elements: the core, the damage zones and the area of host (parent) rocks (Kim et al. 2004; Lin and Yamashita 2013; Liao et al. 2020). The boundaries between the fault core and the damage zones are sharper than those between the damage zone and the host rock. The rocks in the damage zones have been subjected to brittle deformation but the original characteristics of the parent rock have been preserved. The mechanical properties of fault rocks are likely to improve with the distance from the fault plane and the properties of the parent rocks are re-established (Gudmundsson 2004; Faulkner et al. 2006).

The architecture of fault zones in carbonate rocks (i.e., trajectory, material distribution, deformation and slip patterns) differ from other cases. Delle Piane et al. (2017) suggest that these differences are due to the potential of carbonate minerals to sustain ductile deformation under physico-chemical processes and the inherent variability in the lithological content of the parent carbonate rocks. The formations that evolve around a carbonate fault zone are illustrated in Fig. 9.1a. Those include matrix-supported ($M_{S.}$) and grain-supported ($G_{S'}$) breccias and fragmented/fractured parent rocks. In the former cases (M_S and G_S), the Uniaxial Compressive Strength (UCS) of the intact rock is equal to the strength of the breccia which is lower compared to the UCS of the intact parent rock except for rare cases in which the breccia is strongly cemented. Moreover, as shown in Fig. 9.1a, the brecciated types do not have a highly developed fracturing degree and therefore, their mechanical properties are mostly attributed to their intact rock strength. Due to the large deformation near the fault core, the fractures have been filled and replaced by the matrix material. On the other hand, in the fragmented host rock zones, the intact rock strength is that of the host rock but the fracturing degree is higher and comprises the main strength reduction factor.

The carbonates rock masses (limestones and dolomites) found in several hundred meters thick zones along the study area (i.e., the western coastal Lefkada island, Greece) are highly tectonized, almost disintegrated, and at places powdered (Rondoyanni et al. 2012) forming limestone/dolomite fault breccias. Based on the observed fragment sizes, the limestone/dolomite fault breccias can be classified given the Woodcock and Mort (2008) classification system. The majority of the observed breccias are *crackle breccias*. Figure 9.1b presents typical outcrops of the rock masses found in western Lefkada Island. The majority of the investigated rock masses comprise fractured/fragmented limestones and limestone fault breccias cemented in a matrix that spatially varies and may be sandy/calcite or clayey. Both formations belong to extensive damage zones or fault cores developed around the faults in the region.

Despite extensive research on fault rocks and their importance in engineering projects, there is no well-accepted engineering classification that can be easily applied

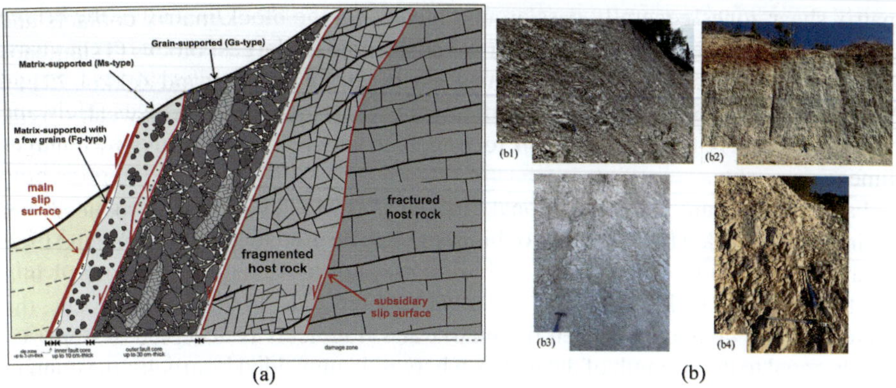

Fig. 9.1 **a** A typical cross-section of a carbonate fault zone in which the distribution of the main types of rock textures are noted (Ferraro et al. 2018); **b** Typical outcrops of fault rocks in western Lefkada island including: (b1) Fine-grained limestone that is highly fractured and fragmented featuring a mixture of angular and rounded shaped fragments; (b2) Limestone fault breccia with calcite/sandy matrix. The fragments range between 2-10 cm; (b3) Limestone fault breccia cemented in a calcite/sandy matrix; (b4) Limestone fault breccia cemented in an incohesive sandy matrix

in the field. The existing classification systems of fault rocks (e.g., Spry 1969, Higgins 1971; Killick 2003; Woodcock and Mort 2008; Mort and Woodcock 2008) approach these materials more from a geological rather than a geotechnical perspective. In addition, laboratory tests to derive the mechanical properties of such materials are halted due to their weak nature and the difficulties associated with sample preparation using the methodologies suggested in the international standards (ISRM 1979; ASTM 2017). Moreover, the existing rock mass classification systems (e.g., RMR, SMR, Q) cannot be successfully applied in such materials since they do not take into account their textural properties that ultimately control their mechanical properties.

9.2 Classification of Fault Breccias Using the “Block-In-Matrix Theory”

In general, the determination of strength and deformation properties of heterogeneous and geologically complex rock masses has been highly challenging and their estimation is increasingly being studied. The literature background (e.g., Medley 1994; Lindquist and Goodman 1994; Sonmez et al. 2006) involves the assessment of the properties of complex formations (i.e., fault rocks, pyroclastic rocks, serpentinites, mélanges); however, evidence regarding the characteristics of fault breccias are limited. Medley (1994) introduced the term “bimrocks”, a contraction of the term “block-in-matrix rocks” and defined a bimrock as “a mixture of rocks, composed of geotechnically significant blocks within a bonded matrix of finer”. The weakest elements in bimrocks are usually the contacts between strong blocks and matrix since matrix shear zones generally pass around blocks via the block/matrix contacts and the most intense shearing is often observed adjacent to the largest blocks (Tsiambaos 2010) but, in some cases, the contacts may also be strong (Medley and Zekkos 2011). Nonetheless, matrix shearing can be dominant when the matrix is of weak origin. Note the block is a term synonym to fragment which is usually used for larger dimensions.

Fault breccias are not conventional bimrocks, as Medley (1994) introduced, but the insights provided by the theory of bimrocks can be used to better understand their behavior. According to the theory of bimrocks, fault breccias can be grouped into two main groups: welded breccias and unwelded breccias. In welded bimrocks, the strength of the contacts between the matrix and fragments is accepted as approximately equal to the strength of the matrix whereas, in unwelded bimrocks, the contact strength between the matrix and fragments is assumed to be less than the strength of the matrix (Riedmüller et al. 2001). Depending on the properties of their components (fragments and matrix) and the contact strength between them, fault breccias may behave as welded or unwelded breccias. Typical examples of each case are provided in Sonmez et al. (2006) and Kahraman and Alber (2006), respectively.

9.3 Mechanical and Textural Properties of the Studied Carbonate Fault Rocks

Rock blocks of fault rocks were retrieved from western Lefkada island and involved two types of fault breccias: A limestone fault breccia (Type E) that behaves as a welded breccia due to the presence of weak calcite matrix and a dolomite tectono-diagenetic breccia (Type P) which features fine crystalline dolomite (strong) matrix and behaves as an unwelded breccia. Twelve (12) and thirteen (13) cubic samples from each rock type were prepared and Uniaxial Compression Tests (UCTs) were performed to derive their UCS. A new sample preparation technique was used as core drilling resulted in extensive disintegration, detachment of the fragments from the matrix and eventually to the material's degradation. The new methodology involved the utilization of a disk cutter with a diamond saw blade (400 mm). The cutting duration was short and the water supply minimum, a fact that prevented fragment-matrix detachment and excess fracturing. Typical samples from each rock type are presented in Fig. 9.2.

In addition, the matrix ratio for each specimen was measured before the UCT. To achieve this, 3D models of each specimen were created based on the Structure from Motion (Sfm) technique and orthophotos of each sample's side were generated. The generated orthophotos were assessed in terms of matrix-fragments distribution. To quantify these two elements an algorithm programmed in MATLAB (2019) that measures the surface that fragments and matrix occupy in a given orthophoto and derive their ratio was established. The final matrix ratio is the average value obtained from all sides of a specimen. The UCS of each specimen and the corresponding

Fig. 9.2 Typical cubic samples prepared from the dolomite tectono-diagenetic breccia (Type P) and the limestone fault breccia (Type E)

matrix ratios for the welded (Type E) and unwelded (Type P) breccias are provided in Table 9.1.

Table 9.1 Uniaxial Compressive Strength (UCS) and matrix ratio (%) for the two fault breccias studied (welded-Type E and unwelded-Type P)

Type E (Welded breccia)		Type P (Unwelded breccia)	
UCS (MPa)	Matrix ratio (%)	UCS (MPa)	Matrix ratio (%)
86.6[a]	0	30.2	57.8
15.4	17	29.9	61.9
13.5	13	14.2	56
24.8	13	32.6	56.4
36.7	11	31.3	61.6
12.2	22	41.0	62.1
20.5	19	28.4	49.6
10.6	29	14.8	47
9.3	31	25.2	55.5
5.8	26	16.9	50.1
3.0	44	33.0	68
3.2	40	58.9	69.9
3.0	100	69.8	75.7

[a]* Corresponds to the UCS of the fragments (UCS_{FR})

9.4 Engineering Classification of the Carbonate Fault Rocks

Coupling the geologic/tectonic regime of fault rocks (Fig. 9.1a), the insights of the theory of bimrocks and the experimental findings, the carbonate rocks formed around a fault zone are classified based on their engineering behavior as described below and illustrated in Fig. 9.3.

- ***Parent Rock (Protolith)***: The intact carbonate rock outside the fault's damage zone that has not been impacted by the fault slippage. Its mechanical properties are not affected by the fault.
- ***Welded carbonate breccia***: The material that has been brecciated due to the fault's slippage forming a fault breccia. The strength of the contact surface between the fragments and the matrix is equal to the matrix (welded). This is indicative of two conditions: 1) Strong cementation between matrix and fragments or 2) The existence of weak matrix. Type E represents a welded limestone breccia.
- ***Unwelded carbonate breccia***: The formation is similar to the welded breccia; however, the contact strength between fragments and matrix is less than the strength of the matrix. This is observed when: (1) The matrix is strong and (2) The cementation between the matrix and fragments is weak. Type P represents an unwelded dolomite breccia.
- ***Matrix***: In matrix-supported breccias (Fig. 9.1a), the matrix may be completely isolated from the fragments and studied individually. This is not a frequent incident and also depends on the scale of the study. Type E includes one specimen solely comprising calcite matrix which is representative of this condition.

It is suggested to classify a material as a welded/unwelded breccia only when the matrix ratio is at least 10%. For a lower matrix ratio, the behavior of the material approximates that of the parent rock or of interlocked fragments therefore, it cannot be classified as a breccia.

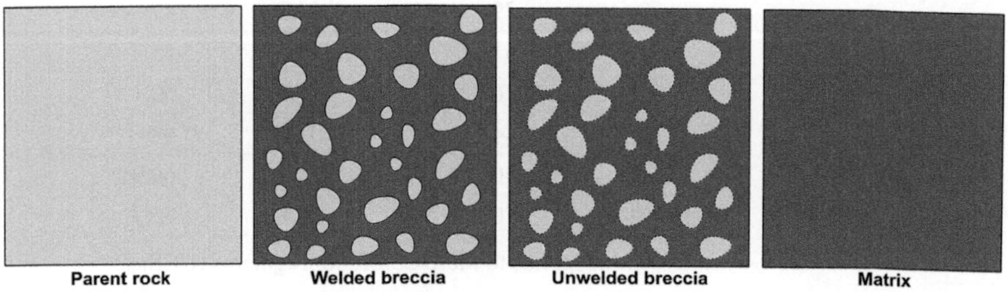

Fig. 9.3 Engineering classification of carbonate fault rocks

9.5 Determination of Uniaxial Compressive Strength

9.5.1 *Parent Rocks (Protolith)*

The case of the parent carbonate rocks, the typical experimental procedures suggested by the ISRM (1979) or ASTM (2017) can be used to directly determine their UCS.

9.5.2 *Welded Carbonate Breccias*

As explained earlier, the strength of a welded breccia depends on the corresponding strength of its fragments and its matrix. More specifically, its UCS is between the UCS of the fragments and the UCS of the matrix (noted as UCS_{FR} and UCS_{MR}, respectively). To account for the intermediate strength behavior, the laboratory findings of Type E are used. Figure 9.4a presents the normalized UCS of the welded rock type, i.e., UCS_{Wbr}/UCS_{FR} (*Wbr*: Welded breccia/*FR*: Fragments). Clearly, the increase of the matrix ratio results in a reduction of the UCS. Interestingly, the data indicate that the matrix proportion controls the overall strength of the material when it exceeds about 40%. The data allow for deriving a relationship between the matrix ratio and the intermediate strength of the welded breccias. The following equation is established:

$$UCS_{Wbr} = UCS_{MR} + UCS_{FR} * \left(\frac{R-1}{R}\right)e^{-1.04M} \tag{9.1}$$

$$R = \frac{UCS_{FR}}{UCS_{MR}} \tag{9.2}$$

where UCS_{Wbr} is the UCS of the welded breccia, UCS_{FR} is the UCS of the fragments, UCS_{MR} is the UCS of the matrix and M (%) is the matrix ratio.

The normalized UCS curves of the welded rock types for various R values are given in Fig. 9.4b. Observe that the trend of each curve qualitatively matches the lab data presented in Fig. 9.4a but the boundary values change depending on R (Eq. 9.2). Consequently, to obtain the intact strength of welded breccias UCS_{Wbr}, the input parameters include the UCS of the fragments UCS_{FR}, the UCS of the matrix UCS_{MR} and the matrix ratio M(%). The former can be derived by performing lab tests on the material that comprises the fragments (the parent rock). The strength of the matrix can also be measured by lab tests or by using a field classification when the material is very weak (see the case of the matrix in 5.4).

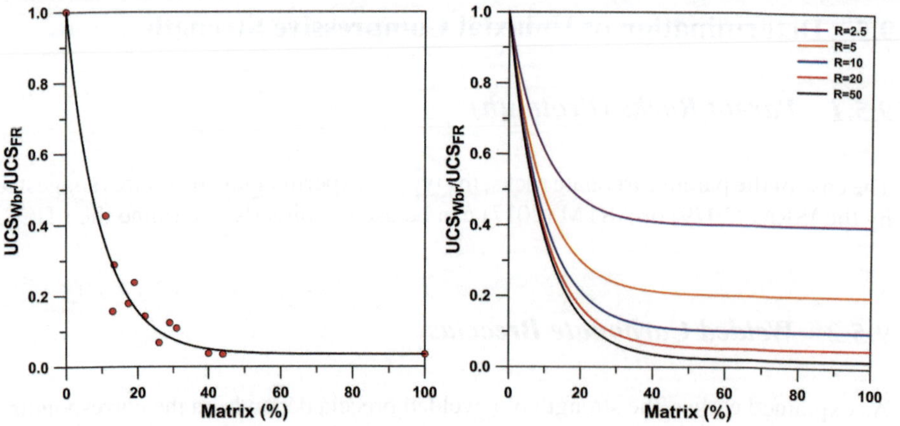

Fig. 9.4 **a** Plot of Normalized UCS of welded breccias (UCS_{Wbr}/UCS_{FR}) with matrix ratio for Type E lab data; **b** plot of normalized UCS of welded breccias (UCS_{Wbr}/UCS_{FR}) with matrix ratio for various $R\left(\frac{UCS_{FR}}{UCS_{MR}}\right)$ values

9.5.3 *Unwelded Carbonate Breccias*

In unwelded breccias, the increase of the matrix ratio also increases the overall strength as the boundaries between the matrix and fragments are reduced. According to the bimrock theory, when fragments replace the matrix, the shear strength of the material is reduced to the corresponding angle of repose (Kalender et al. 2014). Nonetheless, research has shown that the UCS of fault breccias is not nullified once the matrix ratio approximates zero Kahraman and Alber 2006; Afifipour and Moarefvand 2014). The cementation of the fragments and matrix introduces a cohesive strength component; thus interlocking is not the only factor that controls the UCS. These findings are supported by the Type-P experimental results that are used to assess the behavior of unwelded breccias.

To account for the lack of data below ratios of 47% (see Table 9.1-Type P), the corresponding data of Alber and Kahraman (2009) are used. Those are in agreement with the findings of the present study. The plot of the ratio between the UCS of the unwelded rock types and the UCS of the matrix, i.e., UCS_{Ubr}/UCS_{MR} (*Ubr*: Unwelded breccia/*MR*: Matrix) with increasing matrix ratio, based on the assumptions noted earlier, is presented in Fig. 9.5. The UCS of the unwelded breccia can be determined from the following equation:

$$UCS_{Ubr}/UCS_{MR} = 4.6 * 10^{-7} * M\,(\%)^{3.16} + 0.084,\ R^2 = 0.91[3] \tag{9.3}$$

where UCS_{Ubr} is the UCS of the unwelded breccia, UCS_{MR} is the UCS of the matrix and M (%) is the matrix ratio. The intercept (0.084) represents the UCS of the unwelded breccia once the matrix ratio approximates 0. It is equal to 8.4% of the UCS of the matrix. Therefore, the estimation of the unwelded breccia intact strength

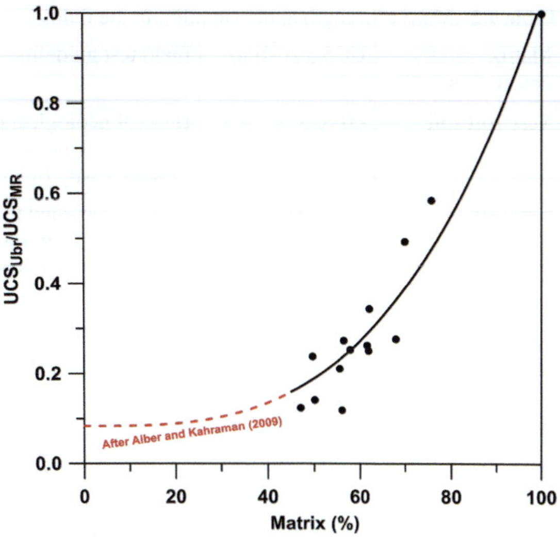

Fig. 9.5 Plot of normalized UCS of unwelded breccias ($\frac{UCS_{Ubr}}{UCS_{MR}}$) with matrix ratio for Type P lab data and the corresponding data of Alber and Kahraman (2009) to account for low matrix ratios (<47%)

requires the derivation of the matrix strength and the matrix ratio (%). Note that the strength of the fragments is not used in this case but it is assumed to be higher than the strength of the boundaries between the matrix and the fragments (which is always the case). Otherwise, the failure surface would traverse the fragments.

9.5.4 Matrix

In cases where the material is found as pure matrix, its intact strength must be derived. This parameter is also required when assessing the behavior of breccias. When the matrix is strong and cylindrical samples can be obtained, the international standard guidelines (ISRM 1979 or ASTM 2017) can be used. If the matrix is very weak (<1.5 MPa), sampling and testing cannot be achieved even if elegant sampling/preparation processes are used. In that case, the field properties suggested by ISRM (2007) and Waltham (2009) can be used to approximate the strength of the matrix (Table 9.2). If the matrix is in the very weak-weak rock range (1.0–12.5 MPa), sampling and testing procedures can be employed but it would still be challenging to measure its mechanical properties. In that case, Point Load tests (PLTs) can be performed in situ using irregular matrix specimens. If those prove to be also infeasible, again, the field properties summarized in Table 9.2 can be used.

Table 9.2 Matrix Strength approximation in the field

Matrix description	UCS_{MR} (MPa)	Field test guidelines	Is laboratory testing enabled?
Very soft soil	<0.025	Head of geological pick can easily be pushed into the shaft of the handle. Molded easily by fingers	No
Soft soil	0.025–0.05	Easily penetrated by thumb. Sharp end of pick can be pushed in to 30–40 mm. Molded by fingers with some pressure	No
Firm soil	0.05–0.10	Indented by thumb with effort. Sharp end of pick can be pushed in to 10 mm. Can just be penetrated with an ordinary spade	No
Stiff soil	0.10–0.20	Penetrated by thumbnail. Slight indentation produced by pushing pick point into soil. Cannot be molded by fingers. Requires hand pick for excavation	No
Very stiff soil/ extremely weak rock	0.20–0.40	Indented by thumbnail with difficulty. Slight indentation produced by blow of a pick point. Requires power tools for excavation	No
Extremely weak rock	0.4–1.0	Break by hand. Requires power tools for excavation	No
Very weak rock	1.0–5.0	Crumbles under firm blows with point of geological hammer. Can be peeled by a putty knife	Yes, but challenging. Alternative method: Point Load tests
Weak rock	5.0–12.5	Cannot cut by hand. Can be peeled by putty knife with difficulty. Indentations made with firm blows with point of a geological hammer	Yes, still problematic. Alternative method: Point Load tests
Moderately weak rock	12.5–25.0	Dent with hammer pick. Shallow indentations made with firm blows with point of a geological hammer	Yes
Medium-strong rock	25.0–50.0	Cannot be scraped or peeled by a putty knife. Specimen can be fractured with a single firm blow of a geological hammer	Yes
Strong rock	50.0–100.0	Specimen requires more than one blow of a geological hammer to fracture it	Yes
Very Strong rock	100.0–250.0	Specimen requires many blows of a geological hammer to fracture it	Yes
Extremely Strong	>250.0	Specimen can only be chipped with a geological hammer	Yes

Modified from ISRM (2007) and Waltham (2009)

Table 9.3 Summary of the methodologies employed to derive the intact rock strength based on the engineering classification of fault carbonate rocks

Description	Methodology
1. Intact carbonate rock	Derive the UCS of the parent rock via laboratory testing on intact specimens
2. Welded breccia: Shear strength of contact points between matrix and fragments is equal to the strength of the matrix	Derive the UCS of the fragments and the matrix and use the corresponding curves/ equations for welded breccias (Eq. 9.1)
3. Unwelded breccia: Shear strength of contact points between matrix and fragments is lower than the strength of the matrix	Derive the UCS of the matrix and use the corresponding curves/equations for unwelded breccias (Eq. 9.3)
4. Matrix	Derive the UCS of the matrix via laboratory testing on intact specimens or use the field estimations (Table 9.2) for rock strength approximation when it is weak (<1.0–5.0 MPa)

9.6 Summary

The determination of the intact rock strength of carbonate fault rocks is of high importance as it controls the quantification of the in situ strength using the existing criteria. For example, in the Hoek–Brown criterion, the UCS of the intact material affects the rock mass strength and the corresponding failure envelopes. The same applies when using the RMR, Q or Q-slope systems. In practice, the UCS of the parent rocks is used in these materials (i.e., intact limestone) without accounting for the different responses of the fault rocks when they are brecciated thus, behaving as welded or unwelded breccias. If the intact strength of the photolith (intact limestone) was used, predicting the shear strength of the material in situ could not be achieved. Therefore, it is essential to divide the carbonate fault rocks into the four (4) categories discussed above and estimate their intact strength as summarized in Table 9.3.

The experimental investigation allowed the establishment of a methodology to assess the textural (distribution of fragments and matrix) properties of welded and unwelded breccias and construct fundamental equations that predict their intact rock strength (noted as UCS_{Wbr} and UCS_{Ubr}, respectively). The input parameters include the uniaxial compressive strength UCS of their fragments (UCS_{FR}), the UCS of the matrix (UCS_{MR}) and the matrix ratio M (%). Moreover, guidelines to measure the strength of the matrix are provided for cases that this is too weak to perform laboratory tests.

Further research should focus on performing more Uniaxial Compression Tests on carbonate fault rocks to verify the proposed equations. Moreover, it would be interesting to study where the transition point between welded and unwelded breccias lies. This could not be investigated in the present study since the studied breccias were clearly either welded or unwelded. The transition between the two discrete behaviors is a function of the strength of the matrix and the strength between the fragments and matrix boundaries.

References

A.S.T.M. (2017) D7012–14e1, Standard test methods for compressive strength and elastic moduli of intact rock core specimens under varying states of stress and temperatures. https://www.astm.org/d7012-14e01.htmls

Afifipour M, Moarefvand P (2014) Mechanical behavior of bimrocks having high rock block proportion. Int J Rock Mech Min Sci 65:40–48. https://doi.org/10.1016/j.ijrmms.2013.11.008

Alber M, Kahraman S (2009) Predicting the uniaxial compressive strength and elastic modulus of a fault breccia from texture coefficient. Rock Mech Rock Eng 42:117–127. https://doi.org/10.1007/s00603-008-0167-x

Delle Piane C, Ben Clennell M, Keller JVA, Giwelli A, Luzin V (2017) Carbonate hosted fault rocks: a review of structural and microstructural characteristic with implications for seismicity in the upper crust. J Struct Geol 103:17–36. ISSN 0191-8141. https://doi.org/10.1016/j.jsg.2017.09.003

Faulkner DR, Mitchell TM, Healy D, Heap MJ (2006) Slip on "weak" faults by the rotation of regional stress in the fracture damage zone. Nature 444:922–925

Ferraro F, Grieco DS, Agosta F, Prosser G (2018) Space-time evolution of cataclasis in carbonate fault zones. J Struct Geol 110:45–64. ISSN 0191-8141. https://doi.org/10.1016/j.jsg.2018.02.007

Gudmundsson A (2004) Effects of Young's modulus on fault displacement. CR Geosci 336:85–92

Higgins MW (1971) Cataclastic rocks. Professional Paper, United States Geological Survey, no 687, p 97

I.S.R.M. (1979) Suggested methods for determining the uniaxial compressive strength and deformability of rock materials. Int J Rock Mech Min Sci Geomech Abstracts 16(2)

I.S.R.M. (2007) The complete ISRM suggested methods for rock characterisation, testing and monitoring: 1974–2007. In: Ulusay R, Hudson JA (eds). ISRM Turkish National Group, Ankara, Turkey

Kahraman S, Alber M (2006) Estimating unconfined compressive strength and elastic modulus of a fault breccia mixture of weak blocks and strong matrix. Int J Rock Mech Min Sci 43(8):1277–1287. ISSN 1365-1609, https://doi.org/10.1016/j.ijrmms.2006.03.017

Kalender A, Sonmez H, Medley EW, Tumultuous C, Kasapoglu KE (2014) An approach to predicting the overall strengths of unwelded bimrocks and bimsoils. Eng Geol 183:65–79. ISSN 0013-7952, https://doi.org/10.1016/j.enggeo.2014.10.007

Killick AM (2003) Fault rock classification: an aid to structural interpretation in mine and exploration geology. S Afr J Geol 106:395–402

Kim Y-S, Peacock DCP, Sanderson DJ (2004) Fault damage zones. J Struct Geol 26(3). ISSN 0191-8141, https://doi.org/10.1016/j.jsg.2003.08.002

Liao Z, Li W, Zou H, Hao F, Marfurt KJ, Reches Z (2020) Composite damage zones in the subsurface. Geophys J Int 222(1):225–230. https://doi.org/10.1093/gji/ggaa158

Lin A, Yamashita K (2013) Spatial variations in damage zone width along strike-slip faults: an example from active faults in southwest Japan. J Struct Geol 57:1–15. ISSN 0191-8141, https://doi.org/10.1016/j.jsg.2013.10.006

Lindquist ES, Goodman RE (1994) Strength and deformation properties of a physical model melange. In: Nelson PP, Laubach SE (eds) Proceedings of the 1st North American rock mechanics symposium. Balkema, Rotterdam, pp 843–850

Matlab (2019) Version R2019a (9.6.0), Natick, Massachusetts: The MathWorks Inc.

Medley EW, Zekkos D (2011). Geopractitioner approaches to working with antisocial mélanges. In: Mélanges: processes of formation and societal significance. John Wakabayashi, Yildirim Dilek. https://doi.org/10.1130/2011.2480(13)

Medley EW (1994). The engineering characterization of melanges and similar block- in-matrix rocks (bimrocks). Ph.D. dissertation, University of California, Berkeley, UMI, Inc., Ann Arbor

Mort K, Woodcock NH (2008) Quantifying fault breccia geometry: Dent Fault, NW England. J Struct Geol 30(6):701–709. ISSN 0191-8141, https://doi.org/10.1016/j.jsg.2008.02.005

Riedmüller G, Brosch F-J, Klima K, Medley EW (2001) Engineering geological characterization of brittle faults and classification of fault rocks. Felsbau 19(4):13–19
Rondoyanni T, Sakellariou M, Baskoutas J, Christodoulou N (2012) Evaluation of active faulting and earthquake secondary effects in Lefkas Island, Ionian Sea, Greece: an overview. Nat Hazards 61:843–860. https://doi.org/10.1007/s11069-011-0080-6
Sibson RH (1977) Fault rocks and fault mechanisms. J Geol Soc London 133:191–213
Sonmez H, Gokceoglu C, Medley EW, Tuncay E, Nefeslioglu HA (2006). Estimating the uniaxial compressive strength of a volcanic bimrock. Int J Rock Mech Min Sci 43(4):554–561. ISSN 1365-1609, https://doi.org/10.1016/j.ijrmms.2005.09.014.
Spry A (1969) Metamorphic textures. Pergamon Press, Oxford, U.K., p 350
Tsiambaos G (2010) Engineering geological behaviour of heterogenous and chaotic rock masses. Bull Geol Soc Greece 43(1):183–195. https://doi.org/10.12681/bgsg.11173
Waltham T (2009) Foundations of engineering geology, 3rd edn. New York, Taylor & Francis, Spon Press, London
Woodcock NH, Mort K (2008) Classification of fault breccias and related fault rocks. Geol Mag 145(3):435–440. ISSN 0016-7568 https://doi.org/10.1017/S0016756808004883

Chapter 10
DEM Simulations of Critical State Behavior of Sand Under Triaxial Compression and Extension

Guobin Gong, Minyi Zhu, Yuanfeng Bao, and Stephen Wilkinson

Abstract This study focuses on examining the behaviors of dense and loose sand subjected to drained triaxial compression and extension using DEM. The research involved performing drained tests on samples, manipulating them under both traditional (constant confining pressure) and constant mean stress conditions using servo-control. The observed drained behavior qualitatively aligns with previously published laboratory findings. It is demonstrated that the void ratio of sand samples remains unaffected by initial packing densities and applied stress paths (CT/CP). The critical state (CS) line is unique regardless of the stress path (CT/CP) conditions, though it shows dependence on whether triaxial compression or extension is applied. Drawing upon the DEM results, a comparison and discussion of various failure criteria are presented. The Matsuoka criterion emerges as the most suitable at the CS. To describe structural anisotropy, a generalized microscopic fabric parameter is proposed, which appears to be unique regardless of the stress paths (CT/CP) conditions.

Keywords DEM · Drained · Triaxial · Sand · Critical state

G. Gong (✉) · M. Zhu
Xi'an Jiaotong-Liverpool University, Xi'an, China
e-mail: Guobin.Gong@xjtlu.edu.cn

M. Zhu
e-mail: 24310033@tongji.edu.cn

M. Zhu
Tongji University, Shanghai, China

Y. Bao
Suzhou Integrated Infrastructure Technology Research Institute, Suzhou, China

S. Wilkinson
University of Wollongong, Dubai, United Arab Emirates
e-mail: Stephenwilkinson@uowdubai.ac.ae

S. Wang et al. (eds.), *Engineering Geology for a Habitable Earth: IAEG XIV Congress 2023 Proceedings, Chengdu, China*, Environmental Science and Engineering,
https://doi.org/10.1007/978-981-99-9057-3_10

10.1 Introduction

The critical state (CS) concept, originated by Roscoe et al. (1958), is commonly employed in the analysis of isotropic elastic or plastic materials exhibiting hardening. The term "critical state" in soils typically denotes a condition where the stress tensor and soil volume remain unchanged despite continuous shearing. Numerous studies have explored the characteristics of soil's CS behaviors through drained triaxial tests, relying on either experimental evidence (Oda et al. 1982; Been et al. 1991; Chu 1995; Ali Rahman et al. 2018), or simulation outcomes (Ng 2009; Thornton and Zhang 2010; Huang et al. 2014; Yang and Wu 2017; Zhou et al. 2017). The uniqueness of the CS line has been a significant topic of discussion (Ng 2009; Zhao and Guo 2013; Yang and Wu 2017). Unlike traditional triaxial tests that assume an axisymmetric stress condition (i.e., $\sigma_2 = \sigma_3 =$ constant), this study introduces a unique condition in axisymmetric triaxial tests where the effective mean stress (p) remains constant, as depicted in Fig. 10.1. Comparing the CS behavior under these two loading conditions, CT and CP, can provide valuable insights, yet such comparisons are seldom found in existing literature.

In this paper, we utilize the discrete element method (DEM) to investigate this question, which is pioneered by Cundall (1971). A variety of drained triaxial compression (AC) and extension (AE) tests with diverse loading paths and confining pressures are simulated for both loose and dense sand using the commercial software PFC 3D. The results are then analyzed and discussed to evaluate the impact of different variables on the CS behavior of sand. Additionally, macroscopic and microscopic parameters are introduced to elucidate the observed soil response.

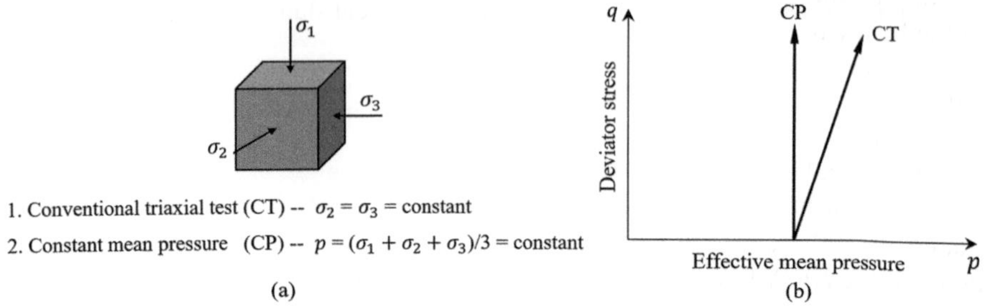

Fig. 10.1 Two types of triaxial tests: **a** Loading conditions; **b** Loading paths

10.2 Overview of DEM Simulation

10.2.1 Model Characteristic

In the DEM simulation, sand, a heterogeneous material, is represented by a collection of dispersed particles interacting at contact points and confined within boundary walls. Original positions for particles and walls, along with their respective contact constitutive models, are established. For every timestep increment, the force–displacement law is employed to each contact point to compute the updated contact force. Subsequently, particle and wall displacements are determined using Newton's second law of motion, updating their positions. This process iterates over each time step (Cundall and Strack 1979; Cundall 2001). In this research, we use the commercial software Particle Flow Code (PFC) with FISH language in its 3D version 5, developed by ITASCA (Itasca 2014), as a platform for conducting numerical simulations.

As depicted in Fig. 10.2, a total number of 7974 particles (diameter ranges from 0.20 to 0.50 mm) are created within a cubic cell measuring 7 mm^3 in PFC 3D. The nominal particle density is set at 2650 kg/m^3. Figure 10.3 displays the grain size versus percentage passing (particle size distribution) of the sample. A density scaling system is employed to enhance simulation efficiency without significantly affecting quasi-static behavior (Thornton 2000).

.

Additionally, a linear contact model is employed for interactions between particle-to-particle and particle-to-wall contacts. This model has been demonstrated to be effective in studying the microscopic behavior of granular media (Di Renzo and Di Maio 2004). Key parameters for particles and contacts are outlined in Table 10.1. The stiffness ratio (normal to shear k_n/k_s) is set at 1.33, which aligns with typical

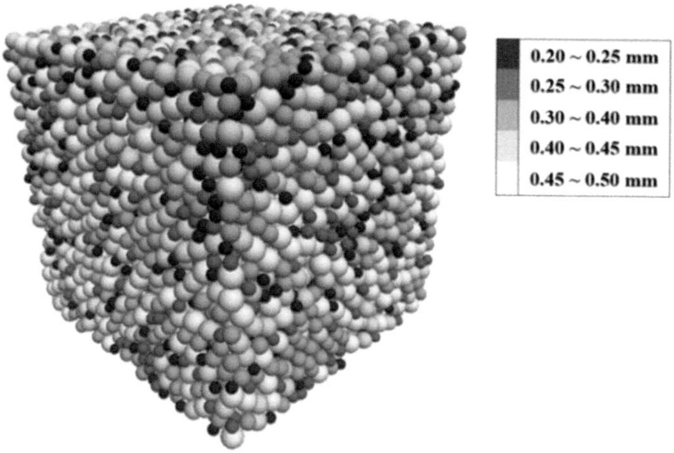

Fig. 10.2 Representative model established in PFC

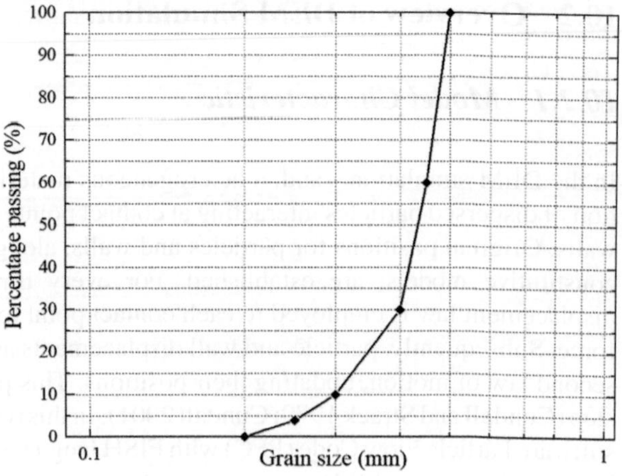

Fig. 10.3 Grain size versus percentage passing

Table 10.1 Micro-scale parameters of sample in the drained tests

Microscopic parameter	Magnitude	Units
Effective modulus E_e	1×10^8	Pa
Particle and wall stiffness (normal-shear) ratio k_n/k_s	1.33	–
Notional density ρ	2650	kg/m^3
Contact damping coefficient ζ	0.7	–
Particle–particle friction μ_p	0.5	–
Wall friction μ_w	0	–

values for granular materials (see e.g., Gong and Liu 2017). A contact damping coefficient $\zeta = 0.7$ is utilized to dissipate the system's excess kinetic energy. It's worth noting that the inter-particle friction μ_p refers to the friction during drained tests, rather than the initial one during particle generation. The gravitational field is not considered in the current models.

Prior to the drained tests, a servo-control system is used to apply isotropic stress on samples until a target stress level is achieved with a maximum tolerance of 0.1%. Once this isotropic stress state is attained, the sample is prepared and the subsequent drained tests (loose/dense, AC/AE, and CT/CP) all commence from this state to ensure consistent initial conditions.

10.2.2 Simulation Details

During the simulation, the opposite walls are allowed to move either towards each other (resulting in an increase in σ_1) for AC, or in opposite directions (leading to a decrease in σ_1) for AE during the shearing process, while the confining stresses σ_2

and σ_3 are regulated by the servo-control system. For CT, the servo-control system can maintain an approximate constant confining stresses and equal to the isotropic stress applied earlier, i.e., $\sigma_2 = \sigma_3 = \sigma_{iso}$ (σ_{iso}—isotropic confining stress). For CP, the servo-control simultaneously adjusts both confining stress values to maintain a constant effective mean pressure, i.e., $\sigma_2 = \sigma_3 = (3\sigma_{iso} - \sigma_1)/2$.

Different samples for the DEM simulation of drained tests are detailed in Table 10.2. A total number of 16 simulations have been conducted, with different confining pressures set at either 100 kPa or 300 kPa. Other variables taken into account include the method of loading (AC/AE), the stress path (CT/CP), and the compactness of the sand (loose/dense). For instance, the sample named CTE_L_100 represents a traditional triaxial test on loose sand under axial extension, with a confining pressure of 100kPa. The initial void ratios for all samples have been noted down. It's noticeable that the void ratio is lower for denser samples subjected to higher confining pressure. The axial strain, denoted as ε_a, is calculated using the formula $\ln(H_0/H)$ (H_0—initial height of sample; H—real-time height of sample). All simulations are halted when ε_a reaches 50%, a point at which the CS behavior of sand becomes significantly evident.

Table 10.2 Prepared samples with key name before tests

Test ID	Loading method	Stress path	Isotropic stress	Sand looseness	Initial void ratio
CTC_D_100	Compression	CT	100 kPa	Dense	0.613
CTC_L_100				Loose	0.756
CPC_D_100		CP		Dense	0.613
CPC_L_100				Loose	0.756
CTC_D_300		CT	300 kPa	Dense	0.594
CTC_L_300				Loose	0.741
CPC_D_300		CP		Dense	0.594
CPC_L_300				Loose	0.741
CTE_D_100	Extension	CT	100 kPa	Dense	0.613
CTE_L_100				Loose	0.756
CPE_D_100		CP		Dense	0.613
CPE_L_100				Loose	0.756
CTE_D_300		CT	300 kPa	Dense	0.594
CTE_L_300				Loose	0.741
CPE_D_300		CP		Dense	0.594
CPE_L_300				Loose	0.741

CT refers to conventional test and CP refers to Constant mean pressure test

10.3 Results & Discussion

10.3.1 Macroscopic Behavior

Figures 10.4a and 10.5a illustrate the progression of deviator stress (q) in relation to axial strain (ε_a) for the two different loading methods, namely AC and AE. Each dense sample firstly exhibits a rapid strain hardening behavior until an initial peak deviator stress is attained and a strain softening behavior occurs subsequently. In contrast, only strain hardening behavior is noticeable in loose samples. The critical state (CS) is achieved for all samples when the axial strain ε_a is about 25%. For the same loading path and confining pressure, the overall deviator stresses (both peak and CS values) of AC tests are apparently higher than those of AE tests. For AC tests, the samples loaded with CT path show a higher overall deviator stress comparing to those loaded with CP path, whereas this phenomenon is inverse for AE tests.

Furthermore, the progression of stress ratio (q/p') for the two loading methods is shown in Figs. 10.4b and 10.5b, respectively, where p' refers to the real-time value of effective mean pressure. The macroscopic stress behavior observed here bears qualitative similarities to the results reported in physical experiments on sands

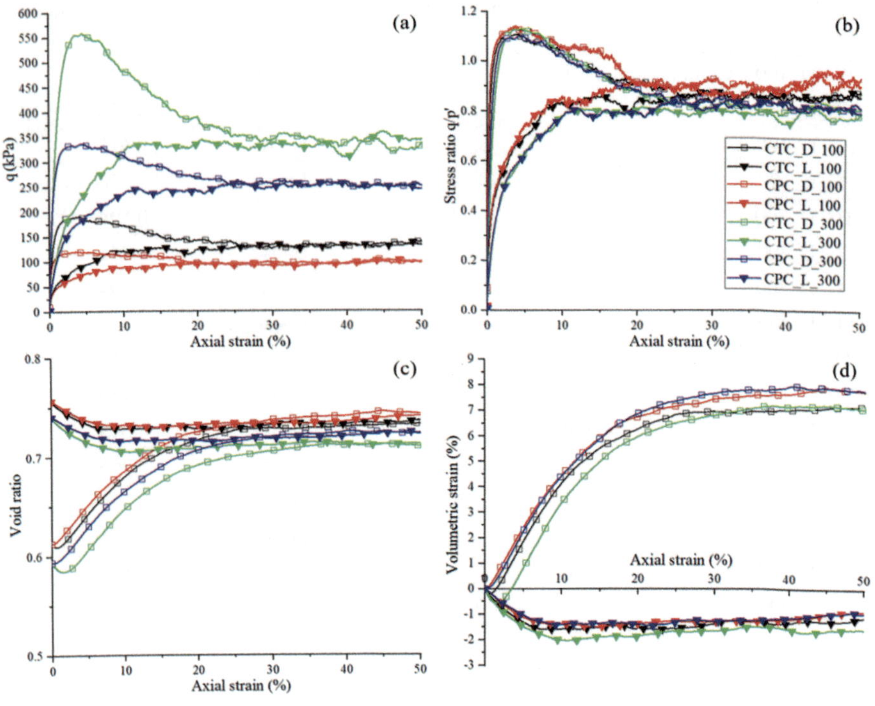

Fig. 10.4 Evolution of macro-parameters in AC tests: **a** Deviator stress; **b** Stress ratio; **c** Void ratio; **d** Volumetric strain

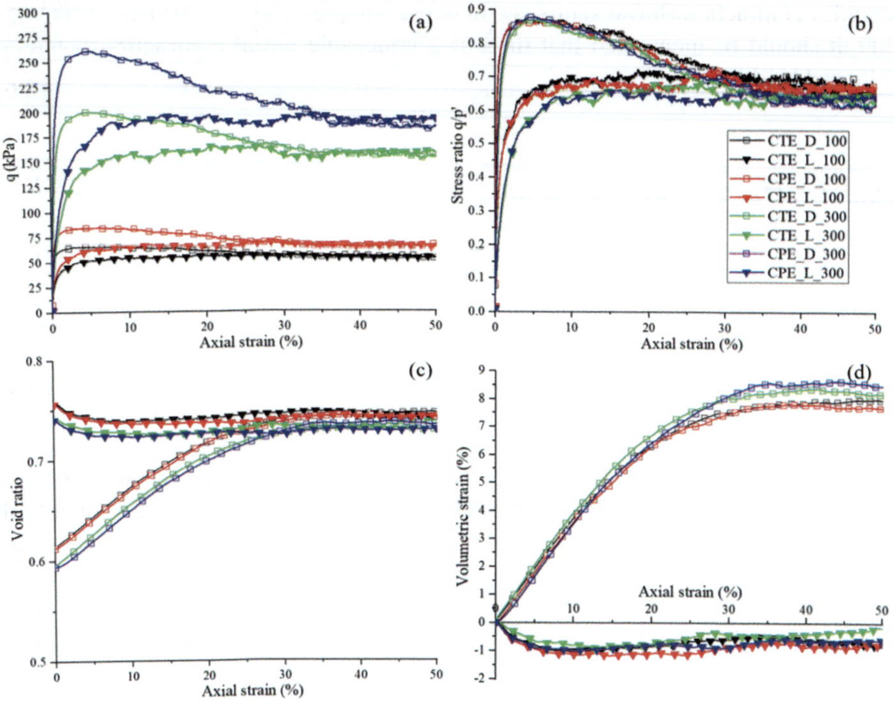

Fig. 10.5 Evolution of macro-parameters in AE tests: **a** Deviator stress; **b** Stress ratio; **c** Void ratio; **d** Volumetric strain

(Yamamuro and Lade 1996). It's noteworthy that regardless of the stress path (CT/CP) and confining pressure, the stress ratios among different samples exhibit a degree of similarity, both at peak values (for dense sand) and at CS values. Additionally, it is evident that at the CS, the stress ratios in AC tests are marginally higher than those in AE tests.

Figures 10.4c and 10.5c depict the progression of void ratio (e) in relation to axial strain (ε_a) for the two loading methods, namely AC and AE. Generally, the void ratio escalates in dense samples and diminishes in loose samples. The CS is attained after an axial strain of approximately 30% for dense sand and 10% for loose sand. Regardless of the initial void ratio (dense/loose) and the stress path (CT/CP), the void ratio at the CS is smaller for samples subjected to higher confining pressure. This, in other words, suggests that the manifestation of a critical density is unique regardless of the initial packing densities and stress path, but it is influenced by the applied confining pressure.

Figures 10.4d and 10.5d illustrate the evolution of volumetric strain in relation to axial strain (ε_a) for the two loading methods, namely AC and AE. The trend of volumetric strain generally mirrors that of the void ratio for each sample. The CS is reached earlier in loose sand than in dense sand. In these figures, a positive value of volumetric strain indicates dilation (dense samples), while a negative value signifies contraction (loose samples). In dense samples CTC_D_100 and CTC_D_300, it should be mentioned that there is a noticeable initial contraction behavior followed by dilation.

10.3.2 Microscopic Behavior

The fabric tensor (ϕ_{ij}), characterizing the structural anisotropic of a granular system, originated by Oda (1982) and Satake (1982) is expressed as follows.

$$\phi_{ij} = \frac{1}{C}\sum_{1}^{C} n_i n_j \tag{10.1}$$

where C represents the total contact number and n_i represents the unit normal vector. Both Thornton (2000) and Kodicherla et al. (2021) quantified the structural anisotropy of a system by introducing a deviator fabric (ϕ_d), where $\phi_d = \phi_1 - \phi_3$ (ϕ_1—major principal value; ϕ_3—minor principal value). Figure 10.6a and b present the progression of deviator fabric (ϕ_d) in relation to axial strain (ε_a) for the two loading methods, namely AC and AE. The development of the deviator fabric follows a consistent trend with the macroscopic behavior of the stress ratio. It is also observed that samples in AC tests reach the CS earlier than those in AE tests. Furthermore, the deviator fabric at CS diminishes as the confining pressure escalates. For a specific confining pressure and loading method, the deviator fabric at the CS appears to be unique regardless of different degrees of looseness and stress paths (CT/CP).

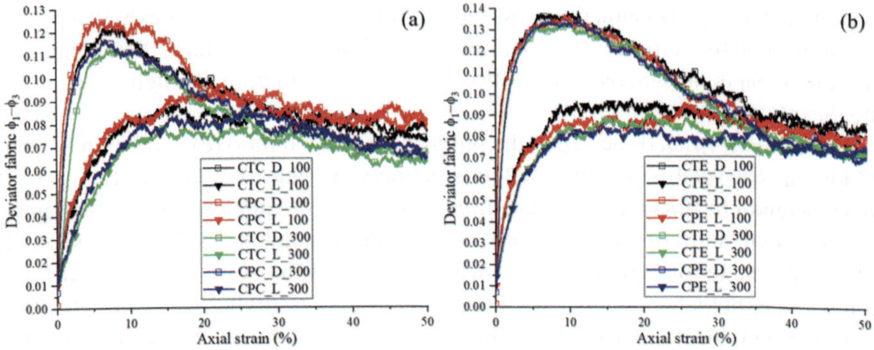

Fig. 10.6 Evolution of deviator fabric: **a** AC; **b** AE

10.3.3 Critical State Analysis

Different failure criterions are illustrated and compared to obtain an appropriate strength criterion for current samples. According to Lade and Duncan (1973), the failure criterion is defined as I_1^3/I_3 (call Lade parameter) (I_1—the first stress invariants; I_3—the third stress invariants. Matsuoka and Nakai (1974) used another failure criterion (call Matsuoka parameter) by taking the second stress invariant (I_2) into account, which is expressed as $(I_1I_2)/I_3$. Figures 10.7 and 10.8 show the evolution of Lade parameter and Matsuoka parameter against axial strain (ε_a) for the different confining pressure, i.e., (a) 100 kPa and (b) 300 kPa, respectively. It is observed that the Lade parameter at peak for dense samples is generally independent of the stress path (CT/CP) applied and loading methods (AC/AE), but it shows great difference for those with different loading methods at CS. The Matsuoka parameter shows good consistency at CS for samples under AC and AE tests. Independent of the stress path, the Matsuoka parameter in AE tests exhibits a higher value than that in AC tests.

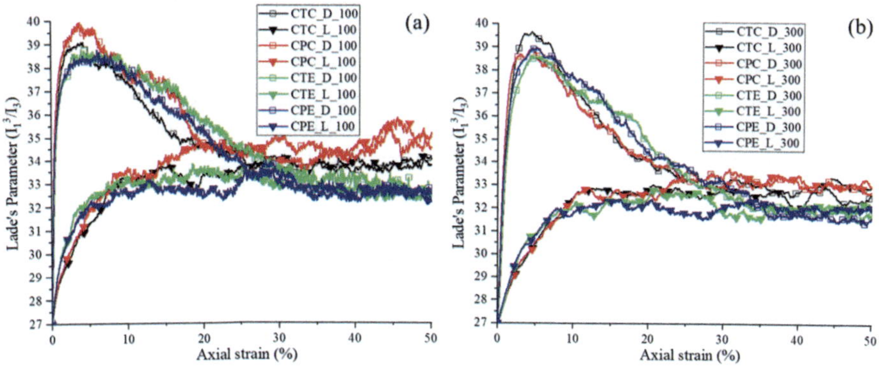

Fig. 10.7 The Lade parameter of samples with confining stresses of **a** 100 kPa; **b** 300 kPa

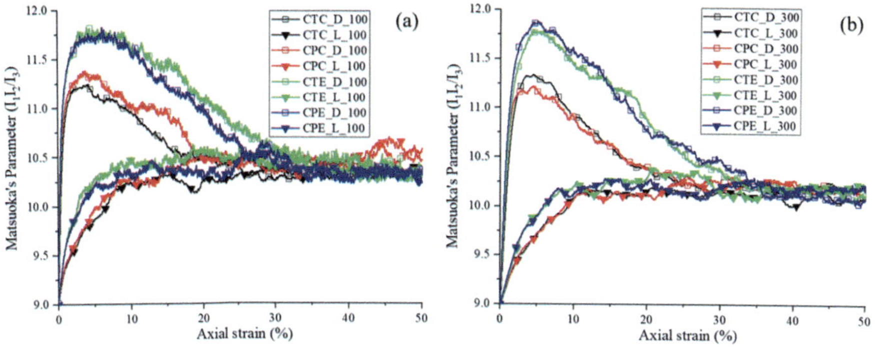

Fig. 10.8 The Matsuoka parameter of samples with confining stresses of **a** 100 kPa; **b** 300 kPa

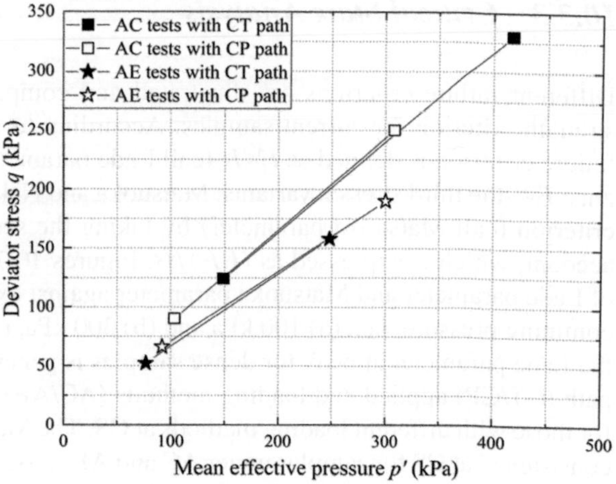

Fig. 10.9 The CS characteristics for dense sand in $q-p'$ space

Figure 10.9 presents the CS characteristics of dense sand in $q-p'$ space. All tests with dense sand are divided into 4 lines by stress path (CT/CP) and loading method (AC/AE). The CS line is shown to be independent of the stress paths, but it exhibits high dependence on whether AC or AE is imposed. Furthermore, the slopes of CS lines for AC tests are obviously larger than those for AE tests. This observation aligns with the findings previously reported by Kodicherla et al. (2021).

10.4 Conclusion

In this study, the critical state (CS) behaviors of both dense and loose sand through DEM simulations of triaxial drained tests under axial compression (AC) and extension (AE) is explored. Taking into account two types of loading conditions—traditional stress condition with constant confining pressure (CT), and constant mean stress condition (CP)—a total of 16 samples are generated for comparison. The discussion revolves around the CS behaviors of sand, demonstrating strong correlations with the evolution of macro- and micro-scale parameters. The main conclusions drawn from the DEM results and subsequent discussions are as follows:

- The peak and CS stress ratios of different samples exhibits the similar values, regardless of the stress path and confining pressure.
- The void ratio (or critical packing density) is indicated to be unique regardless of the initial packing (dense/loose) and stress paths (CT/CP) applied.
- At CS, the deviator fabric is found to be independent of the looseness of samples and the stress path (CT/CP) loaded.
- The Matsuoka parameter is shown to be unique at CS for samples under AC and AE tests. The Matsuoka criterion exhibits to be a more appropriate failure criterion in current simulations.

- The CS line is shown to be unique regardless of the stress path conditions, but it exhibits evident dependence on whether triaxial compression or extension is imposed.

Acknowledgements The authors express their deep gratitude for the funding received from project ID: RDF 18-01-23 and PGRS1906002 at XJTLU. The authors also extend appreciation for the backing provided by the Special Fund ID: KSF-E-19 within XJTLU.

References

Ali Rahman Z, Toll DG, Gallipoli D (2018) Critical state behaviour of weakly bonded soil in drained state. Geomech Geoengin 13(4):233–245

Been K, Jefferies MG, Hachey J (1991) The critical state of sands. Géotechnique 41(3):365–381

Chu J (1995) An experimental examination of the critical state and other similar concepts for granular soils. Can Geotech J 32(6):1065–1075

Cundall PA (2001) A discontinuous future for numerical modelling in geomechanics. P I Civil Eng-Geotec 149(1):41–47

Cundall PA, Strack ODL (1979) A discrete numerical model for granular assemblies. Géotechnique 29:47–65

Cundall PA (1971) A computer model for simulating progressive large-scale movements in blocky rock systems. In: Proceedings of the international symposium on rock mechanics, II-8

Di Renzo A, Di Maio FP (2004) Comparison of contact-force models for the simulation of collisions in DEM-based granular flow codes. Chem Eng Sci 59(3):525–541

Gong J, Liu J (2017) Effect of aspect ratio on triaxial compression of multi-sphere ellipsoid assemblies simulated using a discrete element method. Particuology 32:49–62

Huang X, Hanley KJ, O'Sullivan C, Kwok CY, Wadee MA (2014) DEM analysis of the influence of the intermediate stress ratio on the critical-state behaviour of granular materials. Granul Matter 16:641–655

Itasca (2014) Particle flow code in 3 dimensions (PFC3D) version 5. Itasca Consulting Group Inc., Minneapolis, MN, USA

Kodicherla SPK, Gong G, Fan L, Wilkinson S, Moy CKS (2021) Discrete element modelling of strength and critical state characteristics of granular materials under axial compression and axial extension stress path tests. Particuology 56:152–162

Lade PV, Duncan JM (1973) Cubical triaxial tests on cohesionless soil. Int J Rock Mech Min Sci Geomech Abstracts 11(3):50

Matsuoka H, Nakai T (1974) Stress-deformation and strength characteristics of soil under three different principal stresses. Proc Jpn Soc Civ Eng 1974(232):59–70

Ng TT (2009) Discrete element method simulations of the critical state of a granular material. Int J Geomech 9(5):209–216

Oda M (1982) Fabric tensor for discontinuous geological materials. Soils Found 22(4):96–108

Oda M, Konishi J, Nemat-Nasser S (1982) Experimental micromechanical evaluation of strength of granular materials: effects of particle rolling. Mech Mater 1(4):269–283

Roscoe KH, Schofield AN, Wroth CP (1958) On the yielding of soils. Géotechnique 8(1):22–53

Satake M (1982) Fabric tensor in granular materials. The Netherlands: Balkema, Rotterdam

Thornton C (2000) Numerical simulations of deviatoric shear deformation of granular media. Géotechnique 50(1):43–53

Thornton C, Zhang L (2010) On the evolution of stress and microstructure during general 3D deviatoric straining of granular media. Géotechnique 60(5):333–341

Yamamuro JA, Lade PV (1996) Drained sand behaviour in axisymmetric tests at high pressures. J Geotech Eng 122(2):109–119

Yang Z, Wu Y (2017) Critical state for anisotropic granular materials: a discrete element perspective. Int J Geomech 17(2):04016054

Zhao J, Guo N (2013) Unique critical state characteristics in granular media considering fabric anisotropy. Géotechnique 63(8):695–704

Zhou W, Liu J, Ma G, Chang X (2017) Three-dimensional DEM investigation of critical state and dilatancy behaviors of granular materials. Acta Geotech 12:527–540

Chapter 11
Study on the Strength Characteristics of Granite Residual Soil Under Acidic Wetting–Drying Cycles

Guipeng Luo, Chengming She, Xiuzi Jiang, and Hui Dong

Abstract It is crucial to analyze the long-term stability of granite residual soil slopes by the strength characteristics of granite residual soil under acidic wetting–drying cycles after seasonal acid rain. With this concern in mind, the strength characteristics of granite residual soil and the water-soil interaction under the dual action of acid solution and wetting–drying cycles through direct shear test is studied and analyzed in this work. The results show that the shear strength and parameters are heavily affected by the solution pH value under certain normal stress when considering acidic environment only. The shear strength of granite residual soil decreases rapidly when considering the wetting–drying cycles only. Besides, the internal friction angle and cohesion undergo an increase and decrease trend under the acidic wetting–drying cycles, respectively, in first 2 wetting–drying cycles, and then tend to be stable. A diagonal increase area of shear strength is obtained and becomes more obvious with the increase of normal stress when $pH < 6$ after first 2 cycles. Furthermore, the internal friction angle exhibits a single-peak increase area along the diagonal direction, while the cohesion shows a significant decreasing trend. Based on these results, it is inferred that the shear strength is affected by the dynamic evolution of micro-cracks under wetting–drying cycles, and the change of the electronic double layer thickness caused by ion exchange adsorption and the mineral acid dissolution. The results can provide a useful reference for understanding the mechanical properties and environmental effects of granite residual soil.

Keywords Granite residual soil · Acidic · Wetting–drying cycles · Water-soil interaction

---

G. Luo · X. Jiang (✉) · H. Dong
College of Civil Engineering, Xiangtan University, Xiangtan, China
e-mail: jiangxiuzi@xtu.edu.cn

C. She
Hunan Academy of Building Research Co. Ltd, Changsha, China

S. Wang et al. (eds.), *Engineering Geology for a Habitable Earth: IAEG XIV Congress 2023 Proceedings, Chengdu, China*, Environmental Science and Engineering,
https://doi.org/10.1007/978-981-99-9057-3_11

11.1 Introduction

Granite residual soil is a special soil widely distributed in Hunan Province, China. It is extremely sensitive to the change of water environment. Water infiltration causes granite residual soil easily to collapse and soften result in a loss of shear strength (Nikooee et al. 2013; Bogaard and Greco 2016), which is the main reason for slope instability, ground subsidence and foundation settlement (Zhan et al. 2018). Rainfall is an important inducing factor of slope instability. The wetting–drying process leads to soil structure changes, thus affecting the mechanical properties. Considering the central and eastern parts of Hunan Province are still acid rain areas, and the mechanism of acid rain is not only related to the change of pore water pressure, but also to the change of mineral composition and structure of soil by acid solution (Moore and Brunsden 1996; Jaboyedoff et al. 2004; Bakhshipour et al. 2017). The influence of acidic environment on the basic physical and mechanical properties of granite residual soil has been widely reported. Ji and Li (2017) and Lin et al. (2020) studied the mechanical behavior of granite residual soil under acidic solution, and found that the shear strength and cohesion of granite residual soil decreased with the increase of soaking time. Jin et al. (2017) reported that the change of electronic double layer and cement dissolution were the main reasons for the accelerated disintegration of granite residual soil. Sun and Tang (2018) found that the acidic solution destroyed the cementing material between the granite residual soil particles, made it easier to crack. Li (2020) found that the changes of creep strain and long-term strength of artificially mixed granite residual soil in acidic environment were related to acidic and soaking time, and there was a critical value of acidic solution pH. Obviously, the acidic environment will change the microstructure and mineral composition of granite residual soil, thus change the macroscopic mechanical properties of granite residual soil.

Wetting or drying process during rainy season has a negative impact on the stability of soil and causes engineering accidents (Zhang 2020; Liu et al. 2021; Wang et al. 2020). Scholars have found that the change of mechanical behavior of granite residual soil under wetting–drying cycles is closely related to the generation of micro-cracks and the change of microstructure (Jian et al. 2017; An et al. 2023). The wetting–drying cycles accelerates the development of cracks in shallow granite residual soil, which provides infiltration channel for acid rain infiltration and affects the stability of slope. Therefore, research on the dual effect of acid solution and wetting–drying cycles is of great significance for the mechanical properties of granite residual soil. Chang et al. (2019) reported that the increase of soil expansion rate under the condition of wetting–drying cycles in acid rain environment was more obvious than under acid environment only. Even though, there are few reports on the mechanical behavior of soil under the effect of acidic wetting–drying cycles, and the research objects are mostly focused on rocks (Che et al. 2023; Sun et al. 2023) at present. Therefore, in this paper, the variations of the strength characteristics and the comprehensive effect of water-soil interaction under the dual action of acidic and wetting–drying cycles

Table 11.1 Basic physical properties of granite residual soil

Liquid limit (%)	Plastic limit (%)	Dry density (g/cm^3)	Particle size distribution (%)			
			>2 mm	2–0.075 mm	0.075–0.002 mm	<0.002 mm
46.31	31.04	1.823	7.1	48.3	26.8	17.8

Table 11.2 Mineral component content of granite residual soil

Quartz	Microcline	Sodium feldspar	Illite	Kaolinite
7.4%	16.7%	33.9%	20.3%	38.3%

was evaluated and analyzed, which could provide a theoretical basis for the stability analysis of granite residual soil slope in acid rain areas.

11.2 Materials and Tests

11.2.1 Materials

The granite residual soil samples used in the research were taken from the shallow surface layer of a slope in Xiangxiang City, Hunan Province. The physical parameters and grain size distribution are shown in Table 11.1. The mineral composition of the sample was analyzed by X-ray diffraction and shown in Table 11.2, it is mainly composed of sodium feldspar and clay minerals, of which contains 20.3% illite and 38.3% kaolinite.

11.2.2 Test Procedures

In order to simulate the acidic environment, solutions pH values of 3, 4, 5, 6, and 7 were prepared by diluting concentrated sulfuric acid in the test, and soil samples were mixed with solutions of different pH and compacted at dry density of 1.823 g/cm^3. After that, the samples were subjected to vacuum saturation treatment for 24 h in the same pH acidic solution, which was 0 times of wetting–drying cycles. And then the samples were air-dried and saturated again, that is, 1 wetting–drying cycles was completed. The saturated remolded samples with $N = 0, 1, 2, 5$ and 10 wetting–drying cycles were performed respectively. Then the direct shear test was carried out according to GB/T50123-2019, with a shear rate of 0.03 mm/min under normal stress of 100,200 and 300 kPa. The test instrument was ZJ-type strain-controlled direct shear apparatus produced by Nanjing Soil Instrument Factory.

11.3 Results

11.3.1 Shear Stress-Displacement Curve

The shear stress-displacement curves of granite residual soil under acidic wetting–drying cycles are shown in Fig. 11.1. The curves are strain-hardening type with no obvious peak. And the shear stress increases obviously with the increase of normal stress, and the increase amplification is related to the number of wetting–drying cycles and pH value. According to GB/T50123-2019, the shear stress at the displacement value of 4 mm is taken as the shear strength.

11.3.2 Variation in Shear Strength

Obviously, the shear strength under acidic wetting–drying cycles increases linearly with the increase of normal stress (Fig. 11.2). Figures 11.3 and 11.4 show the variation of shear stress under wetting–drying cycles only (i.e., pH7) and acidic solution pH only (i.e., the number of wetting–drying cycles $N = 0$), respectively. When the normal stress is 100 and 200 kPa, the shear strength is heavily affected by the wetting–drying cycles and solution pH value. When the normal stress is 300 kPa, the shear strength is mainly affected by the normal stress. Specifically, when the solution pH is 7, the shear strength of the granite residual soil shows a downward trend with the increase of wetting–drying cycles. After 1st and 2nd wetting–drying cycles, the strength value decreased sharply. With increasing wetting–drying cycles, the shear strength is rebound slightly (Fig. 11.3). Under the wetting–drying cycles $N = 0$ condition, the shear strength showed a slight decrease with the decrease of solution pH value at low stress (100, 200 kPa), and a slight increase with the decrease of pH value at high stress (300 kPa) (Fig. 11.4).

When considering the dual effect of acidic environment and wetting–drying cycles, the change of shear strength is more complicated under certain normal stress. Overall, when $pH < 6$ after first 2 cycles, shear strength increases in a diagonal area. To be specific, when the normal stress is 100 kPa, the shear strength of the sample decreases greatly after first 2 wetting–drying cycles, and there is a minimum value of shear strength at low pH value (Fig. 11.5a). As the pH value increases, the minimum value of shear strength appears at about 10 times of wetting–drying cycles. When the normal stress is 200 kPa, the shear strength shows a downward trend, even though there are fluctuations after the 2nd wetting–drying cycles at low solution pH, and the fluctuation range increased with the decrease of pH value (Fig. 11.5d, f). When the normal stress is 300 kPa, the shear strength continued to decrease with the increase of wetting–drying cycles, and the shear strength fluctuates heavily and even shows a upward trend, especially at solution pH 4 condition (Fig. 11.5i, j).The reason is that when the wetting–drying cycles is performed 1 or 2 times, the micro-cracks in the soil are in the rapid growth period (Chang et al. 2019), so the decrease of shear stress

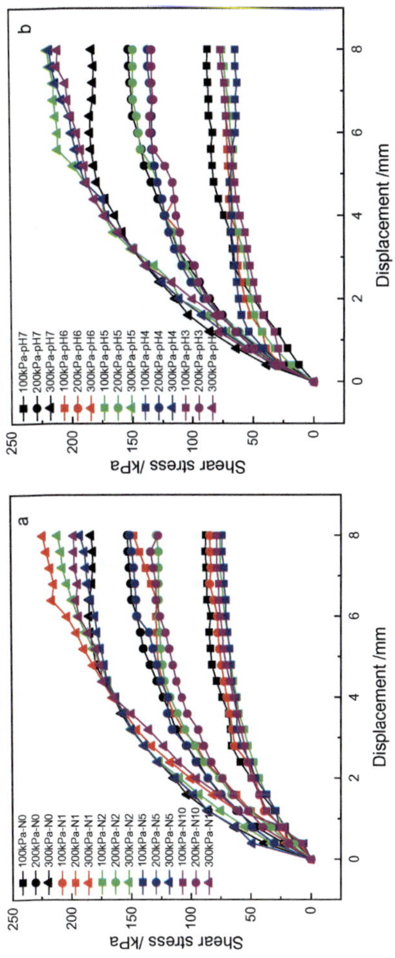

Fig. 11.1 Representative shear stress-displacement curves (in which, **a** is under the solution pH = 7, **b** is under wetting–drying cycles N = 0)

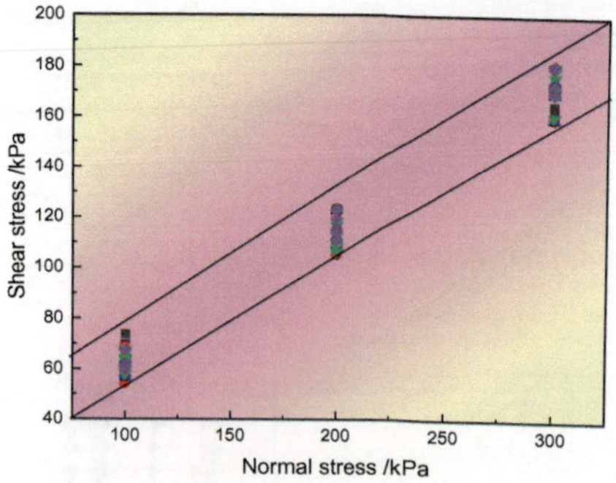

Fig. 11.2 Variation of shear strength with normal stress

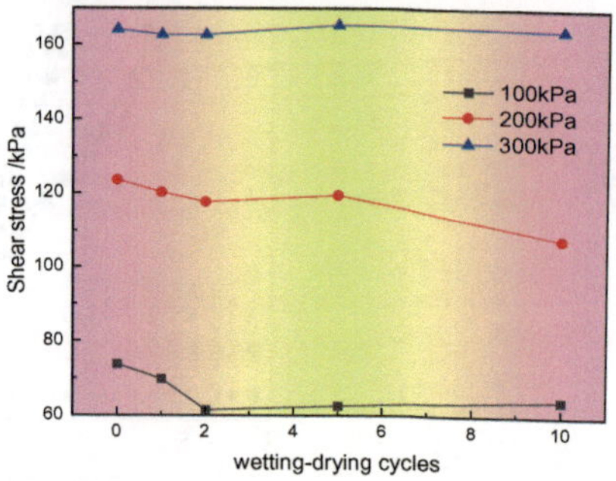

Fig. 11.3 Variation of shear strength with wetting–drying cycles (pH = 7)

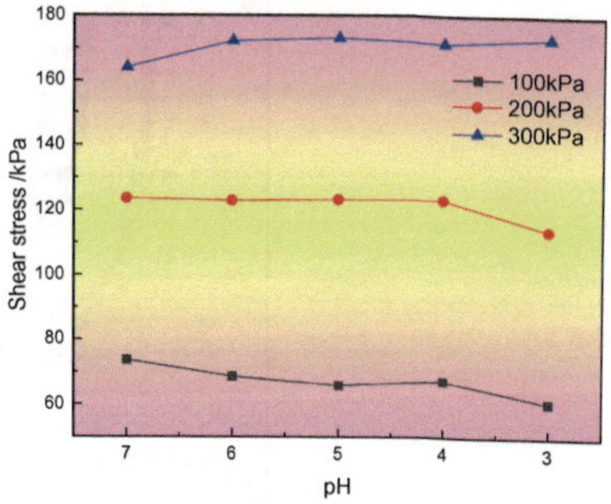

Fig. 11.4 Variation of shear strength with pH value (N = 0)

is very rapid (Fg. 11.5a, d). However, as the normal stress increasing (e.g.300 kPa in this work), the shear strength of the soil is mainly affected by the normal stress. Since the sample is re-saturated and re-compacted due to pore collapse and recoverable deformation under high normal stress during wetting (Wen et al. 2022), it shows an increase in shear strength (Fig. 11.5h).

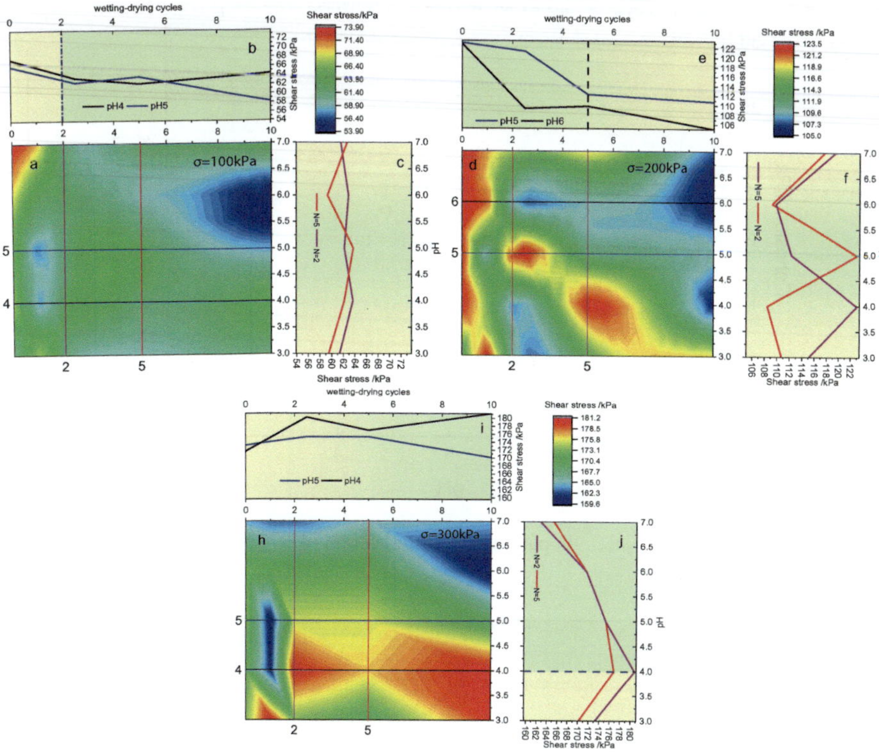

Fig. 11.5 Coutour profiles of shear strength under the dual effect of acidic and wetting–drying cycles

11.3.3 Variations in Shear Strength Parameters

According to the Mohr–Coulomb strength theory, the cohesion and internal friction angle of the sample are determined, and it is found that the cohesion and internal friction angle show the opposite changes (Figs. 11.6 and 11.7). When the influence of solution pH is considered only, the cohesion and internal friction angle show a step-like increase and decrease trend with the pH value decreasing, respectively. And the parameters tend to be stable in the platform range of solution pH from 6 to 4 (Fig. 11.6). Under the wetting–drying cycles only, the cohesion and the internal friction angle separately decreases and increases sharply with the increase of wetting–drying cycles. Meanwhile the parameters tend to be stable after first 2 wetting–drying cycles (Fig. 11.7). When considering the dual effect of wetting–drying cycles and acidic solution, the internal friction angle of the sample has a unimodal increase area along the diagonal direction, which is concentrated in the area interval of wetting–drying cycles > 2 and pH < 6 (Fig. 11.8a). And the cohesion in this area shows a clear decreasing trend (Fig. 11.8b).

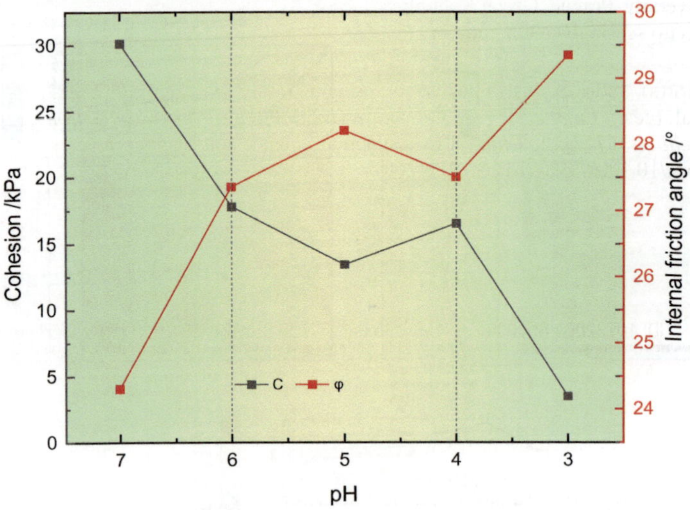

Fig. 11.6 Variation of shear strength parameters with solution pH (N = 0)

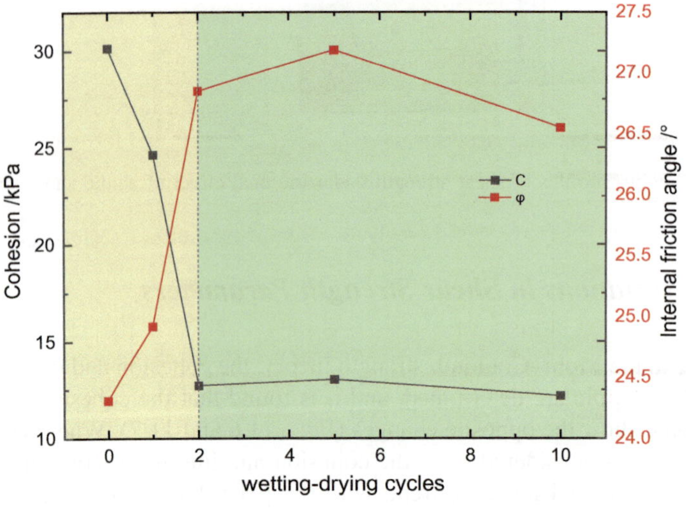

Fig. 11.7 Variation of shear strength parameters with wetting–drying cycles (pH = 7)

11.4 Discussion

Granite residual soil particles are coated by free oxides such as Fe_2O_3 and Al_2O_3. Fe_2O_3 and Al_2O_3 are easy to dissolve in acidic solution (Sun and Tang 2018). Thus, the minerals wrapped in it are exposed, and the contact area with acidic solution is increased. Subsequently, the minerals such as potassium and sodium feldspar and

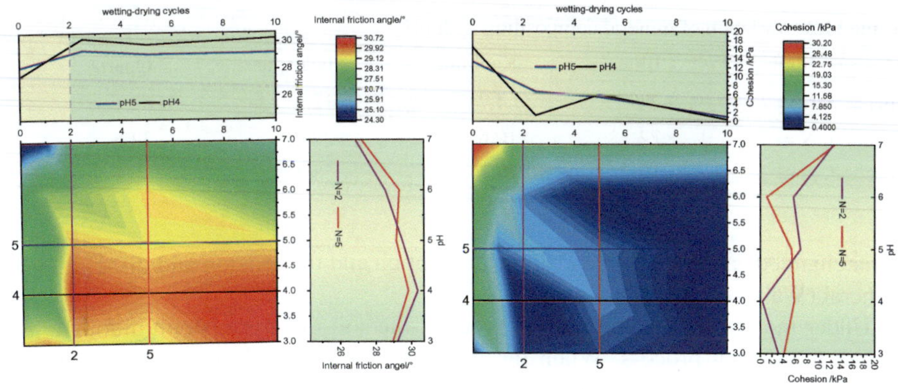

Fig. 11.8 Coutour profiles of shear strength parameters under the effect of acidic wetting–drying cycles

quartz in granite residual soil react with H^+ in acid solution, then the kaolinite and SiO_2 colloid are formed, and the cations K^+ and Na^+ released. Thus, the ion exchange in water-soil interaction is promoted, the crystal structure of the original minerals is changed and the illite is generated. The mineral content at different solution pH of 2 wetting–drying cycles was tested by XRD (as shown in Table 11.3), in which the content of feldspar decreases and the content of illite and SiO_2 increases with the solution pH decreasing. During the water-soil interaction mentioned above, the particle size of the granite residual soil becomes smaller, and the microstructure is destroyed, resulting in a decrease of shear strength and the cohesion of the soil. Previous studies have shown that the chemical change of kaolinite under the action of pore water chemical environment is an important factor for the change of soil mechanical properties (Gratchev and Towhata 2016). The granite residual soil is composed with clay particles, such as kaolinite and illite (Table 11.2). Considering the hydroxyl (OH–) and alumina on the surface and edge of clay particles, the K^+ and Na^+ ions released into the solution are adsorbed on the surface of clay particles, and the closer to the surface, the greater the adsorption force. Due to the H^+ solution erosion continues, the cation content in the solution increases, and the electronic double layer becomes thinner, which result in the flocculation structure of clay particles and the number of large-volume aggregates increasing, also has a significant coarsening effect on the soil particles. While the soil particles size increase, the agglomeration of fine-grained clay minerals is strengthened, resulting in a slight increase in the internal friction angle of the soil (Tang et al. 2023). In addition, Since SiO_2 colloid is generated during the solution erosion, part of the SiO_2 colloid may have formed new cements during the test and reunited the soil particles, thereby the cohesion of the soil samples is improved (Fig. 11.6). So, with the decrease of the solution pH, the fluctuation of the shear strength of granite residual soil and its parameters is affected by the comprehensive influence of the thickness of the electronic double

Table 11.3 Variation of mineral composition with pH

Test condition	Illite	Kaolinite	Feldspar	Quartz
pH = 3, N = 2	24	4.7	12.3	59.1
pH = 5, N = 2	22	10.6	29.9	37.5
pH = 7, N = 2	19.3	5.3	38.4	36.9

layer changed caused by ion exchange adsorption and the erosion and dissolution of mineral (Wahid et al. 2011).

Under the wetting–drying cycles process, the soil moisture is continuously lost during the drying process. When the moisture content is close to dry, the bound water film adsorbed around the clay particles dissipates and becomes thinner. Then the secondary cracks between the particles are developed (An et al. 2023), Which increases the contact area between the acid solution and the soil particles in the next wetting process, aggravates the degree of solution-soil reaction, and causes the granite residual soil to decrease sharply after 1st or 2st wetting–drying cycles, which is similar to the conclusion of Chang et al. (2019). With the increase of wetting–drying cycles, the amount of micro-voids increases, which is more favorable for the infiltration of the acidic solution, accelerated the water-soil interaction mentioned above under the acidic solution condition. Besides, during the process of wetting, small particles may fill the micro-voids due to pore collapse and recoverable deformation (Wen et al. 2022), so that the structural strength of the soil can be recovery, which results in the strength fluctuation in Fig. 11.5. In a word, under the dual effect of acidic wetting–drying cycles, the strength of granite residual soil is affected by water-soil interaction effects, such as the dynamic evolution of micro-cracks under the wetting–drying cycles, the change of the thickness of the electric double layer caused by ion exchange adsorption, and the dissolution of mineral.

11.5 Conclusions

In this work, variations in the shear strength of granite residual soil subjected to acidic wetting–drying cycles were studied. The following conclusions can be drawn:

1. the shear strength and parameters are heavily affected by the solution pH value under certain normal stress under acidic environment only. Under the low normal stress, the shear strength decreases with the pH value decreases. While under the high normal stress, the shear strength has increased slightly. Meanwhile, the cohesion and internal friction angle of granite residual soil decrease and increase by step with the decrease of pH value, respectively, and the pH value platform range is 4 to 6.

2. When considering the wetting–drying cycles only, the shear strength of granite residual soil decreases rapidly after first 2 wetting–drying cycles, and the internal friction angle and cohesion show a trend of decreasing and increasing respectively within first 2 wetting–drying cycles, and then tend to be stable.
3. Considering the dual effect of acidic wetting–drying cycles, the change of shear strength of granite residual soil is not monotonous. There is a diagonal increase area of shear strength, when pH < 6 after first 2 cycles.
4. The shear strength under the dual effect of acidic wetting–drying cycles is affected by the dynamic evolution of micro-cracks under the wetting–drying cycles and the change of the electronic double layer thickness caused by the ion exchange adsorption and the mineral acid dissolution.

Acknowledgements This research was funded by the National Natural Science Foundation of China (No. 42272304), Hunan Provincial Department of Education General Project (18C0090), and Hunan Provincial Department of Natural Resources General Project (2022-18).

References

An R, Kong LW, Zhang XW et al (2023) A multi-scale study on structure damage of granite residual soil under wetting-drying environments. Chin J Rock Mech Eng 42(3):758–767 (in Chinese)

Bakhshipour Z, Asadi A, Sridharan A et al (2017) Acid rain intrusion effects on the compressibility behaviour of residual soils. Environ Geotechn 6(7):460–470

Bogaard TA, Greco R (2016) Landslide hydrology: from hydrology to pore pressure. Wiley Interdiscip Rev Water 3(3):439–459

Chang J, Yang HP, Xiao J et al (2019) Swelling characteristics and microscopical analysis of expansive soil under dry-wet cycles in acid environment. China J Highw Transp 32(03):34–43 (in Chinese)

Che Y, Song Y, Yang H, Chen J, Meng C, Zhang J (2023) Damage behaviour of sandstone induced by combination of dry–wet cycles and acidic environment. Environ Earth Sci 82(1):1–14

Gratchev I, Towhata I (2016) Compressibility of soils containing kaolinite in acidic environments. KSCE J Civ Eng 20:623–630

Jaboyedoff M, Baillifard F, Bardou E et al (2004) The effect of weathering on Alpine rock instability. Q J Eng GeolHydrogeol 37(2):95–103

Ji F.L., Li Q., 2017. Effect of acid solution on physico-mechanical properties of granite residual soil. Chinese Journal of Underground Space and Engineering,13(06):1506–1510+1530. (in Chinese)

Jian WB, Hu HR, Luo YH et al (2017) Experimental study on deterioration of granitic residual soil strength in wetting-drying cycles. J Eng Geol 25(3):592–597 (in Chinese)

Jin X, Xiong CX, Zhong FL (2017) Experimental study on disintegration characteristics of granite residual soil under acid pollution. J Water Resour Archit Eng 15(06):194–198 (in Chinese)

Li QL (2020) Creep characteristics of acid-etched granite residual soil and Time-varying characteristics of saturated water content. Xiangtan University (in Chinese)

Lin QL, Huang ZP, Cao YB et al (2020) Strength evolution characteristics of granite residual soil immersed in distilled water and sulfuric acid solution. J Water Resour Archit Eng 18(02):27–32 (in Chinese)

Liu Y, Chen DX, Wang H et al (2021) Response analysis of residual soil slope considering crack development under drying-wetting cycles. Rock Soil Mech 42(7):1933–1943 (in Chinese)

Moore R, Brunsden D (1996) Physico-chemical effects on the behaviour of a coastal mudslide. Geotechnique 46(2):259–278

Nikooee E, Habibagahi G, Hassanizadeh SM, Ghahramani A (2013) Effective stress in unsaturated soils: a thermodynamic approach based on the interfacial energy and hydromechanical coupling. Transp Porous Media 96(2):369–396

Sun W, Du B, Cheng Q (2023) Experimental study of the dynamic mechanical behavior and degradation mechanism of red sandstone in acid dry-wet cycles. Geofluids

Sun YL, Tang LS (2018) The effect of chemical composition on tensile mechanics of residual granite soils. Acta Sci Rum Naturalium Universitatis Sunyatseni 57(03):7–13 (in Chinese)

Tang LS, Chen Y, Ye YH et al (2023) Influence of Na_2SO_4 solution on physical and mechanical properties of granite residual soil. Chin J Geotechn Eng (in Chinese)

Wahid AS, Gajo A, Di Maggio R (2011) Chemo-mechanical effects in kaolinite. Part 2: exposed samples and chemical and phase analyses. Géotechnique 61(6):449–457

Wang DY, Zeng QJ, Zhou M (2020) Effect of dry-wet cycle on stability of granite residual soil slope. IOP Conf Ser Earth Environ Sci 526(1)

Wen T, Chen X, Shao L (2022) Effect of multiple wetting and drying cycles on the macropore structure of granite residual soil. J Hydrol 614:128583

Zhan L, Zhang Z, Chen Y et al (2018) The 2015 Shenzhen catastrophic landslide in a construction waste dump: reconstitution of dump structure and failure mechanisms via geotechnical investigations. Eng Geol 238:15–26

Zhang PC (2020) Property deterioration and slope stability analysis of granite residual soil under dry wet cycles. Geotech Investig Survey 48(09):19–23 (in Chinese)

Chapter 12
Experimental Study on Meso-Mechanical Behavior of Coarse-Grained Granular Soils

Kaleem Ullah Jan Khan, Wen-Jie Xu, Yang-yang Zhang, and Hong-Bing Wang

Abstract Granular-soils are widely distributed geomaterials in nature which significantly encounter geohazards as landslides and debris flows. The assessment of meso-mechanical behaviours of such geomaterials is quite complex due to the existence of varying conditions of soil compositions. In this study, the soil samples from Yunnan province of China are considered for triaxial tests under different consolidated drained and undrained (saturated and unsaturated) conditions with varying confining stresses. The extensive experimental investigation revealed that different consolidation and draining conditions significantly influence the meso-mechanical behaviour of granular-soils. The results show that the shear strength parameters decrease under different conditions, such as CD (unsaturated), CD (saturated) and CU (consolidated undrained), respectively. The stress–strain mechanism of samples under both CD (saturated and unsaturated) showed an obvious hardening behaviour with no peak stress while strain-softening behavior in CU (consolidated undrained) condition. Furthermore, the Mohr–Coulomb relationship is suggested to obtain and compare the *p-q* linear relationships of shear strength parameters under different conditions. The present study paved the way towards new insights into meso-mechanical behaviour and deformation characteristics in granular-soils existing in nature.

K. U. J. Khan · W.-J. Xu (✉) · Y. Zhang
State Key Laboratory of Hydroscience and Hydraulic Engineering, Department of Hydraulic Engineering, Tsinghua University, Beijing 100084, China
e-mail: wenjiexu@tsinghua.edu.cn

K. U. J. Khan
e-mail: khank10@mails.tsinghua.edu.cn

Y. Zhang
e-mail: zhang-yy21@mails.tsinghua.edu.cn

H.-B. Wang
YCIC Investment Company Limited Yunnan Communications Investment & Constructions Group Co., Ltd, Kunming 650228, Yunnan, China

S. Wang et al. (eds.), *Engineering Geology for a Habitable Earth: IAEG XIV Congress 2023 Proceedings, Chengdu, China*, Environmental Science and Engineering,
https://doi.org/10.1007/978-981-99-9057-3_12

Keywords Meso-mechanics · Granular soils · Consolidated drained (CD) · Consolidated undrained (CU)

12.1 Introduction

Granular soils are extensively distributed geomaterials in the world, which significantly contributes to the occurrence of geohazards as landslides and debris flows. These kinds of materials have been considered in previous studies (Wang et al. 2021; Yu 2017) for triaxial compression to reveal the particle crushing mechanism and its influence on mechanical behavior. The explicit understanding of meso-mechanical behavior of granular soils is needed to demonstrate the deformation mechanism and failure processes of geohazards considering different consolidation and draining conditions. However, the granular materials encounter different mechanical behavior exhibiting both solid and liquid like responses, which are dependent on their different states, cause heterogeneity, non-linearity and possess a serious threat and challenge for researchers and engineers associated with the field of geotechnical engineering.

The efforts have been made over decades to study the granular-soils interfaces to achieve the extensive understanding of strength and deformation characteristics (Potyondy 1961). Existing research have mainly been conducted using direct-shear testing (De Jong and Westgate 2009; Pillai and Gali 2022; Sweeney et al. 2020), simple shearing testing (Martinez et al. 2015; Rui et al. 2021) and ring shearing testing (Fakharian and Vafaei 2021; Kishida and Uesugi 1987). The existing studies have extensively focused on the behavior of granular-soils and the factors which mainly influence the interfaces under both static and quasi-static cyclic loading. The considerable factors include the characteristics of the surface (Ari and Akbulut 2022; DeJong and Westgate 2009), properties of soil (Esmaili et al. 2014; Lashkari and Jamali 2021; Uesugi and Kishida 1986), different conditions of confinement (Afzali-Nejad et al. 2021; Mortara et al. 2007; Pra-ai and Boulon 2017) and the fabric anisotropy (Afzali-Nejad et al. 2018). The variations in the mechanical response of granular-soils under different consolidation and draining conditions and varying confining pressures are ignored in all the aforementioned research findings.

In accordance with existing research findings, an approach in triaxial testing for granular material involvement under different conditions as consolidated drained (saturated and unsaturated) and consolidated undrained (saturated and effective) have been proposed under different confining stresses to extensively reveal the meso-mechanical behaviors of granular soils. The presented study paved the way towards new insights of deformation mechanisms and failure processes in granular materials.

12.2 Description and Preparation of Samples

The granular-soil samples composed of coarse-grained particles from the slag disposal site are mainly maroon, brown and purplish red, slightly wet and loose. It is mainly clay with a small number of granular particles. The granular composition is strong ~ weakly weathered siltstone, mudstone and basalt, among which the particle size of 2–20 mm accounts for about 5%, 20–200 mm accounts for about 18%, >200 mm accounts for about 3% while the rest is fine silt. Thc particle size distribution curve of soil samples was obtained by sieve analysis in the present study, as depicted in Fig. 12.1. The considered sample size in triaxial experiments of the present study was 78 mm and 39 mm in height and diameter, respectively. The natural density and moisture content during sample preparation were kept the same as in the field as control parameters. The natural density of samples from the field was 1.65 gm/cm^3. The prepared samples in this study were composed of four layers compacted with the manner of stratification. The loading rate for soil samples were chosen based on the presence of moisture content in a sample as 0.06 mm/min. The procedures for triaxial testing were conducted at Tsinghua University, China, by keeping the testing apparatus lifting platform mechanically controlled.

The employment of triaxial testing, direct shear tests and plane shear tests are very common to explore the mechanical behaviour and shear strength properties of granular soils. Triaxial testing is comprehensive and more authentic, which is adopted in many studies to reveal the mechanical behaviours of granular soils. The triaxial testing apparatus used in the present study is shown in Fig. 12.2. The testing procedures have adopted four different confining pressures as 100, 200, 400 and 600 kPa to study the meso-mechanical response of the granular soils. Three different scenarios of soil samples are adopted for the triaxial testing system as CD (saturated), CD (unsaturated) and CU (consolidated and undrained. The influence of penetration of

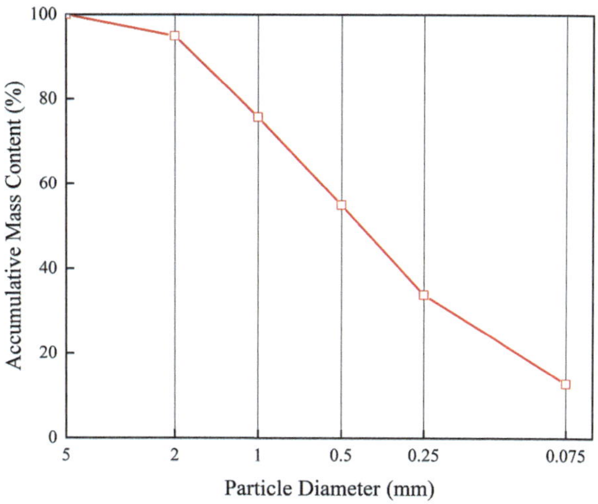

Fig. 12.1 The size distribution curve of soil used in the present study

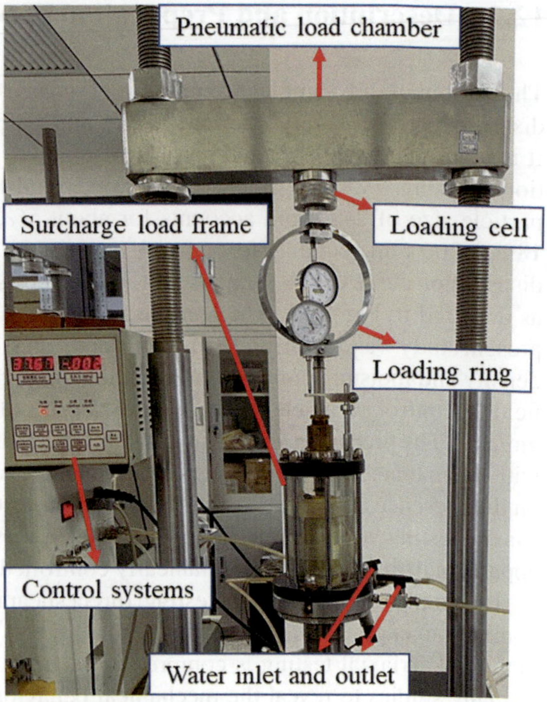

Fig. 12.2 Setup of triaxial apparatus used in the study

rubber membrane would be ignored as the sample afflicted a constant confining pressure during axial compression. The loading rate in the triaxial test was maintained by axial strain until it arrived at 15%. Finally, the volumetric change of the sample was sensitively reflected through this part. In addition, some amount of water was discharged out in the pressure chamber with a downward moment of the shaft. Therefore, the change in volumetric strain of the sample was achieved by measurement of changes in water quality in the closed container as well as with the discounting of axial load shaft effects.

12.3 Results

12.3.1 Meso-Mechanical Response of Granular Soils

It is very crucial to authentically characterise the volumetric-strain relationship with shear stress by constitutive modelling in the field of geotechnical engineering. The series of triaxial tests on granular-soil samples were performed under confining stresses of 100, 200, 400 and 600 kPa. The stress–strain relationship curves of samples

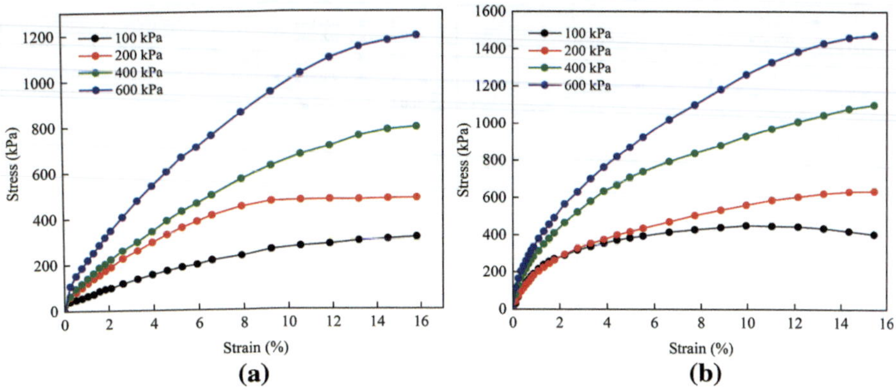

Fig. 12.3 The stress–strain relationship of soil samples under different conditions: **a** CD (saturated); **b** CD (unsaturated)

in CD (saturated and unsaturated) conditions are depicted in Fig. 12.3a, b, respectively, demonstrating the increase in deviatoric stress samples with an increase in confining stress exhibiting varying mechanical behaviors. The increase in deviatoric stress in CD (unsaturated) samples is obvious as the existence of water content within the samples reduces the strength of the soil itself under the CD (saturated) shearing process, which further leads to the reduction in deviation stress when the samples are deformed. The samples demonstrate clear transverse distortion, particularly in their upper-middle parts. The primary factor influencing this phenomenon is predominantly the methodology employed for sample preparation. The increasing rate slows down with the flattening of deviatoric stress, causing an increase in the intensity of sample deformation.

The stress–strain relationship and pore-pressures variations under different confining pressures in CU conditions are depicted in Fig. 12.4a, b, respectively. The samples generate varying positive pore-pressure under the CU shearing process, resulting in the reduction of the deviatoric stress when the sample is deformed, as the generation of high pore-pressure impairs deviatoric stress in the sample. The increasing rate of pore-pressure in granular-soils has obviously influenced the intensity of deviatoric stress in CU condition, which in turn reduced the shear strength of the sample. The elastic stage of the samples in CU condition is larger in comparison to CD (saturated and unsaturated) conditions under same confining pressures showing contractive behaviour. The samples under high stress of 600 kPa show a strain hardening behavior in CU conditions.

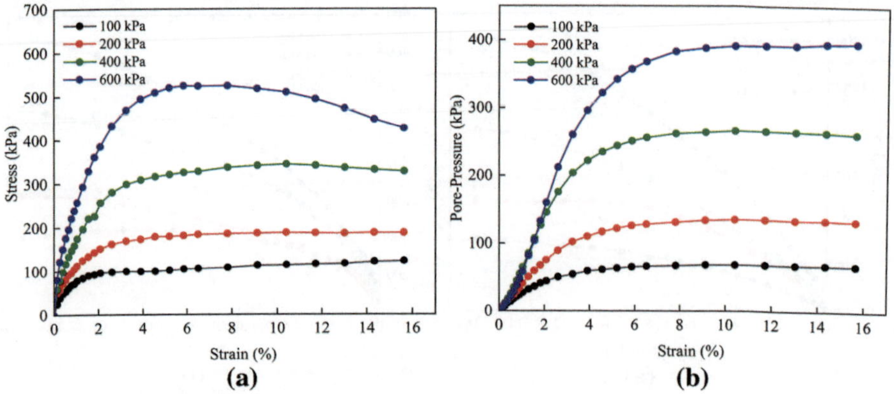

Fig. 12.4 The evolution of meso-mechanical behavior of granular-soils under CU (consolidated undrained) condition: **a** stress–strain mechanism, **b** pore-pressure variations

12.3.2 Volumetric Deformation Response

The evolution of change in volumetric-strain against axial strain and changes under CD saturated and unsaturated conditions are depicted in Fig. 12.5. The deviatoric stress with axial-strain curves of samples in CD (unsaturated) condition shows a strain hardening behaviour due to its loose nature in unsaturated state and thus the sample remains in the regime of bulk shrinkage and typically behaves like a loose geomaterial. The generated volumetric strain in unsaturated condition at the lower confining stress of 100 kPa tend to increase and then decrease with the increase in axial strain showing the tendency of dilation at the end as the volumetric strain shows an increase due to high confining stresses and the samples tend to contract with the increase in axial strain. Referring to the meso-mechanical response, as the applied shear force intensifies, the granular particles undergo a gradual transition towards transverse movement, resulting in a decelerated rate of contraction or expansion within the sample. It is evident that the samples under different confining stresses show larger degrees of shear shrinkage under the CD (saturated) condition in comparison to the CD (unsaturated) condition, where the volumetric shrinkage at all levels of confining stresses with an increase of 6% of shear shrinkage in a saturated condition. The evolution of volumetric-strain under different conditions demonstrates that the samples remain in the regime of shear shrinkage.

12.3.3 Shear Strength Behavior

The relationship of *q*/*p*-axial strain curves showed a linear increase with no peak value observation under CD (saturated) condition. The limit for axial-strain was chosen 15% as a point of benchmark and a stress of critical state to precisely compare

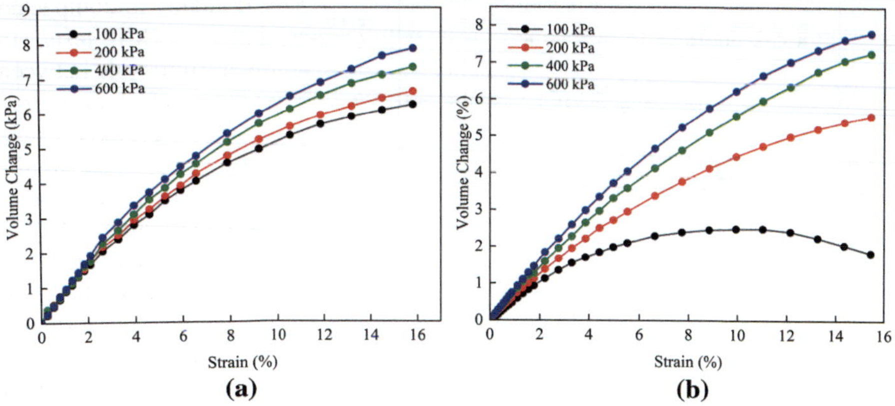

Fig. 12.5 Relationship of volumetric-strain changes of granular soils with shearing stress and CD conditions: **a** CD (saturated); **b** CD (unsaturated)

the correspondence states of both the samples having varying characteristics under different conditions and confining stresses. The primary rationale for selecting this particular value was based on the observation that the volumetric-axial strains exhibited by various soil samples reached a state of equilibrium. Even those samples that initially display an increasing volumetric strain eventually experience a decrease at this juncture. This characteristic is advantageous in providing a concise illustration of the meso-mechanical behavior of granular materials. The p-q linear curve is drawn according to the peak stress under each confining pressure to obtain the strength parameters of the sample under different conditions. Where: $p = \frac{\sigma_1 + 2\sigma_3}{3}$, $q = \frac{\sigma_1 - \sigma_3}{2}$, σ_1 and σ_3 are the axial pressure and confining pressure of the sample respectively. The *p-q* relationships of the samples under CD (saturated and unsaturated) and CU (saturated and effective) conditions in triaxial tests are depicted in Fig. 12.6a, b, respectively. The strength of samples is lower in CU condition due to effective stress generation in the samples, which impairs the deviatoric stress and thus tends the samples to early failure. The effective strength of samples is higher in CU condition as the influence of water content has not been considered in the samples for calculations of shear strength. The second highest strength of samples in CD (saturated) condition is due to the presence of water content in the samples, which decrease the particle-to-particle contacts in the form of cohesion. The highest strength parameters of S-RM are observed in CD (unsaturated) condition as the sample remains in a natural state and retains higher deviatoric stresses upon exposure to external load.

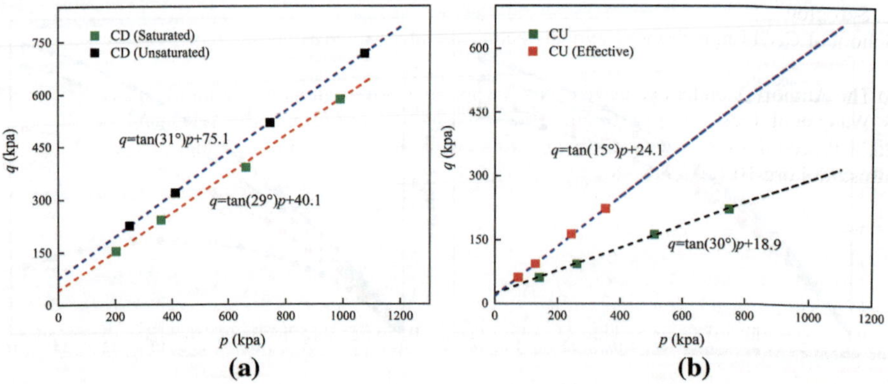

Fig. 12.6 Shear strength curves of S-RM samples under different conditions: **a** CD (saturated and unsaturated) and **b** CU (saturated and effective)

12.4 Conclusions

The granular-soils have different response under varying conditions which significantly influence their meso-mechanical behaviour. Granular-soils are widely distributed in nature, having diversified physical and mechanical characteristic. In this study, a series of triaxial experiments have been conducted considering granular-soils from Yunnan province of China under different consolidation and draining conditions and varying confining stresses. The meso-mechanical response of the soil samples has contrasting behaviours, which can be concluded in major findings as follows:

The deviatoric stress is obviously influenced in CU condition due to pore-pressure generation, which in turn effect the effective stress in the samples and hence decreases the deviatoric stress and causes early deformation in the soil samples.

The volumetric-strain of granular samples showed a contracting behavior under high pressures and dilation under low confining stresses in CD (unsaturated) condition while evidenced shrinkage under CD (saturated) condition.

The consolidation and draining conditions have a significant influence on the shear strength of granular-soils as the shear strength of samples decreased under different conditions as in the sequence of CD (unsaturated), CD (unsaturated) and CU (consolidated undrained), respectively.

Acknowledgements This work was supported by the projects of "Natural Science Foundation of China, China (52079067, 51879142)".

Declaration of Competing Interest The authors declare that they have no known competing financial interests or personal relationships that could have appeared to influence the work reported in this paper.

References

Afzali-Nejad A, Lashkari A, Farhadi B (2018) Role of soil inherent anisotropy in peak friction and maximum dilation angles of four sand-geosynthetic interfaces. Geotext Geomembr 46(6):869–881. https://doi.org/10.1016/j.geotexmem.2018.08.003

Afzali-Nejad A, Lashkari A, Martinez A (2021) Stress-displacement response of sand–geosynthetic interfaces under different volume change boundary conditions. J Geotechn Geoenviron Eng 147(8):04021062

Ari A, Akbulut S (2022) Evaluation of sand-geomembrane interface behavior using discrete element method. Granular Matter 24(1):1–21. https://doi.org/10.1007/s10035-021-01183-6

DeJong JT, Westgate ZJ (2009) Role of initial state, material properties, and confinement condition on local and global soil-structure interface behavior. J Geotech Geoenviron Eng 135(11):1646–1660

Esmaili D, Hatami K, Miller GA (2014) Influence of matric suction on geotextile reinforcement-marginal soil interface strength. Geotext Geomembr 42(2):139–153. https://doi.org/10.1016/j.geotexmem.2014.01.005

Fakharian K, Vafaei N (2021) Effect of density on skin friction response of piles embedded in sand by simple shear interface tests. Can Geotech J 58(5):619–636

Kishida H, Uesugi M (1987) Tests of the interface between sand and steel in the simple shear apparatus. Geotechnique 37(1):45–52

Lashkari A, Jamali V (2021) Global and local sand-geosynthetic interface behaviour. Geotechnique 71(4):346–367

Martinez A, David Frost J, Hebeler GL (2015) Experimental study of shear zones formed at sand/steel interfaces in axial and torsional axisymmetric tests. Geotech Test J 38(4):409–426

Mortara G, Mangiola A, Ghionna VN (2007) Cyclic shear stress degradation and post-cyclic behaviour from sand-steel interface direct shear tests. Can Geotech J 44(7):739–752

Pillai AG, Gali ML (2022) Role of particle shape on the shear strength of sand-GCL interfaces under dry and wet conditions. Geotext Geomembr 50(2):262–281. https://doi.org/10.1016/j.geotexmem.2021.11.004

Potyondy JG (1961) Skin friction between various soils and. Géotechnique 11(4):339–353

Pra-ai S, Boulon M (2017) Soil–structure cyclic direct shear tests: a new interpretation of the direct shear experiment and its application to a series of cyclic tests. Acta Geotech 12(1):107–127

Rui S et al (2021) Cyclic behavior of interface shear between carbonate sand and steel. Acta Geotech 16(1):189–209. https://doi.org/10.1007/s11440-020-01002-x

Sweeney MR et al (2020) Upscaled discrete fracture matrix model (UDFM): an octree-refined continuum representation of fractured porous media. Comput Geosci 24(1):293–310

Uesugi M, Kishida H (1986) Frictional resistance at yield between. Soils Found 26(4):139–149. https://doi.org/10.3208/sandf1972.26.4_139

Wang G, Wang Z, Ye Q, Zha J (2021) Particle breakage evolution of coral sand using triaxial compression tests. J Rock Mech Geotechn Eng 13(2):321–334. https://doi.org/10.1016/j.jrmge.2020.06.010

Yu F (2017) Characteristics of particle breakage of sand in triaxial shear. Powder Technol 320:656–667. https://doi.org/10.1016/j.powtec.2017.08.001

Chapter 13
Analysis on Time-Related Changes of Mechanical and Hydraulic Characteristics of Artificial Fault Core Specimen

Seong-Woo Moon and Yong-Seok Seo

Abstract In this study, a permeability test was conducted on remolded and artificial specimens to examine whether artificial fault core specimens can be applied to permeability tests instead of fault core materials, of which it is difficult to collect an undisturbed specimen in a natural state. In addition, since the physical–mechanical-hydraulic characteristics of fault materials containing clay change with time lapse, the permeability test and the direct shear test were conducted with a time difference in seven steps (immediately after specimen preparation, after 1 day, after 3 days, after 7 days, after 14 days, After 21 days and after 28 days). When the water permeability test was conducted on the remolded and artificial specimens under the same conditions, the hydraulic conductivity was shown in similar ranges for fault breccia and fault gouge, 1.03×10^{-4} to 1.16×10^{-5} cm/s and 4.88×10^{-7} to 2.65×10^{-7} cm/s, respectively, indicating that artificial clay can substitute fault gouge. In addition, as a result of analyzing the change in the permeability and friction characteristics of the artificial specimen, the hydraulic conductivity decreased and the friction angle increased with time laps, clearly showing the hardening phenomenon.

Keywords Fault core · Artificial specimen · Time lapse · Hydraulic conductivity · Friction angle

---

S.-W. Moon · Y.-S. Seo (✉)
Chungbuk National University, Cheongju-Si, South Korea
e-mail: ysseo@cbu.ac.kr

S.-W. Moon
e-mail: msw2080@cbu.ac.kr

S. Wang et al. (eds.), *Engineering Geology for a Habitable Earth: IAEG XIV Congress 2023 Proceedings, Chengdu, China*, Environmental Science and Engineering,
https://doi.org/10.1007/978-981-99-9057-3_13

13.1 Introduction

The fault zone can be largely divided into two parts: the fault core which has undergone a lot of faulting due to concentrated stress, and the fault damage zone in which fracture has occurred due to fault in the periphery (Chester and Logan 1986; Caine et al. 1996; Faulkner et al. 2003; Gudmundsson 2004; Lin and Yamashita 2013). The fault core can be subdivided into the fault breccia and fault gouge according to the content of fragments (Higgins 1971; Sibson 1977; Killick 2003; Woodcock and Mort 2008; Scibek et al. 2016). Fault rock is a material that is difficult to interpret easily because its permeability may increase or decrease compared to the parent rock due to the various movement mechanisms of faults.

The brecciation that occurs during the formation of a fault core makes coarse grains exist in the matrix of fine grains, from which the rock becomes non-cohesive. A precedent study reported that the hydraulic conductivity could increase up to 104–105 cm/s compared to bedrock. (Bense et al. 2013). On the other hand, in the fault gouge, there exist plate-shaped phyllo silicate clay minerals, which play a role in blocking the flow of fluid in the fault zone (Cornelis and Anthony 2016). Although these clay minerals show low permeability by themselves, they also show a greater difference depending on whether or not expandable clay minerals such as smectite are included (Kaproth et al. 2016). In general, the cataclasis action results in an increase in permeability (Blenkinsop 2000), but rather, a large-scale fault gouge zone is formed as the cataclasis action intensifies, sometimes creating impermeability layers (Bense et al. 2013). Due to the above faulting process, fault breccia generally has a higher permeability than fault gouge. In addition, it has been reported that the permeability of fault zones can be changed by external factors other than constituent materials of fault zones, such as fault types (e.g., reverse fault, normal fault, strike slip fault, etc.), alteration or dissolution by surface water and groundwater, cementation, and seismic events (Shi and Wang 2014; Wang et al. 2016; Ingebritsen and Manga 2019). In order to identify the permeability characteristics of the fault zone, it is necessary to analyze various factors such as the particle size distribution, the alignment of breccia, and the content of mineral components that make up the fault rock, however, both the in-situ test and the laboratory test for the fault zone require difficult conditions.

In this study, an artificial fault core specimen was prepared to replace a natural specimen, which is difficult to collect, and the possibility of using them in permeability tests was examined. Through comparison with the fault core specimen in a natural state reshaped to fit the permeability test mold, the applicability of artificial clay was analyzed. In addition, since the artificial fault core specimen contains a large amount of fine-grained clay, the change in permeability materials in accordance with time lapse after specimen preparation was analyzed to understand effects such as hardening and thixotropy of the specimen.

13.2 Types and Characteristics of Fault Core Specimen

13.2.1 Needs for Artificial and Remolding Specimen in Permeability Test

For in-situ tests (permeability test and Lugeon test) for measuring the hydraulic conductivity of a fault zone, there is a problem that the section where the test can be conducted is narrow because the section of the fault zone appears narrow during drilling unless the fault zone is vertically developed. In addition, since the fault core rises as slime during drilling, which involves various problems such as the use of core specimens and the determination of the permeability test section. Even if the fault core is recovered through dry boring, the core wall is often disturbed by the adhesive force of the fault gouge and cannot be used as a permeability test material (Fig. 13.1).

In the case of a laboratory test, as well, disturbance inevitably occurs at the wall of the specimen during sampling due to the components of fault gouge (clay), and when the sampler is inserted, a disturbance occurs due to rotation and crushing of fault breccias (Fig. 13.2). Since the disturbed parts of the fault core specimen may become a water movement path, there is a problem that it is not suitable as a permeability test material.

Fig. 13.1 Fault core materials recovered from dry drilling

(a) Distribution of m3ny breccias within fault core (b) Disturbed specimens due to breccias when it is sampled

Fig. 13.2 The occurrence of disturbance while sampling fault core materials

13.2.2 Manufacturing Method and Characteristics of Artificial Specimen

An artificial specimen was produced by mixing coarse-grained fault breccia, sand, and fine-grained artificial clay. For fault breccia and sand, first, the disturbed specimen collected from the fault zone in the natural state was removed from the fine grains through the soil washing test (ASTM D1140-17, 2017), and then the coarse grains remaining on the sieve were filtered through a No. 4 sieve (4.75 mm) to separate breccia and sand (Fig. 13.3a). Instead of fault gouge whose particle size was too small to recover with a sieve, the test used artificial clay similar to the average value of mineral composition content obtained from 40 non-fault zones in Korea (Yun et al. 2022; Fig. 13.4a, b; Table 13.1).

In this paper, the proportions of breccia, sand, and clay were set at 40%, 40%, and 20%, respectively, in the preparation of the artificial specimen (Fig. 13.3b, c). Since the physical and mechanical properties of the clay changed materials in accordance with time lapse after being disturbed, the tests were conducted with time intervals: immediately after the artificial specimen was prepared, after 1 day, after 3 days, after 7 days, after 14 days, after 21 days, and after 28 days. Since the permeability test

(a) Breccia and sand from the fault core classified using sieve #200 (0.075mm)

(b) Breccia, sand, and clay ready for mixing

(c) Artificial specimen formed using the mold for permeability test

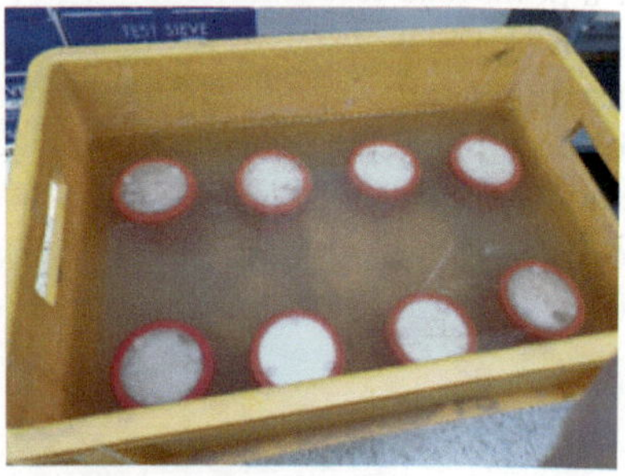

(d) Artificial specimens being submerged in the water

Fig. 13.3 Photographs showing the producing process for artificial specimen

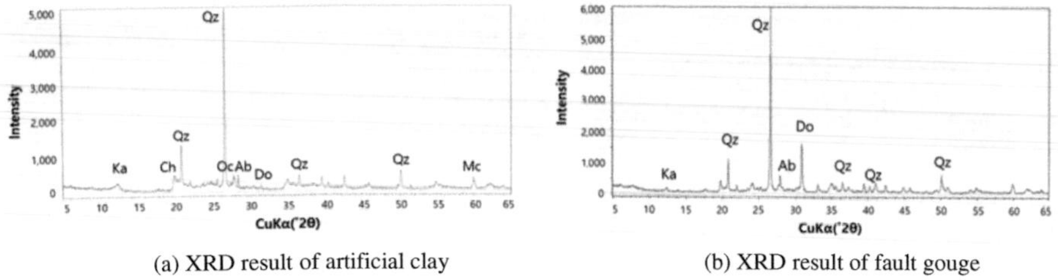

Fig. 13.4 XRD analysis results of artificial clay and fault gouge

Table 13.1 Comparison between mineral components of artificial clay and fault gouge

Type	Mineral components (wt.%)					
	Quartz	Albite	Kaolin	Microcline	Chlorite	Etc
Artificial clay	34.3	18.6	14.7	11.4	8.4	12.6
Fault gouge	39.4	21.6	8.8	–	7.5	22.7

was conducted in a saturated state, the specimen was submerged and stored while preventing the loss of the specimen with a porous plate (Fig. 13.3d).

13.2.3 *Manufacturing Method and Characteristics of Remolding Specimen*

In order to examine the applicability of artificial clay instead of fault gouge, a remolding specimen was prepared in parallel. The remolding specimen means that the fault core specimen was immersed and disturbed, and then remolded to fit the permeability test mold, and is composed of natural fault breccia, sand, and fault gouge (Fig. 13.5).

13.3 Laboratory Test

13.3.1 *Falling Head Permeability Test*

A falling head permeability test was conducted in order to understand the permeability characteristics of the remolded and artificial specimen using the K_{sat} equipment from UMS, Germany, which can perform both constant water head permeability test and falling head permeability test (Fig. 13.6). The tests used a specimen with a diameter of 80 mm and a height of 50 mm, and hydraulic conductivity in the range of $5 \times 10^{-2} \sim 1 \times 10^{-7}$ cm/s was measured.

(a) Immersed fault gouge (b) Remolding specimen of fault gouge

Fig. 13.5 Photographs showing the producing process for remolding specimen

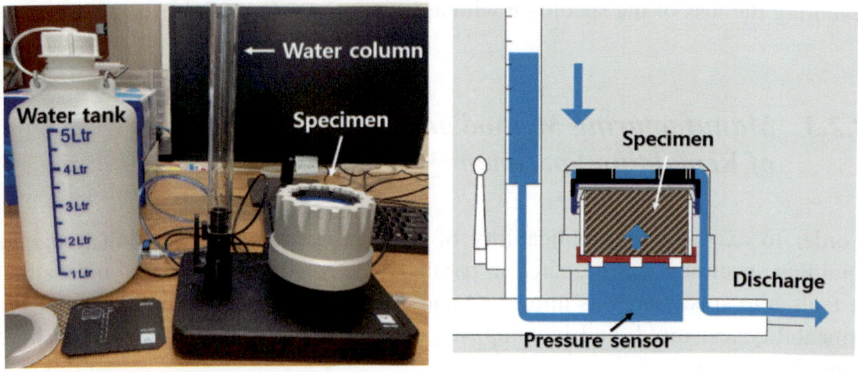

(a) K_{sat} device for permeability test (b) Schematic illustration of permeability test of K_{sat}

Fig. 13.6 The device for measuring the hydraulic conductivity of fault core (UMS 2013)

13.3.2 Direct Shear Test

A direct shear test was performed to determine the friction characteristics of the fault core. For direct comparison with the hydraulic conductivity, the specimen after the permeability test was inserted into a shear box (diameter 80 mm, height 50 mm) in an undisturbed state to conduct a test. The direct shear test was performed in accordance with the ASTM D3080-98 (1998) standard, by setting the shear rate to 0.5–2.0%/min of the sample diameter and the normal stress in three stages (54, 108, and 162 kPa). The stress and displacement generated during shearing are automatically

(a) Apparatus equipped LVDT and loadcell for measurement of displacement and force (b) Artificial specimen of after 7days in shear box

Fig. 13.7 Experimental setup for direct shear test of fault gouge

acquired through the loadcell and the linear variable differential transformer (LVDT) (Fig. 13.7).

13.4 Result Analysis

13.4.1 Comparison of Permeability Test Results of Fault Core Remolding Specimen and Artificial Specimen

In order to examine the applicability of artificial clay as a substitute for fault gouge, the results of the permeability test of the remolding specimen and artificial specimen were compared. A permeability test and particle size distribution analysis were conducted on the remolding specimen, as well as on the artificial specimen that was prepared with a similar particle size to that of the remolding specimen. Both the remolding and artificial specimens were submerged for 4 weeks after preparation so that the changes in physical, mechanical, and hydraulic properties were converged. Table 13.2 is the test result, showing that the fault breccia and fault gouge were 1.03×10^{-4} to 1.16×10^{-5} cm/s and 4.88×10^{-7} to 2.65×10^{-7}, respectively, for both remolding and artificial specimens. It confirms that artificial clay can be used as a substitute for fault gouges in a natural state, when preparing artificial specimens.

Table 13.2 Comparison of the hydraulic conductivity results between the remolding specimens and the artificial specimens

Type of specimen		Particle size distribution			Hydraulic conductivity (cm/s)
		Breccia	Sand	Silt&clay	
Fault breccia	Remolding-1	21.0	29.9	49.1	1.03.E−04
	Remolding-2	19.3	39.5	41.2	1.62.E−05
	Artificial-1	20.0	31.1	48.9	1.16.E−05
	Artificial-2	26.3	29.6	44.1	7.47.E−05
Fault gouge	Remolding-3	3.0	21.5	75.5	2.65.E−07
	Artificial-3	8.1	30.1	61.8	4.88.E−07

13.4.2 Analysis of Change in Permeability Characteristics Materials in Accordance with Time Lapse After Preparation of Artificial Specimen

By analyzing the change in the permeability characteristics of the artificial specimen in accordance with time lapse, it was found that the average hydraulic conductivity was 1.07 × 10–3 ~ 1.54 × 10–4 cm/s from immediately after sample preparation (0 day) to after 7 days, with the standard deviation of $1.70 \times 10^{-3} \sim 2.60 \times 10^{-4}$ cm/s (Fig. 13.8). On the other hand, from 14 to 28 days after sample preparation, the average hydraulic conductivity decreased to 1.64×10^{-6} ~ 8.08×10^{-7} cm/s with the standard deviation of 1.83×10^{-6} ~ 4.06×10^{-7} cm/s. The result shows the hardening of the specimen in which the hydraulic conductivity decreases as time passes after specimen preparation. The drastic decrease in the deviation of the measured values after 14 days means that the hydraulic conductivity stabilizes as the hardening phenomenon is over.

13.4.3 Analysis of Change in Friction Characteristics Materials in Accordance with Time Lapse After Fabrication of Artificial Specimen

Since the permeability test is a non-destructive test, the friction characteristics in accordance with time lapse were analyzed by collecting the specimen after the permeability test was completed and then conducting a direct shear test. Immediately after specimen preparation (0 day), the friction angle showed the lowest value at 20.9°, and after 28 days, it rose to 45.5° showing a clear increasing trend of the friction angle materials in accordance with time lapse. This tendency could be expressed in the form of a logarithmic function of $y = 12.015\ln(x) + 22.321$, of which the determination coefficient was 96.54% indicating a very high correlation. Similar to the decrease in the hydraulic conductivity of the specimen over time after preparation

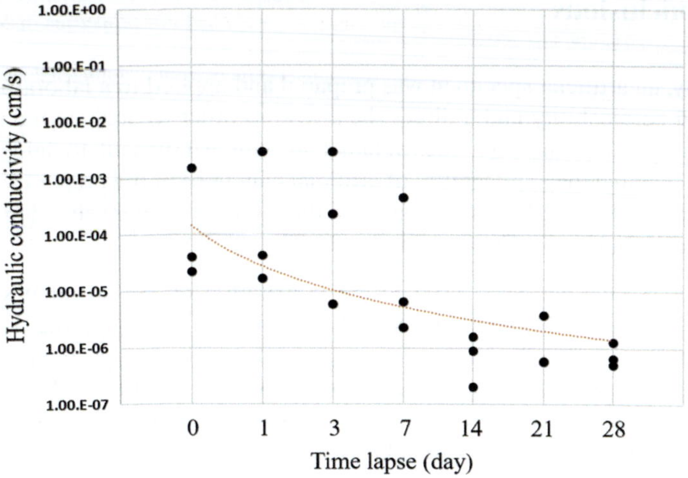

Fig. 13.8 Analysis on the change of hydraulic conductivity of fault core materials in accordance with time lapse

of the aforementioned artificial specimen, the increase in the frictional strength over time means the hardening of the specimen. In addition, the increase range of the friction angle also slowed down after 14 days, showing a similar trend to the results of the permeability test (Fig. 13.9).

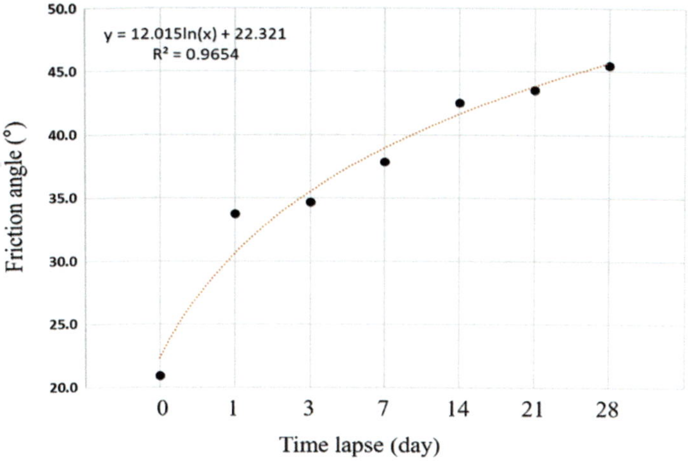

Fig. 13.9 Analysis on the change of friction angle of fault core materials in accordance with time lapse

13.5 Conclusions

In this study, an artificial specimen was prepared and applied to a laboratory test on falling head permeability and a direct shear test, in order to analyze the physical–mechanical-hydraulic characteristics of fault core, which is difficult to field sampling. In order to examine the applicability of artificial clay used in the preparation of artificial specimens, the result was compared to that of a remolding specimen prepared using a natural fault core. The analysis results are as follows.

(1) For the breccia and sand used in the preparation of the artificial specimen, coarse-grained soil filtered by the soil washing test method at the outcrops of the fault core exposed to the outdoors was used, and the fine-grained fault gouge was replaced with artificial clay in the preparation of specimens. The remolding specimen was prepared by adding water to the disturbed fault core specimen and remolding it to fit the permeability test mold, and the particles constituting the remolding specimen were the same as that of the fault core in its natural state.

(2) As a result of a falling head permeability test on the remolding specimen and the artificial specimen under the same conditions (saturation state, 28 days after specimen preparation), the fault breccia was in a range of $\alpha \times 10^{-4} \sim \alpha \times 10^{-5}$ cm/s in both the remolding and artificial specimens, showing similar result with the fault gouge ($\alpha \times 10^{-7}$ cm/s). It indicates that there is no significant difference in the results even when artificial clay is used instead of fault gouge in the permeability test.

(3) Analyzing the permeability characteristics of materials in accordance with time lapse after preparation of the artificial specimen, it was found that the hydraulic conductivity decreased in accordance with time laps, showing that the hardening of the specimen, and it was stabilized and showed similar values after 14 days. In addition, the deviation of the values also decreased from the 14th day onwards, indicating a tendency for the test results to be stabilized.

(4) As a result of analyzing the friction characteristics materials in accordance with time lapse after the preparation of artificial specimens, the friction angle showed the lowest value at 20.9° immediately after sample preparation (0 day), and at 45.5° after 4 weeks (28 days), showing a significant increase with the passage of time. In addition, the increase range in the friction angle slowed down, after the 14 days, showing a similar trend to the results of the permeability test.

The above results presented the applicability of an artificial specimen of fault core. It is expected to be useful in identifying the physical–mechanical-hydraulic characteristics of fault core material, if a heterogeneous fault core material can be replaced with an artificial specimen and laboratory tests are conducted while adjusting various conditions (particle size, content, and size conditions).

Acknowledgements This work was supported by the National Research Foundation of Korea (NRF) grant funded by the Korea government (MSIT) (No. 2021R1F1A1061133).

References

ASTM D1140-17 (2017) Standard test method for determining the amount of material finer than 75-μm (No. 200) sieve in soils by washing, ASTM International, West Conshohocken, PA. https://doi.org/10.1520/D1140-17

ASTM D3080-98 (1998) Standard test method for direct shear test of soils under consolidated drained conditions. ASTM International, West Conshohocken, PA. https://doi.org/10.1520/D3080-98

Bense VF, Gleeson T, Loveless SE, Bour O, Scibek J (2013) Fault zone hydrogeology. Earth Sci Rev 127:171–192

Blenkinsop TG (2000) Deformation microstructures and mechanisms in minerals and rocks. Kluwer Academic Publishers, Dordrecht. ISBN 978-0-412-73480-9

Caine JS, Evans JP, Forster CB (1996) Fault zone architecture and permeability structure. Geology 24(11):1025–1028

Chester FM, Logan JM (1986) Implications for mechanical properties of brittle faults from observations of the Punchbowl Fault Zone, California. Pure Appl Geophys 124:79–106

Cornelis K, Anthony P (2016) Igneous rock-forming minerals. In: Earth materials, 2nd edn. Cambridge University Press, pp 159–196

Faulkner DR, Lewis AC, Rutter EH (2003) On the internal structure and mechanics of large strike-slip fault zones: field observations of the Carboneras fault in southeastern Spain. Tectonophysics 367(3–4):235–251

Gudmundsson A (2004) Effects of Yong's modulus on fault displacement. CR Geosci 336:85–92

Higgins MW (1971) Cataclastic rocks. U.S. Geological Survey Professional, Paper 687: 97

Ingebritsen SE, Manga M (2019) Earthquake hydrogeology. Water Resour Res 55(7):5212–5216

Kaproth BM, Kacewicz M, Muhuri S, Marone C (2016) Permeability and frictional properties of halite-clay-quartz faults in marine-sediment: the role of compaction and shear. Mar Pet Geol 78:222–235

Killick AM (2003) Fault rock classification: an aid to structural interpretation in mine and exploration geology. S Afr J Geol 106:395–402

Lin A, Yamashita K (2013) Spatial variations in damage zone width along strike-slip faults: and example from active faults in southwest Japan. J Struct Geol 57:1–15

Scibek J, Gleeson T, McKenzie JM (2016) The biases and trends in fault zone hydrogeology conceptual models: global compilation and categorical data analysis. Geofluids 16(4):782–798

Shi Z, Wang G (2014) Hydrological response to multiple large distant earthquakes in the Mile well, China. J Geophys Res Earth Surf 119(11):2448–2459

Sibson RH (1977) Fault rocks and fault mechanisms. J Geol Soc 133:191–213

UMS (2013) Operation manual K_{sat}, METER Group AG, p 42

Wang CY, Liao X, Wang LP, Wang CH, Manga M (2016) Large earthquakes create vertical permeability by breaching aquitards. Water Resour Res 52(8):5923–5937

Woodcock NH, Mort K (2008) Classification of fault breccias and related fault rocks. Geol Mag 145(3):435–440

Yun HS, Moon SW, Seo YS (2022) Effects of breccia and water contents on the mechanical properties of fault-core-zone materials. Sci Rep 12:7093

Chapter 14
Evolution of K_0 Coefficient and Deformation Parameters of Coral Sand Under K_0 Consolidation

Weike Peng and Jiru Zhang

Abstract In the design of geotechnical engineering, the coefficient of earth pressure at rest, denoted by K_0, is one of the commonly used parameters. Nonlinear elastic models and their deformation parameters are widely used because of their simplicity and practicality. Coral sand is also increasingly used as backfill or foundation material for marine geotechnical engineering. This study performed a series of K_0 consolidation tests on coral sand to investigate the evolution of K_0 and deformation parameters. The K_0 values of coral sands were measured and their relative breakage index was evaluated. The deformation parameters were calculated based on the test results and generalized Hooke's law. The results show that K_0 value decreases with increasing the effective axial stress and initial relative density, and the relative breakage index increases with the increase of the mean effective stress. The tangent Poisson's ratio decreases as the effective axial stress and initial relative density increase. On the contrary, the tangent elastic modulus increases as the effective axial stress and initial relative density increase. An empirical formula of K_0 and the range of values of deformation parameters are proposed based on the test results. The stress–strain relationship of coral sand under K_0 consolidation is reasonably predicted by the nonlinear elastic model.

Keywords Coral sand · K_0-consolidation · Coefficient of earth pressure at rest · Deformation parameter · Stress–strain model

---

W. Peng · J. Zhang (✉)
School of Civil Engineering and Architecture, Wuhan University of Technology, Wuhan, Hubei 430070, China
e-mail: zhangjr@whut.edu.cn

W. Peng
e-mail: pwk@whut.edu.cn

S. Wang et al. (eds.), *Engineering Geology for a Habitable Earth: IAEG XIV Congress 2023 Proceedings, Chengdu, China*, Environmental Science and Engineering,
https://doi.org/10.1007/978-981-99-9057-3_14

14.1 Introduction

Current scientific research shows that climate change will lead to global sea-level rise through the warming of oceans and melting of ice. Some island countries in the Pacific and Indian Oceans are at risk from sea-level rise. At the same time, they are constantly evolving and face a wide range of natural and human-induced stresses. As a result, these countries have built artificial islands on coral reefs to respond to the threat of sea-level rise and the problem of land demand caused by population growth (Yang 2012; Zhou 2013). In general, the artificial islands are far away from the mainland, and thus dredger fill coral sand is widely used to hydraulic fill the foundation of the artificial island. The coral sand, which is mostly composed of calcareous fragments of marine origin, has low particle strength, irregular shape, high angularity, and abundant internal pores. These characteristics profoundly affect the mechanical properties of coral sand. The stress state of coral sand in hydraulic fill foundation is most likely to correspond to the K_0-consolidation test. Therefore, the coefficient of earth pressure at rest, K_0, defined as the ratio of horizontal to vertical effective stress (Terzaghi 1920; Bishop 1958), plays an important role in the design and construction of hydraulic fill foundation.

Past studies have shown that K_0 value of sands was significantly affected by relative density, stress state and stress history, etc. (Wroth 1973; Mayne and Kulhawy1982; Guo 2010; Zhu et al. 2019; Cai et al. 2020). Chu and Gan (2004) concluded that K_0 value of quartz sand gradually decreases with the increase of cell pressure and relative density through the triaxial instrument. Similar conclusions were obtained by Lee et al. (2013) and Jiang et al. (2020). However, the latest studies showed that K_0 value increased with the increase of relative density (Cai et al. 2021). It is necessary to further study the variations of K_0 with stress level and relative density.

So far, in contrast to quartz sand and rockfill, there is limited information about the effects of K_0 and the stress–strain relationship of coral sand under K_0 consolidation. The mineral composition, particle size, particle texture of rockfill and quartz sand are different from coral sand. Further investigations on the variation of K_0 and deformation parameters of coral sand under K_0 consolidation are needed. In this paper, an experimental technique, using a computer-controlled triaxial test to evaluate K_0 value, is presented. Several triaxial K_0 consolidation tests were carried out on coral sands to investigate the variation of K_0 value and deformation parameters of coral sand and the evolution of particle crushing of coral sand under the K_0 consolidation.

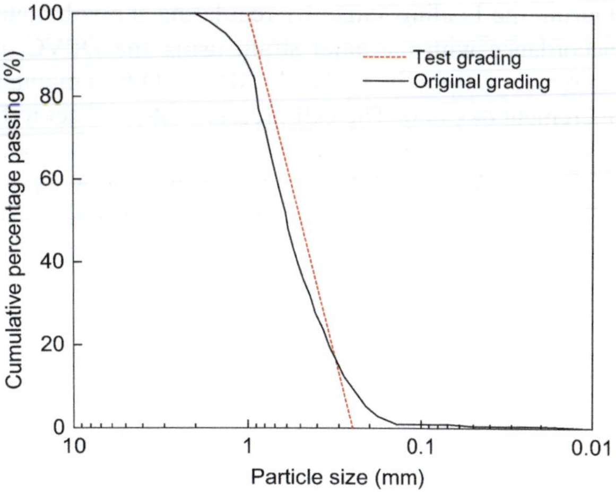

Fig. 14.1 Particle size distribution of coral sand

14.2 Testing Method and Material Tested

14.2.1 Material Tested

Marine-dredged coral sand from the South China Sea was used for this study. The calcium carbonate content of the coral sand is measured by an x-ray fluorescence spectrometer produced by PANalytical B V, Netherlands and the specific gravity was measured by the density bottle method (ISC 1999). The calcium carbonate content of the sand is 96.3%, and its specific gravity is 2.78. A sieving method was used to prepare the sand tested with a particle size ranging from 1 to 0.25 mm, in which the mass percentages of the particles of 1 to 0.5 mm and 0.5 to 0.25 mm are 50%, respectively. The particle size distribution curves of the sand tested and the original sand are shown in Fig. 14.1.

14.2.2 Testing Method

Reconstituted specimens with initial relative densities D_r of 0.6, 0.75, and 0.9 were prepared by the air pluviation method (Ladd 1978) according to the maximum and minimum void ratios of 1.196 and 0.829. The specimens are 39.1 mm in diameter and 80 mm in height. All specimens were fully saturated using back pressure and cell pressure. The B-value of all samples is greater than 0.95 before the K_0 consolidation tests started.

The K_0 consolidation tests were conducted by a GDS triaxial testing system. Three digital pressure/volume controllers (DPVC) were used to apply axial pressure, cell pressure, back pressure, and also to measure the volumetric change of specimen.

During the loading stage, by regulating the volumetric change of the specimen in accordance with the axial strain using the DPVC, the volumetric to axial strain increment ratio to be unity ($d\varepsilon_v/d\varepsilon_1 = 1$) was maintained, to keep the lateral strain increment $d\varepsilon_3 = 0$. The cell pressure rate was set to be equal to 5 kPa/min ($\Delta\sigma_3' =$ 5 kPa/min). The tests were stopped as the cell pressure reached 1000 kPa.

Several K_0 consolidation tests terminated at $\sigma_3' = 100$, 200, 400, 600, and 800 kPa were carried out to investigate the particle crushing of coral sand during the tests. After terminating the K_0 consolidation tests, the specimens were collected to measure the particle size distribution. The relative breakage index B_r, suggested by Hardin (1985), was calculated for quantitative analysis of particle breakage.

14.3 Experimental Results and Analysis

14.3.1 Relationship Between K_0, Stress State, and Initial Relative Density

Figure 14.2 shows the variation of void ratio *e* versus the effective mean stress p'. The effective mean stress is on a logarithmic scale. *e* of the samples decreases with the increase of p' and initial relative density. In the *e*-logp' space, *e* decreases rapidly when p' excess the yield stress σ'_y. The yield stress σ'_y, suggested by McDowell (2002) as the point of maximum curvature in the curves of *e* against the logarithm of p', is shown in Fig. 14.2. The relationship between σ'_y/P_a and the initial relative density is presented in Fig. 14.3. The σ'_y/P_a increases with the increase of initial relative density. The relationship between σ'_y/P_a and the initial relative density is caused by the sample with a larger initial relative density containing more particles in one unit volume, the correspondingly larger stress withstood due to the more contact points of individual particles.

Figure 14.4 shows the relationship of axial strain ε_1 and the effective axial stress σ'_1. σ'_1 increases with the increase of ε_1. The higher the initial relative density, the larger the effective axial stress σ'_1 withstood by the sample under the same axial strain and the relation of curves in Fig. 14.4 all follow a power function relation and can be expressed by Eq. (14.1). As shown in Fig. 14.4, Eq. (14.1) fits well.

$$\frac{\sigma_1'}{P_a} = A\varepsilon_1^B \tag{14.1}$$

where *A* and *B* are material parameters, their values are listed in Table 14.1, P_a is the atmosphere pressure.

The relationship between K_0 and σ'_1 of coral sand is shown in Fig. 14.5. K_0 decreases with the increase of the initial relative density and σ'_1. At the beginning of the tests, K_0 decreases rapidly with the increase of σ'_1. The decreasing trend of K_0 gradually slows down as σ'_1 increases. The same phenomenon was found by

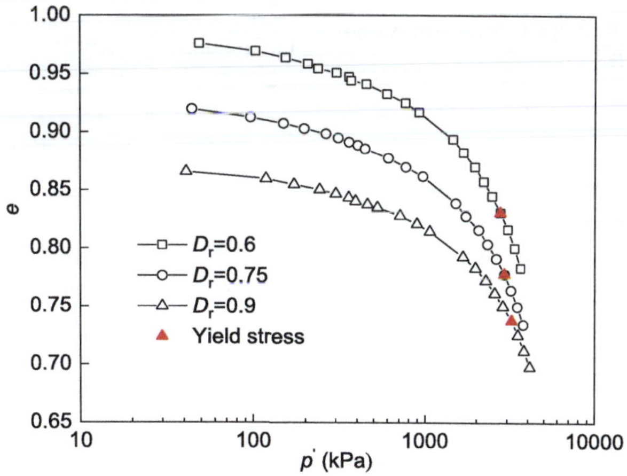

Fig. 14.2 Relationship curve of $e - p'$

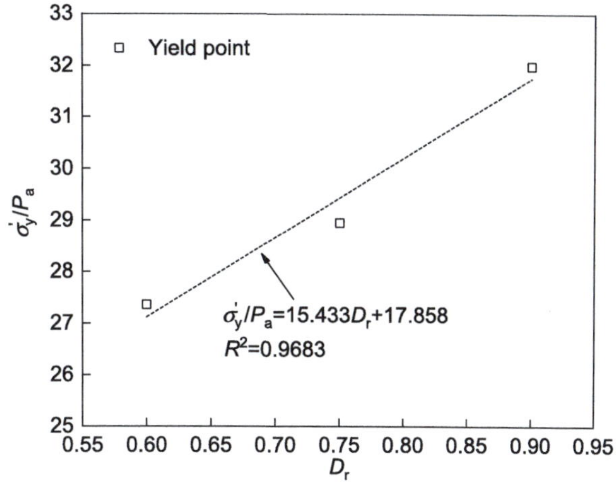

Fig. 14.3 Relationship curve of σ'_y and D_r

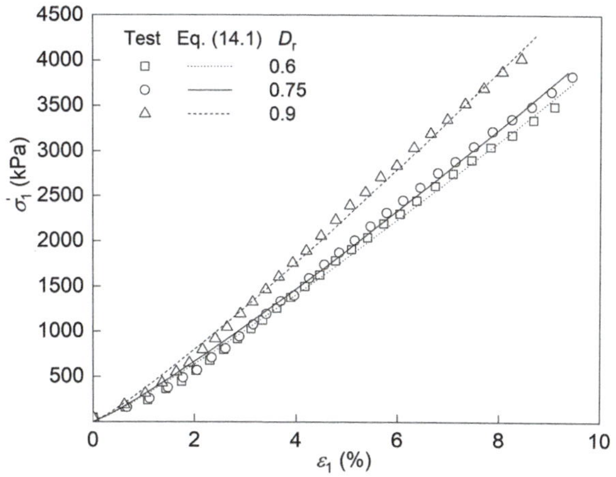

Fig. 14.4 Relationship curve of ε_1 and σ'_1

Table 14.1 Values of A and B

Initial relative density D_r	A	B	R^2
0.6	2.9	1.14	0.9975
0.75	3.03	1.14	0.9974
0.9	3.63	1.14	0.9977

Chu et al. (2004), who conducted K_0 consolidation tests on quartz sand by triaxial instrument. This phenomenon is caused by that, during K_0 consolidation tests, there is a transition process from isotropic to anisotropic in the samples. When σ'_1 exceeds a certain threshold, the change of K_0 tends to level off. With the increase of σ'_1 and initial relative density, the space for coral sand particles in the sample to move is smaller and the structure formed by the particles is more stable. As a result, it becomes difficult for the coral sand particles to get relative motion under the confined consolidation. And the K_0 of the sample tends to decrease with the increase of σ'_1 and initial relative density (Wang et al. 2020).

Through fitting analysis, there is a power function relationship between K_0 and σ'_1 as shown in Fig. 14.5. The relationship between them is well fitted by Eq. (14.2).

$$K_0 = \alpha(\sigma'_1/P_a)^{\beta} \tag{14.2}$$

where α is the value of K_0 corresponding to σ'_1 at one atmosphere. β is mainly related to the rate of the change of K_0. The values of α and β are summarized in Table 14.2.

Figure 14.6 plots the relations of $D_r - \alpha$ and $D_r - \beta$. Both $D_r - \alpha$ and $D_r - \beta$ get a strong linear correlation, which can reflect the influence of the initial relative density on the K_0.

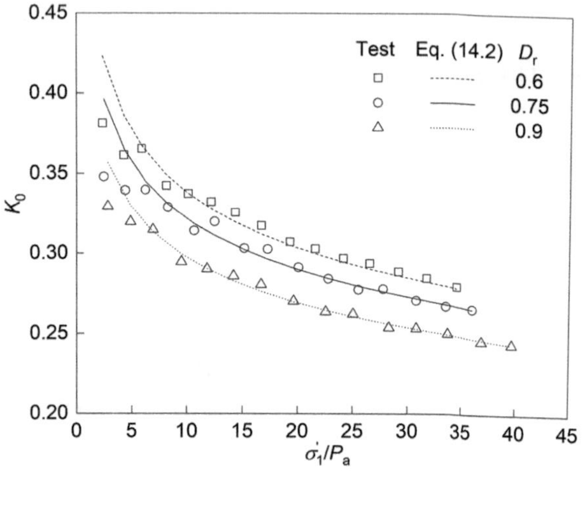

Fig. 14.5 Relationship curve of σ'_1/P_a and K_0

Table 14.2 α, β and correlation coefficient

Initial relative density D_r	α	β	R^2
0.6	0.435	−0.117	0.957
0.75	0.402	−0.108	0.917
0.9	0.356	−0.090	0.972

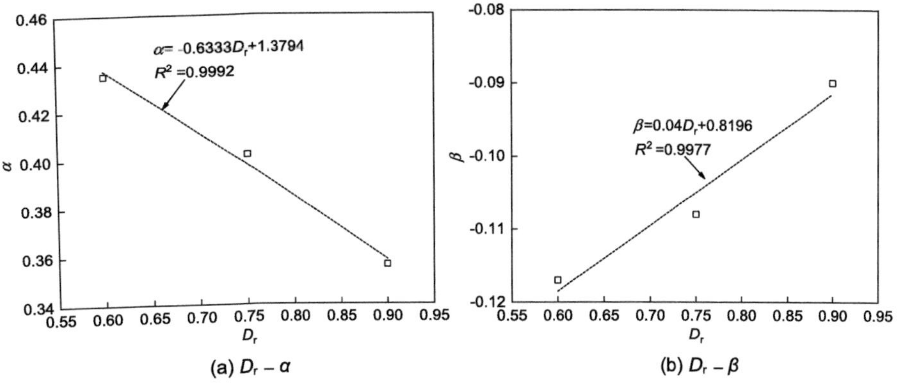

Fig. 14.6 Relationship curve of $D_r - \alpha$ and $D_r - \beta$

14.3.2 Relationship Between Stress State and Particle Crushing

The gradation curves of the specimen after the test terminated at $\sigma'_3 = 100, 200, 400, 600$, and 800kPa under different initial relative densities are plotted in Figs. 14.7, 14.8, and 14.9. The quality of particles smaller than 0.5 mm increases with the increase of σ'_3 and the gradation curves of particles range from 0.075 mm to 0.25 mm approximately parallel from each other in the double logarithmic coordinate system under different σ'_3.

As shown in Fig. 14.10, the relative breakage index B_r increases with the increase of the mean effective stress p'. The increasing trend of B_r increases as p' increases. The larger the initial relative density, the smaller the B_r at the same p' under K_0 consolidation tests. Xiao et al. (2017) and Wang et al. (2021a, b) conducted a series of one-dimensional compression tests on coral sand and found that the breakage index increases with the increase of vertical stress and the sample with a smaller relative density get a larger breakage index, which is similar to the phenomenon in this paper. The possible reason may be that the lower the initial relative density, the smaller the number of particles contained in one unit volume. The correspondingly larger stress at the contact point of individual particles leads to the sample of lower initial relative density geng a larger particle crushing.

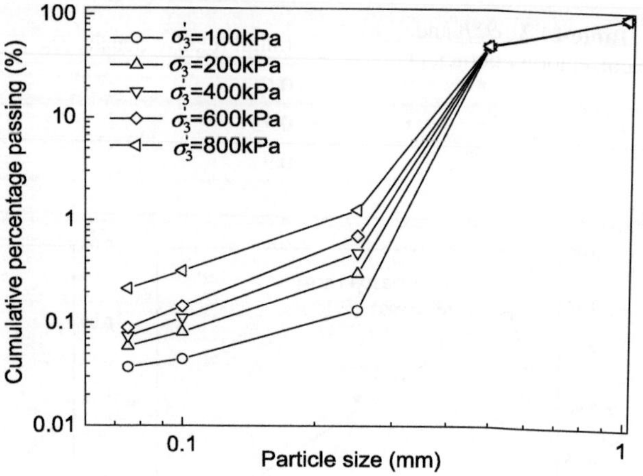

Fig. 14.7 Gradation curves of specimen after test under D_r of 0.6

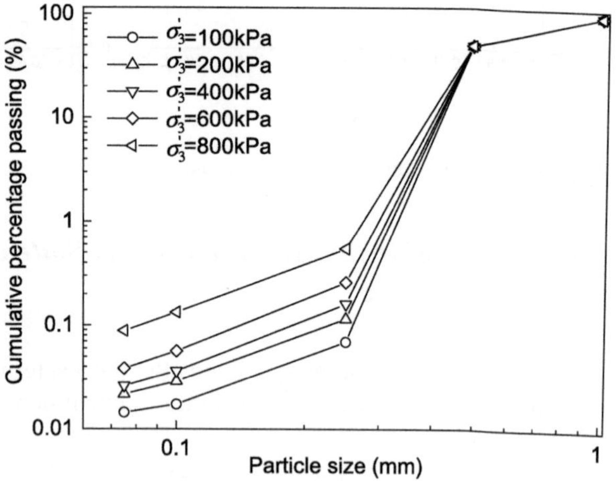

Fig. 14.8 Gradation curves of specimen after test under D_r of 0.75

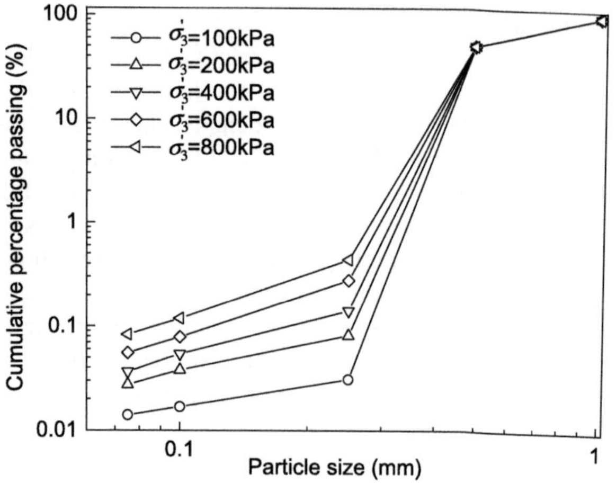

Fig. 14.9 Gradation curves of specimen after test under D_r of 0.9

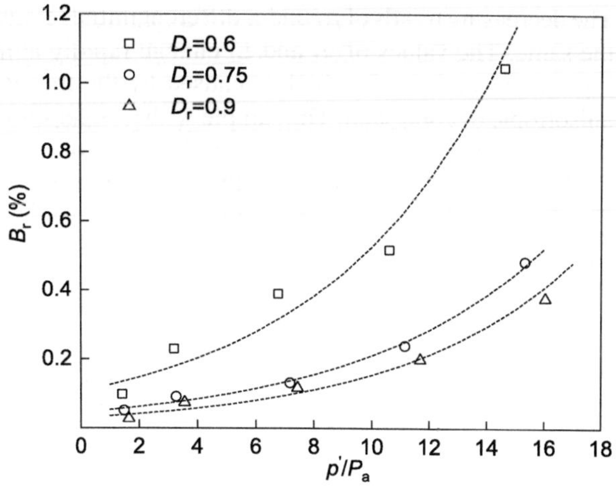

Fig. 14.10 Relationship curve of B_r and p'/P_a

14.4 Evolution of Deformation Parameters and an Incremental Elastic Model

According to the generalized Hooke's law, the axial deformation under the axisymmetric condition is calculated by

$$\varepsilon_1 = \frac{1}{E}(\sigma_1' - 2\mu\sigma_1') \tag{14.3}$$

where E is the elastic modulus, μ is the Poisson's ratio.

By taking $K_0 = \sigma'_3/\sigma'_1$ into Eq. (14.3) and changing it into an incremental form, we can get

$$\mathrm{d}\varepsilon_1 = \frac{(1 - 2\mu_t K_0)}{E_t}\mathrm{d}\sigma_1' \tag{14.4}$$

where μ_t is the tangent Poisson's ratio, and E_t is the tangent elastic modulus.

The tangent Poisson's ratio μ_t and the tangent elastic modulus E_t have been deduced by Zhang et al. (2022), as shown in Eqs. (14.5) and (14.6).

$$\mu_t = \frac{K_0}{1 + K_0} \tag{14.5}$$

$$\frac{E_t}{P_a} = BA^{1/B}(1 - \frac{2K_0^2}{1 + K_0})(\frac{\sigma_1'}{P_a})^{(B-1)/B} \tag{14.6}$$

The values of μ_t and E_t are calculated by Eqs. (14.5) and (14.6). The relationships of $\mu_t - \sigma'_1$ and $E_t - \sigma'_1$ are plotted as shown in Figs. 14.11 and 14.12. μ_t decreases with the increase of σ'_1 and the value of μ_t stabilizes when the σ'_1 exceeds 500kPa.

The decreasing trends of μ_t under different initial relative densities are approximately the same. The values of μ_t and E_t change rapidly at the beginning of the curves in Figs. 14.11 and 14.12, which is caused by the transition process from isotropic to anisotropic of coral sand. The sample with a larger initial relative density gets a lower μ_t under the same σ'_1. In this paper, the range of μ_t varies from 0.2 to 0.3.

As shown in Fig. 14.12, with the increase of σ'_1, the E_t of coral sand increases. The E_t of coral sand stabilizes when the σ'_1 exceeds 500 kPa. The varying trends of E_t under different initial relative densities are approximately the same too and the larger initial relative density gets a higher E_t under the same σ'_1. In this paper, the range of E_t varies from 24 to 50MPa.

To compare the stress–strain response curves of the test and the simulation of the elastic model, the relationship of $\sigma'_1 - \varepsilon_1$ was calculated by the increment elastic model, The calculation method is shown as follows: First setting a small incremental

Fig. 14.11 Relationship curve of σ'_1 and μ_t

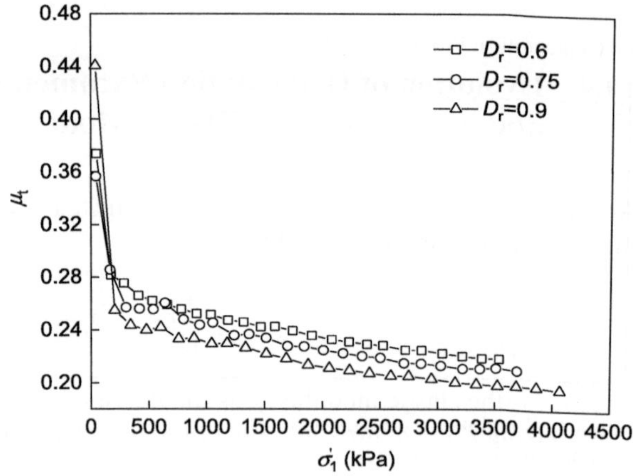

Fig. 14.12 Relationship curve of σ'_1 and E_t

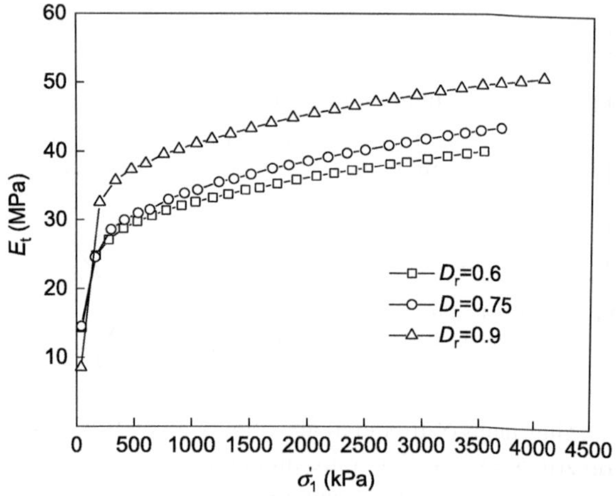

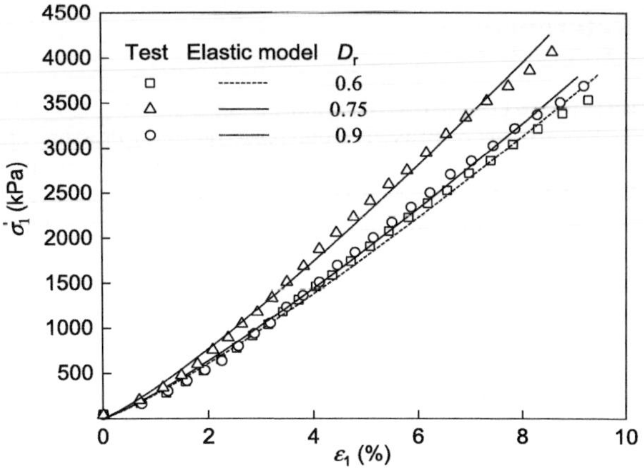

Fig. 14.13 Simulation of the incremental elastic model

step of effective axial stress $d\sigma'_1$, and taking μ_t and E_t calculated by Eqs. (14.5) and (14.6) and the value of K_0 into Eq. (14.4), we can get the increment of axial strain $d\varepsilon_1$. Then the stress increment and strain increment value are accumulated respectively, thus the full $\sigma'_1 - \varepsilon_1$ relationship curve can be obtained. The $\sigma'_1 - \varepsilon_1$ relationship of test results and the simulations by the incremental elastic model are plotted in Fig. 14.13, and the results of the simulation fit well.

14.5 Conclusion

In this paper, K_0 consolidation tests were carried out on coral sand using the GDS triaxial testing system. Based on test results, conclusions are drawn as follows.

(1) The yield stress σ'_y of coral sand increases with the increase of initial relative density.
(2) The coefficient of lateral earth pressure at rest, K_0, is significantly influenced by the initial relative density and stress state. K_0 decreases with the increase of the effective axial stress σ'_1 and initial relative density D_r.
(3) In the triaxial K_0 consolidation tests, particle crushing of coral sand increases with the increase of the mean effective stress p'. The increasing trend of particle crushing of coral sand gradually increases as the mean effective stress p' increases.
(4) The tangent elastic modulus E_t increases with the increase of the effective axial stress σ'_1 and initial relative density D_r. The varying trend of the tangent Poisson's ratio μ_t is the opposite of E_t. The relationship of stress–strain of coral sand under K_0 consolidation is well modeled by the incremental elastic model.

Acknowledgements The authors would like to acknowledge the financial support from the National Natural Science Foundation of China (Grant No. 42172295).

References

Bishop AW (1958) Test requirements for measuring the coefficient of earth pressure at rest. In: Proceedings of Brussels Conference on Earth Pressure Problems, Int. Soc. of Soil Mechanic and Found. Engrg. Brussls, Belgium, Belgium Group, pp 2–14

Cai ZY, Dai ZY, Xu GM, Ren GF (2020) Effect of particle size and compaction on K0 value of sand by centrifugal model test. Rock Soil Mech 41(12):3882–3888. https://doi.org/10.16285/j.rsm.2020.0669

Cai ZY, Zhu X, Dai ZY (2021) Coefficient of earth pressure at rest of sand considering density effect. Chin J Rock Mech Eng 40(8):1664–1671. https://doi.org/10.13722/j.cnki.jrme.2020.0788

Chu J, Gan CL (2004) Effect of void ratio on K0 of loose sand. Géotechnique 54(4):285–288. https://doi.org/10.1680/geot.2004.54.4.285

Guo PJ (2010) Effect of density and compressibility on K0 of cohesionless soils. Acta Geotechnica 5:225–238. https://doi.org/10.1007/s11440-010-0125-0

Hardin BO (1985) Crushing of soil particles. J Geotechn Eng 111(10):1177–1192. https://doi.org/10.1061/(ASCE)0733-9410(1985)111:10(1177)

Jiang MJ, Zhu JG, Mei GX (2020) Experimental study on influence factors of lateral pressure coefficient at rest for coarse-grained soil. Eng Mech 37(03):142–148. https://doi.org/10.6052/j.issn.1000-4750.2019.04.0199

Ladd RS (1978) Preparing test specimens using under compaction. Int J Rock Mech Min Sci Geomech Abstr16(3):50. https://doi.org/10.1016/0148-9062(79)90502-3

Lee J, Yun TS, Lee D et al (2013) Assessment of K0 correlation to strength for granular material. Soils Found 53(4):584–595. https://doi.org/10.1016/j.sandf.2013.06.009

Mayne PW, Kulhawy FH (1982) K0–OCR relationship in soil. J Geotech Eng Div 108(GT6):851–872. https://doi.org/10.1061/AJGEB6.0001306

Mcdowell (2002) On the yielding and plastic compression of sand. Soils Found 42(1):139–145. https://doi.org/10.3208/sandf.42.139

Ministry of Water Resources of China (1999) Industry standards of China: specification of soil test SL237-1999. China Water Resources and Hydropower Publishing House

Terzaghi K (1920) Old earth pressure theories and new test results. Eng News Rec 85(14):632–637

Wang X, Zhu CQ, Wang XZ (2020) Experimental study on the coefficient of lateral pressure at rest for calcareous soils. Mar Georesour Geotechnol 38(8):989–1001. https://doi.org/10.1080/1064119X.2019.1646361

Wang C, Ding X, Xiao Y et al (2021b) Effects of relative densities on particle breaking behavior of non-uniform grading coral sand. Powder Technol 382:524–531. https://doi.org/10.1016/j.powtec.2021.01.015

Wang X, Liu JQ, Cui J, Wang XZ, Shen JH, Zhu CQ. (2021a) Particle breakage characteristics of a foundation filling material on island-reefs in the South China Sea. Constr Build Mater 306:124690. https://doi.org/10.1016/j.conbuildmat.2021.124690

Wroth CP (1973) General theories of earth pressure and deformation. In: Proceedings of the 5th European conference on soil mechanics and foundation engineering, vol 2, Madrid, Spain, pp 33–52

Xiao Y, Liu H, Chen Q et al (2017) Particle breakage and deformation of carbonate sands with a wide range of densities during compression loading process. Acta Geotech 12:1177–1184. https://doi.org/10.1007/s11440-017-0580-y

Yang C (2012) Research on planar shape planning of urban reclamation area based on the sustainable idea. Tianjin University, Tianjin

Zhang JR, Peng WK, Zheng YJ (2022) Stress-strain model and deformation parameters of K0-consolidated coral sand. Chin J Geotech Eng 1-9[2022-09-01]. http://kns.cnki.net/kcms/detail/32.1124.TU.20220614.1340.002.html

Zhou Y (2013) Research and application on reclamation process and the relationship between urban development and the process. Tianjin University, Tianjin

Zhu JG, Jiang MJ, Lu YY et al (2019) Experimental study on the influence of stress state on at-rest earth pressure coefficient for coarse-grained soil. Rock Soil Mech 40(3):827–833+851. https://doi.org/10.16285/j.rsm.2017.2136

Chapter 15
Element Tests Investigating Reinforcement Strain Distribution of Geosynthetic-Reinforced Soil (GRS) Mass Under Vertical Loading

Cheng Liang, Chao Xu, Panpan Shen, and Minmin Luo

Abstract The Geosynthetic-Reinforced Soil (GRS) abutments have the advantages eliminating deferential settlements between the bridge beam and the approaching embankment and reducing the construction costs and difficulties. In recent years, more and more GRS abutments have been constructed throughout the world but the deformation characteristics of the GRS abutments have not been completely understood and further studies need to be carried out. This study conducted 13 two-dimensional element tests to investigate the effects of backfill gradation, reinforcement spacing S_v, and reinforcement strength T_f on the reinforcement strain distribution of the GRS mass under vertical stage loading. Dry silicon sand with three different gradations was used as backfill soil and two biaxial polypropylene geogrids with different ultimate tensile strength were used as reinforcement material. The results showed that the geogrid strain increased with the increase of the backfill strength and modulus, the decrease of the reinforcement spacing, and the increase of the reinforcement strength under the same vertical load. The geogrid strain increased

C. Liang (✉)
Research Center for Hydropower Engineering, Institute of Science and Technology, China Three Gorges Corporation, Beijing 100038, China
e-mail: lcmeijin@163.com

C. Xu
Department of Geotechnical Engineering, College of Civil Engineering, Tongji University, Shanghai 200092, China
e-mail: c_axu@tongji.edu.cn

P. Shen
Shanghai Investigation, Design and Research Institute Co., Ltd., Shanghai 200335, China
e-mail: hermit_shpp@hotmail.com

M. Luo
The Architectural Design and Research Institute of Zhejiang University Co., Ltd., Hangzhou 310028, China
e-mail: zjluominmin@163.com

S. Wang et al. (eds.), *Engineering Geology for a Habitable Earth: IAEG XIV Congress 2023 Proceedings, Chengdu, China*, Environmental Science and Engineering,
https://doi.org/10.1007/978-981-99-9057-3_15

from the edge of the GRS mass to the center and the maximum geogrid strain of all geogrid layers occurred at the mid-height of the GRS mass.

Keywords Geosynthetics · Geosynthetic reinforced soil · Reinforcement strain · Reinforcement spacing · Reinforcement strength

15.1 Introduction

To satisfy the strict deformation requirements in the construction of bridges, rigid deep foundations (e.g., concrete piles) and bearing platforms are usually used to support the bridge beam under its self-weight and traffic loads. However, large modulus difference exists between the rigid deep foundation and the approaching embankments constructed using backfill soil. Under the same load, the deformation of the embankment is usually larger than that of the bridge supported by the rigid deep foundation. This could lead to differential settlement between the bridge beam and the approaching embankment and hence the bump at the end of the bridge beam, which has negative impacts on the driving safety and comfort. In recent years, Geosynthetic-Reinforced Soil (GRS) abutments were chosen as alternatives for rigid deep foundations in the construction of small- and medium-span bridges (Helwany et al. 2007; Pham 2009; Wu et al. 2012). The GRS abutments have the advantages eliminating deferential settlements between the bridge beam and the approaching embankment and reducing the construction costs and difficulties. In recent years, more and more GRS abutments have been constructed throughout the world but the deformation characteristics of the GRS abutments have not been completely understood and further studies need to be carried out.

In this study, two-dimensional (2D) element tests of GRS masses were carried out using a custom-made rigid test box with a height-to width ratio of approximately 2. The test box was connected with two air bags at sides and a hydraulic loading device on the top. The two air bags and the hydraulic loading device were used to provide the confining stress and to apply vertical loads to the GRS mass inside the rigid test box respectively, thus making the boundary condition of the element test similar to that of the triaxial compression test. Three influencing factors were considered in this study, which were backfill gradation, reinforcement spacing S_v, and reinforcement strength T_f. This study aimed to investigate the reinforcement strain distribution of the GRS mass under vertical loading. An appropriate understanding of the reinforcement strain distribution helps to determine the location of the maximum tension in the reinforcement and failure mode of GRS mass, which is important to the design of internal stability of GRS masses under vertical loading.

15.2 Setup of Element Tests

15.2.1 Test Apparatus and Material

The 2D GRS mass, which was constructed inside a rigid test box, had dimensions of 600 mm (long) × 285 mm (wide) × 1000 mm (high). Figure 15.1 shows the cross-section view of the test apparatus and the GRS mass. The height to width ratio of the GRS mass was maintained approximately at 2 to simulate the condition used in triaxial compression tests. This study utilized two air bags, which were located at the left and right sides of the test box and were connected to an air compressor, to provide confining stresses to the GRS mass. A loading plate with dimensions of 580 mm (long) × 270 mm (wide) × 105 mm (thick) was put on top of the GRS mass and a hydraulic jack was connected to the loading plate to apply vertical loads. Detailed information regarding the test apparatus can be found in Xu et al. (2019).

Dry silicon sand with three different gradations but the same maximum and minimum particle diameters ($d_{max} = 3$ mm and $d_{min} = 0.02$ mm) were used as the backfill soil in this study. Table 15.1 shows the properties of the three backfill soil used in this study.

This study utilized two kinds of biaxial polypropylene geogrids as reinforcement material. The aperture size of both PP geogrids was 40 × 40 mm. Table 15.2 shows the

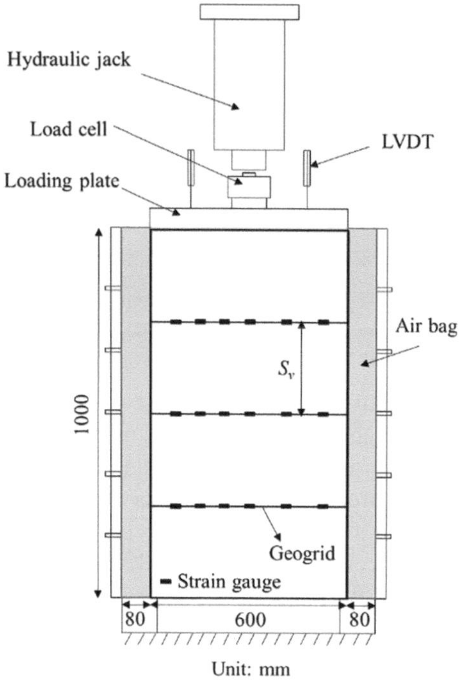

Fig. 15.1 Cross-section view of the test apparatus, the GRS mass, and the instrumentation. Modified from Xu et al. (2019)

Table 15.1 Properties of the backfill soil

Backfill	Properties
Gradation-1 (G1)	$d_{50} = 0.25$ mm, $d_{85} = 0.72$ mm, $C_c = 0.79$, $C_u = 3.79$, $\rho_{d\max} = 1930$ kg/m^3, $\rho_{d\min} = 1510$ kg/m^3, $\rho_d = 1810$ kg/m^3, $D_r = 75\%$, $\phi = 39°$, $c = 0$ kPa
Gradation-2 (G2)	$d_{50} = 0.14$ mm, $d_{85} = 0.24$ mm, $C_c = 0.99$, $C_u = 2.63$, $\rho_{d\max} = 1730$ kg/m^3, $\rho_{d\min} = 1390$ kg/m^3, $\rho_d = 1630$ kg/m^3, $D_r = 75\%$, $\phi = 35°$, $c = 0$ kPa
Gradation-3 (G3)	$d_{50} = 0.78$ mm, $d_{85} = 2.44$ mm, $C_c = 1.55$, $C_u = 8.30$, $\rho_{d\max} = 2050$ kg/m^3, $\rho_{d\min} = 1500$ kg/m^3, $\rho_d = 1880$ kg/m^3, $D_r = 75\%$, $\phi = 42°$, $c = 0$ kPa

Notes d_{50} and d_{85} represent the equivalent particle diameters for which 50 and 85% of soil by weight are finer respectively; C_c and C_u represent the curvature and the uniformity coefficients respectively; $\rho_{d\max}$ and $\rho_{d\min}$ represent the maximum and minimum dry densities respectively; ρ_d and D_r represent the dry density used in construction and the corresponding relative density respectively; ϕ and c represent the friction angle and the cohesion obtained from triaxial tests under confining pressures of 30, 100, and 200 kPa

Table 15.2 Properties of the geogrids

Geogrid	Properties
R20	$T_f = 20$ kN/m, $\varepsilon_f = 9.91\%$, $J_{@2\%} = 380$ kN/m, $J_{@5\%} = 290$ kN/m
R30	$T_f = 30$ kN/m, $\varepsilon_f = 11.56\%$, $J_{@2\%} = 470$ kN/m, $J_{@5\%} = 396$ kN/m

Notes T_f and ε_f represent the ultimate tensile strength and tensile strain at failure respectively; $J_{@2\%}$ and $J_{@5\%}$ represent the tensile stiffness at 2% and 5% tensile strain respectively

detailed properties of the two geogrids. Pullout tests under three normal stresses (i.e., 25, 50, and 75 kPa) were conducted to determine the parameters of reinforcement-soil interaction. Pullout test results showed that the R20-G1, R20-G2, and R20-G3 interfaces had interface friction angles of 38.17°, 38.06°, and 38.27° respectively and zero adhesion. The R30-G1, R30-G2, and R30-G3 interfaces had interface friction angles of 38.24°, 38.13°, and 38.38° respectively and zero adhesion.

15.2.2 Test Plan and Instrumentation

This study conducted 13 element tests to investigate the effects of backfill gradation, reinforcement spacing S_v, and reinforcement strength T_f on the reinforcement strain distribution of the GRS mass under vertical loading. This study also conducted three unreinforced soil element tests for comparison. Detailed test plan could be found in Xu et al. (2019).

The construction of the GRS mass included the compaction of the backfill in ten lifts and the placement of the reinforcement, which is shown in Fig. 15.2. The GRS mass was compacted by a hand tamper in each lift with a lift thickness of 0.1 m using a mass-volume control method. The geogrid was laid inside a GRS mass in a way that its machine direction was parallel to the length direction of the GRS mass.

(a) (b)

Fig. 15.2 Construction of the GRS mass: **a** compaction, and **b** placement of geogrid layers

The area of each geogrid layer was the same as the cross-section area of the GRS mass. During the construction of the GRS mass, the confining stress provided by the two air bags was maintained at 90 kPa in order to simulate a fixed boundary. After the completion of the construction, the confining pressure was slowly decreased to 30 kPa and kept constant during loading. Stage loading was applied on top of the GRS mass with increments of 25 kPa. Each stage was maintained for 10 min to allow the deformation of the GRS mass to be stable. When the vertical deformation of the GRS mass was not stable and the rate of deformation increased rapidly under a certain applied load, the failure was deemed to occur and the loading was terminated. Figure 15.1 also shows the instrumentation layout of the element tests. A load cell was installed between the loading plate and the hydraulic jack to monitor the applied vertical load. Six strain gauges were installed at each geogrid layer at equal distance to monitor the tensile strain of geogrids under loading.

15.3 Test Results

15.3.1 Effects of Backfill Gradation

Figure 15.3 shows the geogrid strain distributions along the geogrid length at three different heights (z = 0.25, 0.50, and 0.75 m) under the vertical load of 75 kPa for the GRS masses with reinforcement spacing $S_v = 0.25$ m and strength $T_f = 20$ kN/m. Figure 15.3 indicates that the geogrid strain increased from the edge of the GRS mass to the center for all three geogrid layers at different heights. However, the location of the maximum geogrid strain was different for different layers. With the increase of the height, the location of the maximum geogrid strain moved towards the left of the GRS mass. In addition, the maximum geogrid strain of all geogrid layers occurred at the height z of 0.5 m, indicating that the maximum lateral deformation of the GRS mass happened at the mid-height and the lateral deformation of the top or bottom of the GRS mass was smaller than that of the mid-height. Comparing geogrid strain distributions of GRS masses with different backfill gradations, Fig. 15.3 indicates that the geogrid layers placed in backfill G3 experienced the smallest deformation (i.e., minimum strain) while the geogrid layers placed in backfill G2 had the largest deformation (i.e., maximum strain). Due to the fact the backfill G3 had higher friction angle and modulus as compared to the backfill G2, under the same vertical load, the GRS mass constructed with backfill G3 showed smaller deformation and consequently lower shear stresses transferred from the backfill to the geogrid layers through reinforcement-soil interaction. As a result, the geogrid layers in the GRS mass constructed with backfill G3 had smaller strain as compared to the other two cases.

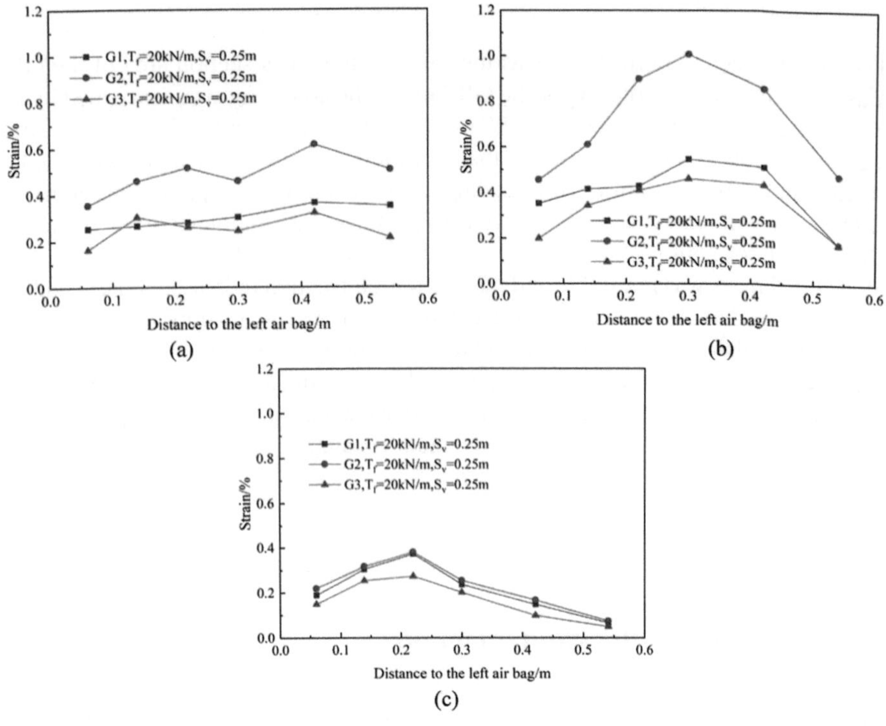

Fig. 15.3 Effects of backfill gradation on the geogrid strain distributions of GRS masses at the height z of: **a** 0.25 m, **b** 0.50 m, and **c** 0.75 m

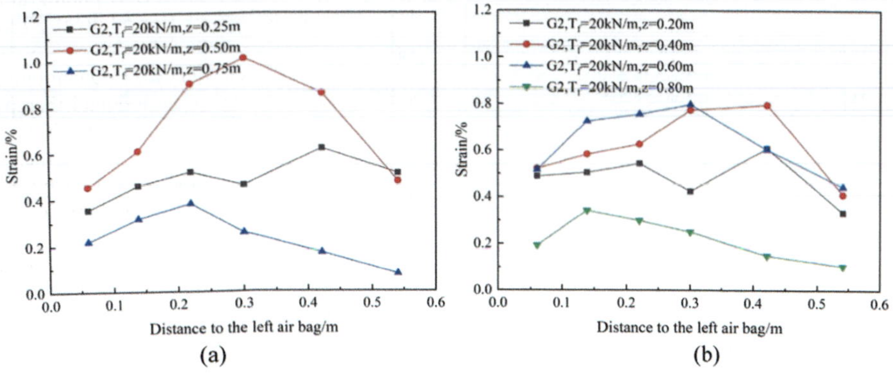

Fig. 15.4 Effects of reinforcement spacing on the geogrid strain distributions of GRS masses at the reinforcement spacing Sv of: **a** 0.25 m, and **b** 0.20 m

15.3.2 Effects of Reinforcement Spacing

Figure 15.4 shows the geogrid strain distributions along the geogrid length under the vertical load of 75 kPa for the GRS masses constructed with backfill G2 and reinforcement strength $T_f = 20$ kN/m. Similar results were found for the GRS masses with backfill G1 and G3 and were not discussed in this manuscript to save space. Comparing Fig. 15.3a and b shows that the geogrid strains in the case of $S_v =$ 0.25 m were larger than those in the case of $S_v = 0.20$ m. In other words, under the same vertical load and other conditions (e.g., backfill gradation and reinforcement strength), the geogrid strain increased with the increase of the reinforcement spacing. According to the equivalent confining stress theory propsed by Yang (1972), the decrease of the reinforcement spacing resulted in the increase of the confining stress applied on the GRS mass, thus decreasing the deformation of the GRS mass under vertical loading. Therefore, the geogrid experienced less deformation and hence smaller strain.

15.3.3 Effects of Reinforcement Strength

Figure 15.5 shows the geogrid strain distributions along the geogrid length under the vertical load of 75 kPa for the GRS masses constructed with backfill G2 and reinforcement spacing $S_v = 0.25$ m. Similar results were found for the GRS masses with backfill G1 and G3 and were not discussed in this manuscript to save space. Comparing Fig. 15.5a–c shows that the GRS mass constructed with the stronger geogrid layers G30 had smaller geogrid strains than that constructed with the weaker geogrid layers G20. This was because that the geogrid G30 had higher tensile strength (or stiffness) than the geogrid G20 and the increase of the geogrid ultimate strength

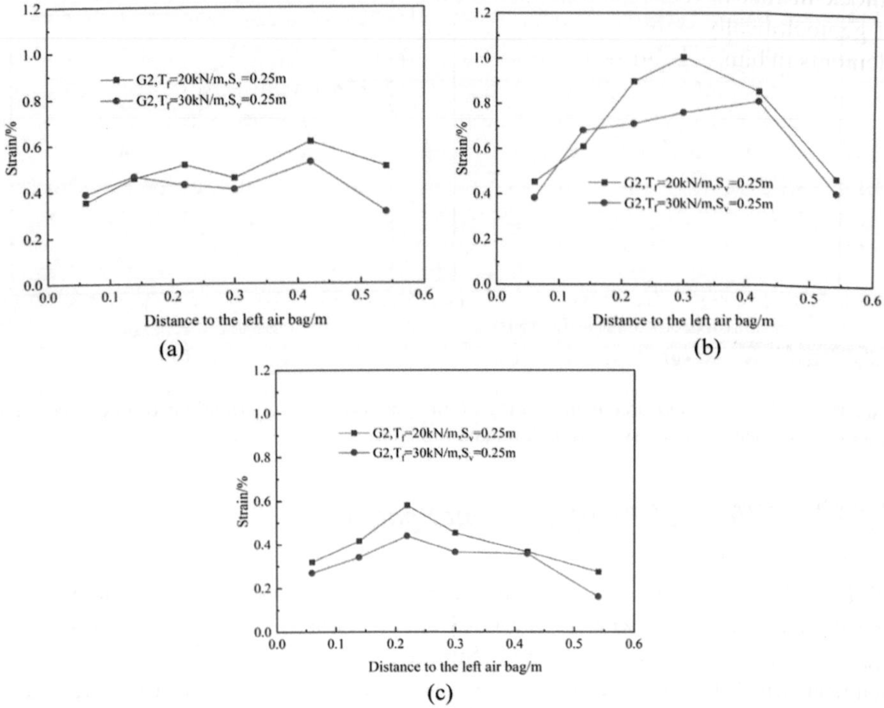

Fig. 15.5 Effects of reinforcement strength on the geogrid strain distributions of GRS masses at the height z of: **a** 0.25 m, **b** 0.50 m, and **c** 0.75 m

(or stiffness) resulted in higher confining stress applied on the GRS mass according to Yang (1972). As a result, the GRS mass constructed with stronger geogrid G30 showed smaller deformation and therefore smaller geogrid strains.

Maximum reinforcement strain distribution of Geosynthetics can be used to determine the failure mode of GRS mass. Failure mode is important to analyze the internal stability and ultimate bearing capacity of GRS mass. Failure mode has been discussed in Xu et al. (2019) through the same element tests and can be described using Ranking failure surface. According to the model test results, the variation of backfill gradation, reinforcement spacing and reinforcement strength has little effects on the location of maximum reinforcement strain. Therefore, it can be concluded that failure mode of GRS mass was not affected by the three influencing factors considered in the element tests.

15.4 Conclusions

This study conducted 13 two-dimensional element tests to investigate the effects of backfill gradation, reinforcement spacing S_v, and reinforcement strength T_f on the reinforcement strain distribution of the GRS mass under vertical stage loading. Dry silicon sand with three different gradations was used as backfill soil and two biaxial polypropylene geogrids with different ultimate tensile strength were used as reinforcement material. The following conclusions could be drawn from this study:

(1) Under the same vertical load, the geogrid strain increased with the increase of the backfill strength and modulus, the decrease of the reinforcement spacing, and the increase of the reinforcement strength.
(2) For each geogrid layer, the geogrid strain increased from the edge of the GRS mass to the center. However, the location of the maximum geogrid strain was different for layers at different heights. The maximum geogrid strain of all geogrid layers occurred at the mid-height of the GRS mass.

References

Helwany SMB, Wu JTH, Kitsabunnarat A (2007) Simulating the behavior of GRS bridge abutments. J Geotech Geoenviron Eng 133(10):1229–1240

Pham TQ (2009) Investigating composite behavior of geosynthetic reinforced soil (GRS) mass. Ph. D. Thesis. University of Colorado, Denver, USA

Wu JTH, Adams M, Pham TQ, Ma CY (2012) A generic soil-geosynthetic composite test. Int J Geotech Eng 6(1):103–116

Xu C, Liang C, Shen P (2019) Experimental and theoretical studies on the ultimate bearing capacity of geogrid-reinforced sand. Geotext Geomembr 47(3):417–428

Yang Z (1972) Strength and deformation characteristics of reinforced sand. Ph.D. dissertation. University of California, Los Angeles

Chapter 16
SPT Interpretation for Engineering Properties of Residual Silt in Vietnam

Kuanjun Wang, Weida Ni, Hao Wu, Lili Ma, Liuyuan Zhao, Yihuai Lou, and Lin Tian

Abstract The strength and stiffness of residual silt, which is extensively dispersed in Vietnam's mountainous regions, are the most crucial design factors for wind power project foundations. Standard penetration test (SPT) is one of the most used in-situ testing techniques in geotechnical investigation and a practical way to assess soil parameters in the absence of soil strength and stiffness data from laboratory test results. To establish SPT correlation for residual silt in Vietnam, this paper carried out laboratory tests, in-situ SPT, and downhole seismic tests based on the Ia Pet wind power project, in Gia Lai province, Vietnam. The correlation between SPT-*N* values and undrained shear strength s_u of residual silt was established based on the results of the direct consolidated quick shear test (DSq) of 31 boreholes. The empirical coefficient '*a*' for strength interpretation of residual silt in Vietnam ranges from 5.25 to 10.72, with an average value of 7.31. For SPT interpretation, it is advised to take

K. Wang · W. Ni
Powerchina Huadong Engineering Corporation Limited, Hangzhou, Zhejiang Province, China
e-mail: wang_kj@yced.com.cn; steven_9090@163.com

W. Ni
e-mail: ni_wd@hdec.com

H. Wu (✉) · L. Zhao · Y. Lou · L. Tian
Zhejiang Engineering Research Center of Marine Geotechnical Investigation Technology and Equipment, Zhejiang Huadong Geotechnical Investigation and Design Institute Corporation Limited, Hangzhou, Zhejiang Province, China
e-mail: wu_h5@hdec.com

L. Zhao
e-mail: zhao_ly@hdec.com

Y. Lou
e-mail: lou_yh@hdec.com

L. Tian
e-mail: tian_l@hdec.com

L. Ma
Ocean College, Zhejiang University, Hangzhou, Zhejiang Province, China
e-mail: malili@zju.edu.cn

S. Wang et al. (eds.), *Engineering Geology for a Habitable Earth: IAEG XIV Congress 2023 Proceedings, Chengdu, China*, Environmental Science and Engineering,
https://doi.org/10.1007/978-981-99-9057-3_16

a conservative approach and use the standard value of 6.158 for the first residual silt layer. The wave velocity and SPT-*N* values of the residual silt are found to have typical power functions when combined with the downhole seismic test results at the indicated three holes. The established SPT correlations for undrained shear strength s_u and shear wave velocity V_s are in good agreement with the measured results, while the interpreted pressure wave velocity V_p has a relatively bad prediction. The undrained shear strength and shear wave velocity of residual silt in Vietnam can be determined using the existing SPT relationships.

Keywords Standard penetration test · Undrained shear strength · Shear wave velocity · SPT correlations

16.1 Introduction

Geotechnical investigation is the most important part of engineering projects, whose primary goal is to provide soil parameters for foundation design. Standard penetration test (SPT) is the most used in situ testing method in geotechnical investigation. Generally, SPT was used to interpret the relative density D_r, and internal friction angle φ of sandy soil (Mulagala and Kadam 2020). The first proposal of a simple linear correlation between undrained shear strength s_u (kPa) and SPT-*N* blow count was presented by Stroud in 1974. SPT is a feasible method to obtain the soil properties of clayey soil (Güllü et al. 2019; Ruge et al. 2018; Viviescas et al. 2019). The SPT-*N* value and undrained shear strength of London clay were linearly correlated by White et al. (2019), who also stressed the significance of analyzing the SPT and s_u data site by site to establish the proper correlation. Small strain shear modulus is also an important parameter for the geotechnical design that can be calculated by shear wave velocity V_s (Anbazhagan et al. 2013; Ataee et al. 2019; Kirar et al. 2016). Based on data from 20 locations in Metro Manila, Daag et al. (2022) established an empirical relationship between V_s and SPT-*N* value for both sandy soil and clayey soil, the interpretation of the empirical coefficient of SPT may vary depending on the site. If the wrong empirical coefficients are used for the SPT interpretation, an unexpectedly substantial deviation may have occurred. Establishing the relationship between SPT and soil characteristics based on site-specific information is crucial. Silt is a common soil type in the mountainous region of central Vietnam. However, the SPT interpretation method and empirical coefficient of silt in Vietnam had not been well studied. Based on the geotechnical investigation of a wind power project in Vietnam, the results of SPT, laboratory tests, and downhole seismic tests have been analyzed as a case study to determine correlations between SPT-*N* values and undrained shear strength s_u, shear wave velocity V_s, and pressure wave velocity V_p, which serve as guidelines for SPT interpretation of silt in the future at Vietnam.

Fig. 16.1 Location of Ia Pet wind power plant, Gia Lai province

16.2 Physical Properties of Residual Silt in Ia Pet

16.2.1 Site Overview

The project of the Ia Pet wind power plant (Fig. 16.1) is in the territory of Ia Pet communes, Dak Doa district, Gia Lai province. The project's location is around 17 km southeast of Pleiku City and about 11 km southwest of the Dak Doa District center. The total installed capacity is 140 MW, which is consisted of 31 turbines of 4.5 MW. Each wind turbine has 31, 40-m-deep boreholes in the middle to expose the engineering geological conditions. SPTs are conducted with 2 m interval at each borehole to determine the engineering geological conditions. To determine the physical and mechanical characteristics of soil samples, drilling samples are subjected to a variety of laboratory tests, including grain size distribution, water content, density, Atterberg limits, one-dimensional consolidation, and direct shear tests. Additionally, three downhole seismic tests are conducted at three chosen representative locations to measure the wave velocity and calculate the shear modulus.

16.2.2 Geological Condition

The eluvial-deluvial zone edQ, the very strongly weathered zone IA1, the strongly weathered zone IA2, and the moderately weathered zone IB are the four primary strata that are highlighted by the borehole logs and laboratory test data.

Table 16.1 provides the geological condition in Ia Pet, where the soil classification is by the requirements of ASTM D2487. Based on the geological condition below the top residual soil, it could be seen that the residual soil is weathered from basalt. In the Casagrande plasticity chart, as illustrated in Fig. 16.2, the top-soil Layer 1 is the reddish-brown residual soil, and Layer 2 beneath it is the reddish brown/bluish-grey residual soil. The physical parameters of Layers 1 and 2 are displayed in Table 16.2.

Table 16.1 Geological conditions

No.	Geological conditions	Type	Top elevation (m)	Thickness (m)
Layer 1	Eluvial-deluvial zone edQ	Clayey soil, soft to stiff	−0.90–5.02	0.50–4.10
Layer 2	Very strongly weathered zone IA1	Clayey soil, stiff	−3.93–1.45	2.00–9.40
Layer 3	Strongly weathered zone IA2	Gravel with clay	−18.48–−4.76	0.10–3.90
Layer 4	Moderately weathered zone IB	Basalt	−18.78–−5.80	1.60–16.40

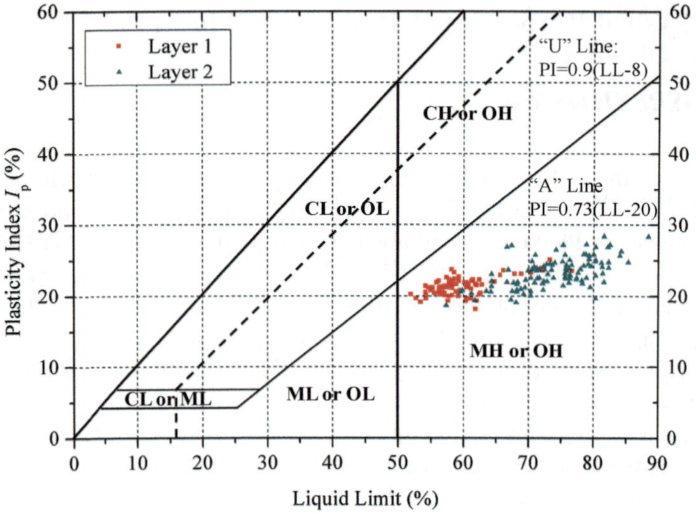

Fig. 16.2 Casagrande plasticity chart for the residual soil

Table 16.2 Physical properties of Ia pet residual silt

	Gravel	Sand				Silt	Clay	
Particle size	4.75–9.5 (mm)	2.0–4.75 (mm)	0.425–2.0 (mm)	0.15–0.425 (mm)	0.075–0.15 (mm)	0.005–0.075 (mm)	0.002–0.005 (mm)	< 0.002 (mm)
Layer 1	0.0	0.1	0.4	0.6	0.8	24.5	28.4	45.2
Layer 2	0.1	0.1	0.4	0.5	1.2	23.5	28.5	45.7
Physical properties	G_s	w (%)	e_0	ρ (g/cm^3)	S_r (%)	w_L (%)	I_p	I_L
Layer 1	2.96	47.5	1.568	1.703	89.7	59.9	21.4	0.418
Layer 2	2.96	53.8	1.692	1.695	94.3	73.9	23.3	0.137

16.3 Strength Properties of Ia Pet Residual Silt

16.3.1 Laboratory Direct Shear Test

The direct consolidated quick shear (DSq) test is used to determine the residual silt strength parameter, with specimen of 61.8 mm in diameter and 20 mm in height. Three specimens are initially consolidated under three distinct overburden pressures of 100, 200, and 300 kPa. After consolidation, rapid shear of 1.0 mm/min is performed on the soil specimens. The cohesion c and internal friction angle φ can be obtained by drawing the Mohr–Coulomb envelope. Figure 16.3 displays the cohesion (c), internal friction angle (φ), and undrained shear strength (s_u) of 31 boreholes. The average values of cohesion c of Layer 1 and Layer 2 are respectively 25.7 kPa and 33.0 kPa, where average values of internal friction angle φ of Layer 1 and Layer 2 are 13.4° and 16.9°, respectively. Although Layer 2's cohesion and internal friction angle are marginally greater than Layer 1's, as depicted in Fig. 16.3c. The following regression equation can be used to fit the undrained shear strength s_u of Layer 1 and Layer 2.

$$s_u = 5.021z + 22.404, \quad R^2 = 0.8928 \tag{16.1}$$

where s_u is the undrained shear strength of silt, and z is the soil depth.

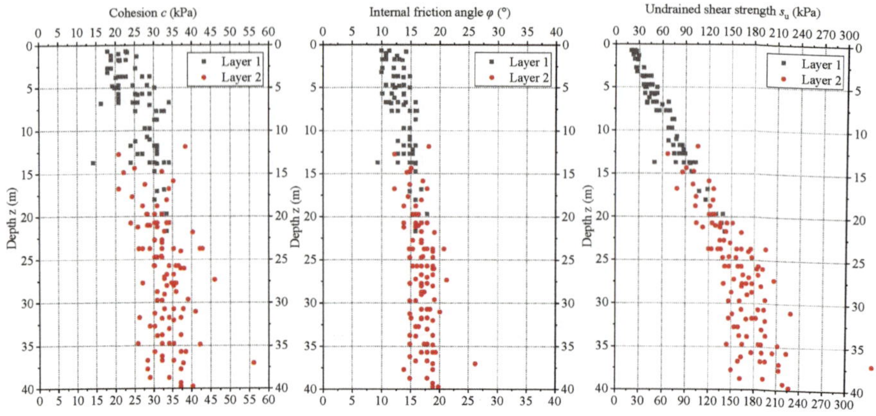

(a) cohesion c (b) internal friction angle φ (c) undrained shear strength s_u

Fig. 16.3 Soil strength parameter versus soil depth

16.3.2 Standard Penetration Test

The standard penetration test (SPT) is carried out by the requirement of Vietnam standard TCVN 9351-2012. Before installing the SPT equipment at the bottom of the borehole and beginning the SPT after drilling to the target depth, the borehole's bottom should be cleaned. SPT equipment consists of a spilt spoon sampler and a hammer. The 63.5 kg hammer is dropped down at the height of 76 cm, where the hammer speed is from 18 to 22 blows per minute. The total penetration depth for each test is 45 cm; the first 15 cm is considered as a seating drive, where the number of blows for the last 30 cm is counted as the SPT-*N* value.

In Fig. 16.4, the SPT data for all 31 boreholes are displayed, and the results for Layers 1 and 2 are represented by square and circle points, respectively. Based on the testing results, linear regression is found between the soil depth z and blow counts

$$N = 0.711z + 3.723, \quad R^2 = 0.8683 \tag{16.2}$$

where N is the blow counts of SPT.

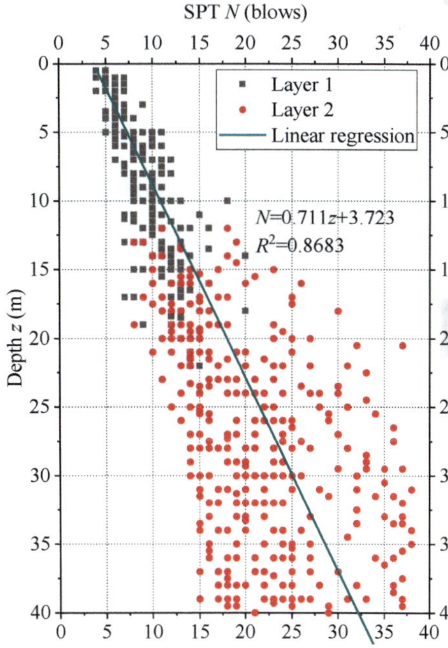

Fig. 16.4 SPT N blows versus soil depth

16.3.3 SPT and Undrained Shear Strength Correlation

As mentioned in Sects. 3.1 and 3.2, the undrained shear strength s_u and SPT-N blows are linearly dependent on the soil depth z. Therefore, the SPT-*N* blows could have a direct relationship with the undrained shear strength, as given by

$$s_u = a \cdot N \tag{16.3}$$

where *a* is the empirical coefficient. The calibrated average values of empirical coefficient '*a*' at all wind turbine locations are listed in Table 16.3, with a minimum of 5.247 and a maximum value of 10.724.

Figure 16.5 shows the distribution of the empirical coefficient *a* of Layers 1 and 2 at different depths. The average empirical coefficient '*a*' values for Layers 1 and 2 are 6.506 and 7.900, respectively, demonstrating that the empirical coefficient '*a*' produced via calibration is relatively discrete. The statistical histogram used to represent the empirical coefficient '*a*' reveals that its overall average value is 7.31, as seen in Fig. 16.6.

Table 16.4 compares some of the correlations between SPT-*N* value and undrained shear strength s_u from earlier studies. Sivrikay and Toğrol (2006) provided the range of the empirical coefficient '*a*' of high and low plastic silts in Turkey, which were 2.0–6.88 and 2.68–6.67, respectively. The average values of the empirical coefficient '*a*' ranged between 3.80 and 4.22, which is a substantial difference from the 7.31 obtained in this study. Hu et al. (2011) carried out research based on the investigation data in Malaysia and obtained the average value of 8.09 of empirical coefficient '*a*' for clayey silt, which was slightly higher than the calibration result in this paper. The empirical coefficient '*a*' for SPT interpretation of silt in various places could be discovered to have significant differences. The strength of the silt may not be accurately estimated if the SPT correlation from other places is used directly.

The standard value calculation method given by the Chinese national standard GB 50021 "Code for Investigation of Geotechnical Engineering" was used to find the recommended empirical coefficient '*a*', for SPT interpretation as given by:

$$\phi_m = \frac{\sum_{i=1}^{N} \phi_i}{n} \tag{16.4}$$

Table 16.3 Calibrated average values of empirical coefficient a at all wind turbine locations

Location	T01	T02	T03	T04	T05	T06	T07	T08	T09	T10	T11
a	6.215	6.937	5.621	5.373	6.885	7.568	6.794	8.002	6.675	7.655	5.711
Location	T12	T13	T14	T15	T16	T17	T18	T19	T20	T21	T22
a	7.352	7.066	5.734	8.094	8.478	5.480	9.464	8.492	8.062	7.570	8.983
Location	T23	T24	T25	T26	T27	T28	T29	T30	T31		
a	10.121	7.367	6.986	6.585	6.715	5.247	5.540	8.873	10.724		

Fig. 16.5 SPT correlation between undrained shear strength

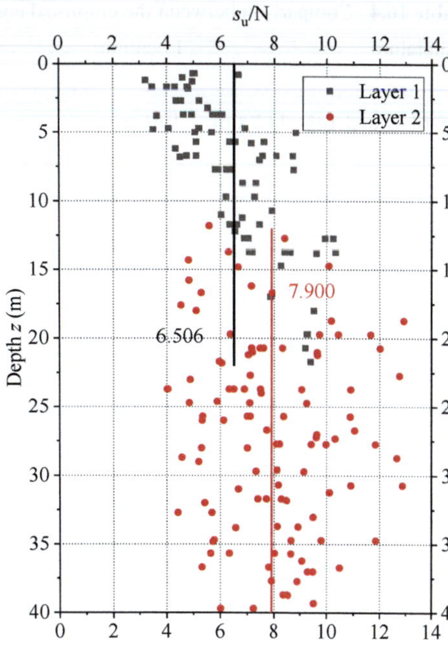

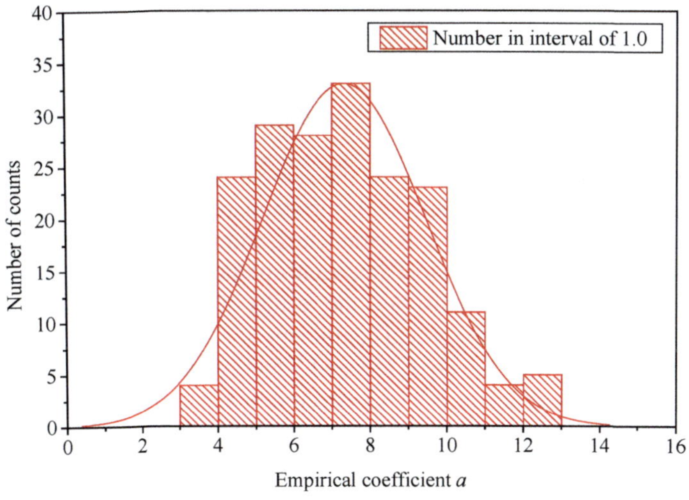

Fig. 16.6 Statistics of calibrated empirical coefficient a values

Table 16.4 Comparison between the empirical coefficient 'a' in this study and existing results

Literature	Location	Soil type	Empirical coefficient *a*	
			Range	Average
Sivrikay and Toğrol (2006)	Turkey	High plastic clays	2.25–17.5	7.52
		Low plastic clays	2.12–13.0	4.98
		High plastic silts	2.0–6.88	3.80
		Low plastic silts	2.68–6.67	4.22
Hu et al. (2011)	Penang, Malaysia	Muddy Silty clay	/	7.95
		Silty clay	/	7.21
		Clayey silt	/	8.09
White et al. (2019)	London, England	Clay	4.50–7.20	5.90
This study	Gia Lai, Vietnam	Silt	5.25–10.72	7.31

$$\sigma_{\mathrm{f}} = \sqrt{\frac{1}{n-1}\left[\sum_{i=1}^{n}\phi_i^2 - \frac{\left(\sum_{i=1}^{n}\phi_i\right)^2}{n}\right]} \tag{16.5}$$

$$\delta = \frac{\sigma_{\mathrm{f}}}{\phi_{\mathrm{m}}} \tag{16.6}$$

$$\gamma_{\mathrm{s}} = 1 \pm \left\{\frac{1.704}{\sqrt{n}} + \frac{4.678}{n^2}\right\}\delta \tag{16.7}$$

$$\phi_{\mathrm{k}} = \gamma_{\mathrm{s}} \cdot \phi_{\mathrm{m}} \tag{16.8}$$

where n is the number of samples, φ_i is the sample value of soil parameters, φ_{m} is the average value of soil parameters, σ_{f} is the standard deviation of soil parameters, δ is the coefficient of variation of soil parameters, and γ_{s} is the correction coefficient of statistics. The standard value φ_{k} of empirical coefficient '*a*' can be determined by the above equation. Table 16.5 displays the results of the intermediate calculations and the average values of the empirical coefficient '*a*'. The estimated standard values of Layers 1 and 2 are respectively 6.158 and 7.544, while the standard value of all calibrated empirical coefficients is 7.043. An empirical parameter of 6.25 is proposed for SPT interpretation of undrained shear strength based on Vietnam's local specification, which differs slightly from the standard value of 6.158 for the empirical coefficient of Layer 1. To interpret the strength of the silt in Vietnam using SPT, it is advised to use the interpretation coefficient of 6.158.

Table 16.5 Standard values of the empirical coefficient '*a*' for each silt layer

Soil layer	Layer 1	Layer 2	All
φ_{m}	6.506	7.900	7.311
n	78	108	185
σ_{f}	1.797	2.163	2.132
δ	0.2762	0.2738	0.2916
γ_{s}	0.9465	0.9550	0.9634
φ_{k}	6.158	7.544	7.043

16.4 Wave Velocity of Ia Pet Residual Silt

16.4.1 Downhole Seismic Test

The downhole seismic test is carried out by ASTM D7400-14 to determine the wave velocity of the soil layers. A vertical borehole should be used for downhole seismic testing, with a PVC casing erected up to the end of the borehole before grouting with a water-cement and bentonite mixture. Set up the three-component geophone next to the borehole wall at the designated testing depth with a 1 m testing interval after detecting the seismic source (hammer and plank) nearby. When a seismic wave is generated by striking a plank with a hammer, the geophone picks up the signal. The wave velocity at each testing depth could then be determined using the formulae below:

$$V_{\mathrm{p}(i,i+1)} = \frac{\Delta H_{(i,i+1)}}{\Delta t_{\mathrm{p}(i,i+1)}} \tag{16.9}$$

$$V_{\mathrm{s}(i,i+1)} = \frac{\Delta H_{(i,i+1)}}{\Delta t_{\mathrm{s}(i,i+1)}} \tag{16.10}$$

where $V_{\mathrm{p}(\copyright,i+1)}$ is the P wave velocity (m/s) at depth of testing point © to testing point $i + 1$, $V_{\mathrm{s}(\copyright,i+1)}$ is the S wave velocity (m/s) at depth of testing point © to testing point $i + 1$, $\Delta H_{(\copyright,i+1)}$ is the difference (m) between the distance of testing point © and distance of testing point $i + 1$, $\Delta t_{\mathrm{p}(\copyright,i+1)}$ is the time difference (ms) between the P wave traveling time of testing point © and testing point $i + 1$, $\Delta t_{\mathrm{s}(\copyright,i+1)}$ is the time difference (ms) between the S wave traveling time of testing point © and testing point $i + 1$. The testing results of P wave velocity and S wave velocity at locations T14, T26, and T27 are presented in Fig. 16.7, where the P wave velocity and S wave velocity are steadily increased with the depth. The velocity data of T26 and T27 are found to be much like each other, while the velocity data of T14 seems to be much larger than that of T26 and T27. The possible reason could be that the location of T26 and T27 is near, but T14 is far away from T26 and T27.

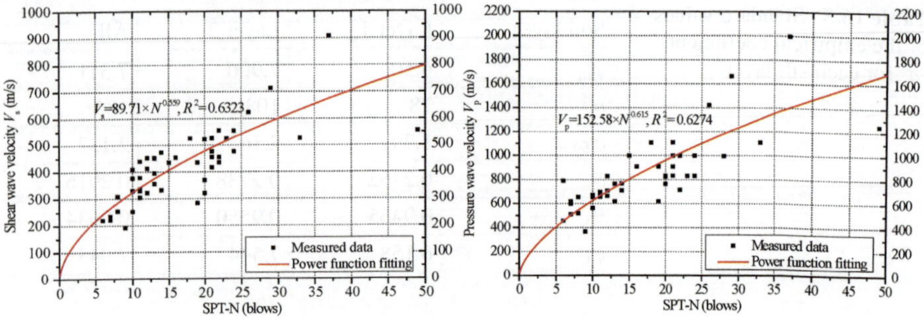

(a) V_s-N (b) V_p-N

Fig. 16.7 Downhole seismic test results at T14, T26 and T27

16.4.2 Velocity and SPT Correlation

The SPT-N values and pertinent velocities V_p and V_s at the same depth are shown in Fig. 16.8 to determine the relationship between the velocities and mechanical characteristics of silt. Power function is employed for shear and pressure wave velocity V_s and V_p fitting analyses. Both V_s and V_p have fitting fair-good expressions, which are displayed as follows:

$$V_s = 89.71 \cdot N^{0.559}, R^2 = 0.6323 \tag{16.11}$$

$$V_p = 152.58 \cdot N^{0.615}, R^2 = 0.6274 \tag{16.12}$$

Most studies concentrated on sand and clay, with only a small number on silt, while examining the relationship between SPT-N value and shear velocity V_s. According to earlier studies, the correlations of silt are displayed in Table 16.6, with the exponent of the power function varying from 0.319 to 0.77 and the coefficient of the regression

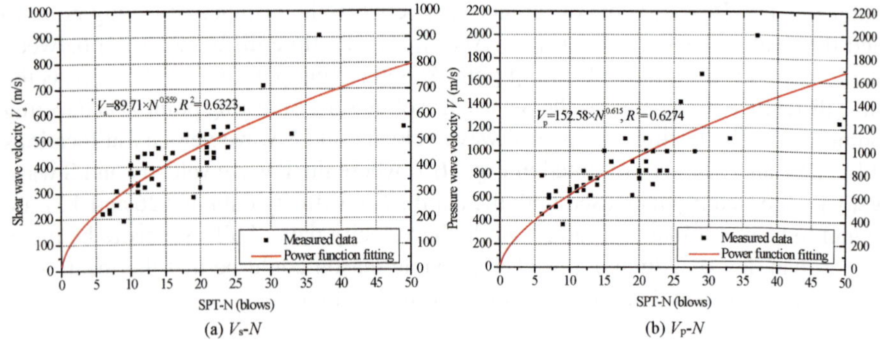

(a) V_s-N (b) V_p-N

Fig. 16.8 Velocities and SPT-N correlations

Table 16.6 Comparison between the N-Vs correlation in this study and previous research findings

Literature	Location	Soil type	Regression equation
Lee (1990)	Taipei, China	Low plastic silt	$V_s = 105.64 \cdot N^{0.319}$
Jafari et al. (2002)	Tehran, Iran	Low plastic silt	$V_s = 22 \cdot N^{0.77}$
Dikmen (2009)	Eskisehir, Turkey	Silt	$V_s = 60 \cdot N^{0.36}$
Daag et al. (2022)	Manila, Philippines	Sandy soil	$V_s = 45.07 \cdot N^{0.5534}$
		Clayey soil	$V_s = 70.26 \cdot N^{0.4220}$
This study	Gia Lai, Vietnam	Silt	$V_s = 89.71 \cdot N^{0.559}$

equation ranging from 22 to 105.64. The coefficient and exponent of the power function of the formula obtained in this study are 89.71 and 0.559, respectively, within a reasonable range. Similar to the correlation of N-s_u, different correlations of N-V_s are established based on the data of different sites. It is very important to assess the SPT and relevant soil parameter data on a site-by-site basis to determine an appropriate correlation.

16.5 SPT Interpretation

The calibrated empirical coefficient a in Eq. (16.3) and regression expressions (16.6) and (16.7) can be used to interpret the undrained shear strength s_u, shear wave velocity $V_{s,}$ and pressure wave velocity V_p. The SPT values of T26 are compared with the obtained findings from measurement to initially validate the interpretation equations. The results of the SPT's interpretation are shown in Fig. 16.9 along with the measurements of the downhole seismic test's pressure and shear wave velocities V_p and $V_{s,}$ respectively, as well as the results of the direct consolidated quick shear test's measurement of undrained shear strength s_u. As can be seen in Fig. 16.9a, the relationship between the undrained shear strength and soil depth fits well with the results of the direct consolidated quick shear test. The SPT interpretation results seem to be slightly overestimated, but with acceptable deviations. In terms of the shear wave velocity V_s and pressure wave velocity V_p interpretation, which are shown in Fig. 16.9b and c, the interpretation findings of V_s are in good agreement with the measured data, but those of V_p are not quite as good. Despite the viability of the SPT interpretation of V_p for the pressure wave velocity, the regression expression's reliance on sparse data may be a contributing factor to the poor prediction of V_p. To perform the regression, which might be more representative, there should be a lot of soil data employed.

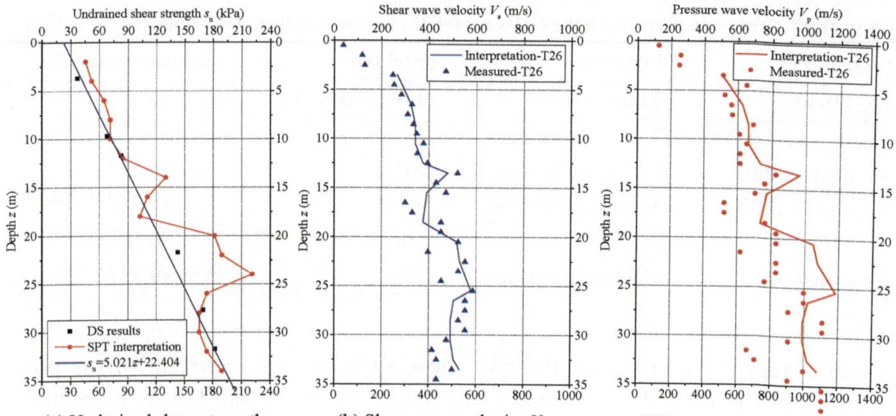

(a) Undrained shear strength s_u (b) Shear wave velocity V_s (c) Pressure wave velocity V_p

Fig. 16.9 SPT interpretation of T26

16.6 Conclusion

This study uses a wind farm in the Gia Lai province of Vietnam to establish correlations between the standard penetration test (SPT) and undrained shear strength s_u, shear wave velocity V_s, and pressure wave velocity V_p. The regression expressions between undrained shear strength s_u, SPT-N values, and soil depth have been established, which are $s_u = 5.021z + 22.404$ and $N = 0.711z + 3.723$, respectively. The empirical coefficients 'a' between the undrained shear strength s_u and SPT-N values are calibrated from 5.247 to 10.724 in 31 wind turbine locations, with an average value of 7.31. For conservative consideration, the standard value of 6.158 for the first residual silt layer is recommended for SPT interpretation. In addition, the correlations between shear wave velocity V_s, pressure wave velocity V_p, and SPT-N values are established as $V_s = 89.71{\cdot}N^{0.559}$ and $V_p = 152.58{\cdot}N^{0.615}$, respectively. At the location of the T26 wind turbine, the preliminary interpretations of the undrained shear strength s_u and shear wave velocity V_s results of the correlations show good agreement with the measured results. However, the interpreted pressure wave velocity V_p has a relatively poor prediction compared to the measured results, which needs necessitating future research with a mass of information. The overall undrained shear strength and shear wave velocity of residual silt in Vietnam can be determined for reference using the existing SPT correlations.

Acknowledgements The authors gratefully acknowledge the supports of the National Natural Science Foundation of China (Grant No. 52108356 and 42102313), the Zhejiang Provincial Natural Science Foundation of China (Grant No. LQ22E090008), and the Science and Technology project of POWERCHINA Huadong Engineering Corporation (Grant No. KY2021-KC-02-02 and KY2020-KC-01).

References

Anbazhagan P, Kumar A, Sitharam TG (2013) Seismic site classification and correlation between standard penetration test N value and shear wave velocity for Lucknow City in Indo-Gangetic Basin. Pure Appl Geophys 170(3):299–318

Ataee O, Moghaddas NH, Lashkaripour GR (2019) Estimating shear wave velocity of soil using standard penetration test (SPT) blow counts in Mashhad city. J Earth Syst Sci 128(3):1–25

Daag AS, Halasan OPC, Magnaye AAT, Grutas RN, Solidum RU Jr (2022) Empirical correlation between standard penetration resistance (SPT-N) and shear wave velocity (Vs) for soils in Metro Manila, Philippines. Appl Sci 12(16):8067

Dikmen Ü (2009) Statistical correlations of shear wave velocity and penetration resistance for soils. J Geophys Eng 6(1):61–72

Güllü H, Canakci H, Alhashemy A (2019) A ranking distance analysis for performance assessment of UCS versus SPT-N correlations. Arab J Sci Eng 44(5):4325–4337

Hu ZH, Li JQ, Li XZ, Zhao XB, Wang P (2011) Determination of undrained shear strength of clayey soils by SPT value. Chin J Underground Space Eng 7(S2):1577–1588

Jafari MK, Shafiee A, Razmkhah A (2002) Dynamic properties of fine grained soils in south of Tehran. J Seismol Earthquake Eng 4(1):25–35

Kirar B, Maheshwari BK, Muley P (2016) Correlation between shear wave velocity (Vs) and SPT resistance (N) for Roorkee region. Int J Geosyn Ground Eng 2(1):1–11

Lee SHH (1990) Regression models of shear wave velocities in Taipei basin. J Chin Inst Eng 13(5):519–532

Mulagala L, Kadam AR (2020) Critical appraisal of shear strength parameters of in situ subsoil by SPT. Indian Geotech J 50(2):261–266

Ruge JC, Mendoza CC, Colmenares JE, Cunha RP, Otalvaro IF (2018) Analysis of the undrained shear strength through the standard penetration test with torque (SPT-T). GEOMATE J 14(41):102–110

Sivrikaya O, Toğrol E (2006) Determination of undrained strength of fine-grained soils by means of SPT and its application in Turkey. Eng Geol 86(1):52–69

Viviescas JC, Osorio JP, Griffiths DV (2019) Cluster analysis for the determination of the undrained strength tendency from SPT in mudflows and residual soils. Bull Eng Geol Env 78(7):5039–5054

White F, Ingram P, Nicholson DP (2019) An update of the SPT-cu relationship proposed by M. Stroud in 1974. In: Proceedings of the XVII European conference of soil mechanics geotechnical engineering, Reykjavik.

Chapter 17
Pore Fluid Composition Impact on the Geotechnical Behaviour of Foundations—A Mini-review

Stephen Wilkinson and Mohanad Talal Alfach

Abstract The composition of pore fluids influences the geotechnical properties and behaviour of soils. If the pore fluid composition is significantly different from that of distilled water, then care must be taken during geotechnical works. Pore fluids vary in composition for a number of reasons, for example, near the coast or in arid environments high levels of salinity mean that porewater densities are higher than those of pure water resulting in an increase in uplift pressures. The presence of saline water is predictable for a given site, and so can be accounted for. However, where sites are contaminated changes in geotechnical properties can also occur due to lubrication, density or viscosity effects for example. Given the increasing number of construction projects on brownfield sites, there will be an increasing need to assess the effects of different pore water compositions on the geotechnical behaviours of soils. The majority of groundwater contamination is caused by either heavy metals or hydrocarbons. Of these two, hydrocarbon contamination is dominant. However, as there are many different types of hydrocarbon, and these can be mixed in different types of soil the impact on soil behaviour is variable. In general, liquid limits decrease and plastic limits increase with hydrocarbon contamination. The maximum dry density and the optimum moisture content decrease. There is a reduction in the angle of friction and an increase in the value of cohesion for pure bentonite or kaolin soils, but for clayey and silty soils with mixed minerology there is an increase in the angle of friction and a decrease in the cohesion. The compression index increases, but the coefficient of volume compressibility decreases, and the permeability decreases. The extent of these effects is dependent both on the type and extent of contamination and the soil properties, it is therefore important that this is quantified during site investigation.

S. Wilkinson (✉)
University of Wollongong in Dubai, Dubai, UAE
e-mail: 2Stephen.wilkinson@gmail.com

M. T. Alfach (✉)
City, University of London, London, UK
e-mail: Mohanad.alfach@yahoo.com

S. Wang et al. (eds.), *Engineering Geology for a Habitable Earth: IAEG XIV Congress 2023 Proceedings, Chengdu, China*, Environmental Science and Engineering,
https://doi.org/10.1007/978-981-99-9057-3_17

Keywords Pore fluid · Geotechnical behaviour · Contamination · Groundwater

17.1 Introduction

Soils are multiphase systems containing a particulate solid phase surrounded by a fluid which can vary widely in its properties e.g., from air to mercury. Classically and typically the fluid phase is understood as a pore-water pressure. However, when carrying out a compaction test, for example, there is a recognition in finding the optimum moisture content that the fluid can affect other soil properties. In this case, the presence of water alters the frictional properties allowing the particles to move closer together than if the fluid phase were air alone. This produces an increase in maximum density as the soil moves towards this optimum moisture content (Partially saturated soil mechanics). This interaction of the fluid, altering the frictional properties of the soil, implies that the fluid phase is important for determining soil strength beyond just the pore water pressure effect. This also implies that the type of fluid will have an impact on soil behaviour, not just the pressure of that fluid. Given that the fluid type can change relatively easily, then the strength parameters on which a geotechnical structure have been designed can be altered by, for example, a spill of contaminant into groundwater. The impact and implications of contamination on the properties of the ground have been studied experimentally (discussed below). However, given the number of possible soil and fluid combinations, identifying the general effects of fluids on soil properties and behaviour is challenging. The primary aim of this paper is the provide an overview of the observed behaviour, identifying the issues in these studies, to broaden the discussion of this topic and highlight the areas of particular challenge.

17.2 Oil Types and Soil Types

Due to their prevalence, the majority of tests on the geotechnical behaviour of contaminated soil, focus on hydrocarbon contamination. As many of the terms used to describe hydrocarbon contamination are non-exact there is uncertainty over the properties of the fluid phase in many geotechnical studies. For example, "crude oil" contaminated soil or "petroleum" contaminated soil, these terms can be applied to pore fluids containing hydrocarbons with a wide range of carbon chain lengths and therefore different fluid properties. Crude oil will change based on the age, and burial/thermal histories of the petroleum reservoir from which the crude oil has been extracted (Clayton 1988). Petroleum is a potentially broader term as it is applied to crude oil as well as to some refined petroleum products. An even broader categorisation used in a wide range of papers is "oil" contaminated soil, which can include for example vegetable oils. Thus, the fluids used in studies of the effects of hydrocarbons on the geotechnical behaviour of soils are very broad. Examples of geotechnical studies on different petroleum fluids are sown in Table 17.1. Given the range of hydrocarbon fluids it is unsurprising that the results of these studies are variable.

In addition to the fluids used, studies of hydrocarbon contamination on soils have equally been carried out on a wide range of soils, including those described as sandy soils (Khamehchiyan et al. 2007), silty soils (Kermani and Ebadi 2012), clayey soils (Karkush and Kareem 2017). Some studies use soils with mixed grades such as silty sand (Hafshejani and Hajiannia 2016) and clayey silts (Resol 1999). Some studies identify the consistency of the soil, e.g., soft clay (Elisha 2012), and some studies provide the origin of the soil e.g., basaltic residual soils (Rahman et al. 2010) and lateritic soils (Ijimdiya 2013). In addition, there are studies on monomineralic soils such as kaolinite soil (Khosravi et al. 2013) and lab made soils such as bentonite-kaolinite mixtures (Daka 2015).

The choices of references in the above section have been made to maximise the overlap of the different oil and different soil groups. It is clear that a large array of possible oils and possible soils exist. This has the potential to make the analysis of the overall effects of hydrocarbon contamination on the geotechnical properties of soils very challenging (Outlining this is the purpose of this paper). In addition to the factors above, time is another important variable. Nazir (2011) identifies that soils which are contaminated with hydrocarbons change properties with time. In a two-year study they identified that the full geotechnical affects of oil contamination occur after approximately 6 months.

Table 17.1 Examples of studies on the geotechnical properties of petroleum and petroleum products the number of Google Scholar results is indicative of the popularity of the terms

Google Scholar search	Number of results	Example geotechnical study
"geotechnical" "oil"	143,000	Kermani and Ebadi (2012)
"geotechnical" "petroleum"	110,000	Tuncan and Pamukcu (1992)
"geotechnical" "hydrocarbon"	26,600	Ota (2010)
"geotechnical" "diesel"	17,600	Safehian et al. (2018)
"geotechnical" "petrol"	12,100	Chaudhary and Singh (2016)
"geotechnical" "crude oil"	9240	Khamehchiyan et al (2007)
"geotechnical" 'fuel oil"	3310	Karkush and Kareem (2017)
"geotechnical" "kerosene"	3190	Karkush and Jihad (2020)
"geotechnical" "gas oil"	2850	Khosravi et al (2013)
"geotechnical" "motor oil"	891	Nazir (2011)
"geotechnical" "engine oil"	774	Hewayde et al (2019)

17.3 Measurements of the Effects of Oil Contamination on Soil Properties

Rajabi and Sharifipour (2018) presents a systematic review which combines data for different oils and soils into single graphs and identifies correlations. This illustrates the extent of the issues of variability outlined above, almost every test graphed and discussed in the paper identifies literature sources which show positive trends and other sources which show negative trends with increasing hydrocarbon contamination. The graphs also show a wide scatter for the presented data. The combination of these results to identify overall trends, allows a broad judgement to be made. However, given different articles show very different trends in results, the broader correlation is an incomplete understanding of the underlying behaviour, and can be misleading in its predictions.

The below section presents the results from commonly carried out laboratory tests with a focus on results which show the effects of increasing concentration. The key focus in the discussion below is the trend of the results not their magnitude. The figures presented in this paper are indicative of the broad scatter and ranging trends presented across the literature, indicating the challenge of studies in this area.

17.3.1 Moisture Content Tests

Both Atterberg limit tests and compaction tests require the measurement of water content. Typically, this test is carried out using an oven set at 105 °C. Where the soil contains both oil and water then, if the oil contains light weight hydrocarbons with a low boiling temperature then most will be driven off. Whereas if the oil contains heavy weight hydrocarbons with high boiling temperatures then less will be driven off. If the oil is a crude oil which contains a mixture of hydrocarbons which will vary by the crude oil used, then only a proportion will be vaporised. This generates an additional factor which makes the comparison of tests in different studies challenging.

Proposals have been made to address this issue in the observed moisture contents of contaminated soils. For example Tong (2008) defined the drying loss of the oil-contaminated porous media as a soils "moisture content" expressed as:

$$W_m = \frac{m_w + m_o}{m_r} = w_w + k_O \tag{17.1}$$

where w_m is moisture content (%), m_o is the oil weight loss caused by drying (i.e., the weight of volatilized oil) (g), w_w is water content (%), k_o is the ratio of m_o and m_r (%). Zheng et al. (2014) derived the water content in porous media sampled from the contaminated site by using:

$$W_w = \frac{(m_t - m_r)\eta - C_O V_O \gamma}{m_r \eta - C_O V_O (1 - \gamma)} \times 100\% \tag{17.2}$$

where m_t is the weight of the contaminated sample used to determine (m_r). m_r and η represent the weight of porous media after oven-drying and the oil-drying loss coefficient, respectively. C_0 is the oil concentration of the extract (g L^{-1}), V_0 is the volume of the extract (L), and γ (−) is recovery coefficient. However, such approaches are rarely implemented, with standard experimental procedures being more commonly followed.

17.3.2 *Atterberg Limits*

The plastic limit (Fig. 17.1a) increases with increasing hydrocarbon contamination for silty soils (Kermani and Ebadi 2012), soft clayey soils (Elisha 2012) lateritic soils (Akinwumi et al. 2014). The plastic limit decreases with increasing hydrocarbon contamination for clayey soils (Khamehchiyan et al. 2007), basaltic residual soils (Rahman et al. 2010), kaolin soils (Khosravi et al. 2013). In general, it appears that the soils with more coarse-grained fraction experience an increase in their plastic limit where as those with a higher fine-grained content experience a decrease in plastic limit. However, given that soft clay soils (Elisha 2012) and clayey soils (Khamehchiyan et al. 2007) produce opposite trends with increasing oil content (both studies "crude oil") another effect is causing this discrepancy. This effect could relate to mineralogy, density or consistency.

The liquid limit (Fig. 17.1b) increases with increasing hydrocarbon content for silty soils (Kermani and Ebadi 2012), soft clayey soils (Elisha 2012) kaolin soils (Khosravi et al. 2013). For kaolin soils the liquid limit starts to decrease at high concentrations (16%). The liquid limit decreases with increasing hydrocarbon content for clayey soils (Khamehchiyan et al. 2007), basaltic residual soils (Rahman et al. 2010), lateritic soils (Akinwumi et al. 2014). The increases and decreases are the same for the liquid limit as for the plastic limit with the exception of that of the kaolin soil (Khosravi et al. 2013), which has the opposite trend. There are a wide range of possible causes for this change in the behaviour of kaolin in comparison to the other samples. For example, kaolin is coarser grained than other clay minerals, it has a smaller specific surface area and it does not experience interlayer expansion in comparison to other clay minerals. It does not generate a diffuse double layer in the fluid phase as is seen in other common clay minerals (Akinwumi et al. 2014). Given the variations in fluid phase type and concentration, the variations in the soils used and the different experimental procedures. Identifying the cause of these differences is challenging.

17.3.3 *Compaction Behaviour*

The observed behaviour with compaction is more mixed than other parameters are due to the variety of approaches used by authors for this test. Some studies use

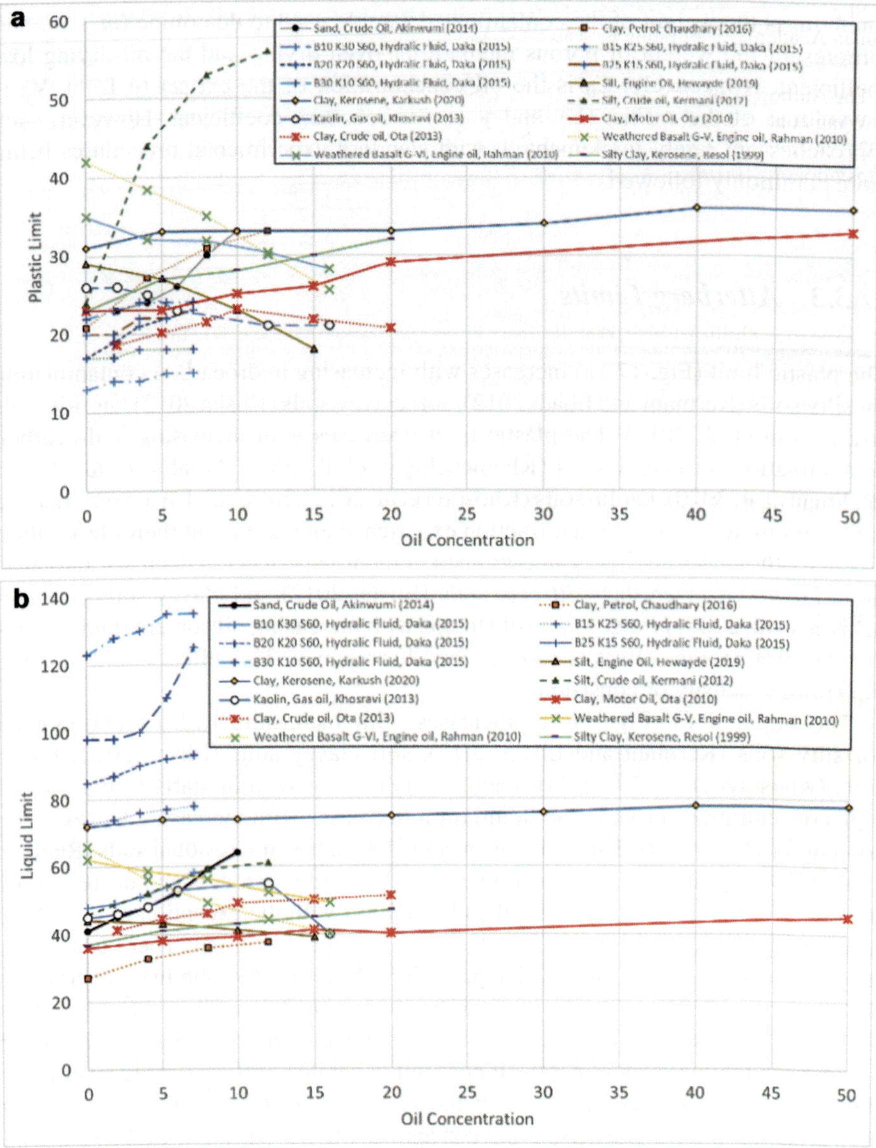

Fig. 17.1 Changes in the plastic (**a**) and liquid limits (**b**) for various types of soil caused by changing the concentration of different types of oil. In the study by Daka (2015) B represents the proportion of bentonite, K represents the proportion of kaolin and S represents the proportion of Sand. In the study by Rahman et al. (2010) V and VI represent the level of weathering of the soil

only oil and change the percentage of oil during compaction, some studies have a starting quantity of soil and oil then mix water, whereas others maintain the ratio of water to oil with increasing fluid volume. In addition, compaction is an unsaturated process and so relates to partially saturated/unsaturated soil mechanics processes, the total amount of saturation at OMC varying between the soil types. Given this, assessing the compaction behaviour with increasing oil content while ignoring the proportion of water produces varying results. However, using total fluid volumes allows for better comparisons. In general, within many studies (e.g., Chaudhary and Singh 2016; Daka 2015; Rahman et al. 2010, etc.) the maximum dry density (Fig. 17.2a) of a soil containing oil is lower than that of a soil containing only water. This may be an artefact of measurement due to some oil being difficult to evaporate at 100 °C. For other studies the maximum dry density (Fig. 17.2a) of a soil containing oil is higher than that of a soil containing only water (Khamehchiyan et al. 2007; Mohammadi et al. 2018), this is possibly due to the better lubrication provided by the oil allowing soil particles to move closer together. While for the optimum moisture content (Fig. 17.2b) looking at just water decreases with increasing oil content, the optimum fluid content in terms of the total sum of fluids in the soil is roughly the same whether the fluid is water, oil or a mix of the two. This pattern observed in sandy soils (Khamehchiyan et al. 2007) and silty soils (Kermani and Ebadi 2012). Two studies report an increase in OMC with oil concentration (Chaudhary and Singh 2016; Resol 1999) indicating that the compaction observations above are not universal for all soils.

17.3.4 Strength Properties

The key strength parameters assessed here are the angle of friction (Fig. 17.3a) and cohesion (Fig. 17.3b) which contribute towards the peak shear strength of the soils studied. The angle of friction increases with increasing hydrocarbon content for clayey soils (Khamehchiyan et al. 2007; Resol 1999), silty soils (Kermani and Ebadi 2012). The angle of friction decreases with increasing hydrocarbon content for sandy soils (Khamehchiyan et al. 2007), bentonite rich soils (Mohammadi et al. 2018), kaolin rich soils (Khosravi et al. 2013) and other clayey soils (Karkush and Kareem 2017). The cohesion increases with increasing hydrocarbon content for sandy soils (Khamehchiyan et al. 2007), bentonite rich soils (Mohammadi et al. 2018), kaolin rich soils (Khosravi et al. 2013). The cohesion decreases with increasing hydrocarbon content for clayey soils (Karkush and Kareem 2017; Khamehchiyan et al. 2007), silty soils (Kermani and Ebadi 2012).

For clayey soils both an increase (Khamehchiyan et al. 2007) and a decrease (Karkush and Kareem 2017) in the angle of friction has been reported. As there is a difference in the hydrocarbon used in these studies (crude oil and fuel oil respectively), and there are a wide variety of minerals and grain sizes that can make up clayey soils it is difficult to identify the reason for this contradiction. In general, for all soils the peak shear strength reduces (Fig. 17.4).

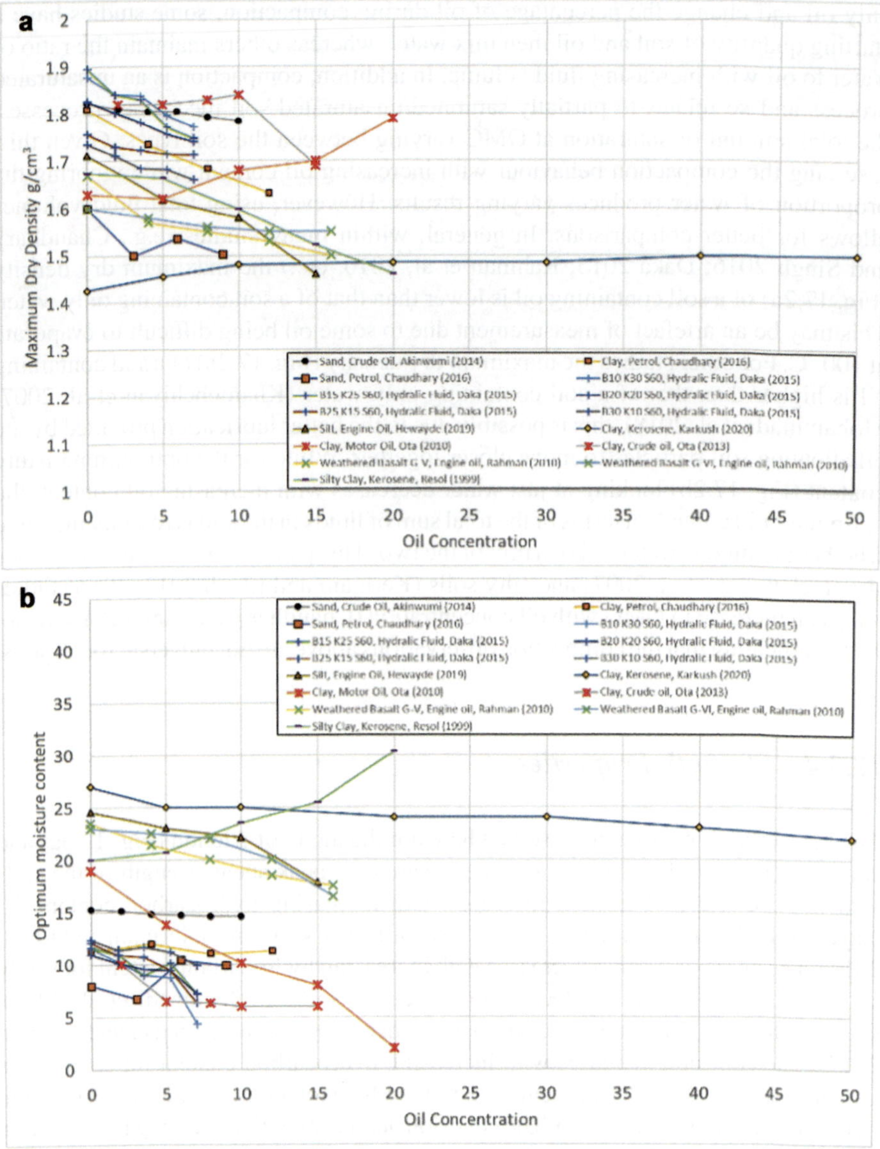

Fig. 17.2 The changes in compaction behavious of different types of soil exposed to different concentrations of several types of oil. Including optimum moisture content (**a**) and maximum dry density (**b**). In the study by Daka (2015) B represents the proportion of bentonite, K represents the proportion of kaolin and S represents the proportion of Sand. In the study by Rahman (2010) V and VI represent the level of weathering of the soil

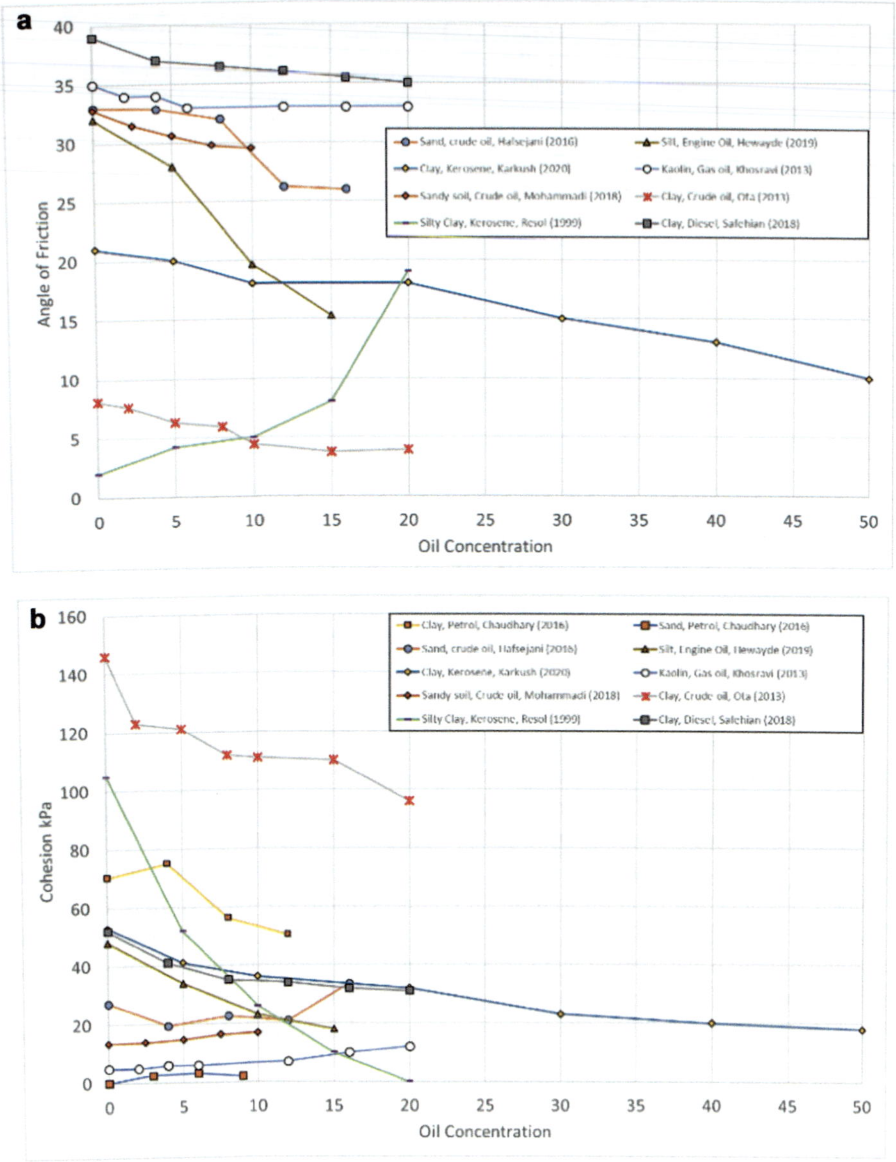

Fig. 17.3 The impact of oil concentrations, for various types of oil on the strength parameters, angle of friction (**a**) and cohesion (**b**), for different types of soil

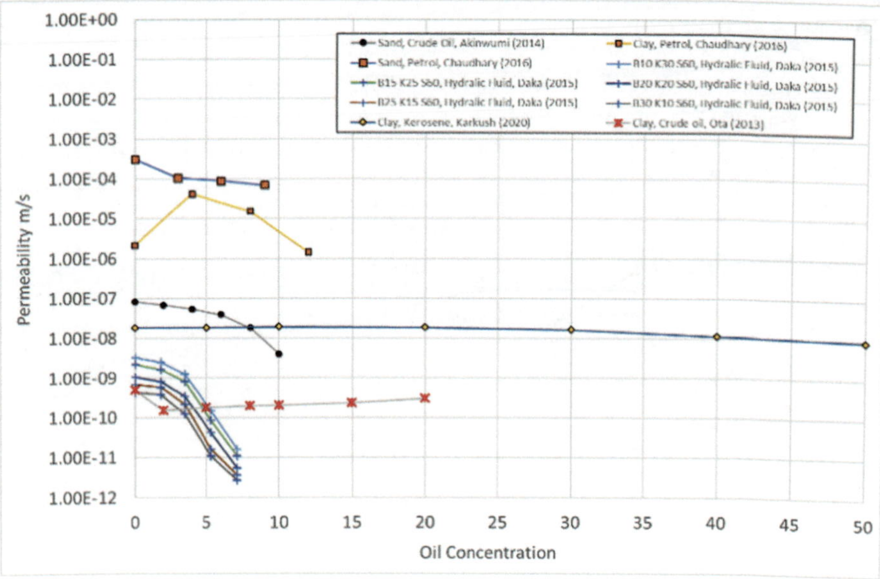

Fig. 17.4 The impact of oil concentration on the permeability of soils. In the study by Daka (2015) B represents the proportion of bentonite, K represents the proportion of kaolin and S represents the proportion of Sand

During unconfined compressive strength tests Ijimdiya (2013) identified an increase in strength with small quantities of oil which decreased rapidly within increasing oil content. The exact mechanism which causes the strength properties reduction associated with the increase in pore fluid viscosity is not entirely understood. For example, as the strength of a granular soil is governed by the mineral to mineral contacts and mineral to mineral contacts in the presence of pore fluids, the observed changes should be associated with particle contacts. Microscopic modelling of granular media has shown that the microscopic friction angle (which is based on the Mohr–Coulomb failure criteria) is a key factor controlling the macroscopic behavior. Therefore, it is reasonable to believe that it is the lubricating effect at microscopic level due to the viscosity of the pore fluid is responsible of the observed macroscopic behavior.

17.3.5 Consolidation Behaviour

There is less uncertainty in relation to consolidation behaviour. With increasing oil content, the compression index increases (Karkush and Kareem 2017; Kermani and Ebadi 2012) and both the coefficient of volume compressibility and the coefficient of consolidation decreases (Ijimdiya 2013). This is likely due to the lubricating nature of the oil, and to its non-polar nature meaning that is does not interact with the diffuse double layer of clay minerals. However, the statistical significance of the observations of the decrease in the coefficient of consolidation have been questioned by Talukdar and Saikia (2013) who state the change is only significant for low compressibility clay soils.

17.3.6 Permeability and Other Properties

There are a few soil properties which show the same trend across all soils with increasing hydrocarbon content. One important one is permeability (Fig. 17.4), which decreases for all soils from the zero oil sample to the maximum oil sample (Daka 2015; Khamehchiyan et al. 2007; Rahman 2010). Where the two phases are miscible this effect can be caused by the increase in viscosity of the resultant fluid. This confirms the importance of viscosity (μ) in determining the permeability in the general statement of Darcy's law:

$$Q = -\frac{k\rho g \Delta HA}{\mu L}$$

where Q is the volumetric flow rate, k is the intrinsic permeability, ρ is the density, g is acceleration due to gravity, ΔH is the change in head, A is the area through which flow is occurring, and L is the distance along the flow path. Where the fluids are not miscible then permeability is reduced due to the blocking of the pores by the non-wetting phase. However Chaudhary and Singh (2016) observed an increase in permeability for intermediate oil contents. And Ota (2013) identified a decrease in permeability for samples containing oil in comparison to the sample with only water, but observed an increasing permeability with increasing oil concentration.

Some additional observations have been made in a few papers. There is an increase in the amount of linear shrinkage with oil content (Ota 2013), this is likely due to lubrication during shrinkage as well as the evaporation of the oil. There is a decrease in specific gravity of the soil with increasing oil content (Akinwumi et al. 2014), primarily as oil has a lower density than water or soil (oil floats on water). There is an increase in the grading modulus with increasing oil content (Daka 2015), this is because the oil causes grains to stick together, or resist dispersion. There is also a decrease in swelling pressures (Elisha 2012), this is probably because oil is a non-polar fluid so has little interaction with the charged surface of the clay particles, but may prevent water from having this interaction.

17.4 Discussion and Conclusions

Given the above observations it is clear that soil-hydrocarbon systems have complex geotechnical behaviour. As the studies in the literature use different soils and different mixtures of hydrocarbons, extracting comparable data and identifying the causes of each difference in behaviour is challenging. This limits the ability to predict how the properties of a soil might change due to contamination. Alfach and Wilkinson (2020) show, using parameters taking from laboratory tests, the significance of changes to the behavioural properties of the ground in relation to pile foundations. They show an increases in piles settlement, piles bending moment, shear forces and normal forces. These all reduce the factors of safety against ultimate and serviceability limit failures. Increasing the risk to these foundations.

While there is value in systematic reviews of this topic, such as that produced by Rajabi and Sharifipour (2018) as it draws together a large quantity of the literature. Simple correlations with complex data do not necessarily enhance the understanding of the behaviour of these materials. Interpreting the differences in results observed by different authors in these complex systems is challenging.

References

Akinwumi I, Diwa D, Obianigwe N (2014) Effects of crude oil contamination on the index properties, strength and permeability of lateritic clay. Int J Appl Sci Eng Res 3(4):816–824. https://doi.org/10.6088/ijaser.030400007

Alfach MT, Wilkinson S (2020) Effect of crude-oil-contaminated soil on the geotechnical behaviour of piles foundation. Geotech Res 7(2):76–89. https://doi.org/10.1680/jgere.19.00017

Chaudhary A, Singh A (2016) Study of behavioural change in the soil contaminated with petrol. J Civ Environ Eng 6:2 3pp https://doi.org/10.4172/2165-784X.1000217

Clayton JL (1988) Classification and composition of crude oil. In: Magoon LB (ed) Petroleum systems of the United States, US Geological Society Bulletin no 1870. Washington, pp 54–56

Daka MR (2015) Geotechnical properties of oil contaminated soil. MPhil thesis, University of Manchester, Manchester, UK

Elisha AT (2012) Effect of crude oil contamination on the geotechnical properties of soft clay soils of Niger Delta region of Nigeria. Electron J Geotech Eng 17(Bundle M):1157–1162

Hafshejani A, Hajiannia A (2016) Contamination dispersion on the load bearing capacity of in-situ concrete piles in SM soils. Electron J Geotech Eng 21(08):2857–2869

Hewayde E, Abbas M, Kubba Z (2019) Influence of engine oil on geotechnical properties of cohesive soil. Int J Eng Res Technol. 12(1):33–41

Ijimdiya TS (2013) The effects of oil contamination on the consolidation properties of lateritic soil. Dev Appl Oceanic Eng 2(2):53–59

Karkush MO, Kareem ZA (2017) Investigation the impacts of fuel oil on the geotechnical properties of cohesive soil. Eng J 21(4):127–137. https://doi.org/10.4186/ej.2017.21.4.127

Karkush MO, Jihad AG (2020) Studying the geotechnical properties of clayey soil contaminated by Kerosene. In: Karkush M (ed) Key engineering materials, vol 857, pp 383–393

Kermani M, Ebadi T (2012) The effect of oil contamination on the geotechnical properties of fine-grained soils. Soil Sediment Contam 21(5):655–671. https://doi.org/10.1080/15320383.2012.672486

Khamehchiyan M, Charkhabi AM, Tajik M (2007) Effects of crude oil contamination on geotechnical properties of clayey and sandy soils. Eng Geol 89:220–229. https://doi.org/10.1016/j.enggeo.2006.10.009

Khosravi E, Ghasemzadeh H, Sabour MR, Yazdani H (2013) Geotechnical properties of gas oil-contaminated kaolinite. Eng Geol 166:11–16. https://doi.org/10.1016/j.enggeo.2013.08.004

Mohammadi A, Ebadi T, Ahmadi M (2018) Effect of bentonite addition on geotechnical properties of oil-contaminated sandy soil. J Civ Eng and Constr 7(4):153–162. https://doi.org/10.32732/jcec.2018.7.4.153

Nazir AK (2011) Effect of motor oil contamination on geotechnical properties of over consolidated clay. Alexandria Eng J 50(4):331–335. https://doi.org/10.1016/j.aej.2011.05.002

Ota J (2010) The effect of hydrocarbon spills on the geotechnical properties of a clay soil. In: Sarsby R, Meggyes T (ed) Construction for a sustainable environment. CRC Press, London 4pp. https://doi.org/10.1201/9780203856918

Ota JO (2013) The effect of light crude oil contamination on the geotechnical properties of Kaolinite Clay Soil. PhD thesis, Anglia Ruskin University, Cambridge, UK

Rahman Z, Hamzah U, Taha M, Ithnain N, Ahmad N (2010) Influence of soil contamination on geotechnical properties of basaltic residual soil. Am J Appl Sci 7(7):954–961. https://doi.org/10.3844/ajassp.2010.954.961

Rajabi H, Sharifipour M (2018) Geotechnical properties of hydrocarbon-contaminated soils: a comprehensive review. Bull Eng Geol Environ 78:3685–3717. https://doi.org/10.1007/s10064-018-1343-1

Resol HA (1999) Some geotechnical properties of oil contaminated soil. MSc thesis, University of Baghdad, Baghdad, Iraq

Safehian H, Rajabi AM, Ghasemzadeh H (2018) Effect of diesel-contamination on geotechnical properties of illite soil. Eng Geol 241:55–63. https://doi.org/10.1016/j.enggeo.2018.04.020

Talukdar DK, Saikia BD (2013) Effect of crude oil on some consolidation properties of clayey soil. Int J Emerg Technol Adv Eng 3(2):117–120

Tuncan A, Pamukcu S (1992) Geotechnical properties of petroleum and sludge contaminated marine sediments. In: The second international offshore and polar engineering conference, San Francisco, California, USA

Tong L (2008) Study on water-physical and mechanical properties of oil-contaminated soils. Dissertation, Ocean University of China

Zheng X, Zhang J, Zheng T, Liang C, Wang H (2014) A developed technique for measuring water content in oil-contaminated porous media. Environ Earth Sci 71:1349–1356. https://doi.org/10.1007/s12665-013-2541-6

Chapter 18
Behavior of Bentonite Under Nuclear Waste Repository Conditions: Insights from a Thermo-hydro-mechanical Model Based on Hypoplasticity

Gianvito Scaringi, David Mašín, Jan Najser, Tomáš Koudelka, Tomáš Krejčí, and Jaroslav Kruis

Abstract Advanced modeling is fundamental for reproducing non-linear and coupled behaviors in expansive soils, especially in unsaturated and non-isothermal conditions. In this contribution, we provide an overview of the recent developments of a modeling approach based on the theory of hypoplasticity, specialized for expansive clays under a double-structure framework and enhanced with thermo-hydraulic and thermo-mechanical couplings at both structural levels. The model, plugged into a single-element driver—in turn introduced in an in-house finite-element solver—has been calibrated on element-volume experiments on MX-80 and B75 bentonites, which are candidate materials for buffer layers in deep geological repositories of radioactive waste. The so-calibrated parameters were then utilized in simulations of laboratory-scale tests, aiming at reproducing the response and homogenization of buffer layers (compacted bentonite blocks and/or pellets) under realistic combinations of hydration and prolonged heating. The relative simplicity of the hypoplastic formulation—featuring key parameters with clear physical meaning—has proven successful in model simulations, which exhibited a reasonable matching between predicted and experimental trends of saturation, volume change, and swelling pressure development under varying temperature. Within the framework of ongoing collaborative projects, further model improvements are envisaged, with the aim to offer better formulations of microstructural behaviors and improved computational efficiency.

Keywords Nuclear waste disposal · Bentonite · Clay · Modeling · Thermo-hydro-mechanical coupling

G. Scaringi (✉) · D. Mašín · J. Najser
Faculty of Science, Institute of Hydrogeology, Engineering Geology and Applied Geophysics, Charles University, Prague, Czech Republic
e-mail: gianvito.scaringi@natur.cuni.cz

T. Koudelka · T. Krejčí · J. Kruis
Faculty of Civil Engineering, Czech Technical University in Prague, Prague, Czech Republic

S. Wang et al. (eds.), *Engineering Geology for a Habitable Earth: IAEG XIV Congress 2023 Proceedings, Chengdu, China*, Environmental Science and Engineering,
https://doi.org/10.1007/978-981-99-9057-3_18

18.1 Introduction

18.1.1 Thermo-hydro-mechanical Coupling in Expansive Clays

The peculiar nature of bentonite, which is mainly composed of montmorillonite (a type of smectite) with various adsorbed cations (Na^+, Ca^{2+}, etc.), is such that it features a very high surface charge, cation exchange capacity, and specific surface (in the order of hundreds of m^2/g). The resulting ability to adsorb large amounts of water confers it a high compressibility even under low confining stresses, but also extremely low hydraulic conductivity and resistance to shearing. The large magnitude of physico-chemical interactions on the surface of the clay particles compared to their mass results in evident coupled behaviours (especially chemo-mechanical and thermo-hydro-mechanical). Because of this, bentonite clays are particularly sought after in engineering applications in which resistance to hydraulic, thermal, and chemical forcing for long periods of time is necessary for the functioning of the infrastructure in question. This is the case in facilities for the final disposal of spent nuclear fuel and other highly radioactive waste, which are typically constructed deep underground and require reliable and maintenance-free insulation from the surrounding host rock and aquifers (e.g., Ahusborde et al. 2020; Chen et al. 2017; Gens et al. 2010).

It is nowadays well known that many aspects of the hydro-mechanical behavior of smectite-rich soils are strongly influenced by the temperature at which processes occur, even though some responses are more apparent than others (Loche and Scaringi 2023; Scaringi and Loche 2022). Water retention capacity and compressibility of compacted bentonite are strongly temperature-dependent (Villar and Lloret 2004a, 2004b; Villar et al. 2010). The water retention capacity, in particular, does not only vary upon changes in water properties (e.g., viscosity, specific volume), but also upon changes in physico-chemical processes on the particle surface—such as interlayer water moving into the macropores and vice versa—which depend on temperature. These processes could be invoked to explain the decrease in suction occurring upon heating in absence of changes in water content (Villar and Lloret 2004a). Similarly, they could explain the lower swelling ability under free-swelling condition and, dually, the lower swelling pressure under constant-volume condition, exhibited as temperature increases (Romero et al. 2005; Tang et al. 2008; Villar and Lloret 2004a). The shear strength of smectite-rich soils is also influenced by temperature. This aspect has, so far, received much less attention but could be critical in engineering geological applications such as slope stability assessments and energy piles (Loche and Scaringi 2023; Loche et al. 2022; Scaringi and Loche 2022).

18.1.2 Thermo-hydro-mechanical Modeling

The complexity of the physico-chemical processes occurring in expansive clays at various scales—from molecular to particle and aggregate levels—is stunning. The most evident coupled hydro-mechanical responses are generally well reproduced by constitutive models (especially phenomenological, macroscopic models) for saturated clays. In unsaturated, compacted expansive clays, additional complexities arise, and more advanced approaches are usually needed in order to achieve accurate results. Some successful models make use of a double-structure framework to successfully reproduce volume change responses and changes in water retention and strength that would otherwise remain unexplained (Ahusborde et al. 2020; D'onza et al. 2011; Gens and Alonso 1992; Mašín 2013a). Even more advanced models are needed in order to incorporate, into this framework, the modeling of stress–strain, fluid pressure, and flow changes originating from chemical, electrical, and thermal gradients and, vice versa, changes in chemical, electrical, and thermal fields induced by changes in mechanical and hydraulic gradients (Deng et al. 2020; Scaringi and Loche 2022; Zhao et al. 2022; Zhu et al. 2022).

In deep (hundreds of meters below ground) geological repositories of radioactive waste, the performance of bentonite barrier layers needs to be ensured for thousands of years. This requirement has promoted a great deal of research, and many modeling solutions have been proposed to reproduce and predict the coupled responses in bentonites under a variety of initial and boundary conditions (Cui and Tang 2013; Gajo et al. 2002; Gens et al. 2010; Ma et al. 2017; Mašín 2017). The work of Mašín, in particular, has focused on modeling the main thermo-hydro-mechanical (THM) responses using the constitutive framework of soil hypoplasticity (Mašín 2013a, 2014, 2017). The performance and capabilities of hypoplasticity in isothermal settings have been well validated in both coarse- and fine-grained soils (e.g., Mašín 2013a, 2014). On the other hand, model development and performance under non-isothermal conditions have lagged behind (Ahusborde et al. 2020). In this contribution, we report on some results of model calibration at the scale of the laboratory sample within the THM (unsaturated and non-isothermal) hypoplasticity framework and discuss outstanding challenges.

18.2 Materials and Methods

18.2.1 The B75 Bentonite

The B75 bentonite is extracted at the *Černý vrch* deposit in the northwest of Czech Republic. It exhibits intermediate properties compared to the extremely active MX80 bentonite and the less active GMZ bentonite (Bernachy-Barbe et al. 2020; Chen et al. 2018). It is mainly composed of Ca-Mg-montmorillonite (~60% in dry weight) and exhibits values of the plastic limit, liquid limit, and specific gravity (G_s) of ~

65%, ~ 230%, and ~ 2.85, respectively, which vary slightly according to the batch. The B75 bentonite has a total surface area of ~ 400 m^2/g, an external surface area (evaluated by BET) of ~ 75 m^2/g (Brázda and Červinka 2016), and a cation exchange capacity (CEC) for the main cations of: 36.9, 26.8, 6.6, 3.0, and < 0.5 mmol/100 g for Ca^{2+}, Mg^{2+}, Na^+, K^+, and H^+, respectively (Sun et al. 2017). A swelling pressure (p) of ~ 3 MPa was evaluated in samples compacted at a dry density $\rho_d = 1.5$ t/m^3 under room temperature (~20 °C) and hygroscopic water content (~7.5%) conditions. A thermal conductivity $\kappa = 0.4$ W/(mK) was evaluated in similar conditions ($\rho_d = 1.45$ t/m^3). Under room temperature conditions and $\rho_d = 1.5$ t/m^3, a saturated hydraulic conductivity $k = 5 \cdot 10^{-12}$ m/s also was evaluated in lab-scale samples (i.e., only considering permeability by porosity, in absence of discontinuities acting as preferential flow pathways). Additional evaluations of the relationships between hydraulic conductivity and dry density and swelling pressure and dry density indicate that k can decrease by less than one order of magnitude with ρ_d increasing from 1.2 to 1.75 t/m^3, whereas p can increase by more than one order of magnitude within the same ρ_d range (Hausmannová and Vašíček 2014). Additional material characterization can be found in Sun et al. (2020, 2022).

Notably, Sun et al. (2018c) found two distinct porosity domains in compacted B75 bentonite samples in porosimetry tests with mercury intrusion (MIP tests). In particular, the authors observed that the pore size (d_t) denoting the threshold between micro- and macro-porosity decreases as the total suction (s_t) increases: d_t decreased from 0.30 to 0.07 μm when s_t increased from 3 to 300 MPa. Interestingly, externally applied compaction (with loads in the order of MPa) did not produce significant effects on the macropore size. Instead, both s_t and compaction were found to affect the macropores. These did not fill with water until low values of total suction (~3 MPa), whereas the micropores became saturated under still large values of s_t (10–60 MPa). Further, volume changes induced during cycles of wetting and drying were completely reversible at the macrostructural level but only partially reversible at the macrostructural level (with net shrinkage or welling depending on initial conditions).The work of Sun et al. (2018a) focused on how the water retention behavior depends on temperature in absence of constraints (free-swelling, unconfined condition). The authors conducted experiments on B75 material at T = 20–80 °C, various ρ_d values, and s_t = 4–300 MPa and employed the vapor equilibrium technique (VET). They demonstrated a decrease in water retention capacity upon temperature increasing, and this was more evident along drying than along wetting paths, and was mostly insensitive to ρ_d.

18.2.2 Hypoplastic Modeling for Expansive Clays

Here, we relied on the constitutive formulation presented in Mašín (2017), which was built by including thermal couplings to an earlier isothermal formulation (Mašín 2013a). The double-structure approach is used, following experimental evidence showing (mainly via porosimetry) two structural levels in compacted expansive soils.

These level are the *macrostructure* (M), that is, the fabric of clay aggregates and their macropores, and the *microstructure* (m), which represents the internal arrangement of the aggregates and their micropores. These aggregates are made of stacked clay platelets. Hydraulic (H) and mechanical (G) behaviors are modelled individually at the two levels, with G–H coupling achieved at both levels and indicated as $G^M H^M$ and $G^m H^m$ for the macrostructure and microstructure, respectively. Connection between the levels is guaranteed by a so-called double-structure coupling function, f_m, which takes the level of compaction into account: $f_m = 1$ means that a volume change of the aggregates results in an equal global volumetric response; $f_m = 0$ implies that volume changes at the aggregate level are realized upon an opposite response of the macropores, so that no volumetric response occurs globally (Mašín 2017). Specifically, the H^M component of the model considers the variable saturation by means of a hysteretic water retention curve expressed in terms of the void ratio (e) (Mašín 2013a). The G^M component includes an explicit formulation of the asymptotic state boundary surface and accounts for stiffness anisotropy (Mašín 2013b, 2014). As for the microstructure, its representation is simplified in that the H^m component considers permanent full saturation, which is reasonable as long as $s_t <$ ~ 100 MPa (Mašín and Khalili 2015), and the G^m component only considers reversible and isotropic behavior.

In the model, temperature can influence the hydraulic as well as the mechanical components of the macrostructure and the microstructure (Mašín 2019). The normal compression line (NCL) can shift its position upon temperature variations. Therefore, the G^M component incorporates a parallel shift of the NCL, which is in the direction of lower void ratios as temperature increases. The effect of temperature on the microstructure is more complex. The combination of changes in hydration condition, dielectric constant, basal spacing, and diffuse double layer (DDL) thickness can produce net swelling or shrinkage (swelling pressure increase or decrease if volume changes are forbidden), or even no effects under the same temperature variation (Romero et al. 2005; Tang et al. 2008; Villar and Lloret 2004a). Furthermore, the effect of temperature on the macroscopic hydraulic response H^M is thought to only depend on changes in the water's surface tension, whereas contact angle variations at the aggregate-water interface are assumed to be negligible. At the microstructural level, volume changes induced by temperature variations can occur (G^m), and these can affect the microstructural water retention capacity (H^m).

The model's general formulation, in the rate form, can be expressed by the following tensorial equation

$$\dot{\boldsymbol{\sigma}}^M = f_s\left[\mathcal{L} : \left(\dot{\boldsymbol{\epsilon}} - f_m \dot{\boldsymbol{\epsilon}}^m\right) + f_d \boldsymbol{N} \left\| \dot{\boldsymbol{\epsilon}} - f_m \dot{\boldsymbol{\epsilon}}^m \right\|\right] + f_u \boldsymbol{H}_s$$

in which $\dot{\boldsymbol{\sigma}}^M$ represent the macrostructural effective stress rate, $\dot{\boldsymbol{\epsilon}}$ is the global Euler stretching tensor, $\dot{\boldsymbol{\epsilon}}^m$ is the microstructural strain rate, f_m is the already mentioned double-structure coupling factor, $\mathcal{L}$ and $\boldsymbol{N}$ are the 4th and 2nd order hypoplastic tensors, $\boldsymbol{H}_s$ defines the wetting-induced collapse, and f_s, f_d, and f_u are scalar factors for barotropy, pyknotropy, and collapsible behaviors (Mašín 2017). The model requires a number of parameters, which are defined in Table 18.1. The model's mathematical formulation is explained in detail in Mašín (2019).

Implementation of the model was conducted in C++ environment and included in the single-element driver TRIAX, which is freely accessible through the *Soilmodels* project (https://soilmodels.com/triax/). The algorithm for solving the model at the single-element level was presented by Janda and Mašín (2017) and subsequently updated for double-structure THM calculations (Mašín 2017). Finally, the model's code was incorporated in the finite-element solver SIFEL (Koudelka et al. 2017, 2018), also freely available (http://mech.fsv.cvut.cz/~sifel/).

Table 18.1 Definition of parameters in the THM hypoplastic model for expansive clays from Mašín (2017)

Symbol	Definition
φ_c	Macrostructural critical state friction angle
λ^*	Slope of the isotropic NCL (INCL) in $\ln(p^M/p_r) - \ln(1+e)$ coordinates, where p^M is the macrostructural effective mean stress, and $p_r = 1$ kPa is a reference stress
κ^*	Macrostructural volume strain upon p^M unloading
N	Position of the INCL in $\ln(p^M/p_r) - \ln(1+e)$ coordinates
υ	Stiffness in shear
n_s	Effect of s on the position of the INCL
l_s	Effect of s on the slope of the INCL
n_T	Effect of T on the position of the INCL
l_T	Effect of T on the slope of the INCL
m	Parameter controlling how wetting-induced and heating-induced compactions are affected by the distance from the state boundary surface, and how wetting–drying and heating–cooling cycles affect e through f_m
α_s	Effect of T on the microstructural volume strain
κ_m	Effect of p^m (the microstructural effective mean stress) on the microstructural volume strain
s_r	Reference value of suction for e^m
e^m_{r0}	Reference value of e^m at zero total stress, reference suction s_r, and reference temperature T_r
c_{sh}	Value of f_m in compression
e^M_0	Reference e^M at the reference air-entry value s_{e0}
s_{e0}	Reference air-entry value of s at e^M_0
T_r	Reference value of T
a	Parameter controlling how T affects s_{e0}
b	Parameter controlling how T affects s_{e0}
a_e	Ratio between air-entry and air-expulsion s values
λ_{p0}	Slope of the main drying-wetting curve in $\ln S^M_r - \ln s$ coordinates, where S^M_r is the macrostructural degree of saturation, considered equal to Bishop's effective stress parameter χ

18.2.3 Experiments

We relied upon a variety of experimental results to calibrate the model. However, since some key conditions had not been investigated using B75 bentonite, we also included data on MX-80 bentonite owing to likely similarities in behaviors. The latter were presented in Mašín (2013a, 2017) and employed in Janda and Mašín (2017) to obtain a calibration of the model for the MX-80 material. The experiments used for model calibration are briefly introduced in the following paragraphs.

Water adsorption under free-swelling condition. Powdered material was placed into 50-mm diameter molds at hygroscopic water content ($w_0 = 10\%$) and submitted to uniaxial compaction. Samples with a height ~ 10 mm were obtained, with dry density values of 1.27, 1.60, and 1.90 t/m^3 (Sun et al. 2018c). The so-obtained samples were then dried in oven for a whole day at 105 °C. VET was utilized to permit water adsorption after the samples were placed in sealed containers in a fixed-temperature environment (20 °C). The conditions in the containers were chosen so that total suction values of 3–300 MPa could be maintained.

Swelling pressure and hydraulic conductivity. In these experiments (Hausmannová and Vašíček 2014), powdered bentonite was placed in molds with 30 mm diameter and compacted uniaxially to achieve sample heights of ~ 10 mm with dry density values in the range of 1.20–1.75 t/m^3. Subsequently, the samples were installed into a device designed ad hoc to allow simultaneous measurement of swelling pressure and hydraulic conductivity. In particular, the samples were installed into a stainless steel cell between porous stones and were kept at constant volume by axial confinement. Through the bottom porous stone, distilled water was provided at a pressure of 1 MPa to displace air and thereby fully saturate the samples thanks to direct contact with the atmosphere through the top porous stone. In the experiments, the injection pressure was increased up to 6 MPa while measuring the water flow and, at the same time, the force of the sample against the (fixed) top plate. The latter corresponded to the swelling pressure after correction for the injection pressure.

Constant-load swelling. Powdered bentonite at hygroscopic water content underwent uniaxial compaction to achieve 8–10 mm high samples in 50 mm diameter molds at values of dry density of 1.24, 1.80, and 1.95 t/m^3 (Sun et al. 2018b). These samples were then submitted to oedometer tests with loading up to 7.8 MPa prior to being exposed to distilled water. Swelling was permitted without altering the confining normal stress until equilibrium conditions were achieved.

Saturated oedometer compression. Some of the samples described in the previous paragraph were subjected to unloading and reloading up to 4–7.5 MPa, cyclically. Each subsequent step was carried out only after attaining complete primary consolidation.

Water adsorption–desorption under free swelling condition. These tests were carried out on MX-80 bentonite samples, prepared at hygroscopic water content ($w_0 = 17\%$), by Tang and Cui (2005). The material was first installed in a sealed desiccator with controlled temperature (20 °C) and left to reach equilibrium with a saturated K_2CO_3 solution (producing a total suction of 110 MPa). The material was subsequently uniaxially compacted and placed again in the desiccator to equilibrate once again. In the end, 20.2 mm diameter samples were obtained, with heights of 8.7 mm and dry density of 1.68 t/m^3. Subsequently, each sample was assigned to a different desiccator in which a set value of total suction was imposed by fixing the relative humidity (range: 26–97%) through vapor transfer with selected saturated salt solutions at T = 20–80 °C.

Heating–cooling isotropic compression. These tests were conducted by Tang et al. (2008). The authors compressed MX-80 bentonite samples isotropically under mean net stress values of 0.1 or 5 MPa and three values of total suction, namely 9, 39, and 139 MPa at temperatures of 25–80 °C. More specifically, the authors used 80 mm diameter samples which had heights of 10–15 mm. The samples were exposed to air with controlled relative humidity and their temperature was controlled via a temperature-controlled bath which was heated and cooled cyclically. Compression was ensured by compressed air imposing stresses via a neoprene interface and was kept constant (together with the total suction) throughout the experiments.

18.2.4 Calibration

For equilibrium states, calibration was done in TRIAX environment, while SIFEL was used to evaluate the temporal evolution in some experiments. Error metrics and optimization algorithm were not used at this stage, preferring expert judgement as a trade-off between the potential (but perhaps not significant) increase in accuracy and the challenge in selecting suitable control points in order to build quantitative metrics. Expert judgement also ensured interpretability and reasonable choices of values.

In fact, for φ_c and υ (see Table 18.1 for all definitions), reasonable values were taken, while for s_r, e_0^M, and T_r, reference values were taken so that they remained within the relevant experimental ranges. Mašín (2017) also noted that these values can be chosen arbitrarily together with e_{r0}^m. The latter can also be fine-tuned to improve the reproduction of the water retention curves. Finally, the value of κ_m was taken so as to suitably model the swelling behavior.

MX-80 bentonite experimental results (isotropic compression tests) were exploited to find suitable values to the basic hypoplastic parameters λ^* and κ^*. Further, corrections were made to the values of N, n_s, n_T, and l_T so as to be able to predict the intrinsic normal-compression line (INCL) appropriately, together with volumetric changes induced by temperature variations. In order to accurately predict the compression behavior of the B75 bentonite (in oedometer confinement conditions), some further corrections were carried out.

The value of α_s was selected to interpret heating tests under high-suction condition, whereas the values of s_{e0} and a_e were just assumed owing to the fact that their influence on the behavior in such condition is very limited. Finally, reasonable values for a and b also were assumed, following the simplifying hypothesis that the role of temperature in the water retention behavior only stems from changes in water's surface tension (Mašín 2017).

18.3 Results

Following calibration—which was in part an iterative process—and reasonable assumptions, a set of values was chosen for the B75 bentonite, as reported in Table 18.2. The table also displays the values obtained for the MX-80 bentonite, for comparison. Notably, similar values of the basic parameters of hypoplasticity (φ_c, λ^*, κ^*, N, υ, n_s, l_s, m) were taken for the two materials because the experiments highlighted similar critical-state friction angles (φ_c) and a similar dependence between the suction (s) and the INCL via the parameters n_s and l_s.. However, the B75 exhibited a steeper INCL (via the parameter λ^*) as well as larger volumetric strains during unloading (as determined by the parameter κ^*). To adapt the MX-80 calibration to the B75 bentonite, important corrections were also made for the parameter N, , which sets the position of the INCL. We also assigned a larger shear stiffness (υ) to the B75 but also a smaller value of m compared to that of the MX-80 bentonite.

The values of the parameters defining the clay's thermal response, namely n_T, l_T, α_s, T_r, a, and b, were taken for the B75 bentonite equal to the values set for the MX-80. This seemed a reasonable choice in absence of results of non-isothermal experiments on the B75, particularly for the values of a and b, following the simplification, discussed above, that thermal effects on the water retention curve are solely resulting from temperature-dependent changes in water surface tension (Villar and Lloret 2004a), neglecting the consequences of the changing solid-water interface contact angle at the aggregate scale and other possible effects. Concerning the reference temperature T_r, , its value could be chosen arbitrarily (similar to s_r, e_0^M , and e_{r0}^m) with the sole objective to match the ranges experienced in the calibration tests. For the temperature-dependent changes to the position and slope of the INCL (parameters n_T and l_T) we did not have sufficient information to establish whether there should be differences between the MX-80 and the B75. Further, the value of the microstructural parameter α_s also was taken equal for both materials owing to an expected similarity in the response of their microstructures, modeled in both cases as isotropic and saturated, and in consideration of their analogous mineralogy. Similarly, the double-structure coupling factor during compression (c_{sh}) was taken equal for both materials. Finally, the parameter κ_m of the microstructure was changed from the default value of 0.10 to improve the prediction of the swelling behavior. The values of the other parameters controlling the hydro-mechanical behavior were chosen as follows: the air-entry value of suction (s_{e0}) was selected along the water retention curve at the reference macrostructural void ratio (e_0^M) evaluated at T_r; the slope of the main drying-wetting curve (λ_{p0}) in the $\ln S_r^M - \ln s$ plane was determined according to water retention tests.

Table 18.2 Parameter values of the THM hypoplastic model calibrated for the MX-80 and B75 bentonites

Parameter	Unit	MX-80	B75
φ_c	°	25	25
λ^*	–	0.230	0.130
κ^*	–	0.055	0.060
N	–	2.70	1.73
υ	–	0.25	0.25
n_s	–	0.020	0.012
l_s	–	0	−0.0050
n_T	–	−0.07	−0.07
l_T	–	0	0
m	–	0.2	1
α_s	1/K	0.00015	0.00015
κ_m	–	0.12	0.07
s_r	kPa	2700	2000
e_{r0}^m	–	0.90	0.45
c_{sh}	–	0.002	0.002
s_{e0}	kPa	1000	2700
e_0^M	–	0.10	0.50
T_r	K	294	294
a	N/m	0.118	0.118
b	N/(mK)	−0.000154	−0.000154
a_e	–	0.50	0.75
λ_{p0}	–	0.55	0.70

Experimental and simulation results relative to free-swelling water adsorption experiments at room temperature are displayed in Fig. 18.1. We can see that the model correctly grasps the role of the initial compaction level, which is noticeable in terms of the S_r against s_t curve (Fig. 18.1a), whereas it is negligible in terms of w (Fig. 18.1b). In particular, the model is able to reproduce the material response when the suction value is rather high (> 20 MPa), that is, for values of the degree of saturation below 50% under high values of dry density ($\rho_d = 1.6, 1.9$ t/m^3). However, the model underestimates the degree of saturation for low-compaction samples, i.e., under $10 < s_t < 100$ MPa, and predicts rapid saturation upon lower s_t values, thereby overestimating the measured values. However, it is worth noting that good predictions were nevertheless achieved for the water content also under very low values of suction (< 500 kPa), as shown in Fig. 18.1b.

Results of swelling pressure experiments are displayed in Fig. 18.2a, while results of swelling tests under constant confining stress are reported in Fig. 18.2b for samples fabricated at different values of initial dry density. We can see (Fig. 18.2a) that the equilibrium value of the swelling pressure is predicted correctly by the model, even

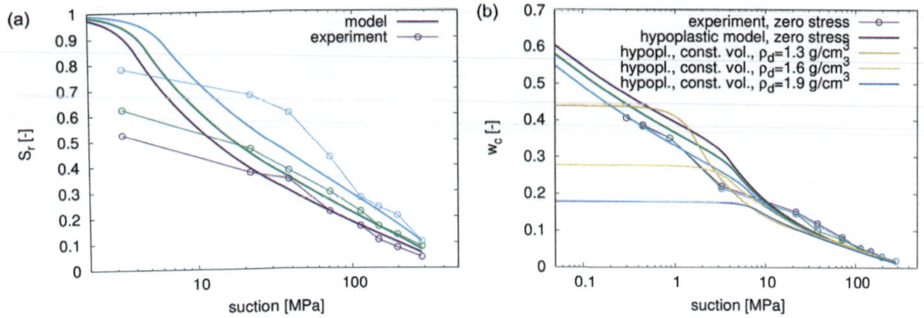

Fig. 18.1 Numerical simulation of free-swelling water adsorption tests performed on B75 bentonite samples compacted to dry density values of 1.2, 1.6, and 1.9 kg/m^3. Degree of saturation (**a**) and water content (**b**) versus total suction

though some overestimation occurs for initial void ratios of $0.7 < e_0 < 1.0$. Under rather high loading (>0.2 MPa), the swelling response is captured well, yet it is quite underestimated under weaker loading (<0.2 MPa). This is not surprising considering that the microstructure was assumed, in the model, to respond as an isotropic elastic material. In reality, an anisotropic response upon anisotropic loading is expected.

Because the hydraulic gradient during the experiments was set and monitored, it was also possible to determine the samples' hydraulic conductivity, which is not a constitutive parameter of the model (which has a single-element formulation) nor it is coupled to any of them. Yet, the hydraulic conductivity is important at the finite-element level as it controls time-dependent responses involving volumetric changes. Thus, in Fig. 18.3, we reported how the swelling pressure develops over time and we modeled the process using SIFEL. In particular, the figure tracks the behavior of samples prepared at three distinct initial dry densities upon injection of water at 1 MPa from one of the sample bases, leaving the other at atmospheric pressure. In order for the simulations to match the experimental results, a dry density-dependent hydraulic conductivity had to be assumed.

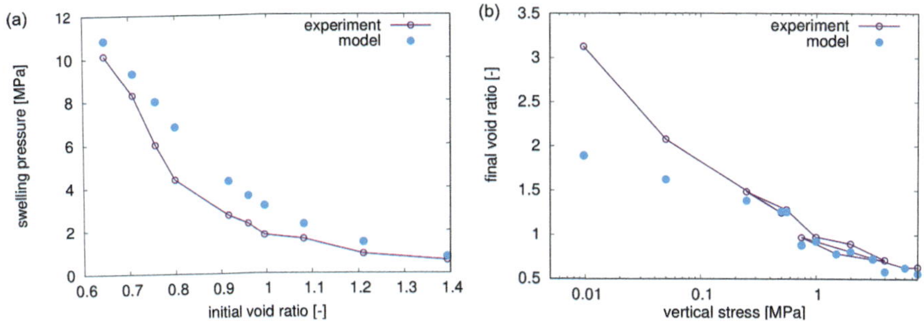

Fig. 18.2 Equilibrium values of swelling pressure (**a**) and swelling (**b**) in experiments on B75 bentonite samples under different compaction levels (initial void ratios)

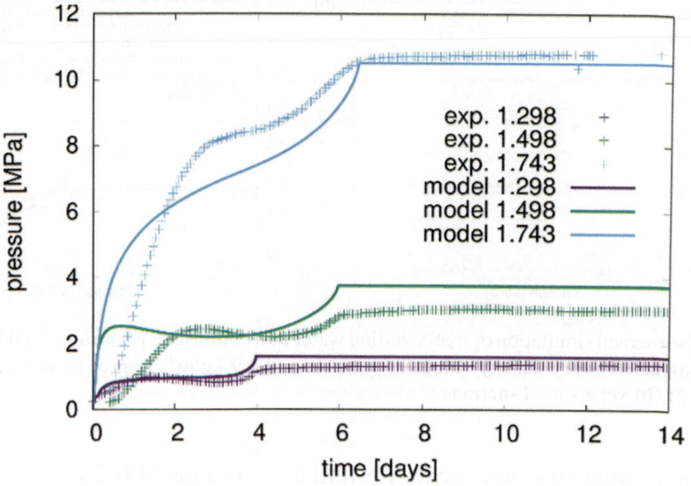

Fig. 18.3 Evolution of swelling pressure in samples compacted at initial dry density values of 1.298, 1.498, and 1.743 t/m^3 under an injection pressure of 1 MPa from the sample bottom (and sample top at atmospheric pressure)

In Fig. 18.4, we reported the results of oedometric loading–unloading experiments performed on saturated materials. Here, model calibration was performed so as to capture the compression line very accurately. Nonetheless, the model did not perform as well upon unloading, likely because of the already described limitations in the microstructural model (elastic and isotropic). Therefore, by assuming oedometric conditions in the model, we underpredict the swelling; however, by assuming isotropic swelling, we can predict the void ratio adequately.

Non-isothermal experiments carried out using MX-80 bentonite material are discussed in detail elsewhere (Tang and Cui 2005; Tang et al. 2008), and so are the results of model calibration (Mašín 2017). Owing to space constraints, here we only report a striking example (Fig. 18.5), in which cyclic heating–cooling under a set isotropic confinement (0.1 MPa) can cause different responses (either reversible swelling or irreversible shrinkage) depending on suction.

18.4 Conclusion and Perspectives

In this paper, we described the basic elements of a hypoplastic model specialized for expansive clays that uses a double-structure framework and accounts for thermo-hydro-mechanical coupling at both structural levels. We performed a model calibration using a variety of experimental results obtained on compacted samples of the Czech B75 bentonite, complemented with selected results obtained on MX-80 bentonite. We showed that the model was capable of reproducing responses

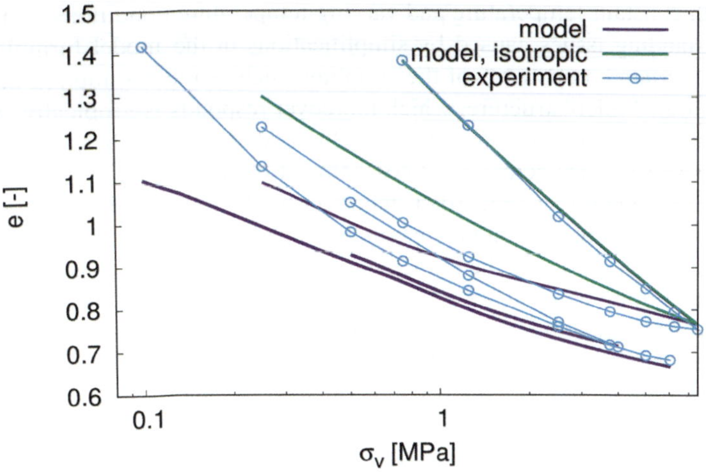

Fig. 18.4 Consolidation tests in oedometer for saturated B75 samples, which were first compacted in ambient moisture, then loaded to a comparative small level, saturated under constant load, and finally loaded-unloaded while maintained saturated

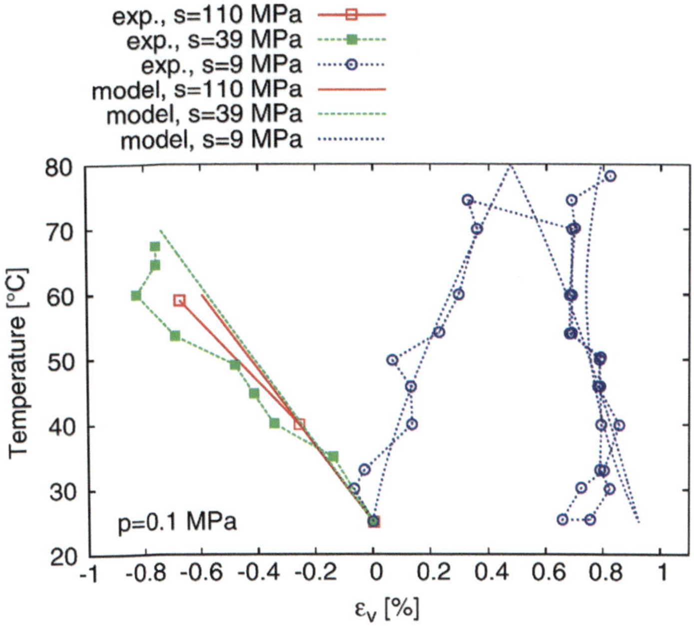

Fig. 18.5 Heating–cooling-induced volumetric strain under a total isotropic stress of 0.1 MPa and constant total suction. Data are from Tang et al. (2008) and simulation results are from Mašín (2017)

under both constant-temperature and varying-temperature conditions. Yet, we identified outstanding issues caused by simplifications in the model formulation with respect to certain components of the coupling, such as the assumption of a permanently saturated microstructure, which moreover responds isotropically even under anisotropic loading, or the assumption of a thermal response only related to changes in water viscosity, neglecting physico-chemical dynamics at the level of the clay surface. Consequently, we recorded imperfect predictions, for instance in terms of an underprediction of swelling under small confining stresses.

Further, we incorporated the implementation in single-element mode into SIFEL, our in-house finite-element solver, in order to predict stress–strain and hydraulic responses of samples during testing (time-dependent response). We are actively working on simulating full-scale tests, such as mock-up tests in which the solicitations experienced by bentonite barriers surrounding cannisters containing radioactive waste are reproduced (BEACON, EURAD, and TAČR projects: see acknowledgements). Additional challenges are brought, however, by this upscaling operation, especially in relation to the complexity of water flows and bentonite homogenization (while heating patterns are typically well captured), and to interfacial friction arising at bimaterial contacts that are hardly accounted for. The use, in some large-scale experiments, of bentonite pellets rather than powder or compacted blocks also introduces challenges owing to the additional structure level above that of the clay aggregates. For this, a triple-structure model formulation could be considered, or the use of equivalent macroscopic parameters accounting for both pellet and macrostructural characteristics. Especially within the framework of the EURAD project, model improvements are currently being studied to overcome the aforementioned limitations.

Acknowledgements The authors acknowledge support by the Euratom research and training programme 2014–2018, grant agreement No. 745942 (project BEACON—Bentonite Mechanical Evolution), by the European Union's Horizon 2020 research and innovation programme, grant agreement No. 847593 (project EURAD), and by grants Nos. TK01010063 and TK01030031 of the Technological Agency of the Czech Republic (TAČR). G. Scaringi also acknowledges support by the Czech Science Foundation (GAČR grant No. GJ 20-28853Y) and the Fund for international mobility of researchers at Charles University (MSCA-IF IV; Project No. CZ.02.2.69/0.0/0.0/20_079/0017987).

References

Ahusborde E, Amaziane B, Baksay A, Bator G, Becker D, Bednar A, Beres M, Blaheta R, Böhti Z, Bracke G (2020) State of the art report in the fields of numerical analysis and scientific computing. Final version as of 16/02/2020 deliverable D4. 1 of the HORIZON 2020 project EURAD

Bernachy-Barbe F, Conil N, Guillot W, Talandier J (2020) Observed heterogeneities after hydration of MX-80 bentonite under pellet/powder form. Appl Clay Sci 189:105542. https://doi.org/10.1016/j.clay.2020.105542

Brázda L, Červinka R (2016) Determination of specific surface area of clay minerals by EGME method. In: Proceedings from the second workshop, final conference of the BELBaR project. Berlin, Germany. http://www.Skb.Se/Belbar/Wp-Content/Uploads/2016/04/D6.14-Proceedings_BELBaR_2016_with_EndUserReviewBoardEvaluation.Pdf

Chen Z-G, Tang C-S, Shen Z, Liu Y-M, Shi B (2017) The geotechnical properties of GMZ buffer/backfill material used in high-level radioactive nuclear waste geological repository: a review. Environ Earth Sci 76:1–16. https://doi.org/10.1007/s12665-017-6580-2

Chen Y-G, Dong X-X, Zhang X-D, Ye W-M, Cui Y-J (2018) Combined thermal and saline effects on the swelling pressure of densely compacted GMZ bentonite. Appl Clay Sci 166:318–326. https://doi.org/10.1016/j.clay.2018.10.001

Cui Y-J, Tang AM (2013) On the chemo-thermo-hydro-mechanical behaviour of geological and engineered barriers. J Rock Mech Geotech Eng 5:169–178

Deng Y, Yan S, Scaringi G, Liu W, He S (2020) An empirical power density-based friction law and its implications for coherent landslide mobility. Geophys Res Lett 47:e2020GL087581. https://doi.org/10.1029/2020GL087581

D'onza F, Gallipoli D, Wheeler S, Casini F, Vaunat J, Khalili N, Laloui L, Mancuso C, MAšíN D, Nuth M, Pereira JM, Vassallo R (2011) Benchmark of constitutive models for unsaturated soils. Géotechnique 61:283–302. https://doi.org/10.1680/geot.2011.61.4.283

Gajo A, Loret B, Hueckel T (2002) Electro-chemo-mechanical couplings in saturated porous media: elastic–plastic behaviour of heteroionic expansive clays. Int J Solids Struct 39:4327–4362

Gens A, Alonso EE (1992) A framework for the behaviour of unsaturated expansive clays. Can Geotech J 29:1013–1032

Gens A, Guimarães do LN, Olivella S, Sánchez M (2010) Modelling thermo-hydro-mechano-chemical interactions for nuclear waste disposal. J Rock Mech Geotech Eng 2:97–102

Hausmannová L, Vašíček R (2014) Measuring hydraulic conductivity and swelling pressure under high hydraulic gradients. Geol Soc Lond Spec Publ 400:293–301. https://doi.org/10.1144/SP400.36

Janda T, Mašín D (2017) General method for simulating laboratory tests with constitutive models for geomechanics. Int J Numer Anal Meth Geomech 41:304–312

Koudelka T, Krejčí T, Kruis J (2017) Coupled hydro-mechanical model for expansive clays. AIP Conf Proc 1863:290008. https://doi.org/10.1063/1.4992445

Koudelka T, Krejčí T, Kruis J (2018) Time integration of hypoplastic model for expansive clays. AIP Conf Proc 1978:080011. https://doi.org/10.1063/1.5043736

Loche M, Scaringi G, Yunus AP, Catani F, Tanyaş H, Frodella W, Fan X, Lombardo L (2022) Surface temperature controls the pattern of post-earthquake landslide activity. Sci Rep 12:988. https://doi.org/10.1038/s41598-022-04992-8

Loche M, Scaringi G (2023) Temperature and shear-rate effects in two pure clays: possible implications for clay landslides. Results Eng 10:10164

Ma QJ, Ng CWW, Mašín D, Zhou C (2017) An approach for modelling volume change of fine-grained soil subjected to thermal cycles. Can Geotech J 54:896–901

Mašín D (2013a) Double structure hydromechanical coupling formalism and a model for unsaturated expansive clays. Eng Geol 165:73–88

Mašín D (2013b) Clay hypoplasticity with explicitly defined asymptotic states. Acta Geotech 8:481–496

Mašín D (2014) Clay hypoplasticity model including stiffness anisotropy. Géotechnique 64:232–238

Mašín D (2017) Coupled thermohydromechanical double-structure model for expansive soils. J Eng Mech 143:04017067

Mašín D (2019) Modelling of soil behaviour with hypoplasticity: another approach to soil constitutive modelling. Springer, Cham

Mašín D, Khalili N (2015) Swelling phenomena and effective stress in compacted expansive clays. Can Geotech J 53:134–147

Romero E, Villar MV, Lloret A (2005) Thermo-hydro-mechanical behaviour of two heavily overconsolidated clays. Eng Geol 81:255–268

Scaringi G, Loche M (2022) A thermo-hydro-mechanical approach to soil slope stability under climate change. Geomorphology 401:108108. https://doi.org/10.1016/j.geomorph.2022.108108

Sun H, Mašín D, Boháč J (2017) Experimental characterization of retention properties and microstructure of the Czech bentonite B75. In: Proceedings of the 19th international conference on soil mechanics and geotechnical engineering, Seoul. pp 1249–1252

Sun H, Mašín D, Najser J (2018a) Thermal water retention characteristics of compacted Bentonite. In: Hu L, Gu X, Tao J, Zhou A (eds) Proceedings of GeoShanghai 2018 international conference: multi-physics processes in soil mechanics and advances in geotechnical testing. Springer, Singapore, pp 71–78. https://doi.org/10.1007/978-981-13-0095-0_8

Sun H, Mašín D, Najser J (2018b) Investigation of the mechanical behavior of compacted bentonite. In: Proceeding 7th international conference unsaturated soils UNSAT2018 3-582018 Hong Kong

Sun H, Mašín D, Najser J, Neděla V, Navrátilová E (2018c) Bentonite microstructure and saturation evolution in wetting–drying cycles evaluated using ESEM, MIP and WRC measurements. Géotechnique 69:713–726. https://doi.org/10.1680/jgeot.17.P.253

Sun H, Mašín D, Najser J, Scaringi G (2020) Water retention of a bentonite for deep geological radioactive waste repositories: high-temperature experiments and thermodynamic modeling. Eng Geol 269:105549. https://doi.org/10.1016/j.enggeo.2020.105549

Sun H, Scaringi G, Mašín D, Najser J (2022) An experimental investigation on the swelling behavior of compacted B75 bentonite. Eng Geol 296:106452. https://doi.org/10.1016/j.enggeo.2021.106452

Tang A-M, Cui Y-J (2005) Controlling suction by the vapour equilibrium technique at different temperatures and its application in determining the water retention properties of MX80 clay. Can Geotech J 42:287–296. https://doi.org/10.1139/t04-082

Tang A-M, Cui Y-J, Barnel N (2008) Thermo-mechanical behaviour of a compacted swelling clay. Géotechnique 58:45–54

Villar MV, Lloret A (2004a) Influence of temperature on the hydro-mechanical behaviour of a compacted bentonite. Appl Clay Sci 26:337–350

Villar MV, Lloret A (2004b) Temperature influence on the mechanical behaviour of a compacted Bentonite. In: Stephanson O (ed) Elsevier geo-engineering book series, coupled thermo-hydro-mechanical-chemical processes in geo-systems. Elsevier, pp 305–310. https://doi.org/10.1016/S1571-9960(04)80058-2

Villar MV, Gómez-Espina R, Lloret A (2010) Experimental investigation into temperature effect on hydro-mechanical behaviours of bentonite. J Rock Mech Geotech Eng 2:71–78

Zhao N, Cheng X, Whittle AJ (2022) Extended TTS model for thermal and mechanical creep of clay and sand. J. Geotech. Geoenvironmental Eng. 148:04022044. https://doi.org/10.1061/(ASCE)GT.1943-5606.0002798

Zhu H, Whittle AJ, Pellenq RJ-M (2022) Potential of mean force for face-face interactions between pairs of 2:1 clay mineral platelets. Langmuir 38:13065–13074. https://doi.org/10.1021/acs.langmuir.2c01632

Chapter 19
Experimental Investigation of Energy Extraction Power from Heated Granite

Dazhao Lu and Wei Wu

Abstract Geothermal energy has emerged as a promising renewable energy source due to its low carbon footprint and high energy output. However, the efficiency and economic viability of geothermal energy harvesting largely depend on the heat extraction power from the subsurface. Here we carried out a series of water flow experiments through a drilled borehole in hot granite to determine the heat extraction power. The outflow temperature higher than the inflow temperature demonstrates successful heat extraction from the hot granite. The experimental results show that the heat extraction power generally increased with a greater flow rate and a higher surrounding temperature. The flow rate and surrounding temperature have an indirect influence on the heat extraction power due to the non-isothermal nature of water flow. The findings provide insights into the design and optimization of geothermal energy harvesting systems, and further numerical studies are needed to investigate the non-linear variation of heat extraction power.

Keywords Geothermal energy · Energy extraction · Bukit Timah granite · Flow rate

19.1 Introduction

The development of geothermal energy in deep earth has gained significant attention as a potential solution to climate change. Hydrothermal systems based on borehole heat exchangers utilizes geothermal fluid to transfer heat energy (Kong et al. 2014). Enhanced geothermal systems are developed to harness geothermal energy in impermeable hot dry rock through hydraulic fracturing. In this process, high pressure water

D. Lu · W. Wu (✉)
Nanyang Technological University, Singapore, Singapore
e-mail: wu.wei@ntu.edu.sg

D. Lu
e-mail: dazhao.lu@ntu.edu.sg

S. Wang et al. (eds.), *Engineering Geology for a Habitable Earth: IAEG XIV Congress 2023 Proceedings, Chengdu, China*, Environmental Science and Engineering,
https://doi.org/10.1007/978-981-99-9057-3_19

is pumped into the targeted layer to create new fractures and widen existing fractures, allowing the injected water to pass through, heat up, and be extracted in another borehole (Lu 2018). The heat transfer between geothermal fluid and heat producing rock is a complex process that involves thermo-hydro-mechanical-chemical coupling mechanisms (Pandey et al. 2018; Rathnaweera et al. 2020; Taron et al. 2009). The hydro-thermal coupling is known as one of the most crucial subprocess that affects the thermal energy production (Gong et al. 2020).

In the hydro-thermal coupling process, the heat convection and the fluid flow are connected by flow rate. Numerous studies have demonstrated the significance of flow rate as an influential factor in geothermal energy extraction. Analytical studies indicate that the flow rate affects the heat exchange between flowing fluid and borehole wall, showing variations in temperature distribution and heat extraction rate (Cimmino 2015). Sensitivity analyses reveal that a gradual increase in fluid flow rate enhances the economic performance and delays the temperature decrease in the production well (Li et al. 2016). Numerical models of heat extraction from fractured hot dry rock highlight flow rate as a critical parameter affecting the thermal extraction performance, and an increase in flow rate leading to reductions in fluid pressure and thermal extraction rate (Ma et al. 2020). However, fewer physical observations exist on heat extraction from hot dry rock, which provides a comprehensive understanding about geothermal energy harvesting.

This study presents a newly developed experimental setup to investigate the energy harvesting efficiency from a granite specimen at a given temperature under varying flow rates and rock temperatures. We first introduced the experimental setup and methodology. We then showed the experimental results, including the inflow and outflow temperatures and the derived heat extraction power, and discussed the future work to study a non-linear variation of heat extraction power using numerical modelling.

19.2 Method

We controlled water flow through a drilled borehole with 6 mm in diameter in a cylindrical specimen with 50 mm in diameter and 100 mm in height to extract heat from surrounding rock at a given temperature. We used Bukit Timah granite sourced from central Singapore as the rock medium.

In the water flow experimental setup (Fig. 19.1), we placed the drilled specimen between two iron coreholders with internal water pipes. We inserted fibre glass between the specimen and the coreholder to minimize heat loss from the specimen. We sealed the specimen using a shrinkable plastic tube and fixed it on the coreholders using steel wire. We covered the sealed specimen using a mica heater, which is capable of heating up to 250 °C and controlled by a closed-loop temperature control box with temperature fluctuation less than 1 °C. We used the mica heater to control the surrounding temperature during the experiments.

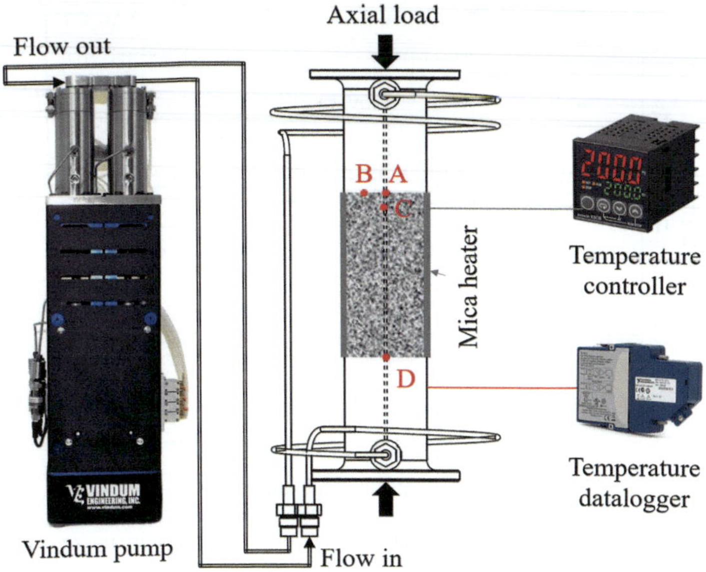

Fig. 19.1 Schematic diagram of water flow experimental setup. Red dots indicate the locations of thermocouples

We then used the Vindum pump to control the continuous flow of distilled water at a given flow rate and four thermocouples to monitor the real-time temperatures of the specimen and water. Two thermocouples were installed next to the borehole ends to monitor the inlet and outlet temperatures of water. The other two thermocouples were attached on the upper end of the specimen and the upper inner wall of the borehole to monitor the temperature of the specimen. The thermocouples with a measurement accuracy of ± 1.5 °C were connected to a LabVIEW data acquisition system at a sampling rate of 10 Hz. During the experiments, the specimen was fixed under an axial pressure of 1 kN, and the room temperature was 22 °C.

We conducted three sets of water flow experiments to investigate the heat extraction power during water flow at 3 flow rates (5, 10 and 15 ml/min) through the heated specimen at 3 surrounding temperatures of 80, 100 and 120 °C (Fig. 19.2). The experimental procedure includes 3 sequential and 3 reverse-sequential steps. For the 3 sequential steps, the specimen was first heated to a desired temperature, and distilled water was pumped into the borehole at a constant flow rate of 5 ml/min. The flow rate was kept for 30 min to ensure the inflow and outflow temperatures stable. The flow rate was subsequently increased to 10 and 15 ml/min and kept for the same duration. After a 30-min break, the specimen temperature was recovered to the initial value. In the 3 reverse-sequential steps, distilled water was first pumped into the borehole at a constant flow rate of 15 ml/min for 30 min, followed by pumping with flow rates of 10 and 5 ml/min and maintained for the same duration. The axial stress, surrounding temperature, and flow rate were fixed during the experiments, and the inflow and outflow temperatures were recorded the data acquisition system.

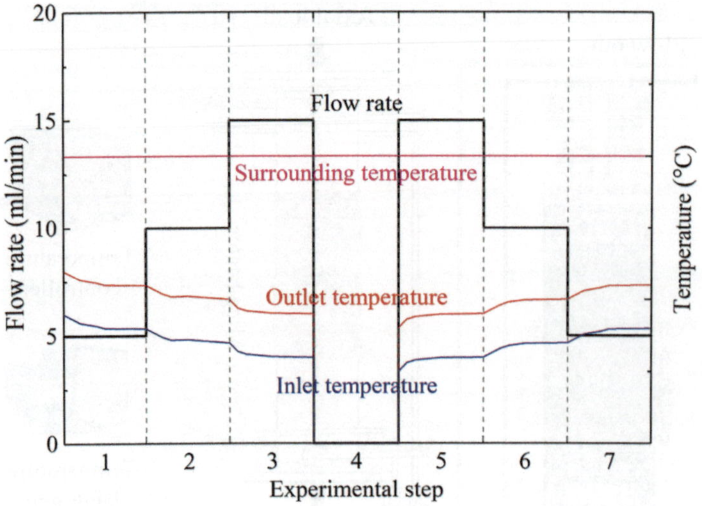

Fig. 19.2 Experimental procedure of water flow experiment. Temperature not to scale

19.3 Results

As shown in Fig. 19.3, the temperature on the upper end of the specimen beyond the borehole (Point A) keeps constant throughout the experiment. The temperature at the inner wall of the borehole (Point B) is recovered between the sequential and reverse-sequential steps. Hence, the sequential and reverse-sequential steps start at the same temperature condition, allowing us to compare the difference between the inflow and outflow temperatures in the sequential and reverse-sequential steps.

Figure 19.4 shows that the outflow temperature higher than the inflow temperature demonstrates successful heat extraction from the hot granite. In the three sets of water flow experiments, the outflow temperature increases with a higher surrounding temperature and decreases with a higher flow rate. As the inflow temperature exhibits similar variations, the difference between the outflow and inlet temperatures should be used to estimate the heat extraction power.

Both the outflow and inlet temperatures approach stable levels at a low surrounding temperature, and the flow rates show minor influences on the temperature fluctuations. The inflow and outflow temperatures are slightly different in the sequential and reverse-sequential steps due to different amounts of extracted heat from previous steps. The average inflow and outflow temperatures of sequential and reverse-sequential steps are used in the calculation of heat extraction power.

We defined the heat extraction power P [W] as the total heat extracted from hot rock ΔH [J] per unit time Δt [s] to evaluate the thermal energy harvesting efficiency:

$$P = \frac{\Delta H}{\Delta t} = c\rho q(T_{out} - T_{in})$$

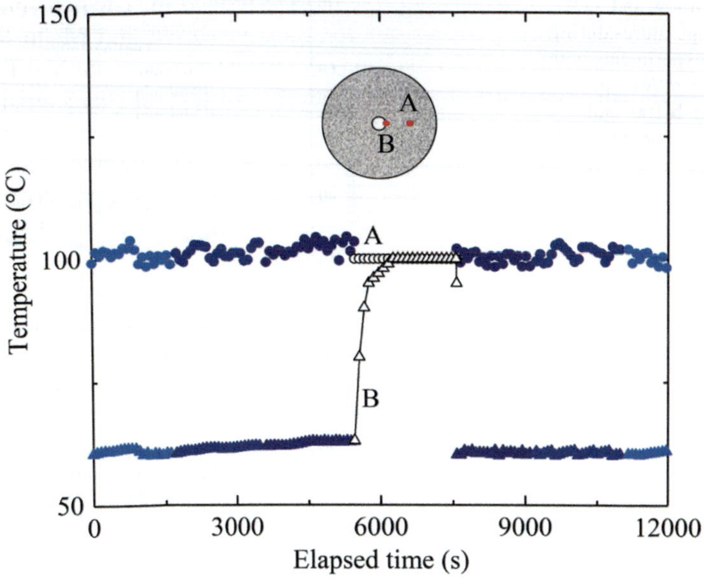

Fig. 19.3 Temperature recovery between sequential and reverse-sequential steps

where c [J/(kg K)] is the specific heat storage of water, ρ [kg/m^3] is the density of water, q [m^3/s] is the volumetric flow rate, T_{out} [K] and T_{in} [K] are the outflow and inlet temperatures, respectively.

Figure 19.5 summarizes the heat extraction power calculated from the experimental data at various combinations of flow rate and surrounding temperature. The heat extraction power is obtained using the mean value of inflow and outflow temperatures, and the error bar is determined using the standard deviation of inflow and outflow temperatures. The heat extraction power generally increases with a greater flow rate and a higher surrounding temperature. The flow rate poses a direct influence on the heat extraction power according to the definition of heat extraction power. The flow rate also affects the difference between the inflow and outflow temperatures (Song et al. 2018) and has an indirect influence on the heat extraction power. The effect of surrounding temperature on the heat extraction power can be complex as water phase changes at 100 °C temperature (Akbar et al. 2016). In this case, the density of water decreases and the difference between inflow and outflow temperatures increases with increasing surrounding temperature (Abdulagatov et al. 2016; Huang et al. 2019). The non-isothermal nature of water flow may lead to a non-linear variation of heat extraction power, which is worthy of further study using numerical models.

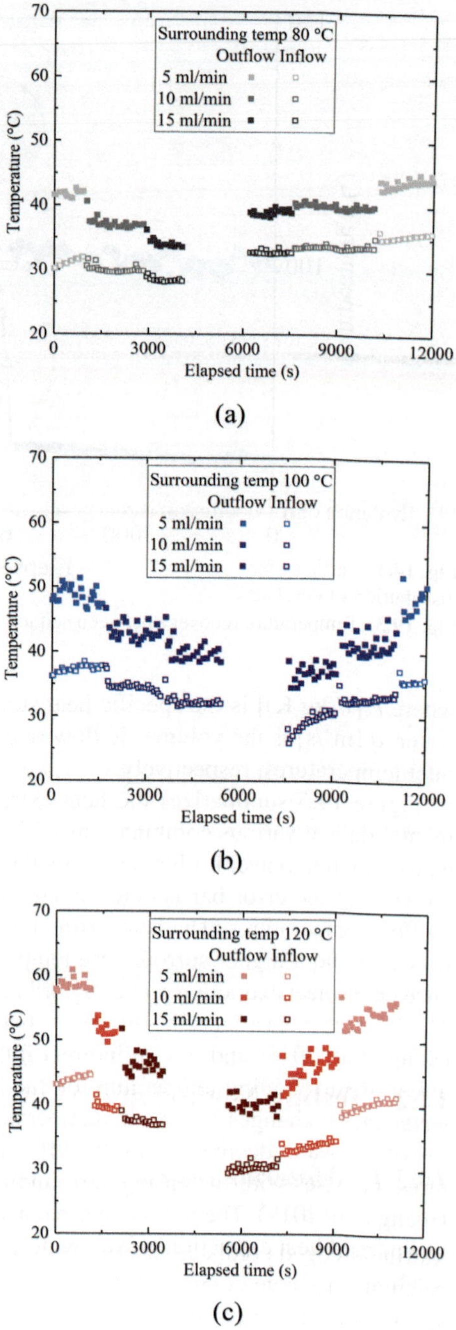

Fig. 19.4 Inflow and outflow temperatures during water flow experiments with 5, 10, and 15 ml/min flow rates at **a** 80, **b** 100, and **c** 120 °C surrounding temperatures

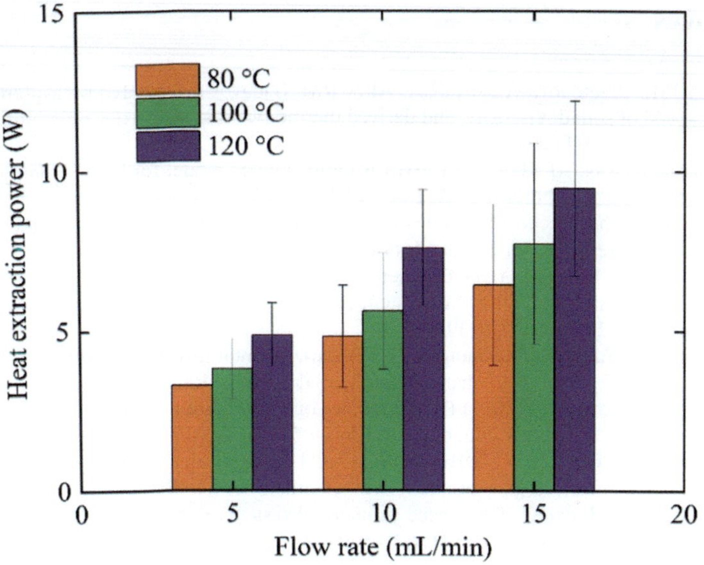

Fig. 19.5 Heat extraction power as a function of flow rate at 80, 100, and 120 °C surrounding temperatures

19.4 Conclusions

We experimentally investigate the heat extraction power using water flow experiments through a drilled borehole in hot granite. The results demonstrate that the outflow temperature higher than the inflow temperature confirms successful heat extraction from the subsurface. The heat extraction power generally increases with a greater flow rate and a higher surrounding temperature. The flow rate and surrounding temperature have an indirect influence on the heat extraction power due to the non-isothermal nature of water flow. The findings suggest that further numerical studies are needed to investigate the non-linear variation of heat extraction power. This experimental system can be regarded as a valuable tool for examining the heat extraction power and investigating other factors that impact energy harvesting. Furthermore, this experimental system can be used to explore other factors that impact heat harvesting. For example, modifying the borehole filler can significantly influence the heat recovery effect, and the roughness of the borehole wall may also be linked to the heat transfer characteristics. Additionally, the effect of in situ stress can be studied by introducing a circumferential pressure to the rock specimen.

Acknowledgements This research is supported by National Research Foundation, Singapore under its Intra-CREATE Thematic Grant (Award No. NRF2019-THE001-0002).

References

Abdulagatov IM, Akhmedova-Azizova LA, Aliev RM, Badavov GB (2016) Measurements of the density, speed of sound, viscosity, and derived thermodynamic properties of geothermal fluids. J Chem Eng Data 61(1):234–246

Akbar S, Fathianpour N, Al Khoury R (2016) A finite element model for high enthalpy two-phase flow in geothermal wellbores. Renew Energy 94:223–236

Cimmino M (2015) The effects of borehole thermal resistances and fluid flow rate on the g-functions of geothermal bore fields. Int J Heat Mass Transf 91:1119–1127

Gong F, Guo T, Sun W, Li Z, Yang B, Chen Y, Qu Z (2020) Evaluation of geothermal energy extraction in Enhanced Geothermal System (EGS) with multiple fracturing horizontal wells (MFHW). Renew Energy 151:1339–1351

Huang Y, Zhang Y, Yu Z, Ma Y, Zhang C (2019) Experimental investigation of seepage and heat transfer in rough fractures for enhanced geothermal systems. Renew Energy 135:846–855

Kong Y, Pang Z, Shao H et al (2014) Recent studies on hydrothermal systems in China: a review. Geotherm Energy 2:19. https://doi.org/10.1186/s40517-014-0019-8

Li T, Shiozawa S, McClure MW (2016) Thermal breakthrough calculations to optimize design of a multiple-stage enhanced geothermal system. Geothermics 64:455–465

Lu SM (2018) A global review of enhanced geothermal system (EGS). Renew Sustain Energy Rev 81:2902–2921. https://doi.org/10.1016/j.rser.2017.06.097

Ma Y, Li S, Zhang L, Liu S, Liu Z, Li H, Zhang H (2020) Numerical simulation study on the heat extraction performance of multi-well injection enhanced geothermal system. Renew Energy 151:782–795

Pandey SN, Vishal V, Chaudhuri A (2018) Geothermal reservoir modeling in a coupled thermo-hydro-mechanical-chemical approach: a review. Earth-Sci Rev 185:1157–69

Rathnaweera TD, Wu W, Ji Y, Gamage RP (2020) Understanding injection-induced seismicity in enhanced geothermal systems: from the coupled thermo-hydro-mechanical-chemical process to anthropogenic earthquake prediction. Earth Sci Rev 205:103182

Song X, Shi Y, Li G, Shen Z, Hu X, Lyu Z, Zheng R, Wang G (2018) Numerical analysis of the heat production performance of a closed loop geothermal system. Renew Energy 120:365–378

Taron J, Elsworth D, Min K-B (2009) Numerical simulation of thermal–hydrologic–mechanical–chemical processes in deformable, fractured porous media. Int J Rock Mech Min Sci 46:842–854. https://doi.org/10.1016/j.ijrmms.2009.1001.1008

Chapter 20
Quantitative Weathering Assessment of Granite Rock Slope Surfaces in Malaysia Using Image Analysis and CIELAB Color Data: A Case Study

Mazlina Razali, Mohd Ashraf Mohamad Ismail, Sharan Kumar Nagendran, Zuraini Zainal, Hayato Tobe, Kensuke Date, Yasuhiro Yokota, and Hamzah Hussin

Abstract This research improved the traditional method of visually assessing macroscopic weathering by using image analysis based on CIELAB color data input. Focusing on the granite rock slope surface assessment in Malaysia, the current study combined the use of CIELAB colour space and image analysis to develop a rapid preliminary quantitative weathering assessment. The obtained results revealed the increase of a* and b* values with the weathering outcomes. The analysis of 50 measurement points of the studied rock slope revealed that a* and b* were positively correlated (R^2 of 0.9027). The results of image analysis further revealed the susceptibility of the major zones (74%) due to structural control and the considerable extent of weathering of the remaining zones (26%) in grades IV–V, which were

M. Razali · M. A. M. Ismail (✉)
School of Civil Engineering, Universiti Sains Malaysia, Engineering Campus, Pulau Pinang, Malaysia
e-mail: ceashraf@usm.my

M. Razali
Civil Engineering Studies, College of Engineering, Universiti Teknologi MARA, Cawangan Pulau Pinang, Permatang Pauh Campus, Pulau Pinang, Malaysia

S. K. Nagendran · Z. Zainal
Public Works Department (JKR), Centre of Excellence for Engineering and Technology (CREaTE), Alor Gajah, Melaka, Malaysia

H. Tobe
Rock Mechanics and Hydro-Geology Group, Kajima Technical Research Institute, Kajima Corporation, Chofu, Japan

K. Date · Y. Yokota
Kajima Technical Research Institute Singapore, Kajima Corporation, Singapore, Singapore

H. Hussin
Department of Geoscience, Faculty of Earth Science, Universiti Malaysia Kelantan, Jeli, 17600 Kelantan, Malaysia

S. Wang et al. (eds.), *Engineering Geology for a Habitable Earth: IAEG XIV Congress 2023 Proceedings, Chengdu, China*, Environmental Science and Engineering,
https://doi.org/10.1007/978-981-99-9057-3_20

validated through geological structural assessment and geomechanical testing. This study successfully developed a rapid preliminary quantitative weathering assessment for geotechnical and technical purposes.

Keywords CIELAB colour · Image analysis · Rock slope instability assessment · UAV photogrammetry · Weathering assessment · Granite

20.1 Introduction

As a pivotal tool in slope stability assessment, quantitative weathering assessment has been a fundamental part of engineering geological studies (Luo et al. 2020). Besides that, image analysis of drone images has recently placed more emphasis on the weathering impact over rock slope instability assessment (Kim et al. 2018). The colour variation of the rock slope surface serves as an indicator of physical weathering and chemical weathering. Geotechnical engineers quantify these changes during the weathering process, classify the weathered rock zones, and estimate the weathering grades. As weathering assessment typically involves qualitative or subjective visual assessments, the interpreted outcomes tend to vary, which can be addressed by quantitatively assessing the rock surface colour through CIELAB colour space (using colorimeter) and image analysis (Tobe et al. 2018). With that, the current study aimed to develop a rapid preliminary quantitative weathering assessment through the use of CIELAB colour space and image analysis. The proposed assessment was expected to determine the rock mass characteristics and the appropriate weathering grade based on the weathered rock zoning percentage. Geological structural assessment and geomechanical testing were also conducted in this study for validation purpose. The susceptibility of rock slope can be determined based on weathering grades, geomechanical input, geological structure, and mineralogical assessment. CIELAB colour space was used to transform and filter 2D orthophoto images for the classification of weathering grade zones, enhancing the digital image reliability. Besides that, the discontinuity orientation for geological structure was extracted using the Compass plugin in CloudCompare, which was then verified through manual compass mapping (Nagendran et al. 2019). Meanwhile, the development of surface 3D models and point-cloud data of the rock slope surface involved the use of unmanned aerial vehicle (UAV) photogrammetry. Using various analytical tools in image analysis generates high-resolution data that provide insights on the geological structure of the rock surface and its weathering condition (Kong et al. 2021; Salvini et al. 2020). These cost-effective and straightforward techniques help to quantify the weathering assessment of granitic rock mass. As illustrated in Fig. 20.1, this study assessed a weathered granite slope in Kuala Kubu Bharu–Teranum–Raub highway, Peninsular Malaysia. The main range S-granites are usually granular porphyritic rocks that contain perthitic alkali feldspar (Jasin 2013). As for the Bentong-Raub suture zone's western border, the main range S-type granite (granitoid belt) periodically intrudes into the suture zone, forming a stitching pluton. It has a covered area of

49,587.4528 m^2, a crest elevation of 327 m, and a circumference of 102 m at latitude 3°33′57.23″N and longitude 101°42′34.97″E.

The mechanical and chemical characteristics of rock masses are mainly altered by rock weathering (Park and Kim 2019). For instance, the weathering of granite rocks leads to certain microstructural changes, including discolouration, micro-fractures, and loosening of grain boundaries (Gokay 2018). There are different key distinguishing elements for rock weathering assessment, such as the colour of the rock surface. Random chemical reactions on the weathered rock surfaces contribute to the variation of colours on these rock surfaces. Mineral composition in rock is the cause of the formation of various coloured rock masses. Based on the visible colour variation, one can determine the extent of weathering (e.g., slightly and moderately weathered, highly weathered, or structurally failed). Visible colour variations and progressive deterioration progress from the open fractures are clearly depicted in Fig. 20.2, which were attributed to physical and chemical weathering of granite (Thuro et al. 2000). These observations indicate the occurrence of iron staining on

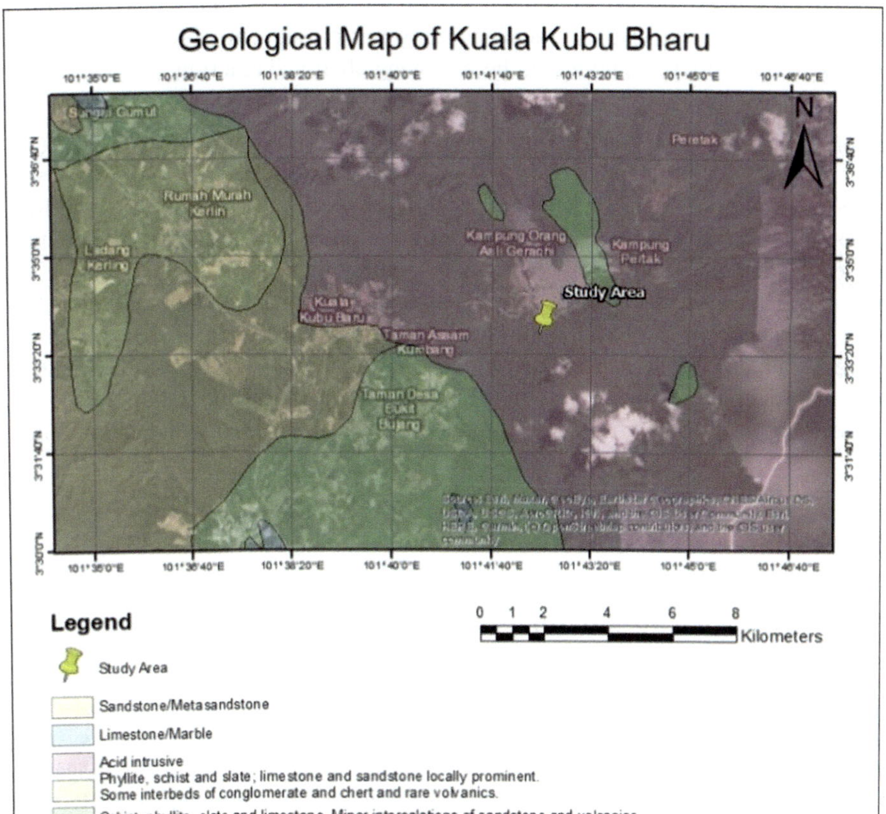

Fig. 20.1 The general lithology of the study area

Fig. 20.2 Physical weathering and chemical weathering of granite (Thuro et al. 2000)

the granite rock slope surface. Furthermore, moderate-scaled oxidation and hydrolysis clearly take place in the discoloured and decayed granite surface. In particular, unweathered granite contains amphibole, biotite, muscovite, sodium potassium feldspar (K feldspar), plagioclase feldspar (Na feldspar), and quartz. The reaction of feldspar in granite with carbonic acid and rainwater subsequently produces kaolinite. Hydrolysis process is expressed in the following chemical reaction:

$$2KAlSi_3O_8 + 2\,H_2CO_3 + 9\,H_2O \rightarrow Al_2Si_2O_5(OH)4 + 4\,H_4SiO_4 + 2K + 2\,HCO_3^-$$

Weathered granitic rock affects the hydrolysis process that involves Na feldspar (Na+) and K feldspar (K+), resulting in the formation of kaolinite (clay). Meanwhile, muscovite and quartz are highly resistant to weathering and often remain as residual minerals. Likewise, the hydrolysis process that involves amphibole and/or biotite also generates clay, while the oxidisation of clay produces iron oxides. As for the current study, the biological weathering of the granite rock slope was rather evident. Minimal vegetation growth was found along the surface runoff and in-between berms, and certain parts of the granite rock slope surface were covered with black mould.

20.2 Materials and Methods

20.2.1 Image Analysis Integration with CIELAB Colour Space

This study used CIELAB colour space to explore the influence of colour systems according to Richard Hunter's prior L*, a*, b* systems. Accordingly, RGB images are converted into CIELAB, which is basically a non-linear transformation of RGB (Shrivastava et al. 2021). In most cases, colour image processing algorithms with CIELAB yield excellent performance. Back in 1976, the International Commission on Illumination (CIE) introduced CIELAB and ΔEab (Eq. 20.1), where the consistency of colour difference shows lacking when compared with varying colours:

$$\text{Colour difference, Delta } E^*_{ab} = \sqrt{(L^*_2 - L^*_1)^2 + (a^*_2 - a^*_1)^2 + (b^*_2 - b^*_1)^2} \quad (20.1)$$

where Delta E^*_{ab} denotes the total colour difference; L* represents lightness; a* represents red/green; and b* represents yellow/blue.

L* denotes the luminance or brightness of the rock's surface and its range values from 0 (black) to 100 (white) (see Fig. 20.3). Meanwhile, a* denotes the proportion of red/magenta (positive value of a*) or green (negative value of a*). There is no specific range of values for a*, but these values typically range from − 100 to 100. On the other hand, b* denotes yellow (positive value of b*) or blue (negative value of b*). Similar to L*, a* and b* do not have specific range of values, which typically range from − 100 to 100. The values of a* and b* are either zero (neutral value) or significantly low (near-neutral value). Red/magenta and yellow (with "+" values) are considered "hot" colours, whereas green and blue (with " − " values) are considered "cold" colours. Most weathered earth materials demonstrate "hot" colours.

Colour variation during oxidation reflects the weathering state of the rock slope surface. The reaction of iron with the grain boundaries in the presence of water or air forms iron oxide/hydroxide (Fe_2O_3/Fe $(OH)_3$), which affects the oxidation of rock minerals. Figure 20.4 depicts different weathering grades based on the colour variation on the rock mass samples. Heavily weathered rock mass sample recorded higher values of a* and b* than unweathered rock mass sample. The oxidation of the

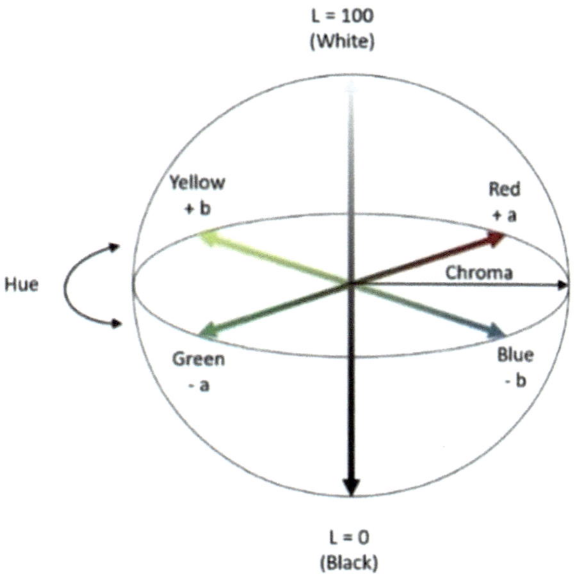

Fig. 20.3 Illustration of CIELAB colour space (Ly et al. 2020)

Fig. 20.4 Samples of CIELAB colour analysis from the study area with different weathering grades

studied rock slope was characterised by different colours from yellowish to reddish-brown, blackish-red, and dark grey along the discontinuities.

The in-situ test involved the use of a colorimeter, specifically the FRU Colorimeter WR18 (see Fig. 20.5). The test specifically served to assess the extent of weathering of rock slope surface in this study; it is the same approach to determine how colour variation reflects weathering grade. In this test, the rock slope surface is illuminated by RGB (red, green, and blue) light sources. These filters are used to distil tri-stimulus (RGB) values, matching how human eyes perceive colour. Transmission or reflection of electromagnetic radiation via a body or solution modifies the strength of the electromagnetic radiation. This equipment can be used to examine highly textured, uneven, irregular, and rough surfaces.

Fig. 20.5 In-situ test with FRU Colorimeter WR18 (40 mm aperture)

Apart from CIELAB colour space, this study incorporated image analysis to further enhance the conventional weathering assessment. As depicted in Fig. 20.6 (Fakunle 2016), JudGeo software was used to analyse the rock slope surface. In order to acquire the best results for the quantitative weathering assessment, the in-situ CIELAB data gathered using the colorimeter served as reference and range input. In the image analysis, each colour reflects the weathering grade of the rock slope surface. Accordingly, image analysis was similarly applied in a prior study to establish geological features, and the results were automatically captured and regenerated via a digital processing technique. Processing 2D orthophoto images of varying colours of rock slope surface into the respective weathering grade zones via the CIELAB colour space improved the accuracy of the digital image analysis. RGB values were converted into L* a* b* values in each pixel. The system also assigned pixels to the weathering grades. Analysing the output results of image analysis considered the orientation of discontinuities and rock strength other parameter or mechanical attribute of the rock mass (Tobe et al. 2018). This approach was deemed less biased than the conventional weathering grade zoning. Besides that, digital camera images were transformed into L* a* b* values to ensure accurate colour evaluation. Thus, similar image processing software was used. However, the task involved complex operations. The following equations were used in the system, and the overall process was automated to reduce the number of complex operations required in the study (Lee Kwan Yee et al. 2022):

$$r = R/255,\ g = G/255,\ b = B/255, \tag{20.2}$$

$$\begin{pmatrix} x \\ y \\ z \end{pmatrix} = \begin{pmatrix} 0.3933 & 0.3651 & 0.1903 \\ 0.2133 & 0.7010 & 0.0858 \\ 0.0182 & 0.11170 & 0.9570 \end{pmatrix} \begin{pmatrix} r^2 \\ g^2 \\ b^2 \end{pmatrix}, \tag{20.3}$$

$$L^{*} = 116\left(\frac{y}{100}\right)^{-\frac{1}{3}} - 16, \tag{20.4}$$

$$a^{*} = 500\left\{\left((x/95.045)^{-1/3}\right) - \left((y/100)^{-1/3}\right)\right\}, \tag{20.5}$$

$$b^{*} = 200\left\{\left((y/100)^{-1/3}\right) - \left((z/100.892)^{-1/3}\right)\right\}, \tag{20.6}$$

where R, G, and B represent to the luminance of the basic colours red, green, and blue (0–255); r, g, and b indicate (0–1); x, y, and z are conversion values; L* values denote lightness; a* values denote redness; b* values denote yellowness.

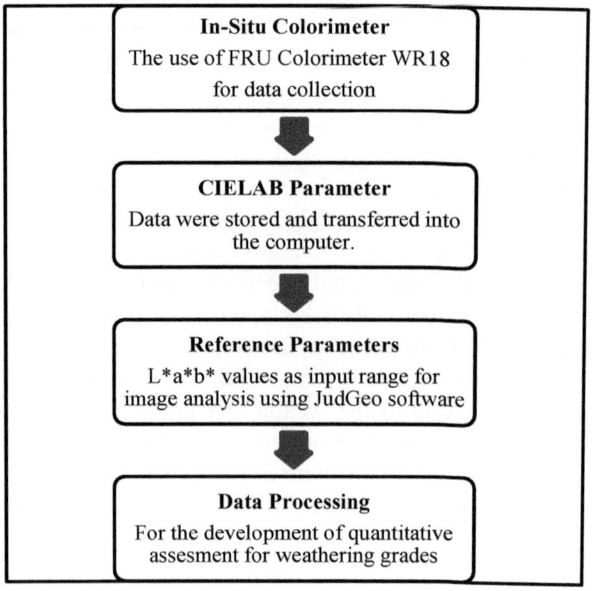

Fig. 20.6 The image processing framework of JudGeo software (Tobe et al. 2018)

20.3 Results and Discussion

The values of L*, a*, and b* and weathering grades were positively correlated. In particular, a* and b* were positively correlated (R2 of 0.9027). Therefore, colour variation can be quantified from unweathered to highly weathered rocks. As depicted in Fig. 20.7, higher values of a* and b* values indicated more weathered rock, suggesting the positive relationship of a* and b* values with the weathering grades. A prior study on the prediction of rockfall from the tunnel face in Japan reported similar results (Tobe et al. 2018). The current results were also compared to the on-site digital photo (see Table 20.1), and its correlation with the weathering description was found significant. The overall assessment based on the gathered samples from 50 measurement points demonstrated strong reliability. The values of a* and b* clearly increased with the extent of weathering grades from slightly weathered to highly weathered (strong to weak rocks).

The obtained results proved the feasibility and reliability of the proposed preliminary quantitative weathering assessment using the CIELAB colour space. Colour variation is a reliable key indicator of the extent of rock weathering. Therefore, on-site digital photos can be an effective tool for the comparison of values gathered using a colorimeter. Figure 20.8 depicts the use of image analysis for this study's weathering assessment. As for the extent of the weathering process, weathering grade VI indicated highly weathered (reddish to orange), weathering grades III (green), IV (yellow), and V (red) indicated the "weathering region", and black indicated unweathered sample. The software automatically calculated the percentage area of these zones to the entire rock slope surface. Areas ranging from weathering grades

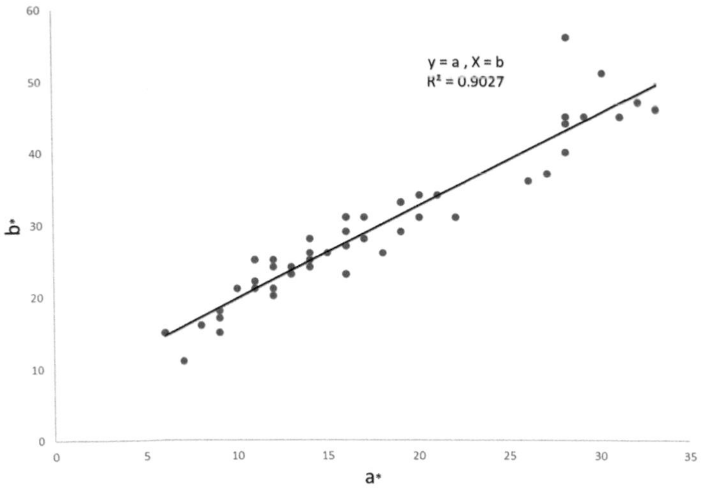

Fig. 20.7 The a* and b* correlation of the study area

I–VI were displayed accordingly. The correlation of the weathering assessment was determined based on the zoning percentage from black to reddish orange.

Based on the results in Fig. 20.8, the colour of rock samples in this study was found to change with the increase in the weathering grade. In particular, the appearance of red and orange colours, which were particularly noticeable near the crest area, indicated the presence of vegetation. In this zone, the rock was deemed weathered, and the rock mass was found highly fracture and loose. There were granites with substantially different texture, even at the outcrop scale, suggesting the various extent of weathering (from small, jointed granites to massive granites). Therefore, the thickness of saprolite may be substantially different as well. The combination of orange and red accounted for 8% of the total zoning, which was utilised for the analysis of the slope vegetation. This zone contained highly weathered granite rocks due to the sun exposure and surface runoff from the top of the slope. Besides that, rotational slides were found common in the red–orange zone, with debris slides in the yellow–brown zone. The yellow–brown zone further indicated the presence of residual soils near certain vegetation types. Meanwhile, the green zone with reddish-yellowish-brown discolouration implied the possibility of rock rolling in the zone. It recorded discolouration and rock degraded into the soil at 12%. The blue zone shared similar possibility of rock failure due to the rolling rocks. Certain rocks disintegrated and/or decomposed into soil, resulting in the discontinuity on the rock slope surface. The blue zoning was the second-most predominant zone, which accounted for 23% of the total zone. In other words, the granite masses had extensive discolouration. The rock texture from medium to coarse grain under moderate weathering was also found to influence the rolling of rocks. On the other hand, the black zone accounted for 51% of the total rock slope (excluding shotcrete), representing a dominant zone as tabulated in Table 20.2. Referring to Fig. 20.8, both left side and right side of the

Table 20.1 Correlations of specific L*, a*, and b* values with the weathering grades description

Point ID	Lab values			Schmidt UCS (MPa)	Photo	Colour simulation	Description
	L*	a*	b*				
1	15	12	25	47.5			Moderately weathered, medium-strong, small and highly fracture, and greenish grey in colour
3	33	6	15	31.0			Weak rock, small to medium joints and fracture, slightly weathered, and whitish-grey colour blend
5	48	11	21	58.5			Moderately weathered and strong, small joints and closed fracture, yellowish-brown colour
40	39	18	26	31.0			Moderately weathered, medium weak rock with small to large-closed joint and fracture, red, yellow, and brown in colour
41	38	12	20	29.5			Weak rock with highly weathered, small to large-closed joint and fracture, grey to brown discolouration

rock slope were highly weathered, which ranged from grades IV to V, whereas the centre of the rock slope appeared slightly to moderately weathered, with grades II and III. The most notable finding was the correlation between the weathering grade zoning and results of image analysis. The results of image analysis (see Fig. 20.8b) revealed more substantial values than that of the macroscopic visual observation (see Fig. 20.8a).

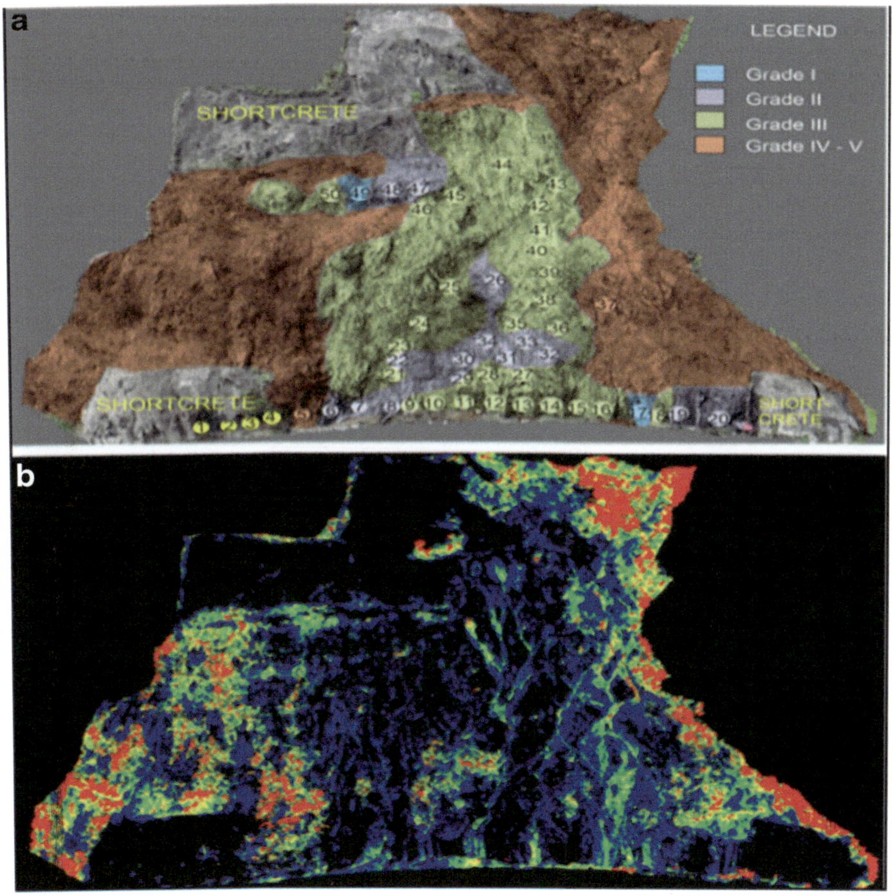

Fig. 20.8 Weathering grade zoning and image analysis for the study area: **a** macroscopic visual observation; **b** image analysis assessment

20.3.1 Geomechanical and Structural Analysis Verification

Referring to Fig. 20.9, the granite rock slope was clearly susceptible to flexural toppling failure (1.72%), wedge sliding (5.51%), and risk of planar sliding (6.90%). Furthermore, a critical joint set, which involved two forms of risk at 12.5%, and a flexural toppling failure risk at 1.72% were observed. The results obtained using the Compass tool from CloudCompare for the determination of rock discontinuity. The field mapping generated 23 visible discontinuity orientations. Planar sliding appeared to have a common trend, whereas failure planes were found subparallel to the slope face. The rock mass also appeared to have slid down along a planar failure surface. The observed failure surfaces were overlain by slightly to completely weathered rock. However, the study was limited to a certain offset of 10° or 10% of the extracted values

Table 20.2 Description of the rock slope possibility failure and weathered rock zoning percentage using image analysis for study area

Colour	Interpretation	Description	Weathering percentage (%)	Rock grade	Estimated schmidt UCS	Possibility of rock failure
	Vegetation/ plants	Weak to very weak rock mass due to some vegetations on the soil	8	R0	5–1	Rotational slides and slumps
	Yellow to brown (residual soils)	Rock discolouration and decomposed to soils	7	R1–R2	25–5	Debris slides
	Green, red, yellow, brown (iron-stained and seepage)	Rock discolouration, decomposed and/or disintegrated into the soil	12	R3	50–25	Rolling rocks
	Grey, brown, bluish (medium to coarse grain)	Rock discolouration at some discontinuities	23	R4	100–50	Rolling rocks
	Light to dark grey, black	Discolouration and stained on the major discontinuities of the rock surface	51	R4–R5	200–100	Rockfalls, block glides, rockslides, and rock slide over sheeting surfaces

from CloudCompare from the actual reading on-site. For example, in the studied region, the extracted Compass plugin slope face orientations recorded 36°/45°, while the manual mapping on-site recorded 35°/40°. There was a difference of 1° for the dip and another difference of 5° for dip directions. Nonetheless, these differences were acceptable. The offset of dip directions was attributed to the misconception of the slope profile viewpoint. The issue of identifying broad slope face orientations was solved by integrating UAV photogrammetry with the measurements of field discontinuity data. Referring to Table 20.3, despite to the presence of strong rock (R4-R5), the rock slope may break due to geological structural failure, including rolling, rockfalls, rockslides, block glides, and sliding across sheeting surfaces. The results of image analysis revealed black zone (51%) and blue zone (23%) as critical zones of slope failure, as these zones were considered structurally controlled failure with the weathering grades II–III. The above results were validated by the results of stereonet (see Fig. 20.9). With the stated type of potential failure, both black zone and blue zone may still slide. Based on these results, the current study demonstrated

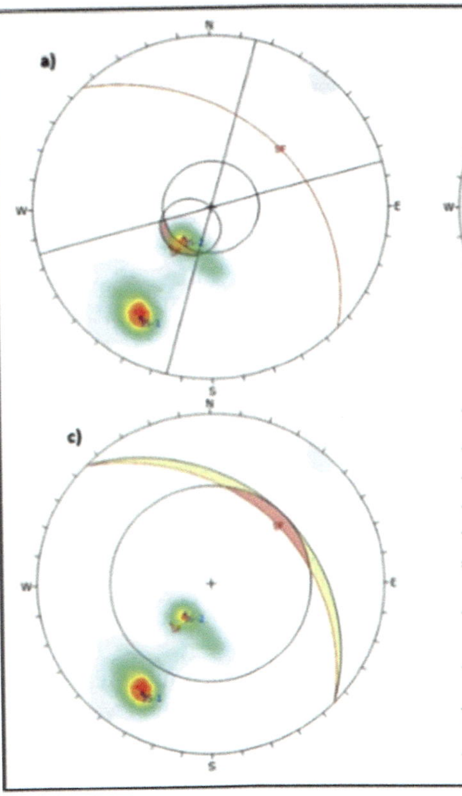

KKB Slope Discontinuity Orientation				
Slope Dip			36	
Slope Dip Direction			15	
Friction Angle			30°	
Lateral Limits			30°	
Kinematic Analysis		**Critical**	**Total**	**%**
a)	Planar Sliding (All)	1	58	6.90
	(Set 2: JS -2	1	8	12.50
b)	Flexural Toppling (All)	1	58	1.72
c)	Wedge sliding	91	1653	5.51

Fig. 20.9 Stereograms illustrating **a** planar sliding, **b** flexural toppling, and **c** wedge sliding

the promising prospects of the proposed quantitative weathering assessment through the combined use of CIELAB colour space and image analysis.

20.3.2 Mineral Composition Verification

This study utilised thin section samples (TS18, TS19, and TS21) from the gathered granite rock sample for mineralogical verification. These samples were classified according to the extent of their weathering process. Overall, the samples revealed weathering grades II and III. As shown in Table 20.3, the mineral evaluation of TS18 revealed 40% quartz, 40% plagioclase, and fragments of rock at 20%, whereas the mineral evaluation of TS19 demonstrated 50% quartz, 40% plagioclase, and 10% matrix/groundmass. On the other hand, the mineral evaluation of TS21 revealed 40% plagioclase, 30% quartz, 20% hornblende chloride, and 20% matrix/groundmass. Mineral variation in granite contributes to the regulation of surface weathering at site. The cooling process and crystallised growth influence the variation in quartz

Table 20.3 Some selected thin section (TS) samples and their associated mineralogy

TS No.	Photographs	Mineral composition percentage (%)		Rock type	Weathering grades
18		Quartz	40		
		Alkali feldspar	–	Granite	III
		Plagioclase	40		
		Matrix/ groundmass	20		
19		Quartz	50		
		Alkali feldspar	–	Granite	III
		Plagioclase	40		
		Matrix/ groundmass	10		
22		Quartz	40		
		Alkali feldspar	20	Granite	II
		Plagioclase	30		
		Matrix/ groundmass	10		

and other mineral products. Therefore, the relationship between mineral variation and surface weathering can be interpreted in various ways. Mineral variation in granite takes place for certain minerals, such as alkaline feldspar, and plagioclase (Duzgoren-Aydin et al. 2002). Plagioclase changed into epidote (whitish stripes) in TS19. The dominance of quartz (40%) in TS19 represented a spike in UCS laboratory readings and weathering grades (II). The low readings of UCS load test suggested heavily fractured sample (weathering grade IV). Based on the results, the current study validated the categorisation of granite rock from the study area as strong rock with weathering grades II and III.

20.4 Conclusion

Mineral composition and variation clearly have direct influence on the resultant weathering grade. The results of the current study showed the rapid advancement of the weathering process. This study successfully achieved its primary objective of developing a rapid preliminary quantitative weathering assessment, specifically for the granite rock slope surface at the study area. This study incorporated CIELAB colour space with the use of a colorimeter to quantify the weathering grades. Besides that, this study used the orthoimages generated from UAV photogrammetry for the image analysis using JudGeo software to generate a spectrum of colours representing the extent of weathering. The colour scheme, which ranges from weathering grade I to weathering grade VI, represents the extent of weathering. There are six colours for different weathering grades: red (weathering grade I), orange (weathering grade II), yellow (weathering grade III), green (weathering grade IV), blue (weathering grade V), and black (weathering grade VI). Based on the obtained results, black colour code was dominant, suggesting the presence of strong rock in the study area. The weathering grades from the image analysis in correspondence to the UCS-Schmidt hammer values (UCS-Schmidt) were categorised via user-defined colours of blue, purple, green, and orange on the orthoimages. The comparison of colour images generated by JudGeo software demonstrated an acceptable trend given the resultant colour codes and the corresponding coverage zones. Unlike UCS-Schmidt (more of a point accumulator), JudGeo images in this study presented more accurate weathering grades. The general geology revealed significant variation under rock formations, which was validated through geological structural assessment and geomechanical testing. This study further compared UCS values obtained from L-type Schmidt hammer test with the standard classification of weathering grades. The results confirmed significant similarity in the range of UCS-Schmidt, regardless of the rock type. The relationship between the strength of intact rock and weathering of the rock slope surface was also emphasised in this study. The accuracy of the obtained results was ensured, and potential uncertainties were also minimised. However, the top area of the studied granite rock slope was excluded from the current study due to technical access limitations. As a result, there was no specific weathering grade or classification for the top area of the rock slope, which may result in interpretation errors. However, this limitation can be addressed and accurately validated through image analysis using the JudGeo software.

Acknowledgements The first author acknowledges Universiti Sains Malaysia for the GRA-Assist Fellowship funding support during the initial year of the PhD study and Universiti Teknologi MARA, Cawangan Pulau Pinang for the congress's financial support. We express our gratitude to Kajima Technical Research Institute Japan and Singapore (KATRIS) for their guidance and provision of tools for this research. We also thank the anonymous reviewers for their insightful suggestions to enhance the manuscript.

Declaration of Funding This study was funded by the Centre of Excellence for Engineering and Technology (CREaTE) "Development of Rapid Quantitative Assessment for Weathered Rock Slope in Tropical Region Using the Integrated Image Analysis Technique (Grant No. 304. PAWAM.6050482.J111)" and AUNSEED-Net/JICA for the fieldwork.

References

Duzgoren-Aydin NS, Aydin A, Malpas J (2002) Re-assessment of chemical weathering indices: case study on pyroclastic rocks of Hong Kong. Eng Geol 63(1–2):99–119. https://doi.org/10.1016/S0013-7952(01)00073-4

Fakunle AA (2016) Detection of weathering signatures using UAV photogrammetry in comparison with ground-based sensors

Gokay MK (2018) Analyses of rock surface colour changes due to weathering. Selcuk Univer J Eng Sci Technol 6(1):1–13. https://doi.org/10.15317/scitech.2018.111

Jasin B (2013) Chert blocks in Bentong-Raub Suture Zone: a heritage of palaeo-tethys

Kim DH, Balasubramaniam AS, Gratchev I (2018) Application of photogrammetry and image analysis for rock slope investigation. Geotech Eng

Kong D, Saroglou C, Wu F, Sha P, Li B (2021) Development and application of UAV-SfM photogrammetry for quantitative characterization of rock mass discontinuities. Int J Rock Mech Min Sci 141(February). https://doi.org/10.1016/j.ijrmms.2021.104729

Lee Kwan Yee A, Razali M, Ismail MAM, Yusoff IN, Nagendran SK, Zainal Z, Tobe H, Date K, Yokota Y (2022) Preliminary analysis of rock mass weathering grade using image analysis of CIELAB color space with the validation of Schmidt hammer: A case study. Phys Chem Earth, Parts A/B/C, 103291. https://doi.org/10.1016/J.PCE.2022.103291

Luo J-Y, Xu Z-M, Ren Z, Wang K, Gao H-Y, Yang K, Tang Y-J, Tian L (2020) Quantitative assessment of weathering degree of the Touzhai rock-avalanche deposit in Southwest China. https://doi.org/10.1016/j.geomorph.2020.107162

Ly BCK, Dyer EB, Feig JL, Chien AL, del Bino S (2020) Research techniques made simple: cutaneous colorimetry: a reliable technique for objective skin color measurement. J Investig Dermatol 140(1):3-12.e1. https://doi.org/10.1016/j.jid.2019.11.003

Nagendran SK, Ismail MAM, Tung WY (2019) Integration of UAV photogrammetry and kinematic analysis for rock slope stability assessment. Bull Geol Soc Malaysia 2019(67):117–123. https://doi.org/10.7186/bgsm67201913

Park J, Kim K (2019) Quantification of rock mass weathering using spectral imaging. J Southern Afr Instit Min Metall 119(12):1039–1046. https://doi.org/10.17159/2411-9717/708/2019

Salvini R, Vanneschi C, Coggan JS, Mastrorocco G (2020) Evaluation of the use of UAV photogrammetry for rock discontinuity roughness characterization. Rock Mech Rock Eng 53:3699–3720. https://doi.org/10.1007/s00603-020-02130-2

Shrivastava D, Srivastava KC, Ganji KK, Alam MK, Zoubi IAl, Sghaireen MG (2021) Quantitative assessment of gingival inflammation in patients undergoing nonsurgical periodontal therapy using photometric CIELab analysis. BioMed Res Int. https://doi.org/10.1155/2021/6615603

Thuro K, Hecht L, Plinninger R, Scholz M, Bierer S (2000) Geotechnical aspects of weathered and hydrothermally decomposed granite in the Königshainer Berge tunnel project. Deutsche Gesellschaft für Geotechnik e.V. (ed) (Issue March, pp. 177–182). In: Proceedings of the Eurock 2000 Symposium, Essen. Glückauf

Tobe H, Miyajima Y, Shirasagi S, Yamamoto T, Kawabata J (2018) A rapid image analyzing method for determining crack distribution and interval on tunnel faces. In: ISRM international symposium—10th Asian rock mechanics symposium, ARMS 2018

Chapter 21
Eancing Rock Mass Stability Through the Efficient Utilization of Backfill

Xin Yu, Shengwen Qi, Yuye Tan, Weidong Song, and John Kemeny

Abstract Backfills have gained significant attention as sustainable materials for recycling waste generated in underground mining processes. However, a comprehensive understanding of the efficient mechanisms and their effects on underground rock mass stability under varying conditions is still necessary. This paper investigates the failure process of two different rock-backfill structures through triaxial compressive experiments, employing acoustic emission (AE) and computed tomography (CT) technology. The study reveals the cooperative behavior of the bi-material system and identifies four key reactions of the backfills on the rock mass: stress redistribution, energy absorption, intact force, and intact friction. The specific reactions and their sensitivities depend on the structure and external confinement applied to the

X. Yu (✉) · S. Qi
Key Laboratory of Shale Gas and Geoengineering, Institute of Geology and Geophysics, Chinese Academy of Sciences, Beijing 100089, China
e-mail: yuxin@mail.iggcas.ac.cn

State Key Laboratory of Lithospheric and Environmental Coevolution, Institute of Geology and Geophysics, Chinese Academy of Sciences, Beijing, China

Innovation Academy for Earth Science, Chinese Academy of Sciences, Beijing 100089, China

University of Chinese Academy of Sciences, Beijing 100049, China

S. Qi
e-mail: qishengwen@mail.iggcas.ac.cn

X. Yu · Y. Tan · W. Song
Key Laboratory of High-Efficient Mining and Safety of Metal Mines (Ministry of Education of China), University of Science and Technology Beijing, Beijing 100083, China
e-mail: tanyuye@ustb.edu.cn

W. Song
e-mail: songwd@ustb.edu.cn

School of Civil and Resources Engineering, University of Science and Technology Beijing, Beijing 100083, China

X. Yu · J. Kemeny
Department of Mining and Geological Engineering, University of Arizona, Tucson, AZ 86721, US
e-mail: kemeny@arizona.edu

S. Wang et al. (eds.), *Engineering Geology for a Habitable Earth: IAEG XIV Congress 2023 Proceedings, Chengdu, China*, Environmental Science and Engineering,
https://doi.org/10.1007/978-981-99-9057-3_21

system. Considering the volume fraction of the materials and the external loading, the study presents technical recommendations for utilizing backfills in diverse mining conditions. These findings provide insights for more efficient and flexible design considerations when implementing the backfilling mining method, contributing to a greener and sustainable future.

Keywords Backfill · Underground mining · Rock mass stability · Response characteristic

21.1 Introduction

As a passive supporting structure within underground mining engineering project, backfill experiences passive forces due to the deformation of the rock mass, which reacts back to the rock mass in a passive manner. This interaction primarily involves three modes of action: stress absorption and transfer, contact support, and stress isolation (Zhao 2021). The research on the interaction between rock and backfill mainly revolves around three aspects: arching stress characteristics of the backfill, strength requirement models for backfill, and the mechanical behaviour of the rock-backfill combination. Many scholars have conducted studies on the interaction mechanisms between the two geomaterials through theoretical calculations, laboratory experiments, numerical simulations, and field measurements.

Since the 1980s, Blight and Clarke, Yamaguchi Umetaro, Oliver and Landriault have started developing physical modeling approaches like the uniaxial compression tests and split-loading physical modelling apparatu on composite specimens and loose backfill materials to analyse the encapsulation characteristics of the backfill on the surrounding rock, the properties of the backfill, and the influence of encapsulation height on the supporting effect on the surrounding rock (Sinclair et al. 1980). The optimization of backfill structure and composition has also been a popular research topic, numerous experiments and studies have been conducted on the mechanical behaviour of rock and backfill during instantaneous static or dynamic evolution, considering different types and contents of cementing materials, porosity, composition ratios (such as cement-tailing ratio, concentration, etc.), and macrostructure. These studies have provided insights into the variation patterns of backfill strength and deformation characteristics, as well as the influence of various factors on backfill effectiveness (Cui and Fall 2016; Du et al. 2018a; Liu et al. 2017a; Zhang et al. 2012). Even though these references data could be utilized in mathematical model establishment, they still simplified or even neglected the interaction with the rock mass. These studies mainly focus on idealized models that primarily consider the stress distribution within the backfill itself.

Then from 2016, scholars have shifted their focus from studying rocks and backfill separately to treating them as a whole system to directly observe the mutual mechanical interactions. During the curing process of the backfilling slurry, it adheres to the rock walls of the mined-out area, generating cohesion and frictional forces. As the backfill hardens, the interface effects (Fang and Fall 2020a), roughness, and shear stress at the interface between the rock and backfill all influence stress transfer (Fang et al. 2021; Guo et al. 2020; Kartawidjaja 2020; Xu et al. 2020). When complex stress conditions near the boundary between the rock and backfill reach a certain level, shear or tensile failure may occur at the contact surface, and even a fracture zone may form nearby (Cui and Fall 2017a; Emad 2017; Fang and Fall 2019a, 2020b; Liu et al. 2017b). However, this can lead to rock damage and failure near the backfill, safety issues in the mined-out area, and the creation of new fluid flow channels. Additionally, backfill properties (such as viscosity and filling ratio), surface characteristics of the mined-out area's rock walls (such as roughness and strength), and curing environmental conditions (such as temperature and drainage) all affect the strength and mechanical properties of the contact surface (Bruning et al. 2018; Gutiérrez-Ch et al. 2018; Koupouli et al. 2017, 2016; Zhao et al. 2018).

Thus, Fall et al. conducted direct shear tests on the contact between rocks and backfill, considering factors such as temperature (Du et al. 2018b), sulphate composition of the backfill, rock type, curing time, humidity, cementation effect, roughness of the contact surface (Cui and Fall 2017b; Fang and Fall 2019b), and other influences on the shear behaviour of the rock-backfill interface. Nabasse J. F. Koupouli et al. numerically studied the mechanical response mechanism of the rock-backfill interface through direct shear tests (refer to Fig. 21.2) (Nabassé 2016). Although these tests were comprehensive, they did not consider the mechanical properties of the rock-backfill interface under complex stress conditions such as triaxial loading, true triaxial loading, dynamic conditions, and fluid-coupled fields (Chang et al. 2018). Therefore, Weilv et al. studied the deformation characteristics and shear strength changes interface inclination angle between the rock and backfill (Weilv et al. 2021). Cai (2009) conducted a study on the coupled fracture of cubic rock enclosing backfill specimens using acoustic emission characteristics and found that the encapsulation of backfill contributed to the enhancement of rock strength. Song and Tan (Song et al. 2017; Tan et al. 2018, 2020), Fu et al. (Wang et al. 2020a, b), and others constructed and conducted tests on mutually encapsulated rock and backfill combination models under uniaxial and triaxial compression, preliminarily exploring the special mechanical properties of different structured rock-backfill composite bodies under combined pressure (Li et al. 2020; Wang et al. 2019; Wang et al. 2020a, b). Wang and Cheng conducted mechanical research on rock-backfill composite load-bearing structures through deformation and acoustic emission characteristics (Cheng et al. 2020). The studies have to some extent revealed the shared load-bearing mechanisms of rocks and backfill under single stress conditions. However, when facing severe water inflow issues in deep mining excavations, the stability of the excavation is closely related to the mechanical properties and deformation failure patterns of the pressurized water inside the backfill, rocks, and their interface under water-pressure coupling (Liu 2020; Zhang 2020). Due to the infiltration process, the generation of water flow fissure channels within the material, the primary pores, fractures (especially the contact fractures between rocks and backfill), and matrix of the rock-backfill composite undergo more complex chemical and physical reactions under changing external mechanical conditions. The continuous fracture process and its patterns under triaxial compression conditions, as well as the long-term coupling load-bearing capacity and the effects on the interface of the two materials under different ratios, are still lacking in research and require further in-depth exploration.

In this study, we summarized our series studies on two different rock and backfill structures, comparing their peak strength, effective mechanical properties, failure mode, damage evolution process considering different confining pressure and volume fraction of rock, revealing the coupling mechanism and response characteristics between rock and backfill for engineering design optimization.

21.2 Material and Method

Two different rock and backfill structures were considered, one is rock-encased by backfills, the other is the backfills-encased by rock. The rock-encased by backfills (RB) structure represent the state that rock pillars surrounded by the backfilling stope in the room and pillar mining method. The backfills-encased by rock (BR) structure represent the backfilling stope surrounded by the rock mass in the sublevel backfilling mining method. The rock in RB and BR structures are granite and marble, separately, based on the preferred occurrence in different mining methods. Then the cemented waste tailing backfills was utilized as it is the most popular backfilling materials in underground mines. For save space, please see the mechanical properties of the granite, marble and backfill, the manufacturing process of the cylindrical RB and BR specimens in our previous papers (Yu et al. 2021a, 2021b, 2021c, 2022a, 2022b, 2023; Yu 2022). Four confining pressures and four volume fractions of rock (VF) in the specimen were tested by triaxial compression test through TAW-2000 system, combined with acoustic emissions and CT scanning.

21.3 Analysis

- Triaxial Peak strength

Figure 21.1a and b present the three-dimensional best-fit surface between peak stress, confining pressure, and VF of BR and RB specimens. The dark black points represent data above the best-fit surface, while the grey points indicate data below the best-fit surface. Table 21.1 shows the model equation of the two specimens, all satisfying the multivariate linear equation with a high reliability to estimate the peak strength of the BR specimen. A comparison of the peak strength models for the BR and RB specimens reveals a substantial difference in sensitivity to confining pressure. The VF coefficient in the BR specimen is only 3.8 times that of the confining pressure, whereas in the RB specimen, the VF coefficient is 27.7 times that of the confining pressure. It demonstrated the mechanical properties of the specimen is not only influenced by their volume fraction, but also is highly related with the lateral contact supporting force between rock and backfill acted inside different structure.

- Equivalent Mechanical Properties

For a bi-geomaterial composite, its mechanical properties can be referred to as equivalent mechanical properties (Long et al. 2016). Figures 21.2 and 21.3 illustrates the

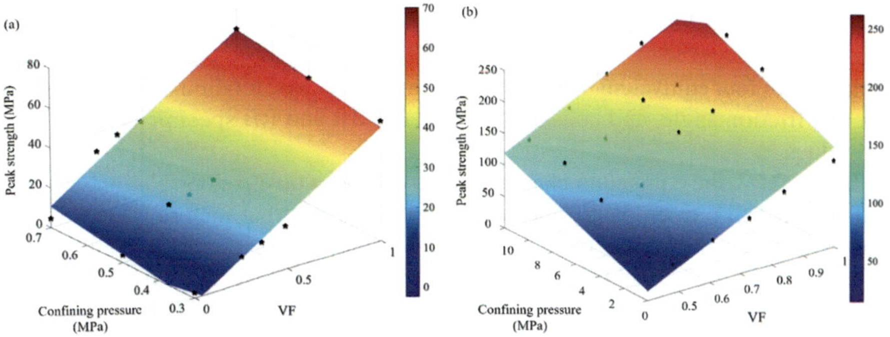

Fig. 21.1 Triaxial peak strength under different confining pressures and VF of the **a** BR, and **b** RB specimen

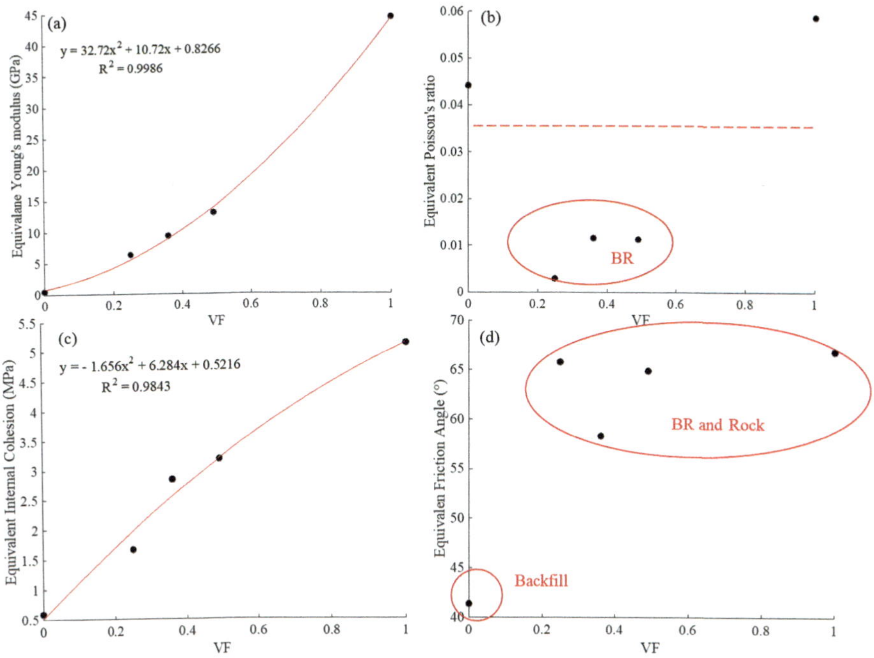

Fig. 21.2 Equivalent mechanical properties under different VF of BR specimen

Table 21.1 The fitting model for the peak stress, confining pressure and VF of BR and RB specimen

Specimen style	Fitting model	Coefficient	Description
RB	$\sigma_{cBR} = -12.057 + 59.037 * VF + 332.1327\sigma_3$	0.9693	σ_{cBR} is the Peak strength of the BR specimen, MPa; σ_3 is the external confining pressure applied to the specimen, MPa
BR	$\sigma_{cRB} = -80.86 + 238.8 * VF + 8.63 * \sigma_3$	0.9608	

variations of four equivalent mechanical parameters (elastic modulus, Poisson's ratio, internal friction angle, and cohesive strength) of BR and RB specimens at different VF values.

In Fig. 21.2, the equivalent elastic modulus shows a positive correlation with VF, and the growth rate gradually increases. This indicates that materials with higher elastic modulus in the composite have a greater influence on the overall values. The Poisson's ratio of the backfill (VF = 0) and the rock (VF = 1) are 0.044 and 0.059, respectively. However, the average equivalent Poisson's ratio of BR specimens is 0.008, which is approximately 1/6 of the two geomaterials, and it does not show a significant correlation with VF. This suggests that the lateral deformation capacity of BR is suppressed, possibly due to the interaction forces between the two geomaterials on the contact surface within the specimen cancelling each other out. The relationship

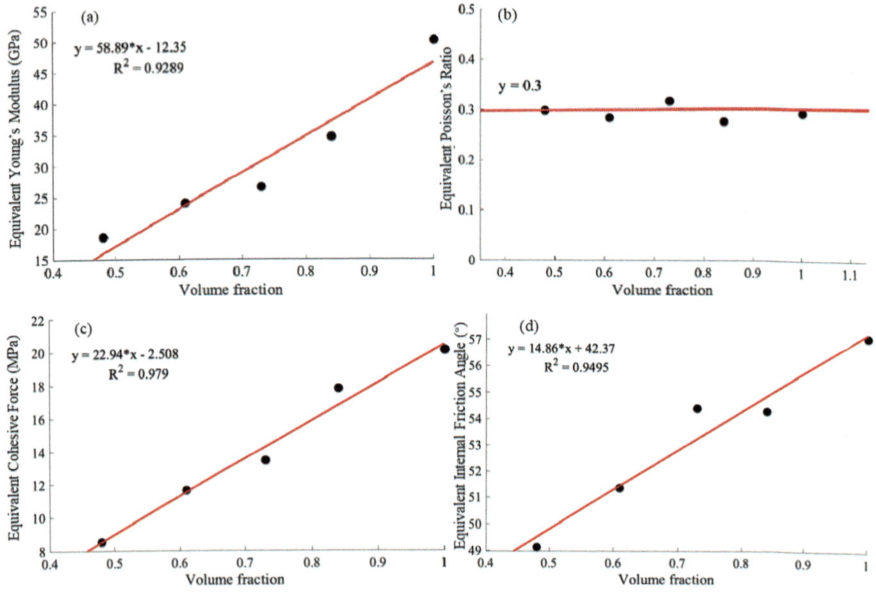

Fig. 21.3 Equivalent mechanical properties under different VF of RB specimen

between equivalent cohesive strength and VF, showing a positive correlation, but the growth rate gradually slows down with increasing VF. The equivalent internal friction angle of BR specimens is like that of the rock, approximately 63.95°, not vary with VF. The internal friction angle of the backfill is significantly smaller than the previous two, approximately 41.37°. Based on the research findings of Wang and Tatsuoka, it is concluded that the adhesion strength of rocks is higher than that of soil materials (such as backfill). Therefore, the internal friction angle of the contact surface between rock and other geomaterials is greatly influenced by the material with stronger cohesion (Tatsuoka 1985; Wang et al. 2018). The Marston non-cohesive model is an important model widely used in engineering practice, and one of its basic assumptions is that the internal friction angle of the backfill is equivalent to the internal friction angle at the filling/rock contact surface. The experimental results validate the rationality of this assumption. In conclusion, the rock in BR specimens plays a positive role in the overall axial deformation capacity and shear strength characteristics, while the lateral deformation and expansion capacity are suppressed due to structural influences.

In Fig. 21.3, the trends of two elastic properties (equivalent elastic modulus and Poisson's ratio) differ significantly with the variation in VF. The equivalent elastic modulus increases with increasing VF, while the equivalent Poisson's ratio remains constant at around 0.3. This indicates that the properties of the external rock have a greater influence on lateral deformation and Poisson's ratio, while VF has a relatively minor effect on these factors. As the content of the backfill increases (VF decreases), both the equivalent cohesion and internal friction angle decrease. The correlation coefficients for the fitting results are 0.979 and 0.9495, respectively, indicating a power function relationship and a linear relationship between the shear strength parameters and volume fraction.

- Failure Mode

Through the analysis of the triaxial compression failure mode of BR (Fig. 21.4a), the following observations can be made: When the confining pressure and VF are small (close to 0), the stress transmitted from the backfill to the rock contact surface is lower under stress isolation. The central rock experiences predominantly uniaxial compression with excessive tensile stress in the longitudinal direction, leading to axial splitting failure (tension). As the confining pressure and VF gradually increase, the influence of contact support causes a significant increase in shear stress within the rock, concentrating in the middle and resulting in multiple shear failures. When the confining pressure and VF exceed a certain value, the large shear angle leads to a significant increase in the probability of pure shear failure, accompanied by significant end effects. Considering the equivalent Poisson's ratio of BR specimens is much smaller than the Poisson's ratio of the two individual media, indicating that when the backfill is subjected to lateral pressure, deformation and stress transfer are prone to occur, exerting stronger lateral contact support force and significantly enhancing the strength of the central rock.

In the case of RB, the influence of the two factors on the failure mode is opposite: as the confining pressure gradually increases, shear failure tends to occur at the ends; as the VF increases, coalescence shear failures often occur in the middle (Fig. 21.4b).

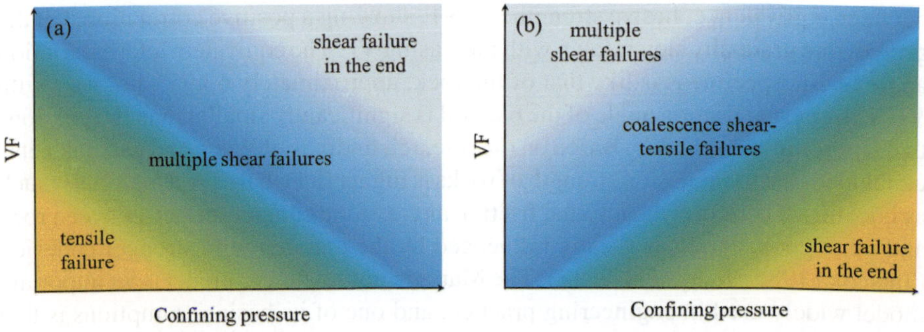

Fig. 21.4 The effect of confining pressure and VF on the failure mode of **a** BR, and **b** RB

According to the research by M. I. Alsayed et al., in the case of thin-walled triaxial compression, the lateral forces on the hollow rock column are uniformly distributed along the wall, and the internal stresses of the cavity cancel each other out and decrease sharply (Alsayed 2002). This results in the suppression of internal lateral deformation, and the external lateral deformation capacity is the same as that of a single rock. Therefore, the equivalent Poisson's ratio of RB is like that of the rock and is not affected by VF. The sensitivity of peak strength to confining pressure is also lower compared to the BR structure.

21.4 Discussion

- Effect of Different Structures on Coupling Mechanism between Rock and Backfills

Here we summarized the fracturing process of BR and RB analysed by the acoustic emission and CT scan technology (Fig. 21.5) and concluded the coupling mechanism between rock and backfills in different structures.

Based on the different primary fracture modes, the stress state at the upper end of the BR specimen can be divided into two stages (Fig. 21.6a): ① The rock bearing stage

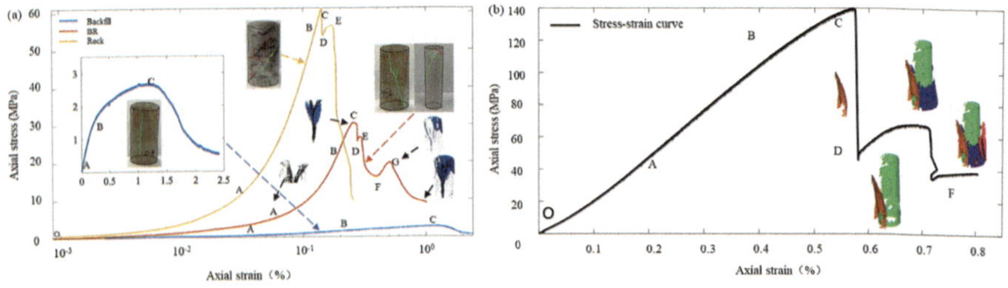

Fig. 21.5 Stress–strain curve of **a** BR, and **b**RB

consists of the initial densification, elastic, yielding, and failure phases. In the early period of this stage, the axial stress is relatively small and mainly acts on the stiffer rock in the middle. Intense end friction occurs between the rock and the loading platens, resulting in significant radial frictional forces between two geomaterials. Excessive shear stress leads to local failure of the rock and backfill near the contact surface. Although the rock undergoes expansion and deformation to squeeze the backfill during the compression process, the backfill has high toughness, allowing it to absorb stress and prevent significant radial deformation of its outer diameter. When the VF decreases or the confining pressure increases, the support force exerted by the backfill on the rock significantly increases, thereby increasing the interaction forces along the contact surface in the normal direction. This effectively reduces the influence of frictional shear stress at the ends, thereby reducing the degree of end failure. ② During the backfill bearing stage, the process can be described as rupture of the rock, secondary rupture, and residual deformation. In the initial stage of this phase, significant internal damage occurs in the rock, leading to complete instability. Intense disturbance and deformation cause the breakage of the contact bond at the ends. As the axial load continues to be applied, the internal stress redistributes. Once the system reaches a new equilibrium, the main stress concentrates within the backfill. If the lateral stress is significant, shear failure occurs within the rock. On the other hand, if the lateral stress is small, axial tensile failure of the backfill occurs around the already damaged contact surfaces at the ends.

Based on the different primary fracture modes, the stress state at the upper end of the RB specimen can be divided into two stages (Fig. 21.6b): ①The initial rock bearing stage is characterized by initial compaction, elasticity, yield, and primary failure. During this stage, the RB specimen experiences relatively uniform and stable

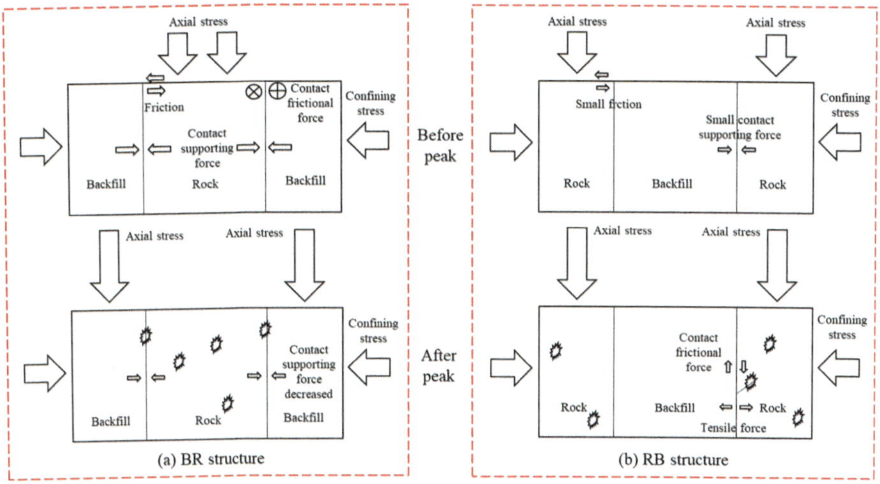

Fig. 21.6 Loading mechanism of rock and backfill under triaxial loading **a** BR, and **b** RB

stress distribution in the process of triaxial compression. The stress mainly concentrates within the hollow thin-walled rock column, while the shear stress at the ends and the support provided by the backfill are relatively small. When the VF increases and the confining pressure decreases, the shear stress decreases, resulting in less damage to the contact surfaces. Conversely, when the volume fraction decreases and the confining pressure increases, the failure becomes more concentrated towards the bottom, leading to increased shear stress and greater rupture energy, thereby enhancing the contact support effect. ② The secondary rock bearing stage is characterized by post-peak step-like failure and residual deformation of the rock. The confining pressure plays a crucial role in the post-peak mechanical behaviour during this stage. When the confining pressure is high and the VF is low, the fractured rock mass expands and slips, and the backfill provides increased support and frictional resistance to resist lateral pressure. However, as the axial stress continues to increase, the contact surfaces gradually deteriorate under the increasing frictional forces and tensile effects. This leads to the expansion of primary failure into secondary shear failure. On the other hand, when the confining pressure is low and the volume fraction is high, the tendency for rock block slippage is reduced, and the support provided by the backfill is sufficient to resist lateral pressure. Additionally, it provides internal resistance for the fractured hollow thin-walled rock mass. The partial cancellation of internal and external lateral pressures leads to significant tensile stresses within the fractured rock mass, triggering intense tensile failure (Yang et al. 2020).

- Response Characteristics between Rock and Backfills and engineering application

According to the transmission directions of the two lateral pressures, a qualitative analysis of the response characteristics of the interaction between rock and backfill can be conducted (Fig. 21.7). In addition to the three main mechanisms of backfill's action on rock (stress absorption and transfer, contact support, and stress isolation, the supporting effect of contact friction (radial and axial) during the mining disturbance process is also considered. This enhances the influence of different mining conditions on the interaction response characteristics between rock and backfill.

Based on this research, adjustments can be made to the proportion and strength of the backfill, and the filling operations can be planned and organized accordingly, considering different depths, mining layouts, and rock mechanics conditions. Strategies and suggestions can be proposed for the practical application of backfilling methods: When the rock strength is low, brittle behaviour is prominent, and the horizontal stress is relatively small, it is advisable to enhance the strength and shear resistance of the backfill material to mitigate frictional failure between the rock and backfill under low stress disturbances. When the rock strength is high, brittle behaviour is prominent, and the horizontal stress is significant, it may be beneficial to add other materials to create a backfill slurry or reduce the particle size of the backfill slurry to enhance the contact stiffness between the rock and backfill, thereby improving the ability of the mining area to withstand high-stress disturbances and large deformations. In scenario b (Fig. 21.7), the role of the backfill material is mainly manifested after the rock failure.

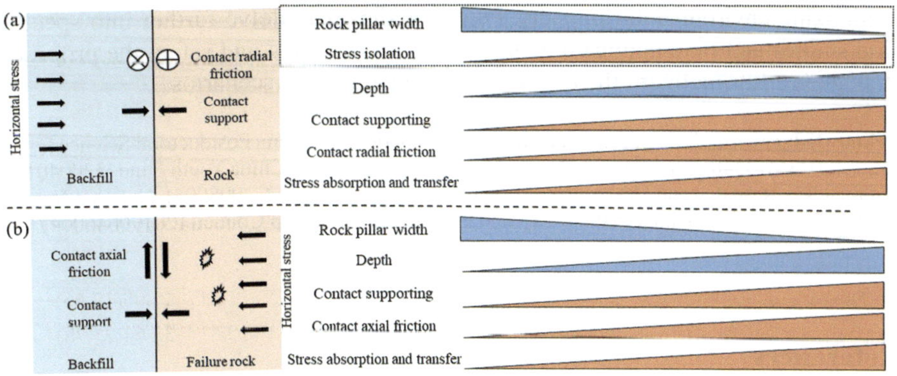

Fig. 21.7 Response characteristics between rock and backfilling **a** horizontal stress comes from the backfill, and **b** horizontal stress comes from the rock

21.5 Conclusions

This study investigated the behaviour and interaction between rock and backfill geomaterials under different stress conditions and mining scenarios. Through the analysis of fracture patterns and stress distribution, it was observed that the rock and backfill system exhibited distinct response characteristics. The presence of backfill played a crucial role in stress absorption, transfer, contact support, and stress isolation. Additionally, the influence of contact friction on the support function during mining disturbances was also highlighted. These findings contribute to a better understanding of the complex mechanics involved in rock-backfill interaction and provide valuable insights for optimizing backfill material composition, strength, and operational planning in various mining environments.

Furthermore, the study offers practical implications for engineering applications. Depending on the rock strength, brittleness, and horizontal stress levels, different strategies can be employed. For weaker rock with lower stress conditions, reinforcing the backfill material's strength and shear resistance can effectively mitigate frictional failures between the rock and backfill under low-stress disturbances. Conversely, in scenarios where rock strength is high and horizontal stress is significant, incorporating additional materials or adjusting the backfill slurry particle size can enhance the contact stiffness between the rock and backfill, enabling the mining area to better withstand high-stress disturbances and large deformations. These insights can guide the design and implementation of backfilling operations, improving the overall stability and performance of mining excavations.

In conclusion, this research provides valuable knowledge for optimizing backfilling strategies in mining operations. By considering the unique characteristics of the rock-backfill system, engineers can tailor the backfill properties and design appropriate mining sequences based on specific geological and mechanical conditions. Implementing these strategies can enhance the safety, efficiency, and sustainability of mining operations by reducing rock failure, supporting excavations, and minimizing environmental impacts. Future studies may delve further into specific case studies and field applications to validate the findings and refine the proposed strategies, ensuring their effectiveness in practical mining scenarios.

Acknowledgements We acknowledge the funding support from China Postdoctoral Science Foundation [2023M733481], the National Natural Science Foundation of China Youth Fund [52004019], National Key Research and Development Program of China [2022YFC2905003], National Natural Science Foundation of China [51974012], and the China Scholarship Council [201906460047].

References

Alsayed MI (2002) Utilising the hoek triaxial cell for multiaxial testing of hollow rock cylinders. Int J Rock Mech Min Sci 39:355–366

Bruning T, Karakus M, Nguyen GD et al (2018) Experimental study on the damage evolution of brittle rock under triaxial confinement with full circumferential strain control. Rock Mech Rock Eng 51:3321–3341. https://doi.org/10.1007/s00603-018-1537-7

Cai S (2009) Basement of backfilling mechanism in mine. Metallurgical Industry Press, Beijing

Chang QL, Tang WJ, Xu Y et al (2018) Research on the width of filling body in gob-side entry retaining with high-water materials. Int J Min Sci Technol 28:519–524. https://doi.org/10.1016/j.ijmst.2017.12.016

Cheng A, Shu P, Zhang Y et al (2020) Acoustic emission characteristics and damage constitution of backfill-surrounding rock combination. J Min Safety Eng 37:1238–1245

Cui L, Fall M (2016) An evolutive elasto-plastic model for cemented paste backfill. Comput Geotech 71:19–29

Cui L, Fall M (2017a) Multiphysics modeling of arching effects in fill mass. Comput Geotech 83:114–131. https://doi.org/10.1016/j.compgeo.2016.10.021

Cui L, Fall M (2017b) Modeling of pressure on retaining structures for underground fill mass. Tunn Undergr Space Technol 69:94–107. https://doi.org/10.1016/j.tust.2017.06.010

Du X, Feng G, Guo Y et al (2018a) Failure analyses of unconfined CCWBM body in uniaxial compression based on central pressure variation. Waste Manag Res 36:159–168. https://doi.org/10.1177/0734242X17748365

Du F, Wang K, Wang G et al (2018b) Investigation of the acoustic emission characteristics during deformation and failure of gas-bearing coal-rock combined bodies. J Loss Prev Process Ind 55:253–266

Emad MZ (2017) Numerical modelling approach for mine backfill. Sādhanā 42:1595–1604. https://doi.org/10.1007/s12046-017-0702-0

Fang K, Fall M (2019a) Shear behavior of the interface between rock and cemented backfill: effect of curing stress, drainage condition and backfilling rate. Rock Mech Rock Eng 53:325–336. https://doi.org/10.1007/s00603-019-01909-2

Fang K, Fall M (2019b) Chemically induced changes in the shear behaviour of interface between rock and tailings backfill undergoing cementation. Rock Mech Rock Eng 52:3047–3062. https://doi.org/10.1007/s00603-019-01757-0

Fang K, Fall M (2020a) Shear behaviour of rock-tailings backfill interface: effect of cementation, rock type, and rock surface roughness. Geotech Geol Eng 39:1753–1770. https://doi.org/10.1007/s10706-020-01586-x

Fang K, Fall M (2020b) Insight into the mode i and mode ii fracture toughness of the cemented backfill-rock interface: effect of time, temperature and sulphate. Constr Build Mater 262:120860–120860. https://doi.org/10.1016/j.conbuildmat.2020.120860

Fang K, Fall M, Cui L (2021) Thermo-chemo-mechanical cohesive zone model for cemented paste backfill-rock interface. Eng Fract Mech 244:107546–107546. https://doi.org/10.1016/j.engfracmech.2021.107546
Guo S, Fall M, Haruna S (2020) Interface shear behavior of cementing underground mine backfill. Int J Geomech 20:1–17. https://doi.org/10.1061/(asce)gm.1943-5622.0001852
Gutiérrez-Ch JG, Senent S, Melentijevic S et al (2018) Distinct element method simulations of rock-concrete interfaces under different boundary conditions. Eng Geol 240:123–139. https://doi.org/10.1016/j.enggeo.2018.04.017
Kartawidjaja J (2020) Experimental study of the effect of passive interface resistance on the multiphysics processes in cemented paste backfill. Dissertation, Lakehead University
Koupouli NJF, Belem T, Rivard P et al (2016) Direct shear tests on cemented paste backfill–rock wall and cemented paste backfill–backfill interfaces. J Rock Mech Geotech Eng 8:472–479. https://doi.org/10.1016/j.jrmge.2016.02.001
Koupouli N, Belem T, Rivard P (2017) Shear strength between cemented paste backfill and natural rock surface replicas. Australian Centre for Geomechanics, Perth, pp 375–385
Li B, Lan J, Si G et al (2020) Nmr-based damage characterisation of backfill material in host rock under dynamic loading. Int J Min Sci Technol 30:329–335. https://doi.org/10.1016/j.ijmst.2020.03.015
Liu E-y, Zhang Q-l, Feng Y et al (2017a) Experimental study of static and dynamic mechanical properties of double-deck backfill body. Environ Earth Sci 76:1–7. https://doi.org/10.1007/s12665-017-7011-0
Liu G, Li L, Yang X et al (2017b) Numerical analysis of stress distribution in backfilled stopes considering interfaces between the backfill and rock walls. Int J Geomech 17:06016014. https://doi.org/10.1061/(asce)gm.1943-5622.0000702
Liu J (2020) Evolution of stress and displacement and stability of surrounding rock mass-backfill combination system in backfilling mining. Dissertation, China University of Mining and Technology
Long K, Du X, Xu S et al (2016) Maximizing the effective young's modulus of a composite material by exploiting the poisson effect. Compos Struct 153:593–600. https://doi.org/10.1016/j.compstruct.2016.06.061
Nabassé JFKT (2016) Direct shear tests on cemented paste backfill-rock wall and cemented paste backfill-backfill interfaces. J Rock Mech Geotech Eng 8:472–479
Sinclair T, Shillabeer J, Herget G (1980) Applications of a computer model to the analysis of rock-backfill interaction in pillar recovery operations
Song W, Zhu P, Qi W et al (2017) Coupling mechanism of rock-backfill system under triaxial compression. J Min Safety Eng 34:573–579
Tan YY, Yu X, Song WD et al (2018) Experimental study on combined pressure-bearing mechanism of filling body and surrounding rock. J Min Safety Eng 35:1071–1076
Tan YY, Davide E, Zhou YC et al (2020) Long-term mechanical behavior and characteristics of cemented tailings backfill through impact loading. Int J Miner Metall Mater 27:140–151. https://doi.org/10.1007/s12613-019-1878-6
Tatsuoka F (1985) On the angle of interface friction for cohesionless soils. Soils Found 25:135–141. https://doi.org/10.3208/sandf1972.25.4_135
Wang Y, Li CH, Hou ZQ (2018) Mechanical behaviors of bimsoils during triaxial deformation revealed using real-time ultrasonic detection and post-test CT image analysis. Arab J Geosci 12:1–16. https://doi.org/10.1007/s12517-018-4179-x
Wang Y, Wang H, Zhou X et al (2019) In situ x-ray CT investigations of meso-damage evolution of cemented waste rock-tailings backfill (CWRTB) during triaxial deformation. Miner 9:52. https://doi.org/10.3390/min9010052
Wang J, Fu J, Song W et al (2020a) Particle flow simulation of mechanical properties and microcrack evolution characteristics of rock-backfill combined model. J China Univ Min Technol 49:453–462

Wang Y, Xiao Y, Hou Z et al (2020b) In situ x-ray computed tomography (CT) investigation of crack damage evolution for cemented paste backfill with marble waste block admixture under uniaxial deformation. Arab J Geosci 13:1018. https://doi.org/10.1007/s12517-020-06003-4

Weilv W, Xu W, Jianpin Z (2021) Effect of inclined interface angle on shear strength and deformation response of cemented paste backfill-rock under triaxial compression. Constr Build Mater 279:122478. https://doi.org/10.1016/j.conbuildmat.2021.122478

Xu X, Fall M, Alainachi I et al (2020) Characterisation of fibre-reinforced backfill/rock interface through direct shear tests. Geotech Res 7:11–25. https://doi.org/10.1680/jgere.19.00029

Yang S-Q, Chen M, Huang Y-H et al (2020) An experimental study on fracture evolution mechanism of a non-persistent jointed rock mass with various anchorage effects by DSCM, ae and x-ray CT observations. Int J Rock Mech Min Sci 134:104469. https://doi.org/10.1016/j.ijrmms.2020.104469

Yu X (2022) Macro-meso mechanical properties and fracture evolution mechanisms of backfill-rock under triaxial effect. Dissertation, University of Science and Technology Beijing

Yu X, Kemeny J, Li J et al (2021a) 3D observations of fracturing in rock-backfill composite specimens under triaxial loading. Rock Mech Rock Eng 54:6009–6022. https://doi.org/10.1007/s00603-021-02498-9

Yu X, Kemeny J, Song WD et al (2021b) Progressive fracturing of rock/backfill specimens under triaxial compression—experiments and modeling. In: 55th U.S. rock mechanics/geomechanics symposium. OnePetro

Yu X, Kemeny J, Tan YY et al (2021c) Mechanical properties and fracturing of rock-backfill composite specimens under triaxial compression. Constr Build Mater, 304:124577. ARTN 12457710.1016/j.conbuildmat.2021.124577

Yu X, Song W, Tan Y et al (2022a) Triaxial cyclic loading fracture of rock-encased-backfill structure by xray and acoustic emission. In: International geomechanics symposium

Yu X, Song WD, Tan YY et al (2022b) Energy dissipation and 3d fracturing of backfill-encased-rock under triaxial compression. Constr Build Mater 341. ARTN 12787710.1016/j.conbuildmat.2022.127877

Yu X, Tan, Y, Song, W et al (2023) Strength and failure mechanism of the rock-encased by backfills structure. J China Univ Min Technol 1–13

Zhang Y (2020) Study on the seepage regularity and water inrush mechanism of fracture rock mass under coupling action of stress field and hydraulic force. Dissertation, University of Science and Technology Beijing

Zhang F, Liu W, Shen L (2012) Damage constitutive model for cemented paste backfill after mixing waste rock. In: World automation congress. IEEE, pp 1–4

Zhao L (2021) Numerical investigation on the mechanical behaviour of combined backfill-rock structure with KCC model. Constr Build Mater 283:122782

Zhao W, Chen W, Zhao K (2018) Laboratory test on foamed concrete-rock joints in direct shear. Constr Build Mater 173:69–80. https://doi.org/10.1016/j.conbuildmat.2018.04.006

Chapter 22
Effect of Initial Freeze–Thaw Damage on Sandstone Failure and Energy Dissipation Mechanisms Under Triaxial Loading

Shanpeng Cao, Caichu Xia, Feng Gao, Keping Zhou, and Wenbo Peng

Abstract To determine the damage nature of cyclic freeze–thaw rock under triaxial stress, the triaxial compression test and freeze–thaw test were performed on sandstone. Based on the energy principle, the energy dissipation mechanism in the process of freeze–thaw rock deformation and failure under the action of surrounding rock is analysed. The experiments show that the confining pressure influences not only promotes the conversion of absorbed energy to elastic energy, but also inhibits the release of elastic energy. The energy storage rate of rocks gradually downscales with freeze and thaw cycle. When the confining pressure increased, the influence level of freeze–thaw on the energy storage capacity increases. The impact of freeze–thaw on the extreme of rock elastic energy represents a changing tendency that first increases and then decreases under low and high confining pressures, and keeps increasing under medium confining pressures. Freeze–thaw reduces the storage efficiency of elastic energy. But the confining pressure promotes the effective conversion of elastic energy. The elastic limit of energy is a comprehensive manifestation of the effects of the freeze-thaw and confining pressure.

Keywords Freeze–thaw · Energy dissipation · Triaxial loading · Confining pressure effect · SEM

S. Cao
College of Civil Engineering, Tongji University, Shanghai, China

C. Xia (✉)
Institute of Rock Mechanics, Ningbo University, Ningbo, China
e-mail: tjxiaccb@126.com

F. Gao · K. Zhou
School of Resource and Safety Engineering, Central South University, Changsha, China

W. Peng
China Communications Second Highway Survey, Design and Research Institute Co., Ltd.,,
Wuhan, China

S. Wang et al. (eds.), *Engineering Geology for a Habitable Earth: IAEG XIV Congress 2023 Proceedings, Chengdu, China*, Environmental Science and Engineering,
https://doi.org/10.1007/978-981-99-9057-3_22

22.1 Introduction

With the smooth progress of the national top-level cooperation initiative "One Belt One Road", the successful practice of "Development of The Western Region in China", "Revitalizing Northeast China" and other strategies, numerous national infrastructure projects have been constructed and applied in cold regions, which have important national defense and international strategic significance (Bartosiewicz 2020; Li et al. 2020; Orakoglu et al. 2016). However, due to the high altitude and cold geographical environment of the region, the problem of rock mass freeze–thaw disasters has become increasingly prominent, which has seriously affected the safety of open-cast mines, slopes, and tunnels in high-cold areas (Bednarik et al. 2022; Niu et al. 2021; Xu et al. 2020; Xu et al. 2022).

In order to reveal the mechanism of Rock cycle freeze–thaw damage in cold regions, some scholars have carried out a lot of research. For example, Wang, et al. (2021) used rock volume strain to describe the freeze and thaw damage variable, which was linearly linked to the number of freeze-thaws, and developed a fatigue damage model that takes into account both mechanical and freeze–thaw damager. Huang, et al. (2022) found that macropore growth is a predominant factor in the development of freeze and thaw damage in stone through microstructure-mechanical property correlation analysis. Ren, et al. (2022) found that crack damage in freeze-thawed red sandstone was mainly located on the sides and edges of the specimens, and that the crack number after rock fracture increased significantly with the number of freeze-thaws. Li et al. (2022) found that new cracks in granite from the Tibetan region developed gradually from the outside to the inside during freeze–thaw cycles. The paths were related to initial defects, including contact boundaries of different materials. Song, et al. (2021) tested freeze–thaw fatigue on Tibetan marble and concluded that both freeze–thaw cycles and fatigue loading accelerated damage to the marble. Mal'a, et al. (2022) monitored the freeze–thaw cracking of andesites as a result of ice crystallification and water pressure generation, which in turn caused the rock's mechanical properties to fail. These research results have effectively improved the understanding of rock freeze–thaw damage mechanisms from the perspectives of fine tube and macroscopic mechanics.

However, the thermodynamic view is that the evolution of the damage magnitude can be seen as an unreversible energy consumption process of the inner structures of the material (JP H 1975; Viafara and Sinatora 2010; Prigogin 1977). When rock gets damaged, microcracks gradually develop from a disorderly distribution to an orderly one, eventually forming macroscopic large cracks that lead to destabilization of the rock (Xie et al. 2004). Thus, the destruction of rocks can be seen in essence as an operation of dissipation, storage and release of energy (Chen et al. 2019a). In the study of damage mechanisms in rocks, the energy dissipation principle also yields many results. Gong et al. (2022) proposed a cyclic unloading damage intrinsic model that takes into account the cumulative effect of plastic strain; Wang et al. (2020a) developed a hard rock elastoplastic damages intrinsic model and a multi-criteria model of stress, energy, stiffness and damage. Zhang et al. (2023) proposed an index of brittle ductility for transversely isotropic rocks, using the energy evolution of the rock. It has also been applied to the study of rock damage mechanisms under dry and wet cycles (Chen et al. 2019b; Liu et al. 2022), high temperatures (Meng et al. 2022; Wang et al. 2020b), chemical corrosion and other conditions (Zhou et al. 2019; Huang et al. 2020).

Currently, new advances have also been made in the energy dissipation principle in freeze–thaw damaged rocks. For example, Wang et al. investigated the behaviors of energy evolution during rock impact failure exposed to freezing and thawing, and divided the rock stress–strain curve into four stages (Wang et al. 2017). Zhang et al. (2019) found through dynamic mechanical experiments that the total strain energy before and after the peak all followed a declining trend as freeze–thaw cycles increased. Gao et al. (2020) investigated the energy dissipation mechanism of freeze-thawed stone in uniaxial pressure, and established a strength development model of freeze-thawed stone by using the energy dissipation ratio. Feng et al. (2022) have developed a freeze–thaw damage model based on relatively variation in the energy dissipation ratio. Su et al. (2023) tested the tensile mechanical properties, energy dissipation patterns, and fracture extension and damage modes of freeze–thaw sandstones through experiments. Jia et al. (2022) used static and dynamic compression experiments to evaluate the effect of freeze–thaw cycles and strain rate on the energy dissipation of sandstones. Zhai et al. (2022) developed a dynamic damage constitutive model that accounts for dissipated energy. At present, the study of energy dissipation in freeze-thawed rocks mostly considers uniaxial action, but the actual engineering rocks in cold areas are mostly in the state of typical triaxial forces and cyclic freeze–thaw. Therefore, based on the energy dissipation principle of rock deformation and failure, triaxial compression tests of sandstone exposed to differing freeze–thaw cycles are performed and the evolution laws of dissipative energy, elastic energy, and total strain energy of typical triaxially loaded rocks are analyzed. The confining pressure influence and the freeze–thaw action of the rock energy evolution law are compared. Furthermore, the energy evolution mechanism resulting from freeze-thawed rock coupling with different confining pressures is revealed.

22.2 Testing Scheme

22.2.1 Materials

Sandstone from an open pit mine in a cold region was used as the research object in this study. The mineral composition of the rock sample as determined by X-ray diffraction is presented in Fig. 22.1. In addition, the rock samples are subject to a degree of screening to reduce errors due to quality, wave velocity and porosity.

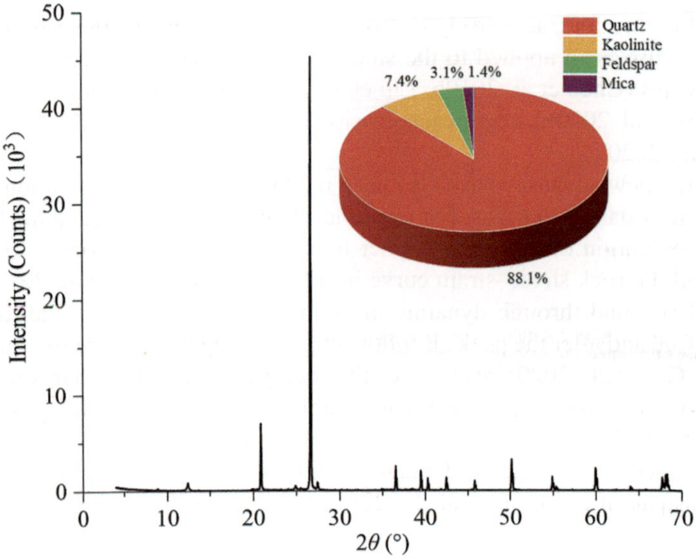

Fig. 22.1 XRD testing

22.2.2 *Equipment*

The TDS-300 freeze–thaw tester, the MesoMr21-060H magnetic resonance imaging system and the MTS815 were the main test instruments used. And the minimum freezing temperature of the TDS-300 is—40 °C, and the maximum melting temperature is + 20 °C, which can be programmed to complete each step of Freeze–Thaw cycles. MTS815's maximum axial pressure is 4600 kN and maximum confining pressure is 140 MPa.

22.2.3 *Experimental Procedure*

The TDS-300 freeze–thaw tester completes the cyclic freeze–thaw test. The freezing temperature and thawing water temperature are − 20 °C and + 20 °C respectively. Both the freezing and thawing times are 4 h. The rock sample was first dried and weighed during the cyclic freeze–thaw test. Then, the sample was fully saturated with water by using a vacuum pump, where dry pumping was carried out for 4 h and wet pumping for 2 h under 0.1 MPa conditions. The rock samples are grouped according to the experimental conditions, where the 0-cycle group underwent compression testing directly after saturation, while the other groups were subjected to 10, 20, 30, and 40 freeze–thaw cycles. Triaxial compression testing was performed using the MTS815 with axial displacement control at 0.1 mm/min. Load tests were applied to

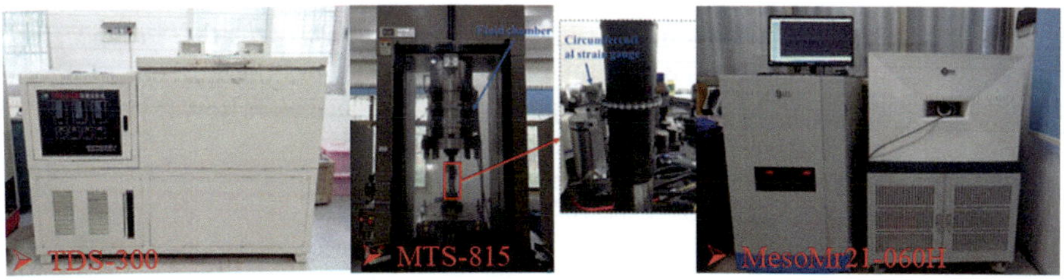

Fig. 22.2 Main test instruments

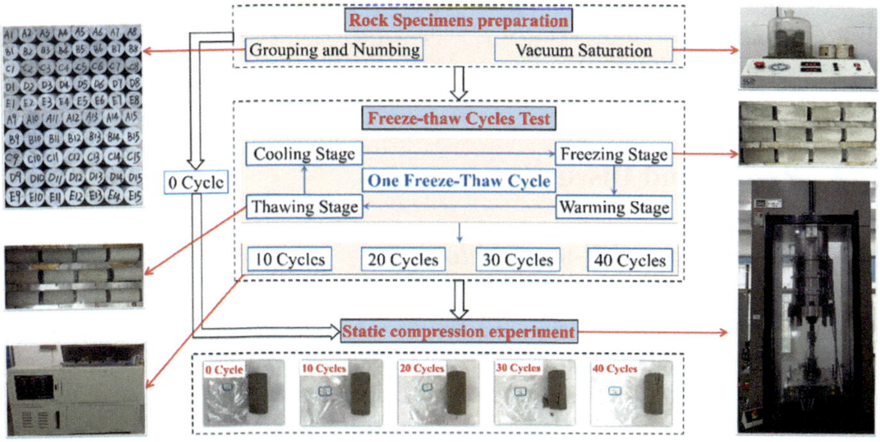

Fig. 22.3 Test procedure

sandstones subjected to various freeze–thaw cycles and confining pressures, which included 0, 3, 6, 9 and 12 MPa. The experimental procedure is depicted in Figs. 22.2 and 22.3.

22.3 Calculation of Strain Energy During Rock Failure

When rocks are deformed by exterior stresses, it is assumed that there is no heat exchange between the rocks and the surrounding environment. In accordance with the first law of thermodynamics, during the process of rock deformation, all external forces are converted into stored and dissipated energy. The total energy produced by the external force is (Xie et al. 2005):

$$U = U^e + U^d \tag{22.1}$$

where U represents the total amount of strain energy, while U^{e} and U^{d} respectively denote the amount of elastic energy that can be released and the amount of energy dissipated by the rock. And the equations for U and U^{e} are as follows:

$$U = \int_0^{\varepsilon_i} \sigma_i d\varepsilon_i \tag{22.2}$$

$$U^e = \frac{1}{2E_0}\left[\sigma_1^2 + \sigma_2^2 + \sigma_3^2 - 2\nu(\sigma_1\sigma_2 + \sigma_2\sigma_3 + \sigma_1\sigma_3)\right]\triangleleft \tag{22.3}$$

where σ_i and ε_i are the stress and strain in each location on the stress/strain graph. σ_1, σ_2 and σ_3 are 1st, 2nd and 3rd principal stresses respectively. E_0 is Young's modulus, ν is Poisson's ratio.

22.4 Results and Discussion

22.4.1 Porosity and P-wave Velocity

The energy released during deformation is not only influenced by the loading conditions but also by the intrinsic structure of the rock. Porosity and P-wave velocity of F1 and F2 subjected to a series of freeze/thaw cycles are plotted in Fig. 22.4. According to Fig. 22.4a, the porosity of sandstone increases exponentially with freezing and thawing cycles, and the growth rate of porosity also grows gradually. The cracks that have already formed will increase the speed at which the cracks develop during the next freeze–thaw cycle. This confirms that freezing and thawing resulted in cumulative damage to the rock. The P-wave velocity gradually reduces with the freeze–thaw cycles, satisfying the quadratic function evolution relationship. Taking F1 as an example, the wave speed decreased by 2.6, 3.6, 7.7 and 10.2% for each additional 10 cycles in the 0–40 freeze–thaw cycles. The cracks in the rock gradually increase under cyclic freeze–thaw, and the number of P-wave refraction and reflection increases during the propagation process. On the other hand, the increase in open pores of frozen-thawed rocks increases the water content, while the speed of longitudinal waves in water is lower than that of rocks, reducing P-wave velocity. The evidence shows that the internal structure of the rock changes under the influence of the freeze–thaw cycles, which will inevitably have an impact on the energy evolution.

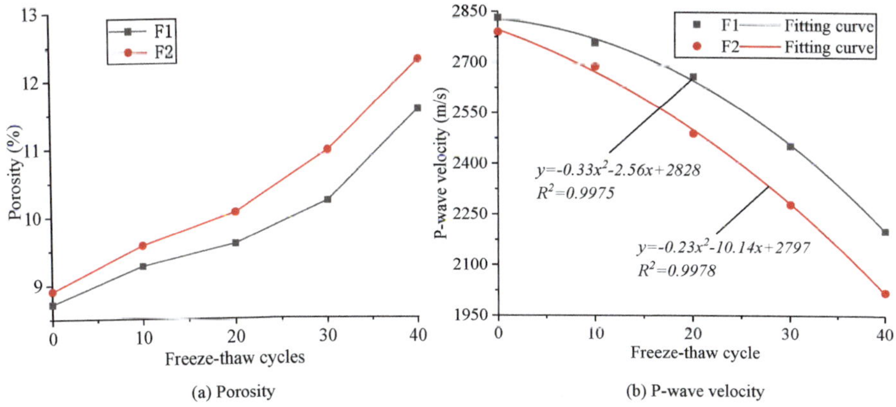

Fig. 22.4 The porosity and P-wave velocity curves of the contrast rock samples

22.4.2 *Mechanical Properties of Freeze–Thaw Sandstone*

Figure 22.5 shows the stress–strain curves of some rock samples under a variety of freeze–thaw cycles and confining pressures. The stress–strain curves of freeze–thaw damaged rock are grouped into compacting stage, elastic stage, plastic yield stage, strain softening stage and failure stage. However, as the number of freeze–thaw cycles increases, the compression phase of the stress–strain curve becomes progressively longer, and the elastic modulus in the elastic phase becomes progressively smaller. Strain softening occurs after freeze–thaw failure of the rock. Macroscopically, the peak strength of the rock specimen gradually declines with increasing number of freeze–thaw cycles. The peak strain, which corresponds to the peak strength, gradually increases. Freeze–thaw has a marked influence on rock strength and strain softening, with a decrease in elastic strain and an increase in plastic strain.

Figure 22.6a shows that the peak strength of the sandstone after freeze–thaw damage is gradually reduced. And the level of freeze–thaw degradation of peak strength at lower confining pressure is significantly lower than at higher. Under

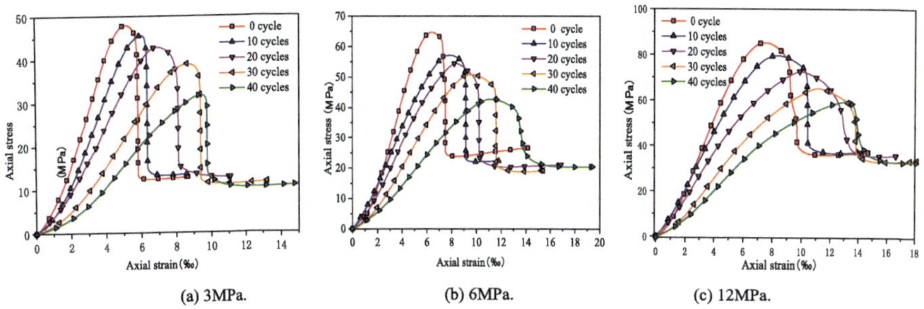

Fig. 22.5 Stress–strain curves for rocks with different freeze–thaw cycles

uniaxial conditions, the rate of peak strength loss was 0.28 MPa/cycle in this experimental study. When the confining pressure is 3 MPa, 6 MPa, 9 MPa and 12 MPa, the strength decay rate per cycle reaches 0.41 MPa, 0.55 MPa, 0.64 MPa and 0.65 MPa, respectively. The findings show that the greater the confining pressure, the faster the peak strength of the frozen-thawed stone decreases. Comparison of the relationship between residual strength and freeze–thaw cycles under various confining pressures in Fig. 22.6b. The freeze–thaw effect of residual strength is not appreciable. The study found that the freeze-thawed rock becomes very dense again as a result of the stress in the remaining distortion phase. The microcracks and micropores caused by freeze–thaw are completely compressed and closed, so the impact of freeze–thaw on the strength of the residual strain is relatively insignificant. Figure 22.6c shows the internal friction angle variation curve of freeze–thaw sandstone. For different confining pressures, the internal friction angle of sandstone showed a lineal decreasing trend with the growing number of freeze–thaw cycles. Freezing and thawing reduces mutual friction between particles in stone, a major cause of loss of strength in frozen and thawed stone. The combination of SEM scans for various states of freeze–thaw damage presented in Fig. 22.10 further supports this conclusion. The degree of weathering of the microscopic rock particles rises significantly as freeze–thaw cycles. Figure 22.6d shows the evolution curve of Young's modulus with freeze–thaw cycles. The Young's modulus decreased rapidly with increasing levels of freeze–thaw cycles, but the rate of decay decreased gradually. On the other hand, the Young's modulus of sandstone shows a tendency to slowly decrease with confining pressure. This also indicates that freeze–thaw and confining pressure are a set of external factors with opposite effects on the elastic deformation behavior of rock before failure (Fig. 22.7).

22.4.3 Mechanisms of Energy Evolution in Unfrozen-Thawed Sandstones

Stress–strain and energy curves. Characteristic stress–strain curves for unfrozen/thawed sandstone at various circumferential pressures, the curves of total strain, elastic and dissipation energy during loading and failure modes are given in Table 22.1. Comparing the rock stress–strain curve and energy curve in Table 22.1, the rock's total strain energy, elastic energy and dissipation energy before peak strength progressively increase as the axial strain progresses. The rate of increase in energy is, in descending order, total strain energy, elastic energy and dissipative energy. For the elastic deformation range, the total strain energy and the elastic energy essentially increase linearly. However, the increment of dissipative energy is small and even remains steady. Energy dissipation begins to increase rapidly during the elastoplastic phase. Under axial load, the servo tester continuously converts mechanical energy into strain energy. Most of the strain energy is held as elastic energy in the rock, and the rest is used for the extension of cracks and pores and friction between

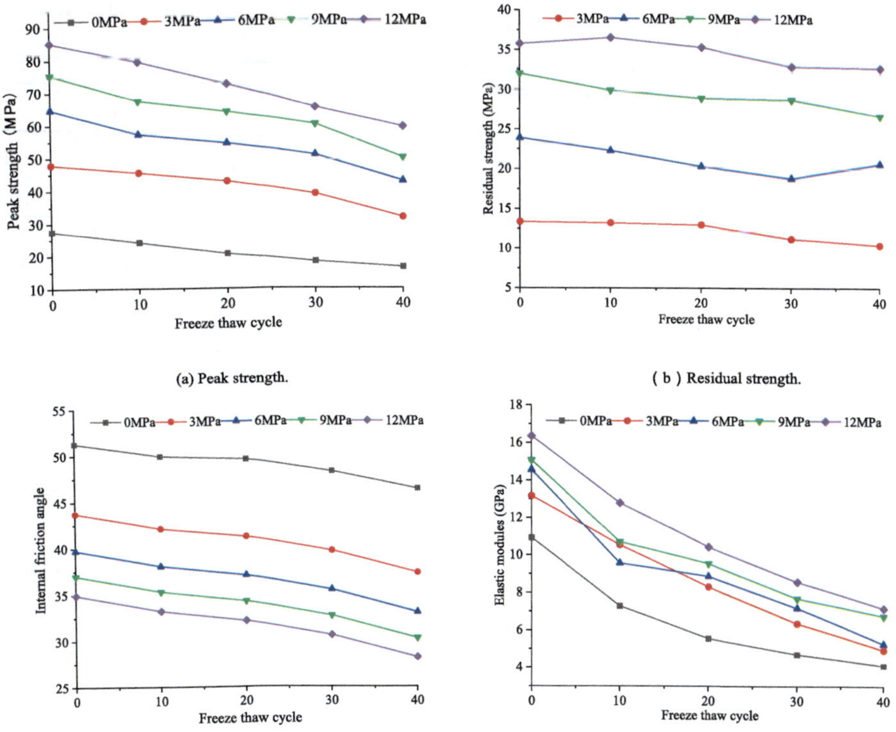

(a) Peak strength.

(b) Residual strength.

(c) Internal friction angle.

(d) Young's modulus.

Fig. 22.6 The relationship between sandstone mechanical parameters and the number of freeze–thaw cycles

particles. The energy evolution in front of the rock peak is a behavior with energy storage as the main body. Moreover, the evolution mode of energy does not change depending on the level of confinement pressure.

In the post-peak failure phase, the compressive pressure has a major influence on the energy development. Once the sandstone has reached its peak strength in uniaxial, the total strain energy still grows at a high speed, and the elastic energy is rapidly released. The reduced elastic energy is applied to expand and penetration of rock macroscopic cracks, and the consequence is the rapid increase of dissipative energy. Finally, the strain energy absorbed in the whole process is converted into dissipative energy. The effect of confining pressure makes the evolution of the total strain energy of rock after the peak strength exhibit a continuous high-speed linear growth, and then a low-speed linear growth trend. Elastic energy presents a trend of slow release first, then high-speed release, and finally maintain a stable level. At the same time, the dissipation energy keeps increasing after the peak intensity, and the speed of growth slowly increases. However, as the confining pressure becomes greater, the growth velocity of dissipative energy in the residual strength stage shows a gradually

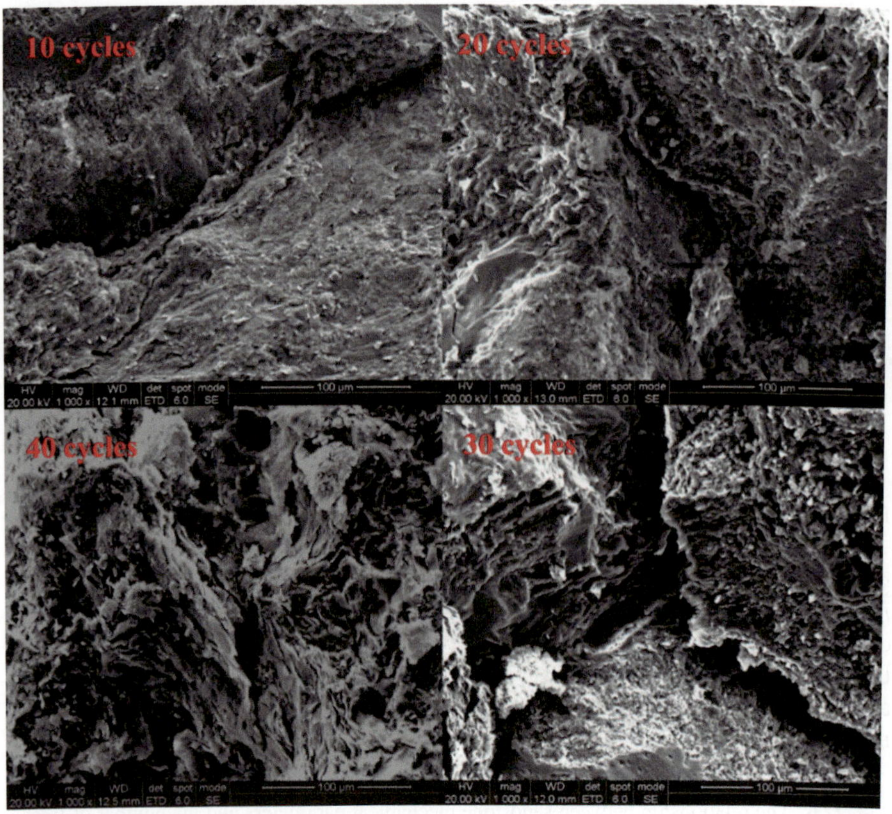

Fig. 22.7 SEM scans of different freeze–thaw damage states (1000x)

decreasing trend. Overall, there is a process of energy releasing and dissipating after peak rock strength. On the one hand, the servo testing machine continues to convert mechanical energy into the strain energy of the rock, but no new energy is stored by the rock, and it is all used for the destruction of the rock. The other side, the elastic energy stored during the pre-peak deformation phase is liberated in large quantities because it exceeds the storage limitation of the rock, resulting in the occurrence of macroscopic damage to rocks. The post-failure model of the sandstone in Table 22.1 indicates that, at low confining pressures, the main cracks of the rock sample penetrate along the vertical. Under the effect of gradually increasing confining pressure, the inclination angle of the main crack gradually decreases, and the fracture surface of the rock gradually evolves from the upper and lower surface through to the side through. The differences in the destruction of rocks are essentially caused by the different characteristics of energy storage and release within the rocks.

Table 22.1 Comparison of energy evolution curves and failure modes of sandstone

Confining pressure	Stress-strain and energy curves	Failure modes
0 MPa	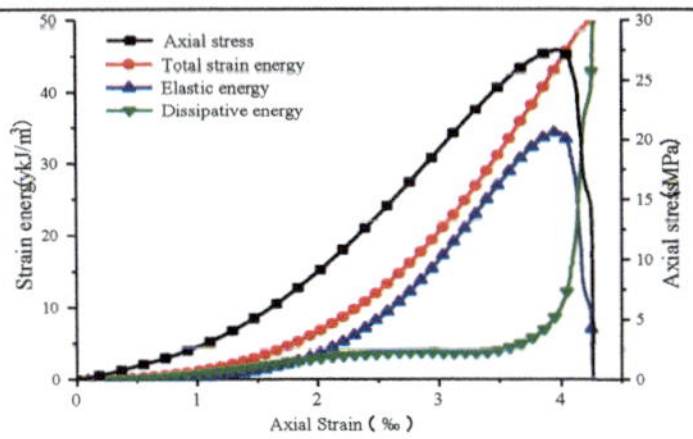	
3 MPa	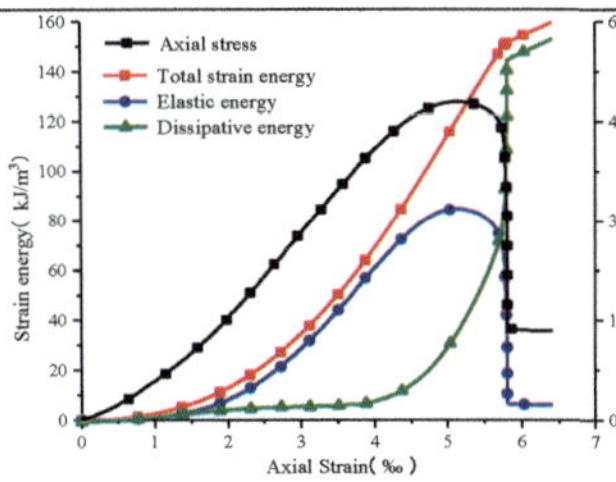	
6 Pa	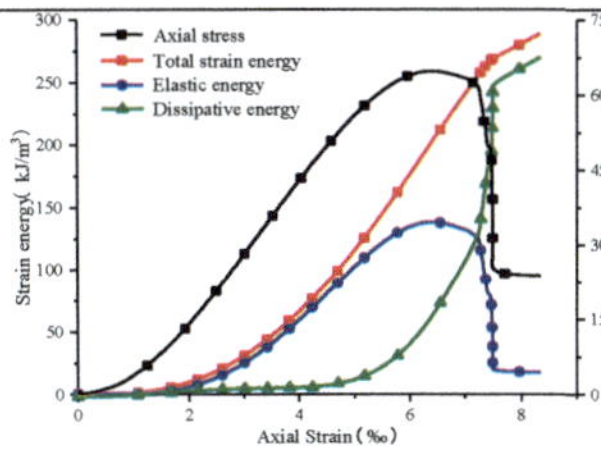	
9 MPa	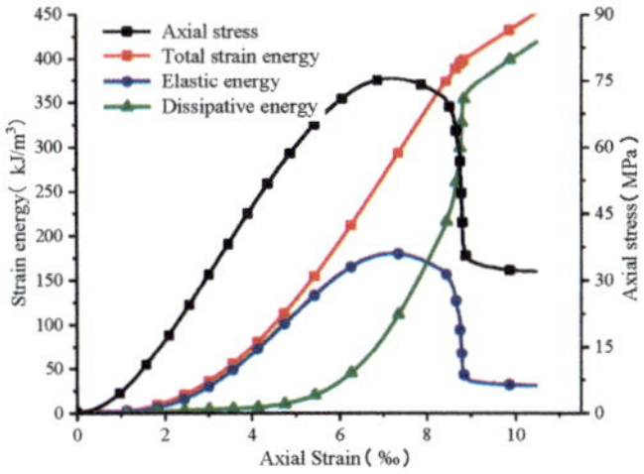	
12 MPa	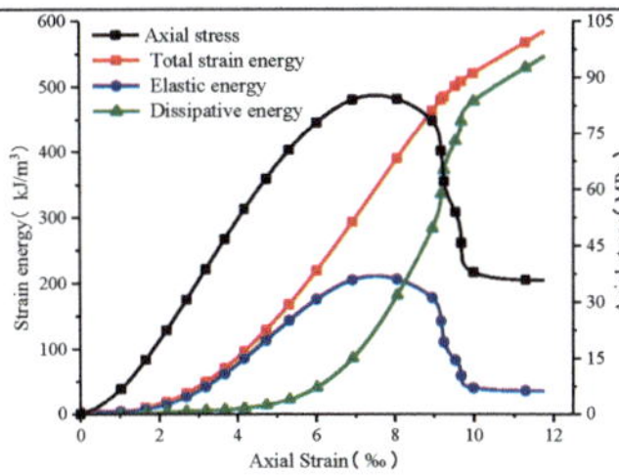	

22.4.4 Confining Pressure Effect of Sandstone Energy Evolution

The comparison of the energy curves of unfrozen and thawed sandstone at differing confining pressures is given in Fig. 22.8. According to Fig. 22.8a, the total strain energy increases as a power function during the deforming and damaging phases of the sandstone. The increase rate of the total strain energy follows an overall increasing tendency with increasing confining pressure. Moreover, the total strain energy level is closely related to the confining pressure. At the residual strength state, At the residual strength state, the linear growth rate of the total strain energy also increased continuously as confining pressure increased, and the growth speed from 3 to 12 Mpa increased from 13.49 to 38.07. In short, the confining pressure greatly enhances the rock's absorption capacity of the mechanical energy transmitted by the servo tester, whether in the forming or failure process or in the residual stage.

The elastic energy evolution curve of sandstone in Fig. 22.8b reflects that the confining pressure influences the elastic energy evolution law only slightly. When the rock sample is not macroscopically damaged, the elastic energy first follows a slow growth trend and then a linear high speed growth trend. After destruction, it first

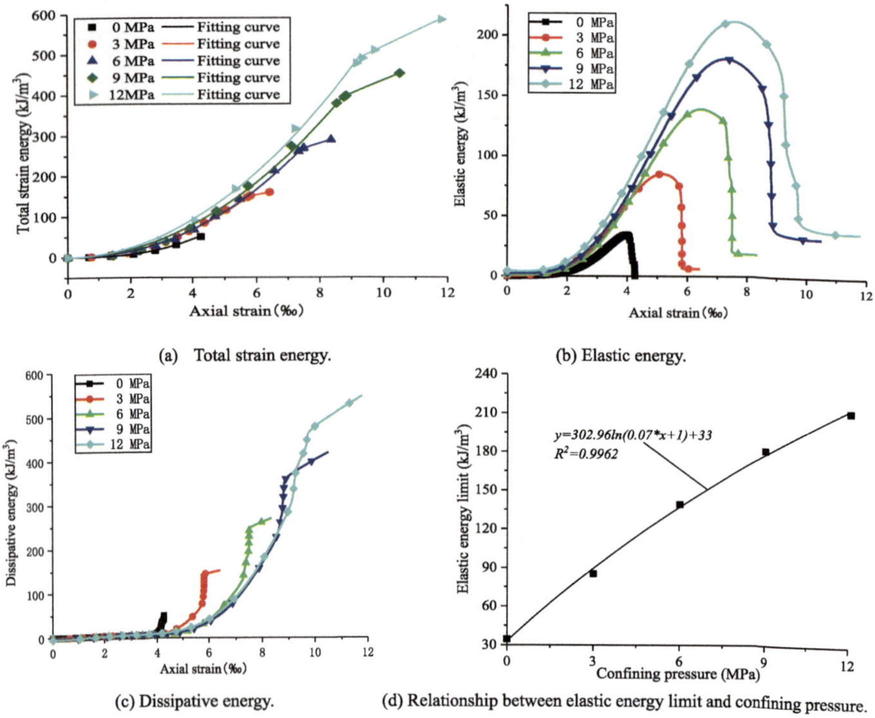

Fig. 22.8 Energy evolution curves of sandstone under different confining pressures

decreases at a moderate rate, then decays at a high rate, and finally maintains a stable trend. As the confinement pressure increases, the efficiency of the pre-peak elastic energy storage gradually increases, i.e. the growth rate of elastic energy will tend to grows with the increase in confinement pressure. When the rock sample is fractured, the rate of release of elastic energy gradually decreases with increasing confining pressure. At the same time, the elastic energy limit of sandstone (i.e. the elastic energy at peak strength) is strongly correlated with the magnitude of the confining pressure. The elastic energy limit grows approximately as a logarithmic function with the confining pressure, and the fitting line between the elastic energy limit and the confining pressure is presented in Fig. 22.8d. Furthermore, the level of rock elastic energy in the residual stage is strongly influenced by the confining pressure, and the residual elastic energy grows as the confining pressure increases. The residual elastic energy is 6.5 kJ/m^3 when the confining pressure is 3 MPa. Residual elastic energy increases 2.9, 5.0 and 5.8 times respectively as confining pressure reaches 6, 9 and 12 MPa. The analysis shows that the confining pressure influence not only effectively promotes the conversion of absorbed strain energy into elastic energy, but also significantly inhibits the liberation of elastic energy.

The evolution of the dissipative energy for various confinement pressures is shown in Fig. 22.8c. Before the elastic energy starts to release rapidly, that is, before the compressive strength is reached, the curves of the dissipative energy of sandstone at differing confining pressures are basically in a state of coincidence. Once the rock has been damaged, the dissipative energy begins to increase rapidly. The total amount of dissipated energy grows significantly with confinement pressure, Although the growth rate with low confinement pressure is obviously larger than that under high. The growth in dissipative energy in the residual phase is in line with the change in total strain energy and continues to increase linearly with the increase in axial strain. This indicates that the evolution behavior of rock energy dissipation is essentially unaffected by confining pressure, but the value of pre-peak energy dissipation and the growth rate after failure are closely related to the magnitude of confining pressure.

22.4.5 *Energy Mechanism of Freeze–Thaw Damaged Sandstone Under Triaxial Loading*

Evolution of total strain energy. In different confinement conditions, the total strain energy progression curves of sandstone undergoing to varying freeze–thaw cycles are given in Fig. 22.9. As the axial strain increases, the total strain energy of the frozen-thaw sandstone before reaching the peak strength exhibits a tendency towards slow growth and then linear high-speed growth. The overall evolution law of total strain energy has not changed because of freeze–thaw conditions. However, if the confining pressure is constant, the energy absorption rate of the rock in the compaction zone will slowly decline as the freeze–thaw cycle increases. This is because the freeze and thaw cycle reduce the storage efficiency of elastic energy in the early stages of compaction. Moreover, the full development of microcracks in the rock makes the closing process longer and consumes less energy under the same amount of deformation. According to Fig. 22.9, cyclic freezing and thawing has a pronounced influence on the energy absorption efficiency in the fast linear growth phase of total strain energy. With increasing freeze–thaw cycles, the rate of linear growth of the total strain energy gradually decreases. The sensitivity of the total strain energy absorbed by sandstone to cyclic freeze–thaw at differing confining pressures is also different. The total strain energy at the peak strength under low confining pressure shows a trend of first increasing and then decreasing with freeze–thaw cycles. Under high confining pressure, the total strain energy displays a general tendency to increase with the number of freeze–thaw cycles. Taking the confining pressure of 12 MPa as an example, the total strain energy of the rock at 0 cycles and 40 cycles is 341.5 kJ/m^3 and 431.6 kJ/m^3, respectively, with an average growth rate of 2.3 kJ/m^3. It is worth noting that the dispersion of the total strain energy taken up by the sandstone in various freeze–thaw cycles also gradually increases as the confining pressure rises. The standard deviation of the peak total strain energy of sandstone with different freeze–thaw cycles under the same confining pressure is shown in Fig. 22.10. This shows that the higher the confining pressure, the more the rock's ability to absorb energy is affected by freeze–thaw cycle.

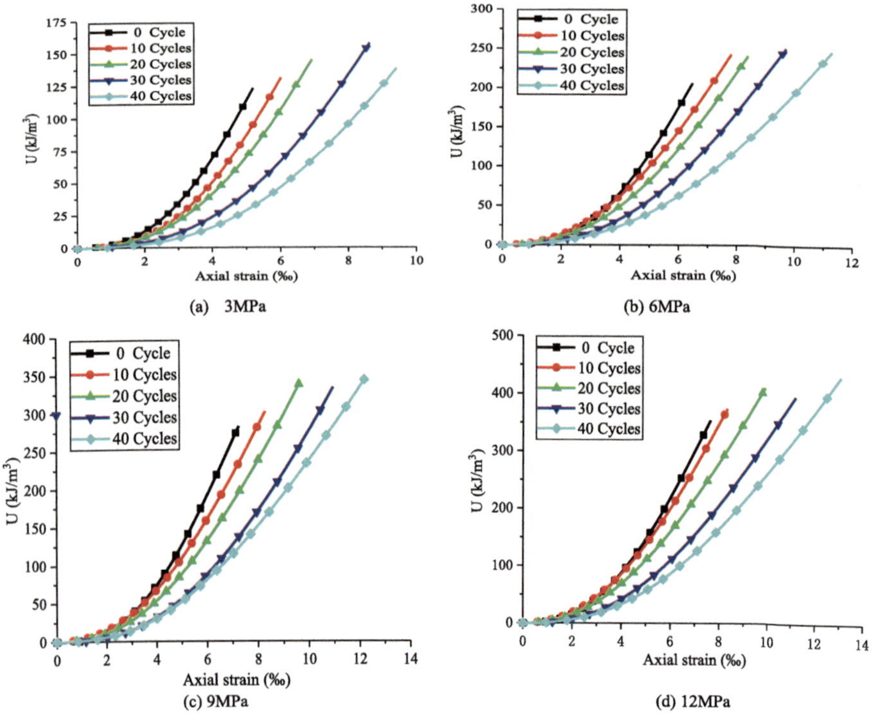

Fig. 22.9 Evolution curve of total strain energy of freeze–thaw sandstone

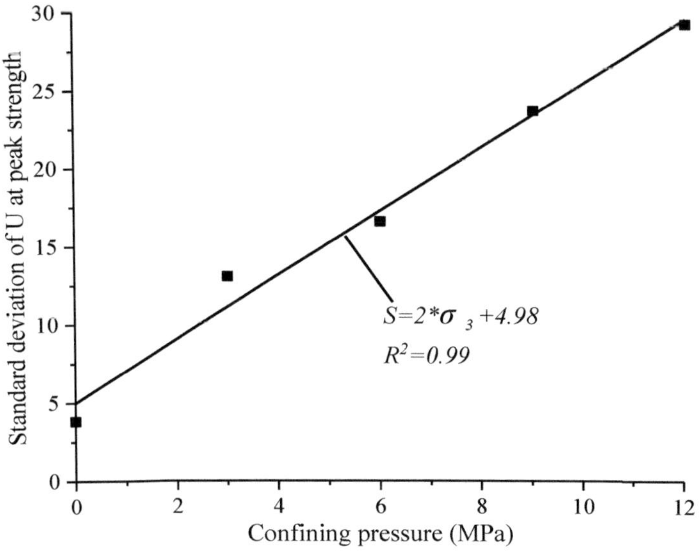

Fig. 22.10 The standard deviation of U with different confining pressure

Energy storage evolution. Energy storage during rock deformation and failure is manifested as the accumulation of elastic energy. The evolution of elastic energy during the deformation process of sandstone at different levels of confinement pressure and cyclic freeze- thaw is shown in Fig. 22.11. According to Fig. 22.11, the cyclic freeze–thaw effect not only reduces the elastic energy conversion efficiency in the compaction phase, which is dominated by energy dissipation, but also slows down the elastic energy storage speed in the elastic deformation stage. For example, under the confining pressure of 3 MPa, the energy storage rates of rocks damaged by 0, 10, 20, 30, and 40 freeze–thaw cycles in the elastic stage are 45.9, 43.6, 36.3, and 30.1 kJ/m^3, respectively. From a microscopic point of view, the freeze–thaw cycle destroys the cement between the internal particles of the sandstone, and the cohesion between the particles gradually decreases. Figure 22.12 shows the correlation between sandstone cohesion and freeze–thaw cycles. This reduces the elastic modulus of the rock, which reduces the elastic energy storage capacity of the frozen-thawed rock.

The elastic energy limit is the expression of the comprehensive mechanical characteristics of the rock, including its peak strength, deformation and Young's modulus, and other parameters. Cyclic freeze–thaw reduces the storage efficiency of elastic energy, and the confining pressure promotes the effective conversion of elastic energy. The elastic energy limit of the rock undergoing the deforming and fracturing stages is the comprehensive performance of its freezing–thawing and confining pressure levels. Combined with the experimental data in this paper, the influence of freeze–thaw cycles on the elastic energy limit of sandstone reveals a tendency to first increase and then decrease at low pressure, continuous growth under medium pressure and first increasing and then slowly decreasing under high confining pressure. Therefore,

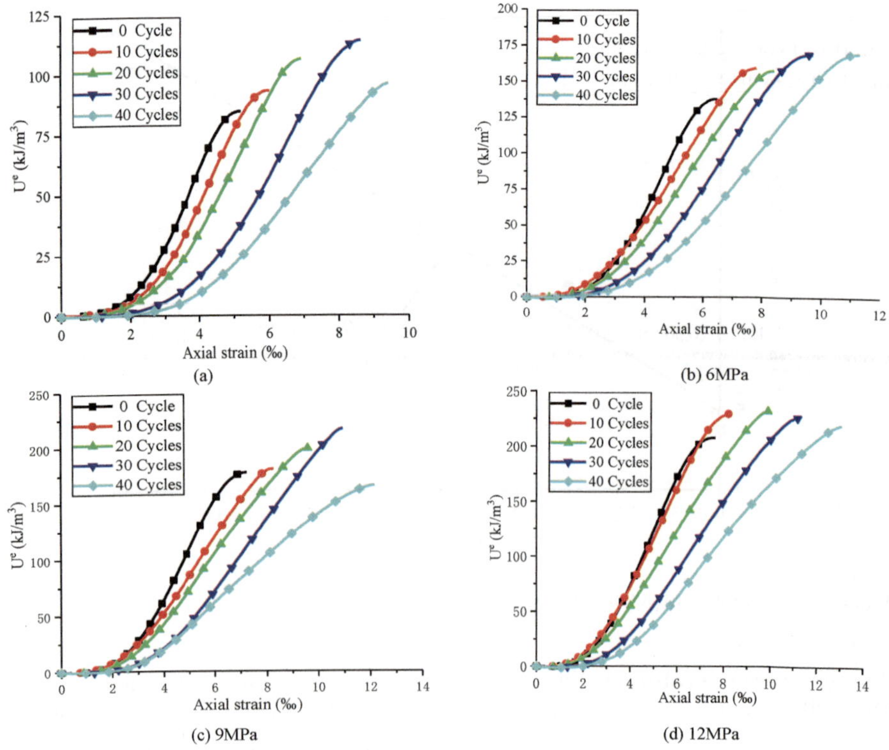

Fig. 22.11 Elastic energy evolution curve

engineering rocks in cold regions need to reasonably estimate their elastic energy limits based on their confining pressure and degree of freeze–thaw damage, and take reasonable support and prevention measures.

Energy dissipation evolution. The process of internal damage development or destruction of rocks affected by external factors is referred to as the process of energy dissipation. Figure 22.13 shows the correlation between the evolution curve of dissipated energy and the freeze–thaw cycle before the overall failure of sandstone at the various confining pressures. The growth in dissipated energy caused by the combined effect of cyclic freeze–thaw and confining pressure can be classified into four stages: low-speed growth stage, stable stage, medium-speed growth stage and high-speed growth stage. There are mainly three aspects of the influence of freeze–thaw cycles on the energy dissipation of sandstone under different stress states.

(A) Cyclic freeze—thaw has little effect on the evolution of dissipated energy before the mutation inflection point. Pressurized to the same degree, the dissipated energy growth curve before the sudden change is almost in a state of coincidence. However, the confining pressure has a significant impact on the dissipated energy before the inflection point, and the dissipated energy at the

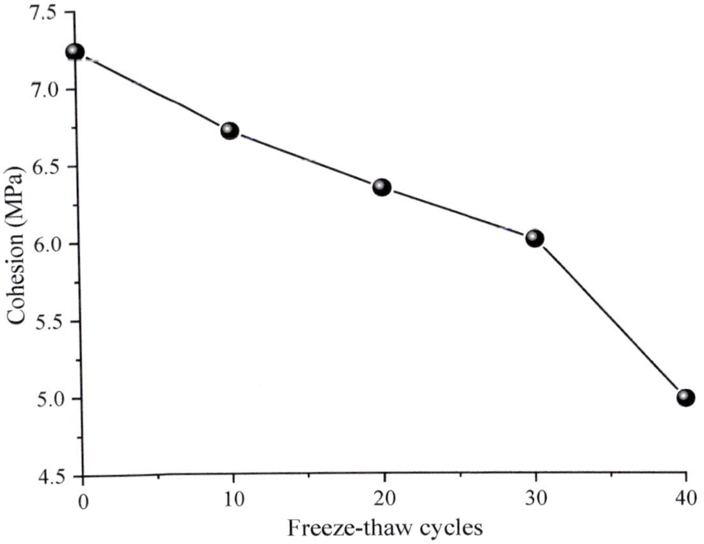

Fig. 22.12 The relationship between the cohesion of sandstone and freeze–thaw cycles

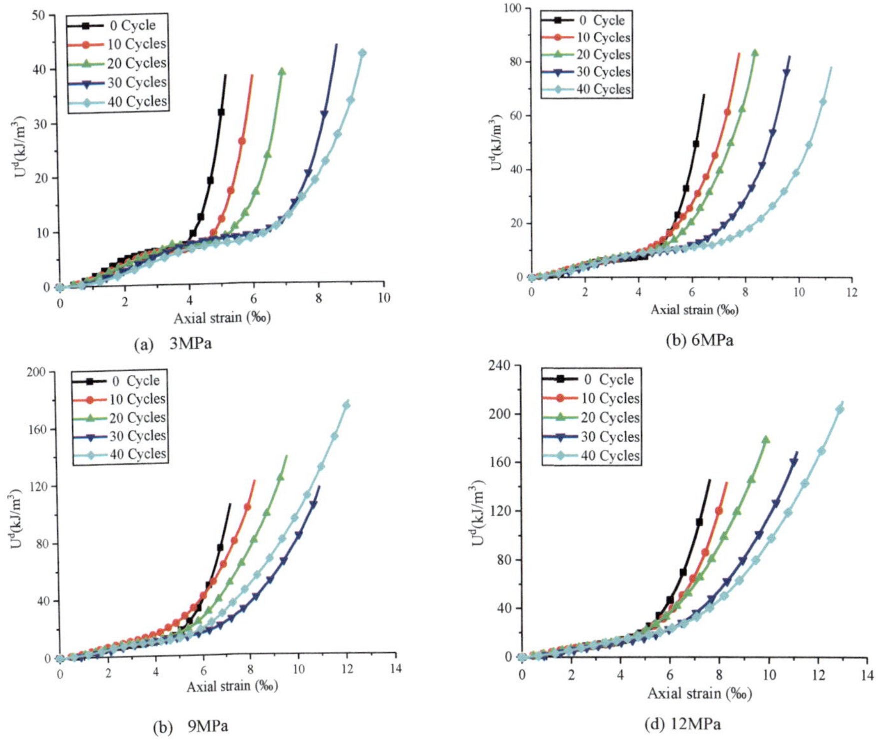

Fig. 22.13 Evolution curve of dissipation energy

inflection point reaches about 7.6, 9.9, 14.5 and 17.4 kJ/m^3 under the conditions of 3, 6, 9 and 12 MPa, respectively.

(B) The strain value at the inflection point of the dissipated energy growth abrupt change shows a tendency to gradually increase with the number of freeze–thaw cycles. Because the freeze–thaw effect reduces the elastic deformation performance of the rock, it can have a greater amount of elastic deformation. Therefore, the more freeze–thaw cycles, the greater the amount of strain before the evolution curve of dissipated energy enters the rapid growth phase. But this trend is gradually restricted as confinement pressure increases.

(C) Before the loading deformation of sandstone reaches the failure limit, the total amount of energy dissipated is affected by the combination of the number of freeze–thaw cycles and the confining pressure. Overall, the medium and low confining pressures first increase and then decrease with the increase of freeze–thaw cycles, while the trend of increasing under high confining pressures is maintained.

22.5 Conclusions

According to the principle of energy dissipation and the triaxial compression experiment of freeze–thaw sandstone, the typical energy evolution law of the deformation process of sandstone under different confining pressures is analyzed. The impact of freeze–thaw on energy storage and dissipation during sandstone loading is then compared.

(1) The total strain energy of the sandstone deformation and failure process increases in a power function. The total strain energy increases linearly in the residual strength stage. The overall increase rate of the total strain energy increases with the increase of the confining pressure, and the confining pressure greatly improves the energy absorption capacity of the rock.

(2) As the confining pressure increases, the efficiency of the storage of elastic energy prior to the peak strength increases gradually, and the release rate of elastic energy after peak strength gradually decreases. The limit level of elastic energy of sandstone is also significantly improved with the increase of confining pressure. The confining pressure effect not only effectively promotes the conversion of absorbed strain energy to elastic energy, but also inhibits its release.

(3) The energy absorption rate of rocks gradually decreases with the increase of freeze–thaw cycles. The higher the confining pressure, the more the ability to absorb energy is affected by the freeze–thaw cycle. The impact of freeze–thaw on the elastic energy limit of sandstone shows a tendency to increase and then decrease under low confining pressure, continuous growth under medium confining pressure, and first increasing and then slowly decreasing under high confining pressure.

(4) Freeze–thaw damage weakens the strength of the rock, but increases its deformation limit. Freeze–thaw overall reduce the storage efficiency of elastic energy. Confining pressure promotes the effective conversion of elastic energy and determines the level of elastic energy limit. The elastic energy limit of rock is a comprehensive manifestation of the effects of freeze–thaw and confining pressure on rock performance.

References

Bartosiewicz A (2020) Poland on the New Silk road: current state and perspectives. Łódź University Press, Łódź

Bednarik M, Holzer R, Pivko D et al (2022) Engineering geological properties of building stones from the north-eastern part of the Vienna basin in the western Carpathians. Geotech Geol Eng 40(6):3287–3305

Chen ZQ, HEC, Ma GY et al (2019a) Energy damage evolution mechanism of rock and its application to brittleness evaluation. Rock Mech Rock Eng 52(4):1265–1274

Chen XX, He P, Qin Z (2019b) Strength weakening and energy mechanism of rocks subjected to wet-dry cycles. Geotech Geol Eng 37(5):3915–3923

Feng Q, Jin JC, Zhang S et al (2022) Study on a damage model and uniaxial compression simulation method of frozen-thawed rock. Rock Mech Rock Eng 55(1):187–211

Gao F, Cao SP, Zhou KP et al (2020) Damage characteristics and energy-dissipation mechanism of frozen-thawed sandstone subjected to loading. Cold Regions Sci Technol 169

Gong FQ, Zhang PL, Du K (2022) A novel staged cyclic damage constitutive model for brittle rock based on linear energy dissipation law: modelling and validation. Rock Mech Rock Eng 55(10):6249–6262

Huang M, Xu CS, Zhan JW (2020) Assessments of the disintegrating behaviour of clay shale considering the effect of pH condition. Eur J Environ Civ Eng 24(1):97–109

Huang SB, Yu SL, Ye YH et al (2022) Pore structure change and physico-mechanical properties deterioration of sandstone suffering freeze-thaw actions. Constr Build Mater 330(5):127200

JP H (1975) Thermodynamics. New York

Jia P, Mao SZ, Qian YJ et al (2022) The dynamic compressive properties and energy dissipation law of sandstone subjected to freeze-thaw damage. Water 14(22):3632

Li X, Wu W, Liu W (2020) Analyzing the highway accessibility in the Belt and Road region based on international highway transport chain. Geogr Res 39(11):2552–2567

Li B, Zhang GH, Wang G et al (2022) Damage evolution of frozen-thawed granite based on high-resolution computed tomographic scanning. Front Earth Sci 10(5):912356

Liang C, Li X, Wang S et al (2012) Experimental investigations on rate-dependent stress-strain characteristics and energy mechanism of rock under uniaixal compression. Chin J Rock Mech Eng 31(9):1830–1838

Liu XX, Li Y, Wang WW (2022) Study on mechanical properties and energy characteristics of carbonaceous shale with different fissure angles under dry-wet cycles. Bull Eng Geol Env 81(8):319

Mal'a M, Greif V, Ondrasik M (2022) Pore structure evolution in andesite rocks induced by freeze-thaw cycles examined by non-destructive methods. Sci Rep 12(1):8390

Meng QB, Liu JF, Huang BX et al (2022) Effects of confining pressure and temperature on the energy evolution of rocks under triaxial cyclic loading and unloading conditions. Rock Mech Rock Eng 55(2):773–798

Niu CY, Zhu ZM, Zhou L et al (2021) Study on the microscopic damage evolution and dynamic fracture properties of sandstone under freeze-thaw cycles [J]. Cold Reg Sci Technol 191:103328

Orakoglu ME, Liu JK, Tutumluer E (2016) Frost depth prediction for seasonal freezing area in eastern turkey. Cold Reg Sci Technol 124:118–126

Prigogine IN G (1977) Self-organization in non-equilibrium systems

Ren JX, Yun MC, Cao XT et al (2022) Study on the mechanical properties of saturated red sandstone under freeze-thaw conditions. Environ Earth Sci 81(14):376

Song ZY, Wang Y, Konietzky H et al (2021) Mechanical behavior of marble exposed to freeze-thaw-fatigue loading. Int J Rock Mech Min Sci 138(6):104648

Su HM, Zhu ZW, Wang L (2023) Experimental and numerical analysis of dynamic splitting mechanical properties and crack propagation for frozen sandstone. Int J Rock Mech Min Sci 163:105325

Viafara C, Sinatora A (2010) Thermodynamic approaches in sliding wear: a review. Int J Mater Prod Technol 38(1):93–116

Wang P, Xu JY, Fang XY et al (2017) Energy dissipation and damage evolution analyses for the dynamic compression failure process of red-sandstone after freeze-thaw cycles. Eng Geol 221:104–113

Wang CL, He BB, Hou XL et al (2020a) Stress-energy mechanism for rock failure evolution based on damage mechanics in hard rock. Rock Mech Rock Eng 53(3):1021–1037

Wang ZL, Hao SY, Zheng J et al (2020b) Study on energy properties and failure behaviors of heat-treated granite under static and dynamic compression. Mech Adv Mater Struct 27(6):462–472

Wang Y, Zhang B, Li B et al (2021) A strain-based fatigue damage model for naturally fractured marble subjected to freeze-thaw and uniaxial cyclic loads. Int J Damage Mech 30(10):1594–1616

Xie HP, Peng RD, Ju Y (2004) Energy dissipation of rock deformation and fracture. Chin J Rock Mech Eng 21:3565–3570

Xie H, Ju Y, Li L (2005) Criteria for strength and structural failure of rocks based on energy dissipation and energy release principles. Chin J Rock Mech Eng 24(17):3003–3010

Xu P, Wu YM, Wang ZJ et al (2020) Distribution laws of freeze-thaw cycles and unsaturated concrete experiments in cold-region tunnels. Cold Reg Sci Technol 172:102985

Xu W, Wang J, Luan M (2022) The effect of freezing and thawing cycles on the mechanical properties and slope stability of the waste dump. J Mining Sci Technolo 7(2):154–165

Zhai Y, Meng FD, Li YB et al (2022) Research on dynamic compression failure characteristics and damage constitutive model of sandstone after freeze-thaw cycles. Eng Fail Anal 140:106577

Zhang J, Deng H, Taheri A et al (2019) Deterioration and strain energy development of sandstones under quasi-static and dynamic loading after freeze-thaw cycles. Cold Reg Sci Technol 160:252–264

Zhang JB, Zhang XH, Huang Z et al (2023) A brittle-ductile index for transversely isotropic rock based on energy evolution of the characteristic stress and its application. Bull Eng Geol Env 82(1):35

Zhou C, Su D, Liu Z (2019) Discussion on energy dissipation mechanism in seepage-chemical damage-softening process of soft rock. J Eng Geol 27(3):477–486

Chapter 23
Applications of Hydro-Mechanical Phase-Field Model: From Laboratory Scale to Natural Fracture Networks

Xing Jinquan, Zhao Cheng, and Niu Jialun

Abstract In recent years, the hydro-mechanical phase-field method has demonstrated its value in studying the geotechnical properties of rock mass. This paper presents a hydro-mechanical phase-field model that couples a length scale insensitive phase-field fracture theory and Biot's theory in poroelastic media. The proposed model exhibits the capability to simulate complex crack behavior, including non-planar propagation and crack coalescence, as well as accurately capturing the pressure drop when induced cracks merge with pre-existing cracks, which indicated the process of fluid filling the pre-existing cracks. Two numerical examples demonstrate the model's ability to simulate hydraulically induced fractures in geological materials across multiple scales and provide insights into fracture behavior. In fracture networks injection, the majority of the time is spent in the process of the dropped pressure gradually increasing to trigger the next fracture event. The proposed model provides an effective tool for investigating fracturing from laboratory-scale to field-scale and has promising application prospects in studying hydro-mechanical fracture behavior in geological materials.

Keywords Phase-field method · Hydraulic fracturing · Hydro-mechanical coupling · Fracture networks

X. Jinquan · Z. Cheng (✉) · N. Jialun
Department of Geotechnical Engineering, Tongji University, Shanghai 200092, China
e-mail: zhaocheng@tongji.edu.cn

X. Jinquan
e-mail: xingjq@tongji.edu.cn

N. Jialun
e-mail: niujialun@tongji.edu.cn

Z. Cheng · N. Jialun
Key Laboratory of Geotechnical and Underground Engineering of Ministry of Education, Tongji University, Shanghai 200092, China

S. Wang et al. (eds.), *Engineering Geology for a Habitable Earth: IAEG XIV Congress 2023 Proceedings, Chengdu, China*, Environmental Science and Engineering,
https://doi.org/10.1007/978-981-99-9057-3_23

23.1 Introduction

Hydro-mechanical numerical methods have become valuable tools in the study of induced fracturing in the field of energy engineering geology and deep earth resource exploitation, and have demonstrated its value in studying the geotechnical properties of rock mass.

Among these methods, the hydro-mechanical phase-field method, which is capable of automatically simulating various aspects of fracture behavior such as nucleation, propagation, kinking, and curvilinear paths, has undergone rapid development in recent years, as demonstrated by numerous studies (Cajuhi et al. 2018; Li et al. 2022; Yu et al. 2021; Zhou et al. 2018). In the related studies, the quasi-static Biot system (Biot 1962) is usually utilized. Miehe and Mauthe (2016), Miehe et al. (2015a, 2015b) developed a macroscopic framework for fracturing in porous media using a continuum phase-field approach. Heider and Markert (2017) applied the macroscopic Theory of Porous Media to investigate the mechanical behavior of multi-field, multi-phase problems in saturated porous media. Ehlers and Luo (2017, 2018) introduced a crack-opening indicator as a variable to model opening, closing, and pre-fractured domains. Zhou et al. (2018) presented a phase-field model for fracture in poroelastic media based on the classical Biot poroelasticity theory. In our latest publication (Xing and Zhao 2023), we successfully coupled the unified phase-field fracture theory (Wu 2017) and Biot's theory, modified the volumetric strain term in the fluid mass conservation equation. Such integrative models have been demonstrated as efficacious in probing into issues pertaining to hydraulic fracturing, and consequently, enhancing our understanding of fracture mechanics. Nonetheless, the application of phase-field models to the domain of engineering is yet to be fully explored and is impeded by numerous challenges. Among these, the scale problem associated with the model stands out as a challenge.

This paper discusses the applications of hydro-mechanical phase-field models, which have been utilized at various scales ranging from several meters (Ehlers and Luo 2017; Heider and Markert 2017; Miehe et al. 2015a, b) to over 100 m (Li et al. 2022; Ni et al. 2020a, b). These models have been particularly useful in investigating hydraulic fracturing-related problems such as crack initiation and propagation in hydraulic fracturing of bedded shale (Liu et al. 2020) or naturally layered rocks (Zhuang et al. 2022a, b), interfacial damage interacting with crack propagation (Nguyen et al. 2016), and crack propagation in fracture networks (Lee et al. 2017; Li et al. 2022; Ni et al. 2020a, b), among others, with significant achievements.

However, it should be noted that the numerical tests above often use different parameters for simulations at different scales, and some may even deviate from the range of geological material parameters. The length scale parameter has a significant impact on the mechanical response in the standard phase-field model (Pham et al. 2011). If the length scale is considered as a material parameter, its value can be estimated using a 1-D analytical solution (Wu et al. 2020), which suggests that the length parameter of rock masses typically falls between several millimeters and hundreds of millimeters. However, when using finite element discretization, the maximum size of the mesh should not exceed half the length parameter (Zhou et al. 2018), as doing so may result in an excessively large number of model elements that cannot be computed. Although adaptive mesh models are available to reduce the number of elements (Lee et al. 2016), handling models with large disparities in length parameters and model dimensions, particularly those with unpredictable crack propagation paths, remains a significant challenge.

This paper introduces a novel hydro-mechanical model that couples length-insensitive phase-field fracture theory with Biot's theory. The model is demonstrated through two numerical examples: laboratory-scale hydraulic fracturing and field-scale fracture networks injection. These examples showcase the model's capacity to effectively simulate fracturing mechanisms in geological materials at multiple scales, and presents an advancement in the understanding of hydro-mechanical fracture behavior.

23.2 A Novel Hydro-Mechanical Phase-Field Model

In the quasi-static Biot system (Biot 1962), with internal sharp-crack boundary Γ, , the total energy functional (Ψ) of poroelastic solid Ω is expressed as follows:

$$\Psi = \int_{\Omega} \psi_{\varepsilon}(\boldsymbol{\varepsilon}) - \alpha p \cdot (\nabla \cdot \boldsymbol{u}) \mathrm{d}\Omega + \int_{\Gamma} G_c \mathrm{dS} - \left(\int_{\Omega} \boldsymbol{b} \cdot \boldsymbol{u} \mathrm{d}\Omega + \int_{\partial\Omega_{hi}} \boldsymbol{f} \cdot \boldsymbol{u} \mathrm{dS} \right) \tag{23.1}$$

where G_c is the critical energy release rate, α is the Biot coefficient and p is the fluid pressure, $\boldsymbol{b}$ is the body force and $\boldsymbol{f}$ is the traction in the von Neumann boundary.

By introducing a scalar field $\phi(\boldsymbol{x})$, and Wu's unified phase-field theory (Wu 2017), which is length scale insensitive, the fracture energy can be expressed as

$$\int_{\Gamma} G_c \mathrm{dS} = \int_{\Omega} \frac{G_c}{\pi} \left(\frac{1}{l_0} (2\phi - \phi^2) + l_0 |\nabla \phi|^2 \right) \mathrm{d}\Omega \tag{23.2}$$

where the internal length scale l_0 determines the width of the transition zone between fractured and intact materials.

As the level of damage increases, the solid material's capacity to store elastic energy decreases. Therefore, the energy density functional can be formulated as

$$\psi_{\varepsilon}(\boldsymbol{\varepsilon}, \phi) = \omega(\phi) \psi_0(\boldsymbol{\varepsilon}), \tag{23.3}$$

where $\omega(\phi) \in [0, 1]$ is the energetic degradation function, expressed as:

$$\omega(\phi) = \frac{(1 - \phi)^2}{(1 - \phi)^2 + a_1 \phi \cdot (1 - \frac{1}{2}\phi)}, \tag{23.4}$$

where $a_1 = \frac{4}{\pi l_0}\frac{E_0 G_c}{\sigma_c^2}$, E_0 is the material elastic modulus and σ_c is the tensile strength.

The reference energy of thermodynamic force, $y = -\psi_0(\boldsymbol{\varepsilon})$, adopts hybrid approaches (Wu 2017), revised by the maximum stress criterion (Rankine Criterion):

$$\overline{y} = \frac{1}{2E}\langle\overline{\sigma}_1\rangle^2, \tag{23.5}$$

where $\overline{\sigma}_1 = \frac{\sigma_1}{\omega(\phi)}$ is the major principal effective stress, and $\langle x\rangle = \max(0, x)$. In order to prevent the cracks recovered, a local strain-history function was introduced

$$H(\boldsymbol{x}, t) = \max_{s\in[0,t]}(\overline{y}) = \max_{s\in[0,t]}\left(\frac{1}{2E}\langle\overline{\sigma}_1(\boldsymbol{x}, t)\rangle^2\right) \text{ in } \Omega \times [0, T] \tag{23.6}$$

For the fluid flow in poroelastic media, the piecewise indicator functions Lee et al. (2016) are used to describe the parameters, such as the porosity, Biot coefficient and permeability coefficient.

$$\chi_R = \begin{cases} 1 & \phi \in [0, \phi_{c1}) \\ \frac{\phi_{c2}-\phi}{\phi_{c2}-\phi_{c1}} & \phi \in [\phi_{c1}, \phi_{c2}) \\ 0 & \phi \in [\phi_{c2}, 1] \end{cases}, \ \chi_F = 1 - \chi_R = \begin{cases} 0 & \phi \in [0, \phi_{c1}) \\ \frac{\phi-\phi_{c1}}{\phi_{c2}-\phi_{c1}} & \phi \in [\phi_{c1}, \phi_{c2}) \\ 1 & \phi \in [\phi_{c2}, 1] \end{cases}, \tag{23.7}$$

where ϕ_{c1} and ϕ_{c2} are two thresholds, and $0 \le \phi_{c1} < \phi_{c2} \le 1$, the empirical values are 0.4 and 0.98 in this paper, respectively. The parameters are modified as $(\cdot) = \chi_R(\cdot)_R + \chi_F(\cdot)_F$, and the $(\cdot)_R$ and $(\cdot)_F$ is the parameters in reservoir domain and fractured domain, respectively.

By classical Biot poroelasticity theory, the mass conservation of the quasi-static system is

$$\rho_f S\frac{\partial p}{\partial t} - \nabla \cdot \frac{\boldsymbol{K}}{\mu}(\nabla p + \rho_f \boldsymbol{g}) = Q_m - \rho_f \alpha \frac{\partial \varepsilon_{vol}}{\partial t}. \tag{23.8}$$

where the storage coefficient $S = \varepsilon_p c + \frac{(\alpha-\varepsilon_p)(1-\alpha)}{K_V}$ derived from the storage model; $\varepsilon_{vol} = \varepsilon_{ii}$ is the volumetric strain of the poroelastic media; p, ρ_f, μ, $\boldsymbol{K}$, and Q_m denote the fluid pressure, the fluid density, the fluid viscosity, the permeability matrix, and fluid source term, respectively.

It is worth emphasizing that the volumetric strain term in Eq. (23.8) is not reduced by the indicator function χ_R, as the physical meaning of the volumetric strain in the fractured domain is the volume change of the space for fluid storage.

The proposed model was implemented numerically using the multifield finite element method in Comsol Multiphysics. To avoid nonphysical oscillations (Cajuhi et al. 2018), quadratic functions were utilized for the displacements, while linear functions were employed for the pore pressure in spatial discretization (Taylor–Hood pair). The discrete functions were solved using an alternate minimization algorithm, and the subproblem was addressed through the damped Newton–Raphson method. Additionally, the second-order backward difference formula was utilized in conjunction with Anderson acceleration. A more detailed exposition of numerical solutions and verification examples can be found in our forthcoming paper to maintain brevity.

23.3 Hydraulic Fracturing Tests of Granite Specimens

In this section, the performance of the proposed model in simulating hydraulic fracturing of small-scale specimens is demonstrated. Specifically, two laboratory-scale specimens with pre-existing cracks are investigated, and the size and boundary conditions are depicted in Fig. 23.1a, b, respectively. In previous studies (Xing et al. 2021; Zhao et al. 2020), we employed Digital Image Correlation to observe the hydraulic fracturing process of granite specimens directly. The specimens were marked by one or two pre-existing cracks with a length of 12 mm, a width of 1 mm, and an inclination angle of at the center, with a rock bridge length measuring 11 mm. A vertical load of 20 MPa was applied, and fluid was injected into the pre-existing cracks at a constant rate of 5 ml/min to drive crack propagation. The parameters used in the simulations are presented in Table 23.1. Triangular finite element meshes were utilized for discretization, with a maximum size that does not exceed $l_0/2$, while the timestep was set to 0.01 s, and the length parameter was fixed at 1 mm.

Figure 23.2 presents the results of the hydraulic fracturing simulation of the laboratory-scale specimens with pre-existing cracks. During the initial stage of fluid injection, the water pressure increased linearly. After reaching the peak, the crack initiated slowly, followed by a decrease in water pressure. At the moment of rapid

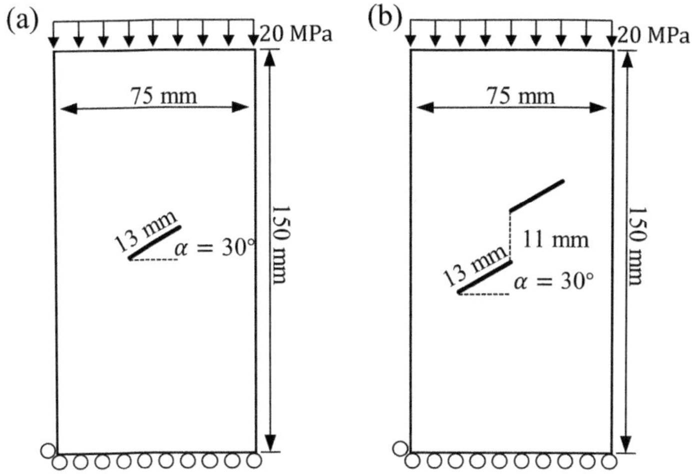

Fig. 23.1 The geometry and boundary conditions, of specimens with **a** single and **b** double pre-existing cracks, which is consistent with the experiments

Table 23.1 Material properties of poroelastic matrix and fluid

Parm	Value	Parm	Value	Parm	Value	Parm	Value
E	77.16 MPa	σ_c	6.49 MPa	ε_{pM}	5e–3	ε_{pF}	1
ν	0.23	G_c	124.55 $N\ m$	k_M	5e–19 m^2	k_F	1e–8 m^2
μ	1e-3 Pa s	c	5.1e–12 1/Pa	α_M	0.3	α_F	1

crack propagation, a cliff-like drop in water pressure was observed, which is consistent with other numerical models (Ni et al. 2020a, b), and experimental observations (Lhomme et al. 2002). This sudden reduction in pressure can be attributed to the increase in volume available for fluid accommodation following the abrupt propagation of the crack. Finally, the fluid-driven cracks propagated steadily as the crack length increased while the fluid pressure decreased.

For the single-flawed specimen, the pressure peaked (at approximately 16.2 MPa) around 14 s, and the water pressure fluctuated around the peak for several seconds. Wing-crack-like damage appeared near the pre-existing crack tips, and at 18 s, the macrocrack began to propagate from the lower crack tips. Until approximately 31 s, the macrocrack expanded rapidly from the upper tips. For the double-flawed specimen, the pressure peaked (at approximately 14 MPa) at approximately 20 s, and

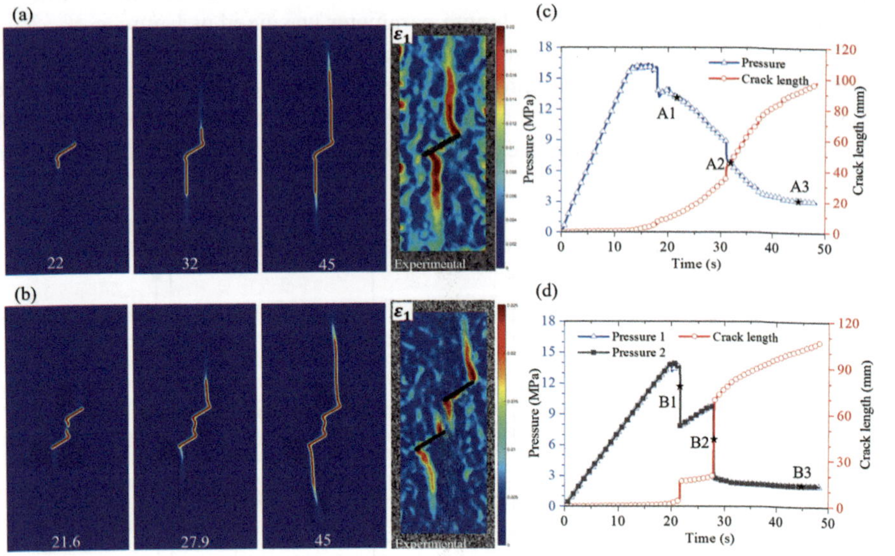

Fig. 23.2 The fracture process of numerical of specimens with **a** single pre-existing crack and **b** double cracks, and the representative moments marked A1–A3 and B1–B3 in (**c**) and (**d**), respectively. The fluid pressures at the pre-existing cracks and the crack length of specimens with **c** single pre-existing crack and **d** double cracks. The crack length is obtained by integrating the crack density function over the entire domain, and the crack length due to the pre-existing cracks is deducted

hydraulically induced fractures began to propagate from the inner tips of the pre-existing cracks at around 21.6 s, resulting in a sharp decrease in pressure to 7.84 MPa. Subsequently, the water pressure gradually increased to 9.74 MPa, and the crack rapidly propagated at 27.8 s. The water pressures in the two pre-existing cracks exhibited a high degree of consistency, becoming completely consistent after the crack coalesced in the rock bridge. During propagation, the damage zones in front of the propagating crack tips were observed.

In general, the simulated crack morphology and fracture process closely resemble those observed in the experiments, suggesting that the proposed model is well-suited for laboratory-scale specimens and capable of accurately simulating crack initiation, non-planar propagation, and coalescence. The founding of the rapid increase in crack length was accompanied by a decrease in water pressure, indicating that the crack propagated rapidly, requiring a fluid amount greater than that injected to produce new fractures.

23.4 Hydraulically Induced Cracks in Fracture Networks

Simulating fracture networks in natural geological bodies that intersect and form complex patterns is a significant challenge, while the phase-field method shows great potential for accurately simulating the intricate crack behavior and multi-field coupling processes in fracture networks (Lepillier et al. 2020; Ni et al. 2020a, b).

The simulated model consists of 15 fractures, as shown in Fig. 23.3a. All external boundaries were fixed as $\boldsymbol{u} = 0$ and $p = 0.2$ MPa, and the initial pressure was 0.2 MPa. Fluids were injected at a constant rate of 1 kg/(m s) into a borehole located at the center of the model. The tensile failure strength was set to 2 MPa; the other material parameters are listed in Table 23.1. The internal length scale was set as $l_0 = 1$ m. The model was discretized using a triangular mesh with a maximum mesh size of $h = l_0/2$, and the time step was $\Delta t = 0.01$ s.

Figure 23.4a illustrates that upon fluid injection, the pressure at the injection point rapidly increases, reaching a maximum value of approximately 1.92 MPa in 0.65 s. This leads to the initiation and propagation of cracks from the borehole, with the crack length increasing rapidly and injection pressure gradually decreasing (Fig. 23.4b). At approximately 4.5 s, the crack gradually coalesces with the existing crack, resulting in a dominant hydraulic channel. The injection pressure then drops sharply to 0.27 MPa, halting further crack propagation. Monitoring point C experiences a sudden pressure increase to the same value. Subsequently, with continuous injection, the pressure and crack length gradually increase until 15 s, when the crack rapidly grows again and connects monitoring points C and D. Similar events occur at 17.9, 29.4, and 38.5 s, increasing the connectivity of the fracture network. The sudden pressure drop followed by a gradual increase reflects the process in which the fluid rapidly fills and gradually opens the natural fractures after inducing the connection of the fractures. While, cracks at monitoring points A and B do not form any macroscopic cracks that connect with the borehole, and the pressure gradually increases by permeation,

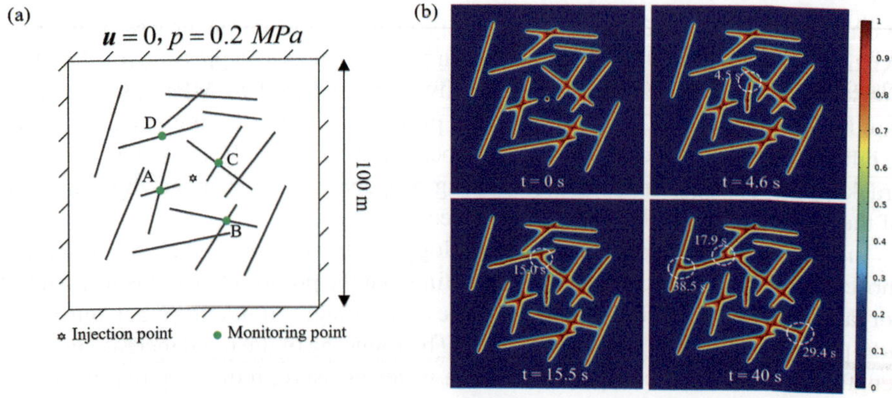

Fig. 23.3 **a** Model setup and boundary conditions. **b** Phase field distribution map at the representative moments. Dashed white lines outline the coalescence parts of the fractures, and the time indicates the moment when the coalescence occurred

eventually inducing crack propagation within 29.4 s at approximately 0.43 MPa. The simulation results demonstrate that rapid crack propagation causes a pressure drop, and when the pressure reaches a critical value, fracture networks propagate again. Moreover, the pressures in the cracks are almost identical after coalescence.

This numerical example demonstrates the capability of the proposed hydromechanical phase-field model to simulate complex crack behaviors in fracture networks, indicating promising prospects for applications. This could provide an effective tool for investigating fracture networks.

The simulated model consists of 15 fractures, as shown in Fig. 23.3a. All external boundaries were fixed as $\boldsymbol{u} = 0$ and $p = 0.2$ MPa, and the initial pressure was 0.2 MPa. Fluids were injected at a constant rate of 1 kg/(m s) into a borehole located at the center of the model. The tensile failure strength was set to 2 MPa; the other

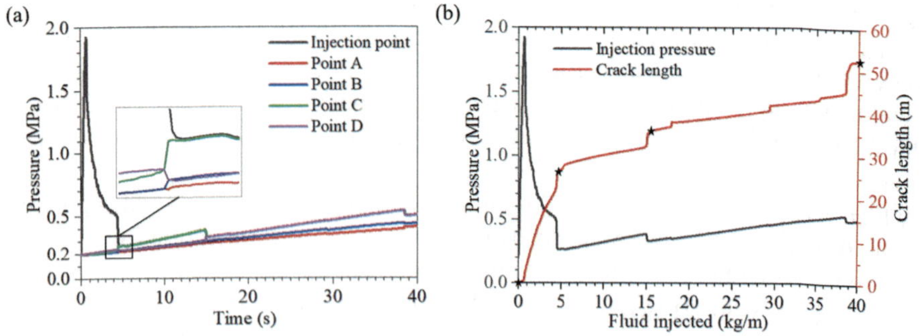

Fig. 23.4 **a** Fluid pressure at the injection point and each monitoring point relative to time. **b** Variation of the injection pressure and crack length with respect to fluid injected; the points marked correspond to Fig. 23.3b

material parameters are listed in Table 23.1. The internal length scale was set as $l_0 = 1$ m. The model was discretized using a triangular mesh with a maximum mesh size of $h = l_0/2$, and the time step was $\Delta t = 0.01$ s.

23.5 Discussion

The previously reported phase-field models (Bourdin et al. 2008; Miehe et al. 2010a, b; Zhuang et al. 2022a, b) have utilized the internal length scale (l_0) as the controlling parameter for the diffuse width of the sharp crack regularization, while l_0 also affected the mechanical response, which may limit the flexibility of the model application. For instance, the peak stress of a 1-D tension bar in Miehe's model (Miehe et al. 2010a; b) is $\sigma_c = \frac{9}{16}\sqrt{\frac{EG_c}{3l_0}}$ (Wu et al. 2020). For a certain material, if parameters E, G_c, and σ_c are determined, l_0 must be a fixed value. In other words, the internal length scale becomes a material parameter instead of a numeric parameter.

In contrast, the proposed model in this study is based on Wu's (2017) unified phase-field model, which optimizes the crack surface density function $\gamma(\phi, \nabla\phi)$ and energetic degradation function $\omega(\phi)$ to exhibit a mechanical response that is independent of the length scale parameter. Thus, the proposed model allows for the incorporation of different l_0 for materials at varying scales, thereby providing assistance in mesh partitioning and reducing computational costs.

As illustrated by the two case studies in this research, the model has demonstrated admirable simulation performance at both laboratory and field scales, which has shown great practical application potential. However, there is still room for improvement. In future studies, the authors plan to extend this method to investigate the effects of material heterogeneity on hydraulic-induced crack propagation at the laboratory scale. At the field scale, the authors plan to further improve the model by simulating the nonlinear flow properties at the constant normal stiffness boundaries of fracture networks, coupling the acoustic emission response with crack behaviour, and conducting numerical experiments to promote understanding.

23.6 Conclusions

This paper proposes a hydro-mechanical phase-field model that couples a length scale insensitive phase-field fracture theory and Biot's theory in poroelastic media. Two numerical examples of laboratory-scale hydraulic fracturing and field-scale fracture networks injection demonstrate the model's ability to simulate hydraulically induced fractures in geological materials across multiple scales and provide insight into fracture behavior. Specifically, the proposed model is capable of simulating complex crack behavior, such as non-planar propagation and crack coalescence. Additionally, the model accurately captures the pressure drop observed when induced cracks merge with pre-existing cracks, indicating that the process of fluid filling the pre-existing cracks and causing crack opens has been captured. In fracture networks injection, the majority of the time is spent in the process of the dropped pressure gradually increasing to trigger the next fracture event.

Acknowledgements This work was supported by the National Natural Science Foundation of China [grant number U22A20597, U1965108]; the Fundamental Research Funds for the Central Universities; and the Shanghai Peak Plateau Discipline (Class I).

References

Biot MA (1962) Mechanics of deformation and acoustic propagation in porous media. J Appl Phys 33(4):1482–1498

Bourdin B, Francfort GA, Marigo J-J (2008) The variational approach to fracture. J Elast 91(1):5–148

Cajuhi T, Sanavia L, De Lorenzis L (2018) Phase-field modeling of fracture in variably saturated porous media. Comput Mech 61(3):299–318. https://doi.org/10.1007/s00466-017-1459-3

Ehlers W, Luo CY (2017) A phase-field approach embedded in the theory of porous media for the description of dynamic hydraulic fracturing. Comput Methods Appl Mech Eng 315:348–368. https://doi.org/10.1016/j.cma.2016.10.045

Ehlers W, Luo CY (2018) A phase-field approach embedded in the theory of porous media for the description of dynamic hydraulic fracturing, part II: The crack-opening indicator. Comput Methods Appl Mech Eng 341:429–442. https://doi.org/10.1016/j.cma.2018.07.006

Heider Y, Markert B (2017) A phase-field modeling approach of hydraulic fracture in saturated porous media. Mech Res Commun 80:38–46. https://doi.org/10.1016/j.mechrescom.2016.07.002

Lee S, Wheeler MF, Wick T (2016) Pressure and fluid-driven fracture propagation in porous media using an adaptive finite element phase field model. Comput Methods Appl Mech Eng 305:111–132. https://doi.org/10.1016/j.cma.2016.02.037

Lee S, Wheeler MF, Wick T, Srinivasan S (2017) Initialization of phase-field fracture propagation in porous media using probability maps of fracture networks. Mech Res Commun 80:16–23. https://doi.org/10.1016/j.mechrescom.2016.04.002

Lepillier B, Yoshioka K, Parisio F, Bakker R, Bruhn D (2020) Variational phase-field modeling of hydraulic fracture interaction with natural fractures and application to enhanced geothermal systems. J Geophys Res-Solid Earth 125(7):20. https://doi.org/10.1029/2020jb019856

Lhomme T, De Pater C, Helfferich P (2002) Experimental study of hydraulic fracture initiation in Colton sandstone. Paper presented at the SPE/ISRM rock mechanics conference

Li H, Lei H, Yang Z, Wu J, Zhang X, Li S (2022) A hydro-mechanical-damage fully coupled cohesive phase field model for complicated fracking simulations in poroelastic media. Comput Methods Appl Mecha Eng 399. https://doi.org/10.1016/j.cma.2022.115451

Liu J, Liang X, Xue Y, Fu Y, Yao K, Dou FK (2020) Investigation on crack initiation and propagation in hydraulic fracturing of bedded shale by hybrid phase-field modeling. Theoret Appl Fract Mech 108:17. https://doi.org/10.1016/j.tafmec.2020.102651

Miehe C, Mauthe S (2016) Phase field modeling of fracture in multi-physics problems. Part III. Crack driving forces in hydro-poro-elasticity and hydraulic fracturing of fluid-saturated porous media. Comput Methods Appl Mech Eng 304:619–655. https://doi.org/10.1016/j.cma.2015.09.021

Miehe C, Hofacker M, Welschinger F (2010a) A phase field model for rate-independent crack propagation: robust algorithmic implementation based on operator splits. Comput Methods Appl Mech Eng 199(45–48):2765–2778. https://doi.org/10.1016/j.cma.2010.04.011

Miehe C, Welschinger F, Hofacker M (2010b) Thermodynamically consistent phase-field models of fracture: variational principles and multi-field FE implementations. Int J Numer Meth Eng 83(10):1273–1311. https://doi.org/10.1002/nme.2861

Miehe C, Mauthe S, Teichtmeister S (2015a) Minimization principles for the coupled problem of Darcy-Biot-type fluid transport in porous media linked to phase field modeling of fracture. J Mech Phys Solids 82:186–217. https://doi.org/10.1016/j.jmps.2015.04.006

Miehe C, Schanzel LM, Ulmer H (2015b) Phase field modeling of fracture in multi-physics problems. Part I. Balance of crack surface and failure criteria for brittle crack propagation in thermo-elastic solids. Comput Methods Appl Mech Eng 294:449–485

Nguyen TT, Yvonnet J, Zhu QZ, Bornert M, Chateau C (2016) A phase-field method for computational modeling of interfacial damage interacting with crack propagation in realistic microstructures obtained by microtomography. Comput Methods Appl Mech Eng 312:567–595. https://doi.org/10.1016/j.cma.2015.10.007

Ni L, Zhang X, Zou LC, Huang JS (2020a) Phase-field modeling of hydraulic fracture network propagation in poroelastic rocks. Comput Geosci 24(5):1767–1782

Ni T, Pesavento F, Zaccariotto M, Galvanetto U, Zhu Q-,Z Schrefler BA (2020) Hybrid FEM and peridynamic simulation of hydraulic fracture propagation in saturated porous media. Comput Methods Appl Mech Eng 366. https://doi.org/10.1016/j.cma.2020.113101

Pham K, Amor H, Marigo JJ, Maurini C (2011) Gradient damage models and their use to approximate brittle fracture. Int J Damage Mech 20(4):618–652

Wu JY (2017) A unified phase-field theory for the mechanics of damage and quasi-brittle failure. J Mech Phys Solids 103:72–99. https://doi.org/10.1016/j.jmps.2017.03.015

Wu JY, Nguyen VP, Nguyen CT, Sutula D, Sinaie S, Bordas SPA (2020) Phase-field modeling of fracture. In: Bordas SPA, Balint DS (eds) Advances in applied mechanics, vol 53. Elsevier Academic Press Inc., San Diego, pp 1–183

Xing JQ, Zhao C (2023) A hydro-mechanical phase field model for hydraulically induced fractures in poroelastic media. Comput Geotech 159. https://doi.org/10.1016/j.compgeo.2023.105418

Xing J, Zhao C, Huang L, Pan H (2021) Direct observations of hydraulic fracturing in rock bridge of granite specimens in grain-scale. Paper presented at the IOP conference series: earth and environmental science

Yu Z, Shao J, Duveau G, Vu M-N, Armand G (2021) Numerical modeling of deformation and damage around underground excavation by phase-field method with hydromechanical coupling. Comput Geotech 138. https://doi.org/10.1016/j.compgeo.2021.104369

Zhao C, Xing JQ, Zhou YM, Shi ZM, Wang G (2020) Experimental investigation on hydraulic fracturing of granite specimens with double flaws based on DIC. Eng Geol 267:105510. https://doi.org/10.1016/j.enggeo.2020.105510

Zhou S, Zhuang X, Rabczuk T (2018a) A phase-field modeling approach of fracture propagation in poroelastic media. Eng Geol 240:189–203. https://doi.org/10.1016/j.enggeo.2018.04.008

Zhuang X, Zhou S, Huynh GD, Areias P, Rabczuk T (2022b) Phase field modeling and computer implementation: a review. Eng Fract Mech 262:40. https://doi.org/10.1016/j.engfracmech.2022.108234

Zhuang X, Li X, Zhou S (2022a) Three-dimensional phase field feature of longitudinal hydraulic fracture propagation in naturally layered rocks under stress boundaries. Eng Comput. https://doi.org/10.1007/s00366-022-01664-z

[illegible] phase-field [illegible] Appl Math Mech (Engl Ed) [illegible]

Miehe C, Welschinger F, Hofacker M (2010b) Thermodynamically consistent phase-field models of fracture: variational principles and multi-field FE implementations. Int J Numer Meth Eng [illegible]

Miehe C, Mauthe S, Teichtmeister S (2015) Minimization principles for the coupled problem of Darcy–Biot-type fluid transport in porous media linked to phase field modeling of fracture. J Mech Phys Solids [illegible] https://doi.org/10.1016/j.jmps.2015.04.006

Miehe C, Schänzel L-M, Ulmer H (2015) Phase field modeling of fracture in multi-physics problems. Part I. Balance of crack surface and failure criteria for brittle crack propagation in thermo-elastic solids. Comput Methods Appl Mech Eng 294:449–485

Nguyen T-T, Yvonnet J, Zhu Q-Z, Bornert M, Chateau C (2016) A phase-field method for computational modeling of interfacial damage interacting with crack propagation in realistic microstructures obtained by microtomography. Comput Methods Appl Mech Eng [illegible] https://doi.org/10.1016/j.cma.2015.10.007

[illegible]

[illegible]

Chapter 24
Study on Compressive Strength Degradation of Hard Rocks Induced by Water Saturation Using Acoustic Emission

Jun Zhu, Jianhui Deng, and Fei Chen

Abstract Water–rock interaction in rock mass practice can lead to significant deterioration of rock strength, posing a substantial threat to the stability and safety of rock engineering projects. By now, the understanding of water weakening properties of hard rocks is still limited, and it remains challenging to accurately quantify and describe the extent of various water weakening effects. In this study, we focused on four representative hard rocks, namely Baoxin marble, Jinping marble, Mahu limestone, and Changtai granite, which posses unique mineral compositions and microstructures. To assess the different water weakening effects, we conducted a series of uniaxial compression tests accompanied by acoustic emission (AE) monitoring. We analyzed certain parameters of both dry and saturated hard rocks, including compressive strength and the dominant frequency of the AE waveform, to evaluate the extent of various water weakening effects. Our findings show that, when the hard rocks are saturated, they exhibited varying degree of compressive strength reductions, e.g., 26% for Baoxin marble, 6% for Jinping marble, 24% for Mahu limestone, and 9% for Changtai granite. A twin-peak feature is observed in the dominant frequency of AE waveforms, regardless of rock types and soaking states. Moreover, water saturation leads to a decrease in the micro shear failures and an increase in micro tensile failures. The degradation of rock strength in hard rocks devoid of clay and siliceous minerals is primarily attributed to the effect of pore water pressure, with the intensity depending on the mineral composition and microstructural characteristics. Stress

J. Zhu (✉)
Key Laboratory of Mountain Hazards and Earth Surface Process, Institute of Mountain Hazards and Environment, Chinese Academy of Sciences (CAS), Chengdu 610299, China
e-mail: zhujun@imde.ac.cn

J. Deng
College of Water Resources and Hydropower, State Key Laboratory of Hydraulics and Mountain River Engineering, Sichuan University, Chengdu 610065, China

F. Chen
College of Architecture and Civil Engineering, Chengdu University, Chengdu 610106, China

S. Wang et al. (eds.), *Engineering Geology for a Habitable Earth: IAEG XIV Congress 2023 Proceedings, Chengdu, China*, Environmental Science and Engineering,
https://doi.org/10.1007/978-981-99-9057-3_24

corrosion may be considered the primary reason for the strength reduction observed in saturated granite.

Keywords Hard rocks · Compressive strength degradation · Water saturation · Acoustic emission · Dominant frequency

24.1 Introduction

Water is widely acknowledged as the most influential environmental factor affecting rock mechanical behavior. Water–rock interaction has the potential to later mineral composition and microstructure of rocks, thereby leading to a decline in strength and deformation of rocks in engineering applications. Hence, the stability and safety of rock mass engineering practice in a water environment would necessitate a complete understanding of water weakening effects.

Previous studies have extensively investigated the water weakening properties in various rock types, including sandstone (Noël et al. 2021; Zhou et al. 2019), limestone (Gajić et al. 2011; Török and Vásárhelyi 2010), mudstone (Erguler and Ulusay 2009; Ewy 2014; Jiang et al. 2014) and tuff (Çelik et al. 2014; López-Doncel et al. 2013). Those studies have generally concluded that when rocks are saturated, their strength, elastic modulus, and friction coefficient exhibit varying degree of reduction, while the Poisson's ratio tends to increase. However, despite the abundant efforts on clay-rich or highly porous rocks, the study on water weakening effects on hard rocks with dense structures remains limited. Further, a plethora of explanations of mechanical degradation induced by water saturation was proposed, including (a) clay mineral softening (Hawkins and McConnell 1992; Zhang et al. 2012), (b) stress corrosion (Brantut et al. 2014; Park et al. 2010), (c) pore water pressure (Huang et al. 2019; Zhou et al. 2016), and (d) friction weakening effect (Kawai et al. 2015; Risnes et al. 2005; Sakuma 2013). Among them, the effects of clay mineral softening and stress corrosion arise in rocks containing clay and siliceous minerals, respectively. However, it is important to note that no universally accepted explanation for water weakening effects exists, as these effects often occur simultaneously. Therefore, it remains unclear which effect predominates in a given rock type. Consequently, there is a pressing need for quantitative descriptions that can elucidate the extent of various water weakening effects.

To this end, we selected four representative hard rocks with special mineral compositions as research objects. Specifically, three rocks (e.g., Baoxing marble, Jinping marble, and Mahu limestone) free of clay and siliceous minerals were used to investigate the effects of pore water pressure and friction weakening, and one granite rock with abundant quartz minerals supported the study of the extra effect of stress corrosion. Systematic uniaxial compression experiments using four rock specimens under dry and saturated conditions were performed, and the acoustic emission (AE) signals released throughout the loading process were recorded. Based on a statistical analysis of the dominant frequency of AE waveforms, the extent of multiple water weakening effects on rock compressive strength was described. This study could provide a new perspective and method to quantify the extent of water weakening effects.

24.2 Experimental Process

24.2.1 Specimen Preparation

In this study, four hard rocks were selected: Baoxin marble from Baoxin County, Jinping marble from Yanyuan County, Mahu limestone from Leibo County, and Changtai granite from Changtai County, all located in China. X-ray Crystallography and Scanning Electron Microscopy (SEM) surveys were first conducted (Fig. 24.1 and Table 24.1). These surveys revealed that Baoxin marble and Mahu limestone consist predominantly of calcite, with a composition exceeding 99.9%. Jinping marble is composed of calcite (12%), muscovite (87%), ferric oxide (0.5%) and potassium chloride (0.5%), and Changtai granite is composed of quartz (25%), albite (50%), ferric mica (15%) and hornblende (10%). It can be concluded that the first three rock types are devoid of clay and siliceous minerals, while Changtai granite does not contain clay minerals. Differences in mineral cementation morphology and micropores among four rocks can be observed by SEM images (refer to Fig. 24.1).

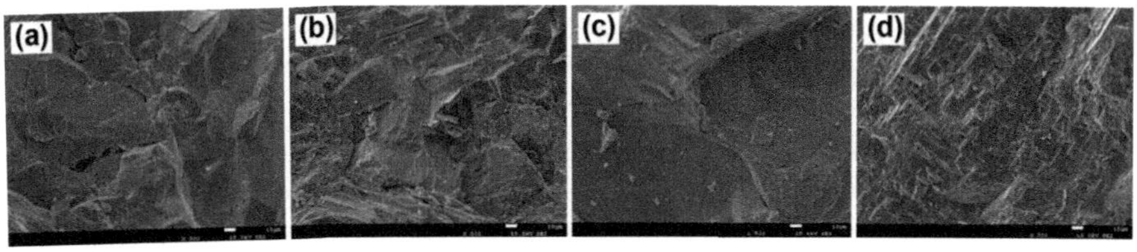

Fig. 24.1 The SEM image of different rocks, **a** Baoxin marble, **b** Jinping marble, **c** Mahu limestone, **d** Changtai granite

Table 24.1 Summary of physical parameters of rock specimens

Rock type		ρ_n (kg/m^3)	P (%)	V_p (m/s)	w (%)
Baoxin marble	$\chi(\pm\sigma)$	2.68(±0.05)	6.16(±0.25)	4262(±30.32)	2.35(±0.21)
Jinping marble	$\chi(\pm\sigma)$	2.98(±0.17)	2.68(±0.13)	4689(±29.32)	0.82(±0.09)
Mahu limestone	$\chi(\pm\sigma)$	2.63(±0.11)	4.80(±0.09)	3826(±18.67)	2.17(±0.12)
Changtai granite	$\chi(\pm\sigma)$	2.78 (±0.14)	2.13(±0.16)	5473(±26.11)	1.03(±0.14)

χ: mean value; σ: standard deviation; ρ_n: density of rocks under natural conditions; P: porosity; w: water content, for saturated samples only; V_p: P-wave velocity of natural samples

The rock samples used in this study were cylindrical in shape, with diameters of 50 mm and lengths of 100 mm, following the experimental standards suggested by the International Society for Rock Mechanics (ISRM). To ensure the reproducibility of experiment, the ultrasonic P-wave propagation velocities of all samples were measured, and any samples that showed significant difference were excluded. The porosity of samples was determined using a Nuclear Magnetic Resonance system (NMR). The basic physical parameters of the rock samples are tabulated in Table 24.1.

Uniaxial compressive tests were conducted on samples prepared under two different conditions: oven-dry and water-saturated. Initially, all the samples underwent a drying process in a 105°C oven for a minimum of 48 h. Afterward, they were transferred to a desiccator at room temperature. Half of the samples were subjected to vacuum saturation using de-aired distilled water for 48 h. Following saturation, the samples were left undisturbed for 4 h, allowing the water to permeate and fill the previously air-removed pores. The water saturation degree of two dry and saturated samples, which were selected from each kind of rock sample randomly, was measured by the atmospheric distillation method. The values of the water saturation degree for dry and saturated samples were close to 0 and 1, respectively.

24.2.2 Experimental Apparatuses and Procedure

The uniaxial compressive experiments were performed using a rock mechanics test system (Model: MTS815 Flex Test GT), alongside the PCI-2 AE monitoring system. The MTS815 has a maximum capacity of 1000 kN. During the sample deformation, axial placements were measured using a linear variable differential transformer (LVDT), with a range of ± 2.5 mm. A constant displacement rate of 0.05 mm/min was maintained during the entire loading process. For AE monitoring, eight micro 30 sensors were symmetrically positioned along the radial direction of cylinder, as illustrated in Fig. 24.2. The sampling rate was set as 1 MHz, and the pre-amplification was 40 dB. Vaseline was used between sensors and samples to enhance the connections.

24.3 Experimental Results and Discussions

24.3.1 Compressive Strength Degradation Induced by Water Saturation

The uniaxial compressive strength (UCS) σ of the four rocks was evaluated under both dry and saturated conditions, as shown in Fig. 24.3. The ratio between the UCS values obtained under saturated and dry conditions (e.g., $\sigma_{sat}/\sigma_{dry}$) was used as an indicator of the water weakening effect, referred to as the softening coefficient f_{σ}. As observed in the figure, there is a noticeable degradation in UCS for all four saturated

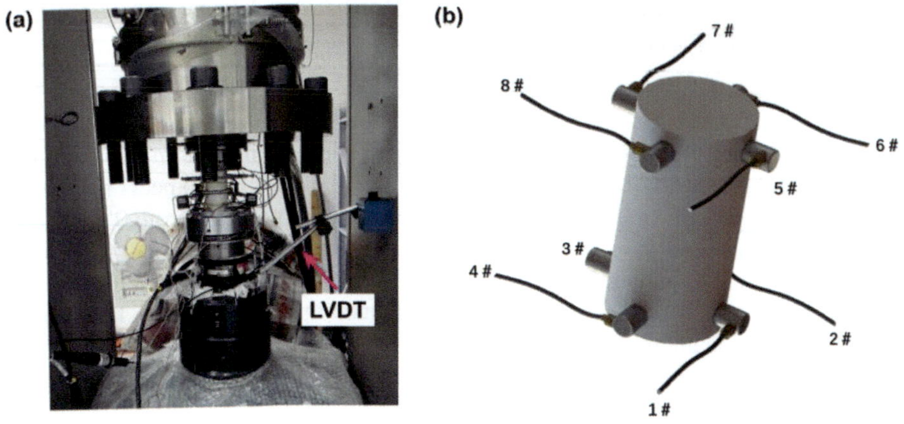

Fig. 24.2 Experimental setup and AE sensors layout

rocks. Water saturation causes a varying reduction in the rock UCS, including 26% for Baoxin marble, 6% for Jinping marble, 24% for Mahu limestone, and 9% for Changtai granite; the corresponding softening coefficients f_σ were calculated as 0.74, 0.94, 0.76, and 0.91, respectively. In comparison, Baoxin marble displayed the most significant water weakening characteristic, whereas Jinping marble exhibited the least pronounced effect.

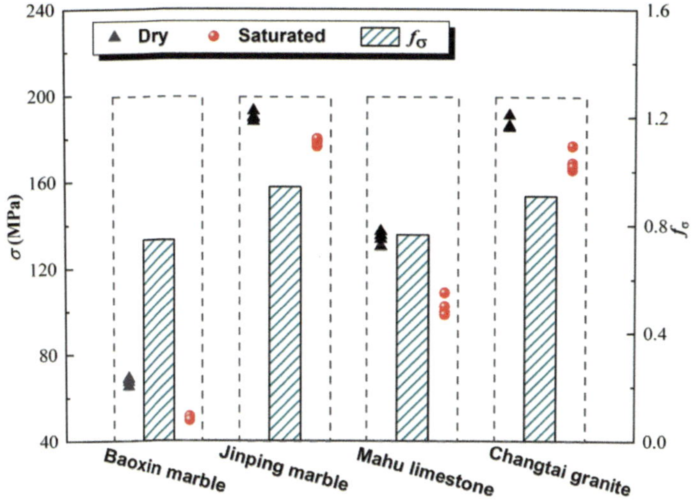

Fig. 24.3 The uniaxial compressive strength and its softening coefficient of different rocks

24.3.2 *Water Weakening Effects on Dominant Frequency of AE Waveforms*

With the aid of MATLAB, the Fast Fourier Transform (FFT) was employed to determine the dominant frequency of AE signals. Subsequently, these dominant frequencies were divided into 46 bands, each with an interval of 10 kHz, expects for the last band. For example, the first band (No. 1) encompassed frequencies from 0 to 10 kHz, while the second band (No. 2) covered frequencies from 11 to 20 kHz, and so on. It is noteworthy that dominant frequencies exceeding 450 kHz were all included in the last band (No. 46), as previous research has indicated that the dominant frequencies of rock AE signals are mostly within the range of 0–450 kHz (Cai et al. 2007). Note that, the accumulated energy of AE waveform signals was computed by squaring the amplitude.

Figure 24.4 illustrates the presence of two concentrations of dominant frequency bands, namely the low dominant frequency band (L-type band) and high dominant frequency band (H-type band), in both dry and saturated rocks. This finding aligns with previous studies (Deng et al. 2018; Li et al. 2017), confirming that twin-peak feature in the dominant frequency is inherent to rocks under compression, irrespective of their rock types and soaking states. Moreover, the existence of water seems to affect the ranges of the H-type band and L-type band.

Figure 24.5 illustrates the amount percentages and corresponding energy of H-type and L-type waveforms for the four rocks. In the case of dry rocks, H-type waveforms, characterized by high energy levels, dominate the AE waveform signals. Among the four dry rocks, the highest and lowest percentages of H-type waveforms are 84.11% and 69.51%, respectively. As rocks become saturated, the percentages of H-type waveforms decrease to varying extents, while the percentages of L-type waveforms correspondingly increase. In comparison, the energy of L-type waveforms for saturated rocks surpasses that of H-type waveforms. Moreover, the L-type waveforms of Baoxing marble, Jinping marble, and Mahu limestone become dominant, with percentages of 73.13%, 58.77%, and 71.15%, respectively. The percentage in L-type bands of saturated Changtai granite is 48.62% compared to that of H-type waveforms of 50.06%. This suggests a significant increase in the occurrence of L-type waveforms with high energy due to the existence of water.

24.4 Analysis of Water Weakening Mechanisms of Different Hard Rocks

The mineral composition of different hard rocks determines the existence of varying water weakening effects. For simplicity, we categorized Baoxing marble, Jinping marble, and Mahu limestone as type *I* rocks, while Changtai granite as type *II* rocks in this study. Type *I* rocks, devoid of clay and siliceous minerals, may exhibit water weakening effects stemming from pore water pressure and fiction weakening.

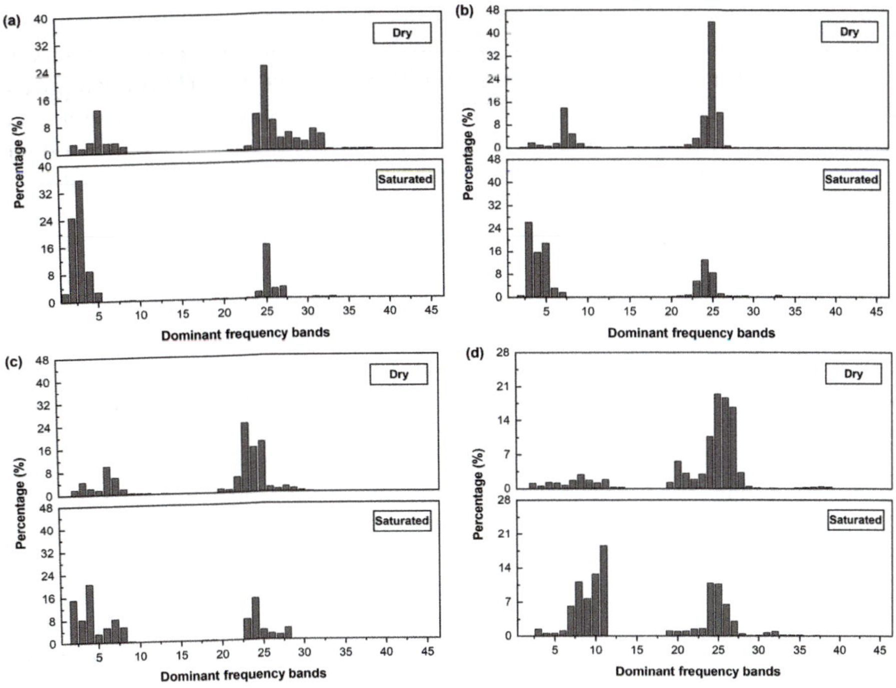

Fig. 24.4 Proportion statistics of the dominant frequency of different rocks, **a** Baoxin marble, **b** Jinping marble, **c** Mahu limestone, **d** Changtai granite

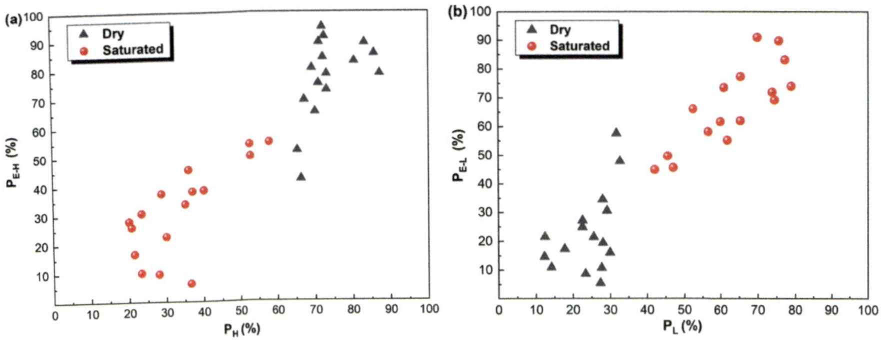

Fig. 24.5 The percentages of H-type and L-type waveforms and their corresponding energy of rocks, **a** H-type waveforms, and **b** L-type waveforms. P_H and P_L indicate the percentages of H-type waveforms and L-type waveforms, and P_{E-H} and P_{E-L} indicate the energy percentages of H-type waveforms and L-type waveforms, respectively

Our experimental results demonstrate that the AE waveforms with low dominant frequency increase significantly and become dominant when the type *I* rocks were saturated. Li et al. (2017) and Deng et al. (2018) reported that H-type waveforms correspond to micro shear failures and L-type waveforms are associated with micro tensile failures. In other words, the presence of water leads to an increase in micro tensile failures in type *I* rocks under compression. Considering the potential water weakening effects in type *I* rocks, such as pore water pressure and friction weakening, it can be deduced that the increase in micro tensile failure may originate from either one or both of them. Notably, the tensile stress around cracks is expected to increase due to the pore pressure effect, resulting in the generation of more micro tensile failures (Zhu et al. 2020). In this connection, the increase in the L-type waveforms for saturated rocks can be explained by the exertion of pore water pressure. In other words, the effect of pore water pressure corresponds to and can be characterized by L-type waveforms.

For type *I* rocks, the f_σ decreases as the percentage increment of L-type waveforms increases (see Fig. 24.6). It emphasizes the negative correlation between UCS reduction and the percentage increment of L-type waveforms, providing further evidence that pore water pressure plays a major role in UCS reduction. The percentage increments of L-type waveforms are greater for Baoxin marble and Mahu limestone, leading to more pronounced UCS reduction compared to Jinping marble. This difference is related to the mineral composition and microstructural characteristics of the three hard rocks. Baoxin marble and Mahu limestone are composed of calcite, with porosity of 6.16% and 4.80%, respectively. However, Jinping marble is composed of various minerals with a porosity of 2.68%, indicating a denser microstructure. As a result, pore water pressure is inhibited and fewer micro-tensile failures (L-type waveforms) are generated. In a word, the effect of pore water pressure depends largely on the mineral composition and microstructural characteristics of hard rocks.

For Changtai granite, the possible reasons for its UCS reduction include pore water pressure, friction weakening effect, and stress corrosion. The greater f_σ of Changtai granite indicates a less UCS reduction. Moreover, the percentages in H-type and L-type waveforms of saturated Changtai granite are almost equivalent. It suggests that the effect of pore water pressure is not dominant any longer. Note that, the UCS reduction of Changtai granite is greater than that of Jinping marble, even though it has the least porosity among the four rocks studied. It can be concluded that other effects, e.g., stress corrosion, may be responsible for the UCS reduction. However, the water weakening effects, including friction weakening and stress corrosion, are complex and future experimental investigations are required to evaluate them quantitatively.

24.5 Conclusions

More than 30 UCS experiments were conducted on samples of four different rock types to better understand the water weakening effects on rock compressive strength. The conclusions can be drawn as follows:

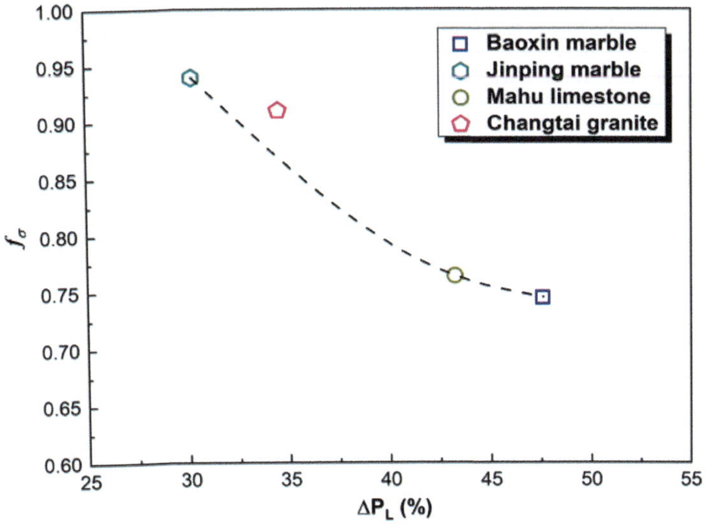

Fig. 24.6 The relationship between the percentage increment of L-type waveforms and UCS softening coefficient

(1) Water saturation of hard rocks leads to a decrease in their uniaxial compressive strength to varying degrees.
(2) A twin-peak feature is consistently observed in the dominant frequency of AE waveforms, regardless of rock types and saturated conditions. The micro shear failures of hard rocks reduce after saturation, while the micro tensile failures increase and become predominant.
(3) The strength degradation of hard rocks that are free of clay and siliceous minerals can be attributed to the effect of pore water pressure. This effect is strongly influenced by the mineral composition and microstructural characteristics of the rocks. The presence of L-type waveforms can serve as an indicator of the pore water pressure. Moreover, stress corrosion may be the primary reason for the strength reduction observed in saturated granite.

References

Brantut N, Heap MJ, Baud P, Meredith PG (2014) Mechanisms of time-dependent deformation in porous limestone. J Geophys Res Solid Earth 119(7):5444–5463

Cai M, Kaiser P, Morioka H, Minami M, Maejima T, Tasaka Y, Kurose H (2007) FLAC/PFC coupled numerical simulation of AE in large-scale underground excavations. Int J Rock Mech Min Sci 44(4):550–564

Çelik MY, Akbulut H, Ergül A (2014) Water absorption process effect on strength of Ayazini tuff, such as the uniaxial compressive strength (UCS), flexural strength and freeze and thaw effect. Environ Earth Sci 71(9):4247–4259

Deng J, Li L, Chen F, Yu J, Liu J (2018) Twin-peak frequencies of acoustic emission due to the fracture of marble and their possible mechanism. Adv Eng Sci 50(5):12–17

Erguler ZA, Ulusay R (2009) Water-induced variations in mechanical properties of clay-bearing rocks. Int J Rock Mech Min Sci 46(2):355–370

Ewy RT (2014) Shale swelling/shrinkage and water content change due to imposed suction and due to direct brine contact. Acta Geotech 9(5):869–886

Gajić V, Matović V, Vasić N, Srećković-Batoćanin D (2011) Petrophysical and mechanical properties of the Struganik limestone (Vardar zone, Western Serbia). Geoloski Anali Balkanskoga Poluostrva 72:87–100

Hawkins A, McConnell B (1992) Sensitivity of sandstone strength and deformability to changes in moisture content. Q J Eng GeolHydrogeol 25(2):115–130

Huang Y, Deng J, Zhu J (2019) An experimental investigation of moisture-induced softening mechanism of marble based on quantitative analysis of acoustic emission waveforms. Appl Sci-Basel 9(3):1–13

Jiang Q, Cui J, Feng X, Jiang Y (2014) Application of computerized tomographic scanning to the study of water-induced weakening of mudstone. Bull Eng Geol Env 73(4):1293–1301

Kawai K, Sakuma H, Katayama I, Tamura K (2015) Frictional characteristics of single and polycrystalline muscovite and influence of fluid chemistry. J Geophys Res Solid Earth 120(9):6209–6218

Li L, Deng J, Zheng L, Liu J (2017) Dominant frequency characteristics of acoustic emissions in white marble during direct tensile tests. Rock Mech Rock Eng 50(5):1337–1346

López-Doncel R, Wedekind W, Dohrmann R, Siegesmund S (2013) Moisture expansion associated to secondary porosity: an example of the Loseros Tuff of Guanajuato Mexico. Environ Earth Sci 69(4):1189–1201

Noël C, Baud P, Violay M (2021) Effect of water on sandstone's fracture toughness and frictional parameters: Brittle strength constraints. Int J Rock Mech Min Sci 147:104916

Park N, Olson JE, Holder J (2010) Stress-corrosion cracking as an alternative time-dependent shale-stability model. SPE Drill Complet 25(02):168–176

Risnes R, Madland M, Hole M, Kwabiah N (2005) Water weakening of chalk—mechanical effects of water–glycol mixtures. J Petrol Sci Eng 48(1–2):21–36

Sakuma H (2013) Adhesion energy between mica surfaces: implications for the frictional coefficient under dry and wet conditions. J Geophys Res Solid Earth 118(12):6066–6075

Török Á, Vásárhelyi B (2010) The influence of fabric and water content on selected rock mechanical parameters of travertine, examples from Hungary. Eng Geol 115(3–4):237–245

Zhang N, He M, Liu P (2012) Water vapor sorption and its mechanical effect on clay-bearing conglomerate selected from China. Eng Geol 141:1–8

Zhou Z, Cai X, Ma D, Du X, Chen L, Wang H, Zang H (2019) Water saturation effects on dynamic fracture behavior of sandstone. Int J Rock Mech Min Sci 114:46–61

Zhou Z, Cai X, Zhao Y, Chen L, Cheng X, Li X (2016) Strength characteristics of dry and saturated rock at different strain rates. Trans Nonferrous Metals Soc China 26(7):1919–1925

Zhu J, Deng J, Chen F, Huang Y, Yu Z (2020) Water saturation effects on mechanical and fracture behavior of marble. Int J Geomech 20(10):04020191

Chapter 25
A Micromechanics-Based Damage Model for Freeze–Thaw Rock Considering Moisture Migration

Huiguan Chen, Cheng Zhao, Jinquan Xing, Lin Huang, and Hongwei Yang

Abstract Under the freeze–thaw cycle, violent moisture migration will occur inside the rock mass, which will change the state of microcracks in the rock mass and ultimately affect its macro-mechanical behaviors. In this paper, the amount of moisture migration was introduced into the thermodynamic system as an intrinsic variable. Based on the friction-damage coupling model, the micro-elastoplastic damage constitutive model considering moisture migration was derived. The proposed model was verified by conventional triaxial tests of rocks under different freeze–thaw cycles and the model can satisfactorily simulate the stress–strain curve of the rocks. The amount of moisture migration under different freeze–thaw cycles is related to the porosity of rock masses. The elastic modulus, compressive strength and plastic strain at the peak strength are all linearly related to the amount of moisture migration. Moisture migration can reduce the strength characteristics of the rock and increase its plastic deformation. The model considers the moisture migration mechanism inside the rock during the freeze–thaw cycle and the micro-physical mechanism of rock failure. Therefore, the parameters in the model have clear physical meanings.

H. Chen · C. Zhao (✉) · J. Xing · L. Huang · H. Yang
Department of Geotechnical Engineering, Tongji University, Shanghai 200092, China
e-mail: zhaocheng@tongji.edu.cn

H. Chen
e-mail: chenhuiguan@tongji.edu.cn

J. Xing
e-mail: xingjq@tongji.edu.cn

L. Huang
e-mail: huanglin206@tongji.edu.cn

H. Yang
e-mail: yanghw@tongji.edu.cn

C. Zhao
Key Laboratory of Geotechnical and Underground Engineering of Ministry of Education, Tongji University, Shanghai 200092, China

S. Wang et al. (eds.), *Engineering Geology for a Habitable Earth: IAEG XIV Congress 2023 Proceedings, Chengdu, China*, Environmental Science and Engineering,
https://doi.org/10.1007/978-981-99-9057-3_25

Keywords Freeze–thaw cycles · Moisture migration · Micro-mechanics · Elastoplastic damage model

25.1 Introduction

The permanent or seasonal cold regions account for more than 50% of the world's land area and more than 70% of the China's land area, mostly distributed in high latitudes or plateaus (Chen et al. 2011). After multiple freeze–thaw cycles, the macroscopic and microscopic physical mechanics and deformation characteristics of rocks will change, which will affect the safety of mining, roads, and water conservancy projects in cold areas (Yang et al. 2018).

In the study of freeze–thaw damage mechanism, the migration of moisture inside the rock mass is an important reason for the deterioration of its characteristics (Cheng et al. 2020). During the freezing process, the in-situ freezing of the water in the pores has a volume increase of 9%. When there is external water supply, the water in the unfrozen area will continuously migrate to the frozen area, and the volume will continue to increase on the basis of 9%. There are pores, microcracks, joints and other defects of different scales inside the rock mass. Moisture migrates inside the rock mass and freezes into ice in these defects. Under the repeated action of frost heave loads, microcracks germinate, expand, and penetrate, which ultimately leads to the deterioration of the macro-mechanics of the rock mass (Yang et al. 2018). Therefore, considering the micro-failure mechanism of defects and the effect of moisture migration, it can better reflect the nature of the effect of freeze–thaw cycles on the macro-mechanical behaviors of rock masses.

The constitutive model is the theoretical basis for studying the mechanical properties of materials. Most scholars considered the influence of the initial damage caused by freeze–thaw and the number of cycles on the parameters, and adopted the widely used failure criterion to establish the phenomenological constitutive model of freeze–thaw rock (Meng et al. 2021). With the clarification of the freeze–thaw damage mechanism, scholars began to use the advantages of thermodynamics to introduce moisture migration to describe changes in rock mechanical properties. Wang et al. (2019a, b, c) analyzed the influence of moisture migration on the parameters of damage mechanics proposed by Lemaitre (1992); Chen et al. (2020) studied the influence of moisture migration on the parameters of the ellipse-parabolic double yield surfaces. The micro-model can describe the nature of deformation and avoid adding unnecessary parameters (Liu et al. 2021). Some scholars used statistical micro-mechanics to study the freeze–thaw damage of rocks, but the accuracy of the model largely depends on the form of the distribution function (Gao et al. 2021). Taking into account the additional stress caused by the freezing of water in the pores, Yan et al. (2015) simulated the freeze–thaw damage of the rock through the deformation and expansion of the microcracks, but the moisture migration was ignored. Zhu (2019) established a series of micro-damage constitutive models considering the friction-damage coupling on the surface of microcracks, which can well describe the multi-scale mechanical properties of rocks. Therefore, the establishment of the micro-mechanics-based constitutive model considering moisture migration has important guiding significance for the safety of engineering in cold regions.

Based on the friction-damage coupling constitutive model, this paper established a micro-damage constitutive model considering moisture migration by introducing moisture migration as an intrinsic variable into the thermodynamic system. The model can reflect the moisture migration mechanism inside the rock mass and the failure mechanism of microcracks, which can provide a reference for engineering in cold regions.

25.2 Model Establishment

25.2.1 Moisture Migration

The internal structural changes and moisture migration during the freezing process of rock masses are very complicated, which are the result of the comprehensive influences of mechanical, physical and physical chemical factors. Most of the analysis on the law of moisture migration inside frozen rocks is improved based on the theory of frozen soil. As shown in Fig. 25.1, the moisture in the cracks in the frozen zone turns into ice, and the moisture in the unfrozen zone will continue to migrate to the frozen zone. The ice crystals in the cracks slowly grow larger, leading to crack propagation and destruction. During the whole process, the frozen fringe moves from the cold end to the warm end, and the freezing area continues to increase.

The amount of moisture migration can be obtained by experimental measurement or theoretical calculation, which can be regarded as the intrinsic variable in the thermodynamic system. In the previous research, based on the moisture migration

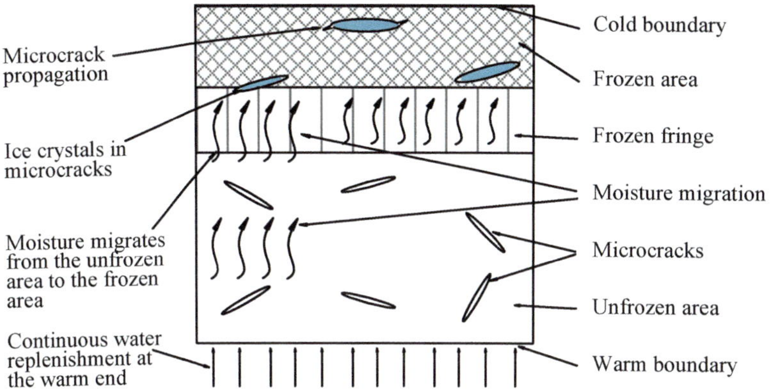

Fig. 25.1 Moisture migration inside the rock (Xu et al. 2013)

mechanism, the macro-mechanical properties of rock (Wang et al. 2019a, b, c) and the volume and shear deformation properties of soil (Chen et al. 2020) were studied. In most compression tests, the rock sample is at a constant temperature, so the influence of temperature can be ignored, and the elastic strain and moisture migration are regarded as the intrinsic variables. The corresponding Helmholtz free energy can be expressed as:

$$\psi = \psi(\boldsymbol{\varepsilon}^{\mathbf{e}}, T, \nu_\alpha) = \psi(\boldsymbol{\varepsilon}^{\mathbf{e}}, V, \nu_\alpha) \tag{25.1}$$

where ψ is the Helmholtz free energy per unit mass, $\boldsymbol{\varepsilon}^{\mathbf{e}}$ is the elastic strain tensor, T is the temperature, ν_α is the α th internal state variable, either a scalar or a tensor, V is the amount of moisture migration. The thermodynamic force considering moisture migration can be expressed as:

$$A_\alpha = \rho \frac{\partial \psi}{\partial \nu_\alpha} = A_\alpha\big(\nu_\alpha(\boldsymbol{\varepsilon}^{\mathbf{p}}), k_\beta(V)\big) \tag{25.2}$$

where A_α is the thermodynamic force conjugated to ν_α, either a scalar or a tensor, $\boldsymbol{\varepsilon}^{\mathbf{p}}$ is the plastic strain tensor, k_β is the β th dissipation parameter of the material.

$$\dot{A}_\alpha = \frac{\partial A_\alpha}{\partial \nu_\alpha}\frac{\partial \nu_\alpha}{\partial \boldsymbol{\varepsilon}^p} : \dot{\varepsilon}^p + \frac{\partial A_\alpha}{\partial k_\beta}\frac{\partial k_\beta}{\partial V}\dot{V} \tag{25.3}$$

By introducing the amount of moisture migration as an intrinsic variable into the thermodynamic system, the moisture migration field is established. In the whole process of the evolution of mechanical properties, moisture migration can be fully considered, which provides a theoretical basis for the study of the mechanical properties of freeze–thaw rocks.

25.2.2 *Micro-Criteria Considering Moisture Migration*

In the theory of microscopic friction-damage coupling proposed by Zhu (2019), the free energy of the system is divided into two parts. The first part is the elastic free energy of the matrix part, and the second part is the free energy caused by the friction and sliding of the microcrack surface. When considering the influence of moisture migration, free energy is related to the elastic strain, the amount of moisture migration, the plastic strain and the damage variable.

$$\psi(\boldsymbol{\varepsilon}^{\mathbf{e}}, V, \boldsymbol{\varepsilon}^p, d) = \frac{1}{2}\big(\boldsymbol{\varepsilon} - \boldsymbol{\varepsilon}^{\mathbf{p}}\big) : C^0 : \big(\boldsymbol{\varepsilon} - \boldsymbol{\varepsilon}^{\mathbf{p}}\big) + \frac{1}{2}\boldsymbol{\varepsilon}^{\mathbf{p}} : C^p : \boldsymbol{\varepsilon}^{\mathbf{p}} \tag{25.4}$$

where d is the damage variable, $\boldsymbol{\varepsilon}$ is the total strain tensor, C_0 is the elastic stiffness tensor of the matrix phase, and C_p is a fourth order tensor related to the friction–damage coupling mechanism of microcracks.

The yield surface represented by the local stress can be expressed as (Hu et al. 2020):

$$f^p(\boldsymbol{\sigma}^{\mathbf{p}}) = \|\mathbf{s}^{\mathbf{p}}\| + k\sigma_m^p \leq 0 \tag{25.5}$$

where $\mathbf{s}^{\mathbf{p}} = \mathbb{K} : \boldsymbol{\sigma}^{\mathbf{p}}$ is the deviatoric stress part of the local stress $\boldsymbol{\sigma}^{\mathbf{p}}$, $\mathbb{K}$ is the introduced fourth-order tensor operator, $\|\mathbf{s}^{\mathbf{p}}\| = \sqrt{\mathbf{s}^{\mathbf{p}} : \mathbf{s}^{\mathbf{p}}}$, k is the friction coefficient, and σ_m^p is the spherical stress part of the local stress. Under the assumption of homogeneity, inelastic deformation can be decomposed into two parts.

$$\boldsymbol{\varepsilon}^{\mathbf{p}} = \boldsymbol{\Gamma} + \frac{1}{3}\theta\boldsymbol{\delta} \tag{25.6}$$

where $\boldsymbol{\Gamma} = K : \boldsymbol{\varepsilon}^{\mathbf{p}}, \theta = tr(\boldsymbol{\varepsilon}^{\mathbf{p}})$. From the relationship between the local stress $\boldsymbol{\sigma}^p$ and the overall stress $\boldsymbol{\sigma}$, that is, $\boldsymbol{\sigma}^p = \boldsymbol{\sigma} - C^p : \boldsymbol{\varepsilon}^p$, the plastic yield criterion expressed by the overall stress can be obtained.

$$f^p = \left\|\mathbf{s} - \frac{2\mu_s\boldsymbol{\Gamma}}{\beta d}\right\| + k\left(\sigma_m - \frac{k_s\theta}{\alpha d}\right) \leq 0 \tag{25.7}$$

where $\mathbf{s}$ is the deviatoric stress part of the overall stress, σ_m is the spherical stress part of the overall stress. k_s and μ_s are the bulk modulus and shear modulus. α and β are parameters related to Poisson's ratio v. Assuming that the rock material complies with the associated flow law.

$$\begin{cases} \dot{\varepsilon}^{\mathbf{p}} = \dot{\gamma}\mathbf{D} \\ \boldsymbol{\varepsilon}^{\mathbf{p}} = \Lambda^p\mathbf{D} \end{cases} \tag{25.8}$$

where $\dot{\gamma}$ is the plasticity multiplier; $\Lambda^p = \int \dot{\gamma}$; and $\mathbf{D}$ is the tensor indicating the direction of plastic strain flow. χ is defined as $\chi = \frac{1}{2}\mathbf{D} : C^d : \mathbf{D} = \frac{k_s k^2}{2\alpha} + \frac{\mu_s}{\beta}$, then the damage criterion can be expressed as (Zhao et al. 2018):

$$f^d = \left(\frac{\Lambda^p}{d}\right)^2 \chi - R(d) \leq 0 \tag{25.9}$$

where $R(d)$ is the damage resistance function, $R(d) = r_c\frac{4d/d_c}{(1+d/d_c)^2}$, d_c is the critical damage value corresponding to the peak stress, and r_c is the maximum value of damage resistance.

In the derivation of the elastoplastic damage constitutive model, there are many expressions describing the evolution process of plasticity or damage. In the microscropic model considering moisture migration, the expression describing the evolution of plasticity and damage can be defined as follows.

$$A_1 = \frac{2\mu_s \boldsymbol{\Gamma}}{\beta d}; A_2 = \frac{k_s \theta}{\alpha d}; A_3 = \left(\frac{\Lambda^p}{d}\right)^2 \chi \tag{25.10}$$

There are five parameters involved in the microscropic model (Zhu 2016), the elastic modulus of the matrix phase E_s, Poisson's ratio v, the generalized friction coefficient k, the critical damage resistance r_c, and the critical damage value d_c. It is assumed that the Poisson's ratio of the rock does not change during the freeze–thaw cycle, and the elastic modulus has a linear relationship with the amount of moisture migration. Other parameters in the model are functions of moisture migration, $k(V)$, $r_c(V)$, $d_c(V)$.

$$E_s = E_s^0 + \varsigma_1 V(N) \tag{25.11}$$

where E_s^0 is the elastic modulus that is not affected by moisture migration, ς_1 is a coefficient reflecting the influence of moisture migration on elastic modulus.

The plastic consistency condition can be expressed as:

$$\frac{\partial f^p}{\partial \boldsymbol{\sigma}} : \dot{\sigma} + \frac{\partial f^p}{\partial \boldsymbol{\varepsilon}^p} : \dot{\varepsilon}^p + \frac{\partial f^p}{\partial d}\dot{d} + \frac{\partial f^p}{\partial V}\dot{V} = 0 \tag{25.12}$$

The damage consistency condition can be expressed as:

$$\frac{\partial f^d}{\partial d}\dot{d} + \frac{\partial f^d}{\partial \boldsymbol{\varepsilon}^p} : \dot{\varepsilon}^p + \frac{\partial f^d}{\partial V}\dot{V} = 0 \tag{25.13}$$

25.2.3 *Micro-Mechanics-Based Model Considering Moisture Migration*

Most compression tests are carried out after freeze–thaw tests. The moisture migration activity in the compression stage has ended, and the moisture migration amount can be regarded as a fixed value. From $\dot{\sigma} = C^0 : (\dot{\varepsilon} - \dot{\varepsilon}^p)$, the tangent modulus of the model in tensor form can be obtained.

$$C^{epd} = C^0 - \frac{C^0 : \mathbf{D} \otimes \mathbf{D} : C^0}{\mathbf{D} : C^0 : \mathbf{D} - \sum_{\alpha=1}^{2} \frac{\partial f^p}{\partial A_\alpha}\left(\frac{\partial A_\alpha}{\partial \boldsymbol{\varepsilon}^p} : \frac{\partial \dot{\varepsilon}^p}{\partial \dot{\gamma}}\right) + \left(\frac{\partial f^p}{\partial d} \Big/ \frac{\partial f^d}{\partial d}\right) \frac{\partial f^d}{\partial A_3} \frac{A_3}{\partial \boldsymbol{\varepsilon}^p} : \frac{\partial \dot{\varepsilon}^p}{\partial \dot{\gamma}}} \tag{25.14}$$

From Eq. (25.7) and Eq. (25.9):

$$f(\boldsymbol{\sigma}) = \|\mathbf{s}\| + k\sigma_m - 2\sqrt{R(d)\chi} \leq 0 \tag{25.15}$$

In the case of triaxial compression, the axial stress can be expressed as:

$$\sigma_1 = \frac{\sqrt{6} + 2k(V)}{\sqrt{6} - k(V)}\sigma_3 + \frac{6}{\sqrt{6} - k(V)}\sqrt{\frac{\chi d r_c(V)/d_c(V)}{(1 + d/d_c(V))^2}} \tag{25.16}$$

From the macroscopic stress–strain relationship (Chaves 2013), $\boldsymbol{\varepsilon} = C^0 : \boldsymbol{\sigma} + \Lambda^p \mathbf{D}$, and the component of the plastic flow direction tensor, $D_{11} = \frac{2}{\sqrt{6}} + \frac{k}{3}$, the axial strain expression can be obtained:

$$\varepsilon_1 = \frac{\sigma_1 - 2\nu\sigma_3}{E_s^0 + \varsigma_1 V} + 2\left(\frac{k(V)}{3} + \frac{2}{\sqrt{6}}\right)d\sqrt{2\frac{r_c(V)}{\chi}\frac{d/d_c(V)}{(1 + d/d_c(V))^2}} \tag{25.17}$$

25.3 Model Verification

25.3.1 *Test Plan*

The proposed model that considers moisture migration was verified by Wang et al.'s (2019a; b, c) experimental data. The samples are sandstone taken from Qinghai, China, and made into a cylindrical shape with a diameter of 50 mm and a height of 100 mm. The freezing temperature and the melting temperature are – 25 °C and 25 °C, respectively. A freeze–thaw cycle totals 16 h, consisting of 8 h of freezing and 8 h of thawing. Samples with freeze–thaw cycles of 0, 30, 60, 90, 120 were selected, the porosity under different cycles were measured, and a series of triaxial compression tests were carried out to obtain stress–strain curves with different freeze–thaw cycles.

25.3.2 *Parameter Identification*

During the test, the amount of moisture migration under different freeze–thaw cycles was not measured, but the change in porosity was measured, and it was found that the porosity increased with the increase in the number of freeze–thaw cycles. Moisture continues to migrate and freeze into ice in the pores, causing the pores to become larger, which is the increase in porosity in the macroscopic view. Therefore, this paper used the increase in porosity after the freeze–thaw cycle to approximately calculate the amount of moisture migration. In the application, the moisture migration measured by the test shall prevail.

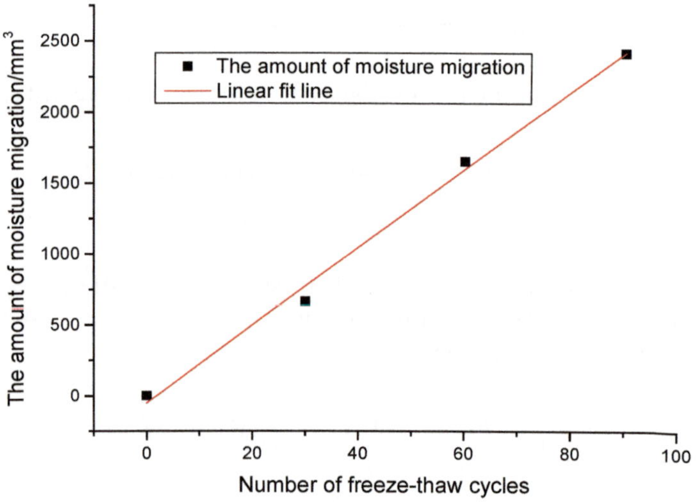

Fig. 25.2 The amount of moisture migration under different freeze–thaw cycles

The relationship between the number of freeze–thaw cycles and the amount of moisture migration can be obtained, and the correlation coefficient is 0.9974, as shown in Fig. 25.2.

$$V(N) = 27.4816N - 53.0302 \quad (25.18)$$

Assuming that the freeze–thaw cycle has no effect on the Poisson's ratio of sandstone, and $v = 0.2$. According to the data in stress–strain curves, the elastic modulus under different cycles can be obtained, which decreases with the increase of moisture migration. The linear fitting expression can be obtained, and the correlation coefficient is – 0.9666, as shown in Fig. 25.3.

$$E(V) = -0.00256V + 13.12576 \quad (25.19)$$

The generalized friction coefficient k is determined according to the ratio of tensile and compression strength of the rock. During the freeze–thaw cycle, the strength of the rock deteriorates. It is assumed that the reduction in tensile strength and compressive strength is proportional. Therefore, the generalized friction coefficient is a constant, and the tension–compression ratio of sandstone is selected to be 7.17, and the generalized friction coefficient is calculated to be 1.6395.

The calculation of the critical damage resistance r_C needs to clarify the peak strength of uniaxial compression under different freeze–thaw cycles. The uniaxial compression test of the specimen was not carried out. The uniaxial compression strength under different freeze–thaw cycles can be obtained by fitting the peak strength under different confining pressures. The correlation coefficient between uniaxial compression strength $\tilde{\sigma}_1$ and moisture migration is – 0.9268, as shown in

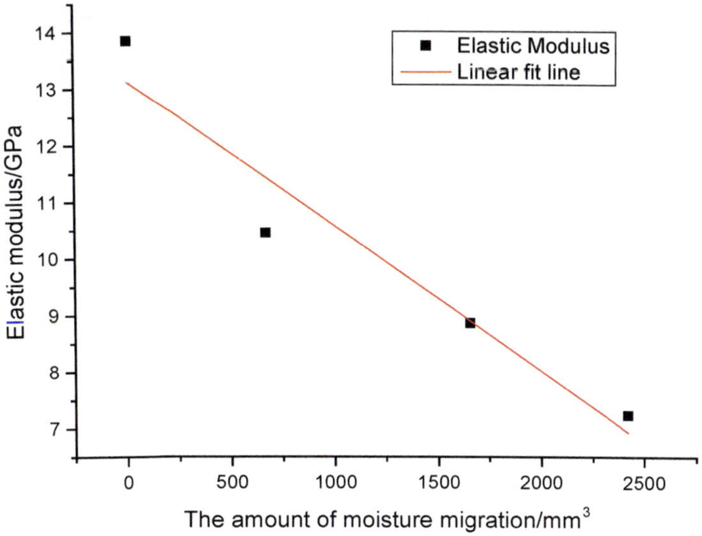

Fig. 25.3 Elastic modulus under different amounts of moisture migration

Fig. 25.4.

$$\tilde{\sigma}_1(V) = -0.0125V + 109.822 \tag{25.20}$$

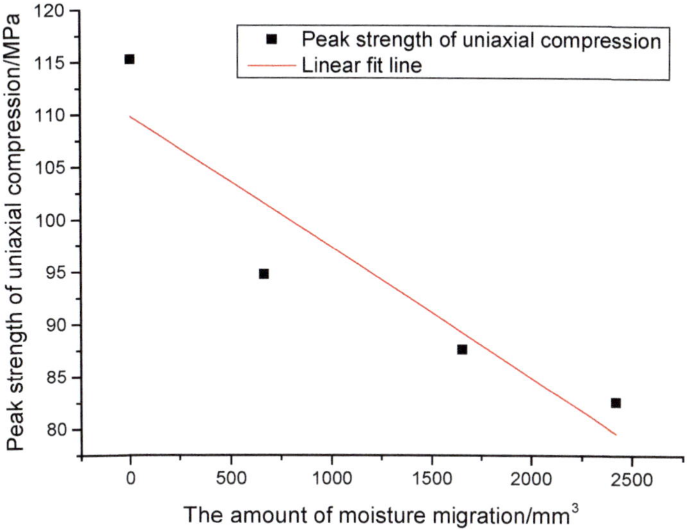

Fig. 25.4 Peak strength under different amounts of moisture migration

Then the relationship between the critical damage resistance and the amount of moisture migration can be obtained.

$$r_c(V) = \frac{[\tilde{\sigma}_1(V)(\sqrt{6} - k)]^2}{36\chi} \tag{25.21}$$

The calculation of the critical damage value d_c requires the plastic strain value at the peak strength $\tilde{\varepsilon}_1^p$. In the initial compaction stage, the rock has a large nonlinear deformation, the influence of this stage should be considered when calculating the plastic strain. The correlation coefficient between the plastic strain value at the peak strength and moisture migration is 0.9974, as shown in Fig. 25.5.

$$\tilde{\varepsilon}_1^p(V) = (0.0002V + 0.7099)/1000 \tag{25.22}$$

The relationship between moisture migration and critical damage value can be obtained

$$d_c(V) = \frac{\tilde{\varepsilon}_1^p(V)}{D_{11}}\sqrt{\frac{\chi}{r_c}} \tag{25.23}$$

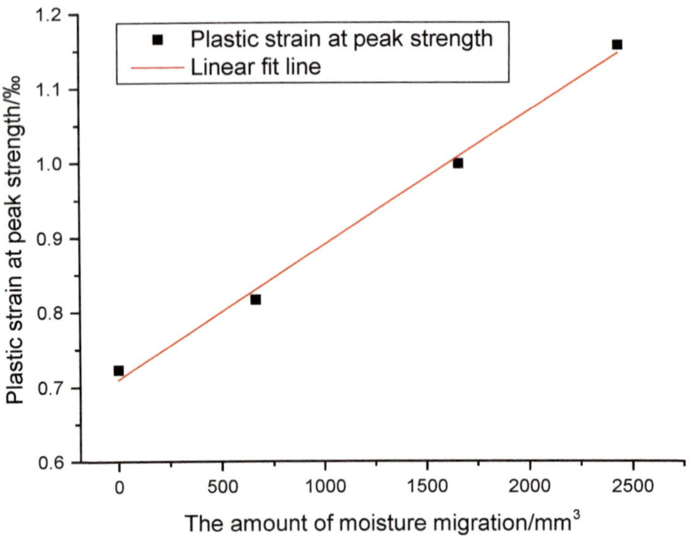

Fig. 25.5 Plastic strain at peak strength under different amounts of moisture migration

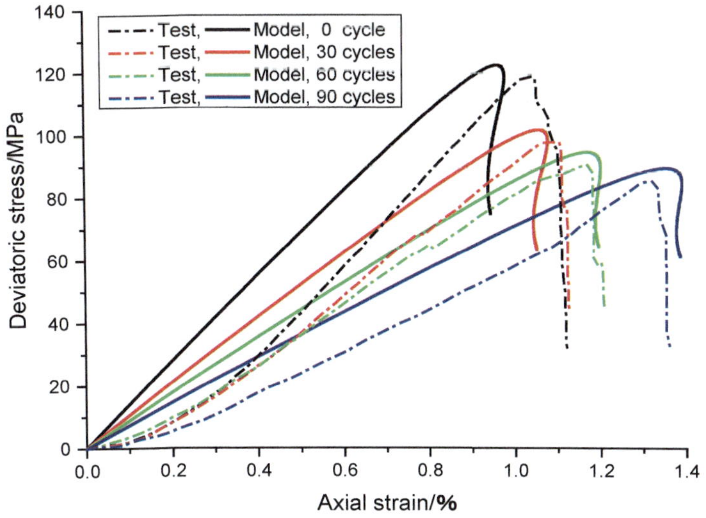

Fig. 25.6 Comparison of model simulation results and test results

25.3.3 Comparison of Calculated and Test Results

The test data with the confining pressure of 1 MPa and the number of freeze–thaw cycles of 0, 30, 60, and 90 are selected to verify the model. In the stress–strain curve of 120 cycles, the elastic modulus at the initial stage and the elastic modulus before the peak strength are both larger than those in the middle. It shows that structural changes may have taken place in the rock after 120 freeze–thaw cycles, so the influence of the 120 times curve is ignored in the simulation. Using the proposed model and parameter determination method, the simulated stress–strain curves can be obtained, as shown in Fig. 25.6.

The proposed model can well simulate stress–strain curves of rock under different freeze–thaw cycles. The role of moisture migration is considered in the model, and specific relationships between moisture migration and the parameters in the model are established. The migration of moisture during the freeze–thaw cycle causes the strength of the rock to decrease and the plasticity to increase. The model can well reflect the degradation of rock mechanical properties under the action of freeze–thaw cycles and the deformation characteristics in elasticity, plasticity and failure stage.

In the initial compression stage, the specimen undergoes large nonlinear deformation. However, the proposed model believes that the sample is in the elastic stage at the beginning, which leads to a gap between the simulated curve and the actual curve. Therefore, future research needs to improve the proposed model for the initial compression stage to better reflect the full stress–strain curve of the rock.

25.4 Conclusions

By introducing the amount of moisture migration into the thermodynamic system, based on the friction-damage coupling model, the micro-mechanics-based elasto-plastic damage constitutive model of rock under freeze–thaw cycles was deduced, and the influence of moisture migration on the parameters in the model was analyzed. Finally, the rationality of the model was verified by conventional triaxial compression tests under different freeze–thaw cycles. The main conclusions are as follows:

The macroscopic and microscopic parameters of the rock are affected by moisture migration. The amount of moisture migration has a linear relationship with the elastic modulus, the uniaxial compression strength, and the plastic strain value at the peak strength, and the correlation coefficients are − 0.9666, − 0.9268, and 0.9974, respectively. It proves that the moisture migration during the freeze–thaw cycle reduces the mechanical properties of the rock and increases its plastic deformation.

The micro-mechanics-based elastoplastic damage constitutive model considering moisture migration can well simulate the triaxial compressive stress–strain curve of rock under different freeze–thaw cycles. The model considers the physical mechanism of the influence of freeze–thaw cycles on rock properties and the coupling mechanism on the surface of microcracks. The parameters in the model have clear physical meanings and calculation methods.

Since the amount of moisture migration will affect the degree of change in rock properties, it is feasible to reduce the damage caused by the freeze–thaw cycle to the engineering by slowing down the migration of moisture in the cold zone.

Acknowledgements This study is supported by the National Natural Science Foundation of China (No. U1965108, No. 42142019, No. U22A20597), the Fundamental Research Funds for the Central Universities, and the Shanghai Peak Plateau Discipline (Class I).

References

Chaves EWV (2013) Notes on continuum mechanics. Springer, Netherlands

Chen H, Zhu Z, Wang Z (2020) Constitutive model with double yield surfaces of freeze-thaw soil considering moisture migration. B Eng Geol Environ 79:2353–2365. https://doi.org/10.1007/s10064-019-01673-1

Chen W, Tan X, Yu H, Yuan KK, Li SC (2011) Advance and review on thermo-hydro-mechanical characteristics of rock mass under condition of low temperature and freeze-thaw cycles. Chin J Rock Mech Eng 30:1318–1336

Cheng H, Chen H, Cao G, Rong C, Yao Z, Cai H (2020) Damage mechanism of porous rock caused by moisture migration during freeze-thaw process and experimental verification. Chin J Rock Mech Eng 39:1739–1749. https://doi.org/10.13722/j.cnki.jrme.2019.1234

Gao F, Xiong X, Xu C, Zhou K (2021) Mechanical property deterioration characteristics and a new constitutive model for rocks subjected to freeze-thaw weathering process. Int J Rock Mech Min 140:104642. https://doi.org/10.1016/j.ijrmms.2021.104642

Hu K, Shao J, Zhu Q, Zhao L, Wang W, Wang R (2020) A micro-mechanics-based elastoplastic friction-damage model for brittle rocks and its application in deformation analysis of the left

bank slope of Jinping I hydropower station. Acta Geotech 15:3443–3460. https://doi.org/10.1007/s11440-020-00977-x
Lemaitre J (1992) A course on damage mechanics. Springer, Berlin Heidel
Liu H, Zhang X, Yan X (2021) A damage constitutive model for a rock under compression after freeze-thaw cycles based on the micromechanics. Geofluids 2021:1–11. https://doi.org/10.1155/2021/3177464
Meng XZ, Zhang HM, Liu XY (2021) Rock damage constitutive model based on the modified logistic equation under freeze-thaw and load conditions. J Cold Reg Eng 35. https://doi.org/10.1061/(ASCE)CR.1943-5495.0000268
Wang L, Li N, Qi J, Tian Y, Xu S (2019a) A study on the physical index change and triaxial compression test of intact hard rock subjected to freeze-thaw cycles. Cold Reg Sci Technol 160:39–47. https://doi.org/10.1016/j.coldregions.2019.01.001
Wang Z, Zhu Z, Chen H, Zhu S (2019) Study on rock stress-temperature-moisture coupled constitutive model under freezing-thawing cycle. Rock Soil Mech 40:2608–2616. https://doi.org/10.16285/j.rsm.2018.0453
Wang Z, Zhu Z, Zhu S (2019c) Thermo-mechanical-water migration coupled plastic constitutive model of rock subjected to freeze-thaw. Cold Reg Sci Technol 161:71–80. https://doi.org/10.1016/j.coldregions.2019.03.001
Xu B, Li N, Li Z, Yan N (2013) Low-temperature LPG and LNG storage caverns and related research review of rock mechanics. Chin J Rock Mech Eng 32:2977–2993
Yan X, Liu H, Xing C, Li C, Wang D (2015) Constitutive model research on freezing-thawing damage of rock based on deformation and propagation of microcracks. Rock Soil Mech 36:3489–3499. https://doi.org/10.16285/j.rsm.2015.12.020
Yang G, Shen Y, Jia H, Wei Y, Zhang H, Liu H (2018) Research progress and tendency in characteristics of multi-scale damage mechanics of rock under freezing-thawing. Chin J Rock Mech Eng 37:545–563. https://doi.org/10.13722/j.cnki.jrme.2017.1295
Zhao L, Zhu Q, Shao J (2018) A micro-mechanics based plastic damage model for quasi-brittle materials under a large range of compressive stress. Int J Plasticity 100:156–176. https://doi.org/10.1016/j.ijplas.2017.10.004
Zhu Q (2016) Strength prediction of dry and saturated brittle rocks by unilateral damage-friction coupling analyses. Comput Geotech 73:16–23. https://doi.org/10.1016/j.compgeo.2015.11.015
Zhu Q (2019) Multiscale rock damage mechanics. China Science Press, Beijing

Chapter 26
Study on Needle Penetration Test for Estimating Mechanical Parameters of Sandy Dolomite

Jia-xing Dong, Lun-shun Zhou, Mei-qian Wang, Jian Mi, Hong-yuan Mu, Yong-hong Wu, Yong-chuan Zhao, and Zhen-luo Shen

Abstract Sandy dolomite is a special engineering geological phenomenon in Yuxi section of Central Yunnan Water Diversion Project. According to the preliminary statistics, it is mainly the completely sandy dolomite and the severely sandy dolomite in the study area, with broken structure, high weathering degree and developed joints. Due to the special engineering characteristics of sandy dolomite, many disasters have occurred in the construction, which seriously hindered the progress of the project. However, there are few references related to the mechanical properties of sandy dolomite. In this paper, we selected the typical sandy dolomite in the Xiaopu tunnel

J. Dong · L. Zhou · Z. Shen
Faculty of Electric Power Engineering, Kunming University of Science and Technology, Kunming 650500, Yunnan, China
e-mail: dong1986@kust.edu.cn

L. Zhou
e-mail: 2019097356@qq.com

Z. Shen
e-mail: 826610323@qq.com

M. Wang · Y. Wu (✉)
Faculty of Civil Engineering and Mechanics, Kunming University of Science and Technology, Kunming 650500, Yunnan, China
e-mail: 549710862@qq.com

M. Wang
e-mail: 496567227@qq.com

J. Mi · H. Mu · Y. Zhao
Yunnan Survey and Design Institute of Water Conservancy and Hydropower, Kunming 650021, Yunnan, China
e-mail: 1429553715@qq.com

H. Mu
e-mail: 34870195@qq.com

Y. Zhao
e-mail: 496567227@qq.com

S. Wang et al. (eds.), *Engineering Geology for a Habitable Earth: IAEG XIV Congress 2023 Proceedings, Chengdu, China*, Environmental Science and Engineering,
https://doi.org/10.1007/978-981-99-9057-3_26

and nearby field outcrops in Yuxi section as the study object, and needle penetration tests were carried out on completely and severely sandy dolomite. The mechanical parameters of sandy dolomite were estimated indirectly by needle penetration index (NPI), and the correlation was compared and analysed. The results show that there was a good correlation between needle penetration index (NPI) and uniaxial compressive strength (UCS). According to the estimation of uniaxial compressive strength (UCS) of needle penetration ratio, the completely and severely sandy dolomite belong to soft rock. The cohesion (c) and internal friction angle (φ) are estimated based on the needle penetration index (NPI), and the value is close to the value recommended by the design institute. Needle penetration test can estimate the mechanical parameters of sandy dolomite simply, quickly and effectively.

Keywords Sandy dolomite · Needle penetrometer test · Mechanical parameters estimation · Water conveyance tunnel

26.1 Introduction

The study of the rock mass strength characteristics is of great significance for the design, construction, and theoretical development of engineering. Rock mass strength includes compressive strength, shear strength, and lion-loaded shear strength, but only the shear strength and the compressive strength are meaningful for engineering rock mass. The uniaxial compressive strength (UCS) refers to the load that a rock specimen can bear per unit area when it is subjected to uniaxial compression to failure (Li and Yang 1994), also known as compressive strength. The uniaxial compressive strength (UCS) is one of the most basic and important mechanical parameters of rock mass, and it is also one of the parameters widely used in various engineering. It is involved in the rock mass rating classification, the rock strength failure criterion (Zhang et al. 2015), the foundation bearing capacity design (Zhang et al. 2015), the underground cavern and slope stability evaluation (Lin et al. 2018). Therefore, it is significant of the rock mechanics parameters study theory and in practice. There are two methods to determine the mechanical parameters of rock mass: direct method and indirect method. The direct method adopts the standard indoor rock mechanics test method, which uses the uniaxial compression test to determine the uniaxial compressive strength of rocks. However, the standard test method usually requires us to make rock samples of standard specimens, which is time-consuming, expensive, demanding and requires large-scale test equipment (Chen et al. 2020). At the same time, the anisotropic characteristics of rock mass have a significant impact on the test results, making it impossible to accurately determine the strength of the rock during actual construction (Lin et al. 2018). In practical work, it is difficult to find hard, intact, non-disintegrating, non-weathering rock masses. Most of the rock masses are soft rocks such as sericite schist, marl, and sandstone, as well as strongly weathered and easily disintegrating rocks. There are significant difficulties in making standard specimens and sampling and it is difficult to carry out standard indoor rock mechanics tests.

Based on the limitations of the above test methods, a series of indirect test methods used such as the point load test, the Schmidt instrument, the longitudinal wave velocity test are proposed (Broch and Franklin 1972; ISRM 2009; Ulusay et al. 2001), it is used for indirect estimation of mechanical parameters of rock mass. However, in the protection project of extremely low strength rock mass and historical relics, we can only determine the mechanical properties of rock mass by in-situ testing. Therefore, Japan has invented a portable, non-destructive, lightweight experimental device named the needle penetrometer, which is suitable for field in-situ testing and laboratory tests without sample preparation and has been published as a recommended method by the Japanese Society of Civil Engineers (JSCE-RMC 1980). And the Japanese Civil Engineering Institute published a draft related to soft rock testing (PWRI 1987). In order to estimate the uniaxial compressive strength of soft rock, Maruto Test Company (Maruto Co. Ltd 1999) has produced a simple penetration test device, which has been widely used in many engineering projects involving soft rock in Japan. According to the recommendations of Japan Society of Civil Engineers (JSCE 1991) and Co, Ltd (JGS 2012), the needle penetration test shall be applied to rock mass with uniaxial compressive strength less than 10 MPa. The international society for rock mechanics (ISRM) approved the needle penetration test as the method recommended by ISRM in 2014 (Ulusay et al. 2014).

In this paper, considered the characteristics of completely and severely sandy dolomite, such as fragmented structure, low strength and easy to disintegration in water, a needle penetrometer is applied. Selected the typical tunnel section and outcrop of sandy dolomite as the study object. The needle penetration index (NPI) is determined, and the uniaxial compressive strength (UCS), the cohesion (c), the internal friction angle (φ) and other relevant mechanical parameters of completely and severely sandy dolomite are indirectly estimated according to the relevant formulas. The rock mass hardness of completely sandy dolomite and severely sandy dolomite is determined, and the estimation results are compared with the in-site direct shear test and estimation results. In addition, studied the suitability of needle penetrometer in completely and severely sandy dolomite.

26.2 Geological Overview

The Central Yunnan Water Diversion Project is a strategic basic project for the sustainable development of Yunnan Province. After the project is completed and put into operation, it can fundamentally solve the problem of water resources shortage in Central Yunnan Province, and has remarkable economic, social and ecological benefits. Sandy dolomite is a typical special adverse geological phenomenon in Yuxi section of Central Yunnan Water Diversion Project (Fig. 26.1). According to preliminary statistics, Yuxi section is 77.069 km long, with eight tunnels, three aqueducts, two concealed culverts and three inverted siphon structures. And there are five large-scale sandy dolomite tunnel sections with a length of 14.406 km, which distributed widely and is rare in the world. The concept of sandy dolomite was first introduced in Pingtou Hydropower Station in Sichuan Province, China (He 2008). Afterwards, it was encountered in some engineering construction in Yunnan, Guizhou and Chongqing. The distribution of sandy dolomite encountered in the process of engineering construction is shown in Fig. 26.2. Relevant researchers have carried out a series of studies on sandy dolomite mechanism and engineering geological characteristics (He 2008; Hu 2009; Li et al. 2018; Tian and Yue 2014; Yu 2013; Zhang et al. 2012; Zhang 2012a, 2012b; Zou and Wang 2012).

Different from ordinary weathered rock mass, the rock mass structure of sandy dolomite is mainly cataclastic structure and granular structure, and the joints and fissures are developed (Fig. 26.3), which seriously threatens the normal development

Fig. 26.1 Typical field outcrop of sandy dolomite

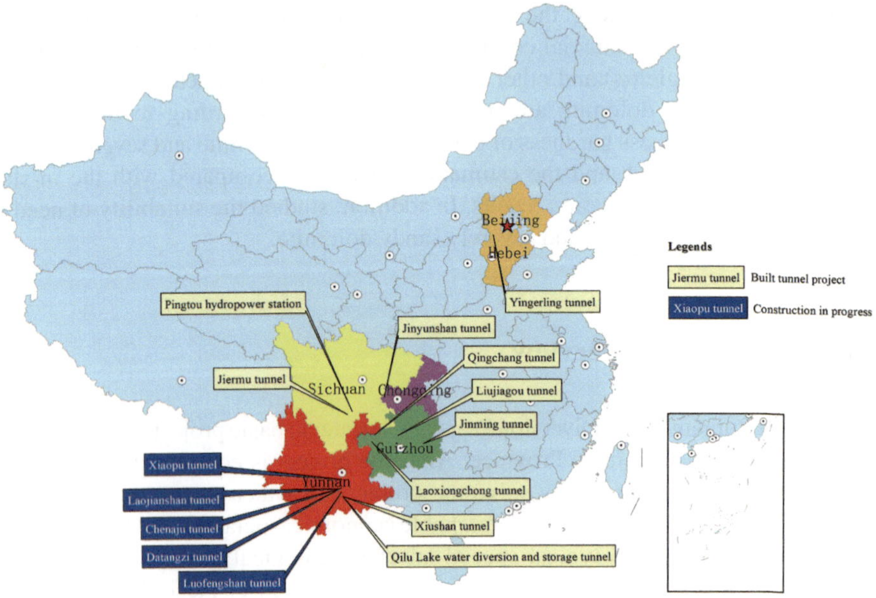

Fig. 26.2 Distribution of tunnels through the sandy dolomite layer

of engineering construction. During the construction of the project, the excavated stratum of the sandy dolomite tunnel is mainly composed of completely and severely sandy dolomite, with few weakly sandy dolomite. There have been several large cave-in disasters, such as water gushing and sand gushing, as well as many small-scale cave-in phenomena (Fig. 26.4).

In this paper, selected the Xiaopu Tunnel and its nearby exposed sandy dolomite as the study object, and conducted relevant experiments on it. The geographical location of the study area is shown in Fig. 26.5, and the layer of the tunnel is mainly composed of completely and severely sandy dolomite. The layer in the study area are mainly composed of the Sinian Doushantuo Formation dolomite and Sinian Dengying Formation dolomite, with a small amount of sandstone, shale, and clay locally. Along the tunnel, the buried depth is shallow, which is located below the underground water level. The tunnel through two large faults. According to the statistics of excavation, there are also many small faults with unknown occurrence in the section of completely and severely sandy dolomite. And the tunnel passes through a large karst water system, and the groundwater is developed. Under the

(a) At the hole depth of 76m, it is the severely sandy dolomite

(b) At the hole depth of 179m, it is the completely sandy dolomite

Fig. 26.3 Dolomite sandy characteristics of core of special part of XQZK203 split core dolomite in Xiaopu tunnel

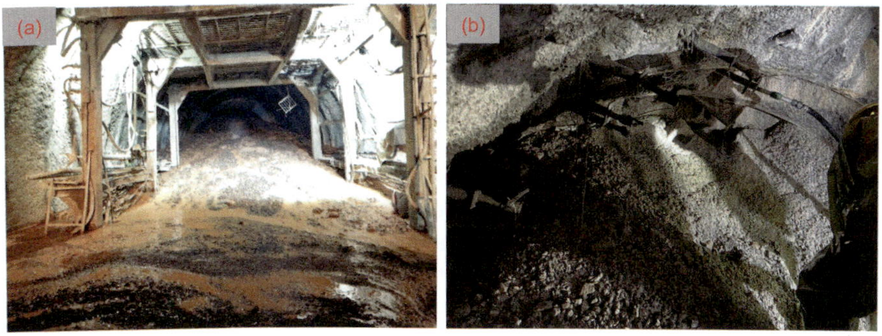

Fig. 26.4 Disaster situation of typical tunnel section

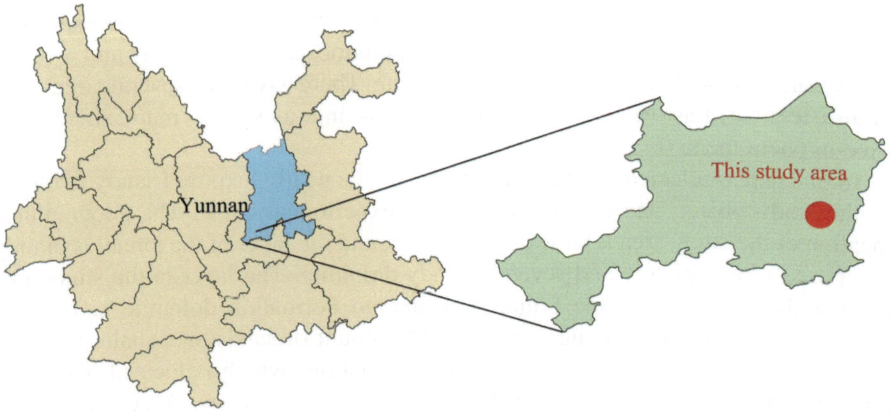

Fig. 26.5 Location of the study area

influence of geological structure, groundwater and other factors, three large collapse and mud phenomena have occurred in study area, and the maximum collapse volume are 3400 m^3 and 2160 m^3/d respectively. In addition, sandy dolomite is widely distributed in the vicinity of the study area, with large outcrop.

26.3 Needle Penetration Test

The needle penetrometer was invented according to the soft rock research in Japan. According to the needle penetration test guidelines, it is necessary to determine the needle penetration gradient (NP) of soft rock first, and then read out the uniaxial compressive strength according to the relationship between needle penetration gradient and uniaxial compressive strength. Needle penetrometer is an instrument that can quickly obtain the needle penetration index (NPI) of soft rock in the field and convert it into uniaxial compressive strength (UCS).

26.3.1 Research Status

The needle penetration test can determine the needle penetration ratio (NPI) of rock mass, and establish the correlation between UCS and NPI ratio through a certain conversion coefficient. Many scholars have studied the relationship between NPI ratio and UCS (Avşar et al. 2015; Aydan 2012; Aydan and Ulusay 2013; Erguler and Ulusay 2007, 2009; Kahraman et al. 2017; Naoto et al. 2004; Okada et al. 1985; Rabat et al. 2020; Rahimi et al. 2020; Takahashi et al. 1998; Uchida et al. 2004; Ulusay and Erguler 2012; Yamaguchi et al. 1997), and they think there is a well connection between the NPI and UCS. Scholars have conducted in-situ tests of various rock masses indoors and outdoors through SH-70 needle penetrometer, and established NPI-UCS empirical formula as shown in Table 26.1, which promotes the engineering practical application of needle penetrometer. According to previous studies, the fitting function of NPI-UCS empirical model is mainly power function and non-zero intercept linear function, and the correlation coefficient is 0.81–0.98, showing a high correlation. The formula given by Maruto Company is the formula recommended for SH-70 needle penetrometer.

The application of needle penetrometer to test the uniaxial compressive strength of rock mass is mainly used in Japan, Turkey, the Netherlands and other countries. In recent years, it has also used to test the strength of soft rocks such as mudstone, bituminous coal, quartz schist and sericite schist in mainland China (Huang et al. 2020; Li 2017; Li et al. 2016; Liu 2018; Yao et al. 2015; Zhang 2020). It is pointed out that the best test range is 0.5–20 MPa.

The needle penetration test is widely used to indirectly estimate the UCS, internal friction angle, Young's modulus, water content, elastic modulus, tensile strength, internal friction angle, cohesion, P-wave velocity, S-wave velocity, as well as the

Table 26.1 Study on the relationship between NPI ratio and UCS

Scholar (Years)	Fitting equation (r)	Types of rock mass
Okada et al. (1985)	$UCS = 0.39719NPI^{0.978}$ (0.92)	Artificial material and mudstone
Yamaguchi et al. (1997)	$UCS = 0.618NPI^{0.982}$ (0.87)	Pyroclastic rocks
Takahashi et al. (1998)	$UCS = 1.5395NPI^{0.9896}$ (0.90)	Sandstone, mudstone, conglomerate, greywacke, tuff
Naoto et al. (2004)	$UCS = 0.273NPI + 0.132$ (0.83)	Sandstone
Naoto et al. (2004)	$UCS = 0.418NPI\text{-}0.004$ (0.90)	Hard claystone
Maruto Co. Ltd (2006)	$UCS = 0.41783NPI^{0.978}$ (0.92)	/
Erguler and Ulusay (2007, 2009)	$UCS = 0.51NPI^{0.8575}$ (0.87)	Marble, siltstone, shale, tuff
Ngan-Tillard et al. (2011)	$UCS = 0.0731NPI$ (0.84)	Calcarenite
Ulusay and Erguler (2012)	$UCS = 0.402NPI^{0.929}$ (0.89)	Marl, tuff, shale, siltstone, sandstone, greywacke, mudstone
Aydan (2012)	$UCS = 0.2NPI$	Tuff, sandstone, pumice, limestone, lignite, mudstone, siltstone, marl, loam
Aydan and Ulusay (2013)	$UCS = 0.3NPI$	Turkish tuffs
Kahraman et al. (2017)	$UCS = 0.35NPI$ (0.81)	Cayirham coal
Rahimi et al. (2020)	$UCS = 0.49NPI$ (0.89)	Siltstones
Rahimi et al. (2020)	$UCS = 0.56NPI$ (0.85)	Gypsiferous rocks
Rabat et al. (2020)	$UCS = 0.13389NPI$ (0.98)	Dry gypsum rocks
Rabat et al. (2020)	$UCS = 0.19405NPI$ (0.97)	Saturated gypsum rocks

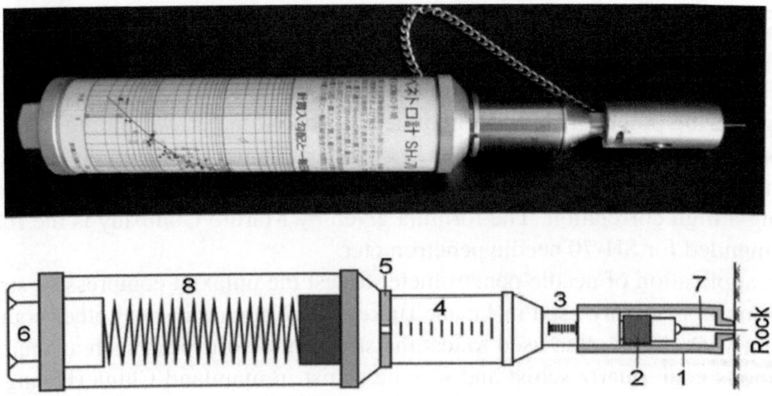

Fig. 26.6 Needle penetrometer, where: (1) presser, (2) chuck, (3) penetration scale, (4) penetration force scale, (5) load-indicating ring, (6) cap, (7) penetration needle, (8) spring

evaluation of anisotropy, weathering state, relaxation and creep characteristics of mechanical properties of rock mass (Aydan 2012; Aydan et al. 2014; Aydan and Ulusay 2013). Therefore, for the completely and severely sandy dolomite with broken structure, it is feasible to indirectly estimate the mechanical parameters of it by needle penetrometer.

26.3.2 Introduction of Needle Penetrometer

The needle penetration test was carried out with SH-70 needle penetrometer instrument produced by Maruto Company of Japan. The instrument consists of needle penetration, chuck (clamping penetration needle), penetration scale, penetration force scale, load-indicating ring (reading penetration force), relationship diagram between needle penetration ratio and uniaxial compressive strength, spring, presser (reading penetration and protecting penetration needle) and cap (including standby penetration needle). The instrument penetration force scale is 10–100 N, and the division value is 10 N. The scale of penetration is 0–10 mm, and the division value is 1 mm. The diameter of the tip of the penetration needle is 0.84 mm and the length is 39.5 mm. The total length of the instrument is about 285 mm and the weight is approximately 700 g. The structure of needle penetrometer is shown in Fig. 26.6.

26.3.3 Test Procedure

(a) Placement of penetration needle. Use the groove of the press to take it out from the penetration scale. Turn the pull lock part counter clockwise slowly, insert the penetration needle, turn it clockwise, tighten and fix it, and then insert the penetration scale piece into the groove of the press.

(b) Scale zero. Adjust the end of the penetration needle at the press end to 0 point of the penetration scale, which can be adjusted through the in and out of the penetration needle. The indicator ring is a proportional type, slide it with your fingers to make it contact the lower end of the hand holding part.

(c) Penetration test. After the penetration depth of the penetration needle reaches 10 mm, pull it out slowly, read out the penetration force and record the penetration force, and amount from the position of the indicator ring at the penetration force scale. If the soft rock is hard, before the penetration needle penetrates to 10 mm, if the penetration force has reached 100 N, slowly pull it out and read the penetration depth from the position of the presser at the penetration scale. Record the penetration force and penetration depth.

It should be noted that the test sequence must be carried out in strict accordance with the operation steps. The use sequence and relationship diagram of the needle penetrometer are provided at the handle. When the needle penetration test is carried out on soft rock with high hardness, the needle penetration may break. In order to prevent the eccentricity of the needle penetration fixed on the test device, the penetration speed shall be appropriately accelerated. To ensure the durability of penetration needle and reduce the dispersion of test data, the optimal UCS range of penetration needle test is 0.5–20 MPa. Repeat the test for the same rock surface data for 3–5 times, the rock thickness shall be greater than 15 mm, and the interval between test points shall be more than 10 mm. When there are production cracks on the rock surface during penetration, the test data shall be discarded, but when the cracks are generated when the penetration needle is pulled out, this data is valid (Fig. 26.7).

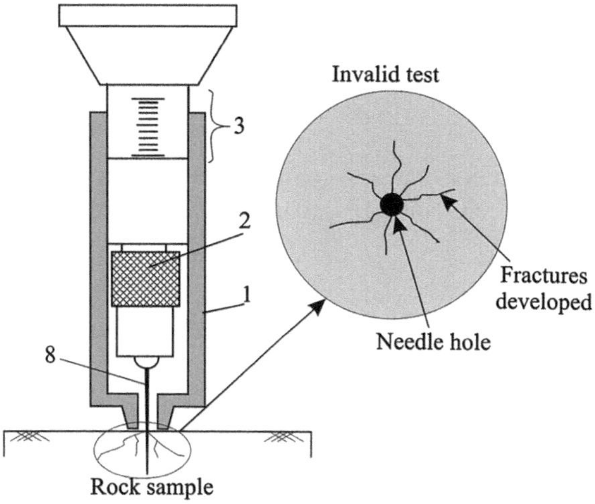

Fig. 26.7 Schematic diagram of invalid needle penetration test

26.3.4 Result Calculation

Needle Penetration Index (NPI). The needle penetration index (NPI) is calculated according to the Formula (26.1, 26.2).

$$For\, D = 10\,\text{mm}\; and\; F \leq 100\,\text{N},\; NPI = 100/D \tag{26.1}$$

$$For\, F = 100\,\text{N}\, and\; D \leq 10\,\text{mm},\; NPI = F/10 \tag{26.2}$$

where D is the penetration depth of the needle (mm), F is the penetration force (N), the unit of the NPI is N/mm.

The value of needle penetration index (NPI) shall be between the 1 N/mm and 100 N/mm, because the minimum graduation of the penetration scale is 1 mm and that of the load scale is 100 N. Relevant studies show that although the impact of the rate of needle penetration into rock mass on the test results can be ignored, the test work of one point shall be completed in about 30 s for each test.

Estimation of Mechanical Parameters of Rock Mass. The UCS is estimated according to the empirical Formula (26.3).

$$\log\, UCS = 0.978 \log\, NPI + 2.621 \tag{26.3}$$

According to the relationship determined by relevant researcher (Aydan et al. 2014), the internal friction angle (c) and cohesion (φ) of rock mass can be estimated, and the empirical formula of c and φ are shown in Formula (26.4, 26.5).

Internal friction angle:

$$c = 0.04 \cdot NPI \tag{26.4}$$

Cohesion:

$$\phi = 13.375\, NPI^{0.25} \tag{26.5}$$

where UCS is uniaxial compressive strength (MPa), C is the internal friction angle (°), φ is cohesion (MPa), 0.04 and 13.375 are empirical coefficients.

26.3.5 Field Needle Penetration Test

The dolomite layer are widely distributed in the study area, and there are many completely and severely sandy dolomite. In order to determine the strength of completely and severely sandy dolomite, needle penetration tests were carried out in the outcrops and tunnels in the study area, and the test points were located at the green mark in Fig. 26.8. And the completely sandy dolomite develops in a large outfall near branch cave No. 2 of Xiaopu tunnel, and its location is shown in Fig. 26.8. The photos of field test work are shown in Fig. 26.9.

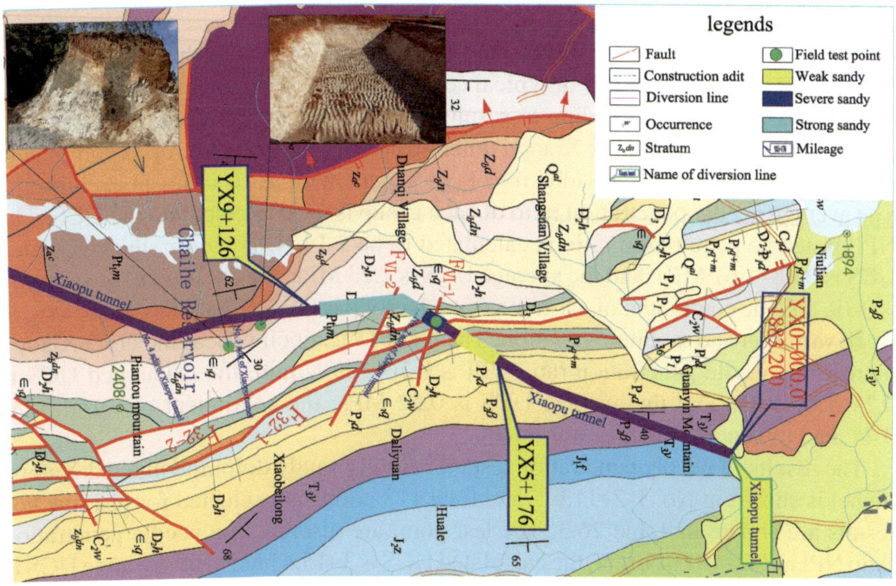

Fig. 26.8 Geographical location of field test point

(a) Field outcrop test (b) Tunnel test

Fig. 26.9 Field needle penetration test

26.4 Results and Discussion

26.4.1 *Test Results*

The calculated needle penetration ratio (NPI) is used to convert the UCS. The field needle penetration test results of typical completely and severely sandy dolomite are shown in Table 26.2. According to results of the needle penetration test, there are 100 test points near the No. 2 Xiaopu tunnel where the natural outcrop is completely sandy dolomite. They are divided into 1–14 groups with 3–10 times in each group. The average needle penetration ratio of each group is used to calculate the UCS. The NPI of needle penetration ratio is calculated to be 2.14–4.13 N/mm, the calculated UCS is 0.88–1.67 MPa, and the average UCS is 1.23 MPa, the coefficient of variation is 0.19, and the coefficient of variation of each group is 0.04–0.21. The data belong to weak variation and medium variation. There are 15 test points near the No.3 Xiaopu tunnel, the needle penetration ratio NPI is 0.70–2.50N/mm, the calculated UCS is 0.29–1.02 MPa, and the average UCS is 0.52 MPa, the coefficient of variation is 0.44, and the data is medium variation.

The results of needle penetration test on severely sandy dolomite in No.2 Xiaopu tunnel is shown in No. 1-2–113-2, and total 113 points are selected for test, the test results are shown in Table 26.3. The test point are discrete, the needle penetration ratio NPI is 7.50–40.0 N/mm, the calculated UCS is 3–15.41 MPa, the average UCS is 5.93 MPa, and the coefficient of variation is medium variation of 0.52. Point number 1-3–7-3 are 7 test points for the severely dolomite sandy exposed from the Xiaopu tunnel (elevation approximately 1876 m, buried depth approximately 115 m). The needle penetration ratio NPI is 3.40–7.8 N/mm, the calculated UCS is 1.38–3.12 MPa, and the coefficient of variation is 0.31.

26.4.2 *Analysis of Needle Penetration Test Results*

Adaptability of Rock Mass Mechanical Parameter Estimation. In the preliminary survey and design stage, Yunnan water conservancy and hydropower survey and design institute (Yunnan water conservancy and hydropower survey and design institute 2018) carried out adit direct shear test and relevant test research on the site, and gave the following recommended values (Table 26.4):

Based on the results of needle penetration index (NPI) obtained from the above field needle penetration test, the internal friction angle and cohesion of completely and severely sandy dolomite are determined. The specific results are shown in the Table 26.5.

In order to judge the suitability of estimating rock mass mechanical parameters with needle penetration index (NPI), the estimation results are compared with the recommended values of the design institute. The analysis results show that the value of the cohesion (c) and internal friction angle (φ) of completely sandy dolomite

Table 26.2 Field needle penetration test results of completely sandy dolomite

Group number	Points	F (N)	D (mm)	NPI (N/mm)	UCS (MPa)
1	10	26.5	10.0	2.65	1.08
2	8	28.0	10.0	2.80	1.14
3	3	32.7	10.0	3.27	1.33
4	5	21.4	10.0	2.14	0.88
5	6	22.2	10.0	2.22	0.91
6	6	36.7	10.0	3.67	1.49
7	8	35.8	10.0	3.58	1.45
8	8	30.0	10.0	3.00	1.22
9	8	29.1	10.0	2.91	1.19
10	6	41.3	10.0	4.13	1.67
11	10	34.7	10.0	3.47	1.41
12	8	24.3	10.0	2.43	0.99
13	8	28.1	10.0	2.81	1.15
14	6	30.3	10.0	3.03	1.24
Maximum	/	41.3	10.0	4.13	1.67
Minimum	/	21.4	10.0	2.14	0.88
Standard deviation	/	/	/	0.6	0.2
Average value	/	/	/	3.01	1.23
RSD	/	/	/	0.19	0.19
Point number	Points	F(N)	D(mm)	NPI(N/mm)	UCS(MPa)
1–1 ~ 15 ~ 1	15	/	/	/	/
Maximum	/	25.0	10.0	1.02	15.41
Minimum	/	7.0	0.70	0.29	3.00
Standard deviation	/	/	/	0.55	0.22
Average value	/	/	/	1.25	0.52
RSD	/	/	/	0.45	0.44

estimated by needle penetration index (NPI) is close to the recommended value. The cohesion (c) and internal friction angle (φ) of severely sandy dolomite estimated by needle penetration index (NPI) is close to and partially equal to the direct shear test data of in-situ rock mass. It indirectly proves the advantage of needle penetration index NPI in estimating rock mass mechanical parameters.

Empirical Model Analysis of Needle Penetration Ratio and Uniaxial Compressive Strength. The empirical formulas of scholars are drawn into a graph (Fig. 26.10). Except for the Takahashi's fitting formula, other equations show consistency when the UCS is less than 5 MPa, and the difference of predicted UCS is small. The UCS estimated by the Takahashi's fitting formula through needle penetration ratio is higher than other empirical formulas. Ulisay believes that it may be related

Table 26.3 Field needle penetration test results of severely sandy dolomite

Point number	Points	F (N)	D (mm)	NPI (N/mm)	UCS (MPa)
1-2–113-2	113	/	/	/	/
Maximum	/	100.0	10.0	40.00	15.41
Minimum	/	50.0	2.5	7.50	3.00
Standard deviation	/	/	/	8.03	3.08
Average value	/	/	/	15.11	5.93
RSD	/	/	/	0.53	0.52
1-3–7-3	7	/	/	/	/
Maximum	/	78.0	10.0	7.80	3.12
Minimum	/	34.0	10.0	3.40	1.38
Standard deviation	/	/	/	1.79	0.71
Average value	/	/	/	5.76	2.31
RSD	/	/	/	0.31	0.31

Table 26.4 Recommended values of rock mass mechanical parameters

Degree of sanding	Cohesion *c* (MPa)	Internal friction angle φ (°)
Severely sandy	0.14–0.41	22.78–28.37
Completely sandy	0.05–0.07	16.7–21.8

Table 26.5 Mechanical parameters of completely and severely sandy dolomite based on NPI

Degree of sanding	NPI (N/mm)	*c* (Pa)	φ (°)
Completely sandy	0.70–4.13(2.10)	0.086–0.17(0.12)	16.17–19.08(17.56)
Severely sandy	7.5–40(15.11)	0.30–1.60(0.604)	22.13–33.64(25.78)

Note The values in brackets are the average of data

to the measured large-grained rock types. When the UCS is greater than 5 MPa, the data estimated by each empirical equation is discrete.

If the empirical formula given by SH-70 needle penetrometer is used as the measured value, and the UCS calculated by the empirical formula of other scholars is used as the predicted value, the NPI obtained from the field completely and severely sandy dolomite needle penetration test is substituted into the UCS calculated in the empirical model in Table 26.1 (except Maruto and Takahashi models), as shown in Fig. 26.11.

The difference between the UCS calculated by the recommended formula of the needle penetrometer and the UCS calculated by other empirical models is expressed by the 1:1 straight line distance. If it is located on the 1:1 straight line, it indicates that the UCS calculated by the recommended formula of the needle penetrometer is consistent with the UCS calculated by other empirical models, while the data far

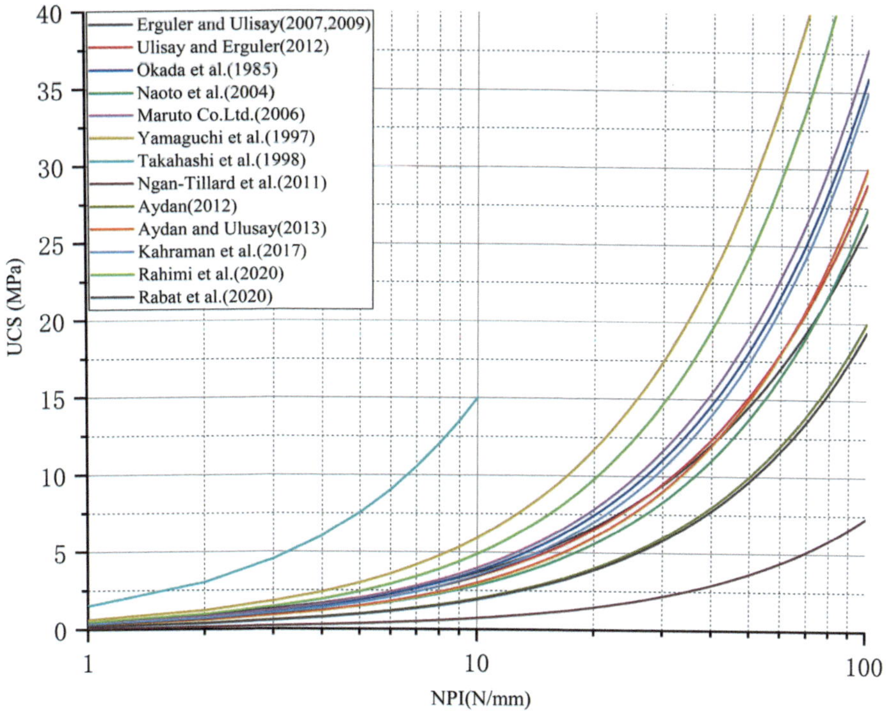

Fig. 26.10 UCS-NPI system diagram of different rock mass needle penetration test by scholars

from the 1:1 straight line is scattered (Avşar et al. 2015; Azadan and Ahangari 2014; Erguler and Ulusay 2009). It indicate that most of the data are located near the 1:1 straight line, but there are still some deviations, especially for UCS calculated greater than 5 MPa, the all data are located in the areas of straight-line 2:1 and straight-line 1:2. The above data shows that the UCS calculated by the recommended formula of needle penetration is close to the UCS calculated by the empirical model of Okada, Naoto and Kahraman, which can be used as a reference formula in future calculations.

The empirical models show the high similarity when the rock strength is less than 5 MPa, and the strength is between the 2:1 and 1:2 straight-line area, which indicate that the needle penetrometer is feasible in predicting the strength of the ordinary soft rock. However, considering the differences of regional and rock type studied by scholars, there may be some differences in the prediction of the completely and severely sandy dolomite UCS-NPI.

Hardness Discrimination of Sandy Dolomite. After analyzing the applicability of needle penetrometer to estimate UCS, the hardness of sandy dolomite can be determined by field test results.

According to the analysis of the field needle penetration test results, the dispersion of the natural outcrop data of severely sandy dolomite is greater than completely sandy dolomite, which may be caused by the special mechanical properties of sandy

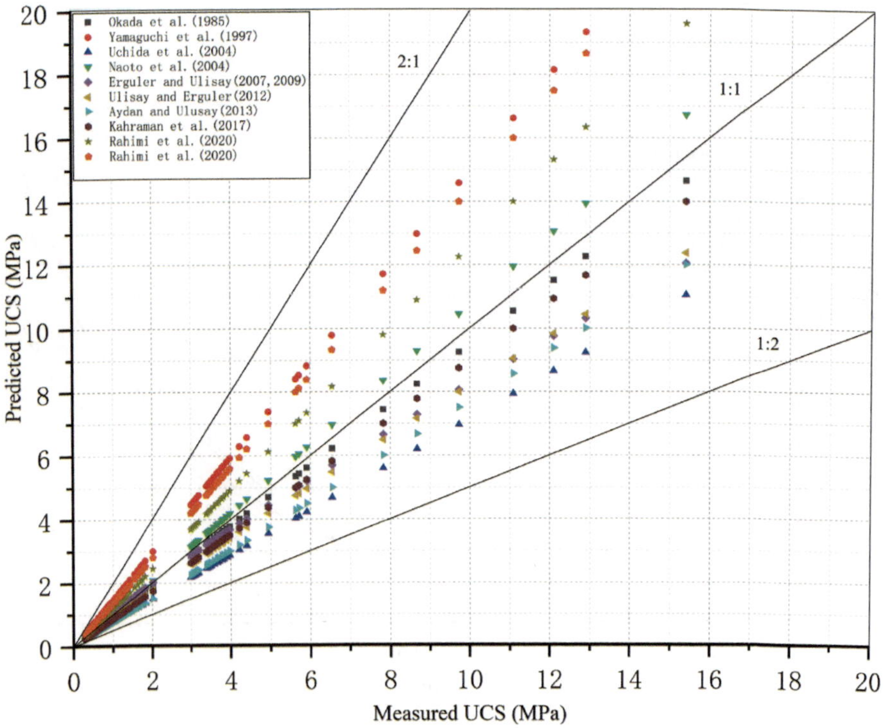

Fig. 26.11 Comparison of UCS calculated by various empirical models

dolomite, such as broken structure, developed joint and fissures, uneven overall rock mass properties and uneven sanding of rock mass. The calculated UCS of the severely sandy dolomite in the tunnel is lower than that in the natural state, and the variation coefficient becomes smaller, which is related to the seepage of the rock mass in the tunnel. The original pores, joints and fissures of the rock are filled with water, affecting the properties and uniformity of the rock mass. In other words, water content is one of the factors affecting strength of rock mass.

First application of field needle penetration test to rock mass strength test of sandy dolomite tunnel, which provides technical parameters for design and construction. The completely and severely sandy dolomite have broken structure and develop joint, it is hard to obtain the test samples for indoor strength test. Therefor, the needle penetration test is a better solution to obtain the UCS of sandy dolomite. According to the classification table of rock hardness (Table 26.6), it indicate that the severely sandy dolomite belongs the soft rock, and the completely sandy dolomite belongs he extremely soft rock.

Table 26.6 Classification of UCS and rock hardness

UCS/MPa	>60	60 ~ 30	30 ~ 15	15 ~ 5	≤ 5
Hardness	Hard rock		Soft rock		
	Hard rock	Relatively hard rock	Softer rock	Soft rock	Extremely soft rock

26.5 Conclusions

This paper takes the completely and severely sandy dolomite in the Xiaopu tunnel of the Yuxi section of the Central Yunnan Water Diversion Project as the study object. According to the needle penetration recommended by the international society of rock mechanics, a large number of needle penetration tests were carried out on the sandy dolomite, the needle penetration ratio NPI was calculated, and the relevant mechanical parameters of sandy dolomite were determined.

(1) The uniaxial compressive strength of severely sandy dolomite is 1.38–3.12 MPa, and the average value is 2.31 MPa. The uniaxial compressive strength of completely sandy dolomite is 0.88–1.67 MPa, and the average value is 1.23 MPa. It can be determined that severely sandy dolomite is soft rock and completely sandy dolomite is extremely soft rock.

(2) The cohesion (c) and internal friction angle (φ) of completely dolomite sandy estimated by needle penetration index (NPI) is close to the recommended value. The Cohesion (c) and internal friction angle (φ) of severely sandy dolomite by needle penetration index (NPI) is close to and partially equal to the field direct shear test data of rock mass.

(3) The needle penetration test can obtain the rock mass mechanical parameters of completely and severely sandy dolomite in the field quickly, economically and conveniently. It can be considered to use needle penetration ratio NPI to characterize the development degree of strongly and severely sandy dolomite.

The needle penetration is a fast, convenient, and accurate indirect method for estimating mechanical parameters for completely sandy dolomite. However, there are errors for severely sandy dolomite, which can be used to confirm its mechanical parameters through methods such as point load tests.

Acknowledgements The author sincerely expresses his gratitude to Dr. Guo Yongfa of Kunming Survey, Design and Research Institute Co., Ltd of CREEC for providing the needle penetration instrument for this test, to Zhao Yiran, Yang Runxue for carrying out the in-site needle penetration test, and to Yunnan Survey and Design Institute of Water Conservancy and Hydropower for providing relevant data. This study was partially sponsored by the National Natural Science Foundation of China (No. 42162026) and Basic Research Program in Yunnan (202201AT070083), Major Science and Technology Project of Yunnan Province (202002AF080003).

References

Avşar E, Ulusay R, Aydan Ö, Mutlutürk M (2015) On the difficulties of geotechnical sampling and practical estimates of the strength of a weakly bonded volcanic soil. Bull Eng Geol Environ 74(4):1375–1394. https://doi.org/10.1007/s10064-014-0710-9

Aydan (2012) The inference of physico-mechanical properties of soft rocks and the evaluation of the effect of water content and weathering on their mechanical properties from needle penetration tests. Paper presented at the 46th U.S. Rock Mechanics/Geomechanics Symposium

Aydan Ö, Sato A, Yagi M (2014) The inference of geo-mechanical properties of soft rocks and their degradation from needle penetration tests. Rock Mech Rock Eng 47(Special I):1867–1890. https://doi.org/10.1007/s00603-013-0477-5

Aydan, Ulusay R (2013) Geomechanical evaluation of Derinkuyu antique underground city and its implications in geoengineering. Rock Mech Rock Eng 46(4):731–754. https://doi.org/10.1007/s00603-012-0301-7

Azadan P, Ahangari K (2014) Evaluation of the new dynamic needle penetrometer in estimating uniaxial compressive strength of weak rocks. Arab J Geosci 7(8):3205–3216

Broch E, Franklin JA (1972) The point-load strength test. Int J Rock Mech Mining Sci Geomech Abstracts 9(6):669–676

Chen YL, Zhang Y, Lin L (2020). Relationship between point load strength and uniaxial compressive strength of sandy gravel. Min Eng Res 35(04):67–73

Erguler ZA, Ulusay R (2007) Estimation of uniaxial compressive strength of clay-bearing weak rocks using needle penetration resistance

Erguler ZA, Ulusay R (2009) Water-induced variations in mechanical properties of clay-bearing rocks. Int J Rock Mech Mining Sci 46(2):355–370. https://doi.org/10.1016/j.ijrmms.2008.07.002

He WX (2008) Study on the causes of dolomite sanding in the site area of Meigu River Pingtou Hydropower Station and its impact on the project (Master). Chengdu University of Technology, Available from Cnki

Hu XB (2009) Study on characteristics and mechanism of dolomite Karst Desertification in deep bank slope of Pingtou Hydropower Station on Meigu River (Master). Chengdu University of Technology, Available from Cnki

Huang YQ, Xu GL, Zhao DW, Li HX (2020) Study on strength deterioration of mica quartz schist under dry wet cycle. China Manganese Industry 38(3):33–37. https://doi.org/10.14101/j.cnki.issn.1002-4336.2020.03.008

ISRM (2009) ISRM Suggested method for determination of the Schmidt hammer rebound hardness: revised version. Int J Rock Mech Min Sci 46(3):627–634

JGS. (2012). Method for needle penetration test (JGS: 3431-2012) Japanese standards and explanations of geotechnical and geoenvironmental investigation methods, no 1. JGS Publication, Tokyo, pp 426–432 (in Japanese)

JSCE-RMC (1980) A suggested method for investigation and testing of soft rocks. In: Japan society of civil engineers, rock mechanics committee, the 4th sub-committee, Tokyo (in Japanese)

JSCE (1991) A suggested method for investigation and testing of soft rocks. Committee on Rock Mechanics of JSCE, p 124 (in Japanese)

Kahraman S, Aloglu AS, Aydin B, Saygin E (2017) The needle penetration test for predicting coal strength(Article). J Southern African Inst Mining Metallurgy 117(6):587–591. https://doi.org/10.17159/2411-9717/2017/v117n6a9

Li JG, Mu HY, Mi J (2018) Preliminary study on engineering geological characteristics of sandy dolomite. Paper presented at the 2nd National rock tunnel boring machine engineering technology seminar, Urumqi, Xinjiang, China

Li ZG (2017) Study on the relationship between engineering characteristics of metamorphic rocks and geological hazards in Duhe Basin. China University of Geosciences

Li ZG, Xu GL, Zhao X, Fu YP, Su C (2016) Applicability of needle penetration test on soft rocks. Electr J Geotech Eng 21(23):7209–7222

Li ZY, Yang YY (1994) Introduction to engineering geology. China University of Geosciences Press, Wuhan, China
Lin J, Sha P, Wu FQ, Chang JY (2018) Study on the relationship between point load index and uniaxial compressive strength of rock like materials based on size effect. J Yangtze River Sci Res Inst 35(03):34–44
Liu XB (2018) Study on strength characteristics and influencing factors of bituminous coal. Chongqing University
Maruto Co. Ltd. (1999) Instructions for SH-70 penetrometer (NANGAN-PENETOROKEI) (in Japanese), Tokyo, Japan, 4 pp
Maruto Co. Ltd. (2006) Penetrometer for soft rock: model SH-70 instruction manual. Tokyo, Japan
Naoto U, Yoshitake E, Hidehiro O, Norihiko M (2004) Strength evaluation of deep mixing soil-cement by needle penetration test. 土と基礎 = Soil mechanics and foundation engineering
Ngan-Tillard D, Verwaal W, Mulder A, Engin HK, Ulusay R (2011) Application of the needle penetration test to a calcarenite, Maastricht, the Netherlands. Eng Geol 123(3):214–224
Okada S, Izumiya Y, Iizuka Y, Horiuchi S (1985) The estimation of soft rock strength around a tunnel by needle penetration test. J Jpn Soc Soil Mech Found Eng 33(2):35–38
PWRI (1987) Draft manual of simple tests for rock mass classification of dam foundation on soft rocks. Technical Note of PWRI, No. 2506:31 (in Japanese)
Rabat Á, Cano M, Tomás R, Tamayo ÁE, Alejano LR (2020) Evaluation of strength and deformability of soft sedimentary rocks in dry and saturated conditions through needle penetration and point load tests: a comparative study. Rock Mech Rock Eng 53(6):2707–2726. https://doi.org/10.1007/s00603-020-02067-6
Rahimi MR, Mohammadi SD, Heidari M, Jalali SH (2020) Evaluation of the needle penetration test to estimate the uniaxial compressive strength of gypsum rocks. Arabian J Geosci 13(1):1–15. https://doi.org/10.1007/s12517-019-4989-5
Takahashi K, Noto K, Yokokawa I (1998) Strength characteristics of Kobe Formation in Akashi Strata (No. 1). Paper presented at the proceedings of 10th Japan national conference on geotechnical engineering. The Japanese Geotechnical Society
Tian QW, Yue CL (2014) Characteristics and surrounding rock classification of karst sandy fine-grained dolomite in Pingtou Hydropower Station. Adv Sci Technol Water Resour 34(5):45–49
Uchida N, Etoh Y, Ono H, Miura N (2004) Strength evaluation of deep mixing soil–cement by needle penetration test. J Jpn Soc Soil Mech Found Eng 52(7):23–25
Ulusay ÖA, Erguler Z, Ngan-Tillard D, Seiki T, Verwaal W, … Sato A (2014) ISRM suggested method for the needle penetration test. Rock Mech Rock Eng 47(Special I):1073–1085. https://doi.org/10.1007/s00603-013-0534-0
Ulusay, Erguler ZA (2012) Needle penetration test: evaluation of its performance and possible uses in predicting strength of weak and soft rocks. Eng Geol 149–150:47–56. https://doi.org/10.1016/j.enggeo.2012.08.007
Ulusay R, Gokceoglu C, Sulukcu S (2001) Draft ISRM suggested method for determining block punch strength index (BPI). Int J Rock Mech Mining Sci
Yamaguchi Y, Ogawa N, Kawasaki M, Nakamura A (1997) Evaluation of seepage failure resistance potential of dam foundation with simplified tests. J Jpn Soc Eng Geol 38(3):130–144
Yao ZQ, Xu GL, Zhu ZH, Hu XL (2015) Application of needle penetration test in mudstone in Wuhan. Construct Technol (A2):66–68
Yu S (2013) Study on engineering characteristics of karst dolomite. West-China Exploration Eng (6):139–140
Yunnan water conservancy and hydropower survey and design institute (2018) Preliminary design and engineering geological survey report of Yuxi section of main water conveyance channel of central Yunnan water diversion project
Zhang, Zhang HQ, Zhao QH, He WX (2012a) Characteristics and development distribution of dolomite sanding in Pingtou Hydropower Station, Sichuan Province. Yangtze River (19):42–44, 78

Zhang CL, Zhang CP, Xu J (2015) Comparative test of rock point load strength and uniaxial compressive strength. Chin J Underground Space Eng 11(S2):447–451

Zhang LX, Zhao QH, Hu XB, Han G, Zhao X (2012b) Laboratory dissolution test and micro dissolution mechanism of dolomite in an area. J Eng Geol (4):576–584

Zhang SK (2020) Study on large deformation prediction and construction control technology of Sericite schist tunnel in western Yunnan. Kunming University of Science and Technology

Zou ZN, Wang XF (2012) On the "sanding" mechanism of fine-grained dolomite and its engineering examples. New Technol New Prod China (08):91–92

Chapter 27
Correlation Between the Shear Strength Parameters of Clastic and Carbonate Rocks with the Brazilian Tensile Strength and the Uniaxial Compressive Strength

Dimitrios Kotsanis and Pavlos Nomikos

Abstract The objective of the current study is to investigate the relation between the Brazilian tensile strength (*BTS*) and the uniaxial compressive strength (*UCS*) with the shear strength parameters (cohesion and internal friction angle) of some clastic and carbonate rocks from the southern part of the East Attica Prefecture, Greece. For this reason, several specimens were prepared and tested for the direct evaluation of *BTS*, *UCS*, and the shear strength parameters of four rock types comprising marbles, limestones, sandstones, and breccias. Applying simple linear regression techniques to the results, it was found that a strong correlation exists between cohesion with *BTS* or *UCS* values, for the studied clastic and the carbonate rocks. On the other hand, the values of the angle of internal friction seem to have no meaningful statistical correlation with *BTS* or *UCS* values. Therefore, a different linear model is used for the correlation of internal friction angle values, considering the ratio between the uniaxial compressive strength with the Brazilian tensile strength.

Keywords Brazilian tensile strength · Uniaxial compressive strength · Shear strength parameters · Clastic rocks · Carbonate rocks

27.1 Introduction

The axisymmetric or conventional triaxial compression test is the most straightforward experimental procedure to simulate the triaxial stress field where rock materials are subjected, under foundations or around excavations. In this test, a cylindrical rock

D. Kotsanis (✉) · P. Nomikos (✉)
School of Mining and Metallurgical Engineering, National Technical University of Athens, Athens, Greece
e-mail: dkotsanis@metal.ntua.gr

P. Nomikos
e-mail: nomikos@metal.ntua.gr

S. Wang et al. (eds.), *Engineering Geology for a Habitable Earth: IAEG XIV Congress 2023 Proceedings, Chengdu, China*, Environmental Science and Engineering,
https://doi.org/10.1007/978-981-99-9057-3_27

specimen is subjected to compression with the intermediate (σ_2) and minor (σ_3) principal stresses equal to the lateral confining pressure (p), while the axially applied major principal stress (σ_1) is constantly increasing. This type of triaxial testing has a long history of usage, and as a result, several failure criteria have been empirically developed to gain insight into the mechanical behaviour of rocks under triaxial stress conditions.

Among the various failure criteria that have been deduced over the years, the simplest and one of the most frequently used in rock mechanics applications is the Mohr–Coulomb failure criterion (MC). Labuz and Zang (2012) emphasize the applicability of this criterion, referring mainly to the simplicity and convenience of its mathematical formulation and the advantage of a clear physical meaning of its shear strength parameters.

According to the Mohr–Coulomb failure criterion, shear failure is assumed to take place along a surface if the shear stress (τ) acting on that plane equals:

$$\tau = S_o + \tan\varphi\sigma_n \tag{27.1}$$

where φ is the internal friction angle and S_o the cohesion of the material, i.e., the shear strength of the material when there is no normal stress (σ_n) on the shearing plane. In terms of the major and the minor principal stresses, σ_1 and σ_3, the linearized form of the Mohr–Coulomb failure criterion can be written as:

$$\sigma_1 = k\sigma_3 + UCS = \frac{1+\sin\varphi}{1-\sin\varphi}\sigma_3 + \frac{2S_o\cos\varphi}{1-\sin\varphi} \tag{27.2}$$

where *UCS* represents the uniaxial compressive strength of the intact rock specimen, while k is the slope of the failure envelope in the $\sigma_3 - \sigma_1$ stress space.

Although it is generally accepted that this criterion is an approximation of the observed failure response of rocks, which is in general nonlinear and can be affected by the intermediate principal stress σ_2 (Paterson and Wong 2005; Mogi 2006), experimental results indicate the applicability of a straight-line failure envelope over a limited range of confining pressures which is of particular interest for civil and mining engineering works (Rummel 1972; Wittke 1990).

Even though the triaxial test is the best straightforward way to determine the MC shear strength parameters (Vutukuri et al. 1974), it cannot be considered an easy and economical experimental procedure. It requires a significant number of high-quality specimens for a series of single-stage triaxial tests and more sophisticated experimental configurations if multiple or continuous failure state tests need to be carried out. With such limitations, empirical relations of the shear strength parameters with other strength or index properties of the intact rock, which may be more easily obtained, are useful at least for an indication of their expected range of values for a particular rock type in a given geographical area.

Empirical relations are common in experimental rock mechanics, where important mechanical properties are estimated through the statistical interpretation of the results of less expensive or simpler experimental procedures. For example, the uniaxial compressive strength can be conveniently determined from physical or dynamical properties, as well as, from the results of index strength tests such as the point load test (Zhang 2016). Nevertheless, this is not the case for the shear strength parameters where the existence of possible relationships of S_o or φ with other rock properties has not been thoroughly studied.

Regarding the research that is based on petrophysical models, it seems that the values of cohesion and internal friction angle of clastic rocks (sandstones and shales) decrease either linear or nonlinear with the increase of porosity (Vernik et al. 1993; Weingarten and Perkins 1995; Crawford et al. 2010). The mineralogical composition also affects the shear strength parameters of rocks. For example, an increase in the clay content results in a nonlinear decrease in cohesion and internal friction angle of some clastic rocks (Crawford et al. 2010). On the other hand, the cohesion of coal measure rocks increases linearly with an increase in quartz content (Smart et al. 1982).

The relation between some basic mechanical properties of rocks and their shear strength parameters has also been investigated, mainly utilizing the results of uniaxial compression and Brazilian tensile strength tests. Karaman et al. (2015) conducted *UCS*, *BTS*, and conventional triaxial tests for various rock types, comprising basalts, dacites, limestones, matabasalts, and volcanic breccias. According to the results of their research, *UCS* and *BTS* are very good indicators of cohesion, where an increase in *UCS* or *BTS* values results in a linear increase in cohesion values. In terms of the internal friction angle and for the whole dataset, the developed linear equations between φ-*UCS* and φ-*BTS* are weak, according to the derived coefficients of determination (R^2). However, improved results were obtained when the regression analysis was carried out for each specific rock type. The developed equations are generally nonlinear with moderate coefficients of determination.

Shen and Jimenez (2018) and Shen et al. (2019) employed previously published datasets and examined the feasibility of *UCS* and *BTS* results to estimate the shear strength parameters for sandstones and coal rocks, respectively. According to the statistical analysis of these results, cohesion values increase linearly with the increase of *BTS* or *UCS* for both rock types. Regarding the correlation between φ-*UCS* and φ-*BTS*, contradictory results are presented. In particular, a weak correlation was observed between the internal friction angle and the *BTS* or *UCS* results for sandstones, while an increase in *UCS* values for coal rocks leads to a linear decrease in cohesion values. However, as shown by Shen and Jimenez (2018), an improvement in the estimation of the shear strength parameters can be achieved through genetic programming.

As for this study, the potential correlations between the shear strength parameters with *UCS* or *BTS* values for clastic and carbonate rocks were investigated, through simple regression analysis techniques. From the results of this analysis, empirical relationships were derived and applicable for preliminary estimations of the cohesion values of these rock types, at least in the study region. Further, for the preliminary estimation of the friction angle values of these rocks, an alternative correlation of φ with the strength ratio *R*, i.e., the ratio of uniaxial compressive strength to the Brazilian tensile strength, was examined.

27.2 Petrographic and Mineralogical Characterization of the Studied Rocks

The rocks tested in the current study were sampled from selected outcrops of the southern part of the East Attica Prefecture, Greece. Block samples of four petrological types from the studied region were collected, distinguished as marbles, limestones, sandstones, and breccias.

According to classification schemes in terms of grain size (Flügel 2010), the sedimentary carbonates are characterized as sparitic and microsparitic limestones (Fig. 27.1a, b). Both types of rocks are characterized by the presence of non-fabric selective pores in the form of vugs, irregular in shape, and randomly distributed. PXRD results revealed only the presence of calcite in all samples.

The metamorphic carbonates are pure calcite marbles. Accessory minerals, such as micas, are present as elongated crystals oriented parallel to the foliation planes (Fig. 27.1c).

The clastic rocks are composed of calcite, quartz, micas, and lithic fragments, with albite and chlorite in lesser amounts. In terms of grain size (Pettijohn et al. 1973), sandstones in this research are classified as very fine-to-medium-grained materials (Fig. 27.1d).

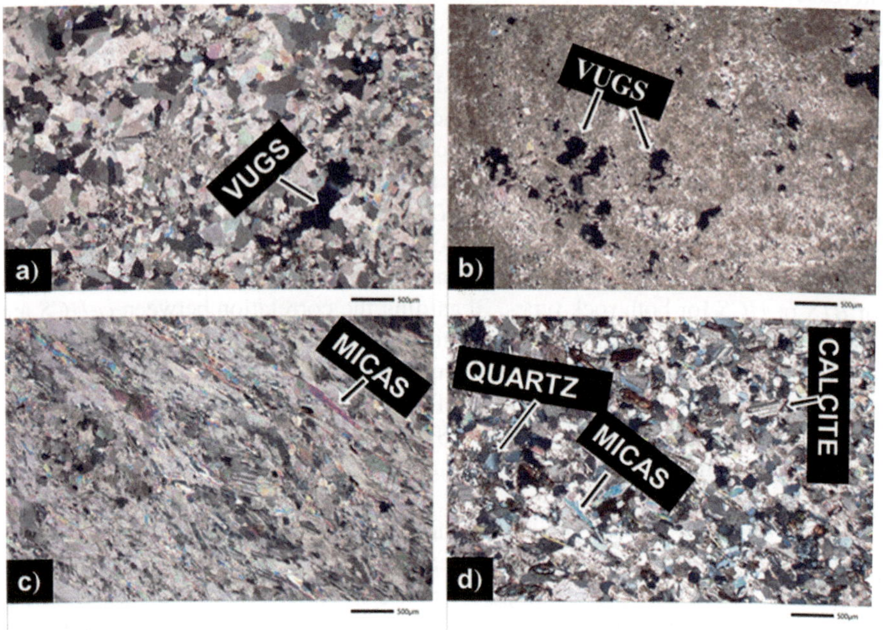

Fig. 27.1 Characteristic photomicrographs of the studied rocks (double polarized transmitted light). **a** Sparitic limestone, **b** Microsparitic limestone, **c** Marble, and **d** Sandstone

27.3 Experimental Program and Results

In the context of this research, eighteen rock blocks were sampled from natural outcrops of the study region. All test results herein refer to right circular cylinders and disks, free from any macroscopic fractures and discontinuities. The specimens were prepared from cores drilled in the laboratory from roughly prismatic rock blocks of sufficient size. All cores were of NX-size, having a diameter of 54.5 mm, and in all cases, their long axis was oriented normal to any visible planes of anisotropy, i.e., the bedding planes of the sedimentary rocks and any foliation of the metamorphic rocks. The final preparation of the specimens was accomplished using a diamond cut-off saw and a lapping machine according to the ISRM suggested methods (ISRM 1978, 1983; Fairhurst and Hudson 1999).

The *BTS* tests (Fig. 27.2a) were performed on solid disks having a thickness approximately equal to their radii. The specimens were placed in an experimental apparatus consisting of two steel loading jaws designed following the ISRM suggested methods (ISRM 1978). To ensure the accurate measurement of the load required to form the primary tensile crack, one linear variable differential transducer (LVDT), capable of measuring the axial displacement with a resolution of 1 μm, was placed between the loading platens (Mellor and Hawkes 1971). All tests were carried out under load control with a constant rate of 200 N/s, in a servo–hydraulic-driven loading frame of 250 kN capacity. The axial load along with the axial displacement were measured and recorded during the test execution. For each rock block, the *BTS* value was taken as the average of the strength values of at least five specimens.

Uniaxial and conventional triaxial compression tests (Fig. 27.2b, c) were performed with a servo-hydraulic loading frame with a 5000 kN capacity. The axial load and the axial displacements were measured and recorded simultaneously throughout the experiments.

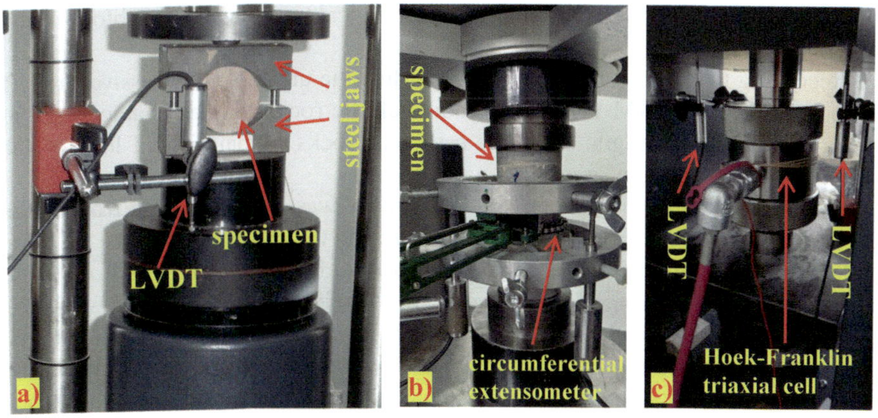

Fig. 27.2 Experimental configuration for **a** the *BTS* tests, **b** the *UCS* tests, and **c** the triaxial tests

The *UCS* tests were conducted on specimens with a slenderness ratio between 2.5 and 3. All tests were performed under lateral strain control, using a circumferential extensometer mounted around the specimen's mid-height. The *UCS* value of each rock block was taken as the average of the strength values of at least three specimens.

The triaxial strength data were obtained by performing one multistage triaxial test for each rock block on specimens with a slenderness ratio of approximately 2:1. The same experimental setup as in Kotsanis et al. (2018) was followed. The experimental procedure was in line with the recommendations of the ISRM (ISRM 1983). The axial displacement was measured through two linear variable differential transducers (LVDTs), placed between the loading platens in opposite directions. To apply the confining pressure, a Hoek-Franklin triaxial cell (Hoek and Franklin 1968) was used, connected to a servo-hydraulic control unit. Four to six loading stages were performed in each multi-stage triaxial test, with a maximum lateral pressure of 24 MPa in the case of hard rocks. For each lateral pressure level, the axial stress was increased until the impending peak strength for the given lateral pressure. Then, the lateral pressure was increased sharply to the next desired level and kept constant. At that time, the axial stress increased again until the corresponding impending peak strength. The same procedure was followed up to the stage of the maximum desired lateral pressure, upon completion of which the peak strength envelope of the test specimen is obtained in the lateral pressure range of the test.

The shear strength parameters of the MC criterion were determined by statistical regression techniques that are implemented in the RocData software (Rocscience Inc. 2017). To this end, a range of confining pressures of $0 < \sigma_3 < 0.42UCS$ along with the average *UCS* value were used. The results of the regression analysis confirm the suitability of the MC criterion for the tested materials since in all cases, the derived coefficient of determination (R^2) was greater than 0.98. Some characteristic failure envelopes, shown with dashed lines, obtained for representative specimens of the four rock types are illustrated in Fig. 27.3.

The mechanical parameters derived from the experimental program are summarized in Table 27.1. In the same table, the petrological type of each rock block is also presented.

According to classification schemes regarding the uniaxial compressive strength of the intact rock (ISRM 1981), the studied clastic rocks may be characterized as rocks of moderate strength, while the carbonate rocks as materials of low to high strength due to variations in grain size and porosity.

The Brazilian tensile strength for clastic rocks was found to range from 3.2 to 6.9 MPa, while in the case of the carbonate rocks, the lowest value was 1.2 MPa, and the highest was 11.5 MPa. Cohesion values for clastic rocks range from 7.6 to 15.3 MPa, while the range for internal friction angle extends from 29.7° to 38.5°. The carbonate rocks yielded cohesion values between 4.3 and 30.1 MPa and internal friction angles between 27.6° to 41.4°. The shear strength parameters obtained by multistage triaxial tests are in line with quoted values in the literature for these rock types (e.g., Farmer 1983).

From the *UCS* and *BTS* values, it was possible to calculate the strength ratio *R* for each rock block sample. In the case of the studied carbonates rocks, this ratio

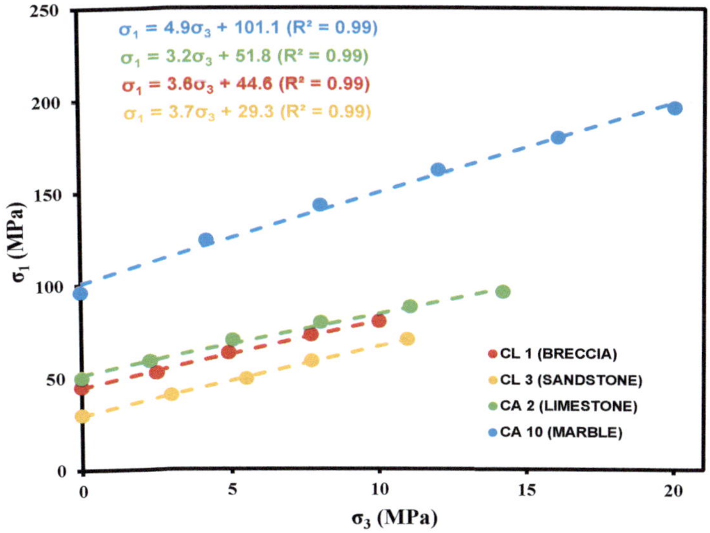

Fig. 27.3 Indicative results of the peak strength data along with their MC fit lines for the studied rocks

Table 27.1 Mechanical properties of the studied rock blocks

Rock block code	Petrological type	S_o (MPa)	φ (°)	UCS (MPa)	BTS (MPa)
CL 1	Breccia	11.8	34.1	44.9	4.9
CL 2	Breccia	13.5	35.4	50.8	5.9
CL 3	Sandstone	7.6	35.2	29.7	3.2
CL 4	Sandstone	15.3	38.5	60.8	6.9
CL 5	Sandstone	14.9	29.7	48.8	6.1
CL 6	Sandstone	9.2	35.6	35.4	3.8
CA 1	Limestone	24.8	32.6	85.0	8.5
CA 2	Limestone	14.4	31.9	50.0	4.6
CA 3	Limestone	8.8	35.5	33.0	3.3
CA 4	Limestone	8.0	28.6	26.7	4.0
CA 5	Limestone	23.3	37.6	88.7	7.9
CA 6	Limestone	4.3	37.6	15.8	1.2
CA 7	Limestone	15.8	27.6	52.8	5.8
CA 8	Limestone	12.4	38.5	51.2	4.8
CA 9	Marble	30.1	38.6	119.5	11.5
CA 10	Marble	22.8	41.4	96.4	8.2
CA 11	Marble	24.4	39.3	101.4	9.4
CA 12	Marble	25.1	37.3	98.1	8.9

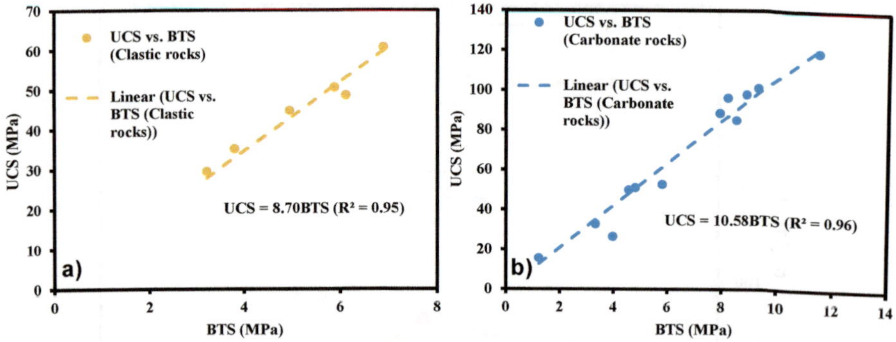

Fig. 27.4 Empirical relationship of *UCS* with *BTS* for **a** the clastic rocks, and **b** the carbonate rocks of this study

varies from 6.72 to 13.08. On the other hand, for the clastic rocks examined, a limited dispersion on these values was observed, where the lowest value was 7.98, and the highest was 9.28.

Results obtained in this research suggest that *UCS* increases linearly with the increase of *BTS* for both rock types, as shown in Fig. 27.4. These empirical relationships are characterized by very strong coefficients of determination ($R^2 = 0.95–0.96$) and expressed by Eq. (27.3) for the clastic rocks and Eq. (27.4) for the carbonate rocks, respectively.

$$\mathrm{UCS} = 8.70\,\mathrm{BTS} \tag{27.3}$$

$$\mathrm{UCS} = 10.58\,\mathrm{BTS} \tag{27.4}$$

27.4 Relationship of Shear Strength Parameters with UCS and BTS

The uniaxial compression strength test is the most common experimental procedure undertaken in rock mechanics laboratories since its results are crucial in many aspects of geotechnical engineering (Pells 1993). On the other hand, the Brazilian tensile strength test is the most popular indirect configuration that is used for the determination of the tensile strength of the intact rock (Mellor and Hawkes 1971). Bearing in mind that *UCS* and *BTS* can be relatively easily determined directly in the laboratory, research concerning their potential correlation with the shear strength parameters seems to be a challenging task in cases where triaxial tests are not available, such as during the early stages of a rock engineering project.

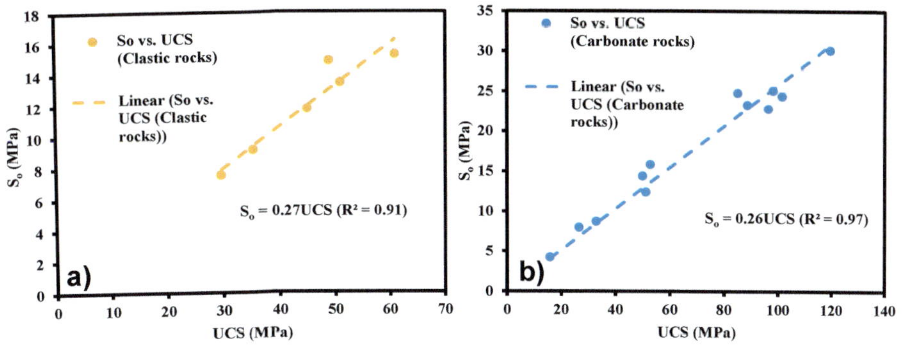

Fig. 27.5 Empirical relationship of cohesion with *UCS* for **a** the clastic rocks, and **b** the carbonate rocks of this study

The statistical analysis of our results reveals a clear positive linear correlation of cohesion with *UCS*, as illustrated in Fig. 27.5. The deduced empirical relationships are characterized by very strong coefficients of determination ($R^2 = 0.91$–0.97) and presented in Eqs. (27.5) and (27.6), for the clastic and the carbonate rocks, respectively.

$$S_o = 0.27\,\mathrm{UCS} \tag{27.5}$$

$$S_o = 0.26\,\mathrm{UCS} \tag{27.6}$$

BTS-based correlations have also been derived in the current study, as presented in Fig. 27.6 and Eqs. (27.7) and (27.8) for the studied clastic and carbonate rocks, respectively.

$$S_o = 2.33\,\mathrm{BTS} \tag{27.7}$$

$$S_o = 2.74\,\mathrm{BTS} \tag{27.8}$$

These relations are also characterized by very strong coefficients of determination ($R^2 = 0.97$–0.98). That is a logical extension of the *UCS* correlation models due to the very strong linear correlation between *UCS* and *BTS*, as described by Eqs. (27.3) and (27.4).

However, our results indicate weak correlations ($R^2 = 0.0001$–0.25) of internal friction angle with *BTS* and *UCS* values for both rock types, as illustrated in Fig. 27.7. Similar results on this topic are also presented by Karaman et al. (2015) and Shen and Jimenez (2018).

In the current study, to achieve a stronger correlation for the internal friction angle, the ratio of the uniaxial compressive strength to the Brazilian tensile strength was also used. The deduced models are presented in Fig. 27.8 and given by Eqs. (27.9)

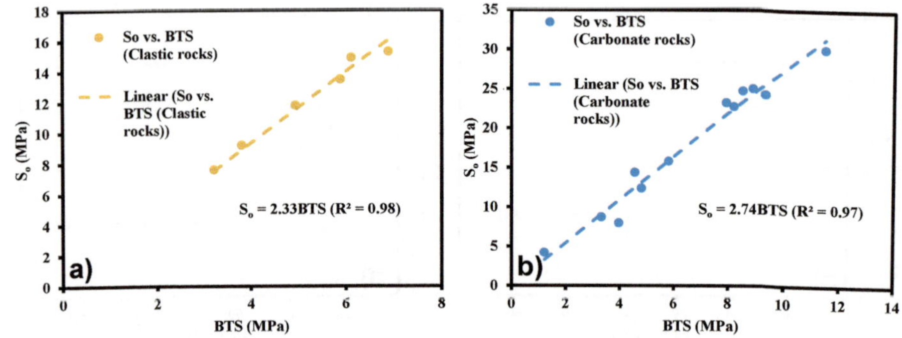

Fig. 27.6 Empirical relationship of cohesion with *BTS* for **a** the clastic rocks, and **b** the carbonate rocks of this study

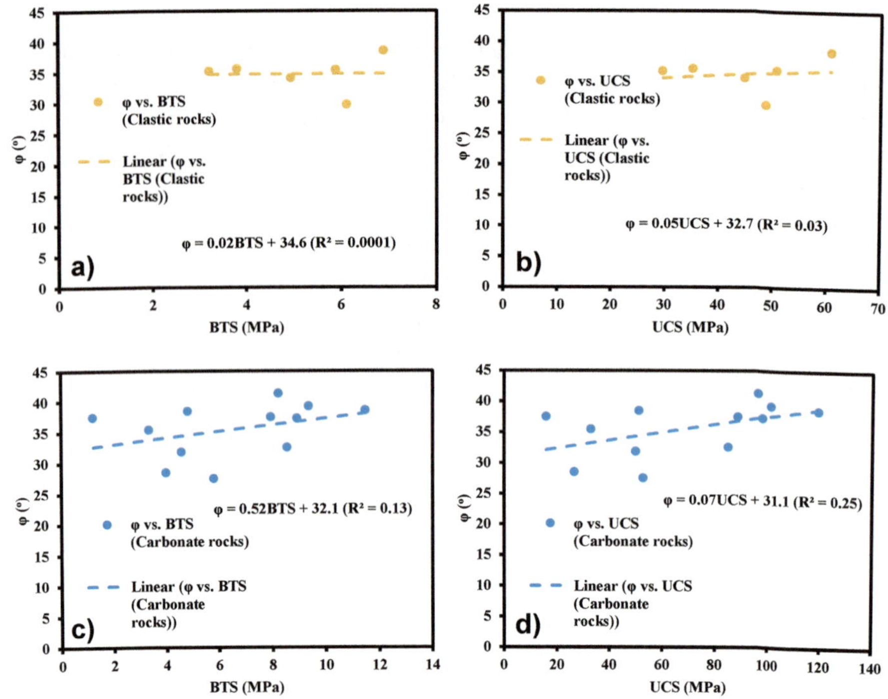

Fig. 27.7 Correlation of internal friction angle with **a** *BTS* for clastic rocks, **b** *UCS* for clastic rocks, **c** *BTS* for carbonate rocks and **d** *UCS* for carbonate rocks

and (27.10) for clastic and carbonate rocks, respectively.

$$\varphi = 3.84\text{R} + 0.80 \tag{27.9}$$

$$\varphi = 2.05\text{R} + 14.08 \tag{27.10}$$

Although the φ-*R* data pairs deviate more or less from the trend line, in both cases, the developed equations are characterized by better coefficients of determination ($R^2 = 0.42$–0.51) from those derived from the predictive models using only *UCS* or *BTS*.

The results presented in this study confirm the findings of previous investigations concerning the reliability of *UCS* and *BTS* predictive models. However, the derived mathematical formulations, as illustrated in Fig. 27.9, differ from each other, and as a result, very different values of cohesion could be obtained. This is a common feature in most of the empirical equations quoted in the literature and emphasizes the need to calibrate a selected equation before its use in a specific site.

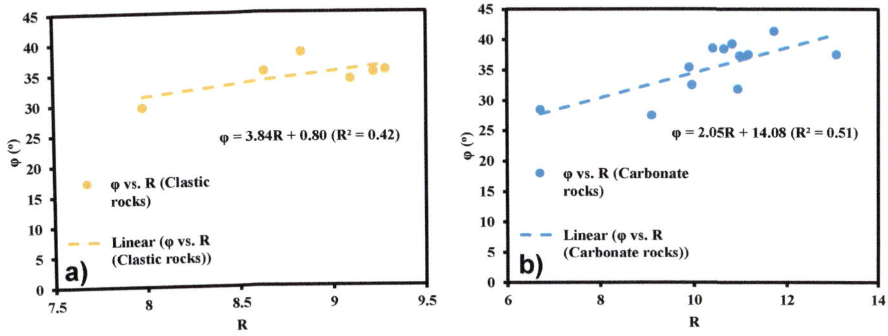

Fig. 27.8 Empirical relationship of internal friction angle with *R* for **a** clastic rocks and **b** carbonate rocks

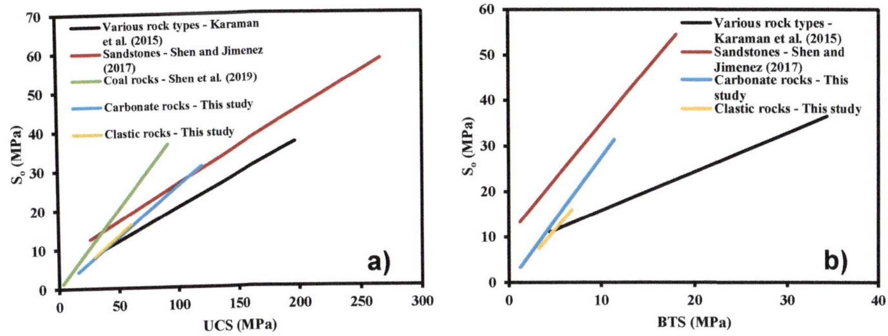

Fig. 27.9 Comparison of previously published equations with the linear relationships derived in this study in terms of **a** S_o-*UCS* and **b** S_o-*BTS*

27.5 Conclusions

As for the content of this study, several specimens of some clastic and carbonate rocks were prepared, from eighteen blocks collected from natural outcrops in the southern part of the East Attica Prefecture, Greece, while a laboratory program was undertaken for the direct determination of *UCS*, *BTS*, cohesion, and the internal friction angle.

From the statistical interpretation of the results, it was found that a strong correlation exists between cohesion and *UCS* or *BTS*. On the other hand, weak correlations were achieved between the internal friction angle and *BTS* or *UCS* for both clastic and carbonate rocks. However, better correlation models were derived based on the strength ratio *R* (=*UCS/BTS*).

Our findings agree with the limited results presented in the literature concerning the applicability of *BTS* or *UCS* to estimate the shear strength parameters of rock. In case triaxial test data are not available, such as during the early stages of a rock engineering project, the derived relations provide an initial estimate of the anticipated range of the shear strength parameters of the studied rocks.

References

Crawford BR, Gaillot PJ, Alramahi B (2010) Petrophysical methodology for predicting compressive strength in siliciclastic "sandstone-to-shale" rocks. In Proceedings of the 44th US rock mechanics symposium, Salt Lake City, 27–30 June 2010

Fairhurst CE, Hudson JA (1999) Draft ISRM suggested method for the complete stress-strain curve for intact rock in uniaxial compression. Int J Rock Mech Min Sci 36:279–289. https://doi.org/10.1016/S0148-9062(99)00006-6

Farmer IW (1983) Engineering behaviour of rocks, 2nd edn. Chapman and Hall, London

Flügel E (2010) Microfacies of carbonate rocks-analysis, interpretation and application, 2nd edn. Springer, Berlin, Heidelberg

Hoek E, Franklin JA (1968) A simple triaxial cell for field and laboratory testing of rock. Trans Instn Min Metall 77:A22–A26

ISRM (1978) Suggested methods for determining tensile strength of rock materials. Int J Rock Mech Min Sci Geomech Abstr 15:99–103. https://doi.org/10.1016/0148-9062(78)90003-7

ISRM (1981) Basic geotechnical description of rock masses (BGD). Int J Rock Mech Min Sci Geomech Abstr 18:87–110. https://doi.org/10.1016/0148-9062(81)90277-1

ISRM (1983) Suggested methods for determining the strength of rock materials in triaxial compression: revised version. Int J Rock Mech Min Sci Geomech Abstr 20:285–290. https://doi.org/10.1016/0148-9062(83)90598-3

Karaman K, Cihangir F, Ercikdi B, Kesimal A, Demirel S (2015) Utilization of the Brazilian test for estimating the uniaxial compressive strength and shear strength parameters. J S Afr Inst Min Metall 115:185–192

Kotsanis D, Nomikos P, Rozos D, Sofianos A (2018) Multistage triaxial testing of various rock types: a case study of East Attica Prefecture, Greece. Procedia Struct Integr 10:112–119. https://doi.org/10.1016/j.prostr.2018.09.017

Labuz JF, Zang A (2012) Mohr-Coulomb failure criterion. Rock Mech Rock Eng 45:975–979. https://doi.org/10.1007/s00603-012-0281-7

Mellor M, Hawkes I (1971) Measurement of tensile strength by diametral compression of discs and annuli. Eng Geol 5:173–225. https://doi.org/10.1016/0013-7952(71)90001-9
Mogi K (2006) Experimental rock mechanics, 1st edn. CRC Press, London
Paterson MS, Wong T-f (2005) Experimental rock deformation-the brittle field, 2nd edn. Springer, Berlin, Heidelberg
Pells PGN (1993) The measurement and estimation of basic rock strength. In: Hudson JA (ed) Comprehensive rock engineering, vol 3. Pergamon Press, Oxford, pp 67–85
Pettijohn FJ, Potter PE, Siever R (1973) Sand and sandstone, 1st edn. Springer-Verlag, New York
Rocscience Inc. (2017). https://www.rocscience.com/
Rummel F (1972) A review of fracture criteria of brittle rock. In: Müller L (ed) Rock mechanics. Springer, Vienna, pp 71–84
Shen J, Jimenez R (2018) Predicting the shear strength parameters of sandstone using genetic programming. Bull Eng Geol Environ 77:1647–1662. https://doi.org/10.1007/s10064-017-1023-6
Shen J, Wan L, Zuo J (2019) Non-linear shear strength model for coal rocks. Rock Mech Rock Eng 52:4123–4132. https://doi.org/10.1007/s00603-019-01775-y
Smart BGD, Rowlands N, Isaac AK (1982) Progress towards establishing relationships between the mineralogy and physical properties of coal measures rocks. Int J Rock Mech Min Sci Geomech Abstr 19:81–89. https://doi.org/10.1016/0148-9062(82)91633-3
Vernik L, Bruno M, Bovberg C (1993) Empirical relations between compressive strength and porosity of siliciclastic rocks. Int J Rock Mech Min Sci Geomech Abstr 30:677–680. https://doi.org/10.1016/0148-9062(93)90004-W
Vutukuri VS, Lama RD, Saluja SS (1974) Handbook on mechanical properties of rocks, vol 1. Trans Tech Publications, Clausthal
Weingarten JS, Perkins TK (1995) Prediction of sand production in gas wells: methods and Gulf of Mexico case studies. J Petrol Tech 47:596–600. https://doi.org/10.2118/24797-PA
Wittke W (1990) Rock mechanics—theory and applications with case histories. Springer, Berlin, Heidelberg
Zhang L (2016) Engineering properties of rocks, 2nd edn. Butterworth-Heinemann, Oxford

Chapter 28
Analysis of Shear Strength of Jointed Rock Masses Based on Connectivity Rate

Yawei Li, Hui Luo, and Xiao Tian

Abstract Irregular joint surfaces in natural rock masses not only weaken their strength, but also cause significant anisotropy in the shear strength across different directions. Assessing the shear strength of rock masses in various directions using on-site survey data is beneficial for engineering design and construction. The connectivity rate indicates the level of connectivity among structural planes in the jointed rock mass, and it can be used to determine the shear strength of the rock mass. This article utilizes the Monte-Carlo random method to construct a fracture network model that aligns with the geometric distribution characteristics of on-site joints, as determined by joint measurement results. The study analyses the connectivity rate and shear strength parameters of joints in various shear directions within engineering rock masses. The findings show a notable variation in joint connectivity and shear strength among different directions within the rock masses. Notably, the direction of occurrence at 146°∠79° exhibits the highest joint connectivity and is the most susceptible to failure.

Keywords Jointed rock · Fracture network · Connectivity rate · Shear strength

28.1 Introduction

Rock mass forms the primary foundation for numerous engineering constructions. In contrast to common engineering materials, engineering rock mass is a distinct material that develops through a lengthy geological evolution process. Due to the presence of joints in natural rock masses, the mechanical strength of the rock mass is significantly reduced compared to intact rock, and extensive engineering practices have demonstrated that joints play a controlling role in the deformation and failure

---

Y. Li (✉) · H. Luo · X. Tian
Beijing Research Institute of Uranium Geology, Beijing, China
e-mail: yawei_cn@163.com

CAEA Innovation Center for Geological Disposal of High-Level Radioactive Waste, Beijing, China

S. Wang et al. (eds.), *Engineering Geology for a Habitable Earth: IAEG XIV Congress 2023 Proceedings, Chengdu, China*, Environmental Science and Engineering,
https://doi.org/10.1007/978-981-99-9057-3_28

strength of rock masses. On one hand, joints can impact the overall engineering quality of the rock mass, and on the other hand, they may also serve as sliding planes during excavation, determining the stability of the rock mass and influencing the design and construction of future projects.

Currently, several commonly used methods exist for determining the shear strength parameters of rock mass. These methods include the indoor small direct shear test method, in-situ large direct shear test method, engineering analogy method, and formula calculation based on empirical strength criterion. The indoor test offers advantages such as standardized operation and controlled test conditions. However, the original state of the test sample differs significantly from the actual engineering environment. On the other hand, in-situ test results are relatively reliable, but they require high test conditions, complex equipment, and are challenging to conduct in large numbers (Liu et al. 2005). The engineering analogy method primarily relies on existing construction experience in similar engineering geological conditions to guide the design and construction of proposed projects. However, its reliability is limited due to regional and empirical differences (Duan et al. 2010). The Hoek–Brown empirical strength criterion considers the strength of intact rock mass, the structural characteristics of the rock mass, and the stress state. Hoek and Brown, for instance, summarized the Hoek–Brown empirical strength criterion based on an extensive number of tests. In this criterion, the rock mass is treated as a homogeneous and isotropic medium, and the shear strength parameters of the rock mass are determined using formulas (Hoek and Brown 1988; Hoek et al. 2002).

In comparison to the aforementioned methods, utilizing the connectivity rate of a fractured rock mass to estimate its shear strength in different directions can more effectively capture the anisotropic characteristics resulting from the occurrence of joints. During the site selection phase of a repository, the analysis of the connectivity rate of structural planes, based on the distribution characteristics of surface joints, enables the determination of the shear strength of the rock mass in various directions. This approach offers a more objective and rational foundation for the azimuth layout design and excavation plan of future repository projects.

28.2 Method for Calculating Connectivity Based on Fracture Network

28.2.1 Definition of Connectivity Rate

Determining the comprehensive shear strength of an engineering rock mass through the assessment of the connectivity rate of fractured rock mass is grounded in the fundamental concept proposed by H. H. Einstein et al. Their hypothesis suggests that during shear failure of a fractured rock mass, the intact rock between structural planes (referred to as "rock bridges") shares the shear force. Accordingly, the comprehensive shear strength of the rock mass can be computed by considering the proportion of the structural planes and rock bridges on the shear plane of the fractured rock mass (Einstein et al. 1983). This approach allows for a holistic evaluation of the shear strength and contributes to a better understanding of the mechanical behavior of rock masses under shear loading conditions.

The connectivity rate of structural plane is an important index to reflect its connectivity in the extension direction. Generally, the connectivity rate of structural plane of jointed rock mass is defined as the ratio of the sum of trace length of structural plane to the length of reference line or the ratio of the sum of area of single structural plane to the total area of a selected datum plane. Due to the limitations of field structural plane measurement, the area of structural plane is difficult to determine. Therefore, the connectivity rate of structural plane is mainly obtained by measuring the trace length of structural plane.

Various calculation methods for determining the connectivity rate have been employed by different researchers in response to diverse practical engineering challenges. Huang et al. approached connectivity rate determination by directly conducting field investigations and measurements (Runqiu and Liuming 2003). Additionally, other methods such as the generalized H–H method, total trace length survey method, and shear path search method have been utilized (Guoming and Runqiu 1999; Xiaogang et al. 1992, 1998). In this study, the bandwidth projection method based on fracture network simulation is adopted. The bandwidth projection method builds upon the concept of the total trace length survey estimation method. It involves defining a statistical measurement band of a certain width and projecting a set of structural planes onto a reference line. The connectivity rate of this set of structural planes is then determined as the proportion of the total projection length on the reference line. The calculation formula for this method is:

$$k = \frac{\sum TL_n}{L},\ n = 1,\ 2,\ 3\ldots \tag{28.1}$$

In Eq. (28.1): k-connectivity; $\sum TL_n$—the total length of the projection of the structural plane in the shear failure direction; L—total length in shear direction; n-Statistics the number of structural plane in the band.

28.2.2 Fracture Network Simulation Principle

The fracture network simulation is based on the probability distribution characteristics of the geometric parameters (occurrence, trace length, spacing, etc.) of the structural plane obtained from the field measurement to derive the joints that conform to these distribution laws. According to the probability statistical model of the geometric parameters of the structural plane on the limited natural outcrop or artificial excavation surface, the Monte-Carlo method is used to simulate the fracture network model consistent with the geometric distribution characteristics of the structural plane on the computer, and the parameter distribution of the structural plane inside the rock mass or in a larger range is further derived (Wang et al. 2016).

28.2.3 *Calculation of Shear Strength Parameters of Rock Mass*

When the rock mass is subjected to shear failure, the comprehensive shear strength of the rock mass is jointly provided by the rock bridge between the structural plane and the structural plane. The shear strength parameters of rock mass can be calculated based on the proportion of rock bridges and structural planes in a certain shear direction, and the calculation formula is as follows:

$$C = C_J \times k + C_R \times (1 - k) \tag{28.2}$$

$$f = f_J \times k + f_R \times (1 - k) \tag{28.3}$$

In Eqs. (28.2) and (28.3).

C_J—Cohesive force of structural plane, C_R—Cohesion of Rock Bridge, f_J—Friction coefficient of structural plane, f_R—Friction coefficient of rock bridge.

28.3 Distribution Characteristics of Surface Structural Plane in Survey Area

The preselected area is an exposed granite rock mass. It is located in the middle and low mountainous area south of Jueluotage in the eastern Tianshan Mountains, which is also commonly known as the Gobi area of the Kumtag Desert. The mountains in the area are low and scattered, the top is round, and the relative height is generally below 200 m. The terrain is high in the east and low in the west, and the overall tilt to the northwest. It is a rare drought and water shortage area in China. Through the fine measurement of more than 4000 fissures within 1 km^2 of the surface rock mass, it is found that the surface joints are mainly distributed in NE and NNW directions, as shown in Fig. 28.1.

According to the isodensity distribution of joints poles, the joints can be roughly divided into four concentrated areas in the stereographic projection plane, including one high concentrated area, two sub high concentrated areas and one low concentrated area. The high density area is distributed near the circumference of the SE domain of the stereographic projection network, representing the most developed group of joints in the area, accounting for approximately 42% of the total number of joints, with a dominant occurrence of 146°∠79°; The two sub dense areas are distributed near the circumference of the NE and NW domains of the stereographic projection

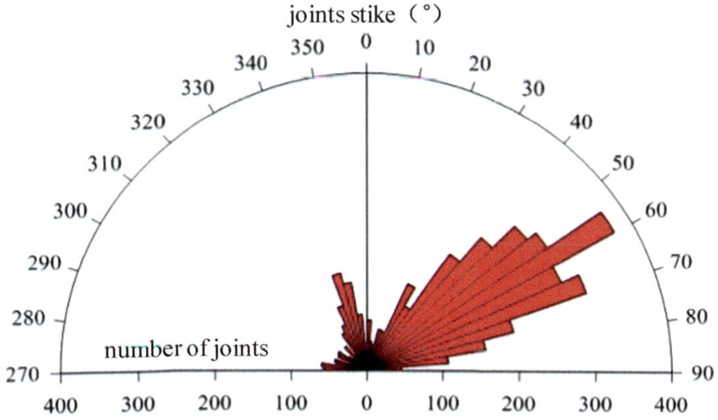

Fig. 28.1 Rose diagram of rock surface joints strike

network, representing the two sets of structural planes developed, accounting for approximately 8% to 15% of the total number of joints. Their dominant occurrences are 74°∠77°and 32°∠78°, respectively; The low density area is distributed north of the SW domain edge of the projection network, accounting for about 5% of the total number of joints. Its dominant occurrence is 25°∠ 80°, as shown in Fig. 28.2. Overall, the four sets of structural planes are mainly composed of medium steep dip joints, accounting for over 75% of the total number of statistical joints, as depicted in Table 28.1.

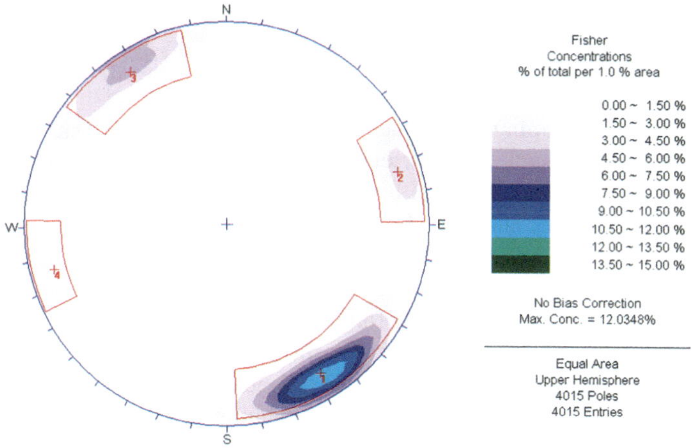

Fig. 28.2 Pole isodensity map of rock surface joints

Table 28.1 The distribution and dominant orientation of joints

The sets	Dip direction range/(°)	Dip range/(°)	Dominant dip direction/(°)	Dominant dip/(°)	Percentage/%
1	122–174	64–90	146	79	42.1
2	56–90	65–90	74	77	8.7
3	303–350	68–90	324	78	15.5
4	244–273	67–90	257	80	5.2

28.4 Simulation of Fracture Network in the Survey Area

The spatial distribution of fractures in rock masses within the same structural zone depends on geometric parameters such as their occurrence, scale, and density. The probability distribution pattern of fitting fracture distribution characteristic parameters from measured data is the foundation of network simulation. The research of relevant scholars at home and abroad shows that the distribution of geometric parameters of discontinuities basically follows four forms: uniform distribution, negative exponential distribution, normal distribution and logarithmic normal distribution. According to the research results other scholars, the fracture tendency is mainly normal distribution and uniform distribution, the dip angle is mainly subject to normal distribution and logarithmic normal distribution, the fracture spacing is mainly negative exponential distribution, and the trace length is mainly logarithmic distribution.

28.4.1 Analysis of Geometric Parameters of Fractures

According to the grouping of surface structural planes in the survey area, the geometric parameters such as occurrence, trace length and spacing of each group of structural planes were analyzed by mathematical statistics, and the frequency statistical characteristics and probability distribution patterns of the representative geometric parameters of each group of structural planes were obtained.

According to statistical analysis, the tendency and dip angle of four sets of structural planes in the rock mass basically follow the normal function curve distribution, while the spacing and trace length mainly follow the negative exponential function curve distribution pattern. See Table 28.2 for statistical parameters such as probability distribution model, mean value and standard deviation of each group of fractures.

Table 28.2 Geometric parametric statistics of fractures of each group

The sets	Statistical index	Probability model	Average	Standard deviation
1	Dip direction/(°)	Normal	148.2	14.636
	Dip/(°)	Normal	77.29	6.398
	Spacing/m	Negative exponential	0.53	0.823
	Trace length/m	Negative exponential	3.11	3.939
2	Dip direction/(°)	Normal	73.12	8.782
	Dip/(°)	Normal	78.27	5.667
	Spacing/m	Negative exponential	1.82	2.318
	Trace length/m	Negative exponential	5.23	5.63
3	Dip direction/(°)	Normal	324.33	13.03
	Dip/(°)	Normal	70.79	5.536
	Spacing/m	Negative exponential	1.03	1.896
	Trace length/m	Negative exponential	2.9	2.639
4	Dip direction/(°)	Normal	256.41	9.973
	Dip/(°)	Normal	80.04	4.887
	Spacing/m	Negative exponential	1.27	1.905
	Trace length/m	Negative exponential	3.3	2.86

28.4.2 Calculation of Structural Surface Connectivity Based on Fracture Network

On the basis of each group of discontinuity plane parameters, the Monte-Carlo method is used to calculate fractures inside the rock mass, and a fracture network model is established. From the analysis of the development and distribution characteristics of general fractures in the survey area, it can be seen that the length of fracture traces is mostly less than 100 m. In order to avoid the influence of the "window effect" of network simulation, a simulation interval of 100 m × 100 m is selected, and at the same time, the simulation interval of fractures is ensured. effectiveness and simplicity. Considering the lack of spatial display of structural surface planar network simulation, multiple section orientations were selected for simulation during simulation, and the number of simulations for each section was 10 times.

Follow the steps below to calculate the structural surface connectivity. (1) Arrange the "reference survey line" on the fracture network diagram according to the orientation of the fractures group whose connectivity rate is obtained. (2) Choose 1 m as the width of the measurement zone, calculate the distance from the midpoint of each crack to the direction line, and take the cracks whose distance is less than 1 m as cracks within the bandwidth, as presented in Fig. 28.3; (3) Calculate the distance between each crack and the direction line in the zone Included angle, project the cracks with an included angle less than 15° onto the direction line; after removing

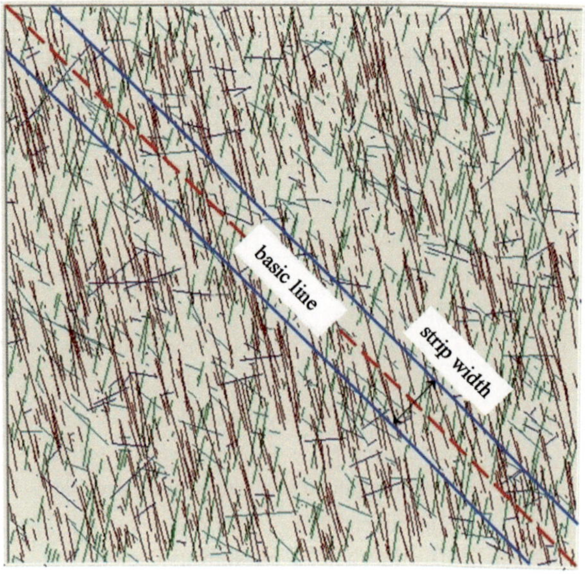

Fig. 28.3 Schematic diagram for calculating the connectivity flux of fractures

the overlapped part, add up the projected length and divide by the length of the direction line as the connectivity rate of the group of fractures.

To comprehensively capture changes in the connectivity of each set of joints within the simulation interval, we employ the center of the fracture network map as the reference point. We modify the shear increment or basic step to 5° and calculate the range from 0° to 180°. As the computer automatically generates random numbers and variables in the simulation, there are slight variations in the simulation outcomes for the same set of fractures. However, the connectivity ratio of each fracture set closely approximates the structural configuration in the shear direction, with the maximum value achieved when the joint dip is attained. By conducting 10 simulations for the four sets of joints within the simulation area, we obtain the statistical eigenvalues for the connectivity ratio of each group of joints in the designated area.

The statistical analysis reveals varying connectivity rates among the four sets of joints within the surveyed area. The first set of joints exhibits the highest connectivity rate, reaching 62.3%. Following this, the third and fourth sets of joints demonstrate connectivity rates ranging from 24.5% to 28.3%. In contrast, the set joint exhibits the lowest connectivity rate at 18.4%.

According to the strength test of intact rock in the survey area and the strength test of rock with joints, the average mechanical strength of intact rock in the survey area is cohesive force $C_R = 31.04$ MPa, friction coefficient $\varphi = 51.44°$. The average mechanical strength of the structural surface has an friction coefficient $C_J = 0.35$ MPa, and the friction coefficient is $\varphi = 40°$. According to the formulas (28.2) and (28.3), the rock cohesive force and friction coefficient values in four sets of dominant occurrence directions are obtained, as shown in Table 28.3.

Table 28.3 The connectivity rate of each group of fractures in the rock mass and corresponding strength parameters in the attitude direction

The sets	Average connectivity rate /%	Cohesive force C/MPa	Friction coefficient φ/(°)
1	62.3	11.92	44.8
2	18.4	25.39	49.6
3	28.3	22.35	53.3
4	24.5	23.52	48.9

28.5 Conclusion

Through on-site measurements of the structural surface at the pre-selected site, we obtained the geometric distribution characteristics of joint trace length, spacing, dip, and dip direction. Utilizing the Monte-Carlo method, we constructed a fracture network model that aligns with the observed geometric distribution characteristics of the fractures on site. Based on this constructed fracture network, we determined the connectivity of structural planes and the strength parameters of the rock mass in different directions. The key findings are as follows:

(1) The joints within the surveyed area predominantly occur in two directions: NE and NNW. They can be categorized into four main dominant joint sets, specifically: 146°∠79°, 74°∠77°, 327°∠78°, and 257°∠80°.
(2) Analyzing the connectivity rate of rock mass joints using the fracture network, we observed that among the four dominant joint sets, the set characterized by an occurrence of 146°∠79° exhibits the highest connectivity rate at 62.3%. Furthermore, this set has a cohesive force of 11.92 MPa, a friction coefficient of 44.8°, and is prone to fracturing. Hence, it requires careful attention in future engineering design and construction processes.

References

Duan Y, Li W, Wu Y (2010) Research status and preliminary exploration on the methods for determining the shear strength parameters of engineering rock masses. Highway Transp Technol (applied Technology Edition) 10:155–161 (in Chinese)
Einstein HH, Venieaiano D, Baecher GB et al (1983) The effect of discontinuity persistence on rock slope stability. Int J Rock Mech Mining Sci Geomech Abstracts 20(2):227–236
Guoming H, Runqiu H (1999) Estimating the connectivity of discontinuous surfaces using the window method. Hydrogeol Eng Geol 1:105–108 (in Chinese)
Hoek E, Brown ET (1988) The Hoek-Brown failure criterion-1988 updating. In: 15th Canadian rock mechanics symposium. University of Toronto, Toronto, pp 31–38
Hoek E, Carranza-Torres C, Corkum B (2002) Hoek-Brown failure criterion. In: Proceeding of NARMS-TAC conference. University of Toronto Press, Toronto, pp 267–273

Liu M, Fu H, Wu J (2005) Current situation of determination methods of shear strength parameters of rock-mass discontinuities and new thoughts. J Chongqing Jiaotong Umoversity 24(5):65–67 (in Chinese)

Runqiu H, Liuming F (2003) Study on probability model of connectivity rate of non-uniformly distributed random discontinuities. Prog Nat Sci 13(6):621–625 (in Chinese)

Wang XG, Jia ZX, Chen ZY et al (2016) Determination of discontinuity persistent ratio by Monte-Carlo simulation and dynamic programming. Eng Geol 203:83–98

Xiaogang W, Zuyu C, Wensong L (1992) Determinate on connectivity ratio and synthetical shear intensity of joint rock mass using the monte Carlo method. Chin J Rock Mech Eng 11(4):345–355 (in Chinese)

Xiaogang W, Zhixin J, Zuyu C (1998) The application on network simulation principle of joint plane developed on rock mass in research of joint rock mass connectivity ratio. Water Resour Hydropower Technol 29(10):43–47 (in Chinese)

Chapter 29
Study on Effective Influencing Factors of Common Landslides Susceptibility Methods

Yu Zou, Shengwen Qi, Xingxing Li, Songfeng Guo, Jiaguo Xia, and Xinyi Guo

Abstract Earthquake-induced landslides are the most common hazard with serious consequences worldwide. Understanding the main controlling factors of the spatial distribution of coseismic landslides is of great significance for future disaster prevention. *Mw* 6.1 Ludian earthquake happened on 3 August 2014, Yunnan province, China. This study evaluated and compared landslide susceptibility maps (LSM) obtained from common landslides susceptibility methods, including frequency ratio (FR), information value (IV), and weights of evidence (WoE) respectively, and find out the most effective influencing factors of coseismic landslides Through remote sensing image interpretations and field surveys, landslide inventory including 1414 coseismic landslides was founded. In the inventory, 1000 landslides (70%) were randomly selected as training landslide dataset for the modeling using FR, IV, and WoE methods, and the remaining 414 landslides (30%) were used for the testing landslide dataset. The factors considered in the landslide susceptibility analysis are slope degree, aspect, elevation, lithology, distance to coseismic fault (BXF), distance to river, distance to road and intensity. Four different combinations of eight factors were taken into consideration in three methods. The results show that all three methods show good predictive capability according to AUCs. However, there is not much improvement in the accuracy of the results in each method. This paper further reveals that slope gradient and distance to coseismic fault are the most important factors

Y. Zou · S. Qi (✉) · S. Guo · X. Guo
Key Laboratory of Shale Gas and Geoengineering (KLSGG), Institute of Geology and Geophysics, Chinese Academy of Sciences, Beijing 100029, China
e-mail: qishengwen@mail.iggcas.ac.cn

Innovation Academy for Earth Science, Chinese Academy of Sciences, Beijing 100029, China

College of Earth and Planetary Sciences, University of Chinese Academy of Sciences, Beijing 100049, China

X. Li
China Academy of Building Research Foundation Engineering Co., Ltd, Beijing 100013, China

J. Xia
China Nuclear Power Engineering Co., Ltd, Beijing 100840, China

S. Wang et al. (eds.), *Engineering Geology for a Habitable Earth: IAEG XIV Congress 2023 Proceedings, Chengdu, China*, Environmental Science and Engineering,
https://doi.org/10.1007/978-981-99-9057-3_29

controlling the spatial distribution of coseismic landslides. In addition, the model, which considers only these two factors, has sufficient accuracy.

Keywords Landslide susceptibility mapping · Coseismic landslides · Frequency ratio · Information value · Weights of evidence

29.1 Introduction

Landslide susceptibility can be defined as the propensity of the occurrence of a landslide based on the relation between the occurrence distribution and a set of predisposing factors (Tseng et al. 2015). For calculating the factor of safety to assess the landslide susceptibility, different stability analysis models are proposed, such as SHALSTAB (Montgomery and Dietrich 1994), SINMAP (Pack et al. 1998), TRIGRS (Baum et al. 2002), YS-Slope model (Kim et al. 2014), etc. Nevertheless, these models require a large amount of detailed input data, which come from laboratory tests. Therefore, they are only applicable to local scale or over a large area homogeneous geological and geomorphological conditions and simple landslide types, such as rainfall-induced shallow landslides.

In contrast to the stability analysis models above, the methods to obtain the landslide susceptibility index (LSI) are much more convenient and only three basic procedures are required, (1) assign the weights to causative factors relative to the landslide occurrence, (2) determine the optimal combination mode of those factors, and (3) validate and compare the results of landslide susceptibility analysis. To determine the weights and the optimal combination of causative factors, different methods are proposed and applied to assess landslide susceptibility by the researchers, such as fuzzy logic (Kayastha 2012), frequency ratio (FR) (Sujatha et al. 2013), weights of evidence (WoE) (Van Westen et al. 2003), information value (IV) (Bijukchhen et al. 2013), certainty factor (Devkota et al. 2013), logistic regression (Su et al. 2010), discriminant analysis (He et al. 2012), cluster analysis (Melchiorre et al. 2008), artificial neural networks (Zare et al. 2013), neuro-fuzzy (Sdao et al. 2013), etc. Among the methods above, FR, IV and WoE are widely used in landslide susceptibility analysis due to explicit weighting function and combination mode of conditional factors.

Earthquake induced LSM in the last two decades has attracted much attention (Sangeeta et al. 2020). A *Mw* 6.1 earthquake occurred in Ludian County, Yunnan province, China on 3 August, 2014. The epicenter was located in Longtoushan Town, with a depth of 12 km and a location of 29.1°N and 103.3°E (Fig. 29.1). The seismogenic fault is the Baogunao-Xiaohe Fault (BXF), which is a previously unknown left-lateral strike-slip fault with a northwest orientation before the earthquake, intersecting with a well-developed northeast-oriented thrust fault, the Zhaotong-Ludian Fault (ZLF) (Xu et al. 2015). This earthquake killed more than 600 people and caused 3143 injured, and huge damage to houses and other infrastructural facilities, causing a tremendous loss in properties (Xu et al. 2008).

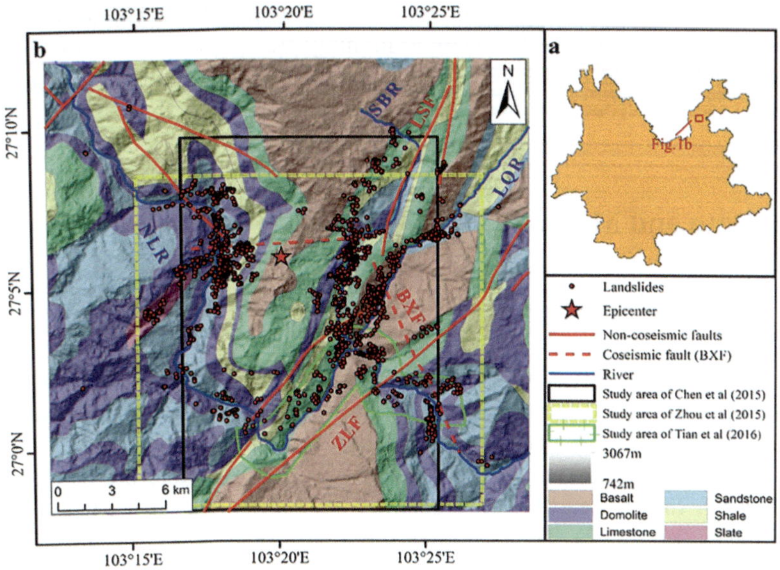

Fig. 29.1 Geological background map (modified from Zou et al. 2021). **a** Map of Yunnan Province; **b** location of the study area and landslide locations

In this study, methods of FR, IV and WoE are applied to assess landslide susceptibility in Ludian earthquake. 4 different combinations of 8 factors are used in each method and 12 separate LSMs are produced. Area Under the Curve (AUC) method is used to implement the model validation process. The goal of this article is to advance general knowledge about which method is more robust for the coseismic landslide susceptibility analysis.

29.2 Geological Setting

The Ludian earthquake occurred between the Chuandian Rhombic Block and the South China Block, east of the Xianshuihe-Xiaojiang Fault Zone and the western section of the ZLF Zone. The ZLF Zone is composed of several NE-oriented faults, including the Lianfeng Fault, ZLF, Shimen fault and Huize faults, all of the faults are right-leteral reverse faults (Xu et al. 2015). The terrain in the area is complex, with geomorphic units such as deeply incised mountains, rift plains, and karst plateaus. The epicenter area is flat, while the eastern and western areas are relatively high in elevation. Most of the mountains are oriented in the NE-SW direction, with the highest peak reaching 3067 m and the lowest point at 742 m above sea level. Three rivers, Niulan River (NLR), Longquan River (LQR) and Shaba River (SBR), are deeply incised with highly developed V-shaped valleys and steep slopes on both sides (Fig. 29.1). The exposed rock types in Ludian County are mainly sandstone, dolomite, basalt, shale, slate and limestone (Zang et al. 2020). The average annual precipitation is 820.3mm, with the largest monthly precipitation occurring during the rainy season from June to August. Due to abundant rainfall, most slopes are covered with short, dense herbaceous plants and sparse tall trees.

29.3 Data and Methods

29.3.1 Inventory: Interpretation and Field Investigation

The first-hand data is difficult to get immediately after the earthquake because of the transportation block and serious secondary disasters. The remote sensing data used in interpreting the landslides in this study includes unmanned aerial vehicle, GeoEye-1, GF-1 and ZY-3 images. The pre-earthquake landslides were interpreted using ZY-3 images, while the post-earthquake landslides were interpreted using GeoEye-1 and GF-1 images (Fig. 29.2). Then, through the field investigation, the information of coseismic landslides was verified, and an inventory of 1414 landslides was established. The inventory contains the information of seismic, topographic, geological of the coseismic landslides.

29.3.2 Co-Seismic Landslides Inventory and the Distribution Regularity

The spatial distribution was analyzed using the landslide concentration index (LC), which is expressed as the number of landslides per unit area for each category (Keefer 2000; Qi et al. 2010). The larger the *LC* value, the more prone the slope is to sliding.

The potential factors selected in the research including slope gradient, aspect, elevation, distance to BXF, distance to river, distance to road, lithology, and intensity. Various factor layers were created using ArcGIS can be seen in Fig. 29.3 (Modified from Zou et al. 2021).

The relationship between the number of coseismic landslides and LC and each influencing factors is shown in Fig. 29.4. Most of the landslides occurred within the slope range of 20°–50°, and the density of coseismic landslides increased with the increase of the slope. Both the number of landslides and LC decreased exponentially with the distance from the coseismic fault or epicenter. The number of coseismic landslides and LC were not significantly correlated with slope aspects, elevation and lithology, but rough negatively correlated with distance from river systems (Zou et al. 2021).

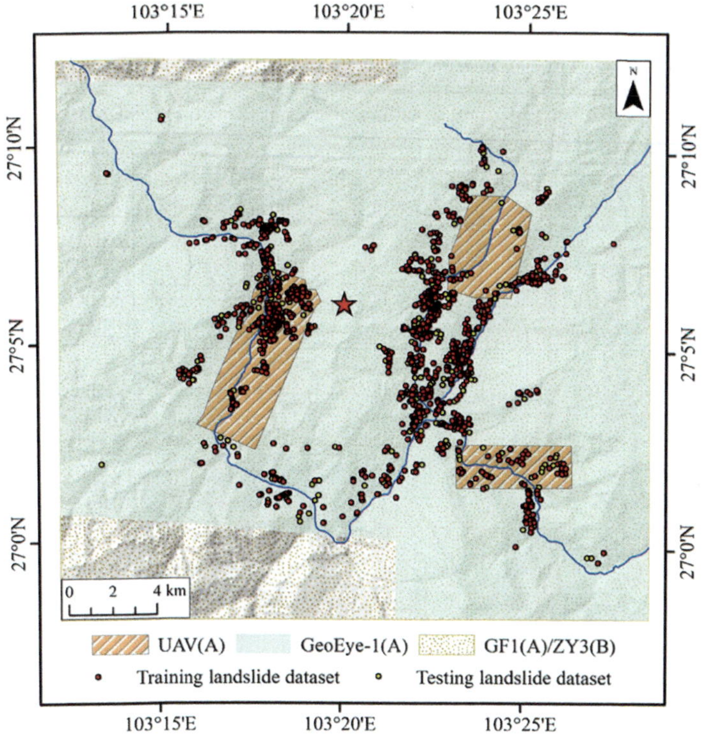

Fig. 29.2 The images used in the interpretation and training–testing landslide dataset. **a** Represents post-earthquake images; **b** represents pre-earthquake images

29.3.3 Training–testing Landslide Dataset

There is no specific limit in determining the ratio of training–testing landslide dataset. The 70% dataset is considered sufficient to conduct analysis and 30% is considered sufficient to validate the model (Silalahi et al. 2019). Hence, the total of 1414 landslide points were randomly subdivided into a training landslide dataset (70%) with 1000 landslides and testing landslide dataset (30%) with 414 landslides. The red points and yellow points represent the training samples and testing samples respectively (Fig. 29.2).

29.3.4 Landslide Susceptibility Mapping (LSM)

Statistical analysis of the relationship between the cause of landslide and the coseismic landslides by using three methods, i.e. to calculate the weights. Then,

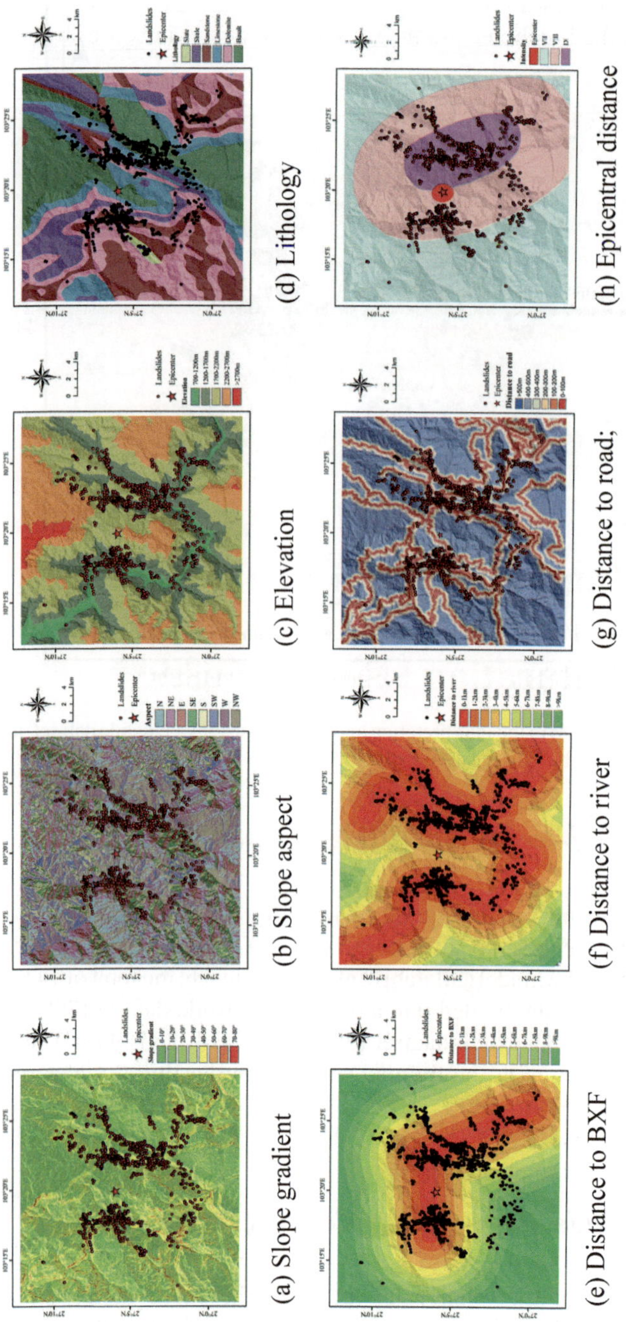

Fig. 29.3 Spatial distribution of coseismic landslides and the factors (Modified from Zou et al. 2021)

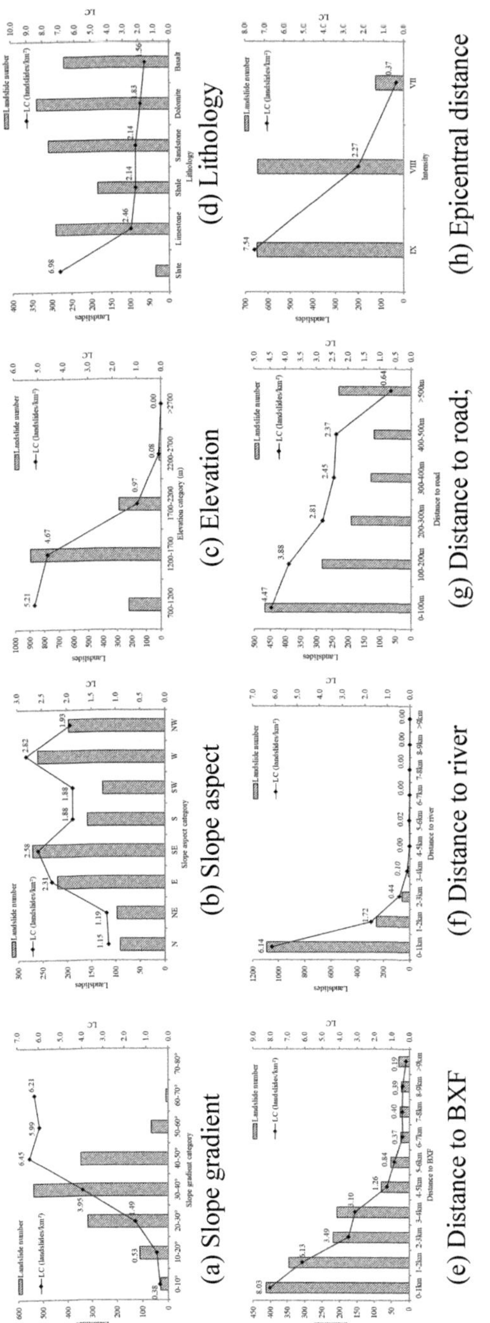

Fig. 29.4 Relation between coseismic landslides and the factors (Modified from Zou et al. 2021)

calculate the landslide susceptibility index of different factor combination models, and use the natural breaks method in ArcGIS to divide them into 5 levels.

29.3.4.1 Application of FR Model

FR model is based on the relationship between the observed spatial distribution of landslides and various factors that affect the occurrence of landslides, revealing the correlation between the location of landslides and various factors (Lee 2005). In the model, the weight of each factor type is assigned as the relationship between the locations of landslides and the value of each factor type, i.e., the ratio of the density of landslides in the factor's type to the density of landslides in the whole study area as follow:

$$FR_{i,j} = \frac{S_{i,j}/N_{i,j}}{S/N} \tag{29.1}$$

$FR_{i,j}$ is the weight of each factor's type; $S_{i,j}$ is the number of landslide cells in the jth type of the ith factor, $N_{i,j}$ is the number of cells in the jth type of the ith factor, S is the number of landslide cells in the entire map, and N is the number of cells in the entire map. A ratio above one indicates the landslide density in the factor's type is larger than average landslide density in the whole area, whereas values less than one imply the landslide density is less than average landslide density.

To obtain the LSI for each cell, each factor's frequency ratio values were summed for each cell in the domain as Eq. (29.2). The larger the LSI value is, the higher the susceptibility to landslides is.

$$LSI_{FR} = \sum_{i=1}^{i=m} FR_{i,j} \tag{29.2}$$

4 combinations of different factors based on FR method were considered for the LSMs (Table 29.1). The abbreviations of FR 2, FR 4, FR 6 and FR 8 represent the combinations with 2 factors, 4 factors, 6 factors and 8 factors respectively.

The natural breaks (Jenks) was confirmed that has the best results than other classification criteria (Chen et al. 2020). After obtaining the LSMs, the LSI is classified into five classes by natural breaks (Jenks) method of ArcGIS for visual interpretation, i.e. very low, low, moderate, high and very high susceptible zone (Fig. 29.5).

29.3.4.2 Application of IV Model

The IV model is a statistical method for spatial prediction of event occurrence based on the probability theory and the relationship between parameters and events. It has proven to be a very useful method for landslide susceptibility analysis. In the practical

Table 29.1 Predictive performance of different factor combinations

Methods	Factors	Abbreviation	AUC (%)
FR	Slope/BXF	FR 2	86.26
	Slope/BXF/Aspect/Elevation	FR 4	87.29
	Slope/BXF/Aspect/Elevation/Lithology/River	FR 6	89.07
	Slope/BXF/Aspect/Elevation/Lithology/River/Road/ Intensity	FR 8	90.14
IV	Slope/BXF	IV 2	85.33
	Slope/BXF/Aspect/Elevation	IV 4	85.91
	Slope/BXF/Aspect/Elevation/Lithology/River	IV 6	86.90
	Slope/BXF/Aspect/Elevation/Lithology/River/Road/ Intensity	IV 8	87.21
WoE	Slope/BXF	WoE 2	85.20
	Slope/BXF/Aspect/Elevation	WoE 4	86.34
	Slope/BXF/Aspect/Elevation/Lithology/River	WoE 6	86.73
	Slope/BXF/Aspect/Elevation/Lithology/River/Road/ Intensity	WoE 8	87.93

application, the information value is calculated as follows (Yin and Yan 1988):

$$I_{i,j} = \ln \frac{S_{i,j}/N_{i,j}}{S/N} \tag{29.3}$$

The information value numerically equals natural logarithm of the frequency ratio, $FR_{i,j}$. When the value is negative, the factor's type is not relevant to landslide spatial distribution and when it is positive, there exists a relevant relationship between the factor's type and landslide distribution, and the stronger the higher the value. The total information value as the LSI for each cell is given by the equation (Yin and Yan 1988):

$$LSI_{IV} = \sum_{m=1}^{i=m} I_{i,j} \tag{29.4}$$

4 combinations of different factors, based on IV method, were considered for the LSMs (Table 29.1). The abbreviations of IV 2, IV 4, IV 6 and IV 8 represent the combinations with 2 factors, 4 factors, 6 factors and 8 factors respectively. Then, the LSIs, classified by the same methods, are applied for frequency ratio model to obtain LSMs, i.e. very low, low, moderate, high and very high susceptible zone (Fig. 29.6).

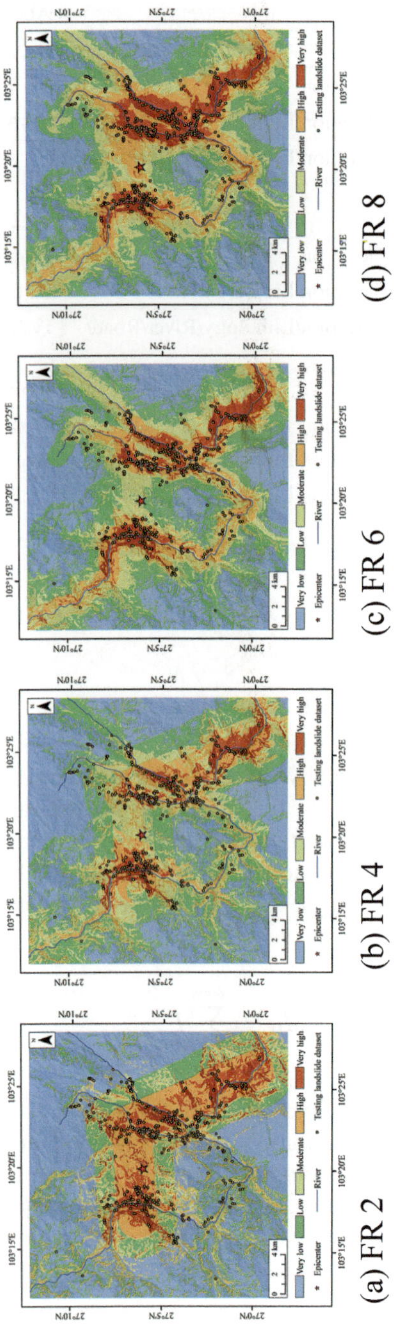

Fig. 29.5 Landslide susceptibility zonation maps based on the FR methods

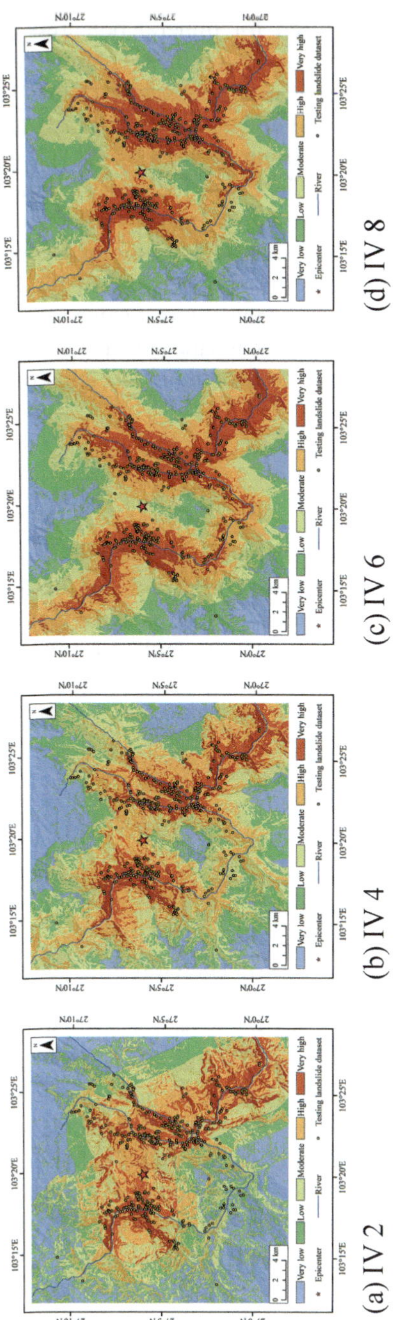

Fig. 29.6 Landslide susceptibility zonation maps based on the IV methods

29.3.4.3 Application of WoE Model

The WoE model was originally used for mineral potential evaluation, based on Bayesian probability theory (Bonham-Carter 1994). Many researchers have used the method in LSM (Van Westen et al. 2003; Kayastha et al. 2012). This method calculates the weight of each landslide causative factor according to presence or absence of landslides in the zone (Bonham-Carter 1994). Weight (W_i^+) indicates that the inducing factor exists at the landslide location, and the value of the weight indicates that there is a positive correlation between the factor and the landslides. Weight (W_i^-) indicate the absence of a causative factor and show the level of negative correlation. The difference between the two weights is called the weight contrast, $W_{ci,j}$ ($W_{ci,j} = W_{i,j}^+ - W_{i,j}^-$). The size of the contrast value reflects the overall spatial association between the triggering factors and the landslides. Therefore, the LSI using weights of evidence methods can be denoted by the weight contrast, as following:

$$LSI_{WOE} = \sum_{i=1}^{i=m} W_{ci,j} \tag{29.5}$$

4 combinations of different factors, based on WoE method, were considered for the LSMs (Table 29.1). The abbreviations of WoE 2, WoE 4, WoE 6 and WoE 8 represent the combinations with 2 factors, 4 factors, 6 factors and 8 factors respectively.Then, the LSIs, classified by the same methods, are applied for frequency ratio model to obtain LSMs (Fig. 29.7).

29.4 Results and Validation

In an ideal LSM, the very high susceptible areas should have the percentage of landslide area and highest landslide density, and from the high susceptibility area to the low susceptibility area, the percentage of landslide area and landslide density gradually decreases (Sarkar et al. 2008). Therefore, the suitability of a LSM can be judged by the above rule. For each landslide susceptibility zones, cumulative percentages of the study area and landslide area are calculated based on the testing landslides dataset. For comparison of LSMs, the approach of the AUC is used.

Comparing the cumulative percentage of landslide area in the susceptibility zone from very high to very low susceptible zone area with the cumulative percentage of the susceptibility zone area, the success rate curve is obtained (Fig. 29.8). Figure 29.8a–c show the success rate curves using FR method, IV method and WoE method respectively. In this study, the values of AUCs are all above 0.85, indicating that the overall success rate of FR, IV, and WoE are above 85%. The results indicate that the LSMs obtained by the three methods have good predictive capability.

As shown in Fig. 29.8, the LSM results with 2 factors have the AUC values of 86.26, 85.33 and 85.20% based on FR method, IV methods and WoE methods. These

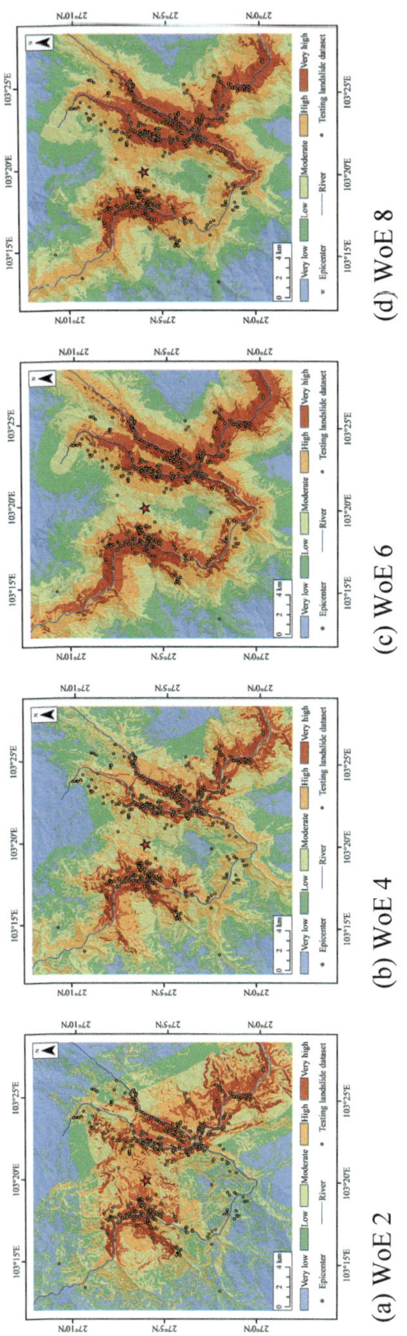

(a) WoE 2 (b) WoE 4 (c) WoE 6 (d) WoE 8

Fig. 29.7 Landslide susceptibility zonation maps based on the WoE methods

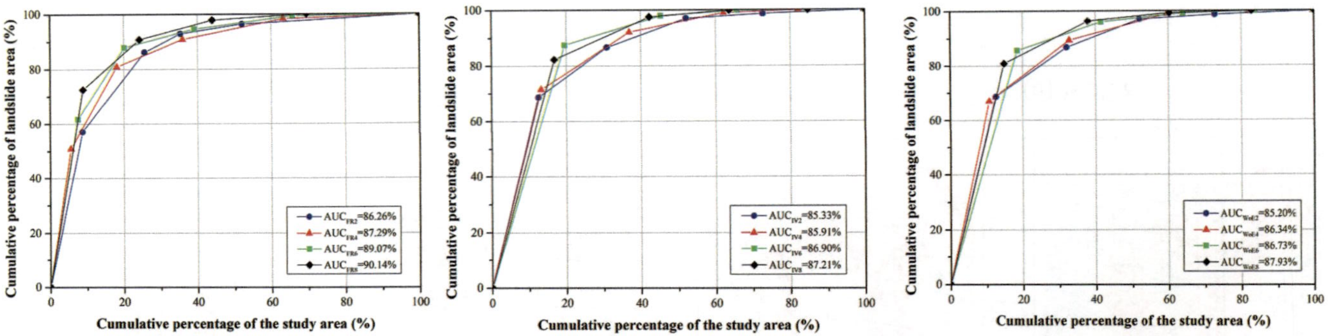

(a) FR method (b) IV method (c) WoE method

Fig. 29.8 AUCs and comparison of the predictive performance between the three methods

AUC values obtained from the method with 2 factors of slope gradient and distance to coseismic fault (BXF) are all above 85%, showing that the model with 2 factor has a high success rate, which is consistent with the research results of Qi et al. (2010) and Zou et al. (2021).

Figure 29.8a shows the LSMs under FR method, the AUC values increase from 86.26% to 90.14%, an increase of 3.88% by using 8 factors compared with 2 factors. Figure 29.8b shows the LSMs under IV method, the AUC values increase from 85.33 to 87.21%, an increase of 1.88%. Figure 29.8c shows the LSMs under WoE method, the AUC values increase from 85.20 to 87.93%, an increase of 2.83%.

29.5 Discussion and Conclusion

In this study, three statistically based models, including FR, IV, and WoE, were used for identifying the areas susceptible to coseismic landslide induced by Ludian earthquake. In order to prepare the LSMs, the landslide inventory of the study area, including 1414 coseismic landslides, was established by remote sensing image interpretations and field surveys. All the three methods show good predictive capability according to AUCs of which are all above 85%. While comparing the LSMs obtained from FR, IV and WoE methods using different combinations of factors, an increase of 1.88% shows a high robustness of IV method. However, overall, there is not much improvement in the accuracy of the results in each method.

It is primarily the interaction between rock mass strength, geometric features, and ground motion that determines the spatial distribution of coseismic landslides (Zou et al. 2021). The presence of faults often results in the fracturing of rock masses on either side of the fault, leading to low strength and an increased susceptibility to landslides in the area. The fault has an impact on the characteristics of rock mass structures, which in turn affects the stability of the slope and ultimately leads to landslides (Wang et al. 2021). Slope gradient, slope relief, and slope aspect are commonly used to describe geometric features. The slope relief is dependent on the slope gradient (Culmann 1875). The influence of slope aspect on geometrical characteristics is related to the impact of climate and vegetation, as well as the direction of coseismic fault movement and seismic wave propagation, but its impact is relatively minor (Qi et al. 2010). As such, the relationship between LC and geometric features is primarily determined by the relationship between LC and slope gradient. The ground motion attenuates gradually with the increase of epicentre distance (for point source rupture earthquake) or coseismic fault distance (for linear source rupture earthquake) (Schnabel et al. 1972). Furthermore, as the slope gradient increases, the peak acceleration also increases (Davis and West 1973; Qi et al. 2003).

In conclusion, as observed in the Ludian earthquake, ground motion weakens gradually as the distance to the coseismic fault increases. Therefore, the occurrence of coseismic landslides decreases exponentially as the distance to the coseismic fault increases. And as the slope gradient increases, the LC also increases. The spatial distribution of coseismic landslides is primarily determined by the slope gradient and distance to the coseismic fault, which are the two most important controlling factors.

Acknowledgements This research was supported by the Second Tibetan Plateau Scientific Expedition and Research Program (STEP), Grant No. 2019QZKK0904, and National Natural Science Foundation of China under Grants Nos. 42207205, 42090051, 42077266, 42002279 and the Youth Innovation Promotion Association of the Chinese Academy of Sciences under grant No. 2022062.

References

Baum RL, Savage WZ, Godt JW (2002) TRIGRS—a Fortran program for transient rainfall infiltration and grid-based regional slope-stability analysis. US Geol Surv Open-File Rep 424:38

Bijukchhen SM, Kayastha P, Dhital MR (2013) A comparative evaluation of heuristic and bivariate statistical modelling for landslide susceptibility mappings in Ghurmi-Dhad Khola, east Nepal. Arab J Geosci 6(8):2727–2743

Bonham-Carter GF (1994) Geographic information systems for geoscientists, modelling with GIS. In: Computer methods in the geosciences, vol 13. Pergamon, p 416

Chen S, Miao Z, Wu L, He Y (2020) Application of an incomplete landslide inventory and one class classifier to earthquake-induced landslide susceptibility mapping. IEEE J Sel Top Appl Earth Observ Remote Sens 13:1649–1660

Culmann C (1875). Die graphische Statik. Leipzig

Davis LL, West LR (1973) Observed effects of topography on ground motion. Bull Seismol Soc Am 63(1):283–298

Devkota KC, Regmi AD, Pourghasemi HR, Yoshida K, Pradhan B, Ryu IC, Dhital MR, Althuwaynee OF (2013) Landslide susceptibility mapping using certainty factor, index of entropy and logistic regression models in GIS and their comparison at Mugling-Narayanghat road section in Nepal Himalaya. Nat Hazards 65(1):135–165

He SW, Pan P, Dai L, Wang HJ, Liu JP (2012) Application of kernel-based Fisher discriminant analysis to map landslide susceptibility in the Qinggan River delta, Three Gorges, China. Geomorphology 171:30–41

Kayastha P (2012) Application of fuzzy logic approach for landslide susceptibility mapping in Garuwa sub-basin, East Nepal. Front Earth Sci 6(4):420–432

Kayastha P, Dhital MR, De Smedt F (2012) Landslide susceptibility mapping using the weight of evidence method in the Tinau watershed, Nepal. Nat Hazards 63(2):479–498

Keefer DK (2000) Statistical analysis of an earthquake-induced landslide distribution—the 1989 Loma Prieta California Event. Eng Geol 58(3):231–249

Kim J, Lee K, Jeong S, Kim G (2014) GIS-based prediction method of landslide susceptibility using a rainfall infiltration-groundwater flow model. Eng Geol 182:63–78

Lee S (2005) Application of logistic regression model and its validation for landslide susceptibility mapping using GIS and remote sensing data. Int J Remote Sens 26(7):1477–1491

Lee S, Choi J (2004) Landslide susceptibility mapping using GIS and the weight-of-evidence model. Int J Geogr Inf Sci 18(8):789–814

Melchiorre C, Matteucci M, Azzoni A, Zanchi A (2008) Artificial neural networks and cluster analysis in landslide susceptibility zonation. Geomorphology 94(3–4):379–400

Montgomery DR, Dietrich WE (1994) A physically based model for the topographic control on shallow landsliding. Water Resour Res 30(4):1153–1171

Pack RT, Tarboton DG, Goodwin CN (1998) The SINMAP approach to terrain stability mapping. In: 8th Congress of the international association of engineering geology, Vancouver, British Columbia, Canada

Qi S, Xu Q, Lan H, Zhang B, Liu JY (2010) Spatial distribution analysis of landslides triggered by 2008.5.12 Wenchuan Earthquake, China. Eng Geol 116(1–2):95–108

Qi SW, Wu FQ, Sun JZ (2003) General regularity of dynamic responses of slopes under dynamic input. Sci China. Series E(46 Supp.):120–132

Sangeeta, Maheshwari BK, Kanungo DP (2020) GIS-based pre- and post-earthquake landslide susceptibility zonation with reference to 1999 Chamoli earthquake. 129:55

Sarkar S, Kanungo DP, Patra AK, Kumar P (2008) GiS based spatial data analysis for landslide susceptibility mapping. J Mt Sci 5(1):52–62

Schnabel P, Seed HB, Lysmer J (1972) Modification of seismograph records for effects of local soil conditions. Bull Seismol Soc Am 62(6):1649–1664

Sdao F, Lioi DS, Pascale S, Caniani D, Mancini IM (2013) Landslide susceptibility assessment by using a neuro-fuzzy model: a case study in the Rupestrian heritage rich area of Matera. Nat Hazard 13(2):395–407

Silalahi F, Pamela, Arifianti Y, Hidayat F (2019) Landslide susceptibility assessment using frequency ratio model in Bogor, West Java, Indonesia. Geosci Lett 6(1)

Su F, Cui P, Zhang J, Xiang LZ (2010) Susceptibility assessment of landslides caused by the Wenchuan earthquake using a logistic regression model. J Mt Sci 7(3):234–245

Sujatha ER, Rajamanickam V, Kumaravel P, Saranathan E (2013) Landslide susceptibility analysis using probabilistic likelihood ratio model—a geospatial-based study. Arab J Geosci 6(2):429–440

Tseng CM, Lin CW, Hsieh WD (2015) Landslide susceptibility analysis by means of event-based multi-temporal landslide inventories. Nat Hazards Earth Syst Sci Discuss 3(2):1137–1173

Van Westen CJ, Rengers N, Soeters R (2003) Use of geomorphological information in indirect landslide susceptibility assessment. Nat Hazards 30(3):399–419

Wang XL, Crosta GB, Clague JJ, Stead D, Sun JJ, Qi SW, Liu HY (2021) Fault controls on spatial variation of fracture density and rock mass strength within the Yarlung Tsangpo Fault damage zone (southeastern Tibet). Eng Geol 106238

Xu C, Xu X W, Yao X, Dai FC (2014) Three (nearly) complete inventories of landslides triggered by the May 12, 2008 Wenchuan Mw 7.9 earthquake of China and their spatial distribution statistical analysis. Landslides 11(3):441–461

Xu XW, Xu C, Yu G, Wu XY, Li X, Zhang JG (2015) Primary surface ruptures of the Ludian Mw 6.2 Earthquake, Southeastern Tibetan Plateau, China. Seismol Res Lett 86(6):1622–1635

Yin KL, Yan TZ (1988) Statistical prediction models for slope instability of metamorphosed rocks. In: Proceedings of the fifth international symposium on landslides, Lausanne, Switzerland

Zang MD, Qi SW, Zou Y, Sheng ZP, Zamora BS (2020) An improved method of Newmark analysis for mapping hazards of coseismic landslides. Nat Hazard 20(3):713–726

Zare M, Pourghasemi HR, Vafakhah M, Pradhan B (2013) Landslide susceptibility mapping at Vaz Watershed (Iran) using an artificial neural network model: a comparison between multilayer perceptron (MLP) and radial basic function (RBF) algorithms. Arab J Geosci 6(8):2873–2888

Zou Y. Coupling between endogenic and exogenic geological mechanisms and spatial distribution prediction model of the traffic corridor in Sichuan-Tibet. University of Chinese Academy of Sciences

Zou Y, Qi SW, Guo SF, Zheng BW, Zhan ZF, He NW, Huang XL, Hou XK, Liu HY (2021) Factors controlling the spatial distribution of coseismic landslides triggered by the Mw 6.1 Ludian Earthquake in China. Eng Geol 296:106477

Chapter 30
Research on Engineering Geological Zoning of Plateau Mountains Traffic Corridors

Yue Yang, Bowen Zheng, Shengwen Qi, Guoxiang Yang, Anna Xu, Lulu Xu, Ruipeng Hao, Wei Lu, Ning Liang, Songfeng Guo, and Shuaihua Song

Abstract The geological conditions of the plateau mountainous traffic corridor project are complex, posing serious challenges to the construction of transportation corridors. To provide systematic engineering geological analysis information for the safe construction of transportation corridor projects in mountainous areas of the plateau, this paper adopts the engineering geological zoning method based on a geographic information system (GIS), selects 10 zoning indicators belonging to different engineering geological conditions, introduces Euclidean distance to establish a hierarchical analysis-entropy value method of subject-objective coupled zoning model, conducts regional engineering geological zoning, and obtains an engineering geological zoning map of the study area. The research results show that the influence indicator of the active fault plays a controlling role in the engineering geological conditions, followed by the terrain undulation and the development density of adverse geological phenomena. The research results can effectively serve the development of an engineering route optimization plan for the traffic corridor, and support the engineering route selection, engineering geological survey design and construction.

Y. Yang · B. Zheng (✉) · S. Qi · R. Hao · W. Lu · N. Liang · S. Guo · S. Song
Key Laboratory of Shale Gas and Geoengineering, Institute of Geology and Geophysics, Chinese Academy of Sciences, Beijing 100029, China
e-mail: zhengbowen@mail.iggcas.ac.cn

B. Zheng · S. Qi · R. Hao · W. Lu · N. Liang · S. Guo · S. Song
College of Earth and Planetary Sciences, University of Chinese Academy of Sciences, Beijing 100029, China

Y. Yang · G. Yang · A. Xu
School of Engineering and Technology, China University of Geosciences (Beijing), Beijing 100083, China

L. Xu
College of Earth Science and Engineering, Shandong University of Science and Technology, Qingdao 266590, Shandong, China

S. Wang et al. (eds.), *Engineering Geology for a Habitable Earth: IAEG XIV Congress 2023 Proceedings, Chengdu, China*, Environmental Science and Engineering,
https://doi.org/10.1007/978-981-99-9057-3_30

Keywords Engineering geological zoning · Engineering geological condition · Traffic corridors in plateau mountainous areas · Hierarchical analysis method · Entropy value method

30.1 Introduction

The transportation corridor in the plateau mountainous region is an important component of the transportation network in the Tibet Autonomous Region of China (Xu et al. 2021a, b). It holds significant significance for areas with varying high and low altitudes (Zhao et al. 2023). The corridor is characterized by active faulting and frequent earthquakes, which result in deformation hazards on shallow slopes and mechanical deterioration of rock masses and construction materials (Wen et al. 2006; Guo et al. 2017; Lu and Cai 2019). The occurrence mechanism of deep-seated engineering disturbance disasters is highly complex, and geological safety risks are extremely severe (Qi et al. 2016; Peng et al. 2020), significantly constraining the construction of transportation corridors in the plateau mountainous region (Pan et al. 2020; Xue et al. 2020; Guo et al. 2021; Qi et al. 2022).

Researchers have carried out some regional geological disaster studies. For example, Li et al. (2017) performed zonation analysis of geological hazard susceptibility along the corridor and adjacent areas using an improved frequency ratio method; Wang (2022) utilized deep learning algorithms such as Convolutional Neural Network (CNN) and Deep Neural Network (DNN) models to assess landslide susceptibility of the transportation corridor; Lu et al. (2023) conducted a hazard assessment of rockslide disasters in the corridor's engineering projects using a frequency ratio-logic regression coupling model. However, these achievements mainly focus on studying the geological safety risks of transportation corridors in the plateau mountainous region from the perspective of geological hazards. There is relatively limited research on the zoning and evaluation of engineering geological conditions.

Engineering geological conditions are a combination of geological elements related to engineering construction; Liu (1961) emphasized the significance of regional engineering geology; Li (1977) proposed to select relatively stable "safety islands" within the active tectonic zone as engineering construction bases and sites; Hu (2001) regarded regional crustal stability as a crucial foundation for regional engineering geological conditions and proposed the "safe island" theory and methodology for evaluating regional crustal stability; Zhang et al. (2022) highlighted the importance of engineering geological assessment in ensuring the safety of major engineering projects and put forth methods and workflow for geological safety evaluation.

In order to provide effective engineering geological analysis information for the safe construction of the traffic corridor project, this paper adopts the hierarchical analysis-entropy method based on geographic information system (GIS) to carry out the engineering geological zoning of the Ya'an–Linzhi section of the traffic corridor in the highland mountainous area with 5 km on each side of the project line as the research area, so as to provide the prediction of engineering geological problems and risk prevention and control for the construction of the project. This will provide a basis for the prediction of engineering geological problems and risk prevention and control of engineering construction.

30.2 General Situation of Regional Engineering Geology

The research area traverses eight high mountains and crosses seven major rivers (Yi 2018; Guo et al. 2020), passing through five geomorphic units (Lu and Cai 2019). The overall terrain is higher in the north and lower in the south, with a west-to-east decline (Li et al. 2020). It exhibits a typical "V"-shaped mountain and gorge topography (Xu et al. 2021a, b) and intersects multiple Quaternary active faults.

The geological strata in the research area are complex. To the west of the Jiali Fault, there are Paleozoic sedimentary rocks, while to the east, there are Mesozoic conglomerates, sandy mudstones, sandy shales, and limestone. Intrusive rocks mainly consist of granite, diorite, and porphyry. Metamorphic rocks are predominantly schist, slate, gneiss, and marble (Xue et al. 2020).

The research area is influenced by the terrain, with long groundwater flow paths, high hydraulic gradients, and high-water pressure. The presence of geological structures provides favorable pathways and storage spaces for groundwater, and karst water is well developed (Zhang et al. 2021).

The research area is located in the Mediterranean-South Asia geothermal anomaly zone and belongs to a high anomaly area of geothermal heat flow (Zhang et al. 2021). There are 612 debris flow gullies and 747 locations of landslide disasters along the corridor, and there are also numerous potential high-risk disaster points (Peng et al. 2020).

30.3 Zoning Method

Analytic Hierarchy Process (AHP) is a subjective analysis method that combines qualitative analysis and quantitative calculations (Harris et al. 2000). The combination of GIS technology and AHP method has improved the application of traditional AHP in geoscience research (Liu 2001; Peng et al. 2005; Shi et al. 2005). Subjective weights can be assigned to the indicators, such as Eq. (30.1).

$$W_a = \frac{\sum_{j=1}^{n} \frac{a_{ij}}{\sum_{k=1}^{n} a_{ki}}}{\sum_{j=1}^{n} \sum_{j=1}^{n} \frac{a_{ij}}{\sum_{k=1}^{n} a_{ki}}}, i, j = 1, 2 \ldots, n \tag{30.1}$$

In the Equation: W_a-feature vector (weight coefficient), n-number of elements, i-element matrix rows, j-element matrix column.

The entropy method is an objective analytical approach that uses the entropy value of indicators to determine their relative importance (Zhang et al. 2010). It measures the significance of indicators based on the differences between their observed values (Wen et al. 2006). Objective weights can be assigned to the indicators, such as Eq. (30.2).

$$W_b = \frac{1 + \frac{1}{\ln(n)} * \sum_{i=1}^{n} \left(\frac{X_{ij}}{\sum_{i=1}^{n} X_{ij}} * \ln \left(\frac{X_{ij}}{\sum_{i=1}^{n} X_{ij}} \right) \right)}{\sum_{j=1}^{m} \left(1 + \frac{1}{\ln(n)} * \sum_{i=1}^{n} \left(\frac{X_{ij}}{\sum_{i=1}^{n} X_{ij}} * \ln \left(\frac{X_{ij}}{\sum_{i=1}^{n} X_{ij}} \right) \right) \right)} \tag{30.2}$$

In the Equation: W_b-weight, n-number of elements, i-sample number, j-indicator number.

Considering that only the subjective evaluation method is used to calculate the indicator weights, the importance of each indicator obtained is generally in line with common sense and experience, but the accuracy of the results will be affected by subjective artificiality; if only the objective assignment method is used, although the zoning process has strong mathematical theoretical support, the knowledge and experience of the evaluators are ignored, and the zoning results may not match the actual situation. This study establishes an AHP-entropy method coupling model for comprehensive evaluation of engineering geological zoning by introducing the Euclidean distance, as shown in Eq. (30.3):

$$\omega = \alpha W_a + \beta W_b \tag{30.3}$$

In the Equation: α and β represent the preference degree of subjective weight and objective weight respectively. According to the Euclidean distance function, the difference degree Equation is established to determine, α, β:

$$D(\omega a, \omega b) = \sqrt{\frac{1}{2} \sum_{1}^{i} (\omega a - \omega b)^2} \tag{30.4}$$

$$D(\omega a, \omega b)^2 = (\alpha - \beta)^2 \tag{30.5}$$

$$\alpha + \beta = 1 \tag{30.6}$$

Thus, the engineering geological zoning indicator is obtained:

$$S_1 = \sum W_i \cdot P_i \tag{30.7}$$

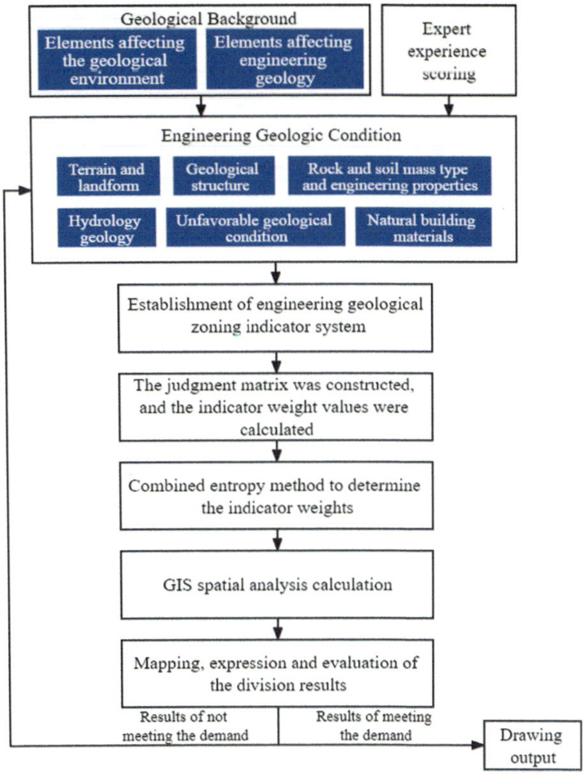

Fig. 30.1 Technical roadmap for regional engineering geological zoning of plateau mountains traffic corridors (Modified after Guo et al. 2009)

In the Equation: S_1-engineering geological zoning indicator, W_i-hierarchy analysis-entropy method coupling weight, P_i-indicator.

Combined with the engineering geological conditions of the study area and the actual engineering planning status, using GIS technology, based on the above model, the engineering geological conditions of the study area were comprehensively divided. The zoning flow chart is shown in Fig. 30.1.

30.4 Determination of Zoning Indicator and Zoning Model

Based on the comprehensive analysis of the regional geological background of the traffic corridor in the plateau mountainous area, the geological environment elements such as geological structure, terrain and landform, stratum lithologic, seismic activity, meteorological and hydrological, adverse geologic actions are selected as the zoning elements of regional engineering geological conditions. The following engineering geological zoning indicator is selected by engineering geological analogy method: active fault impact, tectonic stress field, regional surface deformation, Bouguer

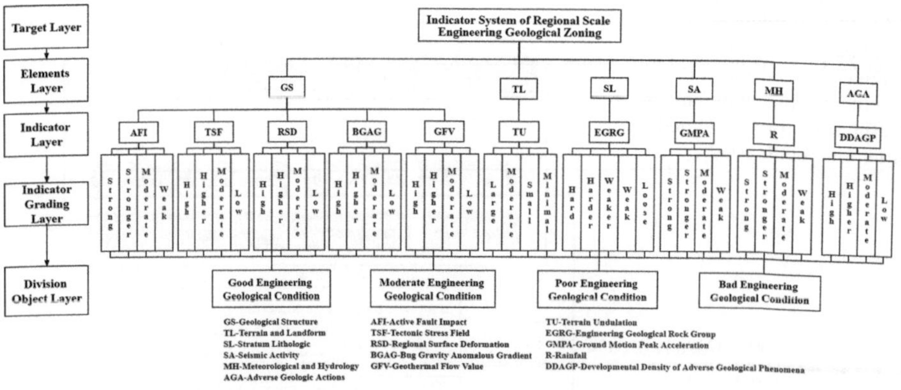

Fig. 30.2 The hierarchical analysis indicator system of engineering geological condition zoning in the study area

gravity anomaly gradient, geothermal flow value, terrain undulation, engineering geological rock group, ground motion peak acceleration, rainfall, developmental density of adverse geological phenomena and so on. According to the basic principles of the establishment and operation of the zoning indicator system, the indicator system of engineering geological zoning in the study area is obtained by using the AHP-entropy method (Fig. 30.2).

\

30.5 Evaluation Process and Zoning Results

30.5.1 Quantitative Approach of Zoning Indicator

After the indicator is determined, the GIS function is fully utilized to form the indicator grid map (Fig. 30.3a–j) on ArcGIS software. According to the scope of the study area, the grid size is set to 30 m × 30 m. For the indicators that can be directly quantified, the indicators can be graded and assigned, and the indicators that cannot be directly quantified can be graded and assigned by scoring comparison.

30.5.2 Determination of Influencing Element and Indicator Weights

The weights and assignment tables of zoning indicators were obtained by hierarchical analysis (Table 30.1). The value of the study area is based on the administrative division. The entropy method is used to obtain the weight of each index, and the two

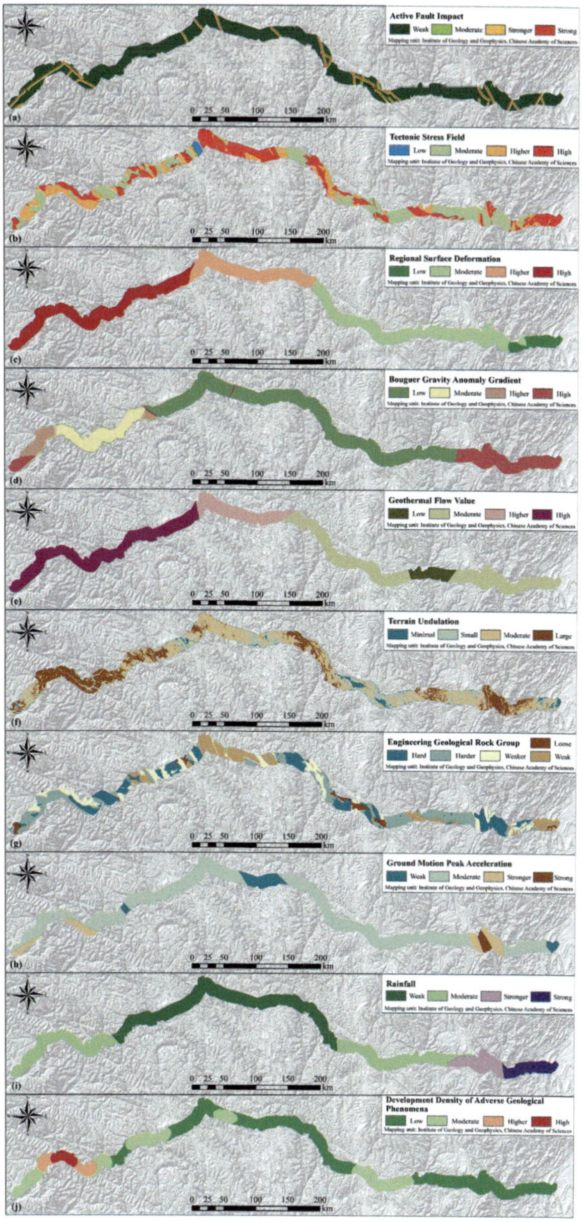

Fig. 30.3 Indicator map of engineering geological zoning in the study area. **a** AFI; **b** TSF; **c** RSD; **d** BGAG; **e** GFV; **f** TU; **g** EGRG; **h** GMPA; **i** R; **j** DDAGP

weights are coupled by the Euclidean distance method to obtain the comprehensive weight table of the engineering geological zoning index in the study area (The results show that the influence of active faults is in the forefront of the weight of each indicator, followed by terrain undulation and the development density of adverse geological phenomena. These three indicators are the basis of engineering geological zoning. Geological structure is a prerequisite for the route selection and construction of traffic corridor projects in plateau mountainous areas. There is little difference in the weights of tectonic stress field, seismic peak acceleration, engineering geological rock group and rainfall. The above analysis shows that the weight of influencing elements determined by the AHP-entropy method is in line with objective reality.)

The results show that the influence of active faults is in the forefront of the weight of each indicator, followed by terrain undulation and the development density of adverse geological phenomena. These three indicators are the basis of engineering geological zoning. Geological structure is a prerequisite for the route selection and construction of traffic corridor projects in plateau mountainous areas. There is little difference in the weights of tectonic stress field, seismic peak acceleration, engineering geological rock group and rainfall. The above analysis shows that the weight of influencing elements determined by the AHP-entropy method is in line with objective reality.

30.5.3 Zoning Results

Based on ArcGIS software, the weights are assigned to each indicator raster for reclassification, and the 10 indicator layers are overlaid and calculated, and the engineering geological condition indicator value is calculated according to Eq. (30.1). Using the natural interruption point grading method, the engineering geological zoning indicator was divided into four grades from low to high, corresponding to good, moderate, poor and bad engineering geological conditions, as the engineering geological zoning results, and the regional scale engineering geological zoning map along the traffic corridor in the mountainous plateau was obtained (Fig. 30.4).

The areas with bad geological conditions are mainly located in the Ya'an-Kangding section, Batang, Chaya-Basu section and Bomi-Linzhi section. Areas within 0–1 km of active faults, especially areas where multiple faults intersect and nearly intersect, areas with intensive geological hazard development and areas with large topographic relief are also areas with bad engineering geological conditions.

The areas with poor engineering geological conditions are mainly located in the periphery of the areas with poor engineering geological conditions, such as Litang, Chaya and Luolong areas. Influenced by active faults, the areas with poor engineering geological conditions are mainly within 1–2 km from active faults, with large topographic relief, dense geological hazard development and located in high tectonic stress areas.

The area with moderate engineering geological conditions is mainly located in the area 2–4 km away from the active fault, mostly in the form of blocks and strips

Table 30.1 Indicator table of engineering geological condition indicator in the study area

Target layer	Element layer		Indicator layer		Indicator grading layer	
	Element	Weight	Indicator	Weight	Indicator classification	Score
Indicator system of regional scale engineering geological zoning	GS	0.395	AFI	0.226	Strong	9
					Stronger	7
					Moderate	5
					Weak	3
			TSF	0.055	High	9
					Higher	7
					Moderate	5
					Low	3
			RSD	0.042	High	9
					Higher	7
					Moderate	5
					Low	3
			BGAG	0.030	High	9
					Higher	7
					Moderate	5
					Low	3
			GFV	0.042	High	9
					Higher	7
					Moderate	5
					Low	3
	TL	0.188	TU	0.188	Large	9
					Moderate	7
					Small	5
					Minimal	3
	SL	0.078	EGRG	0.078	Hard	1
					Harder	3
					Weaker	5
					Weak	7
					Loose	9
	SA	0.124	GMPA	0.124	Strong	9
					Stronger	7
					Moderate	5
					Weak	3
	MH	0.062	R	0.062	Strong	9
					Stronger	7
					Moderate	5

(continued)

Table 30.1 (continued)

Target layer	Element layer		Indicator layer		Indicator grading layer	
	Element	Weight	Indicator	Weight	Indicator classification	Score
					Low	3
	AGA	0.153	DDAGP	0.153	High	9
					Higher	7
					Moderate	5
					Low	3

Table 30.2 The comprehensive weight of engineering geological condition indicator in the study area based on the combination of AHP-entropy method

Indicator	AFI	TSF	GFV	RSD	BGAG	TU	EGRG	GMPA	R	DDAGP
Weight value (%)	21.42	8.28	7.25	4.95	7.51	14.72	7.80	8.08	8.46	11.54

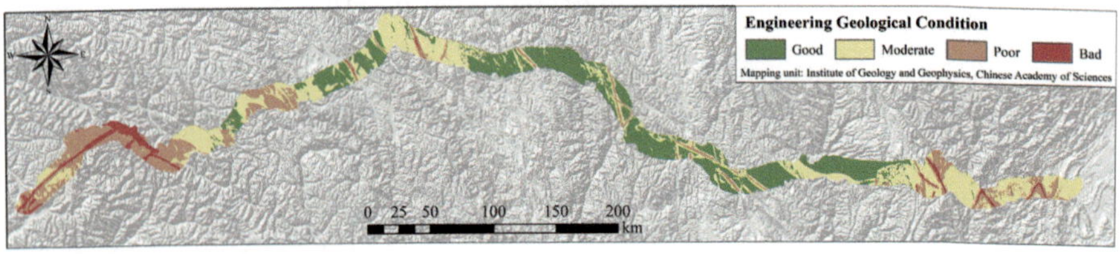

Fig. 30.4 Results of engineering geological zoning in the study area

of scattered distribution, less influenced by the active fault, far from the gully and valley water system, less undulating terrain, and with an average degree of geological hazard development.

The area with good engineering geological conditions is mainly located in the area 4 km away from the active fault, far from the deep canyon, in the form of blocks and strips distributed between the active fault zone and the deep canyon, the terrain is relatively gentle, and geological hazards are not developed.

30.6 Conclusion

Based on ArcGIS platform, this paper establishes the subjective and objective coupling zoning model of AHP-entropy method based on Euclidean distance, divides and evaluates the engineering geological conditions and engineering geological characteristics of traffic corridors in plateau mountainous areas. The influence of active faults is the most influential indicator, followed by the terrain undulation and the development density of adverse geological phenomena. These three indicators are the basis for the zoning of engineering geological conditions in the study area. Geological structure is a prerequisite for the selection and construction of traffic corridor projects. The results of regional engineering geological zoning obtained in this paper provide reliable and intuitive engineering geological data for the safety construction of traffic corridor projects in plateau mountainous areas.

Acknowledgements Thanks to Yongshuang Zhang from China University of Geosciences, Beijing, and Zhihua Yang from Institute of Geomechanics, Chinese Academy of Geological Sciences, for their help during the writing of this article. This research was supported by the Second Tibetan Plateau Scientific Expedition and Research Program (STEP) (grant no. 2019QZKK0904), the National Natural Science Foundation of China (grant no. 41825018) and the Youth Innovation Promotion Association Foundation of the Chinese Academy of Sciences (grant no. 2023073).

References

Guo CB et al (2009) Research on geo-engineering conditions of the Lijiang-Shangri-La Railway Section of the Yunnan–Tibet Railway based on the GIS method. Geoscience 23(3):545–552

Guo CB et al (2017) Discussion on the environmental and engineering geological problems along the Sichuan–Tibet Railway and its adjacent area. Geoscience 31(5):877–889

Guo CB et al (2020) Main progress and achievements of the geological survey project of Sichuan—Tibet Railway traffic corridor. Geological Survey of China 7(6):1–12

Guo CB et al (2021) Typical geohazards and engineering geological problems along the Ya'an-Linzhi Section of the Sichuan-Tibet Railway, China. Geoscience 35(1):1–17

Harris JR et al (2000) Effective use and interpretation of lithogeochemical data in regional mineral exploration programs: application of Geographic Information Systems (GIS) technology. Ore Geol Rev 16(3–4):107–143

Hu HT (2001) The theory and method of evaluation of regional crustal stability based on concept of "safe island. J Geomech 2:97–103

Li LP et al (2017) Geohazard susceptibility assessment along the Sichuan–Tibet Railway and its adjacent area using an improved frequency ratio method. Geoscience 31(5):911–929

Li SG (1977) On earthquakes. Geological Publishing Press, Beijing

Li W et al (2020) Study on route selection of Ya'an–Linzhi Section in Sichuan–Tibet Railway. Subgrade Engineering 5:23–27

Liu GC (1961) Chinese regional engineering geology. China Industry Press, Beijing

Liu ZJ (2001) The application of hierarchical analytical technique to the classification of aeromagnetic anomalies and the ore-prospecting prognosis in central and western qinghai-tibet plateau. Geophys Geochem Explor 3:161–168

Lu CF, Cai CX (2019) Challenges and countermeasures for construction safety during the Sichuan-Tibet Railway Project. Engineering 5(5):6

Lu X et al (2023) Distribution and hazard assessment of collapses and landslides in sichuan-tibet traffic corridor. J Eng Geol

Pan GT et al (2020) Key zones of oceanic plate geology and Sichuan–Tibet Railway Project. Earth Sci J China Univ Geosci 45(7)

Peng JB et al (2020) Challenges to engineering geology of Sichuan–Tibet Railway. Chin J Rock Mech Eng 39(12):2377–2389

Peng SL et al (2005) Application of analytic hierarchy method in large-scale positioning prediction of ore deposits-take the Phoenix Mountain copper mine in Tongling as an example. Geotect Mineral 1:71–77

Qi SW et al (2016) Landslide-risk zonation along mountainous highway considering rock mass classification. In: Proceedings 2015 (15th) annual conference of institute of geology and geophysics. Chinese Academy of Sciences, Beijing, China, pp 286–298

Qi SW et al (2022) Regionalization of engineering geological stability and distribution of engineering disturbance disasters in Tibetan Plateau. J Eng Geol 30(3):599–608

Shi JS et al (2005) GIS-based landslide hazard zonation of the New Badong County Site. Acta Geoscientica Sinica 3:275–282

Wang SB (2022) Evaluation of landslide susceptibility of the Ya'an–Linzhi section of the Sichuan–Tibet Railway based on deep learning. Master, Chang'an University

Wen J et al (2006) Combined weight method for rural drinking water insurance evaluation based on information entropy. J Irrigation Drainage 4:43–47

Xu M et al (2021a) Major engineering hydrogeological problems along the Ya'an–Linzhi section of the Sichuan–Tibet Railway. Hydrogeol Eng Geol 48(5):13–22

Xu ZX et al (2021b) Engineering geological environment and main engineering geological problems of Ya'an–Linzhi Section of the Sichuan–Tibet Railway. Adv Eng Sci 53(3):29–42

Xue YG et al (2020) Main unfavorable geological conditions and engineering geological problems along Sichuan–Tibet railway. Chin J Rock Mech Eng 39(3):445–468

Yi SJ (2018) Engineering geological partition research based on gis for the section of across Suture Zone in Sichuan–Tibet Railway, Master. Chengdu University of Technology

Zhang S et al (2010) The science and technology evaluation model based on entropy weight and empirical research during the 10th five-year of China. Chin J Manage 7(1):34–42

Zhang YS et al (2021) Key problems on hydro-engineering-environmental geology along the Sichuan–Tibet Railway corridor: current status and development direction. Hydrogeol Eng Geol 48(5):1–12

Zhang YS et al (2022) Geological safety evaluation of railway engineering construction inplateau mountainous region: ideas and methods. Acta Geol Sin 96(05):1736–1751

Zhao SY et al (2023) Insights into landslide development and susceptibility in extremely complex alpine geoenvironments along the western Sichuan–Tibet Engineering Corridor, China. CATENA 227:107105

Chapter 31
The Non-dilatant Frictional Resistance of Fresh Tension Fractures in Sandstones and Marbles

Anastasios Tsikrikis, Vassilis P. Marinos, and Theodosios Papaliangas

Abstract The aim of this study is to investigate the influence of origin and mineralogy on the dilation-corrected (non-dilatant) frictional resistance of rock joints. The investigated rocks were two pure, dense, high strength and low porosity calcite marbles and two high strength, low porosity sandstones. Samples with different surface roughness were obtained by artificial splitting of prismatic blocks and tested in direct shear, using a multi-stage testing procedure with sample repositioning at their original position prior to changing the normal load. The tests were conducted at a shear displacement rate of 1.0 mm/min, under constant normal stresses in the range 0.005–2.0 MPa. Detailed shear and normal displacements were recorded and the instantaneous dilation angle over the whole shear displacement was accurately calculated. The results are expressed as dilation-free (non-dilatant) frictional resistance, corresponding to the basic friction angle of essentially flat and naturally textured surfaces, i.e., surfaces with their granular texture exposed but without any macroscopic interlocking and roughness causing dilation during shear. The results demonstrate a clear influence of mineralogy on the textural frictional resistance, which was found to be approximately 35° for sandstones and 40° for marbles. In addition, the dilatant component at peak shear strength was found to decrease logarithmically with normal stress. Also, the influence of grain size on the dilatant and the non-dilatant shear strength is discussed. It is believed that the results of the study provide a better understanding of the origin of shear strength of rock joints.

Keywords Direct shear test of rock joints · Rock discontinuities · Friction angle of rock joints · Dilation · Non-dilatant friction angle

---

A. Tsikrikis (✉) · T. Papaliangas
Department of Environmental Engineering, Hellenic International University, Thessaloniki, Greece
e-mail: tsikrika@geo.auth.gr

V. P. Marinos
Department of Civil Engineering, National Technical University of Athens, Athens, Greece
e-mail: marinosv@civil.ntua.gr

S. Wang et al. (eds.), *Engineering Geology for a Habitable Earth: IAEG XIV Congress 2023 Proceedings, Chengdu, China*, Environmental Science and Engineering,
https://doi.org/10.1007/978-981-99-9057-3_31

31.1 Introduction

The shear strength of rock joints controls the mechanical behaviour of rock masses. The main factors affecting the shear strength of an unfilled joint are the mineral composition, the magnitude of the normal stress on the joint, the roughness and size of the joint, the degree of weathering, the joint wall strength, moisture, and water pressure (Kulatilake et al. 1995).

It is very difficult to theoretically establish a joint shear strength criterion because natural rock joints are very diverse and complex. Apart from the linear Mohr–Coulomb law, a number of other non-linear criteria exist for the determination of the total shear strength of rock joints, most of them based on two components: one constant term, independent of normal stress (i.e., frictional) and one normal stress dependent roughness component. The differences among these criteria arise from the definition and the determination of these two terms.

According to the non-linear criterion proposed by Papaliangas et al. (1995) the peak friction angle of a rock joint can be considered as a purely frictional (independent of normal stress) component, due to sliding between the two naturally textured walls of the rock joint and a dilatant component, due to the surface roughness, i.e., the shear resistance due to dilation. The non-dilatant friction angle corresponds to the frictional resistance of plastically deformed contact areas and is equal the friction angle of the intact rock material at the brittle-ductile transition (Papaliangas et al. 1995).

The shear tests on natural or artificial rough surfaces with instant dilation recording lead to the determination of a dilation-corrected friction angle. This friction angle (non-dilatant) is a material property and is based on the assumption that shearing of a dilating joint is equivalent to a shear movement along an inclined plane (Hencher and Richards 1989, 2015).

The aim of this laboratory study is to confirm experimentally the influence of mineralogy of the non-dilatant friction angle of rough rock surfaces.

31.2 Sample Preparation and Material Properties

The rock samples used in this study involve two marbles and two sandstones from different sites in Northern Greece. Tranovaltos marble (TM) is a white medium crystalline, fine-grained calcite marble, with a mean grain size of 150 μm (range 4–280 μm). Thasos marble (TKM) (commercial name “Krystallina of Thasos”) is a grayish white background and intense gray and black tones, almost pure coarse grained calcite marble with a grain size 1433 μm. Rhodope Sandstone (RS1) have moderately-well sorting and subangular grains with mean size 134 μm. It consists Quartz, Plagioclase, K-Feldspar, Muscovite and quartz fragments and it is classified as Lithic Feldsarenite according Folk’s classification (Folk et al. 1970). Rhodope Sandstone (KI) has poorly sorted and angular-subangular grains with mean size

411 μm. It consists Quartz, Plagioclase, K-Feldspar, Muscovite and quartz fragments and it is classified as Lithic Feldsarenite according Folk's classification (Folk et al. 1970).

All samples from a particular rock were prepared from the same rock block and casted in dental plaster prior to testing.

TM is a low porosity and medium strength rock with a dry density of 2717 kg/m^3, open porosity of 0.15% and uniaxial compressive strength of 93 MPa. TKM is a relatively low-medium strength rock with a dry density of 2709 kg/m^3, open porosity of 0.32% and uniaxial compressive strength of 63 MPa. The main difference of these calcite rocks is that the mean grain size of TKM is 10 time greater than TM. RS1 is a medium strength rock with a dry density of 2517 kg/m^3, open porosity of 3.88% and uniaxial compressive strength of 82.5 MPa. RS2 is a medium strength rock with a dry density of 2513 kg/m^3, open porosity of 4.29% and uniaxial compressive strength of 78.1 MPa. The investigated sandstones have similar mineral composition (11% higher content of quartz for RS2 than for RS1 but the mean grain size of RS2 is 3 times larger than that of RS1 (411 and 134 μm, respectively).

The quantitative mineralogical composition of the investigated rocks was calculated by the X-ray Powder Diffraction (XRPD) method was used. Also, to analyze the micro-structure of investigated rocks, 20–40 mm thin sections were obtained from rock samples and then microstructural observations were performed using a transmitted polarized microscope. The density, the porosity and the uniaxial compressive strength were determined using the test methods suggested by the ASTM. The mineralogical composition and some petrophysical properties are depicted in Table 31.1.

Table 31.1 Material properties of investigated rocks

Property	Units	Rhodope sandstone (RS1)	Rhodope sandstone (RS2)	Tranovaltos white marble (TM)	Thasos "Krystallina" marble (TKM)
Unconfined compressive strength σ_{ci}	MPa	98.7	78.5	98.3	55.2
Mean grain size	μm	134	411	146	1433
Dry density	kg/m^3	2517	2513	2717	2709
Open porosity	%	3.88	4.29	0.15	0.32
Mineral composition[a]	%	Q: 56 K: 18 Ms: 4 Pl: 22	Q: 67 K: 14 Ms: 3 Pl:14 Kln: 2	Cal:95 Do:3 Q:2	Cal: 98 Do: 1 Q: 1

[a] *Q* Quartz, *K* K-feldspar, *Ms* Muscovite, *Pl* Plagioclase, *Kln* Kaolinite, *Cal* Calcite, *Dol* Dolomite

31.3 Experimental Procedure

For each investigated rock 4–5 joint samples were prepared by bending failure of rock prisms and tested in direct shear under 6–7 normal stress each. The total number of samples and direct shear tests were 18 and 126 respectively. The samples size had a length of 8–14 cm, a width of 7–10 cm and a length/width ratio equal to 1.5. The sample preparation and size, and shape were obtained according ISRM guidelines (Muralha et al. 2014).

The direct shear tests were performed under constant normal load (CNL) conditions using the multi-stage shear procedure with repositioning of the joint in its initial natural position before each shearing stage (Muralha et al. 2014). Seven normal stresses were used in the range 0.005–2.0 MPa. All tested rock surfaces, casted in dental plaster, were first tested under their self-weight (equivalent to a normal stress of approximately 5 kPa), so that the maximum dilation angle ψ_o, could be determined. This angle is considered to represent the maximum asperity angle of the sample. Following this "zero normal load" test, the samples were subjected to shearing under a normal stress of 25 kPa for a more detailed record of the dilation at the low normal stress range and then at higher normal stresses up to a maximum of 2.0 MPa. The resulting total number of direct shear runs was 126. The shear displacement rate was constant at 1.0 mm/min and the total relative shear displacement at least 5% of the sample length.

Detailed continuous measurements of direct shear load and shear and normal displacements were recorded. For each pair of shear stress (τ) and normal stress (σ), the corresponding instantaneous dilation angle (ψ) determined for a shear displacement step of 0.2% of the sample length was used to determine the shear and normal stress (τ_1 and σ_1), which correspond to shearing of an effectively planar surface, non-dilating during shearing (Hencher and Richards 2015). The shear displacement step used to determine the dilation angles was 0.2% of the joint length, which according to Patton (1966) represents the second order roughness.

All tests were performed using a purpose-built direct shear box, capable of accommodating samples of different sizes and applying a maximum shear and normal load of 300 kN and 250 kN respectively.

In the case of multi-stage tests, the apparent cohesion can be exaggerated due to accumulation of damage with successive shearing of the same joint specimen (Muralha et al. 2014). In order to reduce this problem but also to get as much information as possible from the available samples, we used different normal stresses in the first-step and also increasing or decreasing level of normal stress among different steps of the same sample. So, each of the available samples from a rock type was tested under different first-step normal stress.

For each data pair of shear (τ) and normal (σ) stress the non-dilatant shear and normal stress (τ_1 and σ_1) was determined using the instantaneous dilation angle (ψ). These stresses represent shearing along a naturally textured plane surface, i.e., shearing without dilation-non-dilating shear. This method is presented by Hencher and Richards (1989, 2015) and is based on the assumption that shearing of a dilating joint is equivalent to a shear movement along an inclined plane, having a slope angle equal to the instantaneous dilation angle. Analysis of stresses along the inclined plane (τ_1) and its perpendicular direction (σ_1) results in the following relations:

$$\tau_1 = (\tau\cos\psi - \sigma\sin\psi)\cos\psi \tag{31.1}$$

$$\sigma_1 = (\sigma\cos\psi + \tau\sin\psi)\cos\psi \tag{31.2}$$

where

ψ is the instantaneous dilation angle at peak shear stress,

τ = measured peak shear stress, σ = measured normal stress,

τ_1 = shear stress along the actual plane of sliding,

σ_1 = normal stress along the actual plane of sliding.

The non-dilatant friction angle is different from that of a saw-cut surface (typically $30° \pm 1.5°$), produced from the same rock material (Hencher and Richards 2015). The non-dilatant shear strength of a rock surface is a shear parameter of a rock surface determined directly from the original surface tested "as it is". Therefore, the non-dilatant friction angle may be used as a characteristic parameter to describe the frictional behaviour of a rock surface.

The values of σ_1 and τ_1 depend upon the shear displacement step used to determine the dilation angles. In our case this step was 0.2% of the joint length, which according to Patton represents the second order roughness (Patton 1966). For a laboratory sample with a nominal length of 100 mm this step is 0.2 mm, which is of the order of the grain size (in this study 134–1433 mm).

The non-dilatant friction angle forms the basic frictional parameter of the non-linear peak shear strength criterion proposed by Papaliangas et al (1995) described by Eqs. (3a, 3b) and (4a, 4b) and used for the analysis of the experimental results of this study:

$$\tau_p = \sigma_n \tan(\varphi_{nd} + \psi) \tag{31.3a}$$

$$\text{or } \varphi_p = \varphi_{nd} + \psi \tag{31.3b}$$

$$\tan\psi = \tan\psi_o\left(\log_{10}\frac{\sigma_{nt}}{\sigma_n}\right) \Big/ \left(\log_{10}\frac{\sigma_{nt}}{\sigma_{no}}\right) \tag{31.4a}$$

$$\psi = ATAN\left[\tan\psi_o\left(\log_{10}\frac{\sigma_{nt}}{\sigma_n}\right) \Big/ \left(\log_{10}\frac{\sigma_{nt}}{\sigma_{no}}\right)\right] \tag{31.4b}$$

where

τ_p is the peak shear stress,

σ_n is the normal stress,

φ_p is the peak friction angle of the rock joint,

φ_{nd} is the sliding friction angle of a flat rock surface and

ψ is the dilation angle at the instant of peak shear strength, under effective normal stress σ_n,

ψ_o is the peak dilation angle under a normal stress σ_{no} which causes negligible asperity deformation (zero load),

σ_{nt} is the transitional effective normal stress under which the normal dilation is completely suppressed.

The friction angle φ_{nd} corresponds to an effectively planar yet naturally textured surface and is different from the friction angle of artificially prepared surfaces such as saw-cut surfaces.

31.4 Results and Discussion

The results are presented in Figs. 31.1, 31.2, 31.3 and 31.4 by means of two diagrams for each rock type: the first diagram shows the variation of dilation rate with normal stress ($\tan\psi - \sigma_n$) and the second diagram the variation of both the peak shear strength and the non-dilatant shear strength with normal stress. The non-dilatant friction angle φ_{nd} is determined from a plot of τ_1 vs. σ_1, as shown in Figs. 31.1, 31.2, 31.3, and 31.4b.

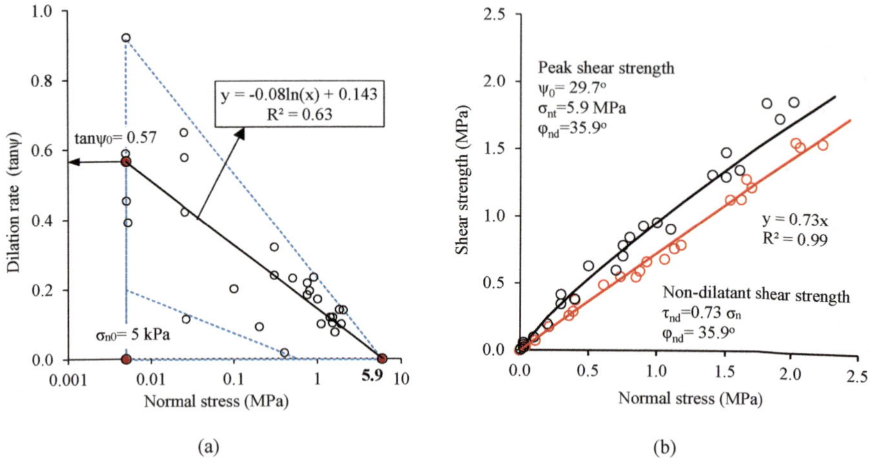

Fig. 31.1 Variation of Dilation rate with normal stress (**a**) and Peak and non-dilatant shear strength (**b**) for Rhodope sandstone 1 (RS1)

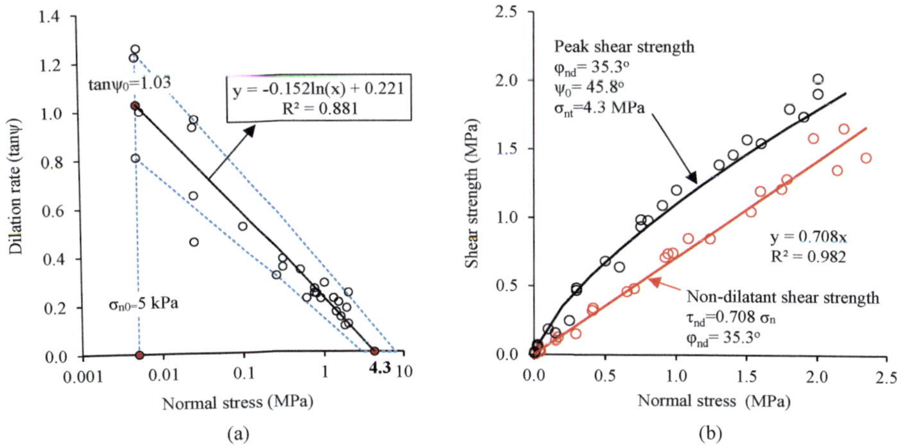

Fig. 31.2 Variation of Dilation rate with normal stress (**a**) and Peak and non-dilatant shear strength (**b**) for Rhodope sandstone 2 (RS2)

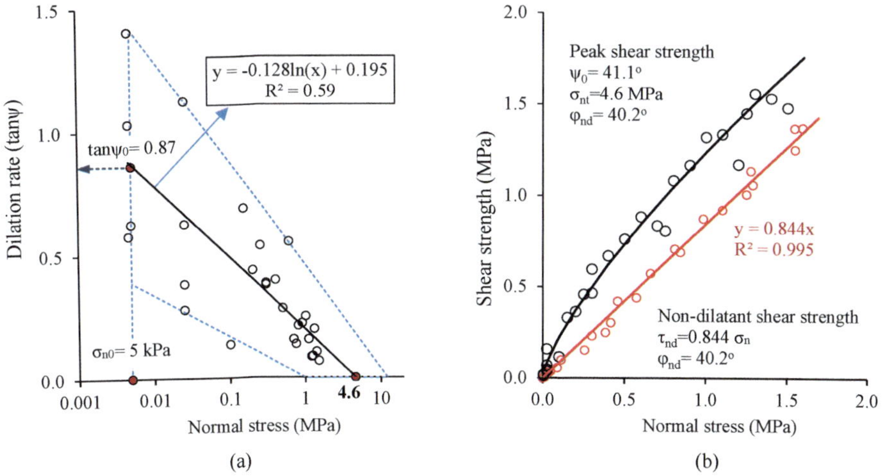

Fig. 31.3 Variation of Dilation rate with normal stress (**a**) and Peak and non-dilatant shear strength (**b**) for Thasos marble (TKM)

Figure 31.1 depicts the results from the tests on Rhodope sandstone 1 (RS1). From the semi-logarithmic regression line fitted to the data of the dilation rate vs. normal stress of Fig. 31.1a, the dilation angle at normal stress 5 kPa ("no load" condition) and the transitional normal stress under which dilation is completely suppressed are determined as $\tan\psi_0 = 0.57$ ($\psi_0 = 29.7°$) and $\sigma_{nt} = 5.9$ MPa respectively. From the linear regression line applied to the non-dilatant shear strength data ($\tau_1 - \sigma_1$) Fig. 31.1b (red symbols) the non-dilatant friction angle is determined ($\varphi_{nd} = 35.9°$).

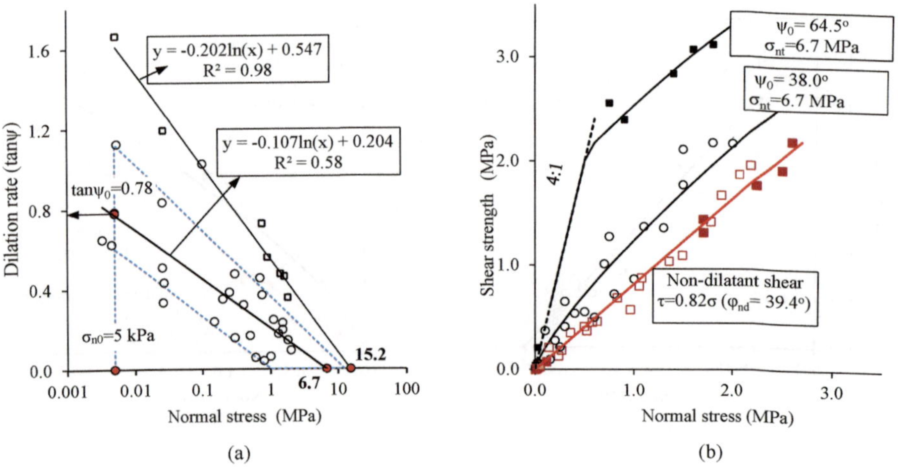

Fig. 31.4 Variation of Dilation rate with normal stress (**a**) and Peak and non-dilatant shear strength (**b**) for Tranovaltos marble (TM)

The above three parameters are used as input parameters in the Papaliangas criterion (Papaliangas et al. 1995) and result in the curve shown by the black curve in Fig. 31.1b. The results from the other three rock types are presented in Figs. 31.2, 31.3 and 31.4 and a summary of the results of the joint shear strength parameters of the studied rock types is given in Table 31.2.

In the case of Tranovaltos marble one (MT4) out of five samples tested, corresponding to a rough undulating surface (measured "no-load" dilation angle equal to $\psi_o = 58°$), the corresponding dilation data in Fig. 31.4a were not used for the overall determination of the values of σ_{nt}, and ψ_o to avoid inordinate divergence from the mean value of the other four samples of the same group. However, the shear strength data were used in the non-dilatant shear strength data (filled red squares) shown in Fig. 31.4b, in close agreement with the data from the other samples. The peak shear

Table 31.2 Summary of shear characteristics of investigated rocks

Property	Units	Rhodope sandstone (RS1)	Rhodope sandstone 2 (RS2)	Tranovaltos white marble (TM)	Thasos "Krystallina" marble (TKM)
Non-dilatant friction angle φ_{nd}	(°)	35.9	35.3	39.4	40.2
Transitional normal stress where dilation is completely suppressed σ_{nt}	MPa	5.9	4.3	6.7	4.6
Peak dilation angle at 5 kPa normal stress ("no-load" condition) ψ_0	(°)	29.7	45.8	38.0	41.1

strength for this sample was much higher than the group of all other samples. The best fitted curve shown corresponds to an increased value of $\psi_o = 64.5°$. The dotted line corresponding to a 4:1 (76°) slope in the σ-τ diagram is the suggested limit of application of the Papaliangas criterion.

The non-dilatant friction angles of the sandstones RS1 and RS2 were 35.9° and 35.3° and for the marbles TM and TKM 39.4° and 40.2° respectively. The average friction angle of the two marbles was 4.2° higher than that of sandstones.

The results indicate a clear influence of mineralogy on the textural frictional resistance. On the other hand, the maximum asperity slope ψ_o peak dilation angle may be positively correlated with the mean grain size. The dilation angle at peak shear strength was found to decrease logarithmically with normal stress. The peak dilation angle ψ_0 under a normal stress $\sigma_{no} = 5$ kPa, was found to be 29.7°, 45.8°, 38.0° and 41.1° for RS1, RS2, TM and TKM respectively. The transitional normal stress corresponding to zero dilation angle (σ_{nt}) is 5.9 MPa, 4.3 MPa, 6.7 MPa and 4.6 MPa for RS2, RS1, TM and TKM respectively. The mean value of 5.38 MPa is approximately three orders of magnitude higher than the normal stress of 5 kPa, which is used to define the "no-load" dilation angle ψ_o. Therefore, it is reasonable to assume that the semi-logarithmic relation between σ_n and $\tan\psi$ holds generally over three orders of magnitude. In this case Eqs. (31.4a, 31.4b) becomes:

$$\tan\psi \approx \frac{1}{3}\tan\psi_o\left(\log_{10}\frac{\sigma_{nt}}{\sigma_n}\right) \tag{31.5a}$$

$$\psi \approx ATAN\left[\frac{1}{3}\tan\psi_o\left(\log_{10}\frac{\sigma_{nt}}{\sigma_n}\right)\right] \tag{31.5b}$$

The grain size seems to have no any effect on the non-dilatant friction angle. More specifically, the mean grain size of RS2 is 3 times larger than that of RS1 (411 and 134 μm, respectively) and that of TKM is 10 times larger than that of TM (1433 and 146 μm, respectively). However, the difference in the non-dilatant friction angle is less than 1°, i.e., 0.6° for sandstones (35.9° and 35.3°) and 0.8° for marbles (39.4° and 40.2°).

The two rock groups, i.e., (marbles and sandstones) with similar minerology per rock type and different mean grain size provide almost same non-dilatant friction angles (average 39.8° and 35.6° respectively). This is in agreement with experimental evidence showing lower friction angles for silicates than those of carbonates (e.g., Barton 1976).

Despite the variation observed in the peak shear strength data for all different rock types used, the best-fit regression lines between normal stress σ_1 and non-dilatant shear strength τ_1 are well-defined (Figs. 31.1b, 31.2b, 31.3b and 31.4b). These linear relations hold all over the normal stress used and irrespective of the surface roughness of the joints. The proportionality between τ_1 and σ_1 clearly indicate that the non-dilatant shear strength is frictional.

The dilation angle ψ_0 is equivalent to the maximum effective roughness under "no-load" conditions. This is not equal to the maximum slope determined from independent linear roughness profiles along the joint surface, but equal to the slope of the steepest plane defined by three contact points at the appropriate shearing direction along which shearing takes place and therefore it is effectively a 3-d parameter.

At normal stresses higher than σ_{nt} the peak shear strength of the joint types is given by the following expressions:

$$\tau_p = \sigma_n \tan(\varphi_{nd})$$

Therefore, the peak shear strength can be interpreted as comprising two components: the first component is due to dilation (dilatant component) and the second due to frictional properties of the rock material (non-dilatant component).

The non-dilatant friction angle, when determined from direct shear tests of natural surfaces, takes into account the effect of various surface conditions like alterations, weathering, surface coatings etc.

The peak shear strength envelopes of Papaliangas non-linear shear strength criterion produced on the basis of the non-dilatant friction angle and the dilation angle fit the experimental data with acceptable accuracy.

31.5 Conclusions

In this paper, several fresh tension fractures with different surface roughness from two carbonates (marbles) and two silicates (sandstones) tested dry in direct shear. The tests resulted in 126 peak shear strength τ_p—normal stress σ_n data pairs with the corresponding instantaneous dilation angle ψ.

From the normal stress—dilation rate ($\tan\psi$) data pairs the peak dilation angle under a normal stress σ_{no} which causes negligible asperity deformation and the transitional normal stress under which the normal dilation is completely suppressed σ_{nt} are estimated.

The main conclusions from this experimental research are:

Over the normal stress range used. The non-dilatant shear strength fits the experimental data all over the normal stress range used irrespective of roughness.

Changes in peak shear strength due to normal stress or roughness can be interpreted in terms of change in the dilatant component only.

The average non-dilatant friction angle of silicate rocks was 4.2° lower than that of carbonate rocks.

The dilation angle at peak shear strength was found to decrease logarithmically with normal stress.

The peak dilation angle ψ_0 under a normal stress σ_{no} which causes negligible asperity deformation depends on mean grain size of rock. The rocks with higher mean grain size provide higher peak dilation angle ψ_0.

The difference of the non-dilatant friction angle for each rock type was lower than 0.8°. This suggests that, for a certain rock type the mean grain size alone may have no effect on the value of the non-dilatant friction angle.

The observed variation in the value of dilation (tanψ) is attributed to the different roughness among samples of the same rock. The results show that the experimental recording of dilation reveals valuable information and therefore is necessary in direct shear tests.

References

Barton N (1976) The shear strength of rock and rock joints. Int J Rock Mech Min Sci Geomech 13:255–279 (Pergamon Press)

Folk RL, Andrews PB, Lewis DW (1970) Detrital sedimentary rock classification and nomenclature for use in New Zealand. NZ J Geol Geophys 13(4):937–968. https://doi.org/10.1080/00288306.1970.10418211

Hencher SR, Richards LR (1989) Laboratory direct shear testing of rock discontinuities. Ground Eng 22:24–31

Hencher SR, Richards LR (2015) Assessing the shear strength of rock discontinuities at laboratory and field scales. Rock Mech Rock Eng 48(3):883–905. https://doi.org/10.1007/s00603-014-0633-6

Kulatilake PHSW, Shou G, Huang TH, Morgan RM (1995) New peak shear strength criteria for anisotropic rock joints. Int Rock Mech Min Sci Geomech Abstr 32(7):673–697. https://doi.org/10.1016/0148-9062(95)00022-9

Muralha J, Grasselli G, Tatone B, Blümel M, Chryssanthakis P, Yujing J (2014) ISRM suggested method for laboratory determination of the shear strength of rock joints: revised version. Rock Mech Rock Eng 47(1):291–302. https://doi.org/10.1007/s00603-013-0519-z

Papaliangas TT, Hencher SR, Lumsden AC (1995) A comprehensive peak shear strength criterion for rock joints. In: Fuji T (ed) Proceedings of 8th international congress ISRM, Tokyo, vol 1. Balkema, Rotterdam, pp 359–366

Patton FD (1966) Multiple modes of shear failure in rock and related materials. Ph.D. thesis, University of Illinois

Chapter 32
Spatial Variability Modelling of Structural Surface Mechanical Parameters for Reliability Assessment of Rock Slopes

Shui-Hua Jiang, Jian-Ping Li, Su Ouyang, and Jinsong Huang

Abstract The spatial variability modelling of mechanical parameters for the highly fractured weathered rock masses and those dominated by a single structural plane has been extensively studied. However, few attempts have been made to characterize the spatial variability of mechanical parameters for the wedge rock masses. In this paper, a method for modelling the spatial variability of the structural surface mechanical parameters for the wedge rock masses is proposed. The function relationship between the factor of safety of the rock slope and uncertain input parameters is constructed using the back-propagation neural network. The non-intrusive stochastic finite difference method is adopted to explore the influence of the spatial variability of the structural surface mechanical parameters on the probability of slope failure. The proposed method can provide an effective and robust means for the characterization of the spatial variability of structural surface mechanical parameters in complex rock slope reliability analysis. Ignoring the spatial variability of the structural surface mechanical parameters can overestimate the probability of slope failure, resulting in conservative slope reinforcement design schemes.

Keywords Rock slope · Reliability assessment · Wedge failure · Structural surface · Spatial variability

---

S.-H. Jiang (✉) · J.-P. Li · S. Ouyang
School of Infrastructure Engineering, Nanchang University, Xuefu Road 999, Nanchang 330031, China
e-mail: sjiangaa@ncu.edu.cn

J.-P. Li
e-mail: jianping_lee@email.ncu.edu.cn

J. Huang
Discipline of Civil, Surveying and Environmental Engineering, Faculty of Engineering and Built Environment, The University of Newcastle, Callaghan, NSW 2308, Australia
e-mail: jinsong.huang@newcastle.edu.au

S. Wang et al. (eds.), *Engineering Geology for a Habitable Earth: IAEG XIV Congress 2023 Proceedings, Chengdu, China*, Environmental Science and Engineering,
https://doi.org/10.1007/978-981-99-9057-3_32

32.1 Introduction

Due to the complex rock structure, environment, loading conditions and geological depositional processes, the geomechanical parameters at distinct positions along different directions of rock masses not only have obvious differences, but also exist certain spatial correlation (e.g., Hsu and Nelson 2006; Renani et al. 2019). This has a significant effect on the deformation and failure mechanism of rock slopes (e.g., Srivastava 2012; Tang et al. 2017; Jiang et al. 2022). Random field theory and geostatistics have been utilized to reasonably model the spatial variability of rock masses mechanical parameters (e.g., Eivazy et al. 2017; Chen et al. 2019; Zhang et al. 2021; Jiang et al. 2023). There are four typical failure modes for rock slopes, including circular failure mode, plane failure mode, wedge failure mode and anti-dip failure mode (Hoek and Londe 1974). Reliability analysis approaches have been emerged as a more reasonable and rigorous way to handle the effects of the spatial variability of rock masses strength properties on the slope stability involving these multiple failure modes. Recently, reliability assessment problems of the highly fractured weathered rock slopes and the rock slopes against a dominated circular failure considering the spatial variability have been extensively studied (e.g., Srivastava 2012; Gravanis et al. 2014; Huang et al. 2016; Tang et al. 2017; Renani et al. 2019; Chen et al. 2019). It was widely proven that ignoring the spatial variability of rock strength properties in the probabilistic rock slope stability analysis could lead to an erroneous estimate of the probability of slope failure.

The abovementioned studies only treated the rock slopes failing along circular slip surfaces or planar sliding surfaces. This was true for the intact rock masses, highly fractured and strongly weathered rock masses, or the rock masses with a predominant weak structural plane. However, it is noted that weak surfaces, joints, bedding planes, faults or discontinuities at critical locations are commonly presented in the heterogeneous rock masses. These joints, bedding planes, faults and discontinuities can cut from each other and/or be combined with each other, leading to complicated failures in the rock slopes, such as wedge failure and toppling failure (Zhou et al. 2017). Few attempts have been made to investigate the reliability of rock slopes against a wedge failure accounting for the spatial variability. Additionally, to take into account the stress–strain relationship of rock masses and ensure the computational accuracy of slope reliability assessment, finite element or finite difference methods are usually employed for deterministic stability analysis of complex rock slopes. It is quite computationally demanding once these methods are combined with the direct Latin hypercube sampling (LHS) for reliability analysis of rock slopes (e.g., Li and Jiang 2016; Chen et al. 2019). Therefore, it is of great necessity to develop an efficient spatial variation modelling method for the reliability analysis of the rock slopes against a wedge failure.

This study proposes a Karhunen-Loève (K-L) expansion-based method for modelling the spatial variability of structural surface mechanical parameters for a wedge rock mass. The non-intrusive stochastic finite difference method is employed for the reliability analysis of rock slope accounting for the spatial variability. A rock slope example against a wedge failure is investigated to illustrate the effectiveness of the proposed method. The influences of the spatial variability of the structural surface shear strength parameters on the probability of slope failure are explored.

32.2 Spatial Variability Modelling and Probabilistic Rock Slope Stability Analysis

32.2.1 *Rand Field Modelling of Spatial Variability*

A large number of structural surfaces including fractures, joints, faults, discontinuities and others typically existed within the rock masses. These structural surfaces intersect with each other, leading to a complex rock mass structure. Generally, the stability and deformation of rock slopes are controlled by the structural surfaces because their mechanical strengths are lower than those of rock blocks. The spatial variability of structural surface mechanical parameters can greatly affect the slope reliability (e.g., Huang et al. 2016; Tang et al. 2017; Chen et al. 2019; Zhang et al. 2021). How to rationally characterize the spatial variability of structural surface mechanical parameters is still an open question. In this study, a K-L expansion-based method is proposed for modelling the spatial variability of structural surface mechanical parameters for the wedge rock masses. The following three-dimensional (3D) isotropic exponential autocorrelation function is utilized to characterize the spatial autocorrelation of structural surface mechanical parameters (e.g., Hsu and Nelson 2006; Srivastava 2012):

$$\rho\left(\tau_x, \tau_y, \tau_z\right) = \exp\left[-2\left(\frac{\tau_x}{\delta_x} + \frac{\tau_y}{\delta_y} + \frac{\tau_z}{\delta_z}\right)\right] \tag{32.1}$$

where τ_x, τ_y and τ_z are the relative distances between any two locations on the structural surface along the x-, y- and z-directions, respectively; δ_x and δ_y are the horizontal scales of fluctuation; δ_z is the vertical scale of fluctuation. It is difficult to directly model the 3D spatial variability of structural surface mechanical parameters since the structural surface is commonly not a plane and its orientation is also not along a specific direction. To this end, a coordinate transformation technique developed by Cheng et al. (2017) is adopted to convert the discretization of a 3D transverse isotropic random field into that of a two-dimensional (2D) rotational random field. A basic rotation matrix is introduced to describe the unit vectors of the new coordinate system using the components of the original coordinate system. Taking a counter-clockwise rotation with an angle of α around the z-axis (see Fig. 32.1), these three unit-vectors of the new coordinate system can be obtained as

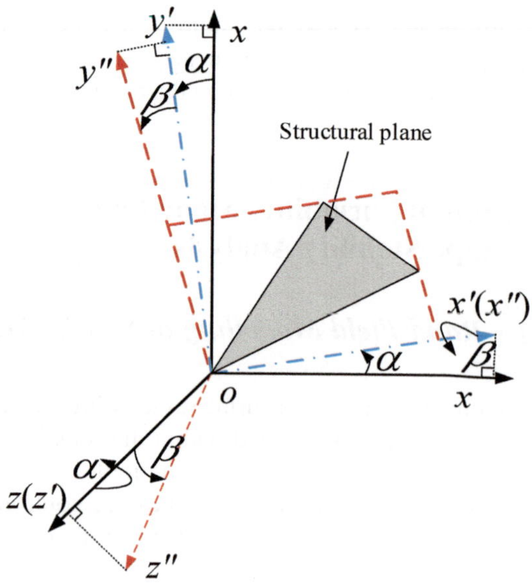

Fig. 32.1 Diagram of different coordinate transformations

$$\boldsymbol{x}' = \begin{bmatrix} \cos\alpha \\ \sin\alpha \\ 0 \end{bmatrix}, \boldsymbol{y}' = \begin{bmatrix} -\sin\alpha \\ \cos\alpha \\ 0 \end{bmatrix}, \boldsymbol{z}' = \begin{bmatrix} 0 \\ 0 \\ 1 \end{bmatrix} \tag{32.2}$$

In this way, the counterclockwise rotation matrix with an angle of α around the z-axis and that with an angle of β around the x-axis can be obtained, respectively, as follows:

$$\boldsymbol{R}_z = \begin{bmatrix} \cos\alpha & -\sin\alpha & 0 \\ \sin\alpha & \cos\alpha & 0 \\ 0 & 0 & 1 \end{bmatrix} \text{ and } \boldsymbol{R}_x = \begin{bmatrix} 1 & 0 & 0 \\ 0 & \cos\beta & -\sin\beta \\ 0 & \sin\beta & \cos\beta \end{bmatrix} \tag{32.3}$$

After these two rotations, the coordinate transformation formula can be obtained as

$$\begin{bmatrix} \tau'_x \\ \tau'_y \\ \tau'_z \end{bmatrix} = \begin{bmatrix} \cos\alpha & -\cos\beta\sin\alpha & \sin\alpha\sin\beta \\ \sin\alpha & \cos\alpha\cos\beta & -\cos\alpha\sin\beta \\ 0 & \sin\beta & \cos\beta \end{bmatrix} \begin{bmatrix} \tau_x \\ \tau_y \\ \tau_z \end{bmatrix} \tag{32.4}$$

where α and β are the orientation and inclination of a structural surface, respectively; τ'_x, τ'_y and τ'_z are the relative distances between any two locations along the x-, y- and z-directions in the new coordinate system, respectively. For simplicity, only the 2D spatial autocorrelation along two mutually perpendicular directions on the structural surface projection plane is considered in this study. Following Cheng et al. (2017), the corresponding 2D autocorrelation function can be estimated as

$$\rho\left(\tau_x', \tau_y'\right) = \exp\left[-2\left(\frac{\tau_x \cos\alpha - \tau_y \cos\beta \sin\alpha + \tau_z \sin\alpha \sin\beta}{\delta_x} + \frac{\tau_x \sin\alpha + \tau_y \cos\alpha \cos\beta - \tau_z \cos\alpha \sin\beta}{\delta_y}\right)\right] \quad (32.5)$$

Moreover, there exists cross-correlation between the structural surface mechanical parameters, such as the cohesion and friction angle (Li and Jiang 2016). Based on the marginal distribution and cross-correlation coefficient matrix of the structural surface mechanical parameters, cross-correlated non-normal random variables in the original space can be obtained as a function of independent standard normal random variables. Here, an equivalent cross-correlation coefficient matrix $\boldsymbol{\rho}_0$ is evaluated based on the cross-correlation coefficient matrix ρ according to Der Kiureghian and Liu (1986). The lower triangular matrix $\boldsymbol{L}$ is obtained by factoring the autocorrelation coefficient matrix $\boldsymbol{\rho}_0$ using Cholesky decomposition algorithm:

$$\boldsymbol{L}\boldsymbol{L}^{\mathrm{T}} = \boldsymbol{\rho}_0 \quad (32.6)$$

Thereafter, the Latin hypercube sampling technique is adopted to generate a set of independent standard normal random samples, $\xi = (\xi_1, \xi_2, \ldots, \xi_n)^{\mathrm{T}}$ (Mckay et al. 1979), which are then multiplied by $\boldsymbol{L}$ to obtain a cross-correlated standard normal random vector $\boldsymbol{\eta}$:

$$\boldsymbol{\eta} = \boldsymbol{L}\xi \quad (32.7)$$

Then, the KL expansion method is utilized to generate the realizations of cross-correlated Gaussian random fields (e.g., Li and Jiang 2016; Chen et al. 2019):

$$\underset{X_i}{\overset{\sim}{\mathrm{H}}}(x, y) = \mu_{X_i}' + \sum_{j=1}^{n} \sigma_{X_i}' \sqrt{\lambda_j} f_j(x, y) \eta_{X_{i,j}} \quad (32.8)$$

where n is the number of truncated K-L expansion terms, which is also the number of discretized random variables for a spatially varying variable; (x, y) are the coordinates in the 2D space; μ_{X_i}' and σ_{X_i}' are the mean and standard deviation of the random field X_i (i.e., cohesion c and friction angle φ), respectively; $\eta_{X_{i,j}}$ is the cross-correlated standard normal random sample; λ_j and $f_j(\cdot)$ are the eigenvalues and eigenfunctions of the autocorrelation function in Eq. (32.5), which is obtained by solving an eigenvalue problem of the homogenous Fredholm integral equation (Li and Jiang 2016). Finally, the realizations of the non-Gaussian random fields of structural surface mechanical parameters in the physical space can be obtained via the following Nataf isoprobabilistic transformation (Kiureghian and Liu 1986):

$$\mathrm{H}_{X_i}^{NG}(x, y) = F_i^{-1}\left(\phi\left[\underset{X_i}{\overset{\sim}{\mathrm{H}}}(x, y)\right]\right) \tag{32.9}$$

where the superscript NG means non-Gaussian random field; $\phi(\cdot)$ is the probability density function of a standard normal random variable; $F_i^{-1}(\cdot)$ is the inverse function of marginal cumulative distribution of the non-Gaussian vector random field X_i. Note that the value of n in Eq. (32.8) directly impacts the modeling accuracy of the spatial variability. According to Jiang et al. (2014), to ensure the discretization accuracy of a random field, 95% of the expected energy preserved is often treated as a criterion for determining the value of n. For a 2D region, the expected energy ratio factor ε is defined as

$$\varepsilon = \sum\nolimits_{j=1}^{n} \lambda_j / \sum\nolimits_{j=1}^{\infty} \lambda_j = \sum\nolimits_{j=1}^{n} \lambda_j / L'_x L'_y \tag{32.10}$$

where L'_x and L'_y are the length and width of the 2D random field region, respectively, which can be estimated as follows based on the coordinate transformation formula in Eq. (32.4):

$$\begin{cases} L'_x = L_y \sin\alpha + L_x \cos\alpha + L_z / \tan\beta \\ L'_y = L_y \cos\alpha + L_x \sin\alpha \end{cases} \tag{32.11}$$

where L_x, L_y and L_z are the length, width and height of the structural surface in the original coordinate system in Fig. 32.1, respectively, which are equal to the distances between the maximum and minimum coordinate values along the three axes on the structural surface. Figure 32.2 shows a brief flow chart for the random field modeling of spatial variability.

32.2.2 Construction of Surrogate Model of Factor of Safety

Having obtained the random field realizations of the structural surface mechanical parameters, they can be assigned to the corresponding finite difference elements in accordance to the coordinates of random field elements. The process is conducted with the aid of the element traversal method and data assignment function built-in FLAC3D. The non-intrusive stochastic finite difference method is then adopted to estimate the probability of slope failure considering the spatial variability of structural surface mechanical parameters (e.g., Jiang et al. 2014; Li and Jiang 2016). The finite-difference strength reduction method is employed for the slope stability analysis, which can well account for the rock mass stress–strain relationship, automatically search for the critical sliding surface, and estimate the factor of safety. However, it is quite time-consuming when the direct LHS method is used to calculate the probability of failure because numerous evaluations of slope stability using the finite element or finite difference methods are required. In addition, the factor of

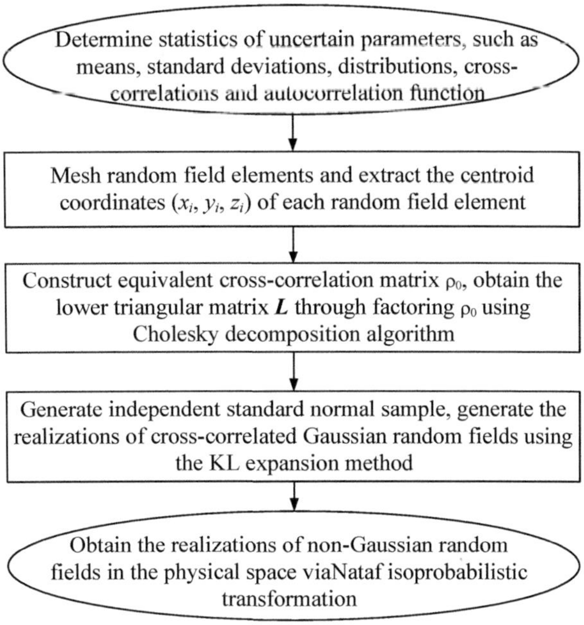

Fig. 32.2 Flow chart for the random field modeling of spatial variability

safety is typically a non-linear implicit function of the structural surface mechanical parameters. To improve the computational efficiency, the back-propagation neural network is adopted to construct the surrogate model of the factor of safety of slope in advance (namely, the function relationship between the factor of safety and uncertain input parameters) (Chen et al. 2022). Interested readers can be referred to Chen et al. (2022) for detailed procedures for the construction of surrogate models using the neural networks.

32.2.3 *Probabilistic Rock Slope Stability Analysis*

Once the surrogate model of factor of safety using the back-propagation neural network is constructed, the LHS technique is adopted to calculate the probability of slope failure p_f with N_t sets of independent standard normal samples as inputs:

$$p_f = \frac{1}{N_t}\sum_{k=1}^{N_t} I\left\{FS_{\min}\left[H_{X_i}^{NG,k}(x,y)\right] < 1.0\right\} \tag{32.12}$$

where $FS_{\min}$ denotes the minimum factor of safety calculated from the surrogate model for a given realization of the random fields of structural surface mechanical parameters (c and φ); $I(\cdot)$ is the indicator function of slope failure. In this way, an interface program between the spatial variability modeling of mechanical parameters and probabilistic rock slope stability analysis using the FLAC3D software is developed.

32.3 Illustrative Example

The rock slope model against a wedge failure in Zhu et al. (2017) is taken as an example to investigate the effect of the spatial variability of the structural surface mechanical parameters on the probability of slope failure. The slope has a height of $H = 33$ m and a rock density of 2600 kg/m^3. Figure 32.3 shows the geometric model of the slope. Table 32.1 summarizes the values of geometric and mechanical parameters of four rock surfaces. The cohesion and friction angle of rock mass are taken as 1.0 MPa and 35°, respectively, which are in accordance with their typical varying ranges of 0.4–2.0 MPa and 31°–53°. The thicknesses of the structural surfaces I and II are both 1.0 m.

The boundary condition of slope model is summarized as follows: (1) Vertical displacement constraints are imposed on the left and right sides and the backside of the model; (2) The top surface III and slope surface IV are free boundaries; (3) The bottom surface is fully constrained by the displacements. The finite-difference strength reduction method is adopted for the slope stability analysis. The rock slope model is discretized into 328 and 206 hexahedral finite difference elements for the structural surfaces I and II, respectively. A total of 22,932 mixed hexahedral, trigonal, tetrahedral, and tetrahedral elements are discretized for the rock mass. A total of 6167 nodes are involved. Based on the means of all rock masses mechanical parameters, the factor of safety of the slope estimated using the finite-difference strength reduction method is 1.296, which is slightly larger than that (1.256) calculated by Zhu et al. (2017) using Swedge software. The slight difference in the factor of safety is because the effect of water pressure on the structural surfaces was considered by Zhu et al. (2017) while that is not considered in this study.

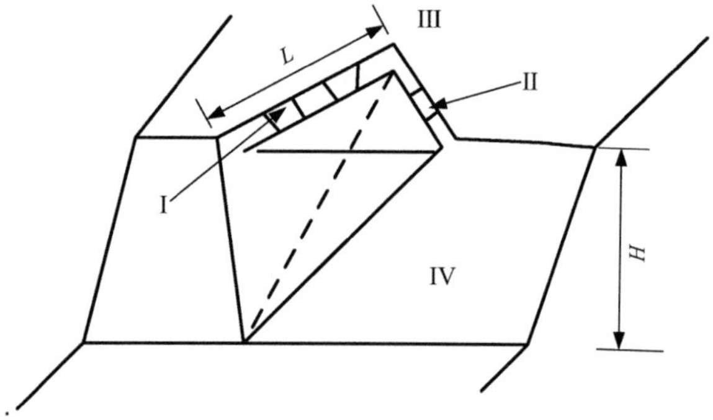

Fig. 32.3 Rock slope model with a wedge failure mode

Table 32.1 Values of the geometric and mechanical parameters of structural surfaces (after Zhu et al. 2017)

Rock surfaces	Parameters	Values
Structural surface I	Dip/°	45
	Inclination/°	105
	Distance of top cracking surface/m	12
	Cohesion /kPa	20
	Friction angle/°	30
Structural surface II	Dip/°	70
	Inclination/°	235
	Cohesion/kPa	50
	Friction angle/°	30
Top surface III	Dip/°	12
	Inclination/°	195
Slope surface IV	Dip/°	65
	Inclination/°	185

Figure 32.4 presents the maximum shear strain increment contour of the slope obtained from the deterministic rock slope stability analysis. It is easily observed from Fig. 32.4 that the stability of this rock slope is mainly dominated by the shear strengths of two weak structural surfaces I and II (namely, lateral sliding surfaces). To incorporate the effect of the spatial variability, the realizations of the cross-correlated lognormal random fields of c and φ of these two weak structural surfaces along their strike directions are generated using the method outlined in Sect. 2.1. The 2D region sizes of the projection planes calculated using Eq. (32.11) are approximately 76 m × 69 m and 70 m × 73 m for the structural surfaces I and II, respectively, as shown in Fig. 32.5. As for the structural surfaces I and II, the random field meshes with a size of 1.9 m are consistent with the respective finite difference meshes. The cohesion and friction angle of the structural surfaces I and II, c_1 and φ_1, c_2 and φ_2, are cross-correlated with a coefficient of $\rho_{c,\varphi} = -0.5$. The exponential autocorrelation function in Eq. (32.5) with $\delta_x = 30$ m and $\delta_y = 30$ m is adopted to depict the spatial autocorrelation of c and φ. The random field mesh size is 0.063 times the horizontal scale of fluctuation for the structure surface shear strength parameters. It indicates that the selected element size can meet the accuracy demands for the discretization of random fields according to Ching and Phoon (2013).

According to Zhu et al. (2017), the coefficients of variations (COVs) of c_1 and φ_1 for the structural surface I are taken as 0.3 and 0.19, respectively, while the COVs of c_2 and φ_2 for the structural surface II are taken as 0.28 and 0.14, respectively. To explore the influences of the COVs of structural surface shear strength parameters on the probability of slope failure, eight groups of COVs in Table 32.2 are considered. Note that the group 4 is the base case chosen in this example. According to the method outlined in Sect. 2.1, a total of 40 independent standard normal random variables

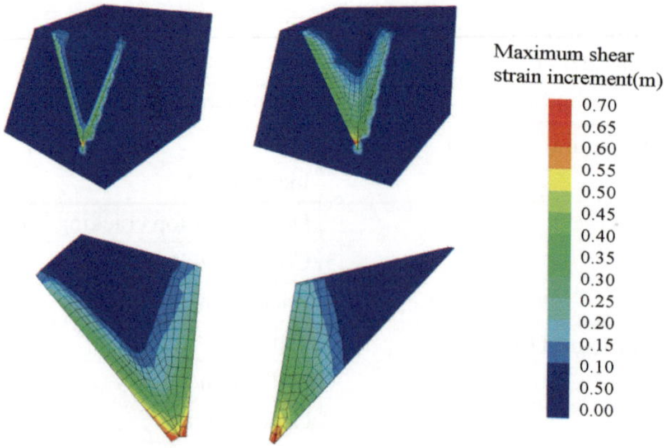

Fig. 32.4 Deterministic slope stability analysis results ($FS_{\min} = 1.296$)

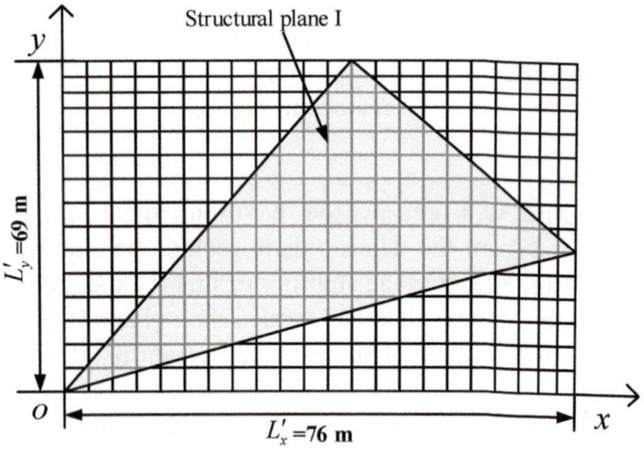

Fig. 32.5 Random field mesh for structural surface I

are needed to discretize the cross-correlated lognormal random fields of c_1 and φ_1 and c_2 and φ_2 because a value of $n = 10$ is required for modeling each spatially variable. The realizations of the random fields of c_1 and φ_1 and c_2 and φ_2 are then assigned to the corresponding finite difference elements. Figure 32.6 presents one typical realization of the random fields of c_1 and φ_1 and c_2 and φ_2 underlying the structural surfaces I and II.

Then, the back-propagation neural network is adopted to construct the surrogate model of factor of safety of the slope, a total of 1000 sets of Latin hypercube samples are generated, 1000 realizations of the random fields of c_1 and φ_1 and c_2 and φ_2 are obtained and assigned to the corresponding finite difference elements in sequence. Then, 1000 evaluations of numerical model using the finite-difference strength reduction method are carried out to evaluate the factors of safety. Herein, the numbers of input layers and output layers for the back-propagation neural network are equal to 40 and 1, respectively. After a lot of trial-and-error calculations, the back-propagation neural network can be best trained when the number of hidden layers equals 18, the double S-type activation function is utilized, the linear activation function is used for

Table 32.2 Comparison of the probabilities of slope failure for different groups of coefficients of variation

No	COVs	The proposed method		Directional simulation method (Zhu et al. 2017)
		Considering the spatial variability	Only considering the variability	Only considering the variability
1	$COV_{c1} = 0.15$, $COV_{\varphi 1} = 0.095$ $COV_{c2} = 0.10$, $COV_{\varphi 2} = 0.095$	1.667×10^{-8}	2.60×10^{-6}	8.733×10^{-6}
2	$COV_{c1} = 0.20$, $COV_{\varphi 1} = 0.129$ $COV_{c2} = 0.16$, $COV_{\varphi 2} = 0.112$	1.00×10^{-6}	0.916×10^{-3}	1.262×10^{-3}
3	$COV_{c1} = 0.25$, $COV_{\varphi 1} = 0.162$ $COV_{c2} = 0.22$, $COV_{\varphi 2} = 0.129$	1.07×10^{-4}	0.77×10^{-2}	1.034×10^{-2}
4	$COV_{c1} = 0.30$, $COV_{\varphi 1} = 0.190$ $COV_{c2} = 0.28$, $COV_{\varphi 2} = 0.140$	2.50×10^{-3}	2.94×10^{-2}	3.114×10^{-2}
5	$COV_{c1} = 0.35$, $COV_{\varphi 1} = 0.230$ $COV_{c2} = 0.34$, $COV_{\varphi 2} = 0.160$	8.90×10^{-3}	5.78×10^{-2}	6.022×10^{-2}
6	$COV_{c1} = 0.40$, $COV_{\varphi 1} = 0.260$ $COV_{c2} = 0.40$, $COV_{\varphi 2} = 0.180$	2.31×10^{-2}	9.43×10^{-2}	9.193×10^{-2}
7	$COV_{c1} = 0.45$, $COV_{\varphi 1} = 0.295$ $COV_{c2} = 0.46$, $COV_{\varphi 2} = 0.195$	0.048	0.133	0.122
8	$COV_{c1} = 0.50$, $COV_{\varphi 1} = 0.330$ $COV_{c2} = 0.52$, $COV_{\varphi 2} = 0.212$	0.083	0.173	0.146

the output layer and the Levenberg–Marquardt algorithm is employed as the training function.

To prove the surrogate model can effectively replace the deterministic analysis of slope stability, the cross-validation method is adopted to compare the factors of safety calculated using the surrogate model and those calculated using the FLAC3D software. The newly generated 1000 samples are divided into five groups (each group accounts for 20%). Each group is successively used as the validation set, and the rest samples are used as the training set. Figure 32.7a–e compare the factors of safety

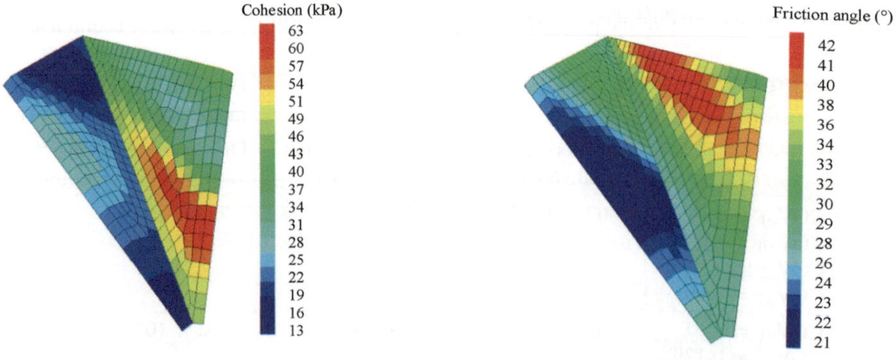

(a) Cohesion (b) Friction angle

Fig. 32.6 One typical realization of random fields of shear strength parameters for the structural surfaces I and II

calculated using the surrogate model versus those calculated using the FLAC3D software for the five validation sets. As observed from Fig. 32.7 that the factors of safety obtained from the two approaches are in good agreements and the determination coefficients R^2 are all larger than 0.99. It indicates the back-propagation neural network-based surrogate model can well replace the finite-difference strength reduction calculations in the FLAC3D software for slope stability analysis.

Finally, the probabilities of slope failure for these eight groups of COVs in Table 32.2 are calculated using the direct LHS with one million samples on the

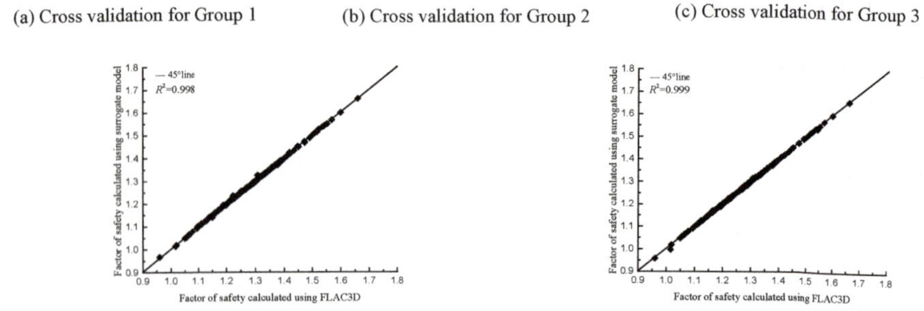

(a) Cross validation for Group 1 (b) Cross validation for Group 2 (c) Cross validation for Group 3

(d) Cross validation for Group 4 (e) Cross validation for Group 5

Fig. 32.7 Cross validation of the back-propagation neural network-based surrogate model back-propagation neural network-based surrogate model. Table 32.2 compares the probabilities of slope failure estimated from the proposed method and directional simulation method reported in Zhu et al. (2017). It can be observed that the probability of slope failure increases as the COVs of c_1, φ_1, c_2 and φ_2 increase. For the scenario that only considering the variability of c_1, φ_1, c_2 and φ_2, the probabilities of slope failure underlying these eight groups of COVs estimated from the proposed method are less than those estimated from the directional simulation method. This is because the latter considers the effect of water pressure on the structure surfaces. Besides, the proposed method can explicitly account for the influence of the spatial variability of structural surface mechanical parameters, whereas the directional simulation method only considers their variabilities. For example, the probability of failure for the group 4 considering the spatial variability calculated from the proposed method is 2.5×10^{-3}, which matches well with that (2.8×10^{-3}) calculated from the direct LHS with 5000 samples. This confirms that the effectiveness of the proposed method for tackling the complex rock slope reliability problems at low probability levels ($p_f < 10^{-4}$). In comparison to the scenario that only considers the variabilities of structural surface mechanical parameters, the probabilities of slope failure considering the spatial variability calculated from the proposed method are reduced significantly. It implies that ignoring the spatial variability of the structural surface mechanical parameters for the rock slope against a wedge failure can overestimate the probability of failure, resulting in conservative slope reinforcement design schemes.

32.4 Conclusions

In this paper, a spatial variability modelling method for the mechanical parameters of wedge rock mass structural surfaces is proposed. The back-propagation neural network is adopted to construct the surrogate model of the factor of safety of the slope, which greatly facilitates the probabilistic rock slope stability analysis. Then, the non-intrusive stochastic finite difference method is adopted to explore the influence of the spatial variability of structural surface mechanical parameters on the probability of slope failure, in which the finite-difference strength reduction method is used for deterministic slope stability analysis. Based on these, an interface program between the spatial variability modelling of the mechanical parameters and probabilistic rock slope stability analysis using the FLAC3D software is developed. Finally, the proposed method is applied to an example of rock slope against a wedge failure. It is found that the proposed method can explicitly account for the influence of the spatial variability and obtain the probability of slope failure with satisfactory accuracy as the directional simulation method. The proposed method provides an effective and robust tool for complex slope reliability assessment at low probability levels ($p_f < 10^{-4}$) considering the spatial variability of rock masses mechanical parameters. Additionally, ignoring the spatial variability of the structural surface mechanical parameters for the rock slope against a wedge failure can overestimate the probability of slope failure, resulting in conservative slope reinforcement design schemes.

Acknowledgements This work was supported by the National Natural Science Foundation of China (Grant Nos. 52222905, 41972280, 52179103 and 42272326) and Jiangxi Provincial Natural Science Foundation (Grant Nos. 20232ACB204031 and 20224ACB204019). The financial supports are gratefully acknowledged.

References

Chen D, Xu D, Ren G, Jiang Q, Liu G, Wan L, Li N (2019) Simulation of cross-correlated non-Gaussian random fields for layered rock mass mechanical parameters. Comput Geotech 112:104–119. https://doi.org/10.1016/j.compgeo.2019.04.012

Chen L, Zhang W, Gao X, Wang L, Li Z, Böhlke T, Perego U (2022) Design charts for reliability assessment of rock bedding slopes stability against Bi-planar sliding: SRLEM and BPNN approaches. Georisk 16(2):360–375. https://doi.org/10.1080/17499518.2020.1815215

Cheng HZ, Chen J, Wang ZS, Hu ZF, Huang YH (2017) Stability analysis of a clay slope accounting for the rotated anisotropy correlation structure. Chin J Rock Mech Eng 36(S2):3965–3973. https://doi.org/10.13722/j.cnki.jrme.2017.0044

Ching J, Phoon KK (2013) Effect of element sizes in random field finite element simulations of soil shear strength. Comput Struct 126(9):120–134. https://doi.org/10.1016/j.compstruc.2012.11.008

Der Kiureghian A, Liu PL (1986) Structural reliability under incomplete probability information. J Eng Mech 112(1):85–104. https://doi.org/10.1061/(ASCE)0733-9399(1986)112:1(85)

Eivazy H, Esmaieli K, Jean R (2017) Modelling geomechanical heterogeneity of rock masses using direct and indirect geostatistical conditional simulation methods. Rock Mech Rock Eng 50(12):3175–3195. https://doi.org/10.1007/s00603-017-1293-0

Gravanis E, Pantelidis L, Griffiths DV (2014) An analytical solution in probabilistic rock slope stability assessment based on random fields. Int J Rock Mech Min 71:19–24. https://doi.org/10.1016/j.ijrmms.2014.06.018

Hoek E, Londe P (1974) The design of rock slopes and foundations. In: General report for third congress of the international society for rock mechanics, vol 53, pp 1689–1699

Huang XC, Zhou XP, Ma W, Niu YW, Wang YT (2016) Two-dimensional stability assessment of rock slopes based on random field. Int J Geomech 17(7):04016155. https://doi.org/10.1061/(ASCE)GM.1943-5622.0000858

Hsu SC, Nelson PP (2006) Material spatial variability and slope stability for weak rock masses. J Geotech Geoenviron 132(2):183–193. https://doi.org/10.1061/(ASCE)1090-0241

Jiang SH, Li DQ, Zhang LM, Zhou CB (2014) Slope reliability analysis considering spatially variable shear strength parameters using a non-intrusive stochastic finite element method. Eng Geol 168:120–128. https://doi.org/10.1016/j.enggeo.2013.11.006

Jiang SH, Huang J, Griffiths DV, Deng ZP (2022) Advances in reliability and risk analyses of slopes in spatially variable soils: A state-of-the-art review. Comput Geotech 141:104498. https://doi.org/10.1016/j.compgeo.2021.104498

Jiang SH, Zhu GY, Wang ZZ, Huang ZT, Huang J (2023) Data augmentation for CNN-based probabilistic slope stability analysis in spatially variable soils. Comput Geotech 160:105501. https://doi.org/10.1016/j.compgeo.2023.105501

Li DQ, Jiang SH (2016) Non-intrusive stochastic analysis method for slope reliability. Science Press, Beijing

Mckay MD, Conover WJ, Beckman RJ (1979) A comparison of three methods for selecting values of input variables in the analysis of output from a computer code. Technometrics 21(2):239–245. https://doi.org/10.1080/00401706.2000.10485979

Renani HR, Martin CD, Varona P, Loring L (2019) Stability analysis of slopes with spatially variable strength properties. Rock Mech Rock Eng 52(10):1–18. https://doi.org/10.1007/s00603-019-01828-2
Srivastava A (2012) Spatial variability modelling of geotechnical parameters and stability of highly weathered rock slope. Indian Geotech J 42(3):179–185. https://doi.org/10.1007/s40098-012-0016-4
Tang H, Yong R, Eldin MAM (2017) Stability analysis of stratified rock slopes with spatially variable strength parameters: the case of Qianjiangping landslide. B Eng Geol Environ 76(3):839–853. https://doi.org/10.1007/s10064-016-0876-4
Zhang W, Meng F, Chen FY, Liu HL (2021) Effects of spatial variability of weak layer and seismic randomness on rock slope stability and reliability analysis. Soil Dyn Earthq Eng 146:106735. https://doi.org/10.1016/j.soildyn.2021.106735
Zhou JW, Jiao MY, Xing HG, Yang XG, Yang YC (2017) A reliability analysis method for rock slope controlled by weak structural surface. Geosci J 21(3):1–15. https://doi.org/10.1007/s12303-016-0058-1
Zhu Y, Zhou H, Feng XT, Zhang CQ, Zhang MQ, Yang FJ (2017) Directional simulation of failure probability of rock slope wedge. Rock Soil Mech 38(S1):157–163. https://doi.org/10.16285/j.rsm.2017.S1.017

Chapter 33
Study on Frictional Slip Behaviors of Rock Discontinuity Using an Impact-Induced Direct Shear Method

Wei Yuan and Jianchun Li

Abstract Understanding the frictional slip behavior of rock discontinuity is of great importance in earthquake engineering, geoengineering and disaster prevention engineering. Due to the inertia effect, the rapid stress change may cause some new slip characteristics of rock discontinuity subjected to dynamic loading. To address this issue, an experimental study on the frictional slip behaviors of rock discontinuity subjected to impact load is conducted. Shear tests on planar saw-cut granite discontinuities are carried out using an impact-induced direct shear method. A near semi-sinusoidal wave with a duration of about 330 μs is used as the dynamic shear load in the test. The slip displacement, velocity, acceleration and frictional resistance during the slip process are obtained. Then, the effects of impact energy on the aforementioned parameters are discussed. Test results show that the frictional slip of the rock discontinuity undergoes two main stages including accelerated slip and decelerated slip. As impact energy increases, the slip acceleration in the decelerated slip stage decreases, and the duration of the slip process increases markedly, even far beyond the duration of the dynamic shear load. It indicates that the inertia effect becomes increasingly important. Throughout the frictional slip process, the maximum slip displacement of the rock discontinuity is proportional to the impact energy. With increasing impact energy, the slip velocity increases, and the maximum frictional resistance of the discontinuity is negatively correlated with the impact energy.

Keywords Rock discontinuity · Frictional slip · Impact-induced direct shear · Impact energy · Inertia effect

---

W. Yuan · J. Li (✉)
School of Civil Engineering, Institute of Future Underground Space, Southeast University, Nanjing, China
e-mail: jcli@seu.edu.cn

W. Yuan
e-mail: yuanw_c@seu.edu.cn

S. Wang et al. (eds.), *Engineering Geology for a Habitable Earth: IAEG XIV Congress 2023 Proceedings, Chengdu, China*, Environmental Science and Engineering,
https://doi.org/10.1007/978-981-99-9057-3_33

33.1 Introduction

Rock discontinuities, e.g., faults, rock structural planes, rock joints, widely exist in natural rock masses. Frictional slip of the rock discontinuity is one of the fundamental mechanical processes during the development of geological and engineering disasters such as rock slope slide and induced earthquake (Brace and Byerlee 1966; Aydan 2016). Understanding the frictional slip behaviors of rock discontinuities may benefit to the prediction and prevention of disasters. Therefore, the frictional slip behavior of rock discontinuities has received increasing attention.

The frictional strength of rock discontinuities is one of the most concerned parameters in engineering practice. Among the methods to test the frictional strength of rock discontinuity, steady-state loading-based shear experiments are most widely used, including direct shear test (Qi et al. 2020), double-plane shear test (Kilgore et al. 2017), biaxial/triaxial compression-shear test (Scholz et al. 1972), and rotary-shear test (Ma et al. 2014). It has been shown that the frictional strength of rock discontinuities was affected by surface morphology, normal pressure, temperature, humidity, mineralogy, and slip velocity (Byerlee and Brace 1968; Barton 1976; Dieterich and Conrad 1984; Di Toro et al. 2004; Tang 2020). In contrast, studies on frictional slip of rock discontinuities subjected to transient loading have been less frequently reported. Compared with steady-state loading, transient loading has the characteristic of real-time variation. The frictional slip process of rock discontinuity under transient loading may be more complicated, especially the inertia effect becomes more significant. Wu and Zhao (2013) designed a dynamic-induced direct-shear model and experimentally explored the frictional slip of simulated granular gouges. The results showed that both slip velocity and maximum slip displacement were negatively linearly related to normal pressure. Yao et al. (2021) developed an impact shear testing system based on split Hopkinson pressure bar (SHPB) and tested the dynamic shear behavior of planar rock discontinuities. Li et al. (2022) proposed a new impact-induced direct shear method based on the biaxial Hopkinson pressure bar (BHPB), which can more accurately measure the dynamic shear and normal stresses in the rock discontinuity during the shear process.

This paper presents an experimental investigation on the frictional slip behavior of rock discontinuity subjected to impact loading. Shear tests on planar saw-cut granite discontinuities are conducted using a self-developed impact-induced direct shear method. In the following sections, the experimental method, and the test results of frictional resistance, slip displacement, slip velocity and slip acceleration are presented. The effects of impact energy on the them are discussed and the inertia effect during the frictional slip process is analyzed.

33.2 Experimental Setups

- Experimental apparatus and loading mechanism

In this study, an impact-induced direct shear method (Li et al. 2022) is adopted. A biaxial Hopkinson pressure bar (BHPB) equipment (see Fig. 33.1) is employed as the loading device. It consists of a dynamic loading unit and a hydraulic loading unit. The dynamic loading unit contains a gas gun, a striker, and 4 pressure bars (i.e., incident bar, transmitted bar, Y-output bars 1 and 2). The striker and bars are made of $60Si_2Mn$ alloy steel (Young's modulus $E = 200$ GPa, longitudinal wave velocity $C_p = 5100$ m/s). The length and diameter of the cylindrical striker are 0.5 m and 40 mm, respectively. All pressure bars have identical cross sections of 50×50 mm^2. The incident bar is 2.5 m long, and the rest three bars are 2.0 m long. The hydraulic loading unit consists of two hydraulic cylinders, two hydraulic pumps, a reaction frame, and the control system. The procedure of loading and unloading are automatically controlled. The initial normal and shear stresses are applied to the specimen using the hydraulic loading unit. For more details about BHPB, please refer to Li et al. (2022).

During the shear test, the specimen is placed among the four bars as shown in Fig. 33.1. Two steel adapters are separately sandwiched between incident/transmitted bar and specimen. The cross section of the adapter is equal to a half of the bar cross section. And the adapters are 5 mm thick. In this way, a direct shear loading configuration is established.

- Data acquisition and processing

The data acquisition unit of the experimental system is displayed in Fig. 33.2, and it contains a stress wave acquisition module and a dynamic displacement measure module. During the test, the dynamic shear and normal stresses were obtained using

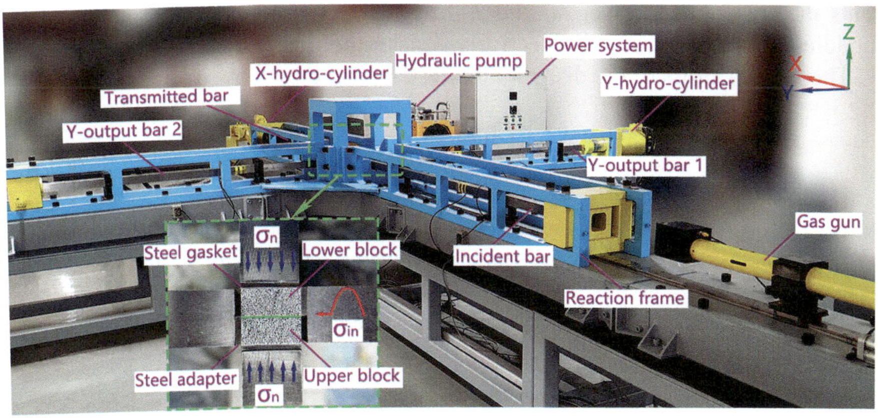

Fig. 33.1 Photos of experimental apparatus (Modified from Li et al. 2022)

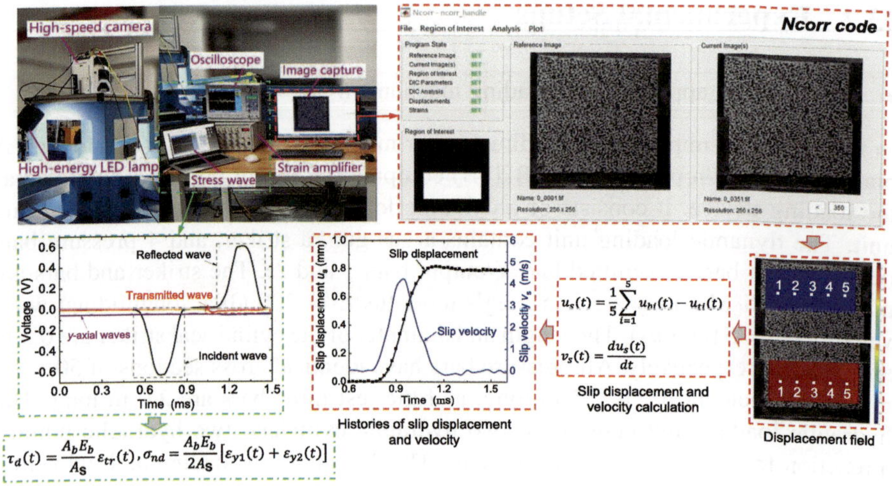

Fig. 33.2 Diagram showing principles of data acquisition and processing

stress wave acquisition module. The stress waves were recorded by the strain gauges pasted on the surface of the four bars. The signals were amplified by the ultra-dynamic strain amplifier, displayed and stored by the oscilloscope. Typical stress wave signals, i.e., incident wave, reflected wave, transmitted wave, and two *y*-axial transmitted waves, obtained in the test are presented in Fig. 33.2. The frictional resistance is calculated from the transmitted wave propagating in the transmitted bar in x-axial direction, that is,

$$\tau_{\mathrm{d}} = A_{\mathrm{b}} E_{\mathrm{b}} \varepsilon_{\mathrm{tr}}(t) \big/ A_{\mathrm{s}} \tag{33.1}$$

where $\tau_{\mathrm{d}}(\mathrm{t})$ represents the frictional resistance of the rock discontinuity; $\varepsilon_{\mathrm{tr}}(\mathrm{t})$ represents the transmitted wave. A_{b} and E_{b} are the cross-sectional area and Young's modulus of the pressure bar, respectively. A_{s} is the nominal area of the shear segment.

The normal stress of the discontinuity is calculated by,

$$\sigma_{\mathrm{nd}}(t) = \frac{A_{\mathrm{b}} E_{\mathrm{b}}}{2A_{\mathrm{s}}}\left[\varepsilon_{\mathrm{y1}}(t) + \varepsilon_{\mathrm{y2}}(t)\right] \tag{33.2}$$

where $\sigma_{\mathrm{nd}}(t)$ represents nominal the dynamic normal stress of the discontinuity; $\varepsilon_{\mathrm{y1}}(t)$ and $\varepsilon_{\mathrm{y2}}(t)$ represent *y*-axial transmitted waves acquired in *Y*-output bars 1 and 2, respectively.

The impact energy is obtained from the incident wave based on Eq. (33.3) (Liu et al. 2020),

$$W_{\mathrm{in}} = A_{\mathrm{b}} C_{\mathrm{p}} E_{\mathrm{b}}(t) \int_{0}^{T} \varepsilon_{\mathrm{in}}^{2}(t)\mathrm{d}t \tag{33.3}$$

where W_{in} denotes the impact energy; ε_{in} represents the incident wave; T is the duration of the incident wave.

The high speed two-dimensional digital image correlation (2D-DIC) technique was adopted to measure dynamic displacements of the rock discontinuity (see Fig. 33.2). The measured surface of the specimen was painted with randomly distributed speckles. A high-speed camera was used to capture the speckle images throughout the test. After the test, an open-source code *Ncorr* was employed to calculate the displacement field of the measured surface. Finally, the slip displacement u_s was obtained from the displacement difference between the two rock blocks. Furthermore, the slip velocity v_s was obtained by the first-order differentiation of the slip displacement. In this test, the image resolution and sampling rate were set to 256 × 256 pixels and 200,000 frames per second, respectively.

- Specimen preparation

The planar rock discontinuity specimen was prepared from a fine-grained granite. Each specimen has two halves (i.e., lower block and upper block) which were saw cut from a cubic rock block. The two halves are identical in size as shown in Table 33.1, and the overall size of the specimen is 50 mm × 50 mm × 50 mm. At a shear rate condition of 0.025 mm/s, the measured friction coefficient and friction angle of the granite are 0.59 and 30.54°, respectively.

- Test conditions

In the test, the amplitude of the impact load was determined by the impact velocity of the striker. Several impact velocities were set to obtain different impact energies. The corresponding impact energies are listed in Table 33.1. The length of the striker is fixed at 0.5 m to produce a constant duration of the incident wave. To obtain an incident wave with a gently rising edge, a rubber disk was used as a pulse shaper. Accordingly, the duration of the incident wave is about 330 μs. The initial normal stress was set at 4 and 12 MPa. To ensure reproducibility, parallel experiments were conducted under the normal stress condition of 12 MPa.

Table 33.1 Details of test condition

Specimen	Normal stress σ_n(MPa)	Impact velocity (m/s)	Impact energy W_{in}(J)
 Size: 50 × 50 × 50 mm^3	4	5.9	37.4
		6.8	62.5
		9.9	162.0
	12	~ 5.9	~ 38
		~ 7.1	~ 72
		~ 8.2	~ 101
		~ 9.6	~ 152
		~ 10.8	~ 195

33.3 Experimental Results and Discussion

- Evolution of frictional resistance

Figure 33.3 illustrates the evolution process of the frictional resistance of rock discontinuities under different impact energy conditions. It can be seen that the evolution of frictional resistance includes two main stages, i.e., stress accumulation and frictional slip. In the stress accumulation stage, the frictional resistance grows continuously. After proceeding to the frictional slip stage, the frictional resistance first undergoes a stress drop and then fluctuates. With the increase in impact energy, the maximum frictional resistance appears a decreasing trend, and the stress drop after the slip initiation is also reduced. The duration of the stress accumulation stage seems to be prolonged, indicating that the growth of frictional resistance is suppressed under conditions of higher impact energy.

- Evolution of slip displacement

Figure 33.4 presents the typical histories of slip displacement and slip velocity under different impact energy conditions. In terms of slip displacement, before the maximum is reached, two stages of accelerated slip and decelerated slip are observed. After the maximum, the slip displacement first decreases slightly and then keeps

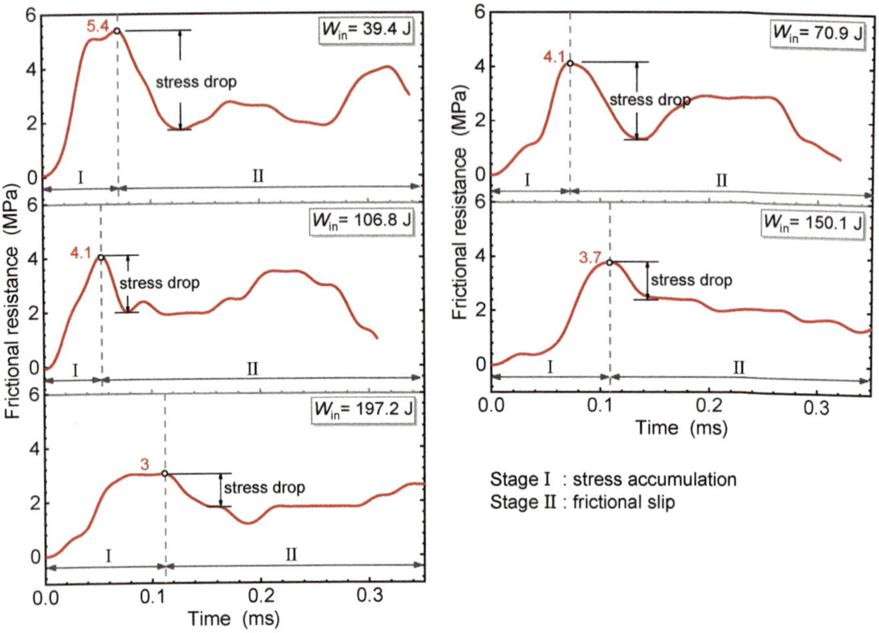

Fig. 33.3 Evolution process of frictional resistance under different impact energy conditions (σ_n = 12 MPa)

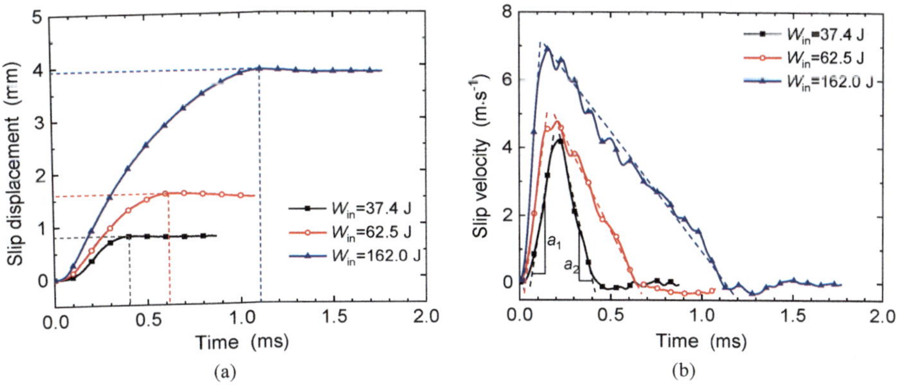

Fig. 33.4 Typical histories of **a** slip displacement and **b** slip velocity under different input energy conditions ($\sigma_{\mathrm{n}} = 4$ MPa)

constant. As the impact energy increases, the maximum slip displacement increases significantly. Additionally, it is noteworthy that the duration of the whole slip process also increases remarkably, even much longer than that of the incident wave. And this phenomenon is mainly related to the increase in duration of the decelerated stage. As shown in Fig. 33.4b, when the impact energy is 37.4 J, the duration of growth segment is almost equal to that of decrease segment; as the impact energy increases, the duration of decrease segment is significantly lengthened. Additionally, according to Fig. 33.4b, the slip velocities are in the order of 0–10^1 m/s, which are much higher than those in most direct shear configurations. During the accelerated and decelerated slip stages, the slip velocity nearly varies linearly.

Figure 33.5 displays the variation of maximum slip displacement with impact energy. As can be seen in Fig. 33.5, within the energy range considered in this study, the maximum slip displacement is nearly proportional to the impact energy. As the normal stress increases, the increase rate of maximum slip displacement versus impact energy decreases, which reflects the constraint effect of normal stress on the frictional slip.

33.4 Discussion

Frictional slip of the rock block along the discontinuity under impact load can be represented by a slider motion model shown in Fig. 33.6. The equation of motion can be expressed as,

$$F(t) = \tau_{\mathrm{d}}(t)A_{\mathrm{s}} + m\ddot{u}_{\mathrm{s}}(t) \tag{33.4}$$

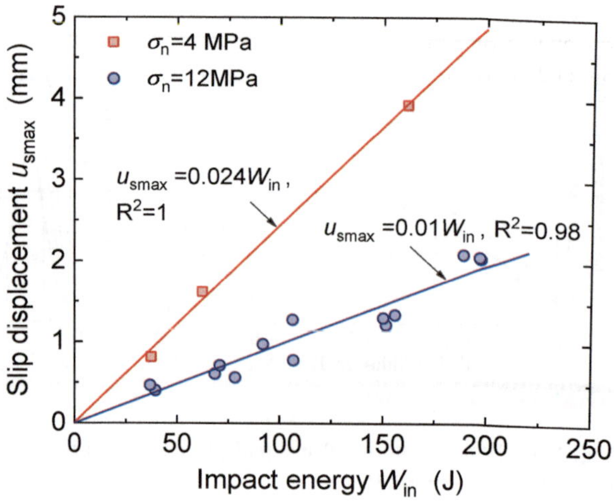

Fig. 33.5 Maximum slip displacement as a function of impact energy

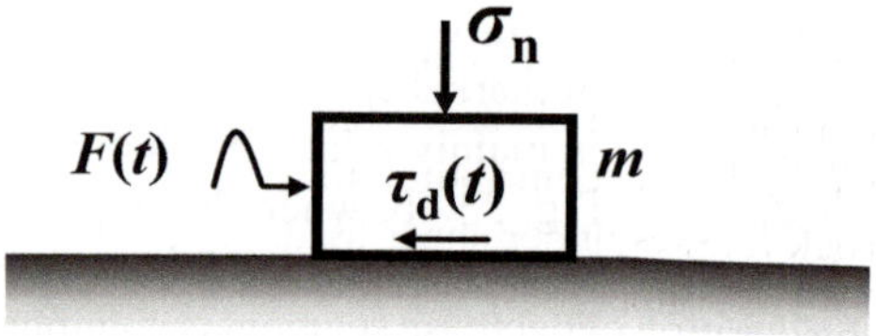

Fig. 33.6 Conceptional model of single degree of freedom slider

where m and $\ddot{u}_s$ are the mass and acceleration of the slider, respectively; $F(t)$ represents the impact load.

Slip initiates when the impact load increases from zero to a value greater than the kinetic frictional resistance. Due to the nonlinear change of the impact load, the slip velocity is not constant and the inertia force cannot be ignored. Once the impact load decreases to a value less than the frictional resistance, it has no effects on the slip; moreover, the frictional slip decelerates due to the interface frictional resistance. In this case, the motion can be described by,

$$\tau_d(t)A_s + m\ddot{u}_s(t) = 0 \tag{33.5}$$

Unlike the displacement-controlled direct shear test, the inertia effect of the slider under impact load is more significant during the frictional slip. Additionally, the magnitude of the inertia force may reflect the residual frictional resistance during the decelerated slip stage.

According to Fig. 33.4b, the slip velocity varies almost linearly during the accelerated and decelerated slip stages. Therefore, two slip accelerations a_1 and a_2 are obtained by averaging the changing rate of slip velocity during accelerated and decelerated slip stages, respectively. Figure 33.7 presents the variation of a_1 and a_2 with

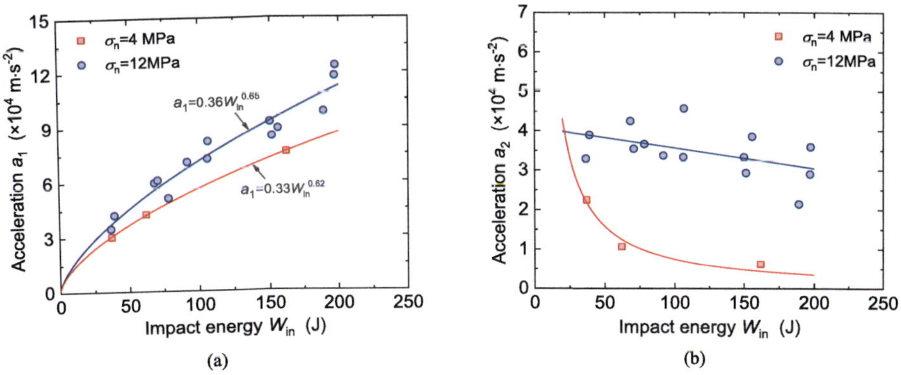

Fig. 33.7 Slip acceleration **a** a_1 and **b** a_2 as a function of impact energy

impact energy. It can be seen that during the accelerated slip stage, the slip acceleration increases nonlinearly with the increasing impact energy. And a power function can well fit the relationship between acceleration a_1 and impact energy. The greater the normal stress, the more significant the increase of acceleration a_1 with impact energy. On the contrary, the slip acceleration a_2 decreases with increasing impact energy. This may be due to the fact that the larger the impact energy, the greater the slip velocity during the accelerated slip stage, the more significant the reduction of frictional resistance of the discontinuity. Therefore, the acceleration of the frictional slip is much lower after entering the decelerated sliding phase. Similarly, the greater the normal stress, the lower the slip velocity during accelerated sliding, and the less significant the reduction of frictional resistance. Consequently, the acceleration a_2 is larger, and its change is less significant.

33.5 Conclusions

To investigate the frictional slip behaviors of rock discontinuity subjected to transient loads, shear tests on planar saw-cut granite discontinuities are conducted using a self-developed impact-induced direct shear method. The effects of impact energy on the frictional slip behavior are analyzed, and the inertia effect on the slip is also discussed.

Test results demonstrates that the frictional slip of the rock discontinuity subjected to the impact load undergoes two main stages, i.e., accelerated slip and decelerated slip. The duration of the accelerated slip stage tends to decrease with increasing impact energy, while that of the decelerated slip stage increases markedly. A positively linear relationship between the maximum slip displacement and the impact energy is obtained. And the lower the normal stress, the greater the increase rate of maximum slip displacement with impact energy. Under the action of impact load, the inertia force is significant, especially during the decelerated slip stage. In this process, the slip acceleration decreases with increasing impact energy; and the lower the normal stress, the greater the decrease of acceleration. This is due to the reduction of kinetic frictional resistance of the rock discontinuity under high-speed slip.

Acknowledgements This research was supported by the National Natural Science Foundation of China (Grant No. 41831281, 42220104007), and the Innovative and Entrepreneurial Team Program of Jiangsu Province, China (JSSCTD202140)

References

Aydan Ö (2016) Large rock slope failures induced by recent earthquakes. Rock Mech Rock Eng 49:2503–2524

Barton N (1976) The shear strength of rock and rock joints. Int J Rock Mech Min Sci Geomech Abstr 13:255–279

Brace WF, Byerlee JD (1966) Stick-slip as a mechanism for earthquakes. Science 153:990–992

Byerlee JD, Brace WF (1968) Stick slip, stable sliding, and earthquakes—Effect of rock type, pressure, strain rate, and stiffness. J Geophys Res 73:6031–6037

Di Toro G, Goldsby DL, Tullis TE (2004) Friction falls towards zero in quartz rock as slip velocity approaches seismic rates. Nature 427(6973):436–439

Dieterich JH, Conrad G (1984) Effect of humidity on time-dependent and velocity-dependent friction in rocks. J Geophys Res 89:4196–4202

Kilgore B, Beeler NM, Lozos J, Oglesby D (2017) Rock friction under variable normal stress. J Geophy Res-Sol Ea 122:7042–7075

Li JC, Yuan W, Li HB, Zou CJ (2022) Study on dynamic shear deformation behaviors and test methodology of sawtooth-shaped rock joints under impact load. Int J Rock Mech Min Sci 158:105210

Liu K, Zhao J, Wu G, Maksimenko A, Haque A, Zhang QB (2020) Dynamic strength and failure modes of sandstone under biaxial compression. Int J Rock Mech Min Sci 128:104260

Ma SL, Shimamoto T, Yao L, Togo T, Kitajima H (2014) A rotary-shear low to high-velocity friction apparatus in Beijing to study rock friction at plate to seismic slip rates. Earthq Sci 27(5):469–497

Qi SW, Zheng BW, Wu FQ, Huang XL, Guo SF, Zhan ZF, Zou Y, Barla G (2020) A new dynamic direct shear testing device on rock joints. Rock Mech Rock Eng 53(10):4787–4798

Scholz C, Molnar P, Johnson T (1972) Detailed studies of frictional sliding of granite and implications for the earthquake mechanism. J Geophys Res 77:6392–6406

Tang ZC (2020) Experimental investigation on temperature-dependent shear behaviors of granite discontinuity. Rock Mech Rock Eng 53:4043–4060

Wu W, Zhao J (2013) A dynamic-induced direct-shear model for dynamic triggering of frictional slip on simulated granular gouges. Exp Mech 54:605–613

Yao W, Wang CL, Xia KW, Zhang X (2021) An experimental system to evaluate impact shear failure of rock discontinuities. Rev Sci Instrum 92:034501

Chapter 34
Numerical Study on the Liquefaction Resistance of Rectangular Closed Diaphragm Walls in Gently Sloping Liquefiable Deposits

Jianlei Zhang, Gengyuan Shi, Qiangong Cheng, and Yan Li

Abstract Rectangular closed diaphragm walls (RCDWs) serve as a new type liquefaction mitigating foundations recently, however, their mechanisms of liquefaction mitigating have been studied scantly. This paper proposed a 2D numerical method to simulate a dynamic centrifuge test with OpenSees, on the basis of Biot's coupled theory and u-p formulation considering the dynamic interactions between fluid and soil particles in the saturated sands. The validity of the method was proved by comparing the results of the dynamic centrifuge test. Furthermore, four different earthquake events with three PGAs were applied to this numerical model to study the capacity to liquefaction mitigating of RCDWs. The results show that RCDWs can mitigate soil liquefaction effectively under all the earthquake events, and the characteristics of excess pore water pressure (EPWP) developed in far field and soil core are different: 1) the development of EPWP in soil core were restrained by RCDWs under small shaking events; 2) liquefaction was restrained in soil core though large EPWP generated under moderate shaking events; and 3) the onset of liquefaction was delayed in soil core under extreme large shaking events.

Keywords RCDWs · Liquefaction-mitigation · Fluid-soil coupling · Dynamic centrifuge test · Numerical simulation

J. Zhang (✉) · G. Shi
China University of Petroleum-Beijing at Karamay, Karamay, China
e-mail: zhangjianlei@cupk.edu.cn

G. Shi
e-mail: 2022216845@st.cupk.edu.cn

Q. Cheng
Southwest Jiaotong University, Chengdu, China
e-mail: chengqiangong@home.swjtu.edu.cn

Y. Li
Sichuan Southwest Jiaoda Civil Engineering Design Co., Ltd., Chengdu, China
e-mail: yanli@my.swjtu.edu.cn

S. Wang et al. (eds.), *Engineering Geology for a Habitable Earth: IAEG XIV Congress 2023 Proceedings, Chengdu, China*, Environmental Science and Engineering,
https://doi.org/10.1007/978-981-99-9057-3_34

34.1 Introduction

Earthquake-induced soil liquefaction and lateral spreading caused the damage of foundation, leading to destruction of bridges in the severe earthquake (Hamada 1992; Cheng et al. 2012). For instance, the liquefaction-induced lateral spreading damaged the pile foundations, causing the serious damage of the Showa bridge in 1964 Niigata earthquake (Hamada 1992), and eight spans of Tubul bridge slid out of wall piers due to the settlement and tilt of the foundations induced by soil liquefaction in 2010 Chile earthquake (Kawashima et al. 2011). Remediation methods are usually required to mitigate the soil shear deformation and resist the lateral force. The diaphragm walls constructed by the deep soil mixing (DSM) can reduce the shear strains of the enclosed soil which can be utilized to mitigate soil liquefaction. Increasing the cross section and stiffness of foundations can enhance the resistance against the lateral force. Few methods can have the both two advantages above. However, RCDWs served as bridge foundation can restrain the liquefaction and have huge stiffness too. RCDWs is a hollow rectangular frame, constructed by 4 pieces of reinforced concrete diaphragm walls, and the walls were connected by rigid joints, as shown in Fig. 34.1. RCDWs foundations have been served as bridge foundation, which are located in deep-water area and the offshore area since 1979, due to its good performance in the liquefiable ground, for example the foundations of Aomori bridge in Japan (Wu et al. 2014).

The performance of the RCDWs as a foundation has been studied by physical models (Wu et al. 2014). The test by Wu et al. (2014) demonstrated that the bearing capacity will be improved and the settlement will be reduced by using RCDWs instead of group piles in soft soils. And a series of dynamic centrifuge tests were performed to investigate the dynamic characteristics of DSM grids in the liquefiable site. Numerical analyses of RCDWs have included the efficiency of mitigating liquefaction of the enclosed soil and the response as foundation during earthquake (Zhang et al. 2021a, b). Namikawa et al. (2007) study the performance of the lattice-shaped walls by the 3D effective stress analyses and the results show the treatment effectiveness depended mainly on the treatment area ratio and the elastic modulus of enclosed soil. The 3D linear elastic dynamic FE analyses by Nguyen et al. (2013) investigated the effect of DSM grids the dynamic shear stress and strain distributions and proposed a method to design the DSM walls. Compared with the pile group, Wu et al. (2015) found that lattice-shaped diaphragm wall is better choice for being used as an earthquake-proof foundation in soft soil by the finite- difference analyses.

Up to now, there have been few centrifuge tests and numerical studies accounted for the seismic performance of RCDWs served as a bridge foundation in the gently sloping liquefiable ground. As a result, this paper presented a dynamic effective stress model to study the seismic performance of RCDWs in liquefiable sites subjected to earthquake ground motions. Compared measured results from the centrifuge test with numerical dates, we verify the accuracy of our numerical code. Subsequently, a parametric study was conducted to understand the effect of wall stiffness, peak

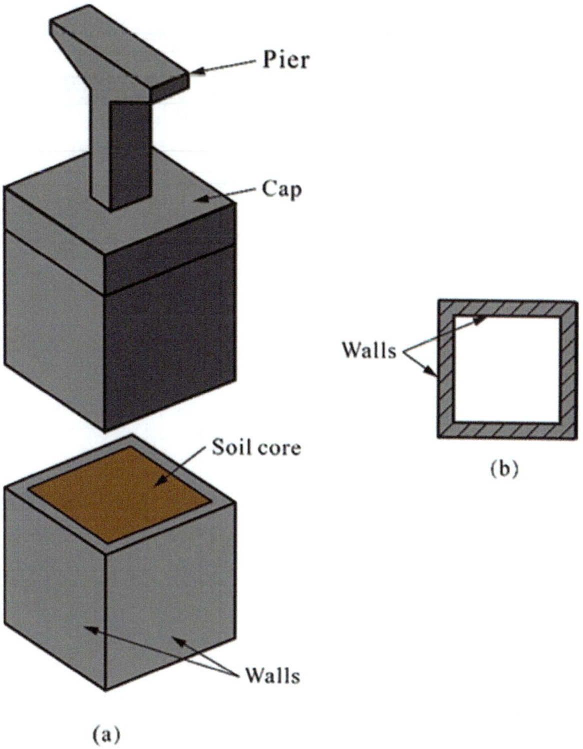

Fig. 34.1 Structure diagram of RCDWs: **a** 3D stereogram; **b** frame cross section

amplitude of acceleration and the sloping angle of the loose sand layer. Finally, several conclusions are drawn from the parametric study.

34.2 Description of Centrifuge Test

The model of RCDWs was constructed in a 1D laminar shear box with internal dimensions of 600 mm × 350 mm × 449 mm (L × W × H) on the ZJU400 4.5-m radius centrifuge device at Zhejiang University. The laminar container was constructed by stacking 20 lightweight aluminium alloy rectangular rings sandwiched with 1-mm roller bearings providing lateral flexibility to allow the frames deform as a shear beam in the shaking direction which is able to absorb the shear wave propagation to decrease the boundary effect.

Hereafter, standard scaling laws are followed and prototype units are presented unless otherwise specified. The soil configuration was a gently sloping deposits consisted of the liquefiable Fujian standard sand with a 13-m- thick medium-dense liquefiable layer (relative density (D_r) ≈ 60%) underlain by a 7-m-thick level dense sand non- liquefiable layer (D_r ≈ 90%) and the water table was as high as the top point

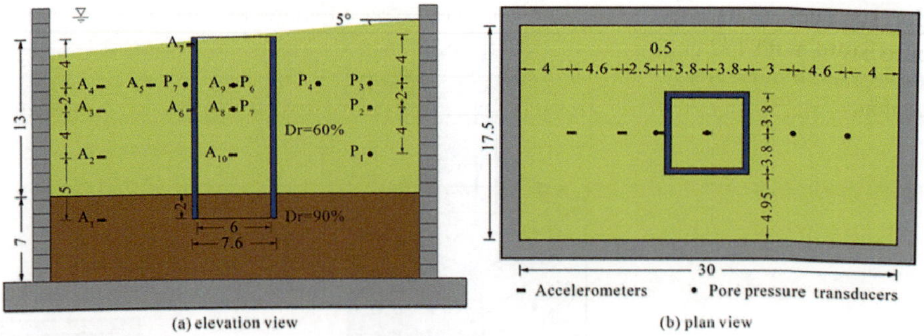

Fig. 34.2 Schematic diagram of the dynamic centrifuge test (unit: m)

Fig. 34.3 Taft wave used in the centrifuge test

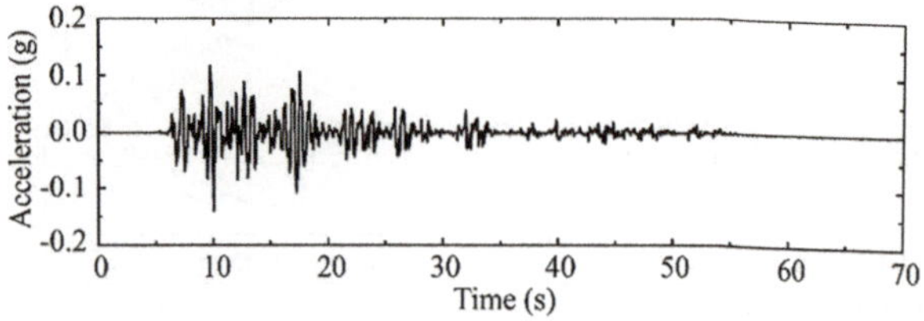

of the soil deposits, as shown in Fig. 34.2 (Li et al. 2019; Zhang et al. 2021a, b). The RCDWs was 15 m high with the bottom 2 m embedded into the underlying dense sand layer, as demonstrated in Fig. 34.2. The frame cross section of the RCDWs was a rectangle with a smaller hollow rectangle, the wall thickness was a 0.8-m. Therefore, the length of sides of hollow rectangle was 6 m (i.e., external dimensions of 7.6 m $\times$ 7.6 m). The more details of the soil profile and the RCDWs can be seen in Table 34.2. The earthquake wave inputted in table base of the dynamic centrifuge test was subjected to the Taft waves with a peak base accelerations (PBAs), a max, base $\approx$ 0.13 g obtained at the No. 1095 station during the 1952 California Earthquake. The recorded time history of acceleration from the base of the centrifuge is presented in Fig. 34.3.

34.3 Numerical Model

The Open System for Earthquake Engineering Simulation (OpenSees) was used to built a 2D FE numerical model in prototype scale. OpenSees is state-of-the-art computational platform developed by UC Berkeley for earthquake engineering simulation. The mech of the model is shown in the Fig. 34.4. The maximum permissible height of soil element was determined by shortest wavelength which can be estimated based on the soil's shear wave velocity (150 m/s) and the maximum frequency (15 Hz of the ground motion). And the height should be smaller than quarter of shortest

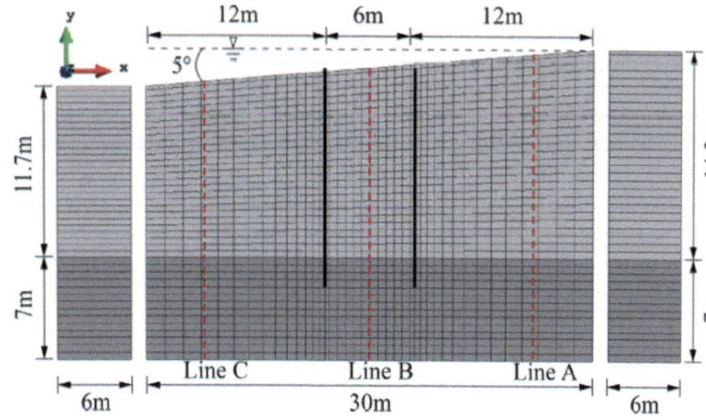

Fig. 34.4 Numerical model of the dynamic centrifuge test

wavelength at different depths to properly absorb the shear wave through the soil mesh in the dynamic analysis. In order to better capture the interaction between soil and RCDWs, soil mech near the interface between soil and RCDWs were divided in smaller size (Zhang et al. 2021a, b).

34.3.1 Soil Models

The saturated Fujian soil was simulated by using a two-phase material in SSPquadUP elements in the numerical simulation (Zhang et al. 2021a, b). The SSPquadUP element is the effective-stress solid–fluid fully coupled quadrilateral elements. These elements are developed based on the porous media theory (u-p formulation), in which u represents the displacement of the soil node and the p was the pore pressure. SSPquadUP elements are not subjected to shear locking and volumetric which lead to bigger coarse mesh accuracy in bending dominated problems and improve accuracy in nearly-incompressible problem. The out-of-plane thickness of the soil elements was 7 m, equal to the width of the centrifuge container in order to capture any pinning effects resulted from the RCDWs in the liquefaction ground. The dynamic behavior of the soil was captured by using the multi-yield-surface plasticity constitutive material developed by Elgamal et al. The material was named PressureDependMultiYield02 (PDMY02) (Elgamal et al. 2008) which was developed in OpenSees to capture the response characteristics of liquefiable sand in dynamic analysis. The parameters of middle dense ($D_r = 60\%$) and dense Fujian sand ($D_r = 90\%$) were different due to the different relative density, and the parameters of PDMY02 material were presented in Table 34.1. Some input parameters were selected based on the recommended values from the constitutive model developer and others were measured from the centrifuge test.

Table 34.1 PDMY02 material parameters for Fujian standard sand (Zhang et al. 2021a, b)

Parameters	$D_r = 60\%$	$D_r = 90\%$
Mass density (kg/m^3)	19.39	19.98
Reference bulk modulus (Pa)	1.62×10^8	2.42×10^8
Reference shear modulus (Pa)	6.51×10^7	1.12×10^8
Friction angle (°)	31.2	33.4
Phase transfer angle (°)	31.2	26.2
Peak shear strain	0.1	0.1
Reference confining pressure (Pa)	8×10^4	8×10^4
The first contraction parameter	0.086	0.281
The second contraction parameter	0.178	0.048
The first dilation parameter	0.001	0.111
The second dilation parameter	0.01	0.051
Void ratio	0.851	0.651

Table 34.2 Arias intensity of earthquake motions (m/s)

PGA	Kobe	Northridge	El-Centro	Hach	Taft
0.07 g	0.069	0.036	0.086	0.085	0.100
0.20 g	0.564	0.298	0.702	0.695	0.813
0.33 g	1.536	0.811	1.912	1.891	2.214

34.4 RCDWs Models and Soil-Pile Interface

In the simulation, the RCDWs was modeled by using beam-column elements as illustrated in the Fig. 34.5. The frame walls were simulated by the columns elements in longitudinal axis with elastic material. And the quasi-rigid beams normal to the walls are applied to account for the actual cross-section of the walls. The stiffness of the quasi-rigid beams are set to very huge, about 1000 times of the wall stiffness, to capture the behavior of the cross-section and reduce the effect of the quasi-rigid elements to the walls (Su et al. 2018).

In the numerical model, the interface between RCDWs and soil elements were simulated using zero-length elements with the nonlinear p-y and t-z materials represented by the PyLiq1 and TzLiq1 materials which accounted for lateral resistance and shaft friction respectively (Fig. 34.5). The PyLiq1 and TzLiq1 materials incorporate liquefaction effects by updating the average mean soil effective stress. The average mean soil effective stress can obtain from a time series of soil effective stress, or from two adjacent soil elements thereby approximately accounting for the effect of the soil liquefaction on the interface. The capacities and backbones of p-y and t-z materials were determined based on American Petroleum Institute formulas and the relation proposed by Mosher respectively as described by Boulanger et al. (1999).

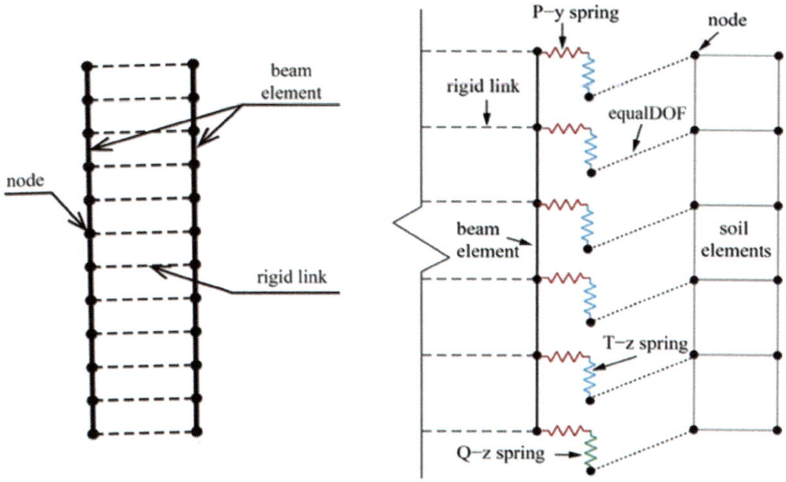

(a) RCDW (b) soil-RCDW interface

Fig. 34.5 Numerical model of the RCDWs and soil-RCDWs interface

The interaction between RCDWs and soil at the bottom was simulated by using zero-length elements with the QzSimple1 material. The ultimate bearing capacity of the material is obtained based on the method proposed by Meyerhof (1976). The bearing capacity factors and the backbone curve of QzSimple1 material was obtained by the method proposed by Vijayvergiya (1977).

34.4.1 Boundary Conditions

The nodes on the bottom of the model had no relative movement in both vertical and horizontal directions by fixing displacements freedom degree in all directions of nodes. Therefore, the numerical boundary was able to simulate a fully reflecting boundary condition of the semi-rigid base of the laminar shear box. In the centrifuge test, the shear box was constructed flexible to deform as one dimensional shear beams condition which can absorb shear wave. In the numerical model, two free-field soil columns were constructed on the left and right sides of the soil mesh to design the lateral boundaries in the direction of shear wave propagation (Fig. 34.6) (Zhang et al. 2021a, b). To ensure that the free-field soil columns behave as flexible shear beam, the movement mode of nodes on either side of these columns are tied together to create shear beam boundary conditions, allowing solely pure shear movement modes (Su et al. 2018). Furthermore, soil columns are given considerably thickness to ensure that soil columns have a significantly mass to simulated infinite boundary conditions for the elements in the interior in the shaking direction. Free-field columns

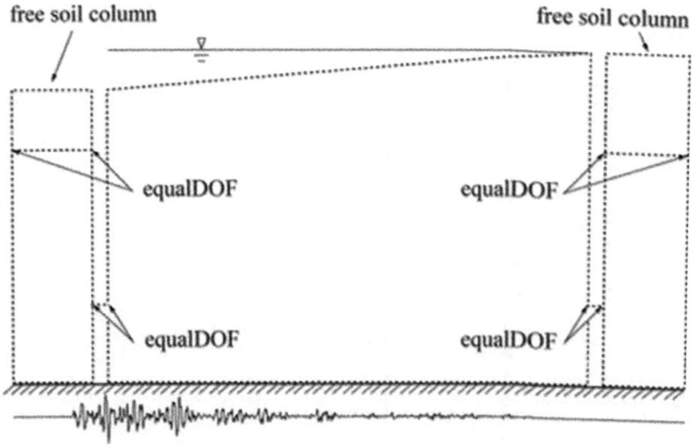

Fig. 34.6 Numerical model boundary

can absorb the shear wave and reduce the impact of free-field columns on the interior soil seismic response.

The free water above ground surface was modelled by applying the distributed hydrostatic pressure and nodal force to the soil nodes at surface (Su et al. 2018; Zhang et al. 2021a, b). Firstly, the hydrostatic pressure generated by the free water was applied to the relevant nodes at surface in the gravity direction to incorporate the dynamic effects of the hydrostatic pressure. Secondly, the specify nodal loads must be applied normal to the top surface to match the effective stresses on the ground surface. The loads caused by the water gravity were divided into vertical and horizontal direction on the soil nodes, while all other soil boundaries were impervious.

34.5 Analysis Results and Discussions

The dynamic response of the soil and RCDWs are analyzed by comparing with the results of the centrifuge test in this section. The main emphasis was putted on the seismic response of soil core, far-field soil and RCDWs consisted of the EPWP, effective stress and the displacement of the RCDWs.

34.5.1 Excess Pore Water Pressure

Figure 34.7 illustrates the EPWP time histories of the soil from the experimental and numerical models, respectively. The numerical EPWP were in good agreement with that of experiment at each location indicating that the numerical model can computed the EPWP well. It was obviously that the EPWP time histories of soil outside of the RCDWs are quite different with that of the soil core which was enclosed by the RCDWs. The EPWP of outside soil reached rapidly to maximum at about 7 s and subsequently kept a nearly constant value during the remaining shaking time.

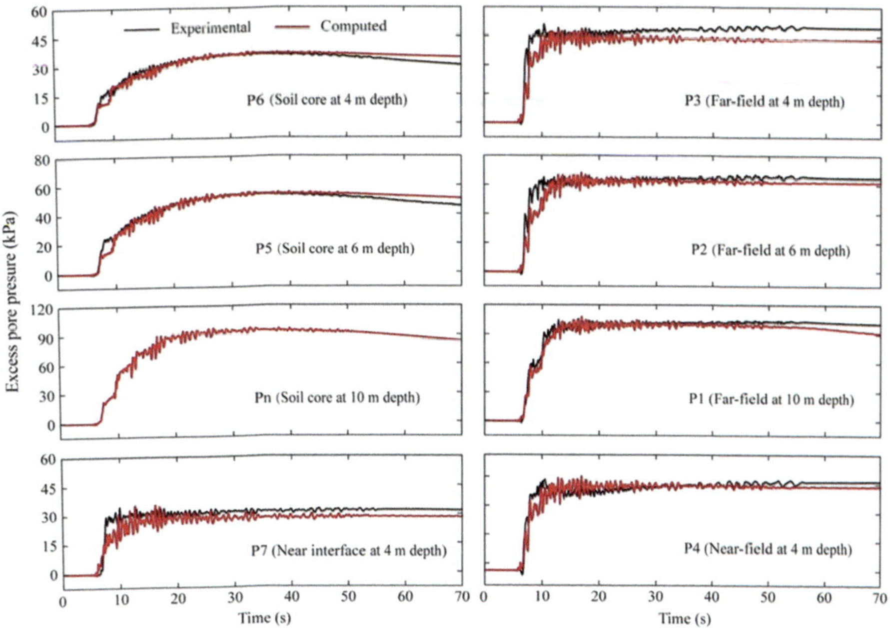

Fig. 34.7 Computed and experimental soil EPWP time histories in soil core and far-field

However, the EPWP of the soil core accumulated slightly to the peak until 35 s and then dissipated slowly to the end of the shaking. It implies that the EPWP generation of the soil core was restrained by the protection of the RCDWs and the soil core was not liquefied even subjected to 0.13 g shake event, while the outside soil without the protection from the RCDWs was liquefied at 7 s quickly and kept that state up to the end of the shaking event.

34.5.2 RCDWs Displacement

The displacement of foundations included the lateral displacement and settlement had very disadvantageous influence on the superstructures. The displacement time histories on top of the RCDWs (Points A and B in the Fig. 34.2) were measured by conventional laser transducers in the dynamic centrifuge test. Figure 34.8 compares these measurements with the datas obtained from the numerical simulation. As shown in Fig. 34.8, the displacement time histories of the RCDWs obtained from the centrifuge test and numerical simulation were in reasonable agreement except the value of fluctuations. It is noteworthy there were some interesting findings that the movement pattern of the uphill side wall was very different from the downhill side wall. The uphill side wall moved up at the beginning of the shaking, after that it shifted the

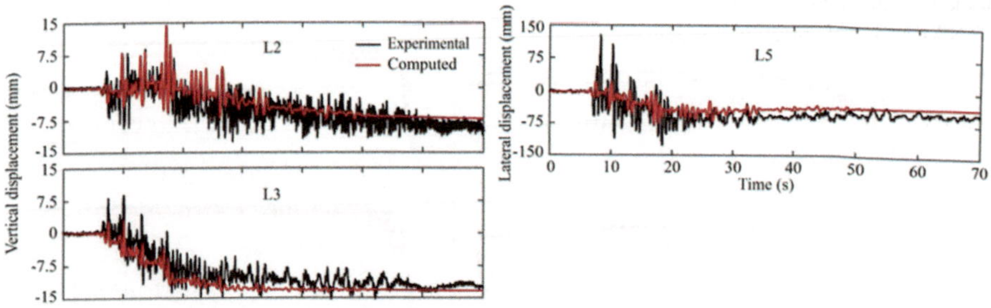

Fig. 34.8 Computed and experimental RCDWs displacement time histories

direction of movement to move down. And settlement quantity of the uphill side wall was smaller than the downhill side wall. These phenomena were possibly caused by the incline of the RCDWs resulted from the lateral spreading which was induced by the liquefaction in the sloping loose sand layer (Zhang et al. 2021a, b).

34.6 Parametric Study

Earthquake ground motions are very complicated, containing different amplitudes, frequency contents, and durations. Therefore, four earthquake ground motions-Kobe, Northridge, El-Centro, and Hach waves (Fig. 34.9) and Taft wave used in the test were selected with three PGAs (0.07, 0.2, 0.33 g) to study the liquefaction-mitigation of RCDWs subjected to different ground motions. The ground motions Arias intensity is shown in Table 34.2.

34.6.1 Liquefaction-Mitigation of RCDWs Under Different PGAs

Figure 34.10 shows the time histories of EPWP ratio (two points are at the same horizontal height) in soil core at a depth of 4 m (Line B) and in far-field at a depth of 4.96 m (Line C), subjected to five ground motions with three PGAs, respectively. Comparing these five graphs, it is found that soil liquefaction was not observed in both soil core and free field subjected to the ground motions with a PGA of 0.07 g. But the time histories of EPWP ratio in soil core and far-field were different. The maximum value of EPWP ratio in soil core was less than 0.6, while the minimum value of EPWP ratio of the far-field soil was greater than 0.7 subjected to the same ground motions. The EPWP ratio in soil core was much smaller than that of far-field soil, so it can be inferred that RCDWs can effectively suppress the soil EPWP in soil core. The EPWP ratio of the far-field soil reached 1.0 within 10 s under the all the ground motions with a PGA of 0.20 g, while the soil EPWP ratio in the soil

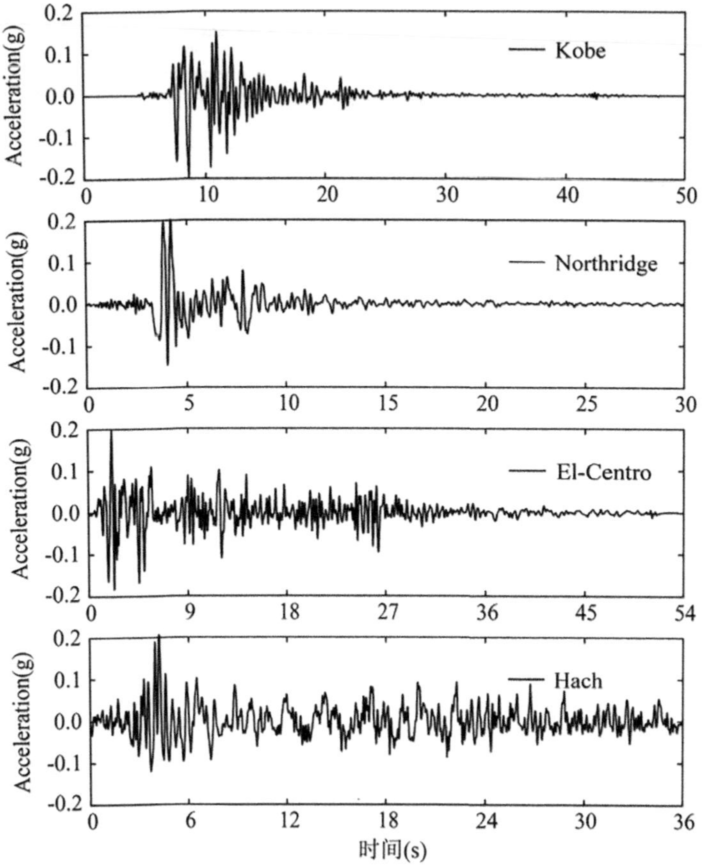

Fig. 34.9 Four different earthquake ground motions

core reached 1.0 only under the Taft ground motions, and the EPWP ratio was less than 0.9 under the other four ground motions. Therefore, the soil liquefaction was not occurred in soil core, while the soil liquefaction was observed in the far-field in the early stage of vibration, indicating that the RCDWs could mitigate liquefaction of soil in soil core under the action of medium earthquake. Soil liquefaction was occurred in both the far-field and soil core under a PGA of 0.33 g, but the initiation time of liquefaction in soil core lagged behind that in the far-field. Except for the Taft wave, the far- field soil was liquefied within 10 s of the vibration, while the soil was liquefied at 25 s in soil core. This indicates that RCDWs can effectively delay the soil liquefaction in the soil core.

When the PGA was 0.07 g, the soil vertical effective stress reduction in soil core was significantly smaller than that of the soil in the far-field, due to RCDWs effectively supress the generation of EPWP in soil core (Fig. 34.10). When the PGAs were 0.20 and 0.33 g, the far-field soil liquefied rapidly under the action of ground

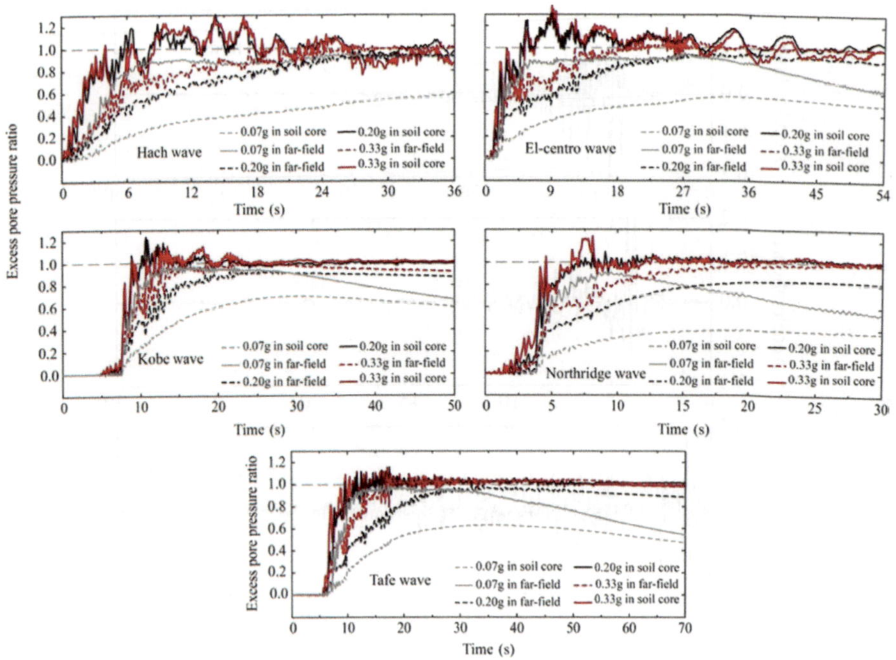

Fig. 34.10 EPWP ratio time history soil core and far-field subjected to Kobe, Northridge, El-Centro, Hach and Taft event

motions, and the soil vertical effective stress decreased to zero rapidly. While the soil vertical effective stress of in soil core decreased relatively slowly, so soil in the soil core can still provide vertical support for the foundation in the early stage of earthquake. Therefore, the soil core soil can still provide vertical support for the RCDWs in the early stage of earthquake, and slow down the damage of the RCDWs due to soil liquefaction.

34.6.2 Liquefaction-Mitigation Performance of RCDWs Under Different Ground Motions

The soil vertical effective stress time histories in the soil core and far-field under different earthquake ground motions with three PGAs (0.07, 0.20, 0.33 g) are shown in Fig. 34.11. It can be seen that the EPWP of soil in far- field rapidly increased to the maximum value, and the effective stress rapidly decreased to zero under different ground motions. The soil in the far-field rapidly liquefied when the PGAs are 0.2 and 0.3 g. While the liquefaction of soil core only occurred at the PGA of 0.33 g. Therefore, it is inferred that the RCDWs can effectively suppress the accumulation of

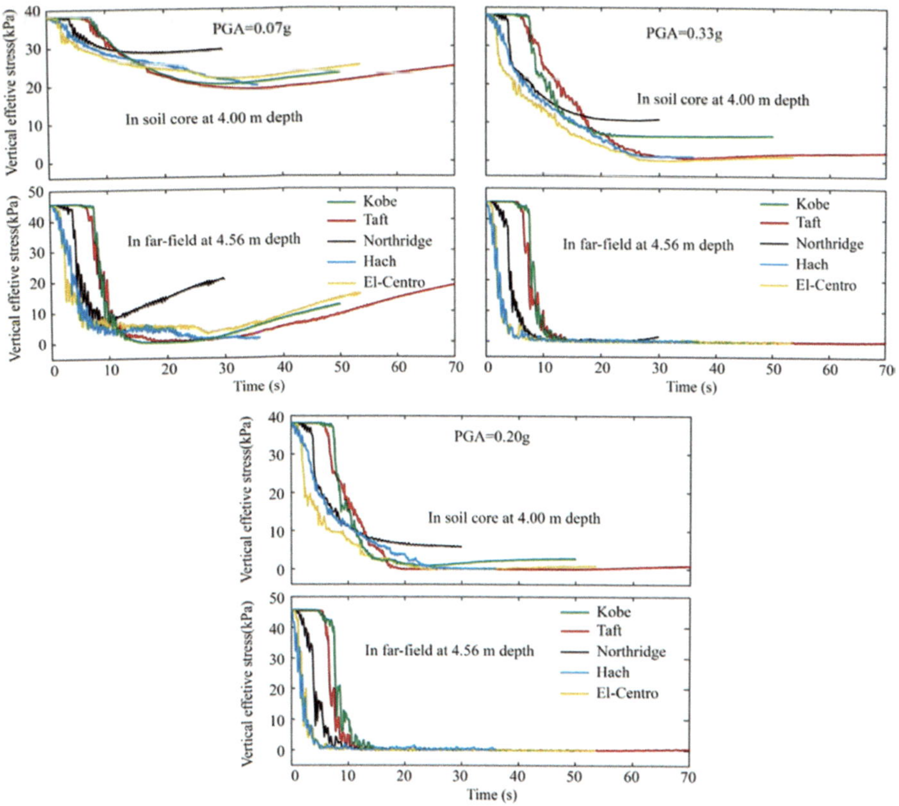

Fig. 34.11 Vertical effective stress time history under PGAs of 0.07 g, 0.20 g, 0.33 g

EPWP in soil core under different ground motions, thus reducing the loss of vertical effective stress and mitigating the soil liquefaction in soil core.

The dynamic response of soil was different under the five different ground motions, due to the different magnitudes of acceleration and appearance times of the five ground motions. The shaking time required for liquefaction of the far-field soil are different under the five different ground motions. The Hach and El-Centro ground motions required the shortest vibration time, followed by the Northridge, the Kobe, and the Taft ground motions required the longest shaking time. Because the Arias intensity of Northridge and Kobe ground motions was smaller (see Table 34.2), soil liquefaction was not occurred in soil core under the Northridge and Kobe ground motions with a PGA of 0.2 g. The soil in the soil cores still did not liquefy under the Northridge ground motion with a PGA of 0.33 g. Therefore, it is necessary to study the liquefaction mitigation of the RCDWs under different ground motions. It can be seen in Fig. 34.10 that the fluctuation of the EPWP ratio in the far field soil is small after soil liquefaction under the Kobe, Northridge and Taft ground motions. This is probably due to the different shearing and swelling effects of the far-field soils under five different ground motions. It can be seen that the soil EPWP ratio of in soil core does not have a large fluctuation, while the EPWP ratio of the far-field soil after liquefaction has a large fluctuation, which indicates that the RCDWs has a great influence on the shear expansion effect of soil in the soil core.

34.7 Conclusions

In this paper, the following conclusions are drawn from the numerical simulations of centrifuge shaking tests on RCDWs and the analysis of relevant parameters:

1. A finite element numerical model is established using the multi-yield surface elastoplastic principal structure. By comparing the results of the centrifuge shaking table test, it is concluded that the numerical model can accurately reproduce the dynamic response of the RCDWs and soil in soil core and far-field in the liquefiable site, which indicated that the numerical model is reasonable and reliable.
2. Comparing the centrifuge shaking test results with numerical simulation results, it can be found that the RCDWs can mitigating the soil liquefaction in soil core very effectively, thus effectively reducing the decrease of vertical bearing capacity of foundation due to liquefaction, which makes the RCDWs have obvious advantages in liquefiable sites.
3. RCDWs can effectively suppress the growth of EPWP in soil core under the ground motions with a PGA of 0.07 g. The soil in the soil core is basically not liquefied during the whole vibration process, while the external far field is liquefied at the early stage of vibration under a PGA of 0.2 g. The soil liquefaction was observed both in soil core and far-field under a PGA of 0.33 g, but soil liquefaction needs more shaking time in soil core.
4. RCDWs can effectively mitigating the soil liquefaction in soil core subjected to five different ground motions. In addition, the Arias intensity of ground motions has a great influence on the initiation of liquefaction of the internal soil cores. When the Arias intensity of ground motions is small, such as the Northridge wave, the soil was still not liquefied in soil cores, even when the PGA is 0.33 g.

Acknowledgements We would like to thank the editor and reviewers for their invaluable and constructive comments. This work was supported by Natural Science Foundation of Xinjiang Uygur Autonomous Region (No. 2022D01F38) and the Research Foundation of China University of Petroleum-Beijing at Karamay (No. XQZX20220007) and "Tianchi Talent" Introduction Plan.

References

Boulanger R, Curras CJ, Kutter LK, Wilson DW, Abghari A (1999) Seismic soil-pile-structure interaction experiments and analyses. J Geotech Geoenviron Eng 125(9):750–759

Cheng QG, Wu JJ, Song Z, Wen H (2012) The behavior of a rectangular closed diaphragm wall when used as a bridge foundation. Front Struct Civ Eng 6(4):398–420

Elgamal A, Yan L, Yang ZH, Conte JP (2008) Three-dimensional seismic response of Humboldt Bay bridge- foundation-ground system. J Struct Eng 134:1165–1176

Hamada M (1992) Large ground deformations and their effects on lifelines: 1964 Niigata earthquake. Case studies of liquefaction and lifelines performance during past earthquake. Technical report NCEER-92-0001. National Centre for Earthquake Engineering Research, Buffalo

Kawashima K, Unjoh S, Hoshikuma JI, Kosa K (2011) Damage of bridges due to the 2010 Maule, Chile, Earthquake. J Earthquake Eng 15(7):1036–1068

Li Y, Cheng QG, Zhang JL, Lyu B, Wang YF, Wu JJ (2019) Seismic behavior of rectangular closed diaphragm walls serving as bridge foundations in gently sloping liquefiable deposit: dynamic centrifuge tests. J Geotech Geoenviron Eng 145(12):04019105

Meyerhof GG (1976) Bearing capacity and settlement of pile foundations. J Geotech Eng Div 102(3):195–228

Namikawa T, Koseki J, Suzuki Y (2007) Finite element analysis of lattice-shaped ground improvement by cement mixing for liquefaction mitigation. Soils Found 47(3):559–576

Nguyen TV, Rayamajhi D, Boulanger RW (2013) Design of DSM grids for liquefaction remediation. J Geotech Geoenviron Eng ASCE 139(11):1923–1933

Vijayvergiya VN (1977) Load-movement characteristics of piles. Proceedings, Ports 77. Am Soc Civil Eng 2:269–284

Su L, Wan HP, Li Y, Ling XZ (2018) Soil-Pile-Quay wall system with liquefaction-induced lateral spreading: experimental investigation, numerical simulation, and global sensitivity analysis. J Geotech Geoenviron Eng 144(11):04018087

Wu JJ, Cheng QG, Wen H, Can JL (2014) Vertical bearing behaviors of lattice shaped diaphragm walls and group piles as bridge foundations in soft soils. Chin J Geotech Eng 36(9):1733–1744 (in Chinese)

Wu JJ, Cheng QG, Wen H, Cao JL (2015) Comparison on the vertical behavior of lattice shaped diaphragm wall and pile group under similar material quantity in soft soil. KSCE J Civ Eng 19(7):2051–2060

Zienkiewicz OC, Shiomi T (1984) Dynamic behavior of saturated porous media: the generalized Biot formulation and its numerical solution. Int J Numer Anal Meth Geomech 8(1):71–96

Zhang JL, Cheng QG, Li Y, Zhang EM, Wang YG, Wu JJ, Xie SY (2021a) Mechanism of liquefaction mitigation by rectangular closed diaphragm walls in sloping liquefiable deposits. Soil Dyn Earthq Eng 142:106582

Zhang JL, Cheng QG, Li Y, Qiu YH, Wang YF, Wu JJ (2021b) Performance of rectangular closed diaphragm walls in gently sloping liquefiable deposits subjected to different earthquake ground motions. Earthq Eng Eng Vib 20:905–923

Chapter 35
Evolution Process of Sand Mesostructure in Liquefaction Using Centrifuge Shaking Table Tests

Xiaoli Xie and Bin Ye

Abstract Previous studies suggested that earthquake liquefaction events would change sand mesostructure, which was also a critical factor affecting the mechanical properties of the reconsolidated sand. However, few studies have directly visualized the evolution of sand mesostructure during liquefaction. In this study, an inflight micro-image acquisition system was designed, developed, and applied to obtain sand particle images of a saturated sand deposit during the whole process of seismic liquefaction based on centrifuge shaking table tests. Excess pore pressure, acceleration and ground displacement were also measured simultaneously. Experimental results show that the beginning small-amplitude shaking increased the interparticle contacts and decreased the voids among particles. The interparticle contacts were reconstructed and the voids were redistributed along with the later larger-amplitude shaking. After liquefaction, the upper seepage of the pore fluid contributed to the expansion of large voids in the approximately vertical and horizontal direction. Sand particles rotated towards vertical under the upward pore fluid seepage. The drastic changes in the sand mesostructure during liquefaction may be the intrinsic mechanism that influenced sand reliquefaction resistance.

Keywords Seismic liquefaction · Mesostructure of sand · Centrifuge shaking table test · Micro-image acquisition system

---

X. Xie (✉) · B. Ye
Tongji University, 1239 Siping Road, Shanghai, China
e-mail: xiexiaoli@tongji.edu.cn

B. Ye
e-mail: yebin@tongji.edu.cn

S. Wang et al. (eds.), *Engineering Geology for a Habitable Earth: IAEG XIV Congress 2023 Proceedings, Chengdu, China*, Environmental Science and Engineering,
https://doi.org/10.1007/978-981-99-9057-3_35

35.1 Introduction

Liquefaction is one of the most devastating seismic geotechnical hazards, which can cause extensive engineering destruction and great economic losses. Previous studies mainly focused on the first liquefaction of soil and an intuitive opinion is that the liquefaction resistance should increase with the increase of density after reconsolidation. Assessing the liquefaction resistance of is the basis of the seismic disaster risk analysis of the sand deposits. After the 2010–2011 Christchurch earthquake Quigley et al. (2013) in New Zealand and the 2011 Great East Japan earthquake Yasuda et al. (2012), researchers have gradually realized that the prior seismic liquefaction events can even decrease the liquefaction resistance of sandy deposits although the soil is denser (Fardad et al. 2021; Yamada et al. 2010; Ye et al. 2018a, b, 2020), which is contrary to the intuition.

Some researchers noted that the unstable mesostructure which was formed after prior liquefaction events were the causes that decreased the liquefaction resistance of reconsolidated soil. Suzuki and Suzuki (1988) firstly observed the formation of sand particles with vertical preferred orientations in cyclic triaxial tests by freezing and silencing the reconsolidated specimens. Based on the 1 g shaking table tests, Ye et al. (2018a) pointed that unstable micro-fabric with vertical preferred particle orientation formed in the first liquefaction events was the main cause of lower reliquefaction resistance. Exploring the dynamic formation and evolution process of these peculiar fabrics has essential significance for studying the internal mechanisms mesoscopically. However, it is difficult to capture particle fabrics throughout the total experimental process by the slicing method (Ibrahim and Kagawa 1991; Suzuki and Suzuki 1988) and static micro-imaging method Ye et al. (2018a) because only mesostructure under a certain state (e.g., before or after tests) can be obtained by these methods.

Centrifuge shaking table tests have advantages of reasonably simulating the real stress–strain behaviours of soil. Few studies on the formation and evolution process of mesostructure of sand particles during liquefaction have been conducted based on centrifuge tests. In view of the lack of research on this filed, an in-flight centrifuge experimental system on macroscopic and mesoscopic liquefaction behaviours were developed in this study, and centrifuge tests were conducted to reveal liquefaction related mesoscopic mechanisms. Sand particle images and liquefaction responses of the sandy deposits were acquired and analysed. The dynamic evolution of sand mesostructure during the total process of liquefaction were revealed. This research results could help understand the origin of the loss in reliquefaction resistance from a mesoscopic view.

35.2 Centrifuge Shaking Table Tests

35.2.1 Development of the Experimental System

Experimental investigations were conducted based on TLJ-150 centrifuge. Photos of the centrifuge and in-flight shaking table are given in Figs. 35.1a, b, respectively. The shaking table was fixed on the centrifuge basket and the model was fixed on the platform when conducting seismic testing. The capacity of the centrifuge was 150 g × ton (i.e., 1500 kg under 100 g centrifugal acceleration) and the maximum payload of the shaking table was 300 kg under 50 g centrifugal field. The output seismic direction was parallel with the rotating axis of centrifuge (i.e., the white arrowed line in Fig. 35.1b) when the centrifuge swung upward.

To successively acquire sand particle images during the dynamic centrifuge tests, remote communication and control method based on local area network (LAN) was employed in this study. Figure 35.2 depicts the principle of the controlling method. First, the personal computer 1 (PC1) placed in the control room was connected to the personal computer 2 (PC2) which was fixed onto the centrifuge arm by the LAN cables and the slip ring of the centrifuge, and the network communication was then correspondingly established. The PC2 and the micro-camera were mutually connected via a USB cable. During the tests, operation orders will be transmitted from the control room to the centrifuge room, and the sand images will be taken. A series of calibration tests were conducted before the actual test, and it was proven that the proposed experimental system could provide favourable performance.

Figure 35.3 depicts the image of the developed experimental system. The experimental model mainly contains a strong model container, a micro-image acquisition system, a steel frame, a piece of transparent synthetic glasses. The steel frame was employed to support the micro-image acquisition system and absolute relative rest between the image acquisition system and the shaking table platform during shaking can be ensured. A circular hole with 110 mm diameter at the position of the objective image area was excavated on the frame, which could provide a window for imaging.

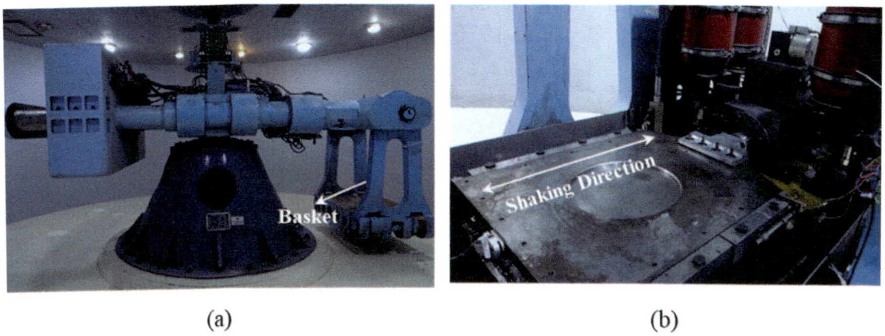

Fig. 35.1 Experiment appratus. **a** TLJ-150 geotechnical centrifuge and **b** in-flight shaking table

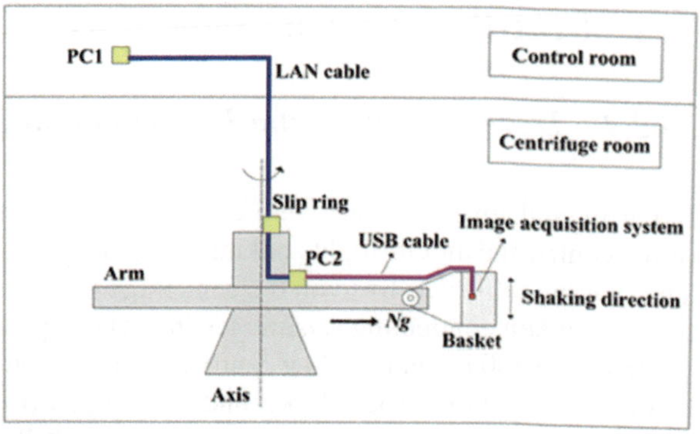

Fig. 35.2 Communication and controlling method of the micro-image acquisition system

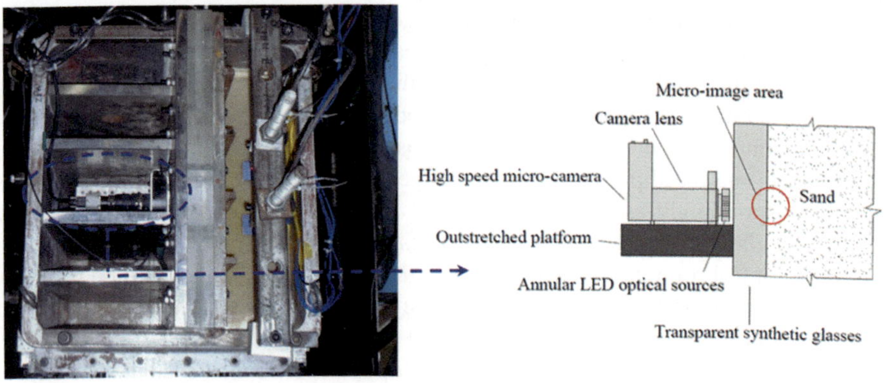

Fig. 35.3 The developed experimental system

An illustration of the micro-image acquisition system is also shown in Fig. 35.3. To decrease the boundary friction on sand particle movement, a thin lubricating grease was applied on the inner side of the model container to decrease the friction. The micro-image acquisition system mainly contains a ring type illuminant, a fast-shutter-speed micro-CCD camera and its lens. The camera can output 150 full frames per second in a continuous mode, with 1920 (horizontal) × 1080 (vertical) active pixel resolution.

35.2.2 *Experiment Procedure*

The Silica sand produced from Anhui was employed in this study. The sand was with sub-angular to angular particle shapes. Some basic properties of the tested sand are summarized in Table 35.1. Note that the d_{10}, d_{30} and d_{60} are defined as the diameter through which 10, 30 and 60% of the total soil mass is passing, respectively.

Table 35.1 Basic properties of the tested sand

Property	Unit	Value
Characteristic particle diameter, d_{10}	mm	0.25
Characteristic particle diameter, d_{30}	mm	0.37
Characteristic particle diameter, d_{60}	mm	0.62
Coefficient of uniformity, C_u	–	2.22
Coefficient of curvature, C_c	–	0.98
Specific gravity of sand particle, G_s	–	2.65
Maximum void ratio, e_{max}	–	1.09
Minimum void ratio, e_{min}	–	0.57

The scaling factors of the dynamic time and the consolidation time were not the same if water was used Adamidis and Madabhushi (2015). To overcome this conflict, the Methylcellulose solution with viscosity of 25 cSt Dewoolkar et al. (1999) was employed as the pore fluid because the tests were conducted at 25 g centrifugal acceleration in this study. Model deposits with initial relative density of 30% was prepared by wet pluviation method as used in the previous studies (Ecemis et al. 2015; Ha et al. 2011), which can well reproduce the natural sedimentary process of hydrogenic sands.

Figure 35.4 depicts the layout of the model deposit and transducers. All data here was presented in prototype scale unless otherwise noted. The model deposit was instrumented with nine pore pressure transducers, ten accelerometers and two laser displacement transducers to capture the excess pore water pressure, horizontal seismic acceleration response and vertical displacement of the ground surface during the sand liquefaction process. To reduce the dynamic boundary effects, two pieces of 40 mm thick buffering boards were attached to the side walls of the model container.

The mesoscopic particle images were taken at the depth of 1.625 m. The depth of the imaged zone was same as the depth of the uppermost pore pressure transducers, which was considered as the depth where liquefaction was most easily triggered in this study. Because the macroscopic seismic responses at different depth were quite different, it can be indicated that there were also significant differences between the microscopic responses at different depth. However, limited by the current experimental conditions, only the microscopic images of sand particles at one depth were captured in this study.

After the data acquisition system was checked and activated before testing, the centrifuge gradually spun to 25 g centrifugal acceleration. The initial static consolidation of the sand deposit cause by the increasing centrifugal acceleration was completed when the values of the pore pressure transducers and the Laser displacement transducers remained at steady levels.

Figure 35.5 shows the time history and frequency components of the input seismic motion. The peak value of the acceleration cycle increased and decreased gradually in the last three seconds. The maximum base acceleration was 0.15 g. The main frequency component was 1 Hz, and there were also some inevitable high-frequency

Fig. 35.4 Layout of the model deposit and the transducers

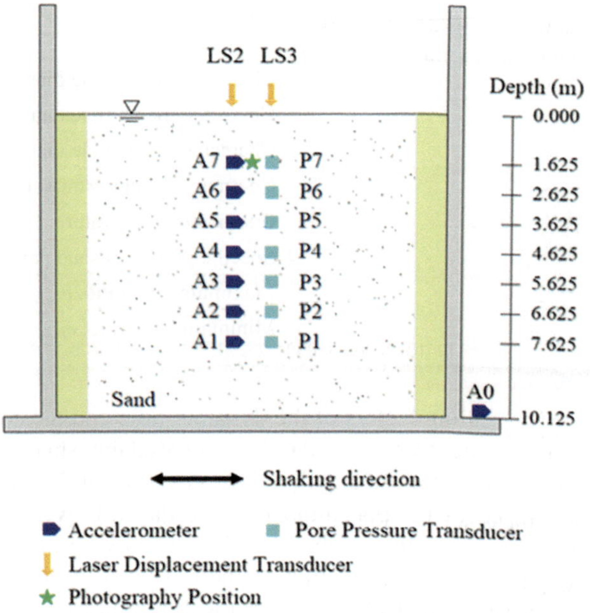

noises with smaller amplitudes. Micro-images of sand particles were captured during the total experimental process.

The macroscopic seismic responses, i.e., the time history of horizontal acceleration and excess pore pressure ratio, of the model deposit are displayed in Figs. 35.6a, b, respectively. It can be seen from Fig. 35.6a that the seismic responses of upper and lower soil deposit were quite different. The spiky acceleration response was observed at the 1.625 m depth, which can be considered as the typical characteristic of seismic liquefaction. The excess pore pressure ratio (r_u), the ratio of the excess pore pressure to the initial vertical effective stress, is an index to evaluate the liquefaction behaviours of soil. The initial liquefaction state was achieved when r_u increased to 1.0. The deposit at the depth of 1.625 m liquefied, while the deposit at the other depths did not liquefy. The initial liquefaction state at the depth of 1.625 m

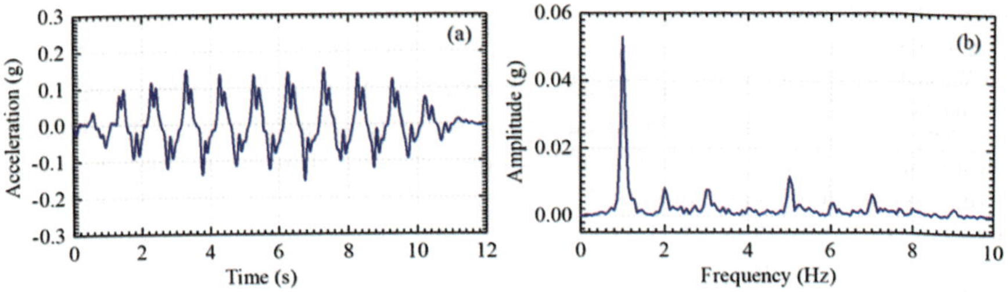

Fig. 35.5 Input seismic motion. **a** Time history, **b** Frequency component

was reached at 20.4 s, indicating the r_u at this depth continuously increased after the end of inputting seismic motion. The r_u at the depth of 3.525 and 5.625 m reached the maximum at the end of seismic motion and then decreased gradually. Because the pore pressure at the lower depth was greater than that at the upper depth, the increase of r_u at the depth of the 1.625 m after the seismic motion could be attributed to the upward seepage of the pore fluid during the post-seismic reconsolidation. With the gradual dissipation of the excess pore pressure, the upward seepage gradually weakened, and the r_u at the depth of 1.625 m also gradually decreased.

The time history of vertical displacement at the ground surface are shown in Fig. 35.7. It can be seen that the ground surface settled gradually with the dissipation of excess pore pressure, which is consistent with prevoius experimental studies Zhou et al. (2013). The final settlement of the ground surface is approximate 325 mm, and this indicates that the settlement of loose sand deposits induced by seismic liquefaction can not be ignored in engineering practice..

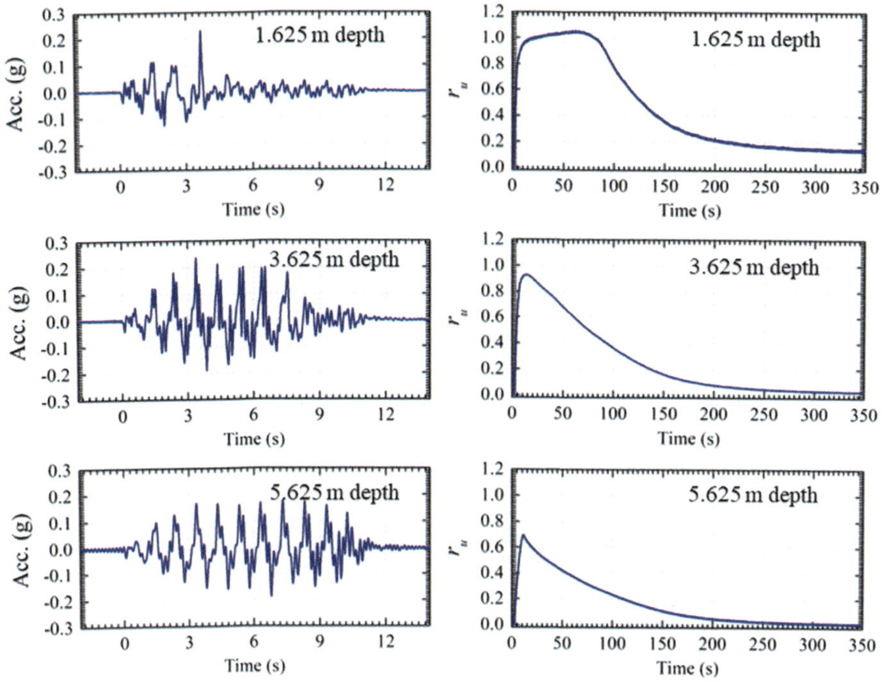

(a) Time history of acceleration (b) Time history of r_u

Fig. 35.6 Macrosopic seismic response of the model deposit

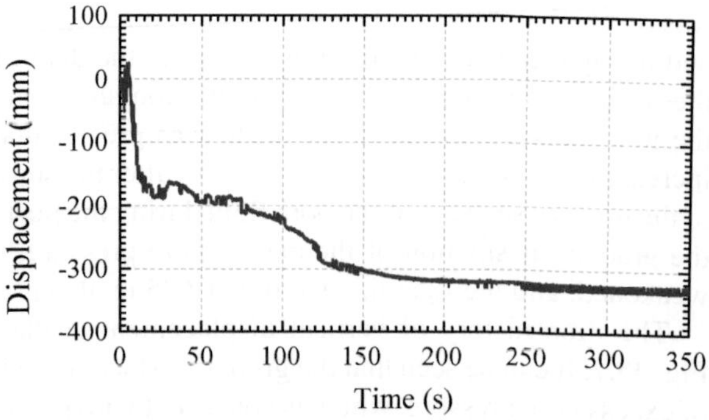

Fig. 35.7 Vertical displcement at the ground surface

35.3 Microscopic Response of the Sand Particles

35.3.1 Seismic Shaking Stage

Figure 35.8 displays the evolution process of sand particle structures during the stage of inputting seismic wave. In this figure, the areas outlined with green dashed line represents interparticle contact conditions. The yellow and white irregular shapes were the outlines of sand particles. As shown in Fig. 35.8a, b, the interparticle voids were compressed, and the distance between sand particles decreased during the beginning 2.78 s. Correspondingly, new interparticle contacts formed, and some particles rotated slightly. With the further increase of the excess pore pressure, the effective stress further decreased. Therefore, the tightness of interparticle contracts decreased gradually. The sand particles with weak contacts were more likely to slide and rotate under the cyclic shear stresses. Consequently, the particles in regions A-C lost contacts with the surrounding particles, and the voids were rearranged, as shown in Fig. 35.8c.

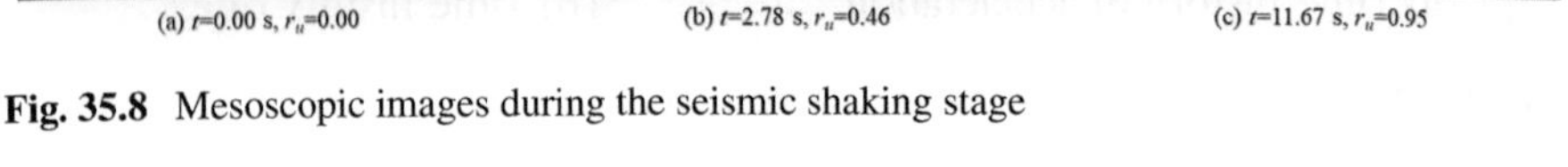

Fig. 35.8 Mesoscopic images during the seismic shaking stage

Fig. 35.9 Mesoscopic images during the liquefaction ($r_u = 1.0$) stage

35.3.2 *Liquefaction Stage ($r_u = 1.0$)*

Figure 35.9 exhibits the evolution of sand particles during the liquefaction (i.e., $r_u = 1.0$) stage. In Fig. 35.9a, the irregular shapes outlined with dashed and solid lines were the positions of sand particles at the moment of 22.22 s and 31.67 s, and the green dashed arrows represent the overall movement direction and distance of sand particles. As shown in Fig. 35.9a, some small particles broke away from the interparticle contacts and move upward rapidly under the pore fluid seepage, while the large particles moved downward simultaneously. These small particles were originally located at the voids between the large particles to maintain the stability of particle structures. Therefore, the migration of the small particles leaded to contact losing between the large particles and the instability of particle structure. As shown in Fig. 35.9b, the original contacts were gradually destroyed under the upward dragging effects of the pore fluid seepage, and the voids among sand particles were expanded. The expanded voids were represented by the green dashed area, and the long-axis direction of these voids were represented by green arrows. Thereafter, these voids gradually connected with each other in approximately vertical direction which was called as open voids. With the further dragging effects from the upward seepage, the long axes of sand particles (represented by yellow irregular shapes) that around the open voids (represented by dashed green arrows) gradually rotated towards vertical, as shown in Fig. 35.9c.

35.3.3 *Post-Liquefaction Stage ($r_u < 1.0$)*

With the pore pressure dissipation, the dragging effects of the upward seepage gradually weakened, thus the sand deposit at the depth of 1.625 m was no longer in liquefaction state (i.e., $r_u < 1.0$). Firstly, most particles moved downward together as indicated by Fig. 35.10a, b. However, the downward moving velocity of different particles varied. For example, the moving velocity of particles 1–3 was smaller than their neighbouring particles. Consequently, the particles 1–3 moved upward relatively to other particles, though the dragging effects of pore fluid were weakened.

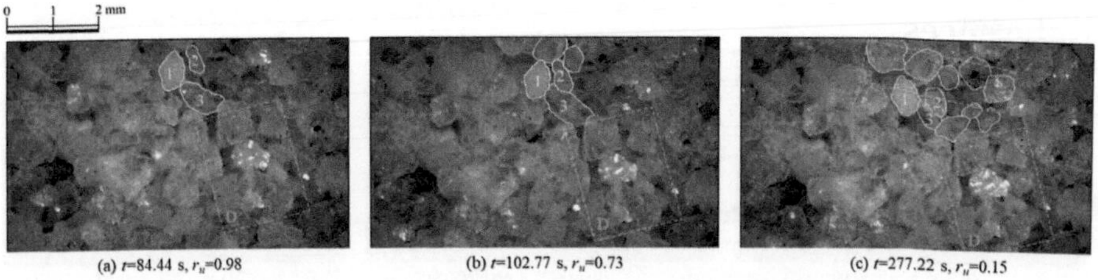

(a) t=84.44 s, r_u=0.98 (b) t=102.77 s, r_u=0.73 (c) t=277.22 s, r_u=0.15

Fig. 35.10 Mesoscopic images during the post-liquefaction ($r_u < 1.0$) stage

The difference between sand particle moving velocity contributed to the formation of new sand mesostructure.

With the further dissipation of the excess pore pressure, the pore fluid was not strong enough to move these particles upward continuously. As shown in Fig. 35.10c, these particles gradually sedimented downward together with other particles to establish new contacts with the particles below them or slightly rotated until stable states were reached. Consequently, slender voids formed in the approximately horizontal direction as shown in zone E. Subsequently, the sand mesostructure remained such a state without obvious changes until the completion of the sand reconsolidation. It should be noted that the open voids and the particle structure with long-axes in the vertical directions (i.e., sand particle structure in zone D) that formed during $r_u = 1.0$ still remained during $r_u < 1.0$, which indicates that the previous liquefaction history changed the sand mesostructure obviously.

35.4 Conclusion

In this study, the evolution process of sand mesostructure during seismic liquefaction was analyzed based on a favorable micro-image acquisition system in centrifuge shaking table tests. Conclusions can be drawn as follows.

1. The interparticle contacts increased gradually, and the voids between particles decreased gradually under the beginning small-amplitude shaking.
2. During sand liquefaction, open voids formed under the upward seepage of the pore fluid with sand particle long axes rotating towards vertical. With the decrease of the excess pore pressure, slender voids formed in the approximately horizontal direction.

Acknowledgements This research was financially supported by the National Natural Science Foundation of China (No. 41977225). All supports are gratefully acknowledged.

References

Adamidis O, Madabhushi GSP (2015) Use of viscous pore fluids in dynamic centrifuge modelling. Int J Phys Model Geotech 15(3):141–149
Dewoolkar MM, Ko HY, Pak RYS (1999) Centrifuge modelling of models of seismic effects on saturated earth structures. Géotechnique 49(2):247–266
Ecemis N, Demirci HE, Karaman M (2015) Influence of consolidation properties on the cyclic re-liquefaction potential of sands. Bull Earthq Eng 13(6):1655–1673
Fardad AP, Huang DN, Wang G et al (2021) Effects of strain history and induced anisotropy on reliquefaction resistance of Toyoura sand. J Geotech Geoenviron Eng 147(9)
Ha IS, Olson SM, Seo MW et al (2011) Evaluation of reliquefaction resistance using shaking table tests. Soil Dyn Earthq Eng 31(4):682–691
Ibrahim AA, Kagawa T (1991) Microscopic measurement of sand fabric from cyclic tests causing liquefaction. Geotech Test J 14(4):371–382
Quigley MC, Bastin S, Bradley BA (2013) Recurrent liquefaction in Christchurch, New Zealand, during the Canterbury earthquake sequence. Geology 41(4):419–422
Suzuki T, Suzuki T (1988) Effects of density and fabric change on reliquefaction resistance of saturated sand. Soils Found 28(2):187–195
Yamada S, Takamori T, Sato K (2010) Effects on reliquefaction resistance produced by changes in anisotropy during liquefaction. Soils Found 50(1):9–25
Yasuda S, Harada K, Ishikawa K et al (2012) Characteristics of liquefaction in Tokyo Bay area by the 2011 Great East Japan Earthquake. Soils Found 52(5):793–810
Ye B, Hu HL, Bao XH et al (2018a) Reliquefaction behavior of sand and its mesoscopic mechanism. Soil Dyn Earthq Eng 114:12–21
Ye B, Zhang L, Wang HH et al (2018b) Centrifuge model testing on reliquefaction characteristics of sand. Bull Earthq Eng 17(1):141–157
Ye B, Xie XL, Zhao T et al (2020) Centrifuge tests of macroscopic and mesoscopic investigation into effects of seismic histories on sand liquefaction resistance. J Earthquake Eng 26(8): 4302-4324
Zhou YG, Liang T, Li YG et al (2013) Dynamic centrifuge tests on liquefaction of clayey sand ground. Chin J Geotech Eng 35(9):1650–1658

Chapter 36
Comparative Analysis of Numerical Simulation Methods for Active Earth Pressure on Rigid Retaining Walls

Yuan Qi and Shiguo Xiao

Abstract There are significant advantages in analysing lateral earth pressure on retaining walls using numerical simulation methods because soil deformation behaviours can be fully involved in the methods. Operational processes as well as advantages and disadvantages of five typical numerical approaches are generally summarized and analysed, including wall active displacement method, wall elastic modulus reduction method, soil weight increase method, wall bottom friction coefficient reduction method and wall weight reduction method. Taking a wall translation mode as a typical example, the numerical analysis of each method is carried out via FLAC3D. The results show that distribution curves of active earth pressure along the height of wall exhibit a parabolic mode with the peak close to the wall bottom. The slip surface in the retained soil is curved, in contrast to the plane mode in the classical theories. The difference of the resultant force of the earth pressure between these numerical methods and the Coulomb theory is about 0.3–7.2%. In the condition that the retained soil reaches the active limit state, the computed wall displacements are about 1.2–2.4‰ of the wall height and they are very close to the test values. The rationality of computation principle, feasibility of simulating wall motion mode, applicability of simulation results and calculation efficiency of the five numerical methods are comprehensively compared, which indicates the calculation principle of the latter two methods is more reasonable than that of the first three methods. It is accordingly suggested to give priority to the wall weight reduction method, which can be followed by the bottom friction coefficient reduction method. The former can present the translational and rotational movement modes of the wall, but the latter can only be used in the translational mode.

Y. Qi · S. Xiao (✉)
School of Civil Engineering, Southwest Jiaotong University, Chengdu, China
e-mail: xiaoshiguo@swjtu.cn

Y. Qi
e-mail: qiyuan7272@163.com

S. Xiao
Key Laboratory of High-Speed Railway Engineering, Ministry of Education, Southwest Jiaotong University, Chengdu, China

S. Wang et al. (eds.), *Engineering Geology for a Habitable Earth: IAEG XIV Congress 2023 Proceedings, Chengdu, China*, Environmental Science and Engineering,
https://doi.org/10.1007/978-981-99-9057-3_36

Keywords Rigid retaining walls · Active earth pressure · Numerical simulation methods · Computational efficiency

36.1 Introduction

Rigid retaining wall is a common engineering retaining structure used to maintain the stability of soils. The rational determination of the active earth pressure on the wall back is very important in the design and calculation of retaining walls. In practical engineering, the Coulomb (1773) and Rankine (1857) theories are generally used to approximate the earth pressure, but they have obvious limitations. Many scholars have improved the classical earth pressure algorithm, including theoretical analysis based on limit equilibrium principle (Bang, 1985; Chang, 1997; Paik and Salgado, 2003; Cai et al. 2017) and model test method (Tsagareli, 1965; Fang and Ishibashi, 1986; Fang, et al., 1994; Khosravi et al. 2013). Limit equilibrium theory is a simple method to analyze earth pressure. However, it is obviously not practical to assume that the soil mass is rigid-plastic without considering the deformation characteristics of the soil mass. The numerical simulation method is more reasonable than the limit equilibrium theory in reflecting the properties of the soil because of fully considering the elastic–plastic deformation characteristics of the soil. At the same time, the active limit state of the soil behind the retaining wall is the result of the displacement of the wall due to its translation, rotation or combined movement under the action of earth pressure. Therefore, the numerical simulation method can fully reflect the process that the soil behind the wall reaches the active limit state caused by the wall displacement.

Potts and Fourie (1986) analyzed the influence of different movement modes of retaining wall on earth pressure under smooth and rough wall backs by finite element method. Matsuzawa and Hzaarika (1996) introduced the interface unit with bilinear stress-displacement relationship between backfill and wall to simulate the friction characteristics of wall, and studied the influence of wall movement modes on active earth pressure. Chen et al. (2004) introduced Goodman unit without thickness between the contact surface of soil and structure to study the influence of different movement modes and displacement conditions on the magnitude and distribution of earth pressure. Fan and Fang (2010) conducted a numerical analysis on the active earth pressure of rigid retaining walls with limited backfill space. The effects of the ratio of length to diameter and internal friction angle on the coefficient of active earth pressure and the application point of resultant force are studied. Ying et al. (2012) studied the active earth pressure of rigid retaining wall with narrow cohesive fill by the finite element analysis software. The critical crack depth of the soil and the development law of the slip surface are discussed. Chen et al. (2019) analyzed the active failure mechanism of narrow backfill on retaining walls in translational mode by the finite element limit analysis software. A new calculation model of active earth pressure is established based on the numerical results. Woerden and Achmus (2013) studied the three-dimensional active earth pressure of the wall in uniformly and normally consolidated sand by finite element method. The effects of wall movement modes, soil width ratio, relative density of sand and wall roughness are discussed. Other scholars have analyzed the earth pressure from different angles by finite element method, such as the interaction between sandy soil and wall (Clough and Duncan, 1971; Ozawa and Duncan, 1976; Nakai, 1985; Bhatia and Bakeer, 1989).

In addition to the finite element method, some scholars also studied the earth pressure by using the discrete element method. Zhou et al. (2008) discretized the soil into rigid strips with sliding joint model. The distribution of active earth pressure under different movement modes and different displacement sizes is simulated by particle flow procedure. Jiang et al. (2014) analyzed the earth pressure of retaining wall under different wall movement modes by using contact model considering rolling resistance between particles. Xiao et al. (2020) and Zhang et al. (2021) studied the influence of soil width on earth pressure in a limited width range by using the discrete element method. There are obvious advantages in solving large deformation and failure problems for discrete element method. However, micro-parameters such as contact stiffness and friction coefficient between particles of granular materials are often difficult to obtain directly by experiment, which needs further exploration.

For the specific operation method of numerical simulation, wall active displacement method (Benmeddour et al., 2012; Benmebarek et al., 2016), wall bottom friction coefficient reduction method (Dai et al., 2009), and soil weight increase method (Dai et al., 2008) can be used to simulate the process that the retained soil gradually reaches the active limit state with the wall displacement. Additionally, when determining the earth pressure by numerical simulation, a series of important problems are involved, such as the rationality of calculation results, calculation astringency (applicability of method), calculation efficiency (time consumption) and so on. However, there were few specific analyses and discussions on these problems in previous studies. Besides, the active displacement method is often used to make the soil behind the wall reach the active limit state in current research which is inconsistent with the physical process of wall-soil movement. The wall displacement is actually caused by the earth pressure, and the soil behind the wall gradually reaches the limit state.

Therefore, it is a key problem in the numerical simulation method to reasonably simulate the process of the backfill behind the wall reaching the active limit state. This paper aims to compare and evaluate the advantages and disadvantages of various numerical simulation methods in terms of scope of application and rationality of the numerical simulation process. In this regard, five numerical simulation methods are elaborated, including wall active displacement method, wall bottom friction coefficient reduction method, wall weight reduction method, wall elastic modulus reduction method and soil weight increase method. The wall translation mode is taken as a typical example for the numerical modelling. Some simulation results, such as the distribution of the active earth pressure, the magnitude and application point of the resultant force of earth pressure, the relationship between the resultant force and the wall displacement, the location of the slip surface in the retained soil, and the calculation efficiency of different methods, are discussed in detail by comparing with the test results. The numerical approach is realized by finite difference software

FLAC3D. Finally, the rationality and operability of the calculation results of different numerical simulation methods are analysed comprehensively.

36.2 Numerical Simulation Method

36.2.1 Wall Active Displacement Method

Among the numerical simulation methods of earth pressure of retaining wall, the wall active displacement method is the most commonly used one at present. Benmeddour et al. (2012) simulated the translation mode by applying equal horizontal velocity at each node of the rigid wall. Benmebarek et al. (2016) also applied a very low velocity on all nodes of the wall actively to realize different wall modes of movements, which led to stable plastic flow of the soil behind the wall gradually. In the simulation, the retaining wall was modelled by structural beam elements connected to the soil grid via interface elements. The main operation process of the wall active displacement method is as follows: Firstly, the numerical model is established. Then, according to different wall modes of movements, the corresponding motion velocity is applied to each node of the wall. And the wall is fixed after reaching the preset displacement. Finally, the solution is solved to iterative convergence.

The wall active displacement method, as a method of moving the wall actively, is easy to operate and realize. However, the shortcomings of this method are as follows: (1) This method is only applicable to the earth pressure simulation process with clear wall motion mode. However, the movement mode of the retaining wall in the actual state is mostly the compound movement of translation and rotation, which is difficult to determine accurately in advance. (2) During the implementation of this method, the velocity applied to the wall needs to be very slow to ensure that the iteration steps of the model are sufficient in the process of the soil following the coordinated changes of the wall. This velocity needs to be estimated by trial calculation in advance. (3) Whether the soil behind the wall can reach the active limit state is determined by the preset displacement of the retaining wall. The preset displacement needs to be decided by the relationship curve between the resultant earth pressure and the wall displacement. The relationship curve needs to be simulated many times to obtain, which leads to low calculation efficiency. (4) From the physical mechanism of wall movement, the actual wall displacement is the effect of earth pressure behind the wall. There is an interaction and adjustment mechanism of "earth pressure-wall displacement—earth pressure—…". The stress path of the soil behind the wall cannot be truly reflected in the simulation method of actively applying wall displacement.

36.2.2 Wall Bottom Friction Coefficient Reduction Method

In this method, the anti-sliding friction of the retaining wall is reduced by decreasing the coefficient of friction between the bottom of the retaining wall and the ground. When the friction coefficient at the bottom of the wall decreases to a certain value, the wall moves forward gradually under the action of earth pressure behind the wall. The shear yield zone in the soil behind the wall runs through the soil surface and approaches the active limit equilibrium state. Dai et al. (2009) used this method to calculate the active earth pressure of L-shaped retaining wall by finite element method. The comparison with the results of classical theoretical methods shows that the distribution and magnitude of active earth pressure on the vertical arm of L-shaped retaining wall calculated by this method are reasonable. For the wall bottom friction coefficient reduction method, the main operation flow is: Firstly, the friction coefficient of wall bottom is estimated based on limit equilibrium theory to determine the reasonable range for the reduction factor. Secondly, typical solution steps of analysis model are extracted through initial stress balance process. Six typical time steps can be used as termination conditions for the subsequent reduction process. Finally, the friction coefficient reduction process is realized by dichotomy. The upper and lower limits of the reduction ratio of friction coefficient are adjusted according to whether the calculation results converge or not. The iteration process is calculated until the accuracy requirements are met. The friction coefficient reduction ratio is defined as the ratio of the actual friction coefficient to the reduced friction coefficient, namely:

$$\delta' = \arctan\left(\frac{1}{F_s}\tan\delta\right) \tag{36.1}$$

where, δ and δ' are the actual and reduced friction angle of the wall bottom, respectively; and F_s is the reduction ratio of the friction coefficient and it is not less than 1.

The wall bottom friction coefficient reduction method achieves the purpose of active wall translation by reducing the wall bottom friction. Therefore, it is suitable for retaining walls with large bottom area. Translation is the main motion mode of this kind of retaining wall, such as cantilever retaining wall. The main shortcomings of this method are as follows: (1) The scope for application of this method is small. Active earth pressure of retaining wall with translation as the main mode can only be simulated in this method. This method is meaningless if the wall rotates. The friction properties of the wall bottom will be ineffective due to the formation of cracks caused by the separation of parts of the wall bottom. Consequently, the vertical velocity of all nodes in the wall needs to be constrained to limit wall rotation throughout the calculation process. (2) Computation time of this method is relatively long. Because multiple friction coefficient reduction calculations are required to meet the accuracy requirements. Meanwhile, non-convergence will lead to a waste of computing time.

36.2.3 Wall Weight Reduction Method

The translation mode or rotation around a point of the retaining wall needs to be assumed in the above two numerical simulation methods so that the active limit state can be reached. However, in practical engineering, the movement mode of retaining wall is more likely to be a combination of translation and rotation. By decreasing the weight of the wall, the anti-slide stability and anti-roll stability of the wall are reduced in the wall weight reduction method, so as to achieve the effect of earth pushing against the wall to produce displacement. The actual movement of the retaining wall can be simulated using this method without assuming the movement mode in advance. Dai et al. (2008) adopted this method to reveal the actual movement mode of rigid retaining wall when the active limit state is reached. The rationality of this method is proved by comparing with classical theory. The numerical simulation flow of the wall weight reduction method is basically the same as that of the wall bottom friction coefficient reduction method. The main differences are as follows: (1) The upper and lower limits of the wall weight reduction factor need to be estimated considering the anti-sliding stability and overturning stability. (2) Form (2) needs to be used to achieve the reduction of wall weight, where the reduction factor of wall weight is defined as the ratio of the actual wall weight to the reduced wall weight. The wall weight reduction coefficient is defined as the ratio of the actual wall weight to the reduced wall weight.

$$\gamma_{\mathrm{w}}' = \frac{\gamma_{\mathrm{w}}}{k_{\mathrm{w}}} \tag{36.2}$$

In the formula: The actual weight of the wall is γ_{w}. The weight reduction coefficient of the wall is k_{w}. The reduced wall weight is γ_{w}'.

Although the wall motion modes need not to be assumed, the disadvantages of this method are as follows: (1) Since the wall weight is reduced in this method, there is a certain difference between the calculated stress field of the soil behind the wall and the real situation. (2) Multiple reduction processes need to be calculated, which leads to low calculation efficiency.

36.2.4 Wall Elastic Modulus Reduction Method

Some measures are used to make the wall move actively in the above numerical simulation methods. However, the active limit state of soil behind the wall can also be reached from another angle in the wall elastic modulus reduction method. In this method, the wall is assumed to be an elastomer with sufficient thickness to make it equivalent to a spring. By reducing the elastic modulus (spring stiffness) of the wall, the soil movement behind the wall is induced by the wall deformation until the active limit state is reached due to sufficient deformation (Zhao et al. 2004). The key to the realization of this method is that sufficient lateral deformation is generated in the soil to make the slip surface in a fully connected state. Therefore, the thickness and elastic modulus of the wall need to be estimated through trial calculation.

The wall elastic modulus reduction method is simple in operation and fast in calculation. It is a relatively efficient numerical simulation method of active earth pressure. However, the main defects of this method are as follows: (1) Retaining wall is assumed to be a spring. And the external boundary of the wall needs to be constrained. Thus, the influence of wall motion mode on active earth pressure cannot be considered. It is also impossible to get the wall motion mode in the real situation. (2) The requirement of advance trial calculation is that the lateral deformation of the soil behind the wall is enough to make the slip surface in the soil pass through. So there are subjective factors in this method. (3) In this method, the wall soil interface is curved due to the reduction of the elastic modulus of the wall, which makes the calculation result different from the actual situation.

36.2.5 Soil Weight Increase Method

In the previous methods, most of them make the wall move away from the soil from the perspective of reducing resistance. On the contrary, the soil thrust is increased in the soil weight increase method. The active limit state of the soil behind the wall is reached under the active movement of the wall. The physical and mechanical parameters such as the weight, elastic modulus and the bottom friction coefficient of the wall have not been changed in this method. Therefore, the actual movement mode of the wall can be truly reflected. The simulation process of the soil weight increase method is basically the same as that of the wall weight reduction method. The increase coefficient of soil is defined as the ratio of the increased soil to the actual soil, namely:

$$\gamma_s' = k_s \gamma_s \tag{36.3}$$

In the formula: The actual weight of soil is γ_s. The increase coefficient of soil is k_s. The increased weight of soil is $\gamma_s{}'$.

The wall motion mode need not be assumed in the soil weight increase method. So it is suitable for the analysis of the actual movement mode of the retaining wall. However, the shortcomings of it mainly include: (1) Due to the change of soil weight, it is difficult to obtain the active earth pressure under the actual physical and mechanical state of soil. (2) The position and shape of the slip surface obtained by numerical simulation are different from the actual situation due to the change of soil weight. Multiple calculations need to be performed, which leads to low calculation efficiency.

36.3 Example Analysis and Comparation

A test model on the active earth pressure on rigid retaining wall with translation mode (Khosravi et al., 2013) is selected herein as a typical example. The numerical model via FLAC3D for the example is shown in Fig. 36.1. In this example, the rigid boundary and retaining wall are linear elastic constitutive materials, and soil is an ideal elastic–plastic Mohr–coulomb material satisfying non-associated flow rule. The interfaces are established between backfill, rigid boundary and retaining wall. There are 18,100 zones and 22,890 nodes in this model. According to the test data, the main physical and mechanical parameters used in the numerical model are shown in Table 36.1. The interface between wall and soil is established by attaching to the wall. According to the test (Khosravi et al., 2013), the external friction angle of the interface is the same as the internal friction angle.

36.3.1 Distribution of Active Earth Pressure

The comparison of distribution of active earth pressure calculated by the above five methods is shown in Fig. 36.2. The depth from the top of the backfill is h. The height of the backfill is H. The horizontal active earth pressure is σ_{ha}. The weight of the backfill is γ. Calculation results of Rankine theory and test results are also shown in

Fig. 36.1 Numerical model via FLAC3D for the example

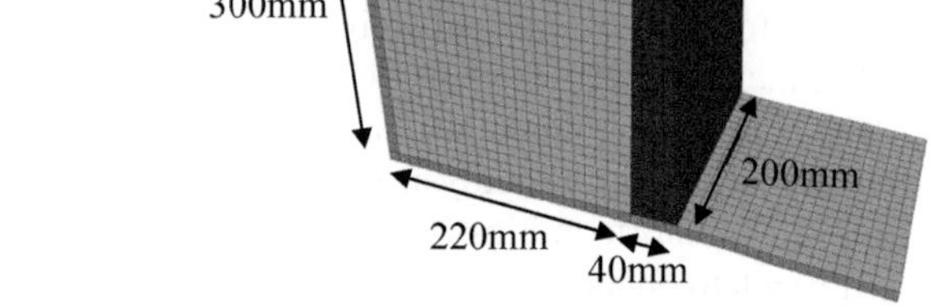

Table 36.1 Main physical and mechanical parameters of the numerical model for the example

Material type	γ /(kN·m^{-3})	c (kPa)	φ (°)	E (MPa)	μ
Rigid boundary	30	/	/	6×10^4	0.2
Backfill	12.7	0	35	3	0.3
Retaining wall	27	/	/	3×10^4	0.22

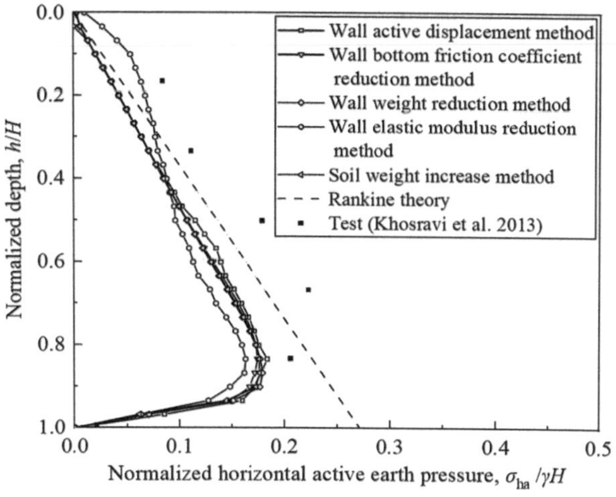

Fig. 36.2 Comparison of distribution of active earth pressure with depth using various methods

Fig. 36.2. It can be seen that the active earth pressure is parabola distributed along the wall height in the numerical simulation and test results. The top of the parabola is close to the bottom of the wall. The smooth curve results can be obtained in the wall bottom friction coefficient reduction method, the wall weight reduction method, and the soil weight increase method. The reason may be that the gradual change process of wall movement can be implemented in these methods and enough precision can be set to meet the calculation requirements. However, the wall moving speed and iteration steps need to be manually set in the wall active displacement method. Wall thickness and elastic modulus also need to be estimated in the wall elastic modulus reduction method. These subjective factors are the reasons why it is difficult to obtain high-precision calculation results in the two methods.

From Fig. 36.2, the distribution law and size of active earth pressure obtained by various numerical simulation methods are very close. The variation trend of earth pressure with depth obtained by simulation is close to the test result, which is different from the linear increase with depth of Rankine theory. In addition, the numerical simulation results are less than the calculation results of Rankine theory and test results. The maximum earth pressure stress of numerical simulation is 17.4–27.0% smaller than the test value.

36.3.2 *Magnitude and Application Point of the Resultant*

Table 36.2 shows the resultant force and its application point of active earth pressure (the action position ratio is the ratio of the height from the application point to the wall bottom to the wall height) obtained by various numerical simulation methods. The results of Rankine theory, Coulomb theory and model test are also included.

Table 36.2 Comparison of resultant force and its application point of active earth pressure using various methods

Numerical simulation methods	Resultant force (N m^{-1})	Height of application point (m)	Action position ratio
Wall active displacement method	117.22	0.113	0.38
Wall bottom friction coefficient reduction method	110.22	0.110	0.37
Wall weight reduction method	108.51	0.110	0.37
Wall elastic modulus reduction method	110.25	0.124	0.41
Soil weight increase method	584.92	0.110	0.37
Rankine theory	154.87	0.100	0.33
Coulomb theory	116.90	0.100	0.33
Test (Khosravi et al. 2013)	134.15	0.107	0.36

It can be seen that, except for the soil weight increase method, the resultant force of active earth pressure obtained by other numerical simulation methods is close to that of Coulomb theory, and the difference is about 0.3–7.2%. However, the result of numerical simulation is 24.3–29.9% smaller than that of Rankine theory. The possible reasons are as follows: (1) According to the test, the wall back is taken as completely rough in the numerical model, while the friction between wall and backfill is ignored in Rankine theory. (2) The ideal elastoplastic constitutive model is adopted in the numerical simulation method, which can fully reflect the soil deformation. However, Rankine and Coulomb theory both belong to the category of limit equilibrium theory, so that soil deformation cannot be considered. The numerical simulation results are 12.6–19.1% smaller than the test results. It may be related to the continuous medium used for soil and grid division.

For the position ratio of resultant force, except for the wall elastic modulus reduction method, results obtained by other numerical simulation methods are close to the test results, which are about 0.37–0.38, and are about 0.04–0.05 higher than that of Rankine theory. The position ratio of the resultant force obtained by the wall elastic modulus reduction method is 0.41, which is about 0.05 higher than the test result.

36.3.3 Relationship Between the Resultant and Wall Displacement

The relationship curve between the normalized resultant force of horizontal earth pressure and the normalized wall displacement obtained by the numerical simulation method is shown in Fig. 36.3, where E_{ha} is the resultant force of horizontal earth pressure, Δx is the translation displacement of the wall. The judgment methods of active limit state are as follows: For the wall active displacement method (Fig. 36.3a), the resultant force under each preset wall displacement is calculated. When the variation of the resultant force decreases to a certain value (1% of the resultant force under small earth pressure), the active limit state is considered to be reached. For the other three methods (Fig. 36.3b–d), typical time step of 6 times (the solution time step of the initial stress equilibrium state of the numerical model) is taken as the calculation termination condition. The difference between two adjacent reduction or increase coefficients is calculated to be less than 0.01, and this state is regarded as the active limit state. The mean value is taken as the calculated value of the reduction coefficient or increase coefficient. It can be seen from Fig. 36.3 that when the soil reaches the active limit state, the wall displacement is about 1.2–2.4‰ of the wall height, which is close to 1.75‰ of test.

36.3.4 Location of the Slip Surface

Figure 36.4 shows the slip surface of soil calculated by numerical simulation method (represented by the contour of max. shear strain increment). Rankine theory and test results are also shown in Fig. 36.4. It can be seen that there are curved characteristics for the slip surfaces obtained by numerical simulation methods, which are close to the model test results. Although the results are different from plane slip surface of Rankine theory, there is little difference between them. The locations of the slip surface obtained by each method are basically consistent.

36.3.5 Comprehensive Comparison

According to the above numerical simulation methods and calculation results of them, the application scope (corresponding to the wall motion mode), the smoothness of the distribution curve of earth pressure, the calculation efficiency, the rationality of the simulation calculation principle, and the advantages and disadvantages of each method are comprehensively analyzed and evaluated as shown in Table 36.3. The calculation efficiency is expressed by the relative quantity (ratio), that is, the number of operation steps of the wall active displacement method is taken as the basic unit,

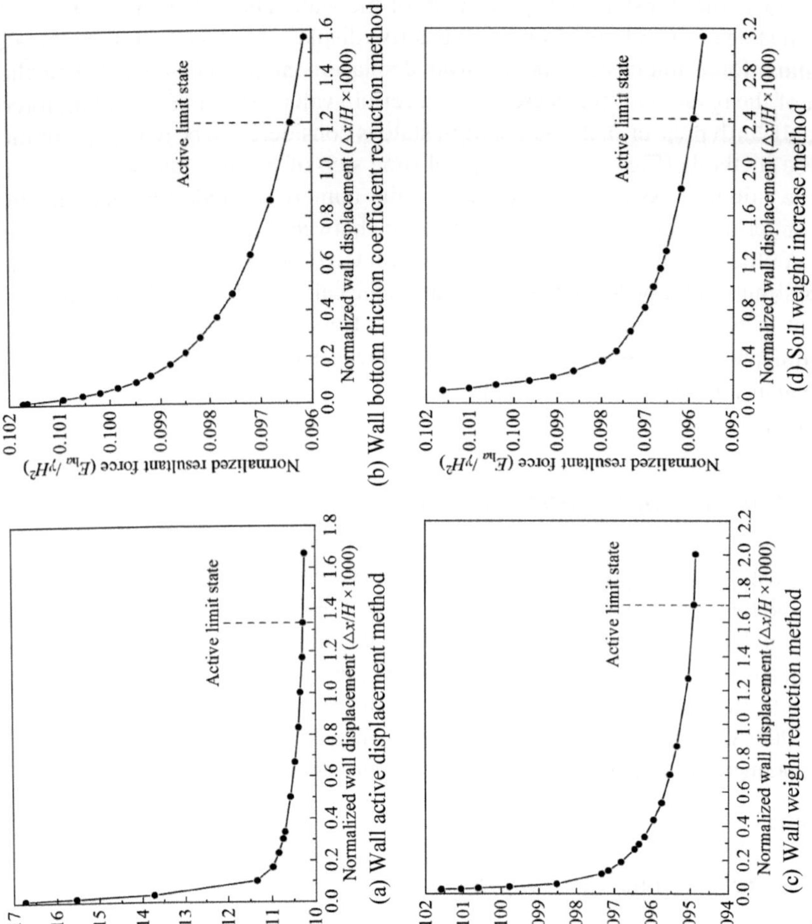

Fig. 36.3 Variation curves of resultant force of earth pressure with wall displacements

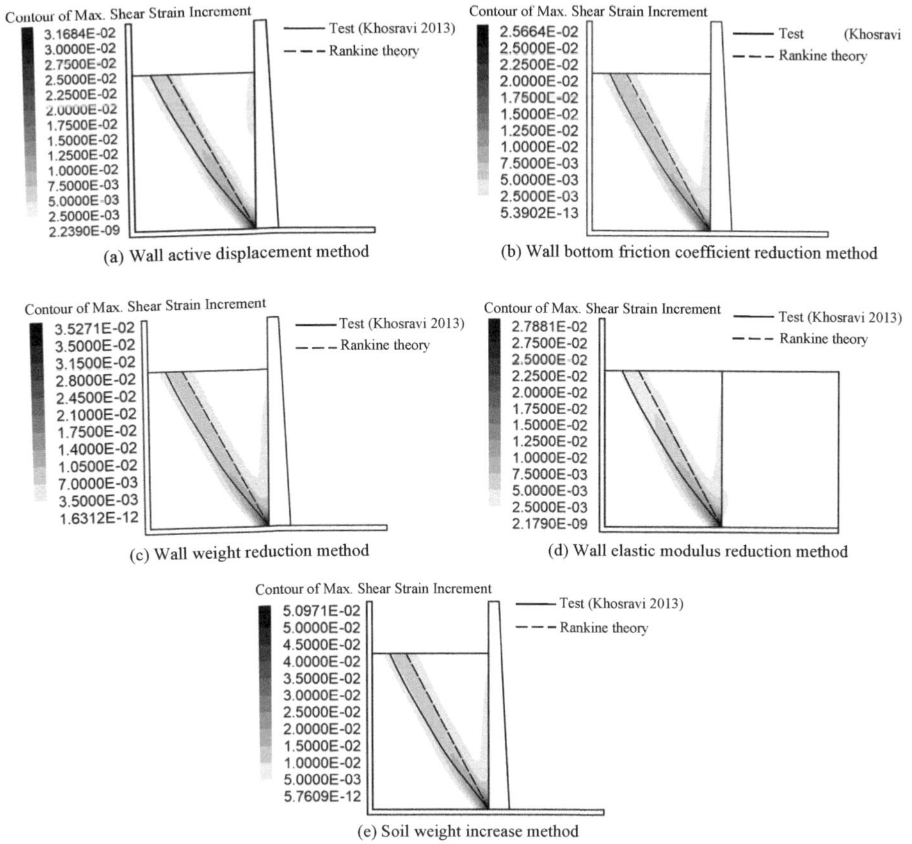

Fig. 36.4 Comparison of sliding surfaces of the retained soil using different methods

and the number of operation steps of other methods is divided by it. All calculations are performed on the same numerical model and on the same computer.

For wall motion mode, the wall active displacement method, the wall weight reduction method and the soil weight increase method can be applied to the motion mode where wall translation and rotation occur simultaneously. The wall bottom friction coefficient reduction method is only applicable to the wall translation mode. The wall motion mode cannot be considered in the wall elastic modulus reduction method.

For the smoothness of the earth pressure distribution curve, the relatively smooth results can be obtained by the wall bottom friction coefficient reduction method, the wall weight reduction method and the soil weight increase method. The distribution curves obtained by the other two methods are relatively bad in smoothness.

In terms of calculation efficiency, the active wall displacement method and the wall elastic modulus reduction method have relatively high calculation efficiency. In particular, the calculation efficiency of latter is higher, and the calculation time of it

Table 36.3 Comprehensive evaluation of different numerical simulation methods

Numerical simulation method	Application scope	Smoothness of earth pressure distribution	Calculation efficiency	Rationality of principle	Advantages	Disadvantages
Wall active displacement method	Translation and rotation	Less smooth	1.00	Defective	Simple and fast	(1) Low calculation accuracy (2) Bad rationality of principle
Wall bottom friction coefficient reduction method	Translation	Smooth	4.35	Reasonable	Conform to the physical mechanism	(1) Only translation mode (2) Low calculation efficiency
Wall weight reduction method	Translation and rotation	Smooth	4.28	Reasonable	Conform to the physical mechanism	(1) Low calculation efficiency
Wall elastic modulus reduction method	Unable to consider	Unsmooth	0.01	Defective	Simple and efficient	(1) Unable to reflect the wall motion mode (2) Bad rationality of principle (3) Low calculation accuracy
Soil weight increase method	Translation and rotation	Smooth	3.87	Defective	Reflect the wall motion mode	(1) Inconsistent with actual state of soil (2) Dimensionless results (3) Low calculation accuracy

is about 0.01 times of the wall active displacement method. The calculation time of other methods is about 3.9–4.4 times of the wall active displacement method.

As for the rationality of the simulation calculation principle, the wall active displacement method belongs to the method of manually applying displacement to the wall. The corresponding physical mechanism of wall soil movement and the mechanism of soil stress change behind the wall (stress path) are inconsistent with the actual situation. In addition, the active limit state of soil is reached by increasing the soil weight in the soil weight increase method. The active earth pressure under the actual physical and mechanical state is hard to be reflected in this method. Furthermore, the influence of the wall motion mode on the active earth pressure cannot be reflected in the wall elastic modulus reduction method. So, these three methods are methods with obvious deficiencies in the calculation principle. However, the reasonable process of wall movement under earth pressure to make soil gradually reach the active limit state can be appropriately reflected in the other two methods. The calculation principles of them are correct.

Considering the rationality of the calculation principles, the feasibility of simulating the wall motion mode, the applicability of the calculation results and the calculation efficiency of these five methods, the wall weight reduction method should be preferred, followed by the wall bottom friction coefficient reduction method.

36.4 Conclusions

Aiming at the numerical simulation methods for the active earth pressure on rigid retaining wall, we summarize and analyse the realization process, advantages and disadvantages of five numerical simulation methods, including wall active displacement method, wall bottom friction coefficient reduction method, wall weight reduction method, wall elastic modulus reduction method, and soil weight increase method. The numerical simulation methods are comprehensively evaluated from wall motion mode, smoothness of pressure distribution curve, calculation efficiency and rationality of the principle. The main conclusions are drawn as follows:

(1) In terms of the realization principles of the five numerical simulation methods, wall active displacement method, soil weight increase method and wall elastic modulus reduction method are of obvious deficiencies. However, the calculation principle is reasonable by the wall bottom friction coefficient reduction method and wall weight reduction method.
(2) The distribution of active earth pressure along the wall height obtained using each method is relatively close, and all of them are in a parabola shape with the apex close to the wall bottom. In addition, the slip surfaces in the retained soil using different methods are curved, which is different from the planar slip surface in the classical theory.
(3) Example analysis results show that except for soil weight increase method, the difference of the resultant force between the numerical methods and the Coulomb theory is about 0.3–7.2%, which is 12.6–19.1% less than the test results. Apart from the wall elastic modulus reduction method, the wall displacement in the numerical methods is about 1.2–2.4‰ of the wall height if the active limit state of the retained soil is reached, which is close to the test value of 1.75‰.
(4) Based on the comprehensive comparison of the rationality of the calculation principle, the feasibility of simulating the wall motion mode, the applicability of the calculation results and the calculation efficiency of the five methods, it is suggested that the wall weight reduction method should be preferred, followed

by the wall bottom friction coefficient reduction method. The former can reflect the translation and rotation modes of the wall, while the latter can only reflect the translation mode.

Acknowledgements This research is supported by the Construction S&T Project of Department of Transportation of Sichuan Province (Grant No. 2020A01) and National Natural Science Foundation of China (Grant No. 51578466).

References

Bang S (1985) Active earth pressure behind retaining walls. J Geotech Eng 111(3):407–412. https://doi.org/10.1061/(ASCE)0733-9410(1985)111:3(407)

Benmebarek N, Labdi H, Benmebarek S (2016) A numerical study of the active earth pressure on a rigid retaining wall for various modes of movements. Soil Mech Found Eng 53(1):39–45. https://doi.org/10.1007/s11204-016-9362-z

Benmeddour D, Mellas M, Frank R, Mabrouki A (2012) Numerical study of passive and active earth pressures of sands. Comput Geotech 40:34–44. https://doi.org/10.1016/j.compgeo.2011.10.002

Bhatia SK, Bakeer RM (1989) Use of the finite element method in modelling a static earth pressure problem. Int J Numer Anal Methods Geomech 13(2):207–213. https://doi.org/10.1002/nag.1610130208

Cai YY, Chen QS, Zhou YT, Nimbalkar S, Yu J (2017) Estimation of passive earth pressure against rigid retaining wall considering arching effect in cohesive-frictional backfill under translation mode. Int J Geomech 17(4):04016093.1–04016093.11 https://doi.org/10.1061/(ASCE)GM.1943-5622.0000786

Chang MF (1997) Lateral earth pressures behind rotating walls. Can Geotech J 34(2):498–509. https://doi.org/10.1139/t97-016

Chen YK, Wang YM, Xu RQ, Gong XN (2004) Numerical analysis of active earth pressure on rigid retaining wall. Chin J Rock Mech Eng 23(6):989–995

Chen FQ, Yang JT, Lin YJ (2019) Active earth pressure of narrow granular backfill against rigid retaining wall near rock face under translation mode. Int J Geomech 19(12):04019133.1–04019133.13. https://doi.org/10.1061/(ASCE)GM.1943-5622.0001525

Clough GW, Duncan JM (1971) Finite element analysis of retaining wall behavior. J Geotech Eng 97(12):1657–1673. https://doi.org/10.1061/JSFEAQ.0001713

Coulomb C A (1773) Essai sur une application des regles des maximis et minimis a quelques problemes de statique relatifs a 1'architecture. Memoires de l'Academie Royale pres Divers Savants 7:343–382. (in French)

Dai ZH, Lin ZY, Zheng YP, Lu JC (2009) Finite element method for computations of active earth pressures acting on L-shaped retaining walls with reduced friction coefficients of base bottoms. Chin J Geotech Eng 31(4):508–514

Dai ZH, Lin ZY, Zou ST, Lu JC (2008) Finite element method for calculations of active earth pressures by reducing densities of rigid retaining walls. In: Proceedings of the 7th Fujian Provincial symposium on geotechnical and foundation engineering. Fujian, China, pp 31–35

Fan CC, Fang YS (2010) Numerical solution of active earth pressures on rigid retaining walls built near rock faces. Comput Geotech 37:1023–1029. https://doi.org/10.1016/j.compgeo.2010.08.004

Fang YS, Ishibashi I (1986) Static earth pressures with various wall movements. J Geotech Eng 112(3):317–333. https://doi.org/10.1061/(ASCE)0733-9410(1986)112:3(317)

Fang YS, Chen TJ, Wu BF (1994) Passive earth pressure with various wall movements. J Geotech Eng 120(8):1307–1323

Jiang MJ, He J, Wang JF, Liu F, Zhang WC (2014) Distinct simulation of earth pressure against a rigid retaining wall considering inter-particle rolling resistance in sandy backfill. Granul Matter 16(5):797–814. https://doi.org/10.1007/s10035-014-0515-3

Khosravi MH, Pipatpongsa T, Takemura J (2013) Experimental analysis of earth pressure against rigid retaining walls under translation mode. Geotechnique 63(12):1020–1028. https://doi.org/10.1680/geot.12.P.021

Matsuzawa H, Hzaarika H (1996) Analyses of active earth pressure against rigid retaining walls subjected to different modes of movement. Soils Found 36(3):51–65. https://doi.org/10.3208/sandf.36.3_51

Nakai T (1985) Analysis of earth pressure problems considering the influence of wall friction and the wall deflection. In: Proceedings of the 5th international conference on numerical and analytical methods in geomechanics. Nagoya, Japan, pp 765–772

Ozawa Y, Duncan J M (1976) Elasto-plastic finite element analyses of sand deformations. In: Proceedings of 2nd international conference on numerical methods in geomechanics. Blacksburg, USA, pp 243–263

Paik KH, Salgado R (2003) Estimation of active earth pressure against rigid retaining walls considering arching effects. Geotechnique 53(7):643–653. https://doi.org/10.1680/geot.2003.53.7.643

Potts DM, Fourie AB (1986) A numerical study of the effects of wall deformation on earth pressures. Int J Numer Anal Methods Geomech 10:383–405. https://doi.org/10.1016/0148-9062(87)91531-2

Rankine WJM (1857) On the stability of loose earth. Scotl Philos Trans R Soc Lond 147(1):9–27

Tsagareli ZV (1965) Experimental investigation of the pressure of a loose medium on retaining walls with a vertical back face and horizontal backfill surface. Soil Mech Found Eng 2(4):197–200. https://doi.org/10.1007/BF01706095

Woerden FT, Achmus M (2013) Numerical modeling of three-dimensional active earth pressure acting on rigid walls. Comput Geotech 51:83–90. https://doi.org/10.1016/j.compgeo.2013.02.004

Xiao XD, Li MG, Wu H (2020) DEM Simulation of earth pressure exerted on rigid retaining wall subjected to confined soil. Chin J Undergr Space Eng 16(1):288–294

Ying HW, Zhu W, Huang D, Xie XY, Li BH (2012) Active earth pressures against rigid retaining walls with narrow cohesive backfill. Chin J Geotech Eng 34(S1):13–18

Zhang HZ, Xu CJ, Liang LJ, Hou SL, Fan RD, Feng GH (2021) Discrete element simulation and theoretical study of active earth pressure against rigid retaining walls under RB mode for finite soils. Rock and Soil Mech 42(10):2895–2907. https://doi.org/10.16285/j.rsm.2021.0458

Zhao SY, Zheng Y, Zhou CF (2004) Study on lateral earth pressure of retaining wall by FEM numerical simulation. In: Proceedings of the 8th national conference on rock mechanics and engineering. Beijing, China, pp 303–309

Zhou J, Peng SQ, Fan L (2008) Particle flow simulation of active earth pressure distribution on rigid retaining wall. Rock and Soil Mech 29(3):629–632+638. https://doi.org/10.16285/j.rsm.2008.03.040

Chapter 37
Durability Assessment of UK and Portuguese Jurassic Mudrocks

J. C. Cripps, F. T. Jeremias, and M. A. Czerewko

Abstract The term mudrock defines a large variety of fine-grained clay-rich sedimentary rocks. As these materials occur extensively they are often encountered in engineering projects, as natural ground on construction and ground engineering sites and as made ground, fill and construction materials. They may display a range of engineering behaviours controlled mostly by their composition and textural characteristics. In some situations, due to relatively rapid breakdown and susceptibility to volume changes they can cause problems both during and after construction. Investigation of mudrock durability can enable engineers to avoid problems, however prior assessment of mudrocks can be hindered if the weaker, less durable materials disintegrate during sampling and testing such that a biased view of the engineering behaviour of these rocks is formed. The paper presents durability assessments for UK Jurassic mudrocks that comprise the Lias, Oxford and Kimmeridge Clays formations. These mudrocks outcrop as a band stretching across England from the Dorset coast in the south west to the Yorkshire coast in the north east. The Portuguese Jurassic mudrocks studied are lower to middle Kimmeridgian in age that outcrop in the Arruda dos Vinhos Basin north/north-east of Lisbon. A review of the engineering properties of these UK and Portuguese Jurassic mudrock formations aims to compare the relationships, of the rocks of similar age, between the mineralogy, structure, strength and in order to investigate their similarities and differences in order to elucidate the controls on the durability of the different materials.

Keywords Mudrock · Durability · United Kingdom · Portugal · Jurassic

J. C. Cripps (✉)
University of Sheffield, Sheffield S1 3JD, UK
e-mail: j.c.cripps@sheffield.ac.uk

F. T. Jeremias
Laboratório Nacional de Engenharia Civil, Lisboa, Portugal
e-mail: ftelmo@lnec.pt

M. A. Czerewko
AECOM Ltd., Chesterfield, UK
e-mail: Mourice.czerewko@aecom.com

S. Wang et al. (eds.), *Engineering Geology for a Habitable Earth: IAEG XIV Congress 2023 Proceedings, Chengdu, China*, Environmental Science and Engineering,
https://doi.org/10.1007/978-981-99-9057-3_37

37.1 Introduction

Both the UK and Portugal possess significant sequences of Jurassic mudrocks in locations where urban development or routeways are being constructed. They form the foundations, occur in natural and cut slopes and are used as construction materials, including embankment fills. Changes in properties during and after the construction process can make it difficult for geotechnical engineers to determine appropriate parameters for design and construction. It is recognised that durability is one of the most important engineering properties of mudrocks, that involve exposure to the weathering environment (Dick et al. 1994; Bell et al. 1997) and that slaking, the process of disintegration of clay-bearing rocks upon interaction with water (Erguler and Shakoor 2009), is one of the principal controls of mudrock durability and breakdown. The paper presents a review of the implications to ground engineering of the variation in durability within the UK Jurassic succession. This is followed by an example of a study into the variation in the durability of the mudrocks of the Arruda dos Vinhos Basin to the north and north-east of Lisbon, Portugal, with regards to the construction of roads, embankments and cuttings in this area.

37.2 Jurassic Mudrock Formations and Engineering Implications in the UK

The Triassic-Jurassic boundary interval was subject to a period of major change responsible for catastrophic mass extinction (Simms 2007). The deposits are the result of a complex pattern of sedimentation over a short time period effected by several minor marine transgressions and regressions. These events were associated with the major world-wide marine transgression brought about by the breakup of the single super continent of Pangaea with rapid development of marine conditions as the epicontinental Rheatic Sea located on the western margin of the Tethys Ocean progressed into Britain (Hesselbo et al. 2004). The full Jurassic sequence is recorded in Britain extending from 105 Ma through to 146 Ma. At this time Britain was located between 30° and 40° north of the equator, with likely temperate climatic conditions. As Fig. 37.1 shows, the sequence outcrops over from the south coast to the north east of England, and in small areas of Scotland and Wales, so the sequence underlies many urban areas and connecting transport routes. Jurassic sediments formed to the west of this have been lost through erosion, while to the east they underlie younger strata.

During the earliest Lower Jurassic, within the expanding and deepening marine environment much of Britain was under water and mostly highly fossiliferous alternating clays and limestones were laid down, which through diagenetic changes have been converted to mudstones with interbedded limestone and argillaceous limestones (Toghill 2000). During the latter part of the Early Jurassic and into the Middle Jurassic a number of active areas of tectonic uplift occurred, resulting in. deposition in three

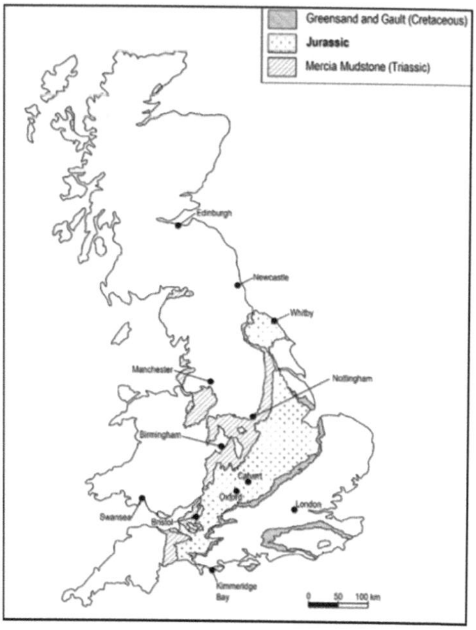

Fig. 37.1 Outcrop of Jurassic strata in the UK, shown bound by outcrop of older Triassic deposits, and younger Cretaceous deposits

interconnected basins separated by highs, (Cox et al. 1999) where there were shelf seas. Thick deposits of predominantly clay and mudstone were formed in the basins, whereas interbedded sequences of mudstone, limestone many of which are oolitic and bioclastic, and sandstone, were deposited in the shelf areas (Simms et al. 2004). This means there are lateral and vertical changes in the materials deposited, and the extent of burial diagenesis also varies. At some stratigraphic levels the same formational sequence can be recognised across large areas of the country. For example the Blue Lias Formation of southwestern England is now recognised as being the stratigraphic equivalent of the Scunthorpe Formation of the midlands and north of the country.

The Middle Jurassic deposits crop out along an almost continuous strip from the south to the north east of England. These predominantly carbonate sediments have been extensively exploited historically for local construction, whereas in eastern England, Middle Jurassic iron rich chamositic oolites were exploited for iron. Extensive areas where strong limestones overlie plastic clays and mudstones suffer from landslides, cambering (Chandler 1974) and undercutting action, leading to rock falls in steep slopes, both natural and man-made.

The Upper Jurassic Series marks a return to marine conditions during which sequences of thick organic clays and impersistent limestones dominate, followed during the close of the Jurassic by a regressive limestone sequence (Toghill 2000). A high input of organic material and volcanic ash deposits occur in Oxford Clay Formation stiff clays and mudrocks. The Kimmeridgian Stage followed this with a period of globally high sea levels in which mudrocks form the predominant lithology

(Wright and Cox 2001). The upper Jurassic of the north east of England is comprised mainly of interbedded units of mudstone, siltstone and sandstone formed under estuarine conditions (Cox and Sumber 2002). The end of the Jurassic is marked by a marine regression and a landmass covered most of Britain.

37.3 Consequences to Ground Engineering of UK Jurassic Mudrocks

The mineralogy of the British Jurassic mudrock deposits is dominated by clay minerals, particularly illite and kaolinite with significant proportions of mixed-layer illite–smectite and clastic minerals (see Table 37.1). But also present throughout the mudrocks and interbedded argillaceous limestones, are significant amounts of pyrite and calcite.

During the early Lower Jurassic, the prevailing marine conditions gave rise to deposition of dominant mudstone sequences in the deeper waters that are strongly affected by diagenesis and burial. These are interbedded with impure argillaceous limestones. Pleistocene glaciations have impacted on the condition of these deposits. In central and northern England repeated surcharging and unloading have structurally weakened the mudrocks when compared with southern counterparts. Periglacial freeze–thaw action and higher rainfall in all areas resulted in landslips, mantles of overlying unstable solifluction deposits and cambering activity. The Cotswold Hills of central England. where limestones overlie mudstone formations were particularly affected by these periglacial processes. Figures 37.2 and 37.3 illustrate lower Jurassic Blue Lias at different locations in England including coastal landsides on the south coast of England, where frost action has caused weakening and de-stressing. Since the Lower Lias sequences were formed in different basins, they experienced different degrees of over-consolidation and induration. Blue Lias mudrocks exposed on the north coast of the southwest peninsular are stronger than their northern England equivalents, seen in Fig. 37.3. This difference is rapidly removed by a few months of exposure, particularly following the onset of pyrite oxidation. Although fresh Lower Jurassic mudstones tend to be weak to moderately strong and rock-like materials, they quickly degrade to low strength, highly compressible plastic clays due to stress relief, weathering of pyrite and leaching of carbonates. Therefore, testing is needed to assess both their durability upon exposure and the potential for the development of chemical aggressivity. (Reid et al. 2017).

The younger Upper Jurassic sequences comprise thick units of pyrite bearing mudrocks that range from stiff over-consolidated clays to very weak mudstones. As with the Lower Jurassic sediments the potential for chemical aggressivity and rapid deterioration following exposure to atmospheric conditions must be assessed and mitigated for in designs, to avoid problems in ground engineering. In central and eastern England the sequences of middle and upper Jurassic Oxford Clay and succeeding Kimmeridge/Ancholme Clay are mostly dark coloured, bitumenous, stiff

Table 37.1 The mineralogy of some British Jurassic mudrocks (from Campbell 1993)

	Non-clay content (%)	Clay content (%)	Mineralogy (%)								
			Quartz	Calcite	Dolomite	Pyrite	A–A[*a]	Chlorite	Kaolinite	Illite	Illite/Smectite[*b]
Upper Jurassic (n = 8)	27.9–48.5	51.5–77.5	8.2–31.3	0.2–30.7	0–3.7	7.5–10.4	0.8–2	1.4–4.3	17.4–24	9–19.6	17–35.7
Mid Jurassic (n = 8)	38.34–55.9	44.1–61.7	10.7–28.2	0.4–31	0–0.6	6.9–10	1.2–1.9	0.8–2.1	9.4–18	15–24.2	13.3–24.3
Lower Jurassic (n = 8)	34.9–63.5	36.5–65.1	10.1–39.6	9.2–44.3	0	3.7–12.8	0.9–2.5	0–2	9–18.5	8.5–27.1	12.2–22

Note Lower Jurassic includes the Blue Lias & Lower Lias formations; mid Jurassic includes Oxford Clay; upper Jurassic includes Ampthill Clay, Kimmeridge Clay and Ancholme Clay formations

[*a] A-A represents apatite & anatase

[*b] Illite/smectite may include discrete smectite

Fig. 37.2 Rapid deterioration consequential of periglacial conditions affecting Lower Jurassic, Charmouth Mudstone on the south coast of England

Fig. 37.3 Variability of Blue Lias strata due to mechanical de-stressing caused by glacial loading and unloading, **a** weak friable mudstone interbedded with argillaceous limestone in Nottinghamshire, central England, **b** medium strong thinly bedded mudstone interbedded with argillaceous limestone at Kilve in southwest England

clay formations with some beds of argillaceous limestone all of which contain pyrite, and which readily degrade on exposure to weathering action, as seen in Fig. 37.4.

If not managed accordingly, rapid degradation in these formations will have adverse impacts in slope and routeway construction. An example of rapid deterioration of prepared road and slope formation levels is shown in Fig. 37.4, where the finished road formation layer deteriorated after a single rain event and cutting side slopes desiccated and ravelled just as quickly.

Weathering action results in hard and soft bands being produced, with selenite, a consequence of pyrite oxidation, occurring in the weathered horizons particularly at former groundwater levels. Weathered horizons occur as orange-brown gypsum rich residual silty clay (see Fig. 37.5a) that overlie fresh extremely weak blue-grey mudstone seen in Fig. 37.5b. The Kimmeridge Clay formation outcrops along the southern coast of England and continues northwards to northern England, where it is known as Ancholme Clay. It consists of organic or calcareous silty and sandy mudstones with thin siltstone and calcareous cementstone beds (see Fig. 37.4a). Planes of weakness are formed along bedding planes, especially if highly pyritic and bituminous horizons are present. Fissures are often present at shallow depth within the upper 3–5 m affected by Pleistocene chemical weathering (Gallois and Horton 2001). Even in this shallow layer, effects (Fig. 37.6) of glacial inter-action, including disturbance due to destressing, and chemical degradation cause routeway construction problems. In excavations, the mudrocks may undergo lateral displacement because

Fig. 37.4 Example of the rapid deterioration in Kimmeridgian Ancholme Formation mudrock. **a** Intact mudrock as very weak dark grey thinly bedded mudstone, **b** Desiccation and ravelling of a slope cut section following a rain event, **c** Deterioration of road formation surface by shrink-swell desiccation, and selenite growth following a single rain event, **d** Rapid growth of selenite crystals in material stockpile resulting from pyrite oxidation

of release of stored strain energy, whereas more rigid limestone beds may become destabilised due to differential movements. Some low durability mudrocks, especially those containing swelling clay minerals typical of Jurassic mudrocks (see Table 37.1), undergo significant swelling when the overburden and lateral restraint are reduced resulting in uplift, lateral displacements, loading of support systems and on re-loading excessive settlement may occur. This process may be exacerbated by the presence of water and it tends to be more rapid in over-consolidated clay deposits such as Oxford Clay than in more durable mudrocks such as Blue Lias. It imposes significant stresses on inter-bedded stronger lithologies within the strata. Examples of geotechnical parameters for a range of Jurassic mudrocks are presented in Table 37.2.

The differences in engineering characteristics generally reflect variations in depositional environment and mineralogy, geological burial history and glacial influences. In view of the rapid deterioration in geotechnical properties that occurs when the mudrocks are exposed to weathering agencies, it is essential that the extent and rate of change in properties are fully characterised so that the engineering works are designed and construction processes carried out in an appropriate manner. The

Fig. 37.5 **a** Weathering profile through Oxford Clay, with the orange-brown horizon possessing gypsiferous enrichment by groundwater. **b** Below the cut slope the sequence becomes fresh extremely weak blue grey pyritic mudstone

Fig. 37.6 Ancholme Mudstone disturbed by glacial action, showing stress fractures and brown ferric-oxide residue from the oxidation of pyrite at the interface with overlying till

presence of smectite and mixer layer clay minerals engenders low strength and high susceptibility to breakdown due to slaking. It also causes the recorded high Ip values in Table 37.2, such that a number of Jurassic clay formations, including Oxford Clay, are prone to troublesome shrink-swell behaviour, both during and after construction, as shown in Fig. 37.7.

Heave may result in distortions to route-way alignments (Fig. 37.7b) and failure of lightly loaded structures. Failure may also occur in excavations such as slopes and highway cuts in which lithological variation within an exposed mixed sequence results in the slaking, ravelling and erosion of mudrocks leaving overlying more competent beds in an unsupported condition (Figs. 37.2 and 37.3b).

37.4 Ground Assessment

Assessments of durability required for optimising construction programming of the construction sequence may be undertaken by carrying out field trials. Figures 37.8 and 37.9 show the progressive degradation following the exposure of samples to natural weathering processes. Relatively short duration exposure led to breakdown of the materials, but this type of test might not provide the necessary data in an acceptable time period, and the weather conditions at the time of the trial may not match those at, during and after the construction period. Testing of ground material should classify the lithologies present in terms of physical properties but most importantly, the mudstones should be characterised in terms of their static and slake durability index strength and plasticity using available test procedures.

Unfortunately this approach is very time-consuming and may need expensive equipment such as large climatic chambers. A more practical approach is to focus on the main weathering mechanisms in mudrock. Studies have shown that mudrocks

Table 37.2 Example of typical geotechnical characteristics of some unweathered Jurassic mudrocks

	1—Lower Jurassic	2—Lower Jurassic	3—Middle to Upper Jurassic	4—Upper Jurassic
Sequence	Blue Lias (Angulata)	Lower Lias (Barnstone member of the Scunthorpe Formation)	Oxford Clay (Peterborough Member)	Ancholme Group (Ancholme Clay)
Locality	Kilve—Somerset	Cotgrave—Nottinghamshire	Calvert—Buckinghamshire	Melton—East Yorks
Description	Moderately weak to medium strong, grey, thinly bedded calcareous mudstone	Extremely weak, thinly laminated to thinly bedded mid grey calcareous silty mudstone	Very stiff to extremely weak, thinly bedded blue grey calcareous clay/mudstone	Extremely weak, thinly to medium bedded, dark grey, calcareous mudstone
UCS (MPa)	7–36 (n = 4)	0.6–35 (n = 16)	0.4–2 (n = 5) (*su = 68–191 kPa, n – = 5*)	0.9–1.9 (n = 3)
Id_2	46–59 (n = 3)	2–4 (n = 4)	2–4 (n = 3)	12–35 (n = 2)
Ij'	3–5 (n = 4)	6–8 (n = 4)	8	8 (n = 2)
mc (%)	6–22 (n = 6)	16–19 (n = 7)	25–36 (n = 4)	10–26 (n = 4)
Ip*[a]	10–38 (n = 6)	23–28 (n = 4)	36–41 (n = 6)	24–28 (n = 3)
Calcite (%)	1–39 (n = 4)	18–22.3 (n = 4)	6–13 (n = 4)	2.2–4.3 (n = 4)
Pyrite (%)	2.1–5.2 (n = 4)	0.1–3 (n = 4)	3.4–7.3 (n = 4)	2–3.9 (n = 4)

*[a] Ip based on < 425 μm fraction material derived from slake durability test. The values are representative of the slaked material only, and generally comprised 4–5% of material total from sample 1 to 58–60% of material for sample 3. See Czerewko and Cripps (2001) for Jar (static) slake classification & Fig. 37.15 for dynamic slake durability classification

Table 37.3 Classification, sampling depth and weathering grade of the Portuguese mudrocks

Sample	Classification/ weathering grade	Depth (m)	Sample	Classification/ weathering grade	Depth (m)
163	Calcareous siltstone (C)	4.1	112	Calcareous siltstone (A)	5.0
81	Mudstone (C)	6.5	114	Calcareous siltstone (A)	9.5
296	Mudstone (B)	10.5	334	Calcareous laminated siltstone (A)	7.0
333	Mudstone (B)	3.5	342	Calcareous laminated siltstone (A)	8.5
162	Calcareous mudstone (C)	3.4	431	Calcareous laminated siltstone (A)	9.0
283	Calcareous mudstone (D)	4.5	336	Mudstone (A)	9.5
285	Calcareous mudstone (C)	6.0	286	Calcareous mudstone (A)	9.5
52	Calcareous laminated mudstone (B)	10.5	436	Calcareous mudstone (A)	17.0
153	Calcareous laminated mudstone (C)	4.5	156	Laminated mudstone (A)	10.5
294	Calcareous claystone (C)	7.5	445	Laminated mudstone (A)	18.0
152	Calcareous laminated claystone (D)	3.0	154	Calcareous laminated mudstone (A)	7.0
			441	Calcareous laminate mudstone (A)	8.5

Weathering grade: (A)—unweathered; (B)—partially weathered; (C)—distinctly weathered; (D)—destructured

are susceptible to slaking in which cyclic wetting and drying results in the degradation of the material. If in the engineering situation the material will also experience mechanical disturbance then the test devised by Franklin and Chandra (1972) should be used, whereas static slaking test described by Czerewko and Cripps (2001) should be used for determining the intrinsic in situ performance of the material (see Fig. 37.10). Their potential to develop an aggressive chemical environment consequential on pyrite oxidation should be assessed though chemical testing (Reid et al. 2017). Shrink-swell behaviour and rapid degradation may be predicted by examination of the material with associated laboratory testing to determine compressive strength, durability by static 'Jar Slake' together with slake durability (Czerewko and Cripps 2001), swell potential and index testing of moisture condition and Atterberg limits.

The intrinsic behaviour of the material will also depend on microstructures, especially any mineral alignment, lamination and micro-discontinuities it contains. More

Fig. 37.7 Shrink-swell behaviour of Oxford Clay: **a** deep desiccation cracks developed in a completed construction layer, **b** undulating deformation and cracking of a road constructed over Oxford Clay

Fig. 37.8 Weathering assessment of Lower Lias Mudstone to determine rate and extent of change enabled management of material for improvement for use as earthwork fill, ensuring complete weathering of pyrite before construction

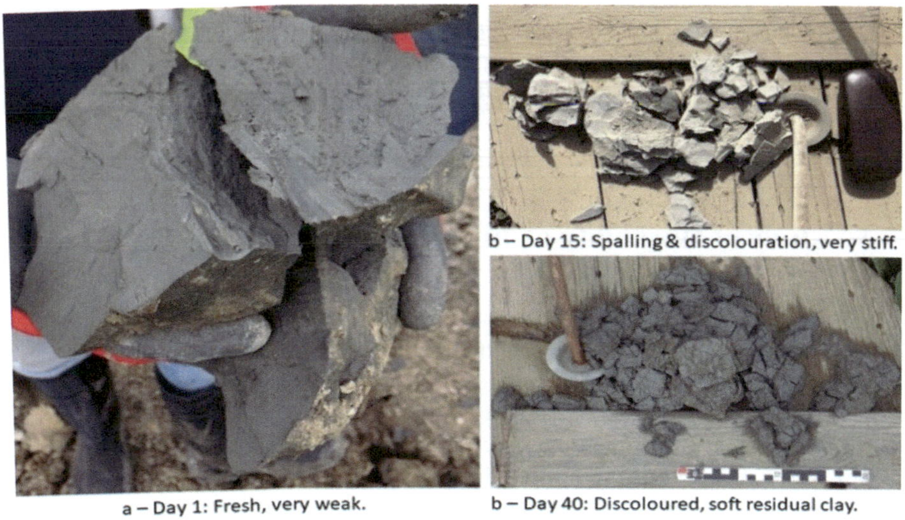

Fig. 37.9 Weathering assessment of Oxford Clay to determine rate and extent of change enabled management of excavations mitigating against avoidable deterioration through staged timing of operations

Fig. 37.10 Durability testing: **a**, **b** Static jar slake tests, showing classification; **c** Slake durability test, where plasticity tests are performed on the slaked residue retained in the water troughs

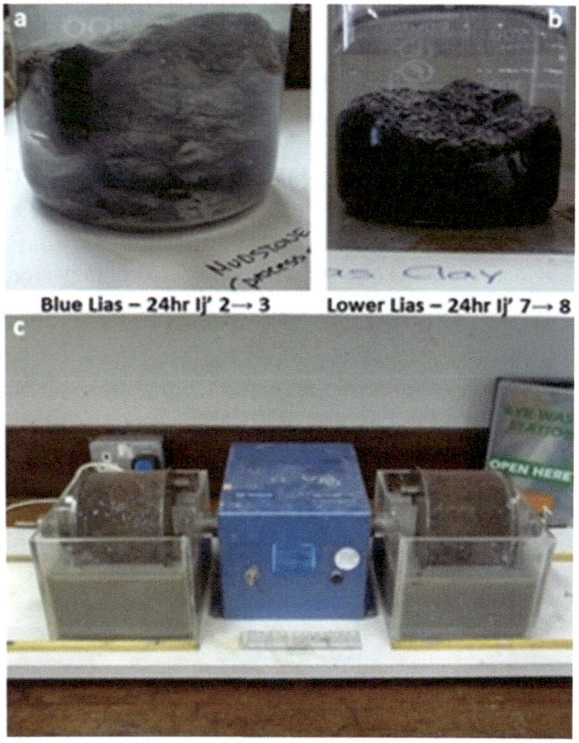

thorough investigation and interpretations may be assisted by determinations of the mineralogy and micro-structure using optical and scanning electron microscopy and X-ray diffraction techniques, although these methods are not usual in routine ground investigations.

37.5 Study of Durability of Jurassic Mudrocks from Portugal

The Portuguese mudrocks studied in this paper occur in the Arruda dos Vinhos basin set in Portuguese western mesocenozoic border in the North Lisbon area—See Fig. 37.11. The geological structure of the area consists of an important domal structure where Jurassic outcrops include lower to middle Kimmeridgian Abadia Beds in the core, overlain by upper Kimmeridgian to lower Tithonian calcareous formations rich in corals (Amaral Beds and Arranhó/Freixial Beds) (Kullberg et al. 2013). Proposed road network improvements in this area provided an opportunity to sample and study the Portuguese late Jurassic mudrocks involving sedimentary sequences of mudrocks interbedded with limestones and sandstones. The engineering problems related to these materials are mostly linked to the stability of excavations and natural slopes and the use of the materials for the construction of embankments. Therefore the durability of these materials was of particular interest. The mudrock samples used in the study were collected from boreholes drilled in the Abadia Beds at the locations shown in Fig. 37.11, mainly in the eastern and southern sectors of the basin. A range of fresh and weathered, cemented and compacted and fissile and massive lithologies was obtained, as listed in Table 37.3.

The mudrocks from the Abadia beds are chiefly constituted of quartz, feldspars (plagioclase and K-feldspar), carbonates and clay minerals, where quartz and feldspars form the coarser detrital fraction of these rocks, occurring either as floating

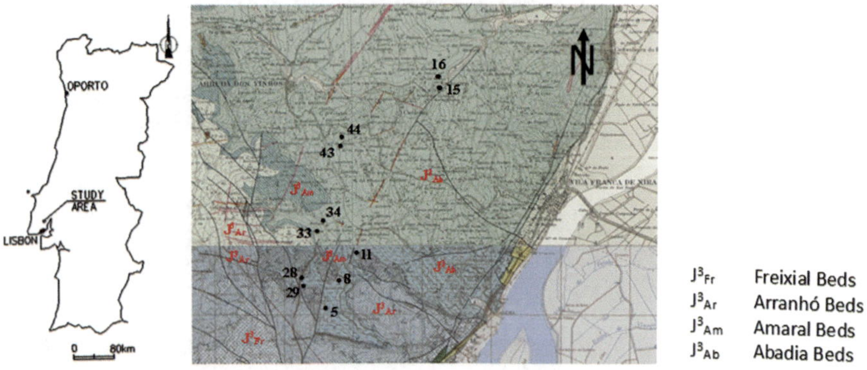

Fig. 37.11 Geological setting of Arruda dos Vinhos basin with the locations of borehole sampling points

grains spread in the matrix of claystone or forming a granular skeleton by grain-to-grain contacts and cementation bonds in siltstones. The carbonate minerals comprise calcite and dolomite that occur as micrite matrix and as grains and sparry cement. The clay minerals, kaolinite, illite, chlorite, smectite and mixed-layer illite–smectite, are the main constituent of the matrix and chlorite and illite also occur as flakes. Laboratory geotechnical characterization of the Portuguese mudrocks included determinations of fundamental physical properties, strength, swelling and slaking, as presented in Table 37.4. In their unweathered state the Jurassic mudrocks studied are classified as weak rocks with a uniaxial compressive strength less than 50 MPa.

Superficial type failures usually involving shallower parts of rock masses, are the cause of many of the instability problems associated with low to medium durability, weak mudrocks, such as the materials studied, especially where they are weathered and less dense (Franklin 1981). Commonly occurring forms of failure in these materials include excessive erosion, slump failures, debris flows and undercutting of stronger interbedded units. Undercutting occurs when mudrocks underlie more resistant bed of limestone or sandstone from which support is removed, or in mudrock formations that possess variable strength and durability characteristics. Although, in high slopes there is the possibility of falls of blocks of the stronger rock formation, there is low risk of a hazard associated with most superficial styles of failures. However, they may require costly remediation and/or routine maintenance (Jeremias et al. 2020).

The suitability of mudrocks as construction materials in embankments is mostly controlled by their durability and breakdown behaviour during and following excavation. For soil-like materials compaction layer thicknesses between 20 and 25 cm are usually adopted, whereas incorporating rock-like mudrocks allows compaction in layers up to 60 cm thick. The design of slopes and the construction of embankments therefore requires the assessment of the durability of the materials during excavation and exposure to weathering agencies (Table 37.4).

37.6 Approaches to Durability Assessment of Portuguese Mudrocks

Various studies point to the importance of the breakdown that occurs due to slaking, in which degradation is caused by cycles of alternate wetting and drying, with and without mechanical disturbance respectively that are evaluated by subjecting the materials to dynamic and static slaking tests. The static slaking test carried out, namely the Jar slake test, is rapid and simple to perform, as explained by Czerewko and Cripps (2001). A dried cube of rock between 40 and 50 mm side length is placed in distilled water and the development of fracturing and extent of breakdown changes (see Fig. 37.10a) are monitored according to their listed criteria at specific elapsed time intervals. Care was taken to ensure the samples were not subjected to vibration or disturbance.

Table 37.4 Geotechnical properties of mudrocks from North Lisbon area

	Weathered mudrocks (n = 11)	Siltstone/ laminated siltstone (n = 5)	Mudstone (n = 3)	Laminated mudstone (n = 4)
G_s	2.68–2.80 (2.71)	2.67–2.71 (2.70)	2.67–2.72 (2.68)	2.70–2.73 (2.71)
ρ_d (Mg m^{-3})	1.93–2.22 (2.13)	2.25–2.37 (2.31)	2.30–2.39 (2.38)	2.25–2.45 (2.39)
N (%)	17.9–23.6 (21.6)	11.7–17.3 (15.8)	12.4–15.8 (15.3)	12.3–16.7 (12.6)
% Clay	17.9–32.7 (23.9)	9.1–21.2 (17.2)	19.2–26.7 (25.8)	13.5–26.6 (20.8)
MBA (g/100 g fines)	3.1–6.0 (4.3)	2.6–4.6 (3.3)	3.5–5.8 (4.0)	2.9–3.3 (3.2)
PI (%)	17.4–29.2 (21.8)	9.6–22.5 (15.3)	15.4–27.4 (18.7)	12.9–15.7 (15.0)
σ_c (MPa)	1.5–5.7 (3.1)	12.9–23.4 (17.6)	13.6–26.0 (17.8)	8.1–27.6 (17.9)
$CI_{(0.23)}$	0.57–1.05 (0.82)	1.22–1.88 (1.81)	1.06–2.30 (1.48)	0.80–2.09 (1.52)
$CI_{(0.635)}$	0.51–0.98 (0.70)	1.03–1.74 (1.37)	0.84–1.51 (1.17)	0.78–1.63 (1.27)
ε_s (%)	11.8–21.2 (14.3)	10.1–18.0 (14.5)	13.2–19.4 (16.4)	12.2–16.0 (14.5)
ε_z (%)	0.71–9.45 (4.34)	1.7–16.7 (9.2)	2.8–11.6 (5.6)	2.0–9.9 (5.7)

G_s—Grain specific gravity; ρ_d—Dry density; N—Effective porosity; % Clay—Percentage of material < 2 μm determined inslake durability test residue; MBA—Methylene blue adsorption value; PI—Plasticity index; σ_c—Uniaxial compressive strength; $CI_{(0.23)}$ and $CI_{(0.635)}$—Cone indenter numbers measured for steel blade deflections, respectively, of 0.23 and 0.635 mm; ε_s—Swelling strain measured on remoulded specimens; ε_z—Swelling strain measured on intact specimens perpendicular to bedding/lamination; Median values shown in brackets ( )

For dynamic slaking tests the apparatus in Fig. 37.10b and the procedure described in ISRM (2007) were used. In this the breakdown caused to 10 No. equidimensional rock lumps with a mass of between 40–60 g each placed in a 2 mm mesh drum that is rotated partly submerged in a water bath is determined. In preliminary tests it was found that although in some samples much material had broken down, it was retained in the drum (i.e. > 2 mm), so the test would over-estimate the durability of the material. To enhance the sensitivity of the test, after the standard two cycles of 200 rotations (i.e. 200 + 200 rotations), the remains of samples were subjected a third cycle of drying and a further 600 rotations in the apparatus, such that all the samples were subjected in total to 1000 rotations through three cycles of 200 + 200 + 600 rotations.

The > 2 mm fraction retained in the drum mesh after the third cycle was photographed and described in terms of the extent of the breakdown and shape, and the grain-size distribution of the < 2 mm fraction was determined by sieving and sedimentation (Fig. 37.12). The slake durability index is determined by expressing the weight of the dry sample retained by the 2 mm mesh as a percentage of the original weight of dry sample. Accordingly, the standard slake durability index I_{d2} after 400 rotations of the apparatus and the modified I_{d3} value after 1000 rotations were determined and the classified according to the values suggested by Dick et al. (1994) indicated on the horizontal axis of Fig. 37.15.

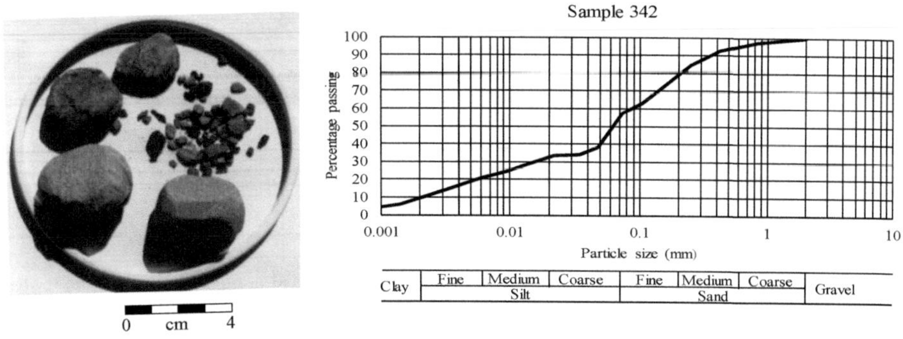

Fig. 37.12 Photographic record of the retained > 2 mm fragments retained in the drum and particle size distribution plot for the < 2 mm fraction from the water bath after 1,000 rotations

37.7 Durability Properties of Portuguese Mudrocks

The results of the jar slake and slake durability tests for Portuguese mudrocks are presented respectively in Figs. 37.13, 37.14 and 37.15 and these results for the different lithological groups are considered below.

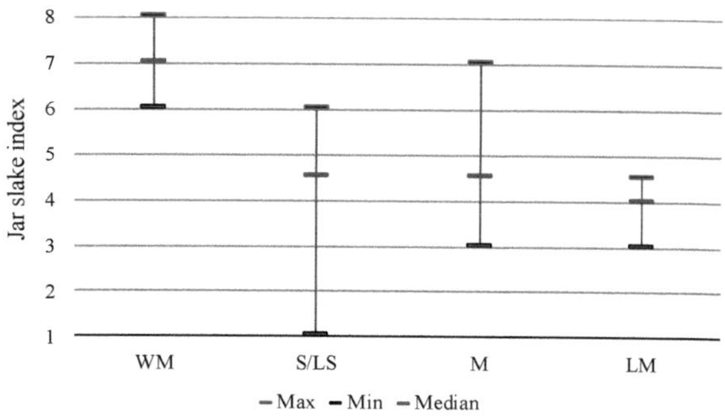

Fig. 37.13 Jar slake test results (WM—weathered mudrocks; S/LS—siltstone/ laminated siltstone; M—mudstone; LM—laminated mudstone)

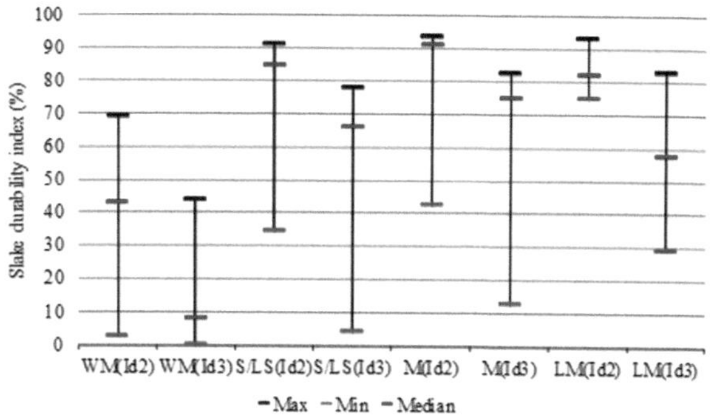

Fig. 37.14 Results of slake durability indexes I_{d2} and I_{d3}

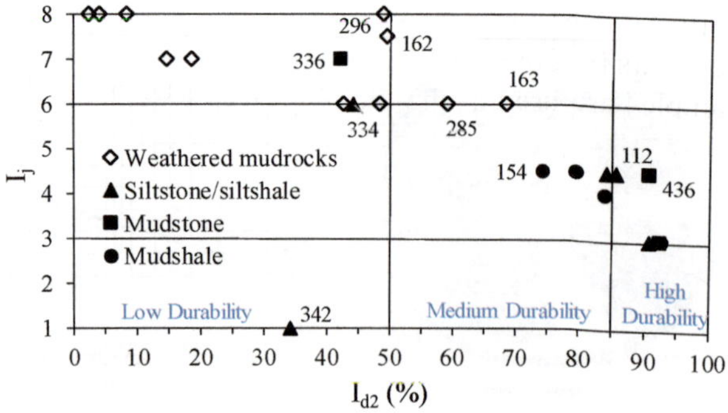

Fig. 37.15 Relationship between Ij and I_{d2} indices

37.7.1 Weathered and Fresh Mudstones

The integrated analysis of the static and dynamic durability tests data show that weathered mudrocks have low durability, where according with the durability classes proposed by Dick et al. (1994), with jar slake values between 6 and 8, which is in line with the results obtained for I_{d2} and I_{d3} in the dynamic tests, values were mostly less than 50%. The values of Id_2 and Id_3 obtained for sample 163 which was the most durable of the weathered mudrocks, come about because it consisted of a mixture of indurated/cemented lumps that did not breakdown and others that totally disintegrated (Fig. 37.15). Most of the fresh mudstones possessed medium to high durability except for sample 336 with rank 7 in jar slake, and Id_2 and Id_3 values of 42.3% and 12.5%, respectively. The least durable mudstones in Fig. 37.14, possessed high expandable clay mineral high contents.

37.7.2 Siltstone and Laminated Siltstone

The siltstone and laminated siltstone samples showed medium to high durability. However the apparently anomalous behaviour of sample 342 (laminated siltstone) with a rank 1 in the jar slake test and an I_{d3} of 26.5%, arose because it consisted of a heterogeneous mix of strong and weak layers, where the latter were rich in swelling clay minerals. Examination of the intact fragments of this sample retained in the slake durability drum after 3-cycles, revealed that they were very strong and similar to the material used in the jar slake test, while the remainder of the sample had totally disintegrated (Fig. 37.12). Sample 334 was thinly laminated such that breakdown into platy shaped fragments that then only slowly disintegrated (Fig. 37.15).

37.7.3 Laminated Mudstones

Laminated mudstones showed medium to high durability, with better median jar slake ranking but the third cycle slake durability data indicated significant breakdown. The effect of laminations over breakdown was not evident after the first and second cycles of the test, but then the fragments developed platy shapes that quickly broke down in the third cycle.

The data obtained from the geotechnical laboratory characterization carried out on Portuguese mudrocks show that the increases of porosity, swelling and plasticity in these materials correspond with lower strength. Susceptibility to breakdown is greater in mudrocks which have higher swelling strain and porosity values.

37.8 Conclusions

The Jurassic sequences of the UK and Portugal provide challenges for geotechnical projects as many members of the sequences possess low strength and high compressibility. Such adverse properties can be catered for if known about early in projects, before construction begins. Many strata may appear strong and incompressible in geotechnical tests, but disturbance, removal of overburden and exposure to weathering agencies then rapidly causes degradation of the materials, such that they no longer perform as anticipated in engineering designs. This paper highlights the variations in the durability behaviour of Jurassic mudrocks of the UK and of the Arruda dos Vinhos basin area of Portugal, and provides explanations for variation in the durability behaviour of the latter. These durability data provide an essential aspect of geotechnical explorations of the formations so that changes in engineering properties during and after construction can be taken into account in designs and in the organisation and management of the construction process.

References

Bell FG, Entwisle DC, Culshaw FG (1997) A geotechnical survey of some British Coal Measures mudstones with particular emphasis on durability. Eng Geol 46:115–129

Campbell IM (1993) Mineralogical and diagenetic controls on the durability of overconsolidated clays. Ph.D. Thesis, University of Sheffield, p 440

Chandler RJ (1974) Lias Clay: the long-term stability of cutting slopes. Geotechnique 24:21–38

Cox BM Sumber MG (2002) British Middle Jurassic Stratigraphy. Geological Conservation Review Series. No. 26, Joint Nature Conservation Committee, Peterborough, p 508

Cox BM, Sumber MG, Ivimey-Cook HC (1999) A formational framework for the Lower Jurassic of England and Wales (on shore area). BGS Research Report, RR/99/01

Czerewko MA, Cripps JC (2001) Assessing the durability of mudrocks using the modified jar slake index test. Q J Eng GeolHydrogeol 34:153–163

Dick JC, Shakoor A, Wells N (1994) A geological approach toward developing a mudrock-durability classification system. Can Geotech J 31:17–27

Erguler ZA, Shakoor A (2009) Quantification of fragment size distribution of clay-bearing rocks after slake durability testing. Environ Eng Geosci 15:81–89

Franklin JA (1981) A shale rating system and tentative applications to shale performance. Transp Res Rec 790:2–12

Franklin JA, Chandra R (1972) The slake durability test. Int J Rock Mech Min Sci 9:325–341

Gallois RW, Horton A (2001) Field investigation of British Mesozoic and Tertiary mudstones. Q J Eng Geol 14:311–323

Hesselbo SP, Robinson SA, Surlyk F (2004) Sea-level changes and facies development across potential Triassic-Jurassic boundary horizons, SW Britain. J Geol Soc 161:365–379

ISRM (2007) Determining water content, porosity, density, and related properties and swelling and slake-durability index properties. In: Ulusay R, Hudson JA (Eds) The complete ISRM suggested methods for rock characterization, testing and monitoring: 1974–2006. ISRM Turkish National Group, Ankara, pp 83–98

Jeremias FT, Olarte JM, Pinho AB, Duarte IMR, Saroglou H, Suárez MCT (2020) Mudrocks as soft rocks—properties and characteristics. In: He M, Sousa LR (eds) Kanji M. Soft rocks mechanics and engineering, Springer, pp 37–107

Kullberg JC, Rocha RB, Soares AF, Rey J, Terrinha P, Azerêdo AC, Callapez P, Duarte LV, Kullberg MC, Martins L, Miranda R, Alves C, Mata J, Madeira J, Mateus O, Moreira M, Nogueira CR (2013) III.3. A Bacia Lusitaniana: Estratigrafia, Paleogeografia e Tectónica. In: Dias R, Araújo A, Terrinha P, Kullberg JC (eds) Geologia de Portugal, Volume II—Geologia Meso-cenozóica de Portugal, Escolar Editora, pp 195–347

Reid JM, Czerewko MA, Longworth I, Cripps JC (2017) Sulfur compounds in soils and rocks—avoiding pitfalls in terminology, test procedures and ground assessment. Gr Eng, 30–36

Simms MJ (2007) Uniquely extensive soft-sediment deformation in Rheatic of the UK: evidence for earthquake or impact? Palaeogeogr Palaeoclimatol Palaeoecol 244:407–423

Simms MJ, Chidlaw N, Morton N, Page KN (2004) British Lower Jurassic Stratigraphy. Geological Conservation Review Series, No 30. Joint Nature Conservation Committee, Peterborough, p 458

Toghill, P. (2000) The geology of Britain. An introduction. Airlife Publishing, p 192

Wright JK, Cox BM (2001) British Upper Jurassic Stratigraphy. Geological Conservation Review Serie, No. 21, Joint Nature Conservation Committee, Peterborough, p 266

Chapter 38
Karst-Sandification Characteristics and Geotechnical Properties of Triassic Dolomite in Yuxi, China

Jian Guo, Wenlian Liu, Mo Xu, Yujie Wang, Sugang Sui, and Hanhua Xu

Abstract As a special rock dissolution phenomenon, dolomite's karst-sandification is widely distributed in the Triassic strata of the Yuxi area of Yunnan Province, China. In order to investigate the influence of karst-sandification on dolomite's geotechnical geological properties, various methods were applied on Triassic dolomite in the study area from macroscope to microscope. Firstly, field investigation, microscopic identification, geochemical analysis and X-ray diffraction tests were used to inquire into the karst-sandification characteristics of dolomites. Furthermore, borehole sonic testing and a series of rock physical and mechanical tests were carried out to quantifying the effect of the karst-sandification on the Triassic dolomite. The results show that the Ca and Mg elements in dolomite were lost during karst-sandification by way of dissolution, whereas SiO_2 was significantly enriched. After the karst-sandification, the integrity of the dolomite was reduced, and the elastic modulus, dry compressive strength, saturated compressive strength and tensile strength of the dolomite were reduced by 64.8%, 68.3%, 75.1% and 78.8%, respectively. Characterizing the karst-sandification and quantifying its influence on the Triassic dolomite can help establishing geological engineering regulations, and has great theoretical and practical significance to engineering construction in other similar areas.

Keywords Dolomite · Karst-sandification · Geotechnical properties · Triassic · Yuxi area

---

J. Guo · M. Xu · Y. Wang
State Key Laboratory of Geo-Hazard Prevention and Geo-Environment Protection, Chengdu University of Technology, Chengdu 610059, China

W. Liu (✉) · S. Sui
Kunming Prospecting Design Institute of China Nonferrous Metals Industry Co., Ltd., Kunming 650214, China
e-mail: lwenl702@sina.com

H. Xu
Yunnan Key Laboratory of Geotechnical Engineering and Geohazards, Kunming 650214, China

S. Wang et al. (eds.), *Engineering Geology for a Habitable Earth: IAEG XIV Congress 2023 Proceedings, Chengdu, China*, Environmental Science and Engineering,
https://doi.org/10.1007/978-981-99-9057-3_38

38.1 Introduction

Karst is a general term for various geological phenomena, forms, and phenomena produced by soluble rocks under the action of water. It is a common natural phenomenon, while the "sandification" of fine-grained limestone is a very special karst phenomenon (Zou and Wang 2012), and there are relatively few studies on the karstification of fine-grained limestone. There is a big difference between limestone and marl in karstification. Under the karstification process, fine-grained limestone areas mainly form fine-grained limestone sand and powder materials (He 2008). Unlike marl areas, special landforms and underground shapes are formed on the surface and underground space, and larger dissolution cavities and underground rivers are less likely to form (Hu 2009). Karstification will reduce rock strength, damage rock mass integrity, and form various voids. There are often many engineering geological problems in karstified areas (Yu 2013). With the rapid development of various types of engineering construction in karstified areas, it is particularly important to understand the engineering geological characteristics of fine-grained limestone sand.

Domestic and foreign scholars have studied the sandification characteristics of fine-grained limestone from different perspectives using various methods. Zhang et al. (2012) and Tian and Yue (2014) both based their research on the Pingtou Hydropower Station in Megui River. The former used a statistical analysis of rock mass integrity, rock structure characteristics, and structural surface status as the basis to qualitatively analyze the impact of karst sandification on rock mass quality classification, proposing a rock mass quality classification model with "karst sandification development degree correction coefficient" as the core. The latter combined excavations and experimental results to determine the characteristics and comprehensive scores of different types of surrounding rocks within the area, establishing the type of surrounding rock being revealed by each chamber excavation. Wu et al. (2014) carried out a large number of immersion tests and conducted direct shear tests to explore the impact of immersion on the anti-shear strength index of fine-grained limestone. Li (2015) used uniaxial compression tests to discuss the impact of primary microcracks on the strength characteristics, deformation characteristics, acoustic emission characteristics, and failure characteristics of fine-grained limestone. Li (2015) analyzed the distribution pattern of fine-grained limestone sand from a plane and spatial perspective using the Yunnan Central Diversion Project as the case study. Based on the degree of sandification, he classified tunnel surrounding rock engineering geology.

Previous research on the engineering geological characteristics of sandified fine-grained limestone has mainly focused on the classification of tunnel surrounding rocks, with little attention given to the physical and mechanical properties of limestone sand. In this study, optical microscopy thin section identification, whole rock mineral composition analysis, and whole rock X-ray diffraction analysis were conducted on fine-grained limestone and limestone sand to investigate their sandification characteristics. Furthermore, drilling acoustic wave testing experiments and a series of indoor tests were carried out to examine the engineering geological properties of these two rock samples in terms of their structural characteristics, integrity, and physical and mechanical properties.

38.2 Study Area and Rock Sampling

The study area is located in Dengloushan town which is between Mile City and Yuxi City in Yunnan Province. The Tunnel Xiaojiang Fault Zone and its secondary faults across the study area formed by the strata formation. During the investigation, several special karst phenomena, dolomite sandification were found in the middle Triassic strata. The strength of the rock mass is low, and it can be easily squeezed into sand by hand, see in Fig. 38.1.

The experimental sandification limestone specimens were all taken from the Dengloushan area of Huaning County, Yuxi City, Yunnan Province. According to the different degree of sandification, a total of 6 rock samples were collected representing un-sandified limestone (4# sample, 12# sample), weakly-sandified limestone (5# sample, 6# sample), and strongly-sandified limestone (8# sample, 13# sample). Each rock sample was processed into 9 cylindrical specimens with a diameter of 50 mm and a height of 100 mm according to the recommended method of the International Society for Rock Mechanics (ISRM) and the actual research needs, which were used for subsequent physical parameter testing, uniaxial compression testing, and tensile strength testing. To reduce the errors caused by the differences between specimens, all specimens were weighed using a high-precision electronic balance and measured with a vernier caliper before testing to ensure that the flatness and perpendicularity of both ends of each sample did not exceed 0.2 mm. Based on the similarity in weight and size, rock samples with large differences were excluded,

Fig. 38.1 Dolomite karst-sandification characteristics in the study area

and the qualified samples were numbered and grouped, with each group containing 3 samples of the same lithology.

38.3 Methods

Microscopic identification, chemical composition analysis and whole-rock X-ray diffraction tests were used to figure out the mineral composition of the sanded dolomite in the study area. The sandification characteristics was determined by material structure then. On top of this, relevant physical and mechanical tests were conducted to obtain the basic physical and mechanical parameters of the site rock samples. To gain the effect of sandification on the physical and mechanical properties of the dolomite, the physical and mechanical properties of the dolomite before and after sandification were compared.

38.3.1 *Sandification Classification*

The formation and evolution of sanded dolomite is not an isolated and static process, dolomites with different degrees of sandification will show different structural characteristics in the rock. By analyzing the evolution of fracture surfaces and the structural zoning characteristics of the sanded material in the dolomite sandification process inside the engineering area of Pingtou Hydropower Station in Sichuan Province, Zhang Liangxi summarized five stages of the macroscopic development and evolution of dolomite sandification, namely, natural sanded stage, moderately sanded stage, strongly sanded stage, intensely sanded stage, and fully sanded stage (Zhang et al. 2012). The identification characteristics of dolomite with various sandification degrees are shown in Table 38.1.

38.3.2 *Geochemical Analysis*

The analytical equipment used for the rock mineralogical testing included: a tube burner, burette, Z-2000 graphite furnace atomic absorption spectrophotometer, DY938 X-ray spectrometer, 5110 Agilent 5110 plasma spectrometer and SRJ4-9 chamber furnace. All test samples were prepared by micro-drilling, grounded into powder by an agate mortar and sieved through a steel sieve to below 200 mesh.

Table 38.1 Identification characteristics of dolomites of different sandification grades

Sandification classification	Sandification strips thickness	Number of sandification strips	Sandification strips features	Rock structure
Non-sanded	< 1	1–2	Off-white dolomite sand	Dense structure
Moderately sanded	1–5	2–4	Grey-yellow dolomite sand and dolomite powder	Dense structure
Strongly sanded	3–8	4–6	Brownish-yellow dolomite sand and dolomite powder	Massive structure
Intensely sanded	5–20	6–8	Yellow–brown dolomitic sand and powdery clay	Block structure
Fully sanded	> 20	> 8	Red-brown powdery clay	Fragmental structure

38.3.3 X-Ray Diffraction Tests

A diffractometer, RINT-Ultima III (Rigaku Corp., Japan), with Cu Kα source was employed. The X-ray scan range was 2.7°–10° with a scan speed of 20°/min under Parallel-Beam geometry. A relative fast scanning speed was chosen for minimizing the possible specimen rebound and water evaporation during X-ray diffraction (XRD) scanning. After XRD tests, water content of each slice was measured.

38.3.4 Rock Physical Tests

Density and porosity were selected to characterize the changes of physical parameters of dolomite sandification process. The rock dry density and particle density were obtained by the volume product method and specific gravity bottle method, respectively, and the rock porosity was calculated by Eq. (38.1).

$$n = \left(1 - \frac{\rho_d}{\rho_s}\right) \times 100 \tag{38.1}$$

where n is the porosity of each rock sample. ρ_d is the dry density of each sample. ρ_s is the particle density of each sample.

38.3.5 Rock Mechanical Tests

Unconfined compressive strength (UCS), ultimate tensile strength (UTS), deformation modulus (E) and Poisson's ratio (μ) were selected to characterize the change pattern of mechanical parameters of dolomite sandification process. The rock was processed into the standard samples with a diameter of 50 mm and a height of 100 mm. Then, the rock samples were subjected to uniaxial compression test and Brazilian splitting test by Shimadzu AG-X250 electronic universal testing machine.

38.4 Results

38.4.1 Sandification Characteristics

The microscopic identification photographs of 6 representative dolomite specimens are shown in Fig. 38.2, and their mineral content, structure, and structural characteristics are shown in Table 38.2.

a. 4# Natural dolomite; b. 12# Natural dolomite; c. 5# Moderately sandy dolomite; d. 6# Moderately sandy dolomite; e. 8# Intensely sandy dolomite; f. 13# Intensely sandy dolomite.

4# and 12# are typical natural dolomites. According to the microscopic identification results, the dolomite in the study area has an overall fine crystal and mud crystal structure with massive structure. The composition of the rocks is mainly dolomite, with a content of over 97%; the content of quartz is low and scattered within the rocks in a punctiform manner, with a content of less than 2.5%. The porosity of the

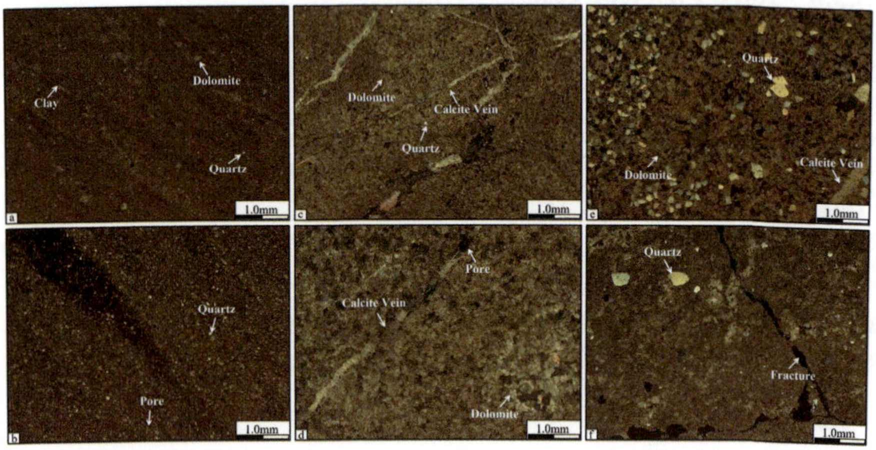

Fig. 38.2 Microscopic features of dolomite flakes

Table 38.2 Table of results for the identification of dolomite mineral fractions in the study area

Rock character	Sample number	Rock texture	Rock structure	Content of main mineral components	Porosity	Designation
Natural dolomite	4#	Argillaceous texture	Parallel bedding structure	Dolomite 99%; Clay, Silty, Quartz 0.5%	$< 0.5\%$	Argillaceous dolomite
	12#	Powder—argillaceous texture	Massive bedding	Dolomite 97%; Quartz 2–2.5%	$< 0.5\%$	Powder—argillaceous dolomite
Moderately sandy dolomite	5#	Aplitic texture	Taxitic structure	Dolomite 89%; Calcite 1%; Quartz 5–10%	1–2%	Aplitic-powder dolomite
	6#	Aplitic texture	Massive bedding	Dolomite 90%; Clay $\leq$ 1%; Calcite $\leq$ 4%; Quartz 5%	1–3%	Aplitic dolomite
Intensely sandy dolomite	8#	Fine-grained structure	Massive bedding, linsen bedding	Dolomite 70%; Calcite $\leq$ 1%; Quartz 20–25%	4–6%	Fine-grained dolomite
	13#	Argillaceous texture	Massive bedding	Dolomite 80%; Calcite 15–20%	5–7%	Argillaceous dolomite

rocks is generally low, all within 0.5%, and the interior of the cascade fractures are mostly filled with clay minerals.

The samples 5#, 6#, 8# and 13# are typical of the moderately and intensely sanded dolomites in the area respectively. The structure and tectonic features of the sanded dolomite, as well as the types of the main minerals within the rock, are basically the same as those before sandification, and the difference is mainly reflected in the relative content of the mineral components. In the weakly sanded dolomite, the dolomite content decreases to 89–90%, while the quartz content increases to 5–10%; in the strongly sanded dolomite, the dolomite content decreases to 70–80%, while the quartz content increases to 15–25%, with the main expression of the sandification process being the dissolution of dolomite and the enrichment of quartz. In addition, the porosity of the rocks tends to increase as the degree of sandification increases, from 0.5 to 1–3% in weakly sanded dolomites and 4–7% in strongly sanded dolomites.

The XRD results of three typical dolomite specimens: #4, #6 and #8, are shown in Figs. 38.3, 38.4 and 38.5. Comparison of the standard PDF card with the diffraction pattern of each specimen shows that the diffraction angle corresponding to each diffraction peak is approximately the same, which indicates that the main mineral species in the three specimens are the same. The relative content of each major mineral component in the rock samples with different degrees of sandification was calculated by the K-value method according to Eq. (38.1). The results of the calculations are shown in Table 38.3.

$$W_X = \frac{I_X}{K_A^X \sum_{i=A}^{N} \frac{I_i}{K_A^i}} \tag{38.2}$$

where, W_X is the relative content of a phase X; I_X is the diffraction intensity of a phase X; A is the name of the phase selected as the internal standard among the N

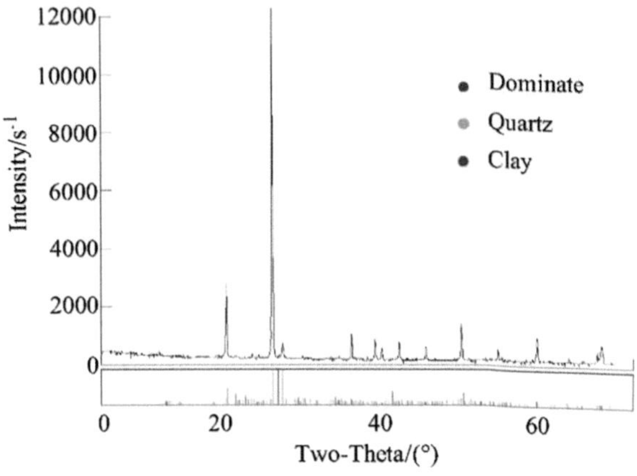

Fig. 38.3 X-ray energy-dispersive spectrum analysis of #4

Fig. 38.4 X-ray energy-dispersive spectrum analysis of #6

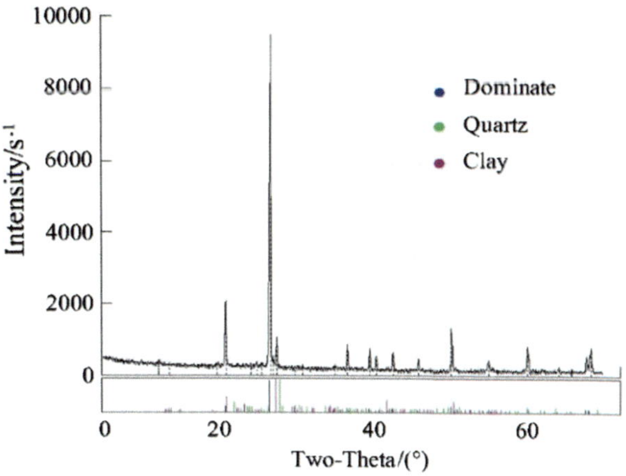

Fig. 38.5 X-ray energy-dispersive spectrum analysis of #8

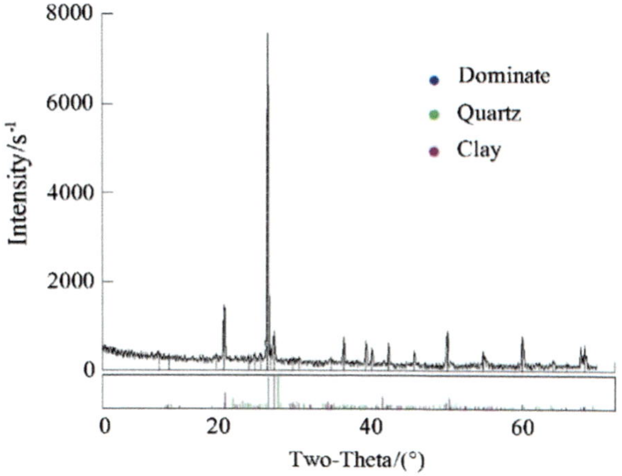

Table 38.3 Dolomite mineral analysis results based on XRD

Sample number	Dolomite	Quartz	Calcite	Clay	Else
4#	97.6	1.3	0	0.4	0.7
5#	85.5	12.3	1.6	0	0.6
8#	68.6	27.9	2.0	0	1.5

phases; and K_A^X is the ratio of the RIR values between a phase X and the internal standard phase A.

As can be seen from Table 38.3, the percentage of dolomite in 4#, 5#, 8# are 97.6%, 85.5% and 68.6% respectively, while the percentage of quartz is 1.3%, 12.3% and 27.9% respectively. The dolomite content decreases and the quartz content increases as the sandification progresses, which is consistent with the results of microscopic identification.

38.4.2 Rock Mineral Components

The chemical composition of the dolomite is closely related to the rock-forming environment, and six representative groups of rock samples were selected for chemical composition analysis to investigate the sandification characteristics. The specific chemical compositions of the six groups of samples and their percentage contents are shown in Table 38.4.

The main chemical components of the dolomite are CaO and MgO, and the sum of these two chemical components in both groups of natural dolomite is greater than 50%. The results of the grinding flake analysis show that the dolomite content is high, both being greater than 90%, and the percentage of burn loss is greater than 45%, thus indicating that the natural dolomite is of high purity and is a typical carbonate rock, and also proving that the climate during the deposition of the strata in this area The climate during the deposition of the strata in this area was arid or semi-arid.

SiO_2, Fe_2O_3, Al_2O_3, K_2O and Na_2O are the minor chemical components, of which SiO_2 is a stable insoluble mineral and the content of this component in the rock sample is less than 1%, so it can be inferred that the dolomite was deposited in a near clear water depositional environment; while the other four chemical components, Al_2O_3, Fe_2O_3, K_2O and Na_2O, can reflect the content of muddy material in the rock The sum of the percentages of the other four components, Al_2O_3, Fe_2O_3, K_2O and Na_2O, can reflect the content of muddy material in the rock, and the sum of the percentages of the four components is less than 2%, which can indicate that the dolomite has a low mud content.

P_2O_5 reflects the biological activity during the diagenesis of an area, and the value is less than 1% in both the two dolomite groups and the two weakly sanded dolomite rock samples, so it can be inferred that the influence of biological activity during the diagenesis of the study area is small.

Table 38.4 Rock mineral composition test results of geochemical analysis

Sample number	CaO	MgO	SiO_2	Al_2O_3	Fe_2O_3	K_2O	Na_2O	TiO_2	P_2O_5	MnO
4#	32.03	20.56	0.6	0.37	0.76	0.008	0.0047	0.0034	0.245	0.038
12#	31.62	20.49	0.68	0.42	0.73	0.0091	0.0042	0.023	0.107	0.043
5#	25.04	19.13	10.84	0.63	0.79	0.43	0.05	0.29	0.59	0.037
6#	26.4	18.71	9.26	0.64	0.85	0.38	0.08	0.26	0.61	0.44
8#	16.94	11.36	32.01	6.03	4.62	1.48	0.2	0.54	1.57	0.15
13#	17.21	11.4	31.92	6.10	4.53	1.55	0.14	0.61	1.62	0.1

Table 38.5 Physical properties of sanded dolomite

Sample number	$\rho_d/(g \cdot cm^{-3})$	$\rho_s/(g \cdot cm^{-3})$	P/%
4#	2.61	2.73	4.35
12#	2.63	2.75	4.26
5#	2.56	2.69	4.54
8#	2.53	2.65	4.46
13#	2.40	2.65	9.32
16#	2.38	2.61	8.77

38.4.3 Rock Physical Properties

The dry density and particle density of 54 dolomite specimens with different degrees of sandification in six groups of rock samples were obtained by the volume product method and the specific gravity bottle method, respectively, and then the porosity of each specimen was calculated by Eq. (38.1). The average values of physical parameters for each group of specimens are shown in Table 38.5.

With the gradual increase of sandification, the dry density and particle density of the specimens have an obvious decreasing trend, while the corresponding porosity gradually increases. From the pattern shown in the previous mineral composition analysis, it is clear that during the sandification process of dolomite, the dissolution of dolomite leads to the increase of porosity inside the rock, while the enrichment of quartz and calcite minerals with lower molar masses causes the decrease of density of the rock. The change of porosity brought by sandification directly changes the water absorption and permeability of the rock. At the same time, the secondary porosity brought about by it reduces the denseness of the rock mass and the strength of the rock mass suffers from deterioration.

38.4.4 Rock Mechanical Properties

Three groups of dolomite, moderately sanded dolomite and intensely sanded dolomite were selected and subjected to uniaxial compression test under dry condition, uniaxial compression test under saturated condition and splitting test respectively, and the relevant mechanical parameters of the specimens were obtained as shown in Table 38.6.

According to the experimental results, the unconfined compressive strength and uniaxial saturated compressive strength have a significant tendency to decrease with the increase of sandification. The average unconfined compressive strength of natural dolomite was 38.73 MPa and the average uniaxial saturated compressive strength was 30.65 MPa, while the average unconfined compressive strength and the average uniaxial saturated compressive strength of moderately and intensely sanded specimens were 35.95 MPa, 26.16 MPa; 14.13 MPa and 8.68 MPa, respectively.

Table 38.6 Mechanical properties of sanded dolomite

Sample number	UCS/ MPa	Sample number	USCS/ MPa	K	E/GPa	μ	Sample number	UTS/ MPa
4#-1	40.4	4#-4	33.2	0.82	31.1	0.28	4#-7	3.9
4#-2	39.1	4#-5	31.8	0.81	30.3	0.3	4#-8	3.4
4#-3	39.7	4#-6	32.7	0.82	31.8	0.27	4#-9	3.6
12#-1	36.7	12#-4	27.9	0.76	27.4	0.31	12#-7	3.2
12#-2	38.6	12#-5	29.5	0.76	28.3	0.31	12#-8	3.5
12#-3	37.9	12#-6	28.8	0.76	28.7	0.29	12#-9	3.3
5#-1	36.6	5#-4	26.7	0.73	21.9	0.33	5#-7	3
5#-2	35.8	5#-5	25.6	0.72	21.1	0.34	5#-8	2.6
5#-3	35.1	5#-6	24.4	0.70	19.6	0.36	5#-9	2.5
8#-1	36.3	8#-4	27.1	0.75	22.5	0.29	8#-7	3.1
8#-2	36.2	8#-5	26.9	0.74	22	0.31	8#-8	3
8#-3	35.7	8#-6	26.3	0.74	21.4	0.32	8#-9	2.8
13#-1	14.1	13#-4	7.5	0.53	10.6	0.32	13#-7	1.2
13#-2	17.5	13#-5	10.8	0.62	9.5	0.38	13#-8	1.5
13#-3	15.3	13#-6	9.6	0.63	9.8	0.36	13#-9	1.3
16#-1	12.8	16#-4	8.1	0.63	9.8	0.36	16#-7	1.2
16#-2	13	16#-5	8.4	0.65	10.3	0.34	16#-8	1.5
16#-3	12.1	16#-6	7.7	0.64	9.2	0.37	16#-9	0.9

UCS—unconfined compressive strength; USCS—uniaxial saturated compressive strength; K—softening coefficient; E—elastic modulus; μ—Poisson's ratio; UTS—ultimate tensile strength

38.5 Discussion

The stage deterioration ΔDi and the cumulative deterioration are introduced to quantitatively characterise the physical and mechanical property damage pattern during dolomite sandification, which is calculated as follows.

$$\Delta D_i(\mathrm{n}) = \frac{T_{n-1} - T_n}{T_{n-1}} \tag{38.3}$$

$$D_i(\mathrm{n}) = \frac{T_0 - T_n}{T_0} \tag{38.4}$$

where $\Delta D_i(\mathrm{n})$ is the degree of deterioration of physical and mechanical parameter i at stage n; $D_i(\mathrm{n})$ is the cumulative degree of deterioration of mechanical parameter i at stage n; T_0 is the physical and mechanical parameter of natural dolomite; T_{n-1} and T_n are the physical and mechanical parameters of dolomite at stage n − 1 and stage n respectively.

Define natural to moderately sanded as the first stage of dolomite sandification and moderately sanded to intensely sanded as the second stage of dolomite sandification, and calculate the stage degradation and cumulative degradation of physical and mechanical properties of the rocks in the two stages respectively. Table 38.7 shows the stages and cumulative deterioration of the relevant physical and mechanical properties of dolomite specimens with different degrees of sandification.

From Fig. 38.6 and Table 38.7, The rock modulus of elasticity has the highest rate of deterioration, reaching 13% and 16% for the first and second stages respectively. Sandification destroys the robustness of the dolomite and reduces its stiffness considerably. The ultimate tensile strength of the rock has a more pronounced deterioration trend compared to the uniaxial compressive strength, with a cumulative deterioration rate of 19.48%, while the cumulative deterioration rates of the dry compressive strength and saturated compressive strength are 7.03% and 15.04% respectively, and the effect of sandification on the tensile strength of dolomite is more significant.

In terms of the compressive strength of the rock, the saturated compressive strength of the sanded dolomite decreases more significantly than the dry compressive strength, with a cumulative deterioration rate of 15.04% and 7.03% respectively. The difference between the dry compressive strength and the saturated compressive strength of the rocks decreases gradually with the sandification process, which is reflected in the decreasing softening coefficient of the rocks.

In addition to this, it can be seen from Fig. 38.6 that the first stage deterioration rate of the dolomite mechanical parameters is generally greater than the second

Table 38.7 Deterioration degree of Rock mechanical and physical properties

Project	UCS	USCS	K	E	UTS	ρ_d	ρ_s
First stage	5.03	11.42	6.34	17.23	13.40	0.95	0.73
Second stage	2.27	3.62	1.13	12.59	6.08	1.93	1.84
Cumulative deterioration	7.03	15.04	7.47	29.82	19.48	2.88	2.57

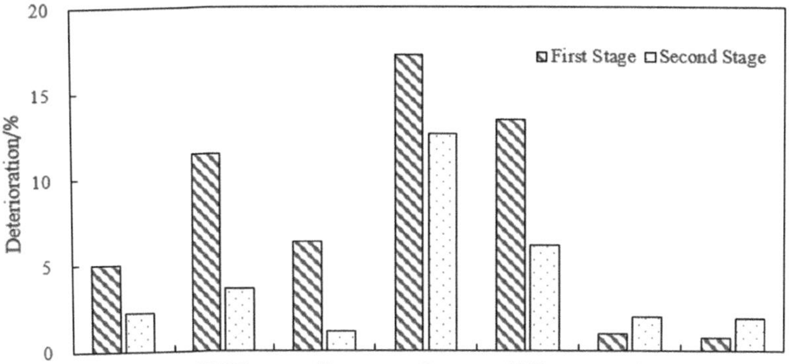

Fig. 38.6 Deterioration degree of Rock mechanical and physical properties

stage deterioration rate. This indicates that during the initial stage of sandification, secondary solution pores develop rapidly within the rock mass and the rock structure is violently damaged, leading to rapid deterioration of the rock strength. With the gradual intensification of sandification, the pores inside the rock gradually developed completely under sandification, and the emergence and expansion of new pores tended to stabilise, and the deterioration of the mechanical parameters of the rock gradually slowed down.

38.6 Conclusion

In this study, we analyzed the karst-sandification characteristics in Triassic dolomite of Yuxi, and compared the geotechnical properties before and after the karst-sandification. The study can be concluded below:

(1) Through microscopic identification XRD, under the sandification process, the content of the main mineral dolomite in dolomite decreases with the strengthening of the sandification process, while the content of quartz shows an increasing trend.
(2) During the process of sandification and dissolution of dolomite, there is a certain negative correlation between the content of CaO and MgO and the content of SiO_2, while the content of Al_2O_3, Fe_2O_3, Na_2O, K_2O, P_2O_5, and MnO shows a positive correlation with the content of SiO_2.
(3) There are significant differences in the damage values corresponding to dolomite with different degrees of sandification, and with the strengthening of the sandification process, its compressive strength, tensile strength, and elastic modulus all show a certain degree of reduction, while the Poisson's ratio shows an increasing trend. This indicates that sandification has a certain damaging effect on both the strength and deformation characteristics of dolomite.

References

He W (2008) Study on the causes of sandification of fine-grained limestone in the Pingtou Hydropower Station Site of Meigu River and its impact on engineering. Chengdu University of Technology. (Chinese)

Hu X (2009) study on the characteristics and mechanism of karst sandification in fine-grained limestone in the deep slopes of Pingtou Hydropower Station in Meigu River. Chengdu University of Technology. (Chinese)

Li X (2015) The influence of primary microcracks on the mechanical properties of dolomite under uniaxial compression. Kunming University of Science and Technology. (Chinese)

Tian Q, Yue C (2014) Characteristics of karstified and "sandified" fine-grained dolomite in Pingtou Hydropower Station and classification of surrounding rocks. Adv Sci Technol Water Resour 34(05):45–49. (Chinese)

Wu Y, Wang J, Liang W, Zhao B (2014) Experimental study on natural water absorption characteristics of dolomite. Geotech Investig Surveying 42(06):1–4+19. (Chinese)

Yu S (2013) Study on engineering characteristics of karstified dolomite. West-China Explor Eng 25(06):139–140. (Chinese)

Zhang L (2012) Study on the formation mechanism and engineering characteristics of karst sandification in fine-grained limestone: a case study of Pingtou Hydropower Station in Meigu River. Chengdu University of Technology. (Chinese)

Zhang H, Zhao Q, Peng S (2012) Mass classifications of karstified dolomite bodies under "sandification" in Pingtou Hydropower Station of Meigu River. China Rural Water Hydropower 07151–155. (Chinese)

Zou Z Wang X (2012) Discussion on the mechanism of "sandification" of fine-grained dolomite and its engineering case. China New Technol Prod 08:91–92. (Chinese)

Chapter 39
Dynamic Mechanical Properties of Unsaturated Red Clay Subjected to Impact Loading

Futian Zhao, Jun Liu, Yue Wang, and Mengyang Zhen

Abstract Red clay, a typical special soil, is frequently utilized as a filling material for engineering operations in southwest and south regions of China. Thus, it is of great importance to study the dynamic properties of red clay for use in engineering applications. However, a detailed characterization of such behaviour for red clay subjected to impact loading has not been reported in the existing literature. In this paper, a Φ50mm Split Hopkinson Pressure Bar (SHPB) apparatus is used to conduct a series of uniaxial impact tests on remolded red clay under passive confining pressure condition. The influence of strain rate and moisture content on the dynamic mechanical properties of red clay are investigated. The experimental results show that the dynamic elastic modulus, ultimate strength and maximum strain of red clay increase with the increase of strain rate. The dynamic elastic modulus and ultimate strength appear as a linear function of strain rate in the range of strain rates from 255 to 564 s^{-1}. Besides, the moisture content has a negligible effect on the strain rate sensitivity of unsaturated red clay dynamic elastic modulus. In contrast, it has a significant effect on the ultimate strength strain rate sensitivity. Eventually, the exponential function relationship between normalized dynamic elastic modulus, strength and strain rate considering the change of moisture content was established, which can be the experimental basis for developing the dynamic constitutive model of unsaturated red clay.

Keywords Unsaturated red clay · Split Hopkinson pressure bar · Impact loading · Dynamic properties

---

F. Zhao
College of Water Conservancy and Hydropower Engineering, Hohai University, Nanjing 210024, Jiangsu, China

J. Liu (✉) · Y. Wang · M. Zhen
College of Civil and Transportation Engineering, Hohai University, Nanjing 210024, Jiangsu, China
e-mail: ljun8@263.net

S. Wang et al. (eds.), *Engineering Geology for a Habitable Earth: IAEG XIV Congress 2023 Proceedings, Chengdu, China*, Environmental Science and Engineering,
https://doi.org/10.1007/978-981-99-9057-3_39

39.1 Introduction

Red clay is widely distributed in southwest and south China, such as Guizhou, Yunnan and Guangxi provinces, accounting for approximately 1/9 of total area (Editorial of Engineering Geology Manual 2007). With the large-scale development of the China's western region, construction projects, which usually take red clay as a filling material, have proliferated over the past few decades. Thus, it is of great significance to study the static and dynamic mechanical behaviors of red clay, which make it possible to obtain a better insight into the its engineering characteristics. Numerous previous studies have examined the static properties of red clay with respect to bearing capacity (Yang and Lu 2011) and deformation mechanism (Long et al. 2009; Bai et al. 2010) and the dynamic properties of red clay induced by a traffic load (Kaynia 2000; Ding and Meng 2014). However, few data exist on the dynamic mechanical properties of red clay subjected to impact loading, representing a gap in light of characterizing the impact responses for red clay in a range of engineering applications, for instance, investigating damage to underground structures in explosive operations, estimating the projectile penetrating-depth in soil during dynamic compaction, and supplying vibration isolation and shock absorption for engineering systems.

Since the Split Hopkinson Pressure Bar (SHPB) was developed and modified by Hopkinson (1914), Kolsky (1949), it has been more commonly used to test the impact properties of hard materials, such as metal, concrete, rock, etc. Because the soil is a typical soft material with characteristics of low wave speed, strain rate sensitivity, inertia effect and low wave impedance (Gaffney et al. 1985), there are many limitations in obtaining the accurate and reliable testing data pertinent to soil impact response via the SHPB. However, researchers have made many useful attempts to improve the conventional SHPB technique, including pulse shaping technique (PST), embedded quartz crystal technique (EQCT) and passive confining pressure technique (PCPT). The PST was introduced by Bragov et al. (2004), Frew et al. (2005) to reduce dispersion effect by placing a pulse sharper between the striker and incident bars. The uniform stress distribution in geotechnical material can be achieved as the rise time of applied stress wave increased and multiple reflections occurred within the specimen. The successful employment of SHPB combined with PST on testing dry sand has been reported by Song et al. (2009), Luo et al. (2011). Due to the huge difference of wave impedance parameters between the SHPB bar and soft material, the transmitted signal is too small to be captured. As the quartz has more sensitivity in detecting stress than the strain gauge, Chen et al. (2000, 2002) proposed the EQCT to directly measure the transmitted force in the SHPB experiment of soft material. Furthermore, Ding et al. (2013) has introduced the EQCT in testing the unsaturated clay dynamic response, where the magnitude of signal recorded by quartz is about three times that of strain gauge. Additionally, most soil SHPB tests are performed with PCPT (Bazhenov et al. 2000; Bragov et al. 1996, 2008; Song et al. 2009), which usually places the soil sample in a rigid cylinder tube and maintains passive lateral confining condition. In order to actively control the lateral confinement, Pierce et al. (1989, 1990) modified the SHPB setup by placing the sample in a rubber membrane and adding a pressurized water layer between the inner deformable membrane and the outer hard cylinder tube.

Based on above improved SHPB techniques, a large number of experiments on dynamic response of sand or clay have been conducted over the last few decades. Song et al. (2009) conducted the passive confining pressure SHPB tests on the fine-grained sand and studied the lateral confinement effects on high strain rate behavior of sand by using three different confining tubes, which are manufactured from steel, polycarbonate and polyolefin respectively. Kabir et al. (2010), Kabir (2010) performed the triaxial SHPB tests on sand by actively controlling the axial and radial constraints. The experimental results indicate that the lateral confinement results in a stiffer soil dynamic response. Murphy et al. (1970), Farr and Woods (1988) investigated the particle breakage characteristics of sand during SHPB test. It can be found that the degree of sand particle breakage significantly increased under the condition of high confining pressure and strain rate. In addition, Charlie et al. (1990), Martin et al. (2009); discussed the moisture content effect on the dynamic compressive behavior and strain-rate dependency of sand. Partially saturated sand was found to be more compressible than dry sand considering the lubrication of pore water. Meanwhile, pore water pressure plays an important role in strain-rate dependency of partially saturated sand. Felice, (1985) pointed that the sand specimen with the optimum moisture content exhibits more significant strain-rate sensitivity than dry sand and saturated sand.

Scholars also investigated the influence of strain rate effect, saturation, compaction, and consolidation on the dynamic responses of clay through SHPB tests. Bragov et al. (1996) conducted a passively confined test on an axial compression clay specimen and revealed the strain rate effect of clay, that is, the higher the strain rate, the higher the dynamic peak strength. Liu et al. (2012) investigated compressive mechanical properties on compacted clays with different projectile speeds by using the SHPB apparatus made of polycarbonate, and introduced dynamic increase factors to analysis the influence of compaction and moisture content. Experimental results show that the effects of strain rate on dynamic strength increase factor can be expressed by exponential approximation, however, the effects of strain rate on peak strain increase factor can be expressed by linear approximation. Luo et al. (2017) studied the dynamic compressive behaviors of clay under three-dimensional stress state. Based on the measurements of the axial responses of soil sample and circular strains of cylindrical confining tube, the mean stress- volumetric strain and compressibility curves can be further determined. The dynamic compressive stress–strain responses of clay were also obtained by Yang et al. (2017) while the consolidation stress state of soil sample was properly considered. The testing results indicate that the ultimate dynamic stress and strain of over-consolidation clay are highly affected by the initial stress history. The bearing capacity of clay are significantly improved as the increasing of previous consolidation pressure.

After reviewing the above literature, it is noted that significant variation with respect to the dynamic impact response of soil bases in concurrence with varying soil component, water content, confinement, strain rate and other factors. However, the dynamic mechanical properties for red clay subjected to impact loading has not been

Table 39.1 Physical properties of remolded red clay

ρ (g/cm^3)	w (%)	G_s	e	w_L (%)	w_p (%)	Grain size (mm)		
						> 0.075	0.075–0.005	< 0.005
1.74	40.31	2.69	1.27	83.9	48.6	0	20.9%	79.1%

well characterized. Therefore, a series of SHPB uniaxial impact tests on remolded red clay under passive confining pressure condition are carried out in this paper. The influence of strain rate effect and moisture content on the high strain rate response of partially saturated granular media are systematically examined. Accordingly, this study aims to establish the quantitative relationship between dynamic parameters of red clay and its associating influencing factors. This act ultimately may offer deeper insight into the dynamic characterization of red clay and be utilized for future engineering applications.

39.2 Dynamic SHPB Tests of Remolded Red Clay

39.2.1 Material Description and Sample Preparation

The red clay investigated was taken from Guizhou area using borehole sampling method according to "Undisturbed Soil Sampling Technical Standard" JGJ89-92. The physical properties of undisturbed red clay were tested and listed in Table 39.1. As shown in Fig. 39.1, the undisturbed red clay sample (Fig. 39.1a) obtained was first dried and ground into a powder, screened through a 2 mm sieve (Fig. 39.1b). Then, the disturbed soil was made into a cylindrical remolded sample (Fig. 39.1d) with 61.8 mm diameter and 80 mm height using a compaction test apparatus (Fig. 39.1c). Finally, the cylindrical remolded sample was divided into three SHPB specimens by a custom cutting ring (Fig. 39.1e). Each SHPB specimen was 50 mm in diameter and 25 mm in thickness (Fig. 39.1f).

39.2.2 Modified SHPB Apparatus and Experimental Design

The experiment would adopt a SHPB apparatus (named ALT100 as shown in Fig. 39.2) in Civil Explosives Engineer Laboratory of Poly Jiu-Lian Co., Ltd. The testing equipment consists of 5 parts, which are axial hydrostatic loading device, confining pressure loading device, punch launcher, stress transfer system, data acquisition and processing program. The incident and transmitted bars are both 2500 mm in length and 50 mm in diameter manufactured from 18Ni, whose density, elastic modulus, and elastic wave velocity are 8 g/cm^3, 190 GPa, and 5000 m/s respectively.

Fig. 39.1 Preparation of remolded red clay sample: **a** undisturbed soil samples; **b** disturbed soil after sifting; **c** compaction test apparatus; **d** remolded cylindrical specimen; **e** custom cutting ring; and (**d**) SHPB specimen

A striker bar composed of the same material with a length of 400 mm is launched by a gas gun, which can control the velocity of the striker bar by adjusting gas pressures.

The specimen was confined within a tube manufactured from 4030 steel with an inner diameter of 50.1 mm and wall thickness of 5 mm to ensure uniaxial behavior of the remolded red clay. To study the effects of moisture content and strain rate on the dynamic behavior of unsaturated red clay, the moisture contents of remolded samples were set as 35%, 39%, 43%, and 47%, corresponding to 80.4%, 82.6%, 91.1%, and 99.8% saturation, respectively. Five gas pressures of 0.05, 0.10, 0.15, 0.20, and 0.25 MPa were used for conducting the impact tests on each moisture content specimen to obtain different strain rates.

Based on the assumption of one-dimensional stress wave theory and dynamic stress equilibrium state, the stress, strain, and strain rate of specimen can be calculated using the strain pulses recorded by strain gauges on the surface of SHPB bars. Equations are as follows:

$$\sigma_1 = \frac{A}{A_S} E(\varepsilon_i + \varepsilon_r) = \sigma_d \tag{39.1}$$

$$\sigma_2 = \frac{A}{A_S} E(\varepsilon_t) = \sigma_d \tag{39.2}$$

$$\dot{\varepsilon} = -\frac{2C}{l_s} \varepsilon_r \tag{39.3}$$

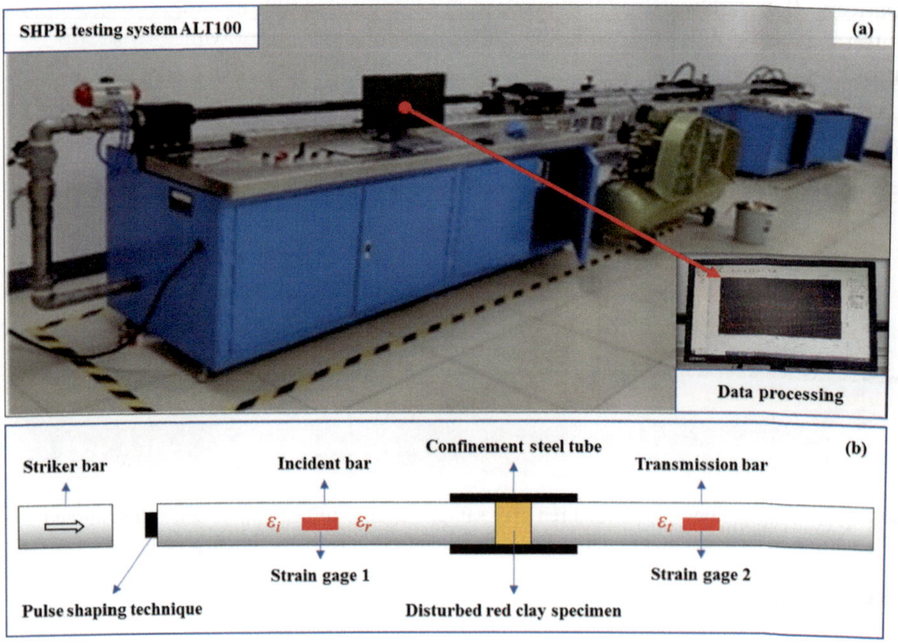

Fig. 39.2 Split Hopkinson pressure bar: **a** product photo; and **b** schematic diagram

$$\varepsilon = -\frac{2C}{l_s}\int_0^t \varepsilon_r dt \tag{39.4}$$

where ε_i, ε_r, and ε_t are the incident, reflected, and transmitted strain pulses, respectively; E is the elastic modulus of the SHPB bar; A/A_S is the ratio of the bar cross sectional area to the specimen cross sectional area; C is the wave propagation velocity of the SHPB bar and l_s is the length of soil specimen.

39.3 Experimental Results and Discussions

Figure 3a shows the typical recorded waveforms of the red clay with 35% moisture content during SHPB test. The incident and reflected waveforms are represented by red solid line, and the transmitted waveform is represented by blue solid line. The original waveforms were imported to data processing program to obtain the stress strain curve. Firstly, the transmitted wave was shifted a certain phase forward (or backward) according to the initial positions of the incident and transmitted waves, ensuring that the rising point of reflected wave coincides with the falling point of transmitted wave. Then, the starting and ending intersections of waveforms were captured, and the shifted waveforms were extracted for calculating.

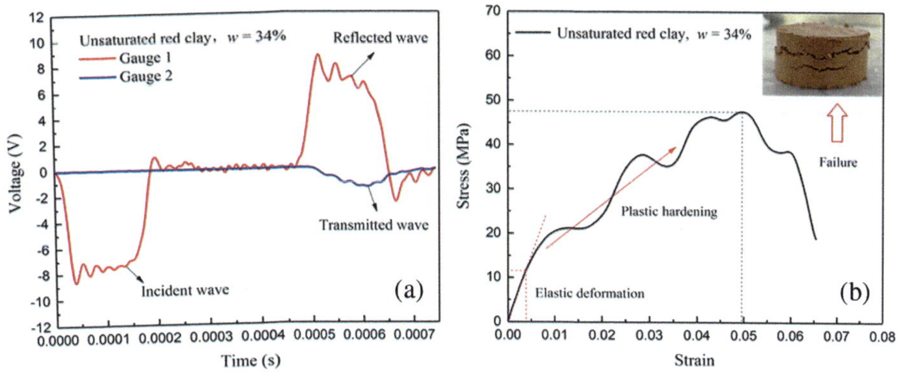

Fig. 39.3 Data processing: **a** original oscilloscope records from gauges in the incident and transmission bar; and **b** calculated dynamic stress strain curve

Figure 3b shows the representative dynamic stress–strain curve of unsaturated red clay. It can be concluded that the stress–strain curve can be divided into three typical stages: (a) Initial elastic compression stage, that is, the stress–strain relationship can be described by a linear function. The yield strength of red clay specimen mainly depends on the structure feature of granular material. Before the red clay reached it yield limit, a tiny strain occurred because the cracks and holes inside were closed soon under the high loading rate condition. (b) Plastic hardening stage, that is, the stress shows a fluctuating growth with strain. The soil particles nearby the incident bar firstly moved along the axis direction under one dimensional impact load, while the soil in the distance still continued to resist deformation. In other words, due to the failure of front-end soil and the resistance of back-end soil, the fluctuation of the stress value in plastic flow occurred considering the clay sample was a unified whole. Similar phenomenon was also reported by Martin et al. (2009) and Yang et al. (2017). (c) Failure stage, that is, the stress was sharply reduced after the strain reached a certain value. The whole structure of the red clay specimen was destroyed and circumferential cracks appeared under under passive confining pressure condition.

39.3.1 *Effect of Strain Rate on Dynamic Response*

Figure 4a shows the dynamic stress–strain responses of unsaturated red clay in the range of strain rates from 255 s^{-1} to 564 s^{-1}. As shown in Fig. 39.4a, the dynamic strength and maximum strain increase with the increase of strain rate. Additionally, the plastic hardening process is obviously enhanced, that is, the ultimate plastic strain and strength limit increase as the strain rate increasing. Figure 4b gives the relationship between dynamic peak stress σ_d and strain rate. The expression of fitting function is $\sigma_d = 0.13\dot{\varepsilon} + 1.46$, $R^2 = 0.9915$.

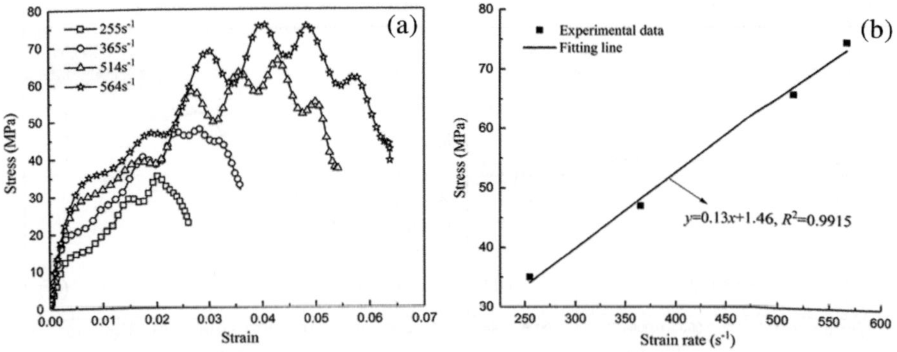

Fig. 39.4 Strain rate effect on the stress–strain responses: **a** stress–strain curves **b** relationship between dynamic strength and strain rate

To further study the relationship between dynamic response parameter and strain rate during the elastic deformation stage, the initial tangent modulus E_d is defined as the slope of stress–strain curve in the elastic compression stage. Figure 5a shows the dynamic responses of red clay specimens in the elastic deformation stage. Experimental results show that the slopes of stress–strain curves and the corresponding yield limits increase as the increasing of strain rate. The relationship between dynamic elastic modulus and strain rate is also fitted by linear function (as shown in Fig. 39.5b). The fitting result is $E_d = 1.07\dot{\varepsilon} + 175.78$, $R^2 = 0.9441$, indicates that there is a good linear positive correlation between dynamic elastic modulus and strain rate.

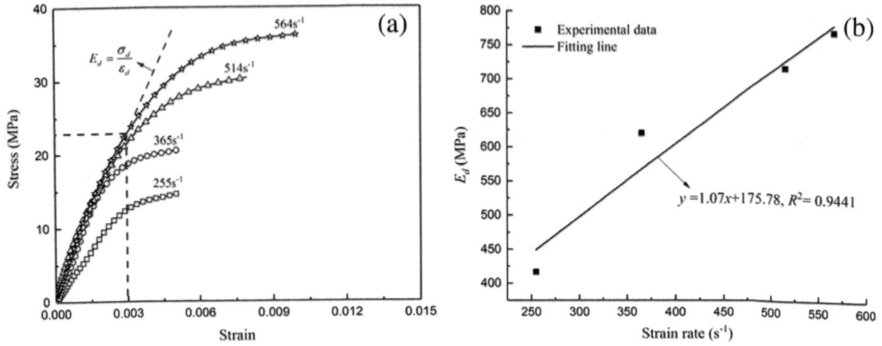

Fig. 39.5 Strain rate effect on the stress–strain responses at the elastic deformation stage: **a** stress–strain curves; **b** relationship between dynamic elastic modulus and strain rate

39.3.2 Effect of Saturation on Dynamic Response

Figure 39.6 illustrates the stress–strain response of clay respectively at saturations from 35 to 47% loaded at a strain rate of 573–579 s^{-1}. It is observed that, the dynamic elastic modulus and peak stress increase when the water content increased from 35 to 43%, but further addition of water would not result in an increase of that. However, the maximum strain of unsaturated red clay decreased as the increasing of moisture content. This behavior is attributed to the increase of the thickness of clay particles capillary film, causing the increase of particles mass at relatively lower degrees of saturation. It is worth noted that free water in soil is generally considered to be incompressible. Therefore, as the increase of moisture content, the applied load on the soil skeleton can be shared by water in soil and the particles movement from original position to void become more difficult. Accordingly, the overall stiffness and resisting capacity of high saturated red clay increased. To summarize, both the multi-phase component and the interaction between phases determine the overall dynamic response of unsaturated soil with solid grains, pore water and closed or semi-closed void.

Further study of the effect of saturation on the strain rate sensitivity of unsaturated red clay is considered next. As observed from Fig. 39.7, linear approximations, for simplicity, were established for the peak dynamic stress and strain rate at each degree of saturation. The degrees of fitting functions are all more than 90%, and the slope of fitting curve k can represent the sensitivity of unsaturated red clay to a certain extent. The slopes of fitting lines in Fig. 39.7a are 0.05, 0.05, 0.06, and 0.06 respectively while the slopes of fitting lines in Fig. 39.7b are 1.81, 0.85, 0.86, and 1.07 respectively, corresponding to 35%, 39%, 43%, and 47% moisture contents of remolded red clay specimens. The experimental results indicate that the strain rate sensitivity of dynamic peak stress is greatly affected by the red clay saturation, while the saturation has little influence on the strain rate sensitivity of dynamic elastic modulus.

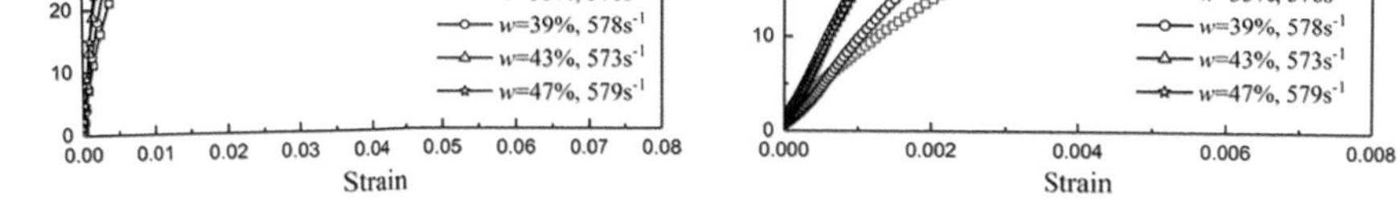

Fig. 39.6 Dynamic responses of different moisture content specimens under same strain rate **a** stress–strain curves, **b** elastic deformation stages

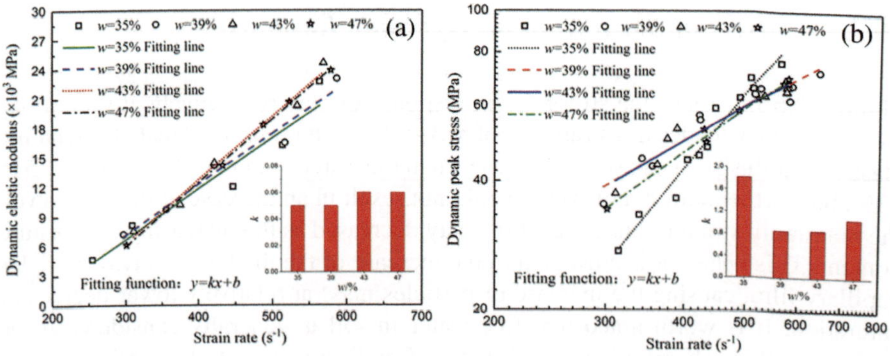

Fig. 39.7 Relationship of strain rate and dynamic parameters: **a** dynamic elastic modulus; **b** dynamic peak stress

Besides, unconsolidated-undrained (UU) tri-axial shear tests on remolded unsaturated soil samples were carried out considering the compression and deformation process are similar to that of SHPB tests in this study. Based on the shear modulus and strength measured by tri-axial test, the normalized elastic modulus and strength were obtained. As shown in Fig. 39.8, the normalized elastic modulus and strength as a function of the strain rate can be expressed in the form of follow equations:

$$E_d^{UU} = 1.5 \times 10^{-3}(\dot{\varepsilon}/\dot{\varepsilon}_0), R^2 = 0.9545 \tag{39.6}$$

$$\sigma_d^{UU} = 7.8 \times 10^{-2}(\dot{\varepsilon}/\dot{\varepsilon}_0), R^2 = 0.9012 \tag{39.7}$$

where E_d^{UU} and σ_d^{UU} are the normalized elastic modulus and strength, $\dot{\varepsilon}_0 = 1 \times 10^{-3}$ is the reference strain rate.

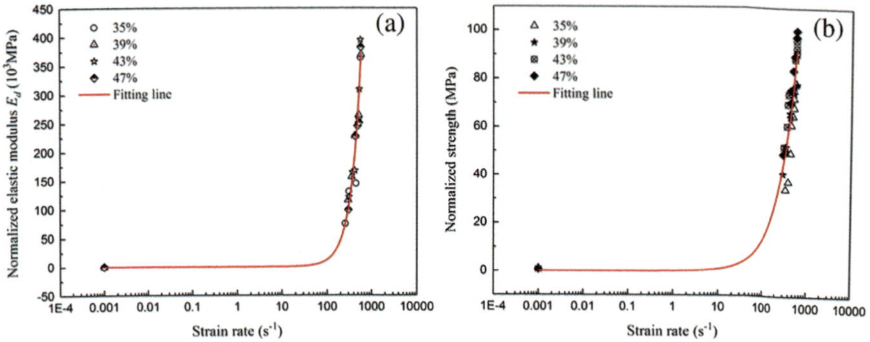

Fig. 39.8 Normalized relationship between strain rate and dynamic parameters: **a** dynamic elastic modulus; **b** dynamic peak stress

39.4 Conclusion

In this study, a series of SHPB tests were conducted to investigate the dynamic behavior of unsaturated red clay subjected to impact loading. The influences of strain rate and moisture content were studied by analyzing the stress–strain behavior. Then, the quantitative relationship between dynamic parameters of red clay and its associating influencing factors were established. The main conclusions are summarized as follows:

(1) Dynamic stress strain curve obtained by SHPB compressive test of unsaturated red clay has three stages in turn: elastic compression, plastic flow, and extreme failure. It is worth noted that the stress has a certain degree of fluctuation as the strain increased over time during the plastic flow stage. The dynamic response shows high strain rate sensitivity, that is, the dynamic elastic modulus and peak stress increase with the increase of strain rate. In addition, there is a good linear positive relationship between the dynamic parameters and the strain rate.

(2) The moisture content has a great influence on the dynamic stress strain response. As the increase of moisture content, the dynamic elastic modulus and strength increases, while the maximum strain decreases. However, the dynamic elastic modulus and strength no longer change when the water content reaches a certain value. There is a significant positive linear correlation between strain rate and dynamic peak stress in the double logarithmic coordinates. The slopes of fitting curves decrease, then increase with the increase of moisture content, indicating that the strain rate sensitivity of unsaturated red clay dynamic peak stress is significantly affected by moisture content.

(3) The exponential function between normalized elastic modulus, ultimate strength and strain rate are established. The expression of fitting function is $y = D(\dot{\varepsilon}/\dot{\varepsilon_0})^{\chi}$, where D and χ denote the fitting parameters based on UU triaxial test and SHPB test. The fitting degrees are more than 90%, indicating that the fitting exponential function can facilitate the development of dynamic constitutive model for red clay.

References

Bai W, Kong L, Guo AG et al (2010) Research on distribution characteristics of bearing capacities and deformation parameters of laterite ground. Rock Soil Mech 31(S2):164–169. https://doi.org/10.16285/j.rsm.2010.s2.021

Bazhenov VG, Bragov AM, Kotov VL et al (2000) Analysis of the applicability of a modified Kolsky's method for dynamic tests of soils in a deformable casing. J Appl Mech Tech Phys 41(3):519–525. https://doi.org/10.1007/BF02465305

Bragov AM, Grushevsky GM, Lomunov AK (1996) Use of the Kolsky method for confined tests of soft soils. Exp Mech 36(3):237–242. https://doi.org/10.1007/BF02318013

Bragov AM, Kotov VL et al (2004) Measurement of the dynamic characteristics of soft soils using the Kolsky method. J Appl Mech Tech Phys 45(4):580–585. https://doi.org/10.1023/B:JAMT.0000030338.66701.e9

Bragov AM, Lomunov AK et al (2008) Determination of physicomechanical properties of soft soils from medium to high strain rates. Int J Impact Eng 35(9):967–976. https://doi.org/10.1016/j.ijimpeng.2007.07.004

Charlie WA, Ross CA, Pierce SJ (1990) Split-Hopkinson pressure bar testing of unsaturated sand. Geotech Test 13(4):291–300. https://doi.org/10.1016/0148-9062(91)90812-z

Chen W, Lu F, Zhou B (2000) A quartz-crystal-embedded split Hopkinson pressure bar for soft materials. Exp Mech 40(1):1–6. https://doi.org/10.1007/s11340-011-9475-2

Chen W, Lu F, Frew DJ et al (2002) Dynamic compression testing of soft materials. J Appl Mech 69(3):214–223. https://doi.org/10.1115/1.1464871

Ding YQ, Tang WH, Xu X et al (2013) Experimental study of dynamic mechanical behaviors of unsaturated clay subjected to uniaxial loading. Rock Soil Mech 34(9):2546–2550. CNKI:SUN:YTLX.0.2013-09-018

Ding JX, Meng YW (2014) Review of applied study for red clay in highway subgrades. AMR 919–921:1144–1149. https://doi.org/10.4028/www.scientific.net/amr.919-921.1144

Editorial of Engineering Geology Manual (2007) Engineering geology manual. Architecture and Building Press, Beijing, China

Farr JV, Woods RD (1988) A device for evaluating one-dimensional compressive loading rate effects. Geotech Test J 11(4):269–275. https://doi.org/10.1520/GTJ10658J

Felice CW (1985) The response of soils to impulse loads using the split-Hopkinson pressure bar technique. University of Utah

Frew DJ, Forrestal MJ, Chen W (2005) Pulse shaping techniques for testing elastic-plastic materials with a split Hopkinson pressure bar. Exp Mech 45(2):186–195. https://doi.org/10.1007/BF02428192

Gaffney ES, Brown JA, Felice CW (1985) Soils as samples for the split Hopkinson bar. In: 2nd symposium on the interaction of non-nuclear munitions with structures. https://www.osti.gov/servlets/purl/6164579

Hopkinson B (1914) A method of measuring the pressure produced in the detonation of high explosives or by the impact of bullets. Proc R Soc Lond Ser A 213:437–456. https://doi.org/10.1098/rsta.1914.0010

Kabir ME, Song B, Martin BE et al (2010) Compressive behavior of fine sand. Terminal Ballistics Technology Department, Sandia Laboratories, New Mexico

Kabir ME (2010) Dynamic behavior of granular materials. Purdue University

Kaynia AM (2000) Ground vibration from high-speed trains: prediction and countermeasure. J Geotech Geoenviron Eng 126(6):531–537. https://doi.org/10.1061/(ASCE)1090-0241(2000)126:6(531)

Kolsky H (1949) An investigation of the mechanical properties of materials at very high rates of loading. Proc R Soc Lond, Ser. B 62:676–700. https://doi.org/10.1088/0370-1301/62/11/302

Liu JX, Chen ZF, Xu WF et al (2012) Experimental study on dynamic mechanical properties of compacted clay under different compaction degrees and water contents. Rock Soil Mech 33(6):1631–1639. https://doi.org/10.3969/j.issn.1000-7598.2012.06.006

Long WX, Chen KS, Xiao T, et al (2009) Research of general triaxial test for unsaturated red clay. Rock Soil Mech 30(S2):28–33. https://doi.org/10.16285/j.rsm.2009.s2.095

Luo H, Lu H, Cooper WL et al (2011) Effect of mass density on the compressive behavior of dry sand under confinement at high strain rates. Exp Mech 51(9):1499–1510. https://doi.org/10.1007/s11340-011-9475-2

Luo H, Hu Z, Xu T et al (2017) High-strain rate compressive behavior of a "natural soil" under uniaxial strain state. In: Conference proceedings of the society for experimental mechanics series, Springer, Cham. https://doi.org/10.1007/978-3-319-62956-8_15

Martin BE, Chen W, Bo S et al (2009) Moisture effects on the high strain-rate behavior of sand. Mech Mater 41(6):786–798. https://doi.org/10.1016/j.mechmat.2009.01.014

Murphy DJ (1970) Soils and rocks: composition, confining level and strength. Duke University

Pierce SJ, Charlie WA (1990) High-intensity compressive stress wave propagation through unsaturated sands. ESL-TR-90-12. Tyndall AFB, Fla: Air Force Engineering and Services Laboratory

Pierce SJ (1989) High intensity compressive stress wave propagation through unsaturated sands. M.Sc. thesis, Colorado State University; Fort Collins, Co, 1989

Song B, Chen W, Luk V (2009) Impact compressive response of dry sand. Mech Mater 41(6):777–785. https://doi.org/10.1016/j.mechmat.2009.01.003

Yang R, Chen J, Yang L et al (2017) An experimental study of high strain-rate properties of clay under high consolidation stress. Soil Dyn Earthq Eng 92:46–51. https://doi.org/10.1016/j.soildyn.2016.09.036

Yang S, Lu TH (2011) Research on unventilated-undrained triaxial shear test of red cohesive soil. Rock Soil Mech 32(S1):356–359. https://doi.org/10.16285/j.rsm.2011.s1.038

Chapter 40
Influence of Grading on the Undrained Flow Instability of Saturated Loess Soil

Xu Ji, Qiang Xu, and Lanting Wei

Abstract Loess slope failure caused by static liquefaction can result in a great number of fatalities. However, the influence of grading on the mechanical responses of saturated loess, especially its critical states and liquefaction potential, remains unclear. Loess collected from a typical silty loess zone on the northwestern part of the Chinese Loess Plateau where loess flowslides have widely occurred was studied. Detailed experimental research using undrained standard triaxial tests was carried out on artificial and natural loess soils with various gradings under saturated states. The results reveal that during undrained shearing, the critical state lines moved downward and rotated synchronously from the natural grading position as the fines content decreased, until reversed shift occurred at a fines content of around 25%. For a similar global void ratio and lower confining stresses, the potential for flow liquefaction of loess soils showed a reverse pattern with increased fines content. While they became stronger monotonously with the addition of fines under higher stresses. Additionally, the applicability of the intergranular and interfine state variables in analyzing the undrained fragility of the loess soils was discussed. The patterns obtained from these laboratory tests provide some new insights into the influence of grading on the static liquefaction of loess soils, as well as new information for risk mitigation of loess flowslides.

Keywords Loess · Soil grading · Shearing behaviour · Critical states · Liquefaction

X. Ji · Q. Xu (✉)
State Key Laboratory of Geohazard Prevention and Geoenvironment Protection, Chengdu, China
e-mail: xq@cdut.edu.cn

L. Wei
Herriot-Watt University, Edinburgh, UK
e-mail: w.lanting@hw.ac.uk

S. Wang et al. (eds.), *Engineering Geology for a Habitable Earth: IAEG XIV Congress 2023 Proceedings, Chengdu, China*, Environmental Science and Engineering,
https://doi.org/10.1007/978-981-99-9057-3_40

40.1 Introduction

Loess with low moisture contents provides earth slopes in northwestern China with sufficient shear strength to resist failure (Lin and Wang 1988). However, due to loess's metastable structure, these stable slopes may transform into catastrophic flowslides with high speeds and long run-out distances upon wetting or saturation (Liu et al. 2016; Peng et al. 2019). The rising groundwater level generated by intense agricultural irrigation has triggered numerous landslides with such flow failure patterns on the edges of the Loess Plateau, especially on the Heifangtai terrace in Gansu Province, China (Gu et al. 2019; Xu et al. 2012, 2014). By introducing the critical state soil mechanics framework, it is believed that loess flowslides can be attributed to the static liquefaction of saturated loess (Feng et al. 2021; Qi et al. 2018; Zhou et al. 2014).

The influence of fines content (FC) or grading on the static liquefaction of saturated soils have been thoroughly investigated through laboratory experiments (Lade and Yamamuro 1997; Monkul and Yamamuro 2011). However, the evolution of the shape and location of the critical state lines (CSLs) in the e-lnp' plane as fines were added to change the grading of the sand seems contradictory. The steeper part of the CSLs became flatter as fines were added to carbonate sand (Coop and Atkinson 1993), while the gradient of the CSLs of quartzitic sand increased with increasing fines content (Been and Jefferies 1985). Another important finding was that the CSLs of sand–silt mixtures moved downward from the clean sand position as fines were added until a reverse trend was observed (Yang et al. 2006). The grading that gave the lowest location was defined as the transitional fines content (TFC) (Thevanayagam et al. 2002).

Loess is a typical silt-sized soil with a silt fraction of 75–100%. Due to the different movement patterns of coarser and finer particles during wind-driven deposition, the Chinese Loess Plateau can be divided into three zones based on soil grading: a sandy loess zone, a silty loess zone, and a clayey loess zone (Xu et al. 2019). By carrying out one-dimensional oedometer compression tests, a recent study determined that particle breakage occurred in loess with different natural gradings (Gao et al. 2021), implying that there is a possible change in the loess grading in geotechnical mega-projects such as mountain excavation and city construction (MECC) in these regions. It is generally accepted that soils with their in situ states above the CSL in the e-lnp' plane, i.e., with positive state parameters, are more prone to static liquefaction (Been and Jefferies 1985; Carrera et al. 2011). This raises the question of how the variation in the loess gradings with location and engineering practice influences loess's mechanical properties, especially its critical states and liquefaction potential. However, few studies have focused on this issue.

In this work, loess from a tableland where flowslides are widely distributed was split into two gradings of sand and fines and was remixed in different proportions to analyze the influence of grading on the potential for undrained flow instability. Series of undrained triaxial tests were conducted to identify the CSLs of loess soils in order to examine the liquefaction potential relative to the location of these lines.

Intergranular and interfine state variables (Rahman and Lo 2008; Rahman et al. 2008, 2011; Thevanayagam 1998; Thevanayagam and Mohan 2000; Thevanayagam et al. 2002) were also used to study the undrained fragility.

40.2 Materials and Methodology

The loess used in this study was collected from the backscarp of a flowslide on the margin of the Heifangtai terrace, Lanzhou City, Gansu Province. Given the small fraction of clay in the finer grains, the term fines is used to represent particles smaller than 75 μm, while sand denotes particles larger than 75 μm but smaller than 2 mm. The particle size distributions of the sand and fines were analyzed using the standard dry sieving test and hydrometer test respectively (Fig. 40.1) (ASTM 2006). The geotechnical index properties of the two materials are listed in Table 40.1.

Series of standard triaxial tests were conducted on loess soils prepared by mixing different percentages of sand and fines. The label FC (fines content) denotes the percentage of the total weight of the dry solids accounted for by the fines. Four mixtures were prepared, i.e., FC0, FC25, FC50, and FC75. FC87.4 represents the natural grading of the Heifangtai silty loess. The soil mixtures with fines contents of 75–100% are representative of the natural gradings of loess on the Chinese Loess Plateau.

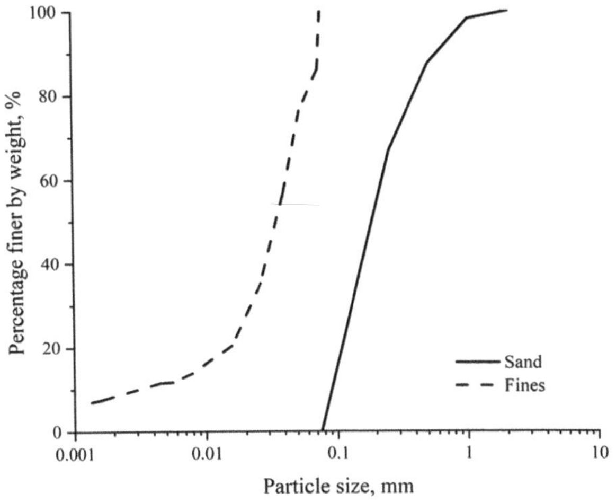

Fig. 40.1 Particle size distributions of the sand and fines

Table 40.1 Geotechnical index properties of the tested materials

Sample	Specific gravity-G_s	Liquid limit	Plastic limit	Plasticity index
Sand	2.705	–	–	–
Fines	2.718	32.6%	16.7%	15.9%

The triaxial specimens were 50 mm in diameter and 100 mm in height. All of the samples were prepared using the wet compaction method, which allowed the loosest samples to be created to investigate their liquefaction potential. The under-compaction method of Ladd (1978) was adopted to avoid the development of macro-pores. The specimens were flushed with CO_2 and distilled water first, and then, they were saturated under a back pressure until a B value of at least 0.95 was achieved. In the consolidation stage, the samples were all isotropically compressed to a variety of stress levels and then sheared under undrained conditions at 0.1 mm/min, particularly with regard to static liquefaction. All of the samples were tested with initial void ratios of 0.83–0.85 (Table 40.2), which is similar to thc in situ dry density of 1.46–1.48 g/ cm^3 of the loess flowslide. Given the small particle sizes of the loess soils, there was no need to consider the membrane penetration effect.

40.3 Results and Interpretation

40.3.1 Stress–Strain Response

Figure 40.2a presents the typical stress–strain data under undrained conditions. Almost all of the stress–strain curves had similar shapes, except for U00. The initial deviatoric stresses reached peak points at 1–2.5% of the axial strain, and subsequently significant strain-softening behaviors were observed, indicating the potential for flow instability. Carrera et al. (2011) defined that true liquefaction occurred when the confining stress was equal to the pore pressure for soil specimens (i.e., $q \approx 0$ and $p' \approx 0$), which incorporates the definitions given by different researchers (Poulos et al. 1985; Yamamuro and Covert 2001). Although severe strain-softening behaviors were observed in samples U50-1 and U87.4-3 (Table 40.2) (the q and p' at the end of the tests came quite close to the original point), none of the tests in this study reached true liquefaction, which is consistent with the results of previous studies that focused on the static liquefaction of loess (Xu et al. 2012; Zhou et al. 2014). It should be noted that the samples with FC $= 0\%$ exhibited the undrained response called reverse behavior, which has been widely observed in silty sand (Lade and Yamamuro 1997; Yamamuro and Lade 1998). Flow instability was discovered under lower stress (i.e., $p' = 200$ kPa), and the stability as well as dilatancy increased with increasing confining pressures. The undrained stress–strain responses of U00-2 and U00-4 changed from strain-softening to phase transformation points, followed by strain hardening. Accordingly, the stress paths of U00 in the q-p' plane exhibited an elbow shape (Fig. 40.3). This unusual mechanical response may be due to the change in the particle structure during loading. The small proportion of the finer silt derived from the mechanical disruption or particle breakage would slide into the empty pore spaces between the larger particles as the sand was compressed in the initial stage. As shearing continued, the coarser grains would move into better contact with each

Table 40.2 Details of the triaxial tests

Sample	Preparation method	Specific volume after saturation	Confining effective stress, kPa	Specific volume after consolidation	End of test		
					Deviatoric stress, q_{cs}: kPa	Mean effective stress, p'_{cs}: kPa	Specific volume
U00*-1	WC*	1.838	200	1.714	76.364	58.815	1.714
U00-2	WC	1.840	300	1.679	213.462	148.464	1.679
U00-3	WC	1.845	400	1.640	187.865	142.812	1.640
U00-4	WC	1.836	500	1.650	368.677	250.022	1.650
U25-1	WC	1.843	100	1.730	30.926	18.119	1.730
U25-2	WC	1.844	200	1.718	76.193	54.598	1.718
U25-3	WC	1.839	300	1.694	121.044	92.268	1.694
U25-4	WC	1.839	500	1.652	222.713	174.228	1.652
U50-1	WC	1.846	50	1.811	23.673	5.632	1.811
U50-2	WC	1.843	100	1.782	21.839	10.500	1.782
U50-3	WC	1.850	200	1.727	64.871	41.574	1.727
U50-4	WC	1.842	300	1.707	100.749	73.243	1.707
U50-5	WC	1.840	500	1.682	196.407	156.359	1.682
U75-1	WC	1.852	100	1.792	29.105	14.422	1.792
U75-2	WC	1.850	200	1.749	55.347	34.649	1.749
U75-3	WC	1.855	300	1.732	98.576	73.019	1.732
U75-4	WC	1.858	400	1.714	136.589	103.210	1.714
U75-5	WC	1.853	500	1.681	202.040	156.267	1.681
U87.4-1	WC	1.882	50	1.844	20.913	15.681	1.844
U87.4-2	WC	1.849	100	1.799	27.858	11.456	1.799
U87.4-3	WC	1.855	100	1.795	19.976	9.402	1.795
U87.4-4	WC	1.849	200	1.786	145.570	36.510	1.786
U87.4-5	WC	1.841	300	1.721	114.261	98.167	1.721
U87.4-6	WC	1.848	500	1.717	186.696	156.102	1.717

Note WC—wet compaction, U00—isotropic consolidated undrained tests for samples with FC = 0%

other, which caused the soil to exhibit increasing dilatancy as the confining stresses increased.

A constant deviatoric stress was still not achieved in some of the tests (Fig. 40.2a), which means they did not quite reach the critical states (CS) even at larger strains. However, for the majority of the specimens, it could be assumed that an axial strain of 20% was sufficient to reach their CS. Figure 40.2b shows the effective stress paths of the excess pore water pressures, which were normalized using the mean effective stress at the start of shearing (p'_0) to eliminate the effect of the different confining

Fig. 40.2 Typical results of triaxial tests: **a** Undrained deviatoric stress–axial strain curves, **b** normalized pore pressure responses

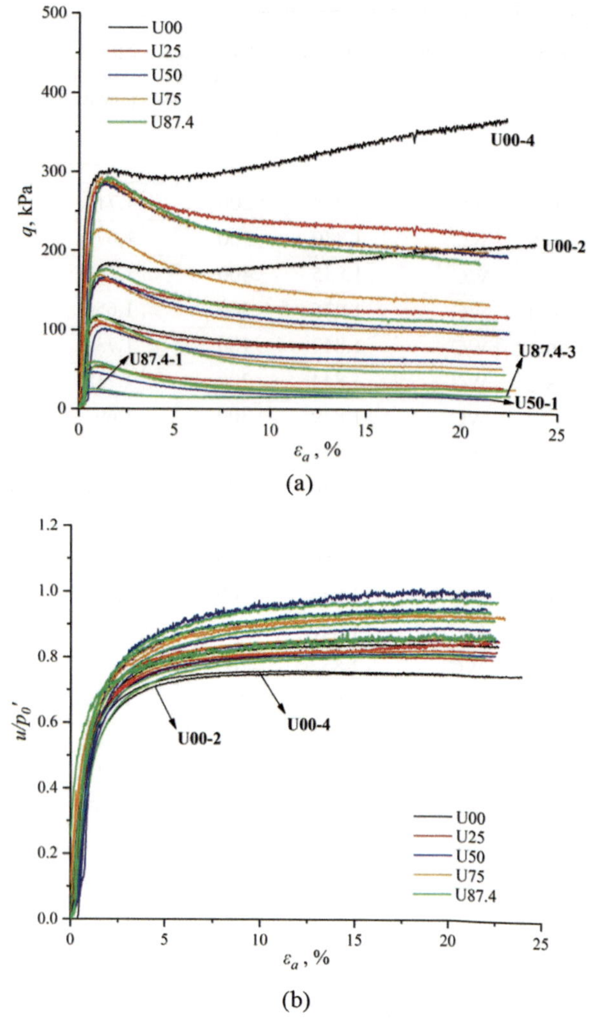

pressures. Almost all of the samples contracted at the start of shearing with a sharp increase in pore water pressure, then they reached peak points and remained stable. Samples U00-2 and U00-4 exhibited a dilative trend after u_{max}/p'_0 was reached, and their pore water pressures subsequently began to decrease, causing an increase in the deviatoric stresses.

As can be seen from Fig. 40.3, the shape of the undrained stress paths of almost all of the tests were quite similar, i.e., *C* shaped, which indicates that contraction occurred at nearly all of the stress levels. The much looser samples, U50-1 and U87.4-1, reached liquefaction when sheared at 50 kPa, but their stress paths exhibited a small inversion at the end of the tests, which was probably due to the lateral restraint of the membrane at high strains. A more *C* shaped stress path was observed for the specimen

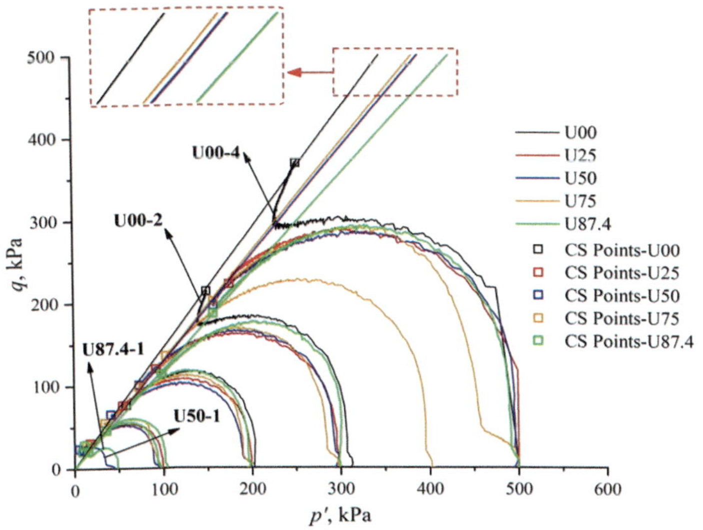

Fig. 40.3 Stress paths and critical state points of loess soils

with an FC of 0% at 200 kPa. However, at higher confining stresses, the stress paths were more *S* shaped, and the phase transformation points could be clearly identified. This trend is consistent with the stress–strain and pore water pressure responses of pure sand. The end points of the triaxial tests defined different CSLs in the q-p' plane for each grading. The gradient of the different CSLs changed with the particle size distributions (Fig. 40.3) and the angle of the shearing resistance at critical states (i.e., φ'_{cs}) exhibited a reverse pattern with increasing FC, indicating that the soil grading had an impact on the friction angle of the loess soils. This was probably due to the angular to sub-angular particle shapes of all the sizes.

40.3.2 Critical States and Liquefaction Potential

As is shown in Fig. 40.4a, a unique CSL could be determined for each grading. As FC decreased from 87.4% (i.e., natural grading), the CSLs moved downward and rotated counterclockwise synchronously until a reverse trend was observed when the FC reached ~ 25%. The continuous decrease in the FC resulted in the rotation of the CSLs with steeper gradients, which suggests that when everything else is the same, the state parameter increases at higher stresses (FC = 0%) compared to the other groups. For the loess soils, the influence of grading on the location of the CSLs was more pronounced under lower stresses, exhibiting reversed shift pattern. While under higher stresses, the CSLs moved downward monotonously with decreasing FC. The rotation of the CSLs in the v-lnp' plane was similar to the findings of Been and Jefferies (1985) for sand–fines mixtures.

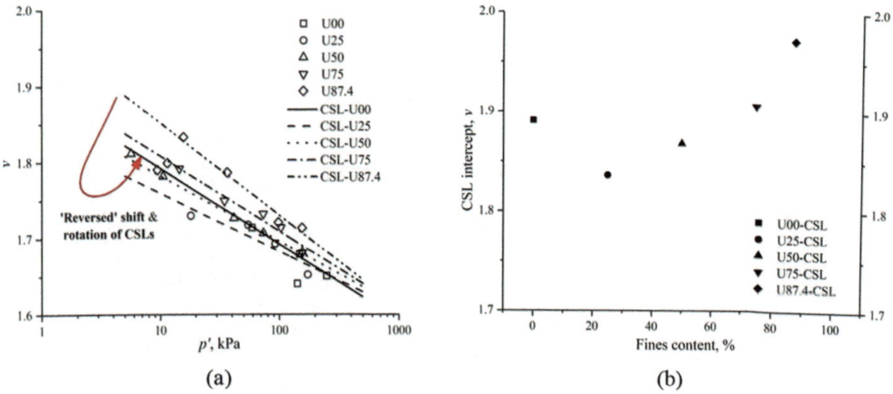

Fig. 40.4 Critical state data for all tests: **a** Critical state lines, **b** variations in intercepts of CSLs

Previous studies have indicated that the relative locations of CSLs are closely linked to the soil grading (Thevanayagam et al. 2002; Yang et al. 2006). Zuo and Baudet (2015) defined the TFC at which the intercept of the CSLs reached a minimum value under monotonic loading conditions. Although the gradings of loess soils were mostly original and not artificial, and was noted above, the CSLs in this paper actually tended to translate more than rotate, so this criterion was adopted. Figure 40.4b shows that the intercept of the CSLs reached a trough at an FC of ~ 25%, defining the lower bound of the possible value. As a result, an FC of ~ 25% was determined to be the TFC for the Heifangtai silty loess. The TFC for the loess soils in this paper was different from those of Thevanayagam et al. (2002) (40%) and Carrera et al. (2011) (50%). Fourie and Papageorgiou (2001) did not observe a change in the downward movement of the CSLs with a FC even of 60%. This gap can be ascribed to the different origins of the soils. Although the term TFC seems to be inapplicable at higher confining stresses, it is of great interest in analyzing the mechanical responses of loess soils at low stress levels in order to identify the risk of static liquefaction.

The collapse potential (CP) was proposed by Thevanayagam et al. (2002) to describe the magnitude of the maximum positive shear induced pore pressure ratio during monotonic shearing:

$$CP = \frac{p'_0 - p'_{cs}}{p'_0} \tag{40.1}$$

where p'_0 and p'_{cs} are the mean effective stresses at the start of shearing and in the critical state respectively. For specimens that exhibit initial collapse followed by dilation, p'_{cs} is equal to the mean effective stress in the quasi-steady-state. Figure 40.5 shows that at similar specific volumes (after consolidation) (circled by the red lines), CP decreased with increasing FC when FC > TFC, while the opposite trend occurred when FC < TFC. Consequently, when FC > TFC, an increase in the finer particles in loess could restrain the magnitude of the pore water pressure induced by undrained

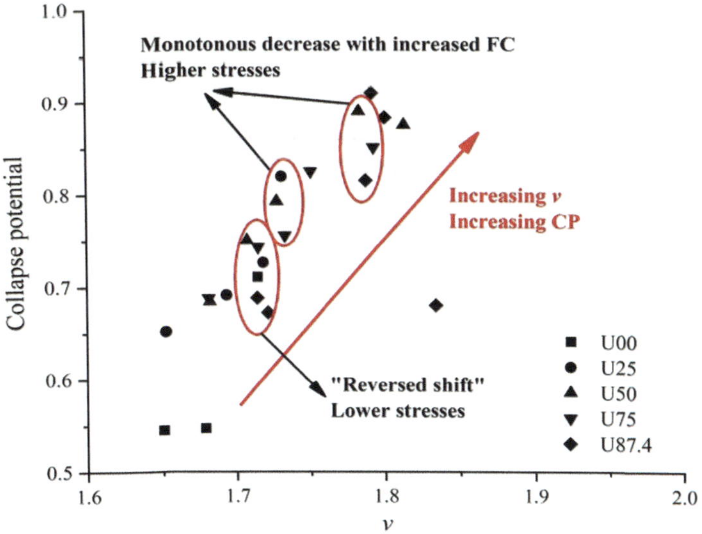

Fig. 40.5 Collapse potential of all tests

shearing, thus lowering the risk of flow instability. This clearly reflected the relative locations of CSLs in both lower and higher stresses. In addition, CP increased concurrently with the void ratio regardless of the soil grading, indicating that static liquefaction was most likely to occur in saturated loess soils in loose states.

40.4 Intergrain Contact Friction Effect on Critical States

The coarse grain contacts play a significant role in the shearing responses, and the fine grains offer a secondary effect when FC < TFC. In contrast, the interfine grain contact becomes dominant when FC > TFC. Since the global void ratio (e) is not suitable to describe the mechanical behavior of sand containing fines or fines containing sand, previous studies have successfully identified unique mechanical behaviors in sand–fines mixtures, regardless of the FC, by using the intergrain state variables (Ni et al. 2004). Except for the first-order indices (i.e., the intergranular and interfine void ratios, e_c and e_f), the equivalent contact density indices (i.e., the equivalent intergranular and interfine void ratios, $(e_c)_{eq}$ and $(e_f)_{eq}$) considering to the secondary effects of the soil particles serve as better approaches to investigating this issue (Thevanayagam et al. 2002):

$$(e_c)_{eq} = \frac{e + (1 - b)fc}{1 - (1 - b)fc} \tag{40.2}$$

$$(e_f)_{eq} = \frac{e}{fc + \frac{1-fc}{R_d{}^m}} \tag{40.3}$$

where $0 < b < 1$, and b denotes the proportion of the fine grains that contributes to the active intergrain contacts, fc is the fines content (as a decimal); R_d is the size disparity ratio, $0 < m < 1$, and m is a coefficient that depends on the grain characteristics and fine grain packing. The proper calculation method for R_d may vary for different studies, and no strict rule has yet been proposed. Yang et al. (2006) took R_d as $D_{50\text{-}sand}/d_{50\text{-}fines}$, but $R_d = D_{10\text{-}sand}/d_{50\text{-fines}}$ has been extensively used. The second criteria was used in this study ($R_d = 2.605$). Analyses were conducted to determine whether the undrained shearing response of the loess soils could be normalized. Because the TFC for the Heifangtai silty loess was determined to be 25% in this study, the samples with FCs between 0 and 25%, 50% and 87.4% were investigated separately.

As is shown in Figure 40.6b, c, when $(e_c)_{eq}$ and $(e_f)_{eq}$ were used, the CSLs with different shapes could be normalized by one unique line if FC < TFC. The CS points of the samples with FC of > 25% also exhibited perfect convergence to a unique equivalent interfine CSL, regardless of the FC. Both fitting curves had $R^2 > 0.98$. It should be noted that b was determined using a back-analysis procedure that was independent of the FC, and $b = 0.96$ was found to be optimum, indicating that although the FC was below the TFC, the finer grains in the loess soils made large contribution to the formation of the force chain. Rahman et al. (2008) examined the factors affecting b and proposed a semi-empirical equation:

$$b = \{1 - \exp[-2.5(fc)^2/k]\}(rfc/f_{thre})^r \tag{40.4}$$

where $r = d_{50\text{-}fines}/D_{10\text{-}sand}$, $k = 1 - r^{0.25}$, and f_{thre} is the TFC (in decimal form). Equation (40.4) is based on the grain size of the fines relative to that of the host sand. Figure 40.6a shows that a unique CSL can hardly be defined ($b = 0.523$ according to Eq. (40.4) in this study). So far there has been no uniform standard for the calculation of possible values of b. Consequently, while b indicates to what extent the fines and sand skeleton support the external stress, it actually has no physical meaning. Furthermore, there is no clear evidence from which to determine the accurate value of m. The choice of m was also based on back-analysis procedures, so that perfect convergence was achieved ($m = 0.1$ in this study).

40.5 Conclusions

Experimental studies have been conducted on loess soils with various gradings. Their shearing behaviors, as well as the potential for undrained flow liquefaction, were found to be significantly affected by the soil grading in the saturated state. In the undrained shearing tests all of the samples exhibited contractive behavior under lower stresses, indicating the potential for flow instability. Under lower stresses, the CSLs moved downward and rotated simultaneously as the fines content decreased,

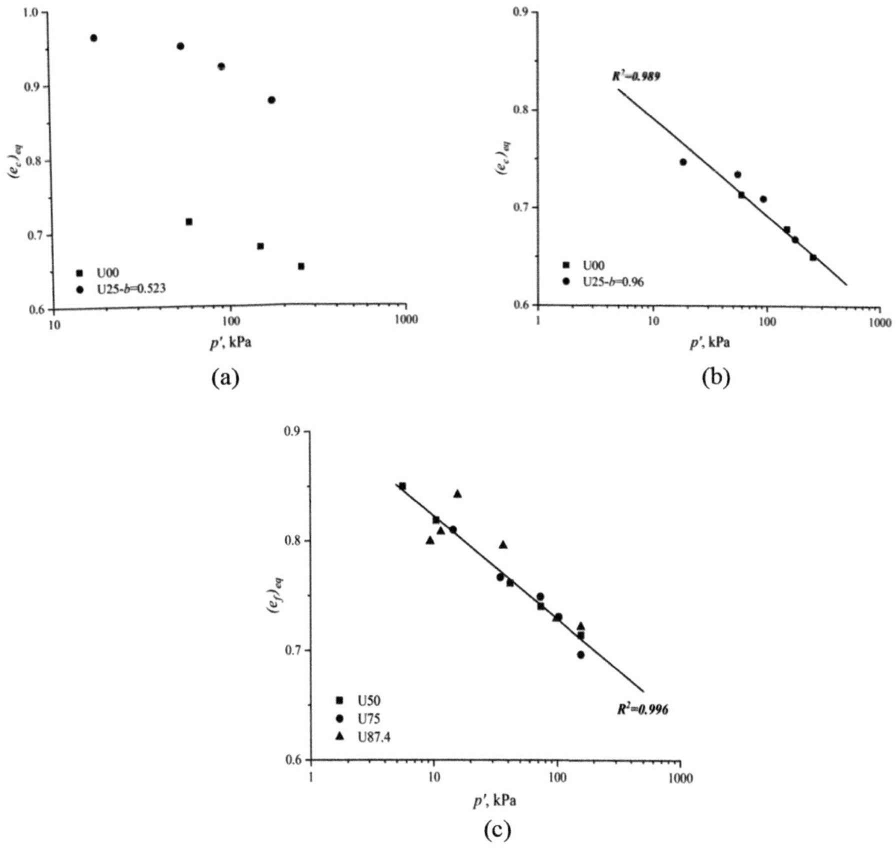

Fig. 40.6 Critical state data in terms of the intergrain void ratios: **a** $(e_c)_{eq}$ V.S. p'_{cs} ($b = 0.523$), **b** $(e_c)_{eq}$.V.S. p'_{cs} ($b = 0.96$), **c** $(e_f)_{eq}$.V.S. p'_{cs} ($R_d = 2.605$, $m = 0.1$)

until reversed shift was observed at a fines content of ~ 25%. Conversely, the CSLs moved downward monotonously with decreasing fines content under higher stresses. The transitional fines content of the Heifangtai loess was determined to be 25%. In general, the intergrain state variables can be used as better indices to capture the undrained collapse of finer loess.

On this basis, we conclude that loess soils originating from the sandy loess zone, with lower fines contents, may be more prone to static liquefaction compared with silty and clayey loess. Hence, great care must be taken when carrying out agricultural irrigation in sandy loess regions. Additionally, some geotechnical mega-projects such as the removal of the tops of mountains and infilling of valleys by compaction trigger particle breakage in loess, causing an increase in the fines content. In some ways, this type of project restrains the liquefaction potential of loess in light of the soil grading. Future work needs to be conducted to determine whether this study can be extended to loess soils with different natural gradings on the Chinese Loess Plateau.

Acknowledgements This work was supported by the National Natural Science Foundation of China (grant numbers 41630640, 41790445). Prof. Fangzhou Liu from University of Alberta is greatly thanked for his valuable suggestions in the conceptual phase.

References

ASTM (2006) Annual book of ASTM standards. ASTM International, West Conshohocken, PA, United States

Been K, Jefferies MG (1985) A state parameter for sands. Geotechnique 35:99–112. https://doi.org/10.1680/geot.1985.35.2.99

Carrera A, Coop M, Lancellotta R (2011) Influence of grading on the mechanical behaviour of Stava tailings. Geotechnique 61:935–946. https://doi.org/10.1680/geot.9.P.009

Coop M, Atkinson J (1993) The mechanics of cemented carbonate sands. Geotechnique 43:53–67. https://doi.org/10.1680/geot.1993.43.1.53

Feng L, Zhang S, Jin Z, Zhang M, Sun P, Jia J, Chu G, Hu W (2021) The critical mechanics of the initiation of loess flow failure and implications for landslides. Eng Geol 288:106165. https://doi.org/10.1016/j.enggeo.2021.106165

Fourie A, Papageorgiou G (2001) Defining an appropriate steady state line for Merriespruit gold tailings. Can Geotech J 38:695–706. https://doi.org/10.1139/t00-111

Gao C, Ling X, Coop MR, Huang C, Zuo L (2021) An investigation of particle breakage in loess. Eng Geol 286:106083. https://doi.org/10.1016/j.enggeo.2021.106083

Gu T, Zhang M, Wang J, Wang C, Xu Y, Wang X (2019) The effect of irrigation on slope stability in the Heifangtai platform, Gansu Province, China. Eng Geol 248:346–356. https://doi.org/10.1016/j.enggeo.2018.10.026

Ladd R (1978) Preparing test specimens using undercompaction. Geotech Test J 1:16–23. https://doi.org/10.1520/GTJ10364J

Lade PV, Yamamuro JA (1997) Effects of nonplastic fines on static liquefaction of sands. Can Geotech J 34:918–928. https://doi.org/10.1139/t97-052

Lin Z, Wang S (1988) Collapsibility and deformation characteristics of deep-seated loess in China. Eng Geol 25:271–282. https://doi.org/10.1016/0013-7952(88)90032-4

Liu Z, Liu F, Ma F, Wang M, Bai X, Zheng Y, Yin H, Zhang G (2016) Collapsibility, composition, and microstructure of loess in China. Can Geotech J 53:673–686. https://doi.org/10.1139/cgj-2015-0285

Monkul MM, Yamamuro JA (2011) Influence of silt size and content on liquefaction behavior of sands. Can Geotech J 48:931–942. https://doi.org/10.1139/t11-001

Ni Q, Tan T, Dasari G, Hight D (2004) Contribution of fines to the compressive strength of mixed soils. Geotechnique 54:561–569. https://doi.org/10.1680/geot.2004.54.9.561

Peng J, Wang S, Wang Q, Zhuang J, Huang W, Zhu X, Leng Y, Ma P (2019) Distribution and genetic types of loess landslides in China. J Asian Earth Sci 170:329–350. https://doi.org/10.1016/j.jseaes.2018.11.015

Poulos SJ, Castro G, France JW (1985) Liquefaction evaluation procedure. J Geotech Eng 111:772–792. https://doi.org/10.1061/(ASCE)0733-9410(1985)111:6(772)

Qi X, Xu Q, Liu F (2018) Analysis of retrogressive loess flowslides in Heifangtai, China. Eng Geol 236:119–128. https://doi.org/10.1016/j.enggeo.2017.08.028

Rahman MM, Lo S (2008) The prediction of equivalent granular steady state line of loose sand with fines. Geomech Geoeng: Int J 3:179–190. https://doi.org/10.1080/17486020802206867

Rahman MM, Lo SR, Gnanendran CT (2008) On equivalent granular void ratio and steady state behaviour of loose sand with fines. Can Geotech J 45:1439–1456. https://doi.org/10.1139/T08-064.doi:10.1139/T08-064

Rahman MM, Lo S, Baki MAL (2011) Equivalent granular state parameter and undrained behaviour of sand–fines mixtures. Acta Geotech 6:183–194. https://doi.org/10.1007/s11440-011-0145-4

Thevanayagam S (1998) Effect of fines and confining stress on undrained shear strength of silty sands. J Geotech Geoenviron Eng 124:479–491. https://doi.org/10.1061/(ASCE)1090-0241(1998)124:6(479)

Thevanayagam S, Mohan S (2000) Intergranular state variables and stress–strain behaviour of silty sands. Geotechnique 50:1–23. https://doi.org/10.1680/geot.2000.50.1.1

Thevanayagam S, Shenthan T, Mohan S, Liang J (2002) Undrained fragility of clean sands, silty sands, and sandy silts. J Geotech Geoenviron Eng 128:849–859. https://doi.org/10.1061/(ASCE)1090-0241(2002)128:10(849)

Xu L, Dai F, Gong Q, Tham L, Min H (2012) Irrigation-induced loess flow failure in Heifangtai platform, north-west China. Environ Earth Sci 66:1707–1713. https://doi.org/10.1007/s12665-011-0950-y

Xu L, Dai F, Tu X, Tham LG, Zhou Y, Iqbal J (2014) Landslides in a loess platform, North-West China. Landslides 11:993–1005. https://doi.org/10.1007/s10346-013-0445-x

Xu L, Gao C, Lan T, Lei J, Zuo L (2019) The influence of grading on the compressibility of saturated loess soils. Géotech Lett 10:198–204. https://doi.org/10.1680/jgele.19.00097

Yamamuro JA, Covert KM (2001) Monotonic and cyclic liquefaction of very loose sands with high silt content. J Geotech Geoenviron Eng 127:314–324. https://doi.org/10.1061/(ASCE)1090-0241(2001)127:4(314)

Yamamuro JA, Lade PV (1998) Steady-state concepts and static liquefaction of silty sands. J Geotech Geoenviron Eng 124:868–877. https://doi.org/10.1061/(ASCE)1090-0241(1998)124:9(868)

Yang S, Sandven R, Grande L (2006) Steady-state lines of sand–silt mixtures. Can Geotech J 43:1213–1219. https://doi.org/10.1139/t06-069

Zhou YF, Tham LG, Yan WM, Dai FC, Xu L (2014) Laboratory study on soil behavior in loess slope subjected to infiltration. Eng Geol 183:31–38. https://doi.org/10.1016/j.enggeo.2014.09.010

Zuo L, Baudet BA (2015) Determination of the transitional fines content of sand-non plastic fines mixtures. Soils Found 55:213–219. https://doi.org/10.1016/j.sandf.2014.12.017

Chapter 41
Experimental Study on Static and Dynamic Characteristics of Tunnel Slag as Road-Base Material

Zhiwei Xie, Yiyue Ma, Zhongjie Zhang, Sheng Zeng, Xianlin Liu, and Xilin Lü

Abstract Growing body of research focused on utilizing tunnel slag as a substitute for natural aggregate in road-base fillers, aiming to mitigate the environment impact of slag deposits. However, as a load-bearing material, the tunnel slag could have considerably permanent deformations due to long-term traffic loads during the whole service life which eventually leads to very large settlement on road. To reveal the long-term deformation mechanism of the tunnel slag, a series of static and cyclic large-scale triaxial tests were conducted under three levels of confining pressures mimicking practices. Most importantly, special attention was placed on investigating the effect of particle gradation and cyclic stress amplitude. Experimental results demonstrated that the particle gradation had various effects on both static and cyclic characteristics since different spatial skeleton-void structures. More specifically, under static loadings, specimens with an uneven skeleton-void structure exhibited superior strength than ones with pure or suspended skeleton structures, while in dynamic tests the pure skeleton structure leads to the highest modulus with a minimal long-term deformation. Cyclic cumulative deformation of specimens

Z. Xie · X. Liu · X. Lü (✉)
Department of Geotechnical Engineering, Tongji University, Shanghai, China
e-mail: xilinlu@tongji.edu.cn

Z. Xie
e-mail: xiezhiwei@tongji.edu.cn

Y. Ma
Zhejiang Communications Construction Group Co., Ltd., Hangzhou, China

Z. Zhang
Shanghai Urban Construction Design and Research Institute (Group) Co., Ltd, Shanghai, China
e-mail: zhangzhongjie@sucdri.com

S. Zeng
Department of Civil Engineering, University of British Columbia, Vancouver, Canada
e-mail: shengzch@mail.ubc.ca

X. Liu
Guangxi Communications Design Group Co., Ltd., Nanning, China

S. Wang et al. (eds.), *Engineering Geology for a Habitable Earth: IAEG XIV Congress 2023 Proceedings, Chengdu, China*, Environmental Science and Engineering,
https://doi.org/10.1007/978-981-99-9057-3_41

overall increases as the increase of coarse content and cyclic stress amplitude, but their relationship gradually changes from linear pattern following the mechanism of plastic shakedown, to non-linear patterns following plastic creep and incremental collapse. Finally, for a given gradation, confining pressure was found to display a positive correlation with resilient modulus so negative with cumulative strains.

Keywords Tunnel slag · Large-scale triaxial tests · Static and cyclic loading · Resilient modulus

41.1 Introduction

Environmental problems arising from mining mountain rocks have long been a topic of concern. With the increasing emphasis on sustainable development and environmental protection, the potential of reusing tunnel slag waste material (TSM) as road filler warrants further investigation. TSM, a by-product of tunnel construction, exhibits properties that depend heavily on the geological condition of the excavation area. This dependency has resulted in insufficient laboratory tests and field analysis. Over the past few decades, only a few researchers have dedicated their efforts to investigating the road performance of the TSM under various conditions (Leventhal and Ambrosis 1985; Wang et al. 2019; Tasalloti et al. 2015; Indraratna et al. 1994, 2020). Despite these efforts, relevant technical standards are still absent.

The road fillers serve to bear and transmit the static and traffic dynamic loads in the base or subbase layer of road structures. Previous investigations found that the long-term effect of traffic load can lead to gradual deterioration of the road base and subbase, which shortens service life (Hong and Prozzi 2006; Aasim et al. 2012). Consequently, studying the characteristics of TSM is crucial for increasing the long-term deformation stability of roads. TSM generally has large particle size; to mitigate the particle size impact on testing results, researchers employ large-scale triaxial apparatus to minimize particle size discrepancies between test specimens and actual field materials (Anhdan and Junichi 2004; Lenart et al. 2014; Thakur et al. 2013; Wichtmann et al. 2010; Sun et al. 2014a, b; Yi and Wang 2017). It has been found that the deformation of coarse material is influenced not only by external loads but also by internal factors such as composition and gradation (Cai et al. 2018). Vallejo and Mawby (2000), Vallejo (2011) highlighted that rock content is one of the major factors affecting the mechanical properties of coarse aggregate. Similarly, the content of block stone is associated with the Poisson's ratio and compression performance (Guo 1984; Donaghe and Torrey 1994). It has been demonstrated that the stressed skeleton formed by rock aggregate can bear significantly external loads, and is particularly resistant to large rotations or relative displacements (Xu et al. 2016; Zhang et al. 2016. Additionally, Suiker et al. (2005) investigated the effect of particle size and gradation on permanent deformation of ballast and subballast materials. In a large triaxial test, Cai et al. (2018) examined the impact of gradation on strain calculation of tuffs under long-term cyclic loading. Sun et al. (2014a, b) explored how cyclic loads influence the particle shape, grain size distribution, and rearrangement of crushed rock aggregates in large specimens with three different grain sizes. Moreover, it was found that the particle shape significantly affects the permeability parameters of the aggregate material (Ghabchi et al. 2015). However, most research has primarily focused on the characteristics of natural rock aggregates as railway ballast, with few studies conducted on TSM as road fillers.

In this paper, a series of static and cyclic large-scale triaxial tests were conducted on TSM specimens under relatively low confining pressure. Starting from an analysis of particle size distribution characteristics, this study examined the stress and strain development law of TSM. Experimental results include the relationship between peak stress and different confining pressure conditions, as well as static mechanical indexes across various gradations. Furthermore, the study analyzed the impact of particle gradation, cyclic stress amplitude, and confining pressure on the dynamic responses of TSM. The final axial strain and resilient modulus were determined, and their variations with cyclic stress ratio and confining pressure were established. Those findings not only support the utilization of TSM as road-base material but also provide fundamental data for further constitutive and numerical modelling.

41.2 Materials and Testing Program

Test materials for this study were sourced from a tunnel excavation construction site of RuiAn-CangNan Expressway in Zhejiang Province, China. The primary components of the material included tuff, mudstone, and sandstone, along with some clay. Specifically, tuff constituted approximately 70% of the material, mudstone about 20%, and sandstone and other soil approximately about 10%. The rock slag particles predominantly exhibited a polygonal shape. For this study, three sets of TSM specimens with different gradations were prepared, as shown in Fig. 41.1. According to the definition that the critical point of coarse and fine particle size is 5 mm (Guo 1998), it can be found that the coarse particle content of the three sets of TSMs increases gradually from G1 to G3. The optimum moisture content and maximum dry density of each gradation set were determined through modified Proctor compaction tests, as shown in Fig. 41.2. Physical properties of TSMs material are summarized in Table 41.1.

The schematic of the large-scaled triaxial apparatus used in the tests is shown in Fig. 41.3. Cylindrical specimens with a height of H = 300 mm and a diameter of D = 150 mm were employed in this triaxial tests. According to Fig. 41.1, the maximum grain size d_{max} of prepared specimens is approximately 30 mm, which satisfies the ratio of specimen diameter to grain size of $D/d_{max} > 5$. To ensure the production of high-quality specimens, 3% kaolin was utilized as a binding agent during sample preparation. In specimen preparation, the initial compaction degree of specimens was verified to exceed 95% to comply with the Chinese road standards. This was achieved by preparing the specimens at the optimal moisture content, and then placing and compacting them in layers of equal weight to ensure precision in

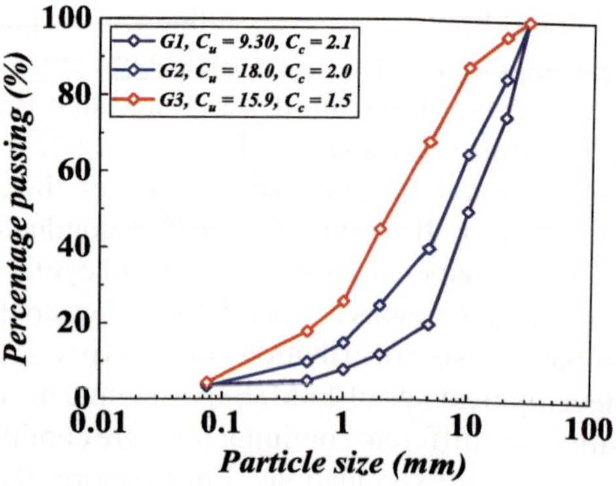

Fig. 41.1 Gradation curves of tested TSMs

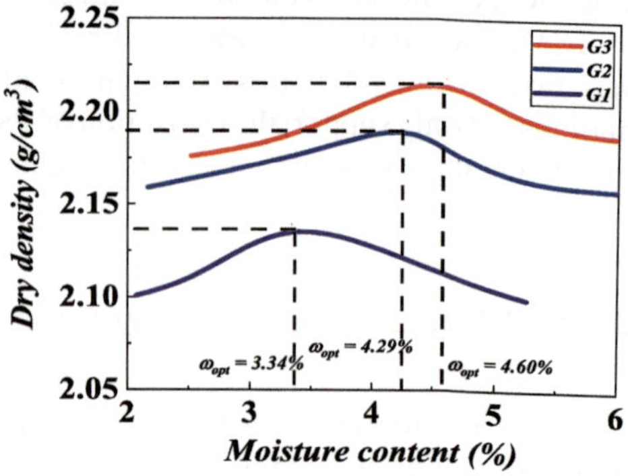

Fig. 41.2 Modified proctor compaction curve of TSMs

Table 41.1 Properties of tested materials

Index property	G1	G2	G3
Maximum grain size, d_{max} (mm)	30	30	30
Mean grain size, d_{50} (mm)	10.0	7.0	2.6
Fine particle content (%)	20	40	68
Maximum void ratio, $e_{\max}$	0.532	0.469	0.407
Minimum void ratio, $e_{\min}$	0.309	0.236	0.219
Maximum dry density, $\rho_{d,\max}$ (g/cm^3)	2.14	2.20	2.22
Specific gravitys	2.21	2.29	2.32
Optimum moisture content, ω_{opt} (%)	3.34	4.29	4.60
Compaction degree, D_c (%)	95.1	95.3	95.3
Grain shape	Subangular	Subangular	Subangular

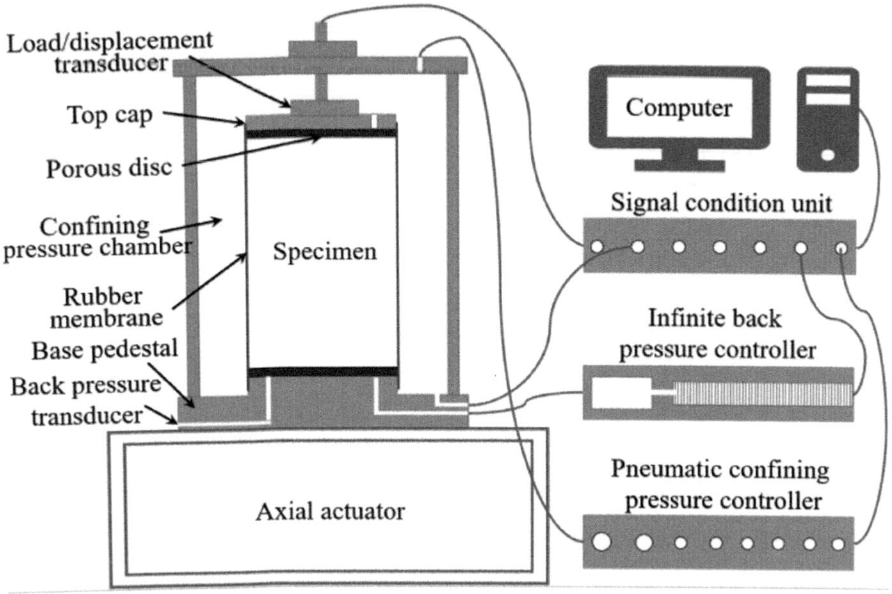

Fig. 41.3 Schematic of triaxial apparatus

compaction. The compacting hammer was set at the same initial relative height in layers to ensure uniformity of specimens. After installing the cylinder specimens into the loading chamber, all static test specimens were saturated by a specific saturation steps: injecting CO_2 into the specimens for more than 2 h to exhaust the air inside the specimens; filling de-air water into the specimen for more than 5 h; applying a back pressure of 200 kPa for 12 h to dissolve the remaining CO_2. The dynamic tests were conducted on soil samples at the optimal moisture content.

After preparation of the TSM specimens, all specimens were isotropically consolidated by low confining pressures to mimic the low-stress level in road base. This study considered the effect of particle gradation, initial confining stress, static and cyclic loading type, and cyclic stress amplitude (q^{ampl}) on the behaviour of TSM. Totally 33 tests were conducted, which can be divided into 3 groups: (1) nine static shearing tests (TS1-9) to study the effect of gradation on the static shear strength; (2) fifteen cyclic shearing tests (TC1-15) to study the coupling effect of particle gradation and cyclic stress amplitude on the irreversible strain accumulation and resilient modulus; (3) nine cyclic shearing tests (TC16–TC24) to study the effect of particle gradation and confining pressure. The plan is listed in Table 41.2. In the static shearing tests (group 1), the loading was controlled by axial strain until it reached 15%, where the shear rate was chosen as 0.1%/min. The cyclic tests were applied a load-controlled mode with a half-sinusoid compressive waveform at a frequency of 1 Hz. Totally 10,000 cycles were applied for each specimen. Considering the superior drainage properties of coarse-grained soil subgrade in practical engineering applications, all tests were conducted under drainage condition.

Table 41.2 Summary of test programs

Test series and ID		Gradation	Confining pressure, σ_3 (kPa)	Cyclic stress amplitude, q^{ampl} (kPa)
Static tests	TS1-TS3	G1	20, 40, 60	/
	TS4-TS6	G2		
	TS7-TS9	G3		
Cyclic tests	TC1-TC5	G1	20	20, 60, 100, 140, 180
	TC6-TC10	G2		
	TC11-TC15	G3		
	TC16-TC18	G1	20, 40, 60	60
	TC19-TC21	G2		
	TC22-TC24	G3		

41.3 Result of Static Shearing Tests

The effect of moisture content on the strength characteristics of coarse aggregate is considerable, as material strength continuously decrease as the increase of moisture content (Duong et al. 2013; Su et al. 2020). Hence, to investigate the static strength of TSM material under worst-case scenarios, shear tests were performed on fully saturated TSM specimens. The experimental results of three gradations under different confining pressure are presented in Fig. 41.4. Overall, the deviatoric stress of the three gradation specimens show an increasing trend as axial strain until it reaches the peak value and then start to decrease, as shown in Fig. 41.4a. The confining pressure exhibits a positive proportional relationship with the peak deviatoric stress among them. In this study, the deviatoric stress in post-peak decreases very rapidly, a behaviour known as "strain-softening" (Giessen and Borst 1998; Suiker et al. 2005), which is a common phenomenon in granular material. This "strain-softening" of granular materials is usually accompanied by large deformations, which is observed in this study. From Fig. 41.4b, it can be observed that the volume strain exhibits a trend of transient contraction before lasting dilatation with the progression of axial strain. As the axial strain increases, the movement of particles transitions from tight interlocking to flipping and sliding. This phenomenon is also the primary reason for the rapid decay of deviatoric stress in the later stages of specimen deformation. The dilatancy behaviour was weaken by the increase of confining pressure. This is mainly because the higher confining pressure provides greater restraint and further limit the deformation of the specimen.

The effect of particle gradation on the mechanical behavior of materials cannot be ignored, as shown in Fig. 41.4. It is known that the specimen inherent structure formed by different particle gradation is variant. As an illustration shown in Fig. 41.5, when the coarse particles are used as the skeleton of the material, there are three main structures of material with the increase of the content of fine particles: skeleton structure, uneven skeleton-void structure, and suspended skeleton structure.

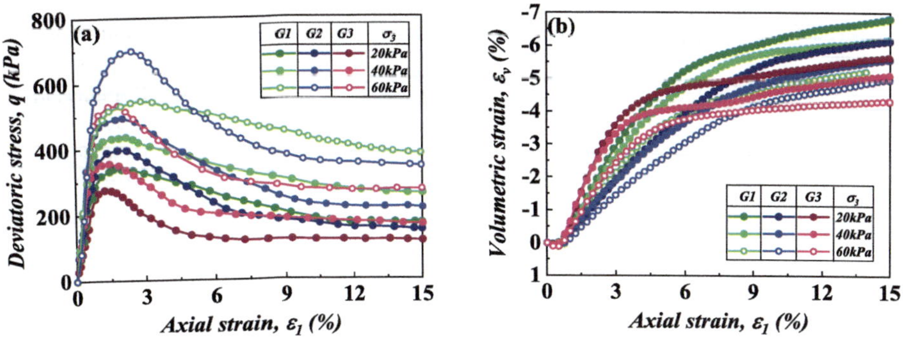

Fig. 41.4 Results of drained triaxial tests of TSM under static shearing

The particle gradations of G1, G2, G3 in this study approximately correspond to the three structures. It can be found that the peak deviatoric stress of the three grading specimens under same confining pressure is G2, G1, G3 from high to low. This trend can be explained in terms of the material structure: (1) the G2 specimens form an interlocking structure among skeleton particles to transmit external forces, and the fine particles filling the pores provides extra resistance to the rotation of the coarse particles, resulting in the maximum peak deviatoric stress in comparison with the two other cases; (2) the G1 specimens lacks skeleton particles to form a effective force transfer, and also lacks fine particles to fill the largest void ratio that can avoid relative movement of large particles, resulting in a relatively weaker shearing strength; (3) the G3 specimens, mainly composed of cohesionless fine particles with a suspended skeleton structure, have difficulty in forming effective interlocking among particles and, thus, was rapidly destroyed by static shearing. The effect of particle gradation on material properties is also observed in volumetric strain in the earlier stage of shearing. It can be observed from Fig. 41.4b that the rate of volumetric strain at the onset of dilatancy is G3, G1 and G2 in a descending order, which reflects the same trend of strength. After post-dilatancy large deformation and volume change, the final volumetric strain (15% axial strain) is positively related to the content of coarse particles in the same confining pressure.

In this paper, the peak deviatoric stress in the curve of axial strain—deviatoric stress is defined as the failure stress of TSM. Figure 41.6 illustrates the evolution of failure and residual shear stress with respect to confining pressure. It can be concluded that the deviatoric stresses of various grading samples increase with the increase of confining pressure. Under identical confining pressures, sample G2 exhibits the highest peak shear stress, followed by G1 and G3. However, the residual stresses exhibit a different development law, in which the stress values increase with the increase of coarse particle content. It is explained that even though the axial strain develops to a significant level, coarse grains can still produce considerable interlocking effects. The envelopes of the failure stress with different particle gradation under low confining pressure are shown in Fig. 41.7. According to the failure envelopes, the shear strength parameters of the specimens with three kinds

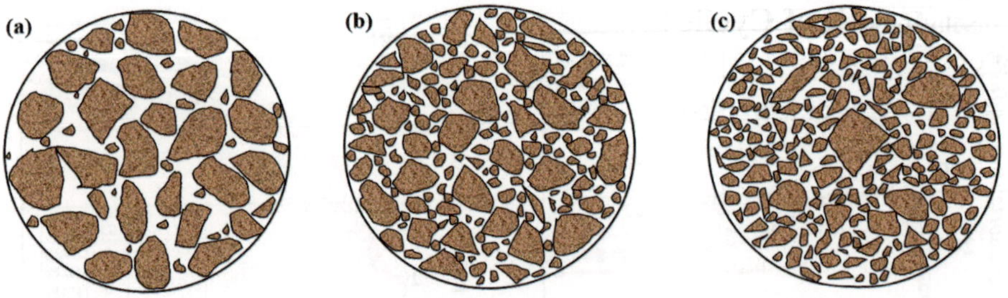

Fig. 41.5 Graded material structure: **a** skeleton structure; **b** uneven skeleton-void structure; **c** suspended skeleton structure

of particle gradation were determined, i.e., friction angle $\varphi = 46°$, $52°$, $50°$, and apparent cohesion $c = 47$, 37 and 22 kPa for G1, G2, G3 respectively.

Fig. 41.6 Deviatoric stress versus confining pressure with different particle gradation

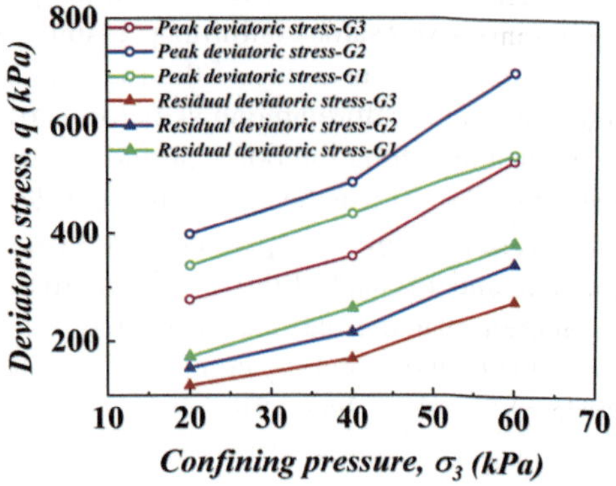

Fig. 41.7 Strength envelope of TSM with different particle gradation

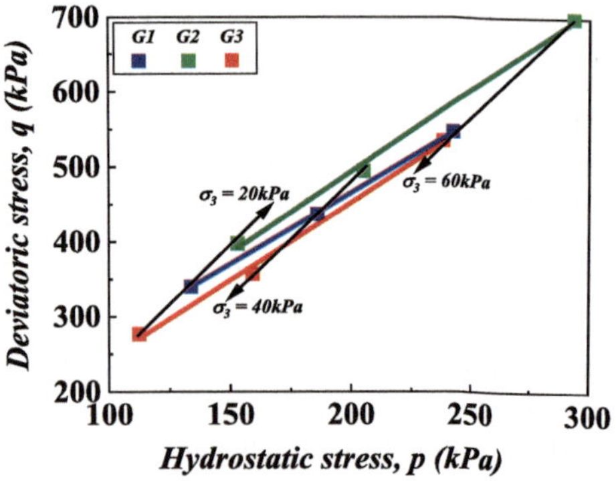

41.4 Result of Cyclic Shearing Tests

41.4.1 Stress and Strain

The effect of traffic cyclic load on road structure decayed with soil depth (Picoux and Houédec 2005; Lu et al. 2009). Here, a series of tests (group 2) were conducted to analyze the mechanical behaviour of the three sets of gradations specimens with a confining pressure of 20 kPa and various cyclic loads. Generally, the total strain ε_1 is composed of a resilient strain ε_1^{ampl} and a residual strain ε_1^{acc}, as shown in Fig. 41.7, where the residual strain corresponds to the difference between the recorded total strain and resilient strain. Figure 41.8 presents the change of the cumulative axial residual strain (ε_1^{acc}) with the number of cycles. It can be observed that the cyclic response of the ε_1 for the three sets of particle gradation specimens were similar under low cyclic stress amplitude ($q^{ampl} < 100$ kPa), following a mechanism called plastic shakedown. Overall, the ε_1^{acc} increases with the q^{ampl} for the same particle gradation. The ε_1^{acc} for various particle gradations show different involving patterns when cyclic stress amplitude increases ($q^{ampl} > 140$ kPa). For instance, when the specimens were subjected to a stress amplitude of 140 and 180 kPa (the data of $\varepsilon_1^{acc} - N$ was incomplete due to failure of the G3 specimen under $q^{ampl} = 180$ kPa), the ε_1 of G1 specimen initially increased rapidly and then stabilized, which was similar to the trend observed on low amplitude cases. For G2 specimen, after the initially increasing stage, the constant rate of the lasting strain until failure implied a plastic creep behavior. Finally, the G3 specimen showed a deformation behavior that can be interpreted as incremental collapse (Werkmeister et al. 2005), corresponding to rapid failure of specimen after a certain number of cycles.

The gradation of TSM was proved to be able to affect the inherent internal structures in static shearing tests. Following the same logic and mechanism, the gradation can also affect the cyclic behaviour of TSM. In experiments, it can be seen from

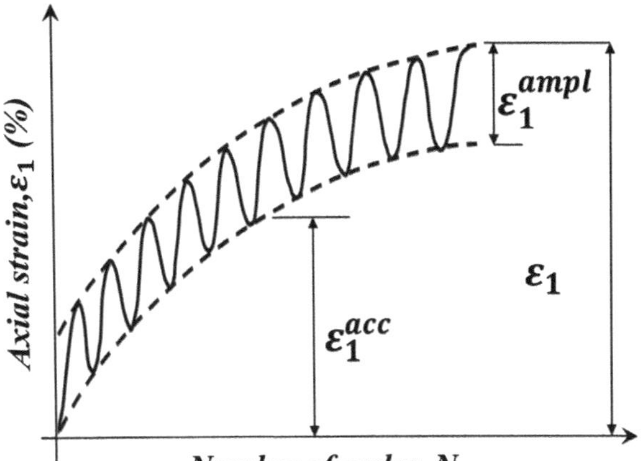

Fig. 41.8 Schematic of the resilient and plastic cumulative axial strain

Fig. 41.9a that the accumulation of ε_1^{acc} in the first few cycles in G1 is slower than others under the same q^{ampl}. This trend is different from the trend of volumetric strain versus coarse particle content in the static test since the cyclic tests were completed in relatively small strain ranges with mainly contraction instead of dilation responses as the discussion in static shearing tests. The initial contraction behaviour was attributed mainly to the initial void ratio and inherent structures. As listed in Table 41.1, the minimum void ratio of the three kinds of particle gradation specimens increases as coarse particle content. Under the same cyclic stress, G3 specimens with the smallest voids and lowest permeability is more likely to generate excess pore water pressure, which weakens specimen strength and then increases axial strain. The G1 specimens has the maximum content of coarse particles and the largest void volume, leading to a faster dissipation of the excess pore water pressure generated during loading than others. Besides, the particle shape is also partially responsible to the behavior. Under cyclic loading, the edges and corners of mudstone particles are prone to be grinded and fractured, which increases roundness and change particle gradations. The fractured mudstone particles fill the specimen pores and densify the specimens, to reduce the ε_1^{acc}. The G1 specimen has the highest content of large-grained mudstone, implying a smaller axial strain accumulation. Thus, it's reasonable to adequately consider both initial gradation and gradation change during cyclic shearing on TSM.

To study the effect of confining pressure on deformation properties of TSM, the experimental result of group 3 is plotted in Fig. 41.8b. It can be observed that the axial strain at three confining pressures have a similar pattern under the same particle gradation. At the same cyclic stress, the axial strain has a negative correlation with the confining pressure, which is mainly because the larger confining pressure provides greater constraints to limit the lateral and axial deformation of the specimen. In addition, the change of ε_1^{acc} in different particle gradation under three confining pressures is consistent with the trend: the ε_1^{acc} decreases with the increase of coarse aggregate content.

The long-term cyclic response can be quantitively analyze in terms of the final accumulated axial strain ε_1^{acc} as shown in Fig. 41.9. The results indicated that there

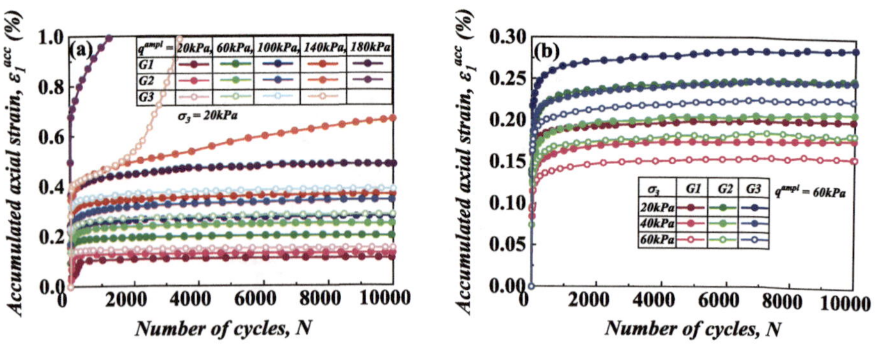

Fig. 41.9 Cyclic response of TSM considering effect of: **a** gradation and confining stress; **b** confining pressure and cyclic stress amplitude

were certain differences in the development trend of the final $\varepsilon_1^{acc} - q^{ampl}$ curves in the three particle gradation specimens. It reveals that the final ε_1^{acc} is mainly affected by three parameters: cyclic stress amplitude, particle gradation, and confining pressure. Figure 41.9a shows the q^{ampl} has a substantial influence on the final ε_1^{acc}, whereby an increase in amplitude leads to a continuous augmentation of the strain. Additionally, changes in particle gradation also results in differences in the final ε_1^{acc}. The final ε_1^{acc} of G1 specimens shows a linear increase with the increase of the q^{ampl}, while G2 and G3 specimens exhibit exponential growth. Accordingly, a key stress amplitude q^{ampl} = 60kPa is considered as a threshold for different patterns of the final $\varepsilon_1^{acc} - q^{ampl}$ curves: (1) when $q^{ampl} \leq 60$ kPa, the development trend of final $\varepsilon_1^{acc} - q^{ampl}$ curves of G1 and G2 is consistent with that of G3 specimens; (2) when $q^{ampl} > 60$kPa, the final ε_1^{acc} of the two kinds of specimens starts to increase exponentially, and the growth rate of final ε_1^{acc} in G3 specimens is much higher than that of G2. In other words, the final ε_1^{acc} of specimens with different particle gradation decreases with the increase of coarse grain content, and the strain difference between two kind of particle gradation specimens becomes larger with the increase of q^{ampl}. Figure 41.9b presents the relationship between the final ε_1^{acc} and the confining pressure under different particle gradation. It can be observed that the final ε_1^{acc} of specimens in all gradation scenarios decreases linearly with the increase of confining pressure for $q^{ampl} = 60$ kPa, which implies the decraesing effect of traffic load on soil along depth. In practice, it was suggested that the content of coarse particle in TSM should be as large as possible.

41.4.2 Resilient Modulus

As an important index to evaluate the stiffness of road fillers, the resilient modulus of materials, $M_r = q^{ampl}/\varepsilon_1^{ampl}$, has always been the focus of study. To analyze the effect of the particle gradation, q^{ampl}, confining pressure and other parameters on the M_r of TSM, Fig. 41.10 was plotted to show the resilient modulus under different factors. The result indicates that the M_r stabilizes after a few cycles, and only few factors are able to cause continuous fluctuation of the M_r, such as the continuous or rapid failure of the specimen caused by large q^{ampl}. Except for two specimens (G2-$q^{ampl} = 180$ kPa and G3-$q^{ampl} = 140$ kPa), the M_r of the same gradation specimens generally increases with the increase of the q^{ampl}. The primary reason for this trend is that large q^{ampl} results in the continuous failure in the structure, leading to unstable development of M_r. Therefore, the continuous fluctuation of M_r can also be used as one of the indicators for determining whether the specimen reach failure state.

The effect of particle gradation on M_r is also considerable. The particle gradation positively correlated with M_r while negatively correlated with ε_1^{acc}. Specifically, the M_r increases with the increase of coarse particle content under same q^{ampl}, except for failed specimens. It indicates that the increase of coarse particle content will reduce the ε_1^{ampl}. It is worth noting that the effect of particle gradation on M_r varies

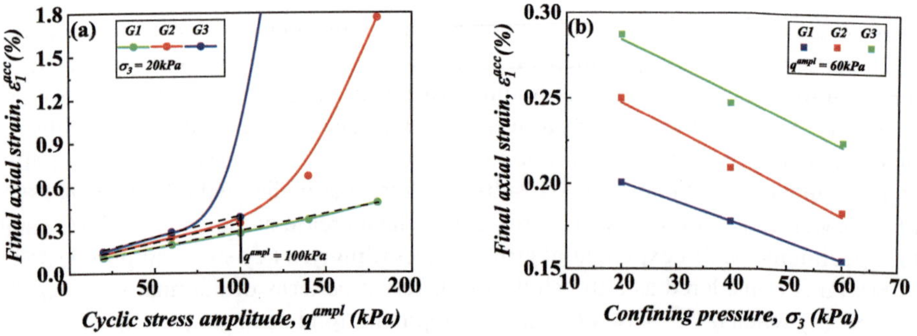

Fig. 41.10 Relationships between the final ε_1^{acc} at $N = 10{,}000$ and various factors: **a** Final ε_1^{acc} versus q^{ampl} and particle gradation; **b** Final ε_1^{acc} versus confining pressure and particle gradation

with q^{ampl}. When $q^{ampl} = 60$ kPa and 100 kPa, the M_r of G1, G2 and G3 specimens approximately forms an arithmetic progression. However, when $q^{ampl} = 20$ kPa, the M_r of G1 specimen with the highest coarse content is significantly greater than that of the other two particle gradations. It implies that the coarse particle content has a constraint effect on the development of the ε_1^{ampl} at lower q^{ampl}.

To elaborate the effect of confining pressure, the curve of the resilient modulus versus loading cycles was plotted in Fig. 41.11b. It is observed that the M_r increases with the confining pressure under the same particle gradation. This can be attributed to the fact that higher confining pressures provide greater lateral confinement, limiting the further growth of elastic strain. It cannot be ignored that more breakage was observed in higher confining pressure cases under cyclic loading. The breakage shortens the difference of resilient modulus between the G2 and G3 specimens. Moreover, the development of M_r is influenced by the coupling effect of particle gradation and confining pressure. As shown in Fig. 41.11b, the M_r of G1-σ_3 = 20 kPa and 40 kPa are almost the same as that of G2–σ_3 = 40 kPa and 60 kPa respectively.

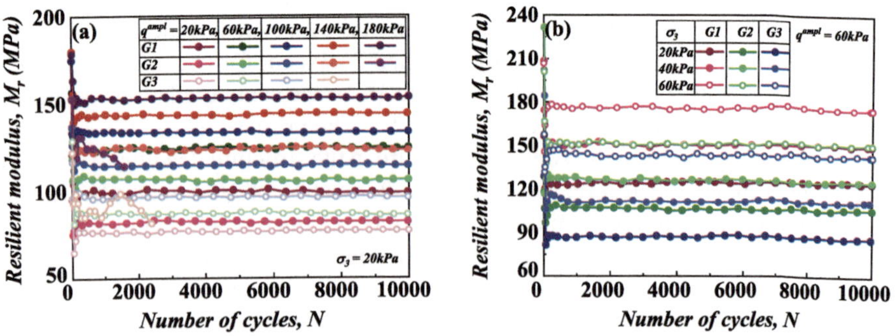

Fig. 41.11 Effect of different factors on M_r of TSM in cyclic tests: **a** M_r versus q^{ampl} and particle gradation; **b** M_r versus confining pressure and particle gradation

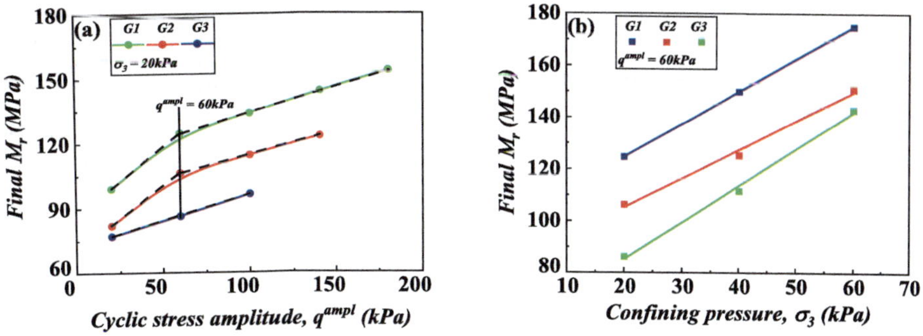

Fig. 41.12 Relationships between the final M_r. at $N = 10{,}000$ and various factors: **a** Final M_r versus q^{ampl} and particle gradation; **b** Final M_r versus confining pressure and particle gradation

The final M_r at $N = 10{,}000$ against cyclic stress amplitude and confining pressure was plotted in Fig. 41.12a, b, respectively. In Fig. 41.11a, the $q^{ampl} = 60$kPa was regarded as a critical point of the final M_r curves of G1 and G2, where an obvious rate change in the final M_r was observed. Overall, the content of coarse particles has a positive correlation with the final M_r. It is noticed that the data of G2 and G3 in Fig. 41.12a is incomplete as the specimen failed at a higher cyclic stress amplitude. The date of G1 specimens is complete mainly because it has the largest number of coarse particles that can ensure the structure's stability under large q^{ampl}. As well as axial strain response, the confining pressure has a nearly linear relationship with the value of the final M_r.

41.5 Conclusions

In order to investigate the mechanical behaviour of TSM as road-base filler, a series of large-scale static and cyclic triaxial tests were carried out under saturated and unsaturated conditions. The effects of particle gradation, cyclic stress amplitude and confining pressure on the cumulative axial strain and resilient modulus of TSM were analysed. The conclusions are summarized as follows:

1. In TSM materials with varying gradations, the highest peak strength is attained as a result of the optimal ratio of coarse to fine pores within the uneven skeleton-void structure. The structure characterized by the highest content of fine particles, lacking interlocking action between larger particles, exhibits the lowest strength. Nevertheless, the final volumetric strain exhibits a positive proportional development relationship with coarse particle content.
2. Under cyclic shearing, particle gradation has a significant effect on the long-term strain accumulation of TSM. The cumulative axial residual strain ε_1^{acc} of TSM negatively correlated with the coarse particle content, while positively correlated with the cyclic stress amplitude q^{ampl}.

3. The failure mode in high cyclic stress amplitude cases shows three patterns: plastic shakedown (e.g., G2-$q^{ampl} \leq 100$ kPa), plastic creep (e.g., G2-$q^{ampl} = 140$ kPa), and incremental collapse (e.g., G2-$q^{ampl} = 180$ kPa).
4. Resilient modulus increases with the coarse particle content, q^{ampl} and confining pressure.

Acknowledgements The financial support provided by the Key R&D projects in Guangxi (AB24010246, AB22080032) are gratefully acknowledged.

References

Aasim I, Tatsuya I, Seiichi M (2012) Effect of principal stress axis rotation on cyclic plastic deformation characteristics of unsaturated base course material. Soils Found 52(3):465–480. https://doi.org/10.1016/j.sandf.2012.05.006

Anhdan L, Junichi K (2004) Effects of large number of cyclic loadings on deformation characteristics of dense granular materials. Soils Found 44(3):115–123. https://doi.org/10.3208/sandf.44.3_115

Cai YQ, Chen JY, Cao ZG et al (2018) Influence of grain gradation on permanent strain of unbound granular materials under low confining pressure and high-cycle loading. J Geotech Geoenviron Eng 18(3):04017156.1–04017156.11. https://doi.org/10.1061/(ASCE)GM.1943-5622.0001054

Donaghe RT, Torrey VH (1994) A compaction test method for soil-rock mixtures in which equipment size effects are minimized. Geotech Test J 17(3):363–370. https://doi.org/10.1520/GTJ10111J

Duong TV, Tang AM, Cui YJ et al (2013) Effects of fines and water contents on the mechanical behavior of interlayer soil in ancient railway sub-structure. Soils Found 53(6):868–878. https://doi.org/10.1016/j.sandf.2013.10.006

Ghabchi R, Zaman M, Kazmee H et al (2015) Effect of shape parameters and gradation on laboratory-measured permeability of aggregate bases. Int J Geomech 15(4): 04014070.1–04014070.11. https://doi.org/10.1061/(ASCE)GM.1943-5622.0000397

Giessen E, Borst R (1998) Introduction to material instabillities in solids. In: Borst DR, Giessen VDE (eds) Material instabilities in solids (eng) 241(1–2):1–13

Guo QG (1984) Compaction performance and compaction data for non-cohesive coarse-grained soils. Dam Obs Geotech Tests 1:41–49. https://doi.org/10.1002/jcp.20384

Guo QG (1998) The engineering properties and application of coarse-grained soil. The Yellow River Water Conservancy Press, Henan, China (in Chinese)

Hong F, Prozzi JA (2006) Estimation of pavement performance deterioration using Bayesian approach. J Infrastruct Syst 12(2):77–86. https://doi.org/10.1061/(ASCE)1076-0342(2006)12:2(77)

Indraratna B (1994) Geotechnical characterization of blended coal tailings for construction and rehabilitation work. Q J Eng Geol 27(4):353–361. https://doi.org/10.1144/GSL.QJEGH.1994.027.P4.06

Indraratna B, Qi Y, Tawk M et al (2020) Advances in ground improvement using waste materials for transportation infrastructure. Proc Inst Civ Eng Ground Improv 19:1–44. https://doi.org/10.1680/jgrim.20.00007

Lenart S, Koseki J, Miyashita Y et al (2014) Large-scale triaxial tests of dense gravel material at low confining pressures. Soils Found 54(1):45–55. https://doi.org/10.1016/j.sandf.2013.12.005

Leventhal AR, Ambrosis L (1985) Waste disposal in coal mining-a geotechnical analysis. Eng Geol 22(1):83–96. https://doi.org/10.1016/0013-7952(85)90040-7

Lu Z, Yao HL, Luo XW et al (2009) 3D dynamic responses of layered ground under vehicle loads. Rock and Soil Mech 30(10):2965–2970. (in Chinese). https://doi.org/10.16285/j.rsm.2009.10.010

Picoux B, Houédec DL (2005) Diagnosis and prediction of vibration from railway trains. Soil Dyn Earthquake Eng 25(12):905–921. https://doi.org/10.1016/j.soildyn.2005.07.002

Suiker ASJ, Selig ET, Frenkel R (2005) Static and cyclic triaxial testing of ballast and subballast. J Geotech Geoenviron Eng 131(6):771–782. https://doi.org/10.1061/(ASCE)1090-0241(2005)131:6(771)

Sun B, Yang L, Bai W et al (2014a) Experimental investigation on porosity reduction of a coarsely crushed rock layer subject to vertically cyclic loading. Cold Reg Sci Technol 104–105:88–96. https://doi.org/10.1016/j.coldregions.2014.05.004

Su Y, Cui YJ, Dupla JC et al (2020) Investigation of the effect of water content on the mechanical behavior of track-bed materials under various coarse grain contents. Constr Build Mater 263(10):120206. https://doi.org/10.1016/j.conbuildmat.2020.120206

Sun QD, Indraratna B, Nimbalkar S (2014b) Effect of cyclic loading frequency on the permanent deformation and degradation of railway ballast. Géotechnique 64(9):1–6. https://doi.org/10.1680/geot.14.T.015

Tasalloti SA, Indraratna B, Rujikiatkamjorn A et al (2015) A laboratory study on the shear behavior of mixtures of coal wash and steel furnace slag as potential structural fill. Geotech Test J 38(4):361–372. https://doi.org/10.1520/GTJ20140047

Thakur PK, Vinod JS, Indraratna B (2013) Effect of confining pressure and frequency on the deformation of ballast. Géotechnique 63(9):786–790. https://doi.org/10.1680/geot.12.T.001

Vallejo LE, Mawby R (2000) Porosity influence on the shear strength of granular material-clay mixtures. Eng Geol 58(2):125–136. https://doi.org/10.1016/S0013-7952(00)00051-X

Vallejo LE (2011) Interpretation of the limits in shear strength in binary granular mixtures. Can Geotech J 38(5):1097–1104. https://doi.org/10.1139/t01-029

Wang D, Tawk M, Indraratna B et al (2019) A mixture of coal wash and fly ash as a pavement substructure material. Transp Geotech 21:100265. https://doi.org/10.1016/j.trgeo.2019.100265

Werkmeister S, Dawson AR, Wellner F (2005) Permanent deformation behavior of granular materials. Road Mater Pavement Des 6(1):31–51. https://doi.org/10.1080/14680629.2005.9689998

Wichtmann T, Rondon HA, Niemunis A et al (2010) Prediction of permanent deformations in pavements using a high-cycle accumulation model. J Geotech Geoenviron 136(5):728–740. https://doi.org/10.1061/(ASCE)GT.1943-5606.0000275

Xu WJ, Hu LM, Gao W (2016) Random generation of the meso-structure of a soil-rock mixture and its application in the study of the mechanical behaviors in a landslide dam. Int J Rock Mech Min Sci 86:166–178. https://doi.org/10.1016/j.ijrmms.2016.04.007

Yi W, Wang YH (2017) An experimental study on engineering characteristics of improved weathered granite soil for a high-speed railway subgrade in south china. J Test Eval 0090-3973. https://doi.org/10.1520/JTE20140358

Zhang HY, Xu WJ, Yu YZ (2016) Triaxial tests of soil-rock mixture with different rock block distribution. Soils Found 56(1):44–56. https://doi.org/10.1016/j.sandf.2016.01.004

Chapter 42
Investigation on the Permeability Coefficient of Sand Soil Impacted by the Seepage Erosion via Laboratory Test

Xuwei Wang, Yeshuang Xu, Xiaowei Li, and Rui Hao

Abstract Seepage erosion inside sand soil occurs when the hydraulic gradient exists, and geology disasters are likely to happen when the seepage erosion develops to a certain degree. The seepage erosion leads the migration of soil particle, which can reflect for the change of the permeability coefficient of the soil. A laboratory test on horizontal seepage erosion in sand soil is conducted to discuss the influence of seepage erosion on the permeability coefficient. The hydraulic head difference (Δh) ranges from 300 to 900 mm in the laboratory test, and seven monitoring points for the groundwater level along the seepage path are set on the horizontal tubular experimental apparatus. The result indicates that the approximately average permeability coefficient of the whole experimental sand soil (kav) presented rising tendency generally with the increase of Δh, and the variation tendency of the permeability coefficient of the soil between two monitoring points (kij) was different along the seepage path. When Δh increased from 0.3 to 0.9 m, kij near the start part and the end part increased while fluctuated around the middle. The kij close to the start and the end of the seepage path was more easily to be affected by seepage erosion than that around the middle.

X. Wang · Y. Xu (✉) · X. Li · R. Hao
State Key Laboratory of Ocean Engineering, Shanghai Jiao Tong University, Shanghai, China
e-mail: xuyeshuang@sjtu.edu.cn

Shanghai Key Laboratory for Digital Maintenance of Buildings and Infrastructure, Department of Civil Engineering, Shanghai Jiao Tong University, Shanghai, China

X. Wang
e-mail: 392538143@sjtu.edu.cn

X. Li
e-mail: lixiaowei07@sjtu.edu.cn

R. Hao
e-mail: hao-rui@sjtu.edu.cn

S. Wang et al. (eds.), *Engineering Geology for a Habitable Earth: IAEG XIV Congress 2023 Proceedings, Chengdu, China*, Environmental Science and Engineering,
https://doi.org/10.1007/978-981-99-9057-3_42

Keywords Permeability coefficient · Seepage erosion · Laboratory test · Hydraulic head difference

42.1 Introduction

Seepage erosion refers to the migration and loss of fine particles between coarse particles in soil under water-head difference (Richards and Reddy 2007; Li et al. 2010). Seepage erosion develops inside soil and results in seepage compaction phenomenon at first (Li et al. 2010). Only when the water-head difference reaches a certain degree, seepage erosion may cause the failure of engineering or geological disaster, such as water inrush disaster (Ma et al. 2019), collapse of river bank (Karmaker and Dutta 2013) and land subsidence (Xu et al. 2017). Actually, because migration of fine particles caused by seepage erosion results in occupation and blockage of voids (Ke and Takahashi 2014a, b), the change of soil parameters due to seepage erosion inevitably exists inside the soil even if the water-head difference is small. The permeability coefficient of soil is one of the soil parameters influenced by seepage erosion (Thaisiam et al. 2018; Yang et al. 2019a). The permeability of soil is related to many factors, such as compaction degree (Xiao and Shwiyhat 2012), particle size (Yang et al. 2019b) and hydraulic gradient (Moffat and Fannin 2006). Laboratory test with convenient operation is an effective way to conduct seepage erosion test under different conditions (Hunter and Bowman 2018; Benamar et al. 2019), which can be applied to analyze the permeability coefficient of sand soil impacted by the seepage erosion.

The objective of this paper is to analyze the change of permeability coefficient at different position of soil under horizontal seepage erosion by laboratory test. The hydraulic head difference (Δh) varied in the test. Permeability coefficient of whole and partial experimental soil under different Δh was discussed.

42.2 Laboratory Test

42.2.1 Test Equipment

Figure 42.1 shows the schematic of laboratory test system. The laboratory system includes water supply-collection system, test tube system, and monitoring system, which are introduced as follows:

(1) Water supply-collection system: this system includes water-supply box, hosepipe and water-collect box. The hydraulic head difference (Δh) can be changed by changing the height of water-supply box. The hosepipe connects the water-supply box and the test tube system. The water-collect box is located at the tail of test tube system.

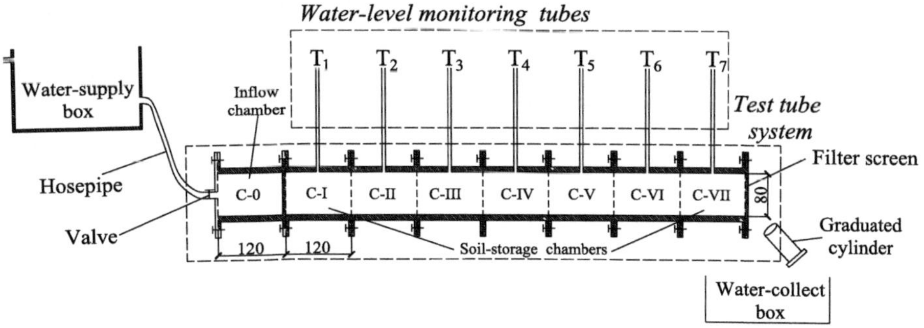

Fig. 42.1 Schematic of laboratory test system

(2) Test tube system: the test tube is a horizontal tubular apparatus, with a diameter of 80 mm and an overall length of 960 mm. The test tube is connected by eight short-tubes, each of which is 120 mm in length. The first short-tube (labelled as C-0) is the inflow chamber to buffer the water pressure from hosepipe and the other seven separated short-tubes (labelled as C-I to C-VII) are soil-storage chambers. The seven short-tubes are connected in turn by waterproof cushion and bolts. Filter screen with void of 0.038 mm is installed on the tail end of C-VII to prevent soil loss.

(3) Monitoring system: this system includes water-level monitoring and flow rate monitoring. Seven water-level monitoring tubes (labelled as T_1 to T_7) are set at the middle point of each single soil-storage chamber for water-level monitoring. The graduated cylinder is set at the tail of test tube system to monitor the flow rate.

42.2.2 *Test Material*

Sandy soil mixed with coarse sand and fine sand are chosen as the test material. The diameter of the coarse sand ranges from 1 to 5 mm, and that of the fine sand ranges from 0.063 to 0.3 mm. The weight percentage of fine sand is 15%, and the grading curve of sandy soil is shown in Fig. 42.2.

42.2.3 *Test Procedure*

Assembling Test Tube System. Firstly, mixing coarse sand of 8.5 kg and fine sand of 1.5 kg and string the mixture about 2 h; Then, filling the mixed sandy soil into each soil-storage chamber, and connecting all the seven soil-storage chambers together in turn; Next, assembling test tube system by connecting the soil-storage chambers with the inflow chamber.

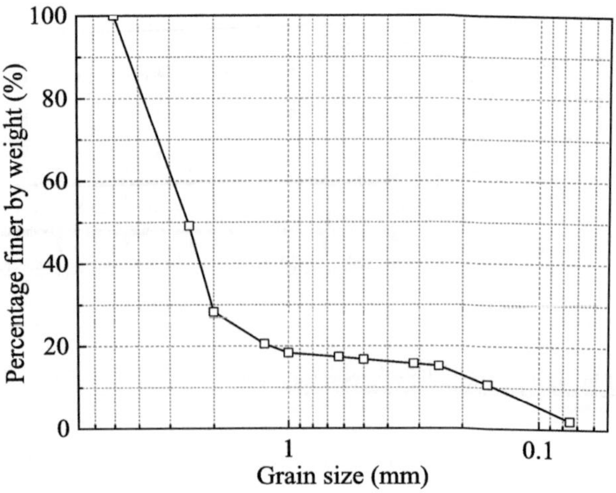

Fig. 42.2 Grading curve of sandy soil

Saturating Soils. Connecting the water-supply box and horizontally arranged test tube system at the same level, and saturating the soil for 2 h.

Erosion Test and Recording Data. Changing the height of water-supply box and setting the hydraulic head difference (Δh) which is set as 300, 450, 600, 750, and 900 mm, respectively; Starting the erosion test under the action of different hydraulic gradient, then recording the water-level data of seven monitoring tubes and the flow rate in one minute every 15 min. Each erosion test process lasted 3 h.

42.3 Results

Figure 42.3 shows the change of water level (h) of water-level monitoring point during the case that Δh is 750 mm. At the point T_1, h kept the same almost, and the lowest water-level occurred at 30th minute. The fluctuated tendency of h on other six monitoring points (T_2 to T_7) was similar, which decreased and increased alternately. During the seepage process, there were two troughs occurred at 30th minute and 105th minute because migration of fine particles under the action of seepage erosion. For example, as shown in Fig. 42.3, h of point T_7 decreased from 509 to 420 mm in the first 30 min, and it recovered to 452 mm at 90th minute; then h decreased again to 425 mm at 105th minute, and increased to 465 mm at 165th minute; and h decreased to 458 mm in the last 15 min.

Figure 42.4 presents the flow rate per minute (Q) with time. As shown in Fig. 42.4, when Δh was 300 or 450 mm, Q decreased with time, and that kept increasing with time when Δh was 750 or 900 mm. When Δh was 600 mm, Q changed fluctuating with time. Q decreased in the first 15 min from 102.7 ml/min to 96.9 ml/min, and Q increased to 113.9 ml/min at 90th minute; then it slightly decreased to 109.2 ml/min at 120th minute, and eventually increased slowly to 114.6 ml/min at 180th minute.

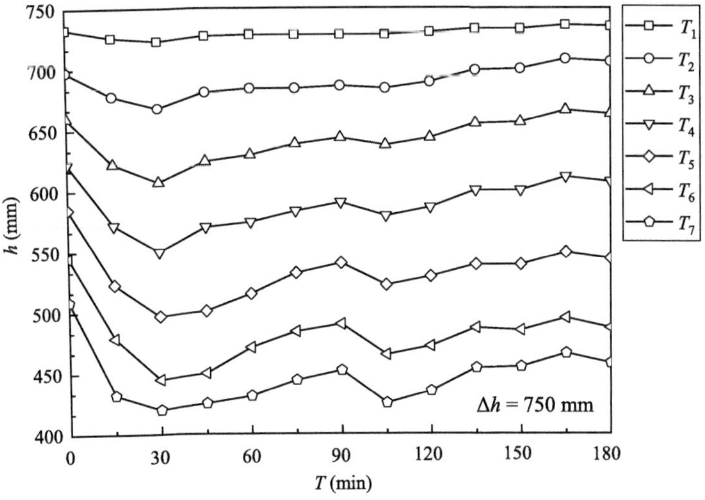

Fig. 42.3 Change of water-level with time ($\Delta h = 750$ mm)

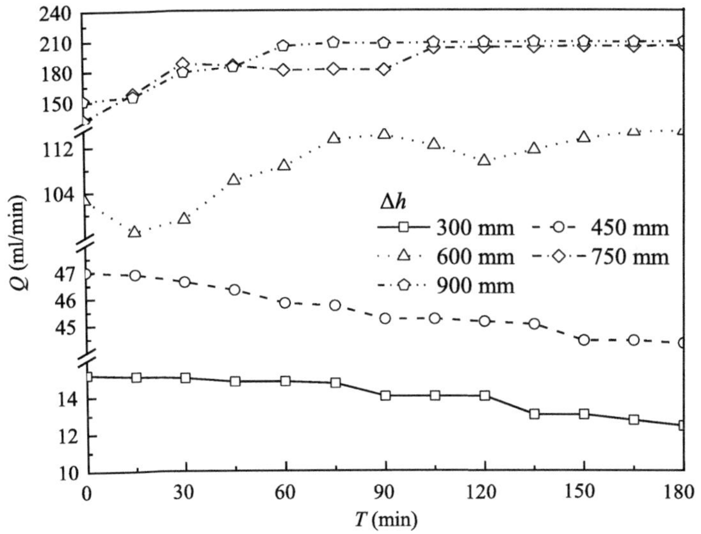

Fig. 42.4 Change of velocity of outflow water (Q) with time

42.4 Discussion

According to the Darcy's law, the permeability of sandy soil can be calculated as follows:

$$\begin{cases} v = \frac{Q}{A} \\ I_{ij} = \frac{\Delta h_{ij}}{\Delta L} \\ k_{ij} = \frac{v}{I_{ij}} = \frac{Q\Delta L}{A\Delta h_{ij}} \end{cases}$$

where v is the seepage velocity; Q is the flow rate; A is the area of seepage section; k_{ij} is the permeability coefficient of soils between two monitoring tubes; I_{ij} is the hydraulic gradient between two monitoring tubes; Δh_{ij} is the difference of water-level between two monitoring tubes; ΔL is the length of seepage between two monitoring tubes.

Figure 42.5 shows the change of k_{ij} and k_{av} with time when Δh is 750 mm. k_{av} means the approximately average permeability coefficient of whole experimental sand soil, which is equal to k_{17}. As shown in Fig. 42.5, k_{av} increased with time generally, indicating the seepage ability was enhanced macroscopically. k_{av} increased from 0.141 to 0.177 cm/s and the increasing rate is 25.5%. Due to the migration of fine particles under water pressure, original voids may be occupied and even blocked, which resulted in the decrease of permeability coefficient. The accumulation of fine particles at different position caused different change of permeability coefficient.

Fine particles in the start part ran away and resulted in the increase of k_{12}, from 0.150 to 0.281 cm/s, and the increasing rate was 87.3%. Also, fine particles of C-II chamber also moved to downstream, and the increasing rate of k_{23} was 37.3%. The increase of k_{ij} at the middle part, including k_{34}, k_{45} and k_{56}, was smaller than that of k_{av}, because of the accumulation of fine particles. The increasing rate of k_{34} and k_{56} was 5.43% and 9.04%, respectively, and k_{45} decreased approximately 8.75%, which

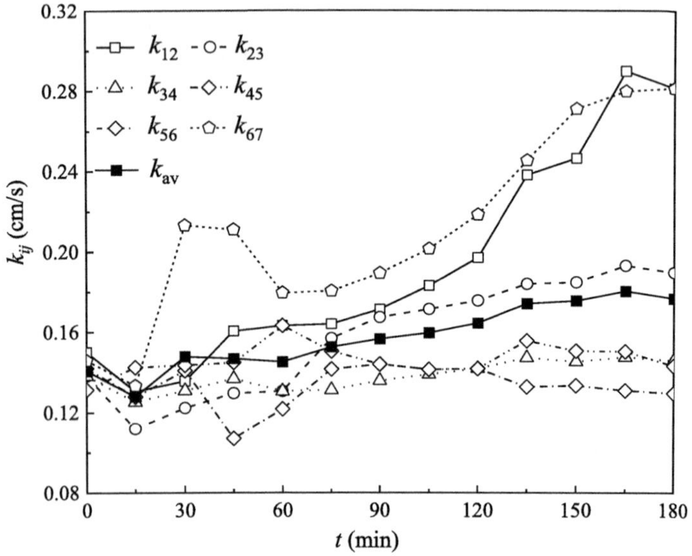

Fig. 42.5 Change of $k$ij and $k$av with time ($\Delta h = 750$ mm)

presented severe blockage of voids occurred at C-IV and C-V chamber. Moreover, the increasing rate of k_{67} was 92.9%, indicating the permeability at the end part was easy to be influenced by seepage erosion.

The difference of k_{av} between initial time and specific time is defined as Δk_{av}, and the relationship between Δk_{av} with time under different Δh is shown in Fig. 42.6. The value of Δk_{av} in the final time increased with the increase of Δh. When Δh was 300 mm and 450 mm, Δk_{av} is −0.0031 cm/s and −0.0019 cm/s at 180th min, respectively. Δk_{av} at the final time was a little smaller than that at the initial time. When Δh was 600, 750 and 900 mm, Δk_{av} at 180th min was 0.028 cm/s, 0.036 cm/s and 0.056 cm/s, respectively. Comparing with k_{av} in the initial time, k_{av} decreased firstly and then increased with time. k_{av} decreased to the minimum value at 15th to 30th min due to blockage of fine particles. In general, Δkav in the final time was larger than that in the initial time when Δh was 600, 750 and 900 mm.

Change rate (R) of k_{ij} along seepage path under different Δh is diverse, which is shown in Fig. 42.7. It is obvious that k_{ij} close to the start and end part of seepage path was more easily to be affected by seepage erosion than that in the middle part. When Δh was 300 and 450 mm, the variation of R presented inverted funnel-shaped and that presented funnel-shaped when Δh was 750 and 900 mm, whereas the tendency of R fluctuated when Δh was 600 mm. R close to the start and end part increased with Δh. For example, as Δh increased from 300 to 900 mm, R at the start part (L = 120 mm) increased from −11.0% to 144.6%. There should exist a critical value of Δh between 450 and 600 mm, indicating R changed from negative value to positive value. However, R at the middle part (L = 480 mm) was 1.98%, −5.74%, 11.6%, −8.75% and −2.79%, respectively, as Δh was 300, 450, 600, 750 and 900 mm. This shows the variation of R in the start and end part was much larger than that at the

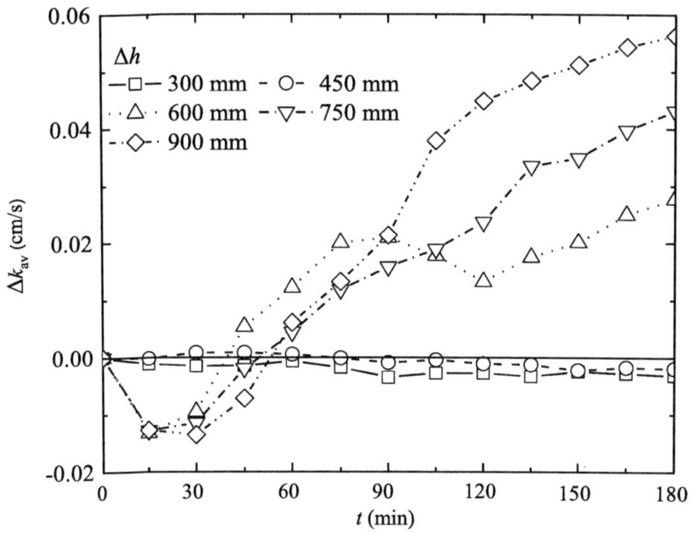

Fig. 42.6 Δk_{av} with time under different Δh

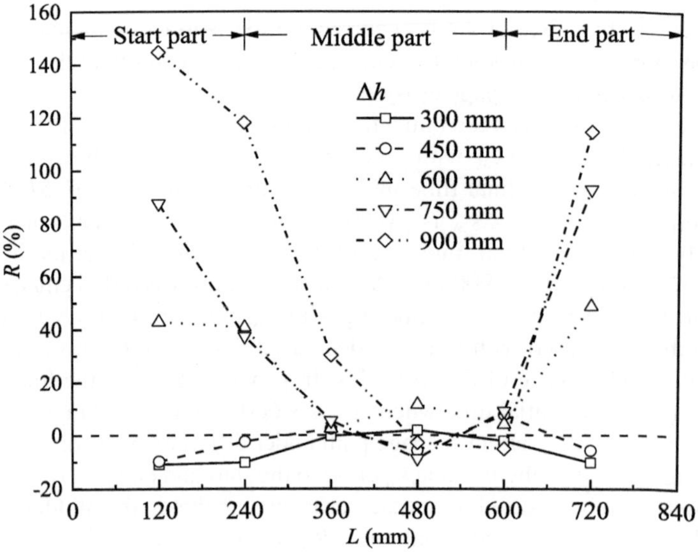

Fig. 42.7 Change rate of k_{ij} along seepage path under different Δh

middle part. The influence degree of seepage erosion on the permeability coefficient was related to seepage path, which should be considered in the actual engineering.

42.5 Conclusion

(1) A laboratory test of horizontal seepage erosion is conducted to analyze the change of permeability coefficient of sandy soil along the seepage path. The test equipment includes water-supply-collection system, test tube system and monitoring system. The test material was sandy soil mixed with coarse and fine sand.

(2) The hydraulic head difference (Δh) varied from 300 to 900 mm in the test. Approximate average permeability coefficient of the whole experimental sand soil (k_{av}) presented rising tendency generally with the increase of Δh. When Δh was 750 mm, k_{av} increased from 0.141 to 0.177 cm/s, while the permeability coefficient of soils between two monitoring tubes (k_{ij}) differed in different position.

(3) Change rate (R) of k_{ij} along seepage path under different Δh was diverse. k_{ij} close to the start part (k_{12} and k_{23}) and end part (k_{67}) of seepage path was more easily to be affected by seepage erosion than that in the middle part (k_{34}, k_{45} and k_{56}). R close to the start and end part increased with Δh, while it fluctuated around the middle. There should exist a critical value of Δh between 450 and 600 mm, which made R change from negative to positive value.

Acknowledgements The research work described herein was funded by the National Nature Science Foundation of China (NSFC) (Grant No. 41877213). This financial support is gratefully acknowledged.

References

Benamar A., Correia-dos-Santos RN, Bennabi A, Karoui T (2019) Suffusion evaluation of coarse-graded soils from Rhine dikes. Acta Geotech 14:815–823. https://doi.org/10.1007/s11440-019-00782-1

Hunter RP, Bowman ET (2018) Visualisation of seepage-induced suffusion and suffosion within internally erodible granular media. Geotechnique 68(10):918–930. https://doi.org/10.1680/jgeot.17.P.161

Karmaker T, Dutta S (2013) Modeling seepage erosion and bank retreat in a composite river bank. J Hydrol 476:178–187. https://doi.org/10.1016/j.jhydrol.2012.10.032

Ke L, Takahashi A (2014a) Experimental investigations on suffusion characteristics and its mechanical consequences on saturated cohesionless soil. Soils Found 54(4):713–730. https://doi.org/10.1016/j.sandf.2014.06.024

Ke L, Takahashi A (2014b) Triaxial erosion test for evaluation of mechanical consequences of internal erosion. Geotech Test J 37(2):347–364. https://doi.org/10.1520/GTJ20130049

Li XA, Huang RQ, Peng JB, Chen ZX (2010) Establishment of conceptual models of physical sub-ground erosion. J Eng Geol 18(6):880–886 (in Chinese)

Ma D, Duan HY, Li XB, Li ZH, Zhou ZL, Li TB (2019) Effects of seepage-induced erosion on nonlinear hydraulic properties of broken red sandstones. Tunn Undergr Space Technol 91:102993. https://doi.org/10.1016/j.tust.2019.102993

Moffat RA, Fannin RJ (2006) A large permeameter for study of internal stability in cohesionless soils. Geotech Test J 29(4):273–279. https://doi.org/10.1520/GTJ100021

Richards KS, Reddy KR (2007) Critical appraisal of piping phenomena in earth dams. Bull Eng Geol Env 66(4):381–402. https://doi.org/10.1007/s10064-007-0095-0

Thaisiam W, Kaewnon P, Pornprommin A (2018) Experiment of channelization due to seepage erosion. Int J GEOMATE 14(46):137–142. https://doi.org/10.21660/2018.46.WRE41

Xiao M, Shwiyhat N (2012) Experimental investigation of the effects of suffusion on physical and geomechanic characteristics of sandy soils. Geotech Test J 35(6):890–900. https://doi.org/10.1520/GTJ104594

Xu YS, Wu HN, Wang ZF, Yang TL (2017) Dewatering induced subsidence during excavation in a Shanghai deposit. Environ Earth Sci 76(9):351. https://doi.org/10.1007/s12665-017-6685-7

Yang J, Yin ZY, Laouafa F, Hicher PY (2019a) Analysis of suffusion in cohesionless soils with randomly distributed porosity and fines content. Comput Geotech 111:157–171. https://doi.org/10.1016/j.compgeo.2019.03.011

Yang J, Yin ZY, Laouafa F, Hicher PY (2019b) Modeling coupled erosion and filtration of fine particles in granular media. Acta Geotech 14:1615–1627. https://doi.org/10.1007/s11440-019-00808-8

Chapter 43
Research Status and Future Research Direction of Soil Sample Disturbance

An Liu, Zhigang Shan, and Weida Ni

Abstract In the process of engineering geological investigation, test of soil sample is a key link to provide accurate calculation index for designers. However, soil disturbance is an unavoidable problem while sampling, which may result in poor estimates of geotechnical parameters, leading to potentially significant and costly design errors. Comprehensive research of the soil sample disturbance is essential and has been going on for many years. This paper firstly made a summary on the research of soil disturbance, which mainly includes the influence of disturbance on soil parameters, the quantitative evaluation index and corresponding classification of disturbance, the pretreatment method of disturbed samples. Based on the review, the paper furthermore raised some considerations and provided new ideas to hopefully advance the research of sample disturbance, which promisingly promote the development of engineering geological investigation.

Keywords Geotechnical investigation · Undisturbed soil · Sample disturbance · Disturbance evaluation · Pretreatment method

43.1 Introduction

The main objective of the engineering geological investigation is to obtain the mechanical parameters of soil by geotechnical test, so as to carry out the design and calculation of various geotechnical buildings. It can be said that the disturbance of soil samples has been paid attention to at the beginning of the development of soil mechanics, and has been continuously studied and improved. Although the in-situ test can basically avoid or weaken the sampling disturbance problem, up to now, it can not completely replace the methods of drilling, sampling and indoor test, mainly because the in-situ test instruments are complex, especially when the test depth is very deep, the current equipment can not meet the requirements. In addition, it is difficult to

A. Liu (✉) · Z. Shan · W. Ni
Power China Huadong Engineering Corporation Limited, Hangzhou, China
e-mail: liu_a2@hdec.com

S. Wang et al. (eds.), *Engineering Geology for a Habitable Earth: IAEG XIV Congress 2023 Proceedings, Chengdu, China*, Environmental Science and Engineering,
https://doi.org/10.1007/978-981-99-9057-3_43

master the boundary conditions of in-situ test, and the indoor test has the advantages of flexibility, economy and easy control. Hvurslev (1949) studied the disturbance factors of soil sampling as early as 1949, mainly including the release of in-situ stress, the change of water content and void ratio, the destruction of soil structure, chemical changes. Wei (1987) studied the influence of soft clay sampling disturbance and believed that the disturbance action was mainly caused by the following factors: (1) The effective stress acting on the soil changes due to the removal of the overlying and surrounding pressure by taking the sample out of the foundation; (2) The above removed stress part is replaced by the surface tension of the pore water of the soil, so that the pore water bears negative pressure, and some gas dissolved in the water escapes, thus reducing the residual effective stress in the soil sample and slightly expanding the volume; (3) Additional disturbance caused by soil sampling, transportation, storage and processing. Hu (1997) summarized the causes of soil sample disturbance and different disturbance situations, namely stress relief, artificial disturbance (including drilling disturbance, sampling disturbance, transportation disturbance, soil sample preparation disturbance, etc.), environmental disturbance (including storage conditions, laboratory medium and environment, laboratory humidity and temperature, etc.), biological disturbance, etc.

Considerable effort have been made to study the sample disturbance, which can be divided into three general parts:

a. Influence of soil sample disturbance on soil parameters;
b. Quantitative evaluation index and classification of soil sample disturbance;
c. Pretreatment test method of disturbed samples.

This paper firstly made a comprehensive summary on the research of soil disturbance. Then based on the review, the paper furthermore raised some considerations and provided new ideas to hopefully advance the research of sample disturbance.

43.2 Research Status

For the key issues mentioned in the above, some literature has been collected and the summary of the research status of these issues is made.

43.2.1 Influence of Soil Sample Disturbance on Soil Parameters

The influence of soil sample disturbance on soil parameters is mainly divided into three parts: the influence on residual stress and saturation, including saturation, pore pressure coefficient and average effective stress reduction; the influence on strength characteristics, including undrained strength, failure strain and undrained deformation modulus; Influence on consolidation characteristics, including compressibility, coefficient of consolidation, etc.

Residual Stress and Saturation. Skempton and Sowa (1963) studied the change of effective stress caused by soil sample disturbance in 1963. In the study, only the influence of stress removal is considered, that is, the influence of "ideal" soil sampling or the change of stress in "ideal" soil sample (Noorany and Seed 1965; Davis and Poulos 1966). Ladd and Lambe (1963) qualitatively described the effective stress change caused by the disturbance in the whole process of drilling and soil borrowing, as shown in Fig. 43.1.

Hvurslev (1949) according to the sampling observation and research, they believe that the expansion occurs under the condition of no water absorption and is caused by the precipitation of air dissolved in pore water during stress removal or the increase of bubble volume. They analyze the expansion characteristics of soil samples after they are taken out of situ.

Strength Characteristics

Undrained strength

Many scholars' experimental studies on "ideal" soil sampling show that "ideal" soil sampling only removing the overlying stress also leads to the reduction of undrained strength of saturated soft clay samples with normal compaction. Skemptone and

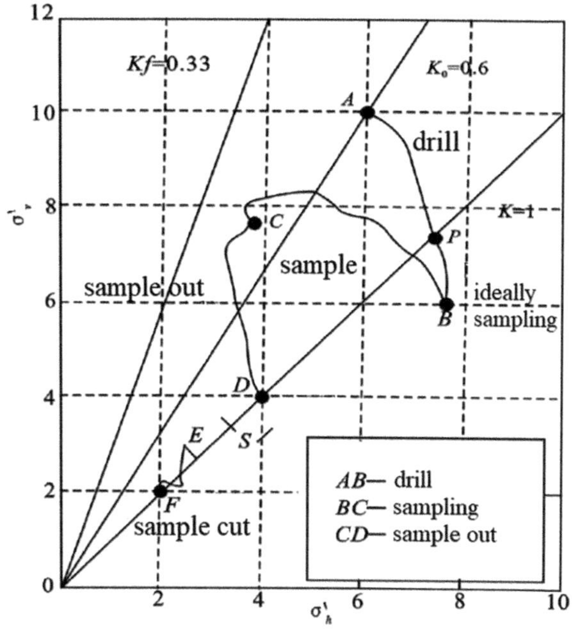

Fig. 43.1 Schematic diagram of stress change in soil process

Sowa (1963) proposed that the strength reduction of soft soil caused by unloading stress is small, generally less than 10%, and its disturbance is much smaller than mechanical disturbance.

Stress strain curve and undrained deformation modulus

The test shows that the undrained stress–strain blue line of remolded soil sample becomes gentle and its position decreases, and its deformation modulus is seriously affected by soil sample disturbance. Rufledge (1944) summarized the influence of soil sample disturbance on unconfined compressive strength and initial tangent modulus of stress–strain curve. The rcsults show that the initial tangent modulus of remolded soil samples is about 20% less than the undisturbed value, and some are even only 3–4%. Broms (1980) pointed out that the influence of soil sample disturbance on stress–strain curve is much greater in brittle soil than in plastic soil, as shown in Fig. 43.2.

Pore pressure coefficient

The pore pressure coefficient decreases with the increase of disturbance ratio, and different consolidation pressures or test methods do not cause great differences.

Remain strength parameters

Calhoon (1954) found that the consolidated undrained strength under in-situ stress measured in laboratory is greater than that in-situ due to soil sample disturbance. The effective stress intensity parameter c and φ are much less affected than the total stress intensity property (Osterberg and Murphy 1979). Ladd and Lambe (1963) believed that when the failure standard is adopted, the soil sample disturbance may slightly reduce the measured c and φ; However, if the failure criterion is adopted, the φ remains unchanged.

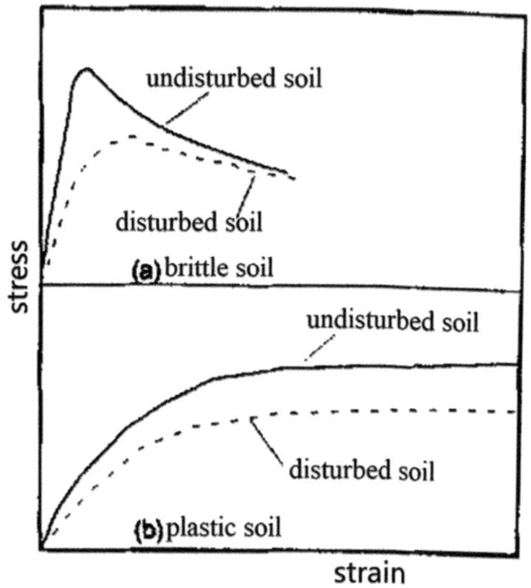

Fig. 43.2 Influence of soil sample disturbance on stress–strain curve

Consolidation Characteristics. Rutledge (1944) summarized the influence of soil sample disturbance on the results of indoor consolidation test: the reduced void ratio of soil sample under any given pressure results in the downward movement of the position of indoor compression curve; The slope of indoor compression curve decreases, which means the compressibility of soil decreases.

43.2.2 *Quantitative Evaluation Index and Classification of Soil Sample Disturbance*

The second part is about the quantitative evaluation index and classification of soil sample disturbance. Different soil sample parameters are affected differently, and the parameters with greater influence can be used as the index to evaluate soil sample disturbance.

Firstly, the residual pore water pressure method is given, and the ratio of pore water pressure of disturbed sample to pore water pressure of undisturbed soil is used as the disturbance index. Samples are undisturbed when λ_d is 1, while totally disturbed when λ_d is 0.

$$\lambda_d = \frac{u_r}{u_p} \tag{43.1}$$

λ_d disturbance degree;
u_r residual pore water pressure, it can be obtained by using the triaxial undrained shear test and a additional calculation (Hu and Chen 1987);
u_p pore water pressure;

$$\mu_p = -\gamma z[K_0 - A(1 - K_0)] \tag{43.2}$$

γ soil bulk density;
K_0 static side pressure coefficient;
z sampling depth;
A pore water pressure coefficient;

Schmertmann (1955) proposed the volume compression method, which uses the pore ratio difference ratio of ideal compression curve, actual compression curve and complete disturbance curve under the action of pre pressure to quantitatively evaluate the disturbance degree (λ_d). It is mainly obtained according to the compression curve (Fig. 43.3).

$$\lambda_d = \frac{\Delta e}{\Delta e_0} \tag{43.3}$$

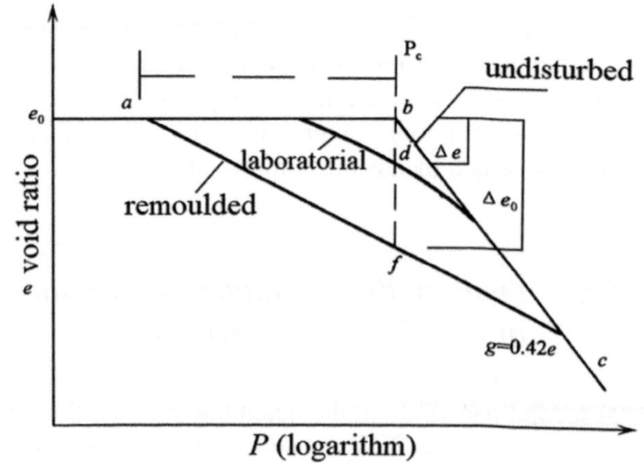

Fig. 43.3 Volume compression curve

Δe difference of void ratio between ideal compression curve and actual compression curve;

Δe_0 difference of void ratio between ideal compression curve and actual compression curve under the action of pressure P_c (Table 43.1);

Ladd and Lambe (1963) raised the undrained modulus comparison method. The degree of disturbance is estimated by the undrained modulus of saturated soil sample, and the calculation formula is as follows:

$$\lambda_d = \frac{[E_u] - E_{50}}{[E_u] - [E_{50}]} \tag{43.4}$$

E_{50} undrained modulus when strain of undisturbed soil is 50%;

$[E_{50}]$ undrained modulus when strain of remolded soil is 50%;

$[E_u]$ undrained modulus of undisturbed soil sample;

The undrained modulus of saturated soil sample is most sensitive to the disturbance, so it is possible to use the change of undrained modulus to evaluate the degree of disturbance.

Anderson and Kolstad (1979) put forwarded a evaluation method, which using the void ratio under self-weight stress as the evaluation index (Table 43.2).

$$\varepsilon_V = \frac{\Delta V}{V} \tag{43.5}$$

Table 43.1 Disturbance index classification and evaluation

λ_d	< 0.15	0.15–0.30	0.30–0.50	0.50–0.70	> 0.70
Disturbed degree	Undisturbed	Lightly	Moderately	Largely	Remould

$$\varepsilon_V = \frac{\Delta e_0}{1 + e_0} \tag{43.6}$$

e_0 initial void ratio of soil sample;
Δe_0 difference of void ratio from loading to self-weight stress;

Lunne et al. (1997) proposed a new index I_e to evaluate the disturbance of soil samples. Table 43.3 indicated the evaluation of sampling quality by the disturbance index I_e.

$$I_e = \frac{\Delta e_0}{e_0} \tag{43.7}$$

e_0 initial void ratio of soil sample;
Δe_0 difference of void ratio from loading to self-weight stress;

Hong and Onitsuka (1998) proposed the improvement of volume compression method, which uses the slope ratio of different curves in $ln(1 + e){-}lgp$ curve to define the degree of disturbance (Fig. 43.4).

$$\text{SD} = \frac{C_{CLB}}{C_{CLR}} \times 100\% \tag{43.8}$$

C_{CLB} and C_{CLR} are the slope of compression curve in $ln(1 + e){-}lgp$ coordinate before yield of disturbed sample and remolded sample respectively.

When SD is 100%, the soil sample is completely disturbed and the compression curve is the compression curve of remolded sample; When SD is between 0 and 100%, the compression curve is between the compression curve of in-situ and completely remolded soil; When SD is 0, the soil sample is undisturbed and the compression curve is in-situ compression curve.

Nagaraj et al. (1990) suggested using the consolidation yield stress to calculate the consolidation yield stress of undisturbed and completely remolded samples by using the consolidation yield stress of disturbed soil samples in different degrees on

Table 43.2 Evaluation standard of soil sample disturbance

ε_V (%)	< 1	1–2	2–4	4–10	> 10
Quality	Excellent	Good	Medium	Poor	Bad

Table 43.3 Evaluation standard of soil sample disturbance

OCR	I_e (%)			
	Excellent	Good	Poor	Bad
1–2	< 4	4–7	7–14	> 14
2–4	< 3	3–5	5–10	> 10

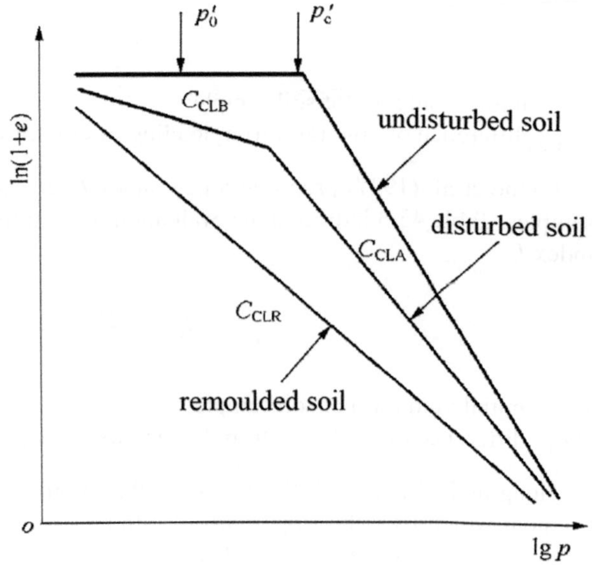

Fig. 43.4 Definition of disturbance degree by modified volume compression method

a straight line, and uses this formula to define the degree of disturbance (Fig. 43.5).

$$\mathrm{SD} = \left(\frac{\sigma_y - \sigma_{y(a,b,\ldots)}}{\sigma_y - \sigma_R}\right) \times 100\% \tag{43.9}$$

σ_y Consolidated yield pressure of undisturbed sample, kPa;
$\sigma_{y(a,b,\ldots)}$ consolidation yield pressure of disturbed samples with different degrees, kPa;
σ_R Consolidation yield pressure of remolded sample, kPa;

When SD is changing from 1 to 0, the sample was totally disturbed and undisturbed.

Landon et al. (2007) suggested using the shear wave velocity of soil samples to evaluate disturbance, which to measure the shear wave velocity V_{vh} and in-situ shear wave velocity V_{SCPTU} of soil samples, and then evaluate them by their ratio (Table 43.4).

$$a = \frac{V_{vh}}{V_{SCPTU}} \tag{43.10}$$

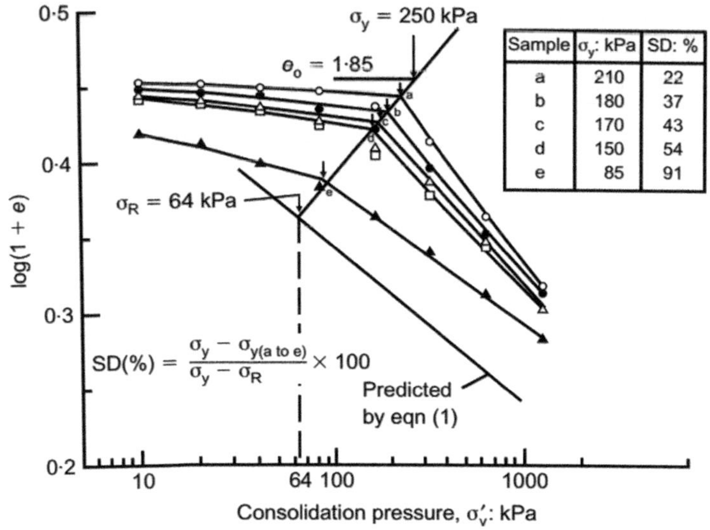

Fig. 43.5 Schematic diagram of consolidation yield stress method

Table 43.4 Classification and evaluation criteria of disturbance index

a	> 0.60	0.35–0.60	< 0.35
Quality	Good	Poor	Bad

43.2.3 *Pretreatment Test Method of Disturbed Samples*

During the survey and sampling, due to the stress release, the initial stress state of the sample is different from that of the natural soil layer. During the triaxial test in the laboratory, the initial stress state must be consistent with the in-situ stress state of the natural soil layer, so as to make the test results consistent with the natural strength of the soil. Soil sample pretreatment is to simulate the in-situ state of soil and restore the soil sample to the in-situ consolidation state by applying a certain pressure, so as to determine the mechanical properties of soil. Recompression method can effectively reduce the impact of soil sample disturbance, and soil sample pretreatment is to more carefully restore the in-situ stress of the soil before the indoor test, so as to more truly determine the properties of the in-situ soil. It is generally believed that soil sample disturbance mainly affects the cohesion of soil, and it is difficult to recover after the loss of cohesion, while the shear strength of soil can be achieved by applying confining pressure. Through the consolidation pretreatment of soil samples, the strength indexes measured after returning to the in-situ consolidation stress state can truly reflect the properties of in-situ soil. Therefore, the consolidation pretreatment before the test has high practical significance for measuring the in-situ stress state of soil samples.

At present, there are two main methods for soil sample pretreatment, they are respectively K_0 consolidation pretreatment and isotropic consolidation pretreatment. K_0 consolidation pretreatment mainly includes reconsolidation method, which means the soil sample is consolidated to the effective overlying stress state of in-situ soil. The condition of K_0 consolidation pretreatment is that the vertical consolidation pressure is γh and the lateral consolidation pressure is $K_0\gamma h$. At consolidation stage, The K_0 consolidation method adopts graded consolidation, and the consolidation time is determined according to the consolidation speed and consolidation drainage rate of soil samples, conventional undrained shear test shall be conducted after consolidation.

The expression of isotropic consolidation pretreatment method is as follows:

$$\sigma_3 = \frac{\gamma h(1 + 2K_0)}{3} \tag{43.11}$$

h burial depth of soil sample, m;
K_0 static side pressure coefficient, which is determined by test;
γ effective gravity of soil, kN/m^3;

Generally, the indoor soil sample pretreatment should adopt K_0 pretreatment method whose vertical and lateral consolidation pressure are equal to the in-situ effective consolidation pressure, restoring the soil sample to the state close to the in-situ stress as far as possible, so that the measured mechanical property indexes of the soil are more in line with the reality. However, K_0 consolidation pretreatment method has destroyed some soft clay before pretreatment, which is difficult to deal with in practice. Therefore, in general, the laboratory adopts isotropic consolidation pretreatment instead of K_0 consolidation pretreatment (Takeshi and Akil 1989; Li and Qian 1987). The Bjerrum method has been widely used in the laboratory consolidated undrained triaxial test, which means the undrained shear method is carried out after the soil sample is consolidated to the effective overlying stress state of the field soil (Bjerrum 1973). Ladd and Fott (1974) proposed the SHANSEP (Stress History And Normalized Soil Engineering Parameters) method. They thought that only when the consolidation pressure is equal to 1.5–2.5 times the site maximum preconsolidation pressure, the consolidated undrained shear strength of the soil sample can more truly reflect the original strength characteristics of the site soil. The advantage of SHANSEP method is that it can obtain the relationship curve between normalized undrained shear strength and over consolidation ratio with less soil samples (generally 5–6). If the soil conditions of the site are uniform and the stress history is clear, the strength profiles of the site soil at different boreholes and depths can be calculated, so as to greatly reduce the cost of borehole sampling and the workload of laboratory test.

43.3 Future Research Direction

The pretreatment to the soil sample can effectively reduce the influence of disturbance on the strength index of soil sample in laboratory test, and more truly reflect the natural strength of soil, but it can not completely restore the original stress state of the sample. It is difficult to restore the properties of undisturbed soil through the treatment of disturbed samples; Moreover, although domestic and foreign scholars have put forward different quantitative evaluation indexes for different indexes of soil, some need complex calculation, and some take "ideal soil" as undisturbed soil for evaluation, and the establishment of quantitative indexes should be based on the real in-situ soil properties. In view of the above thinking, the future research direction are proposed.

First, It is important to determine the quantitative evaluation index of soil sample disturbance based on the real in-situ test parameters. Hight and Leroueil (2003) pointed out that most quantitative evaluation method are usually laboratory-based and destructive, moreover require expensive and time-consuming tests, which indicating that samples are not suitable for determining engineering properties.

The shear wave velocity method is selected here. Its advantage is that it can take the in-situ test parameters as the index, does not need to destroy the soil sample, and the measurement is fast and simple. The calculation formula adopted is to divide the shear wave velocity difference between the in-situ soil and the soil sample by the shear wave velocity difference between the in-situ soil and the remolded soil. The shear wave velocity of the in-situ soil can be measured by the seismic cone test, and the shear wave velocity of the soil sample can be measured by the shear wave velocity tester (Fig. 43.6), which is a simple and fast tester designed abroad. The pattern is fixed with an acrylic plate, and then the shear wave velocity is measured by the inserted piezoelectric bending element.

Then, according to the evaluation index, the relationship between the two kinds of soil sample strength from the indoor test and in-situ test should be established, based on which the pretreatment test method of disturbed samples is finally studied to restore the strength of soil to the full extent. The specific procedure is as follow:

a. Sampling the soil in certain area, simulating the disturbance of soil samples by vibration test, and preparing soil samples with different disturbance degrees (*SD*) based on index;
b. The strength parameter q of soil samples with different *SD* is obtained through laboratory test, data analysis and formula fitting are carried out, and the mathematical model $q = f\ (SD)$ is established (Fig. 43.7);
c. Modifying the mathematical model for different depth soil layers in the area to improve the mathematical model;
d. Based on the experimental data of various regions, a unified mathematical model is tried to be established.

The same procedure can also be used in the method using the resistivity to evaluate the degree of soil disturbance, which is also non-destructive and straightforward.

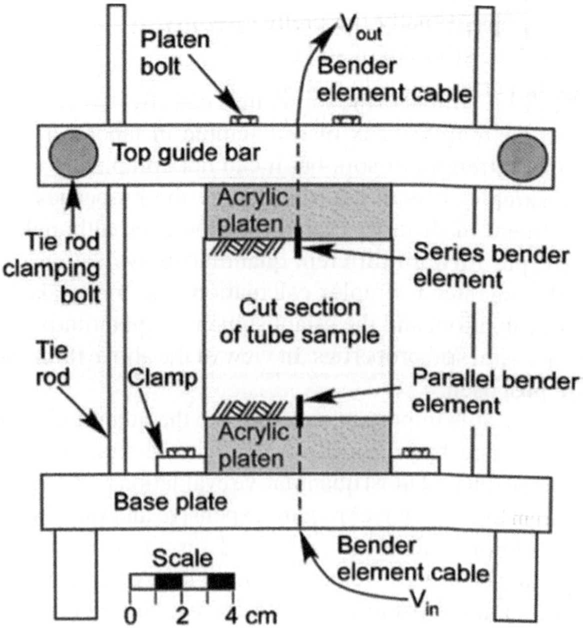

Fig. 43.6 Schematic for bender element measurement of shear wave velocity on a cut section of a tube sample (Landon et al. 2007)

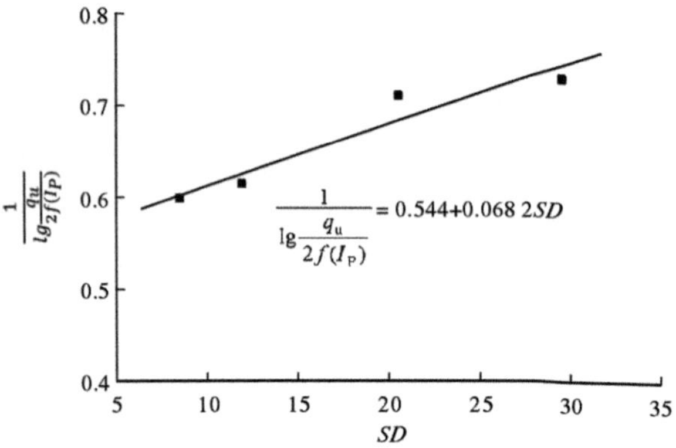

Fig. 43.7 Relationship between SD and $\frac{1}{lg\frac{q_u}{2f(I_P)}}$ (Deng and Liu 2007)

43.4 Conclusions

The paper made a summary on the research of soil disturbance and based on the review, a reasonable way of research was furthermore raised.

Albeit different quantitative evaluation indexes of soil sample disturbance were raised, the establishment of quantitative indexes should be based on the properties of really undisturbed soil. Then based on the evaluation index, the relationship between the two kinds of soil sample strength from the indoor test and in-situ test should be established. Finally the pretreatment test method of SHANSEP should be used to restore the strength of soil to the full extent.

The result can help researchers who are still beginners to understand soil sample disturbance in a short time and realize what they should do in future.

Acknowledgements This research was supported by Major Science and Technology Project of PowerChina Huadong Engineering Co., Ltd. (KY2021-ZD-03).

References

Anderson A, Kolstad P (1979) The NGI 54mm samplers for undisturbed sampling of clays and representative sampling of coarser materials. Proceeding of the International Symposium on Soil Sampling, Singapore, July 28th. Japanese Society of Soil Mechanics and Foundation Engineering, Tokyo, Japan, pp 12–13
Bjerrum L (1973) Problems of soil mechanics and construction on soft clays. In: Proceedings 8th ICSMFE, Vol 3, pp 111–160
Broms BB (1980) Soil sampling in Europe: state of the art, Proe. ASCE, No. GTl, P65
Calhoon ML (1954) Effect of sample disturbance on the strength of a clay. ASCE 80(570)
Davis EH, Poulos HG (1966) Laboratory investigation of the effects of sampling, CE Trans, Vol CE9, No 1
Deng YF, Liu SY (2007) Effect of sample disturbance on soft soil strength. Chin J Rock Mech Eng 09:1940–1944
Hight DW, Leroueil S (2003) Characterization of soils for engineering purposes. In: Proceedings. International Workshop on Characterization and Engineering Properties of Natural Soils, Singapore 1:255–360
Hong Z, Onitsuka K (1998) A method of correcting yield stress and compression index of Ariake clays for sample disturbance. Soils Found 38:211–222
Hu ZX (1997) Soil mechanics and environmental engineering. Tongji University Press
Hu ZX, Chen WH (1987) Evaluation of disturbance degree of soil samples. Port Eng 05:46–52
Hvurslev MJ (1949) Subsurface exploration and sampling of soils for civil engineering purposes. Vicksburg, MI
Ladd CC, Fott R (1974) New design procedure for stability of soft soil. ASCE J Geotech Eng Div 100(GT7):736–786
Ladd CC, Lambe TW (1963) The strength of undisturbed clay determined from undrained tests. Symposium on laboratory shear testing of soil. ASTM Special Tech Publ (STP) 361:342–371
Landon MM, Degroot DJ, Sheahan TC (2007) Nondestructive sample quality assessment of a soft clay using shear wave velocity. J Geotech Geoenviron Eng, ASCE 133(4):424–432
Li T, Qian SY (1987) Evaluation of soil sample disturbance and determination of its preconsolidation pressure. Chinese J Geotech Eng 9(5):21–29
Lunne T, Berre T, Strandvik S (1997) Sample disturbance in soft low plasticity Norwegian clay. In: Almeida A (ed) Symposium on recent developments in soil and pavement mechanics. Balkema, Rotterdam, pp 81–92
Nagaraj TS, Murthy BRS, Vatala A et al (1990) Analysis of compressibility of sensitive soils. J Geotech Eng, ASCE 116(1):105–118
Noorany I, Seed HB (1965) In-situ strength characteristics of soR clays. In: Proceedings ASCE, No. SM2
Osterberg JO, Murphy WP (1979) State of the art of undisturbed sampling of cohesive soils. In: Proceedings on International Syrup. Soil Sampling, Singapore
Rutledge PC (1944) Relation of undisturbed sampling to laboratory test. Trans. ASCE, Vol 109

Schmertmann JH (1955) The undisturbed consolidation on behaviour of clay. Trans ASCE 120:1201
Skempton AW, Sowa VA (1963) The behaviour of saturated clays during sampling and testing. Geotechnique 13(4):269
Takeshi K, Akil N (1989) Undrained shear strength anisotropy of K0-overconsolidated cohesive soils. Soils Found 29(3):145–151
Wei RL (1987) Strength and deformation of soft clay. People's Communications Press

Chapter 44
Characteristics of Sandification Dolomite and Effect on In-Situ Bearing Capacity of Pile Foundation

Yun Mo, Xinli Hu, Deshan Cui, Guangming Yuan, Zonghai Xiong, and Bo Liu

Abstract A group of Carboniferous and Permian dolomites are widely developed in the Tianxingzhou belt of carbonate rock in Wuhan. Drilling revealed that the dolomite has different degrees of sandification, which has caused many difficulties to the exploration, design and construction. By means of in-situ drilling sampling, saturated uniaxial compression test, thin section identification, etc., the physical and mechanical properties of sandification dolomite were analyzed, and the classification table of dolomite sandification degree was compiled. Combined with the in-situ load test data of 14 sets of bored test piles, the characteristics of pile foundation bearing capacity and the effect of pile side and pile tip resistance in sandification dolomite stratum are analyzed. The research results show that when layer ⑤$_1$ is used as the supporting layer, the ratio of length to diameter is inversely proportional to the end resistance. When layer ⑤$_2$ is used as the supporting layer, the end resistance ratio has a strong correlation with the depth-diameter ratio, but has little correlation with the length-diameter ratio. In general, in the sandification dolomite formation, the effect of the pile tip resistance is low. In comparison, the pile type with $d = 800$ mm in layer ⑤$_1$ fully sandification dolomite stratum, the pile tip resistance is better. The pile foundation can achieve greater bearing capacity and is more suitable for the proposed construction in this stratum.

Keywords Sandification dolomite · Sandification level · Bearing capacity of pile foundation · Side resistance · Tip resistance

Y. Mo · X. Hu (✉) · D. Cui
Faculty of Engineering, China University of Geosciences, Wuhan 430074, Hubei, China
e-mail: Huxinli@cug.edu.cn

Y. Mo · G. Yuan · Z. Xiong
WuHan Fengda Geology Engineering Co., Ltd., Wuhan 430074, Hubei, China

B. Liu
Zall Smart City Group Co., Ltd., Wuhan 430074, Hubei, China

S. Wang et al. (eds.), *Engineering Geology for a Habitable Earth: IAEG XIV Congress 2023 Proceedings, Chengdu, China*, Environmental Science and Engineering,
https://doi.org/10.1007/978-981-99-9057-3_44

44.1 Introduction

The Tianxingzhou belt in the northern part of the carbonate rocks in Wuhan is about 49.8 km long in the east–west direction, about 5.27 km wide in the north–south direction, and the narrowest is about 0.35 km, covering an area of about 110 km^2. A group of Carboniferous Huanglong Formation (C_2h) and Permian Qixia Formation (P_2q) dolomites are widely developed in the strata. Drilling results of multiple projects have found that the dolomite has different degrees of sandification. After dissolution, dolomite sand and powdery substances are often formed (Touir et al. 2009; Bischoff et al. 2011). The development of karst causes the quality of rock mass to decrease (Avinoam et al. 1982; Galvão et al. 2015; Vondráčková et al. 2015) which has caused many difficulties to the exploration, design and construction (Hallal et al. 2013). The "dolomite sandification" refers to the dolomite that is strongly affected by tectonic and dissolution. Under the action of karstification, the dolomite is dissolved into a gravel and sandy mixture. During the construction of domestic Pingtou Hydropower Station (Wenxiu 2008; Xiangbo 2009), Yingerling Tunnel (Zinan and Xiangfeng 2012), Jinyun Mountain Tunnel (Hao et al. 2020) and other projects, they all encountered dolomite sanding problems, which led to slow building construction progress and increased project costs. Wang et al. (Pengpeng et al. 2019; Xiliang et al. 2012; Qiwen and Chaolin 2014; Zhongtang et al. 2011; Xiaozhuan et al. 2013; Qi et al. 2015; Jianqiang and Zhihong 2016; Caijin et al. 2020) studied the sandification mechanism of dolomite and its physical and mechanical properties such as rock structure characteristics, integrity, compression, tensile, shear strength, and elastic modulus from the macro–micro and geochemical levels. The problem of dolomite sanding has been studied in Yunnan, Guizhou, Guangxi, Guangdong Province, and other regions. Most of them are concentrated on dolomite sandification for surrounding rock chambers, tunnels or slope support. However, there are few research on the impact of sandification dolomite formation on engineering issues such as the selection of the foundation of a proposed building and the bearing capacity parameters of the pile foundation.

Over the years, Gao, Gong, Zhou (Jingxing 2013; Dazhao 2010; Xiaonan 2016; Weiping and Jiansan 2004; Meng et al. 2013; Ming et al. 2013) and other scholars have conducted theoretical studies on the bearing capacity of pile foundations, the distribution and laws of pile side friction and pile tip resistance. Researches have formed some general understandings: (1) In general, the pile side friction resistance takes effect before the pile tip resistance, the test load and the corresponding pile top displacement required for the pile side friction resistance and the pile tip resistance are also different; (2) For piles with larger l/d ratio, even if the bearing layer of pile tip is rock or hard soil, pile tip resistance can hardly be exerted under working load due to the compression of pile itself. When $l/d \geq 100$, pile tip resistance can be ignored; (3) There is a critical depth problem in the effect of pile side friction and pile tip resistance. After the pile tip enters a certain depth of the rock and soil layer, the pile side friction and pile tip resistance no longer increase with the increase in depth but tends to be a fixed value. In-situ tests can more truly reflect the performance of side friction and end resistance of a single pile under load. It has become the most direct means of theoretical research and engineering practice. Many scholars (Qiuyan et al. 2004; Zhongju et al. 2005; Wenfeng et al. 2009; Yunxiu et al. 2018) used this method to test the bearing capacity characteristics value of pile foundations in Tertiary mudstone areas, "western Yunnan red beds" areas, volcanic ash deposition areas, and karst areas. Weidong et al. (2016, 2015) carried out a study of ultra-long rock piles under the condition of the first-level terrace in Wuhan and obtained valuable engineering experience.

However, there were few research on the influence of the distribution, origin, mechanism, physical and mechanical properties of dolomite sandification in Wuhan area and its degree of sandification on engineering construction, as well as in-situ test research on the bearing capacity of bored piles in sandification dolomite formations. Therefore, it is of great engineering significance to carry out in-situ experimental research on the characteristics of Wuhan sandification dolomite and its influence on the bearing capacity of bored piles.

44.2 Project Summary

44.2.1 Project Introduction

The Wuhan Airport International Sports Center project is located in the south of Hangcheng Avenue, west of Yunhu Road and north of Maijiahu Road in Huangpi District, Wuhan City, Hubei Province. It is located in the central and northern core area of tianxingzhou belt and Jiangbei section in the northern carbonate rocks of Wuhan City. The project has a land area of about 503,000 m^2 and a construction area of about 1.35 million m^2. At present, T5, T7, and T8 parts have completed detailed exploration, construction exploration and partial pile foundation construction. Among them, T5 and T7 are multi-storey, high-rise commercial and residential buildings. Six multi-storey buildings are planned to be built, and the maximum design load of the columns is 2700–13,000 kN. It is planned to build 10 high-rise residential buildings and hotels, and the maximum design load of the column is generally 26,000 kN. The foundation buried depth is 9.0–11.0 m. T8 is a large-scale stadium, consisting of a 60,000-seat professional football game field and 2 outdoor standard football training fields. The football stadium is 57.0 m high, with a maximum design load of 30,000 kN of the column.

44.2.2 Formation

The landform in the study area is relatively flat. During the investigation, the site was not leveled, and the terrain was slightly undulating. The surface layer of the site is filled with soil. The layers are alluvial-diluvial clay, red clay, residual soil and Carboniferous to Permian dolomite. According to the investigation results, relevant mechanical parameters recommended for each layer are shown in Table 44.1.

Table 44.1 Basic mechanical parameters of strata at construction site

Name	Natural density γ	Cohesion c	Internal friction angle φ	Bearing capacity f_{ak} (f_a)	Modulus of compressibility $E_{S1\text{-}2}$ (E_0)	Side resistance q_{sia}	End bearing q_{pk}
	$kN \cdot m^{-3}$	kPa	(°)	kPa	MPa	kPa	kPa
② silty clay	19.4	30	15	180	7.0	36	–
③$_1$ clay	19.7	42	16	400	16.0	44	–
③$_2$ gravel soil	[a]20.0	[a]15	[a]35	450	(30.0)	70	–
③$_3$ silty clay mixed with gravel	19.6	45	17	420	17.0	45	–
④$_1$ red clay	18.5	35	15	260	10.0	30	–
④$_2$ residual soil	18.6	25	12	150	(15.0)	40	–
⑤$_1$ fully sandification dolomite	–	–	–	800	(52.0)	100	1100
⑤$_{1a}$ strongly sandification dolomite	–	–	–	(2000)	–	150	2000
⑤$_{2a}$ medium sandification dolomite	–	–	–	(4000)	–	250	4000
⑤$_2$ weakly sandification dolomite	–	–	–	(5000)	–	300	5000
⑤$_{2b}$ slightly sandification dolomite	–	–	–	(7000)	–	400	7000

Note (1) As there is no relevant value for sandification rock in the "Code for design of building foundation" (DB 42/242–2014) in Hubei Province, according to actual conditions such as core and uniaxial compression strength, the standard value of the layer ⑤$_1$ refers to the standard value of the strong weathering of the softer rock, the standard value of the layer ⑤$_{1a}$ refers to the standard value of the medium weathering in the soft rock, the standard value of the layer ⑤$_{2a}$ refers to the standard value of the medium weathering in the softer rock, the standard value of the layer ⑤$_2$ refers to the standard value of the medium weathering in the harder rock, and the standard value of the layer ⑤$_{2b}$ refers to the standard value of medium weathering in the hard rock. (2) [a] is the empirical value. (3) The listed data is the total stress index

44.2.3 Structure

The study area is mainly controlled by three large NNE faults, and the latest strata are the continental red clastic rocks of the Cretaceous-Paleogene Anjianzhai Formation, which significantly reformed the earlier structure of the area. It is speculated that the formation was formed by the Late Yanshan Movement. NWW trending structures are developed between the NNE trending faults, which are cut by the NNE trending faults. Therefore, it is inferred that the NNE trending faults formed earlier than the NNE trending faults. The NW trending structure is mainly formed by compression, among which the synclinal core is Triassic, so it can be inferred that it was formed after Triassic, and associated with the development of compressional faults in the same strike. A small NE trending normal fault was developed in the northwest of the site, dipping to the SE with a dip angle of 84°, indicating that it was formed by a large tensile stress and was limited to the NWW trending fault at the bottom. Therefore, it was speculated that the small normal fault was an associated fault formed in the same period of the large NNE trending fault.

In general, the proposed site is located between three faults. Among them, a small NWW trending syncline is developed, with Devonian strata exposed at the core and Carboniferous and Permian strata exposed at the two wings. The inversion of the strata between the three faults indicates that the NNE-SSW compressive stress was very strong in this area during the early Yanshan Period. At present, no neotectonic movement has been found in the Holocene within the proposed construction site, which is a relatively stable area (Fig. 44.1).

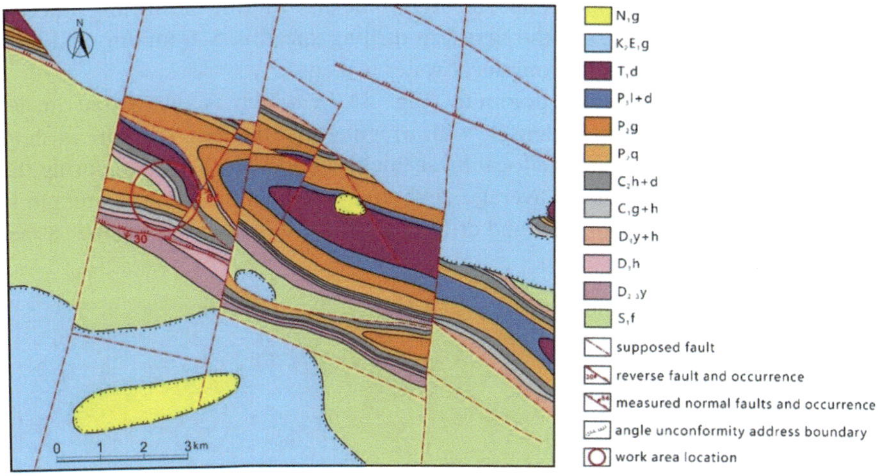

Fig. 44.1 Regional geological and structural outline of the proposed construction site

44.2.4 Hydrogeological Condition

There is no surface water in the study area, but small puddles are formed after the rain in the local area. The west and south sides of the site are adjacent to Mojia Lake. The lake has long-term infiltration, which can supply karst fissure water in a certain extent. Within the scope of the exposed stratum, the groundwater types of the site are mainly upper unconfined water and karst fissure water. The upper unconfined water is mainly stored in the artificial fill on the upper part of the site. During the survey period, the static groundwater depth of the upper unconfined water on the site was about 0.30–2.80 m (elevation of 20.16–30.23 m). Karst fissure water occurs in the dissolution fissures and structural fissures of the bedrock at the lower part of the site. Drilling shows that the underlying sandification dolomite is in a broken to extremely broken state, with well-developed fissures and moderately developed karsts.

44.3 Characteristics of Sandification Dolomite

44.3.1 Drilling Process and Core Characteristics

Drilling Process. $⑤_1$ Fully sandification dolomite (Fig. 44.2), distributed throughout the site, the overlying soil is red clay or residual soil; the rock mass is extremely broken, and the core is completely sugar-like fine particles, which cannot be sampled; the core rate is low, about 45% around, the RQD index is 0; during the drilling process, the footage is faster, generally the first 3 m drilling speed is 5–6 min/m, and then slower; slight water leakage, strong local water leakage.

$⑤_{1a}$ Strongly sandification dolomite (Fig. 44.3), which is distributed in $⑤_1$ fully sandification dolomite as lenses, with irregular distribution. The core is short columnar or fragmented, and the local layer thickNess can reach 1–2 m; during the drilling process, the footage is average, and when the local rock mass strength is high, the drilling tool is worn out and drilling is difficult; the core rate is low, about

Fig. 44.2 Typical core photo of fully sandification dolomite

Fig. 44.3 Typical core photo of strongly sandification dolomite

60%, and the RQD index is about 10%; the general drilling speed is about 7 min/m, and the local can reach 15 min/m above. This phenomenon should be a typical "sand-packed stone" phenomenon in sandification dolomite, that is, the unevenness of sanding.

⑤$_{2a}$ Medium sandy dolomite (Fig. 44.4) is distributed in the weakly sandification dolomite as lenses, with irregular distribution, and the core is massive or short columnar; the core rate is slightly higher, about 65%, and the RQD index is 50%. During the drilling process, the weak sandy dolomite is slightly faster, the local strength is higher, the drilling tool wear is more serious, and the drilling is difficult. The general drilling speed is about 15 min/m.

⑤$_2$ weakly sandification dolomite (Fig. 44.5), distributed throughout the site, affected by uneven sanding, the burial depth varies greatly, the core is short or long columnar; the core rate is higher, about 80%, RQD index is about 70%; during the drilling process, the footage is slow, the drilling tool wear is serious, and the drilling is difficult. The general drilling speed is more than 20 min/m.

⑤$_{2b}$ Micro-sandification dolomite (Fig. 44.6), distributed in weakly sandification dolomite in the form of lenses, with irregular distribution and columnar core; the

Fig. 44.4 Typical core photos of mid-sandification dolomite

Fig. 44.5 Typical core photo of weakly sandification dolomite

Fig. 44.6 Typical core photo of micro-sandification dolomite

core rate is high, about 85%, RQD index is about 75%; during drilling, the footage is slow, the local strength is extremely high, the drilling tool wear is extremely serious, and the drilling is difficult. The general drilling speed is more than 25 min/m.

Core Characteristics. It can be seen from the drilled core that dolomite is affected by dissolution sandification, and its sandification phenomenon mainly has three states: (1) Karst sanding occurs only on some structural surface of core, and a small amount of silty sand is filled; (2) There is karst sandification on the main structural surface of the core, filled with silt sand; (3) The core as a whole has the phenomenon of sand, almost completely sand into gravel, granular structure of sandstone; (4) The sandification of dolomite is mainly full sandification and weak sandification, and the other degrees of sandification (strong, medium, and micro) rock formations cannot be formed separately.

On the whole, the "sandification" effect in the same site shows a gradual weakening trend as the buried depth increases, but the thickNess of full and weak sanding varies greatly, reaching more than 20 m. Regardless of the degree of sanding, unevenness is always accompanied by the entire drilling process, which causes great difficulties for drilling construction, exploration layering and pile foundation selection.

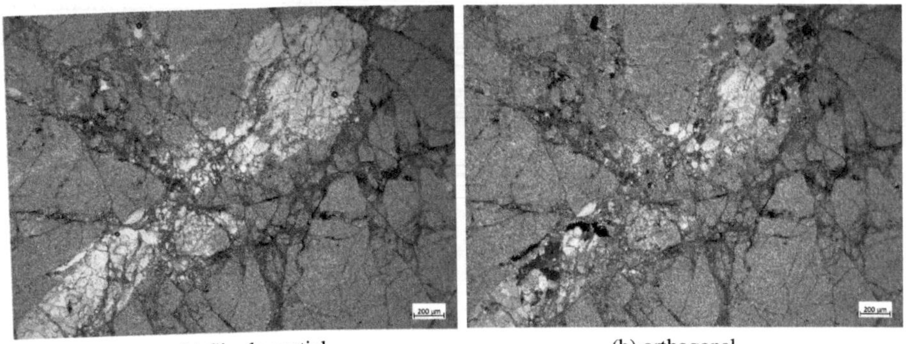

(a) Single partial (b) orthogonal

Fig. 44.7 Results of thin section identification of sandification dolomite

44.3.2 *Thin Sections Identification*

The thin section identification test results show (Fig. 44.7) that the rock is mainly composed of dolomite clastics (65–70%), crushed particles (20–25%) and calcite (5–15%), and the mineral crushed particles play a role in cementing. The long axis of the rock fragments is slightly oriented, indicating that the rock has undergone shallow metamorphism. The dolomite has a microcrystalline-fine-crystalline structure, and the internal cleavage and fissures of the breccia are relatively developed. The crushed grains and powders include quartz, dolomite, etc., which are filled between the dolomite breccias and have a cementing effect. The cracks are filled with splendid calcite, which is irregular and granular, which also plays a certain role in cementing.

44.3.3 *Saturated Uniaxial Compression Test*

A large number of drilling and sampling tests were carried out on the sandy dolomite in the proposed site, and 60 sets of saturated uniaxial compression tests were statistically analyzed (Fig. 44.8), and referring to the relevant specifications "according to the saturated uniaxial compressive strength (f_{rk}) of the rock) According to the standard of "classification of rocks", a statistical analysis was carried out, and the results are shown in Table 44.2.

From Table 44.2 and Fig. 44.8 combined with the coring results of a large number of exploration holes, it can be seen that: (1) Layers ⑤$_{1a}$, ⑤$_{2a}$, and ⑤$_{2b}$ are mainly interlayers of ⑤$_1$ and ⑤$_2$. They are mainly affected by the uneven degree of sandification. The above performance is the difference in uniaxial compressive strength; (2) The saturated uniaxial compressive strength of strong sanding and micro-sanding is relatively discrete compared with medium sanding and weak sanding, and in terms of distribution range and quantity, ⑤$_{1a}$ and ⑤$_{2b}$ is a intercalation or lens. If its f_{rk} index is used to participate in the parameter statistics of the ⑤$_1$ or ⑤$_2$ layer and used

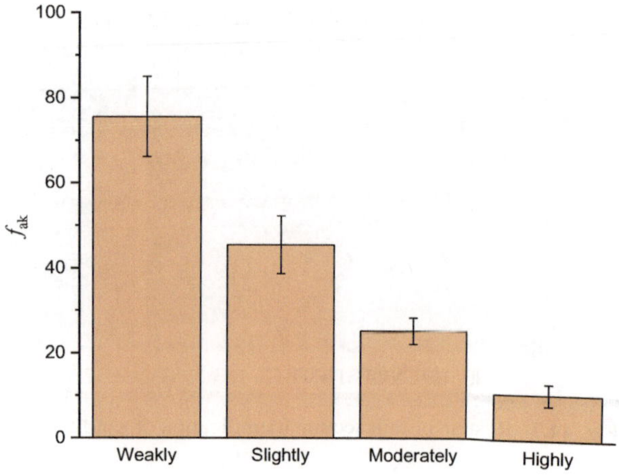

Fig. 44.8 Distribution diagram of saturated uniaxial compressive strength test values of dolomite with different sandification degrees

as the basis for the value of the pile foundation parameters, it will inevitably cause the parameter value of the ⑤$_1$ or ⑤$_2$ layer to be too large, which will affect the safety of the project.

44.4 Pile Foundation in Situ Test

In order to study the effect of bored pile foundation bearing capacity and its influencing factors in sandfication dolomite area, 14 sets of single-pile vertical static loading tests and pile internal force tests were carried out in strict accordance with the relevant requirements of the Code for Testing of Building Foundation Piles (JGJ 106-2014).

44.4.1 Static Load Test

Static Load Test Device. In static loading test, the stack weight platform reaction device was used. The counterforce device and protective stack material are channel steel. The loading beam with threaded steel as a U-shaped hanger installed under the reactor platform. The test reference beam is made of steel with large stiffness and consistent properties. In order to ensure sufficient accuracy, the two ends of the reference beam are respectively fixed. The deformation of test pile is measured by four dial gauges placed symmetrically on the pile head. The resolution of the dial indicator used is better than or equal to 0.01 mm.

The bearing capacity of the loading device is not less than 1.2 times the estimated maximum load. An electric oil pump is applied to drive three parallel 6000 kN oil

Table 44.2 Classification list of dolomite with different sandification degrees

Degree of sandification	Full sandification	Strong sandification	Medium sanding	Weak sandification	Micro-sandification
Corresponding layer number	⑤$_1$	⑤$_{1a}$	⑤$_{2a}$	⑤$_2$	⑤$_{2b}$
f_{rk}	$f_{rk} \leq 5$ MPa	$5 < f_{rk} \leq 15$ MPa	$15 < f_{rk} \leq 30$ MPa	$30 < f_{rk} \leq 60$ MPa	$f_{rk} > 60$ MPa
Test average	Unable to sample	9.43	24.20	44.62	73.48
Rock category	Extremely soft rock	Soft rock	Moderate soft rock	Moderate hard rock	Hard rock
Characteristics	Large-scale distribution within the site, the core characteristics are completely sugary fine particles, can not be sampled	Scattered contained in fully sandy dolomite, the core characteristics are short column or block	Lenticular body contained in weak sandy dolomite, the core characteristics are long column, can not be alone into layers	Full site distribution, but the buried depth varies greatly, the core characteristics is long column	The lenticular body contained in weakly sandy dolomite, the core characteristics is long column, can not be alone into layers

pressure jacks. The pressure value is given by a calibrated pressure gauge, and the calibration curve of the jack is then converted into a load value. The pressure gauge accuracy is better than or equal to 0.4 levels.

The Internal Force Test of the Pile Shaft. The distribution of pile shaft axial force and resistance were tested during the single vertical static loading test by installing sensors in the pile shaft (Fig. 44.9). The original pile shaft force was tested using rebar force gauge, which is installed on the main reinforcement of bored pile. According to the drilling data of piles, the test section is set at the interface of different soil layers, and each section is symmetrically installed with 2 steel dynamometers. The pile top is set with sensor calibration section, and 4 sensors should be buried. When the test point is buried, the main reinforcement is disconnected at the position of the test section, and the welding is replaced by the rebar force gauge. The force measuring line is connected to the ground with the main reinforcement cage, protected by a soft plastic tube wrapped at the top.

The reinforcing bar stress sensors are calibrated in the room, and a special instrument (CTY-202 sensor recorder) is used to measure the frequency and convert it into axial force according to the calibrated curve. 4 times of reading under each load, and the interval time is 30 min.

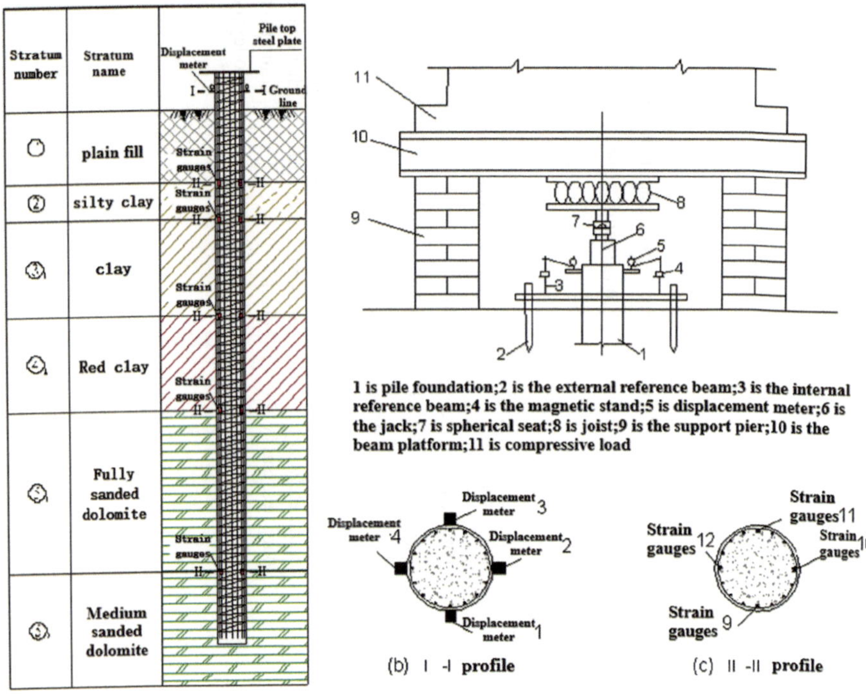

Fig. 44.9 A diagram of the installation and burial of the test device

Loading Method. The experiment was carried out by using the loading method of slow maintenance load: (1) Loading was carried out in stages, and equal quantities step by step. The loading of the first stage was 1/10 of the maximum estimated ultimate bearing capacity. The loading of the first stage was twice of the load of the first stage. (2) After each level of loading, data were read at 5 min, 15 min and 30 min, and then every 15 min. After 1 h, data were read every 30 min. Pile top settlement and other data were recorded. (3) The relatively stable settlement standard of test pile is that the settlement of pile top does not exceed 0.1 mm every 1 h and occurs twice in succession. When the settlement rate of pile top reaches the relatively stable standard, the next level of loading can be applied.

44.4.2 *The Overall of Test Results*

According to the pile foundation theory, there are many factors that affect the bearing capacity of pile foundation. But in this experiment, the factors that affect the bearing capacity of pile foundation are mainly pile length, pile diameter and bearing layer. According to the test information, this test is divided into four categories for discussion (Table 44.3): Class I—pile diameter 700 mm, bearing layer $⑤_1$; Class II—pile diameter 800 mm, bearing layer $⑤_1$; Class III—pile diameter 900 mm, bearing layer $⑤_1$; Class IV—pile diameter 900 mm, bearing layer $⑤_2$.

44.4.3 *Axial Force Distribution of Pile*

The results of a large number of test piles show that the pile shaft axial force of bored pile generally increases with the increase of test load and decreases with the increase of pile length. A total of 14 groups of tests conducted in this study also have similar rules.

According to the distribution curves of pile shaft axial force of each test pile under the maximum test load (Fig. 44.10), it can be seen that: in the soil layer, pile shaft axial force basically increases linearly, indicating that the change of pile shaft axial force in the soil layer is relatively uniform. After entering layer $⑤_1$ or layer $⑤_2$, the pile shaft axial force appears inflection point due to the large increase of lateral friction resistance, indicating that the pile shaft axial force in soil layer decreases with the increase of buried depth.

Table 44.3 List of basic conditions of test piles

Test pile number		Pile length l (m)	Pile diameter d (mm)	Bearing layer	Maximum test load Q(kN)	Ultimate bearing capacity Q_u(kN)	Maximum stable settlement of pile top S_u (mm)
I	A5	32.2	700	⑤$_1$	12,100	11,000	23.23
	A6	33.3	700	⑤$_1$	12,100	11,000	22.47
	A7	32.4	700	⑤$_1$	12,100	11,000	19.87
II	A4	32.1	800	⑤$_1$	14,300	13,000	22.31
	B2	28.8	800	⑤$_1$	13,200	12,000	24.73
	A8	33.3	800	⑤$_1$	14,300	13,000	22.99
	B8	30.0	800	⑤$_1$	13,200	12,000	7.28
III	A1	32.2	900	⑤$_1$	14,300	13,000	17.36
	A2	32.8	900	⑤$_1$	14,300	13,000	14.27
	A3	32.2	900	⑤$_1$	14,300	13,000	16.25
	B6	24.7	900	⑤$_1$	11,200	9600	13.75
IV	B3	34.3	900	⑤$_2$	16,000	14,400	17.74
	B4	28.8	900	⑤$_2$	14,400	12,800	25.53
	B5	29.1	900	⑤$_2$	16,000	14,400	20.54

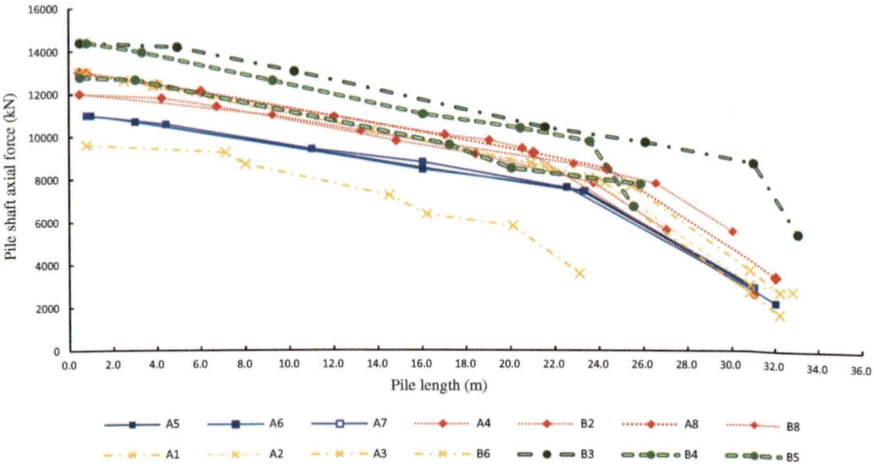

Fig. 44.10 Distribution curve of measured pile shaft axial force under maximum test load

44.5 Characteristics Analysis of Pile Side Friction and Pile Tip Resistance

The vertical ultimate bearing capacity (Q_u) acting on the pile top is borne by the total ultimate friction resistance (Q_s) acting on the pile side and the total ultimate tip resistance (Q_p) acting on the pile tip, respectively, $Q_u = Q_s + Q_p$. The development process of pile side friction resistance and pile tip resistance is the load transfer process of pile-soil system. When the top of pile is subjected to vertical pressure, the pile compresses and moves downward, and then the relative movement between the pile side surface and the adjacent soil occurs. The pile side surface begins to be subjected to the upward friction resistance of the soil. The load is transferred to the soil around the pile through the side friction, so that the axial force and the compressive deformation of the pile decreases with the depth. As the load increases, the side friction of the lower part of the pile also gradually plays an important role. When the load increases to a certain value, the vertical displacement of the pile tip begins to occur, and the reaction force of the pile tip also begins to play.

44.5.1 Pile Side Friction

Distribution of Pile Side Friction. By embedding sensors at the interfaces of different soil layers, the distribution of pile shaft axial force along single pile is tested in the process of vertical compressive static loading test. According to the change, the distribution curve of pile lateral friction resistance and its exertion can be calculated.

Table 44.4 is the statistical analysis table of measured characteristic values of pile side friction resistance (q_{sia}). The analysis shows that: (1) In the soil layer around the pile (except the filled soil layer), the distribution law of side friction resistance of each test pile is similar. When the pile is loaded to the maximum load, the average characteristic value of side friction resistance (q_{sia}) is about from 29.5 to 47.2 kPa. Except the measured value of gravel soil layer is lower than the standard value, the measured value of residual soil layer is basically consistent with the standard value. (2) Dolomite sanding phenomenon has significant influence on the characteristic value of pile side friction resistance. For layer ⑤$_1$, it becomes granulated sugar particles after completely sandification, but it has a certain cementation force in situ. The measured average value of characteristic value of pile side friction resistance is 129.6 kPa. This index can basically reach the upper limit (130 kPa) of the recommended value of hard rock and strongly weathered rock in Table 44.1. (3) For layer ⑤$_2$, the measured average value of characteristic value of pile side friction resistance is only 260.7 kPa, and this index cannot reach 270 kPa, which is the lower limit of recommended weathering value of hard rock, as shown in Fig. 44.11.

Table 44.4 Statistical analysis of characteristic values of pile side friction resistance

Stratum information		The characteristic values of pile side friction resistance q_{sia} (kPa)														Average value
Stratum number	Stratum name	Class I			Class II				Class III				Class IV			q_{sia}
		A5	A6	A7	A4	B2	A8	B8	A1	A2	A3	B6	B3	B4	B5	(kPa)
①	plain fill					9.7						5.1	7.4	8.3		7.6
②	silty clay	34	35	27	35	31	33	23.2	34	33	34	32	32.8		32	32
③1	clay	38	39	40	38	40	38	38	37.6	38.6	37	57.2	39.3	38.7	40	39.6
③2	broken stone				34											37.3
③3	Clay with gravel											45	41	68.6		47.2
④1	Red clay			28	25	33.6		31.2					27	42.3		29.5
④2	Residual soil	34.4	28.8	41.7	49	50.4	43.6	49	46.2	47.6	38.9	44.8	34		35	40.8
⑤1	Fully sandification dolomite	131.6	128.8	131.6	128.8	131.6	130.3	126	128.8	127.4	128.8	134.2				129.6
⑤2	Medium sandification dolomite												275	271.3	268	260.7

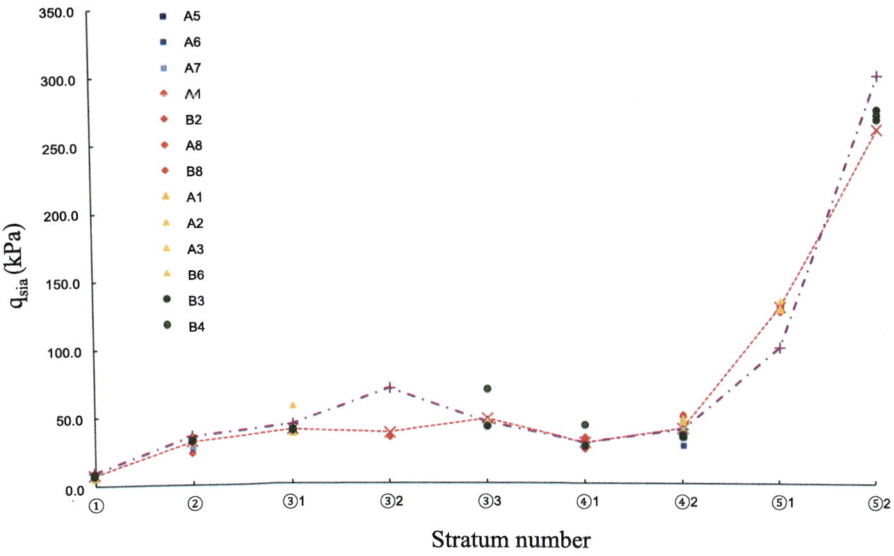

Fig. 44.11 Distribution of characteristic values of pile side friction in each soil layer

The Exertion of Pile Side Friction Resistance. The exertion process of side friction resistance of 13 groups of piles under static loading test was analyzed in detail, and the diagram of maximum value of pile side friction resistance at different depths under the action of ultimate bearing capacity was drawn in Fig. 44.12. The following rules were obtained: (1) Pile side friction increases with the increase of depth, and it basically increases linearly in the soil layer. After entering layer ⑤$_1$ or ⑤$_2$, pile side friction increases greatly, resulting in an inflection point. (2) The pile side friction of rock-socketed section is higher than that of the overlying soil and increases with the increase of buried depth. The maximum value of pile side friction occurs near the pile tip. (3) In the process of loading, the side friction plays a role on the pile shaft before the bottom. Regardless of the bearing layer is layer ⑤$_1$ or ⑤$_2$, the side friction plays a role before the end resistance, and the proportion of the side friction is greater than the that of the end resistance. However, compared with the layer ⑤$_1$, when the bearing layer is layer ⑤$_2$, the pile tip resistance develops significantly earlier and the required loading is smaller. (4) No matter the thickNess of surface fill is 0.5 m or 4.9 m, the side friction resistance of pile is very small or even zero.

44.5.2 *Pile Tip Resistance*

The Exertion Process of Pile Tip Resistance. By embedding sensors at the interfaces of different soil layers, the distribution changes of pile shaft axial force along

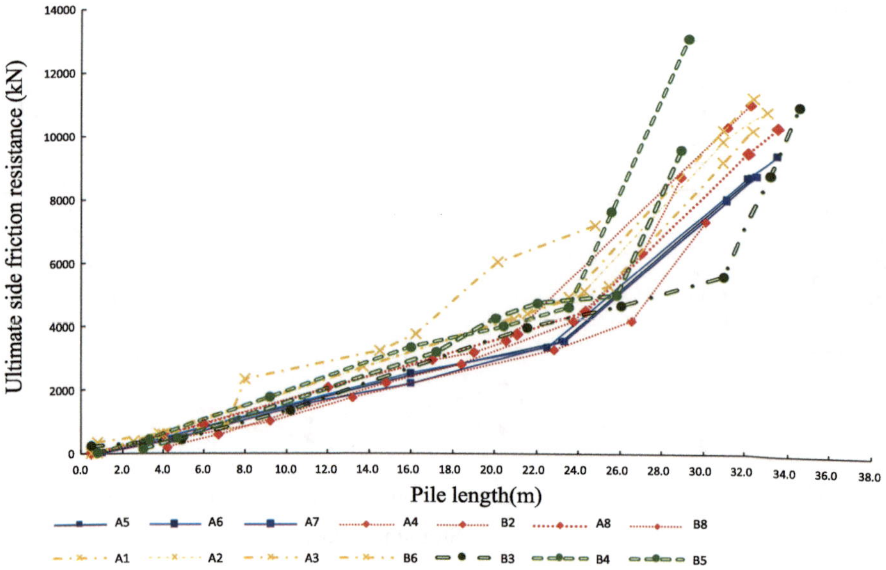

Fig. 44.12 Exertion of the limit value of pile side friction resistance at different depths under ultimate bearing capacity load

pile length in the process of vertical compressive static load test of single pile are tested. According to this change, the value and development of pile tip resistance can also be calculated.

Many test pile data show that the pile side friction began to play a role at the beginning of the loading, and the pile tip resistance is required after the load reaches a certain level. As long as the load began to apply, the displacement of pile head began to play.

Table 44.5 shows the exertion conditions of pile tip resistance under the five categories. It can be seen that: (1) under Class I, when the load is between 7700 and 8800 kN the pile tip resistance comes into play (pile tip displacement is between 13.16 and 15.59 mm). (2) In Class II, when the load is between 6000 and 10,400 kN the pile tip resistance plays a role (pile tip displacement is between 7.25 and 16.15 mm). (3) In Class III, when the load is between 6400 and 10,400 kN the pile tip resistance plays a role (pile tip displacement is between 8.37 and 12.64 mm). (4) In Class IV, when the load is between 4800 and 8000 kN the pile tip resistance plays a role (pile tip displacement is between 4.45 and 5.98 mm).

It can be seen that when the pile diameter is the same and the harder layer ⑤$_2$ is used as the bearing layer, the pile tip resistance takes effect earlier and the pile top displacement is smaller. When the bearing layer is the same, the pile tip resistance with larger diameter takes effect earlier than that of the pile with smaller diameter, and the pile top displacement is smaller.

Table 44.5 Exertion conditions of pile tip resistance under different categories

Test pile type	Class I			Class II				Class III				Class IV		
Test pile number	A5	A6	A7	A4	B2	A8	B8	A1	A2	A3	B6	B3	B4	B5
Bearing stratum	⑤$_1$			⑤$_1$				⑤$_1$				⑤$_2$		
s (mm)	13.65	15.59	13.16	16.15	7.25	12.28		12.64	9.71	10.95	8.37	5.48	4.45	5.98
s_u (mm)	23.23	22.47	19.87	22.31	24.73	22.99	7.28	17.36	14.27	16.25	13.75	17.74	25.53	20.54
s/s_u (%)	58.8	69.4	66.2	72.4	29.3	53.4		72.8	68.0	67.4	60.9	30.9	17.4	29.1
Q (kN)	7700	8800	7700	10,400	6000	9100	6000	10,400	9100	9100	6400	8000	4800	6400

Note s (mm)—Pile tip displacement when pile tip resistance comes into play; s_u (mm)—Total displacement of pile top; Q (kN)—Test load when pile tip resistance comes into play

The Exertion Degree of Pile Tip Resistance. For the ultimate bearing capacity of piles, pile side friction and tip resistance were analyzed and compared. The ratio of pile length to pile diameter (*l*/*d*) and the ratio of pile depth to pile diameter (*h*/*d*), are shown in Table 44.6 and Fig. 44.13.

The statistical analysis shows that: (1) The exertion degree of pile tip resistance (ultimate tip resistance and its proportion) has a certain correlation with bearing layer, length-diameter ratio, and depth-diameter ratio; (2) When layer ⑤$_1$ is taken as the bearing layer, the average ultimate tip resistance of pile with pile diameter of 700 mm is $Q_p = 1984$ kN and the average value of tip resistance ratio is 18.04%. The average tip resistance of 800 mm pile diameter is $Q_p = 3411$ kN, and the average tip resistance ratio is 26.85%. The average value of ultimate tip resistance of 900 mm pile is $Q_p = 2233$ kN and the mean value of tip resistance ratio is 18.78%. When layer ⑤$_2$ is taken as the bearing layer, the average ultimate tip resistance of pile with pile diameter of 900 mm is $Q_p = 2613$ kN and the average value of tip resistance ratio is 19.07%. (3) When layer ⑤$_1$ is taken as the bearing layer, there is a certain inverse relationship between the length diameter ratio and the tip resistance ratio. According to the linear formula in Fig. 44.14, no matter what pile diameter, when *l*/*d* is from 46.2 to 54.4, the tip resistance ratio tends to zero. Accordingly, it is speculated that when $d = 700$ mm and pile length reaches about 36.6 m, when $d = 800$ mm and pile length reaches about 37.0 m, or when $d = 900$ mm and pile length reaches about 49.0 m, the tip resistance almost does not play a role. (4) When layer ⑤$_1$ is used as the bearing layer, there is a more obvious inverse relationship between the depth to diameter ratio and the tip resistance ratio. According to the linear formula in Fig. 44.15, when $h/d \approx 16.2$–21.0, the tip resistance ratio tends to zero. According to this, when d = 700 mm, the depth of rock-socketed reaches about 14.7 m, when d = 800 mm, the depth of rock-socketed reaches about 15.5 m, or when d = 900 mm, the pile length of rock-socketed reaches about 14.6 m, the tip resistance almost does not play a role. (5) When the harder layer ⑤$_2$ is used as the bearing layer, the tip resistance ratio has a greater correlation with the depth to diameter ratio, but not with the length to diameter ratio. According to the linear formula in Fig. 44.15, when d = 900 mm and the depth of rock-socketed reaches about 7.1 m, the tip resistance almost does not play a role.

Therefore, for foundation in sandification dolomite, it is not enough to increase the pile length alone to obtain a large pile bearing capacity, and unreasonable pile length will affect the exertion of pile tip resistance.

It is evident that (1) According to the rules shown in Figs. 44.14 and 44.15, the maximum pile tip resistance and tip resistance ratio can be achieved for pile diameter 800 mm in⑤$_1$ fully sandification dolomite stratum. That is to say, when the pile side friction resistance of pile diameter is almost the same, the pile tip resistance can be better played and larger pile bearing capacity can be achieved. It should be relatively suitable for the construction of heavy buildings in this stratum. (2) The linear relationship between depth to diameter ratio and tip resistance ratio is far superior to that between length to diameter ratio and end resistance ratio, so it can be seen that depth to diameter ratio can better reflect the characteristics of pile tip bearing capacity in rock-socketed section. (3) In sandification dolomite stratum, the

Table 44.6 Statistical analysis of test values of ultimate bearing capacity, side friction resistance and tip resistance

Test pile type	Class I			Class II				Class III				Class IV		
Test pile number	A5	A6	A7	A4	B2	A8	B8	A1	A2	A3	B6	B3	B4	B5
Bearing stratum	⑤$_1$			⑤$_1$				⑤$_1$				⑤$_2$		
h (m)	9.0	10.8	9.1	11.6	5.1	9.0	5.0	8.7	8.6	6.8	4.6	3.4	3.0	5.6
l (m)	32.2	33.3	32.4	32.1	28.8	33.3	30.0	32.2	32.8	32.2	24.7	34.3	28.8	29.1
d (mm)	700	700	700	800	800	800	800	900	900	900	900	900	900	900
Q_u	11,000	11,000	11,000	13,000	13,200	13,000	12,000	13,000	13,000	13,000	9600	14,400	12,800	14,400
Q_s (kN)	8766	9474	8807	11,070	8743	10,387	7358	11,281	10,901	10,246	7241	11,050	9607	13,103
Q_p (kN)	2234	1526	2193	1930	4457	2613	4642	1719	2099	2754	2359	3350	3193	1297
$\underline{Q_p}$ (kN)	$\underline{1984}$			$\underline{3411}$				$\underline{2233}$				$\underline{2613}$		
q_{pa} (kPa)	2904	1984	2851	1921	4436	2601	4620	1352	1651	2166	1855	2634	2511	1020
Q_p/Q_u (%)	20.3	13.9	19.9	14.8	33.8	20.1	38.7	13.2	16.1	21.2	24.6	23.3	24.9	9.0
$\underline{Q_p/Q_u}$ (%)	$\underline{18.04}$			$\underline{26.85}$				$\underline{18.78}$				$\underline{19.07}$		
l/d	46.0	47.6	46.3	40.1	36.0	41.6	37.5	35.8	36.4	35.8	27.4	38.1	32.0	32.3
$\underline{l/d}$	$\underline{46.6}$			$\underline{38.8}$				$\underline{33.9}$				$\underline{34.1}$		
h/d	12.9	15.4	13.0	14.5	6.4	11.3	6.3	9.7	9.6	7.6	5.1	3.8	3.3	6.2
$\underline{h/d}$	$\underline{13.8}$			$\underline{9.6}$				$\underline{8.0}$				$\underline{4.4}$		

Note: h—Depth of pile foundation into bearing layer (m); l—Pile length (m); d—Pile diameter (mm); Q_u—Ultimate bearing capacity of pile (kN); Q_s—Ultimate pile side friction resistance (kN); Q_p (kN)—Ultimate pile tip resistance; $\underline{Q_p}$(kN)—The average limit pile tip resistance; q_{pa}(kPa)—Characteristic value of pile tip resistance; Q_p/Q_u—Tip resistance ratio (ratio of pile tip resistance to ultimate bearing capacity of pile); $\underline{Q_p/Q_u}$—The average tip resistance ratio; l/d—Length-diameter ratio (ratio of pile length to pile diameter); $\underline{l/d}$—The average length to diameter ratio; h/d—Depth to diameter ratio (the ratio of pile depth into bearing layer to pile diameter); $\underline{h/d}$—The average depth to diameter ratio

The significance of underline means the average value

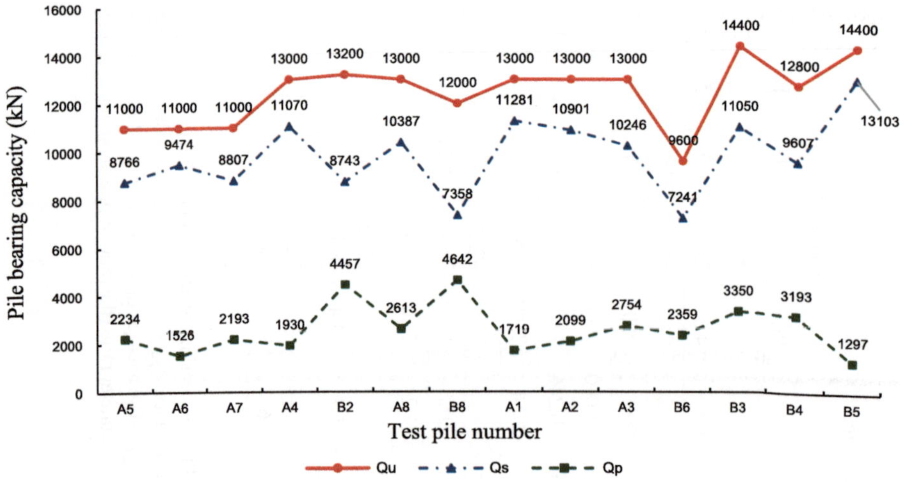

Fig. 44.13 Comparison of ultimate tip resistance with ultimate bearing capacity and ultimate side friction resistance of pile

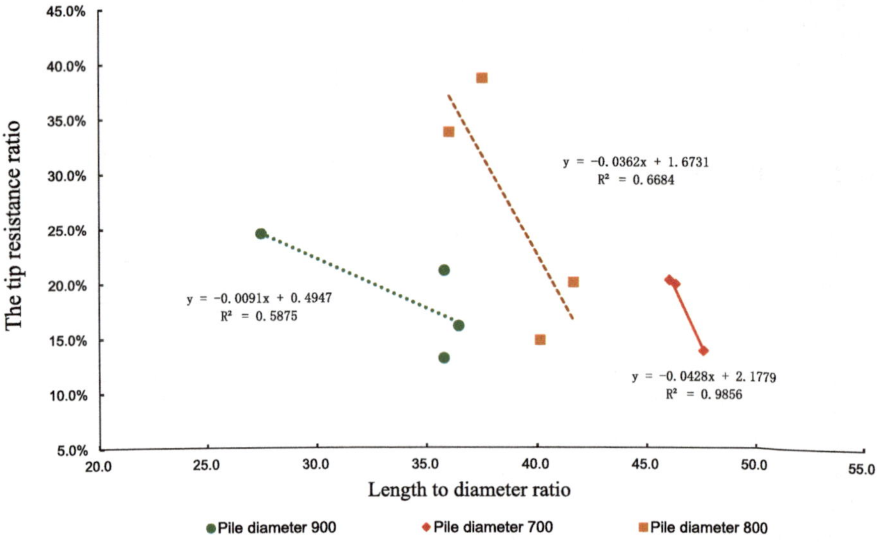

Fig. 44.14 The relationship between length diameter ratio and tip resistance ratio in layer ⑤1

exertion of pile tip resistance is relatively low. The average tip resistance ratio of pile in layer $⑤_1$ and layer $⑤_2$ is generally less than 30%, which belongs to end-bearing friction pile.

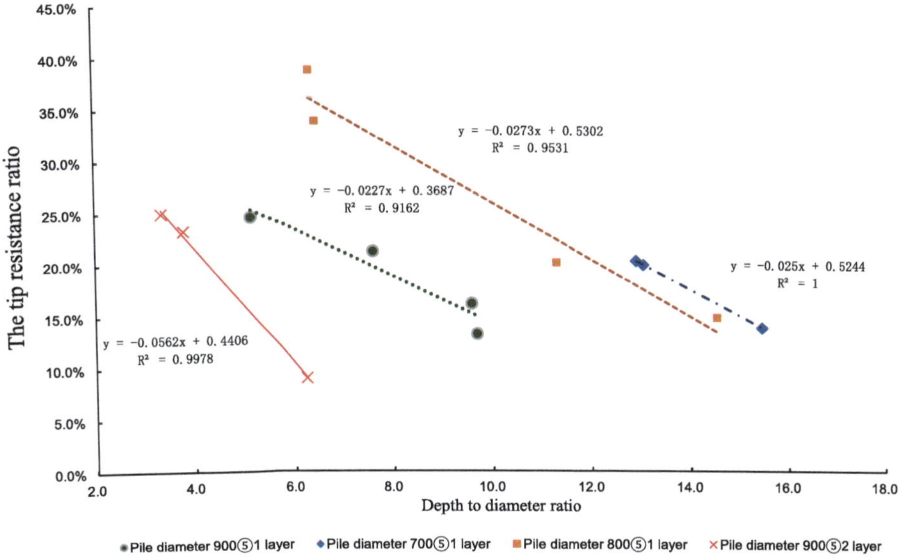

Fig. 44.15 Relationship between depth-diameter ratio and tip resistance ratio

44.6 Discussion

According to the measured results in Sect. 44.5, under the same conditions, it is safe to use layer $⑤_1$ ($q_{sia} = 100$ kPa) according to the index of strongly weathered soft rock in the code, but it is too conservative. However, it is dangerous to use layer $⑤_2$ ($q_{sia} = 100$ kPa) according to the index of moderately weathered hard rock. Therefore, the current local standards in Table 44.1 need to be further studied and improved.

Considering that although $⑤_{2a}$ is a relatively weak interlayer of $⑤_2$, its distribution is more extensive than $⑤_{1a}$ and $⑤_{2b}$, it seems more appropriate to calculate the f_{rk} values of $⑤_{2a}$ and $⑤_2$ together. The standard value after comprehensive statistics is 32.5 MPa, which belongs to the lower value in the hard rock class. Thus, according to the relevant standards in Table 44.1, the value of layer $⑤_2$ should be $q_{sia} = 270$ kPa. In this way, the value of characteristic value of pile side friction resistance is closer to the measured index. However, for pile tip resistance, whether it is layers $⑤_1$ or $⑤_2$, there is a problem of the exertion degree.

Dolomite sandification phenomenon has a significant effect on the characteristic values of pile tip resistance. For layer $⑤_1$, the average q_{pa} values are 2579 kPa (d = 700 mm), 3394 kPa (d = 800 mm), and 1755 kPa (d = 900 mm), respectively. For layer $⑤_2$, the measured average value of pile tip resistance q_{pa} is 2055 kPa (d = 900 mm). This is very different from the recommended value in the code. Because the side friction changes linearly in the soil layer, it plays a stable role. Therefore, the main reason for the difference is the different bearing capacity characteristics of pile foundation embedded in rock segment.

A large number of pile test results show that the side friction of pile is mainly affected by the physical and mechanical properties of soil around the pile. However, the conditions under which pile tip resistance plays a role are more complicated, which are related to the type and properties of pile tip soil, pile length, pile diameter, pile forming technology and construction quality.

Notice that some sensors are not installed at the bottom of the pile and there is a certain distance from the pile tip, the pile axial force measured at the bottom sensor is not the actual pile tip resistance. Therefore, when analyzing pile tip resistance, the relatively accurate pile tip resistance can be obtained only after deducting this part of pile side friction resistance. Although the items of Technical Code for Building Pile Foundations (JGJ 94–2008) explain the relevant generalized model in 5.3.9, it is considered that for soft rock, pile side friction of rock-socketed section is distributed in the form of single hump. For hard rock, pile side friction of rock-socketed section presents a single hump distribution.

44.7 Conclusion

(1) When layer ⑤$_1$ is used as bearing layer, there is a certain inverse relationship between length-diameter ratio and tip resistance ratio. When $l/d \approx 46.2$–54.4, the tip resistance ratio tends to zero. The inverse relationship between depth-diameter ratio and tip resistance ratio is more obvious. When $h/d \approx 16.2$–21.0, the tip resistance ratio tends to zero. When the harder layer ⑤$_2$ is used as the bearing layer, the tip resistance ratio has a strong correlation with the depth to diameter ratio, but not with the length to diameter ratio.

(2) In general, the exertion of pile tip resistance is lower in sandification dolomite stratum. The average tip resistance ratio is generally less than 30% for piles in either layer ⑤$_1$ or layer ⑤$_2$ bearing layer, which should belong to the end-bearing friction pile. When using the standard value, it is suggested that the characteristic value of pile side friction of layer ⑤$_1$ should be taken as economic value, while that of layer ⑤$_2$ should be taken as safe value to avoid engineering problems.

(3) In comparison, the pile type with pile diameter of 800 mm has the largest pile tip resistance and the largest proportion of tip resistance in ⑤$_1$ fully sandification dolomite stratum. That is to say, the pile type with pile diameter of 800 mm has a better exertion of tip resistance and can achieve a larger pile bearing capacity when the pile side friction resistance is almost the same. It should be relatively suitable for the construction of heavy load buildings in this stratum.

(4) Even for rock-socketed piles, it is very uneconomical to increase the pile length unilaterally and support the pile end on the deep hard stratum to obtain high tip resistance, which increases the cost of the project but cannot improve the bearing capacity a lot.

Acknowledgements The work was financially supported by the NSFC Key Projects of International Cooperation and Exchanges (42020104006), the National Key Research and Development Program of China (2017YFC1501302), and the Key Program of National Natural Science Foundation of China (41630643).

References

Avinoam, Danin, and, et al. Patterns of limestone and dolomite weathering by lichens and blue-green algae and their palaeoclimatic significance[J]. Palaeogeography, Palaeoclimatology, Palaeoecology, 1982, 37(2–4): 221–233.

Bischoff JL, Juliá R, Shanks I et al (2011) Karstification without carbonic acid: bedrock dissolution by gypsum-driven dedolomitization. Geol 22(11):995

Caijin X, Junying R, Dengkai L et al (2020) Experimental study on isotropic elastic mechanics of layered argillaceous dolomite. J Guangxi Univ (Natural Science Edition) 45(5):1033–1044

Dazhao G (2010) Answers to difficult questions in geotechnical engineering. China Communications Press

Galvão P, Halihan T, Hirata R (2015) Evaluating karst geotechnical risk in the urbanized area of Sete Lagoas, Minas Gerais, Brazil. Hydrogeol J 23(7)

Hallal N, Bougdal R, Bensafia W (2013) Dédoublement du tunnel d'Aokas (BEJAIA): geologie et caractérisation géotechnique du massif traversé. Stabilité de l'ouvrage. Bull Eng Geol Environ 72(3–4)

Hao L, Fei D, Gang H et al (2020) Curtain grouting technique in dolomite sandification zone in tunnel. Sci Tech Engi 20(18):7441–7450

Jianqiang G, Zhihong H (2016) Physical and mechanical characteristics of muddy dolostone under different experimental condition. J Guizhou Univ (Natural Science) 33(3):110–113

Jingxing Z (2013) Foundation engineering, 3rd edn. Tsinghua University Press

Liu Qi G, Yaoru ZL (2015) The experimental study of dolomite dissolution and pore characteristics in Shibing, Guizhou. Acta Geosci Sin 36(4):413–418

Meng G, Ying W, Guang-yun G et al (2013) A method for determining tip and side resistances of large diameter pedestal pile. Rock Soil Mech 34(3):797–801

Ming Q, Zhihao Z, Ziren Z (2013) Analysis of bearing capacity of bored pile. J Guilin Univ Tech 33(3):493–506

Pengpeng W, Jing Y, Li J (2019) Sandification characteristics of Guizhou Dolomite and the influence on tunnel support structure. J Guizhou Univ (Natural Sciences) 36(03):44–48

Qiuyan F, Qixing W, Guogui Z (2004) Assessment method for the pile-tip bearing capacity of tertiary mudstone in Guangxi. J Eng Geol 12(4):408–411

Qiwen T, Chaolin Y (2014) Characteristics of karst sanded fine-crystalline dolomite and wall rock classification for Pingtou Hydropower Station. Adv Sci Tech Water Res 34(5):45–49

Touir J, Soussi M, Troudi H (2009) Polyphased dolomitization of a shoal-rimmed carbonate platform: example from the middle Turonian Bireno dolomites of central Tunisia. Cretac Res 30(3):785–804

Vondráčková T, Jozef V, Bulko R (2015) The effect of exogenous factors on building construction. Proced Earth Planetary Sci 15

Weidong W, Jiangbin W, Shubo N (2015) Field loading tests on large-diameter rock-socketed bored piles of Wuhan Greenland Center Tower. Chinese J Geotech Eng 37(11):1945–1954

Weidong W, Jiangbin W, Shubo N (2016) Field loading tests on large-diameter super-long rock-socketed bored piles of the Wuhan Center Tower. J Build Struct 37(6):196–203

Weiping P, Jiansan L (2004) Analysis and selection of building pile design parameter in Guangzhou. Chinese J Rock Mech Eng 23(6):1032–1036

Wenfeng R, Xinghua W, Xiaofei H (2009) Test on bearing capacity of bored pile in pozzolana sediment area. J Chang'an Univ (Natural Science Edition) 29(6):59–63

Wenxiu HE (2008) Study of genetic mechanism about dolomite sandification in the site area of Pingtou Hydroelectric Station of Meigu river and its influence on the project. Chengdu University of Technology

Xiangbo H (2009) Study of Dolostone's Karstic sandification characteristic and mechanism in the deep bank slope of Pintou hydropower station in Meigu River. Chengdu University of Technology

Xiaonan G (2016) Handbook of pile foundation engineering, 2nd ed. China Architecture & Building Press

Xiaozhuan C, Pingping L, Huayao Z et al (2013) Geochemical characteristics and fluid origin of the Changxing formation dolomitic rock in the Xinglongchang area of East Sichuan Basin. Acta Geol Sin 7:1031–1040

Xiliang Z, Haiquan Z, Qihua Z et al (2012) Characteristics and distribution of dolomite sandification at Pingtou Hydropower Station. Yangtze River 43(19):42–44

Yunxiu D, Zhongju F, Yumeng H et al (2018) Experiment on bearing capacity of bridge pile foundations in karst areas and reasonable rock-socketed depth. J Traffic Transp Eng 18(6):27–36

Zhongju F, Yongli X, Hongguang Z et al (2005) Comprehensive analysis on influencing factors of bearing capacity of large diameter pile foundation for red bed in West Yunnan. Chinese J Geotech Eng 27(5):540–544

Zhongtang S, Hongde C, Fenyan X et al (2011) Geochemistry and dolomitization mechanism of Majiagou dolomites in Ordovician, Ordos, China. Acta Petrologica Sinica 27(8):2230–2238

Zinan Z, Xiangfeng W (2012) Discussion on "sandification" mechanism of fine crystal dolomite and engineering examples. New Tech New Prod China 08:91–92

Chapter 45
Study on the Dynamic Development Law of Cracks in Expansive Soil Under Dry Wet Cycling

Guo Jianhua, Zhong Kun, Wang Hanhui, Li Shichang, and Dai Zhangjun

Abstract In the middle route project of the South to North Water Diversion Project, the slope of the expansive soil channel is in a dry wet alternating environment. Studying the crack propagation law of expansive soil under the effect of dry wet cycles is of great significance for revealing the instability mechanism of expansive soil slopes. A crack propagation experimental device is designed and used to conduct crack propagation experiments on expansive soil under different dry wet cycles. The water evaporation process and the dynamic development process of apparent cracks over time are systematically measured, and the following conclusions are drawn: (1) The evaporation rate over time under different dry wet cycles can be divided into stable evaporation stage, deceleration rate evaporation stage, and residual evaporation stage. (2) As the number of wet and dry cycles increases, the width of the crack gradually decreases, while the total length of the crack gradually increases. (3) Under the action of dry wet cycles, the surface area crack rate of the soil gradually increases and tends to stabilize. (4) Under multiple dry wet cycles, the particle distribution always develops towards a stable direction, and the humidity stress distribution also develops towards a more advantageous state.

Keywords Expansive soil · Crack development · Dry wet cycle · Humidity stress

---

G. Jianhua (✉) · Z. Kun · W. Hanhui
Changjiang Institute of Survey, Planning, Design and Research Corporation, Wuhan 430010, China
e-mail: guojianhua17@mails.ucas.edu.cn

L. Shichang · D. Zhangjun
State Key Laboratory of Geomechanics and Geotechnical Engineering, Institute of Rock and Soil Mechanics, Chinese Academy of Sciences, Wuhan 430071, China
e-mail: lishichang17@mails.ucas.edu.cn

D. Zhangjun
e-mail: zjdai@whrsm.ac.cn

S. Wang et al. (eds.), *Engineering Geology for a Habitable Earth: IAEG XIV Congress 2023 Proceedings, Chengdu, China*, Environmental Science and Engineering,
https://doi.org/10.1007/978-981-99-9057-3_45

45.1 Introduction

The total length of the main canal of the Middle Route Project of China's South to North Water Diversion Project is 1432 km, of which the cumulative length of the canal crossing the expansive soil is about 340 km. The cracks in expansive soil are an important factor inducing the instability and failure of channel slopes (Krisdani et al. 2008; An et al. 2018; Wang et al. 2018). These cracks divide the soil and damage its integrity. In arid climates, the development of cracks greatly weakens the soil structure, leading to a decrease in soil strength and stability (Kodikara and Costa 2013; Hewitt and Philip 1999; Rayhani 2007; Nelson and Miller 1997; Lozada et al. 2015). Under rainfall conditions, cracks provide an advantageous channel for water intrusion, causing expansive soil cracks to further expand into the deep soil. In nature, the slope of expansive soil channels is always in a dry wet alternating environment. During the process of repeated rainfall evaporation, the cracks in expansive soil are always dynamically changing, and the cracks in expansive soil have undergone a cyclic development process from cracking to gradually expanding to healing. Studying the development law of cracks in expansive soil under the effect of dry wet cycles is an important prerequisite for revealing the disaster mechanism and instability mode of expansive soil slopes in the Middle Route Project of the South to North Water Diversion Project.

There has been rapid progress in the study of crack propagation laws. Based on different internal conditions, scholars mainly studied the influence of soil size and clay content on the propagation law of soil cracks. Yesiller et al. (2000) studied the effect of different fine particle contents on surface crack rate. Through conducting three sets of dry tests on cohesive soil with different fine particle contents, it was concluded that the surface crack rate is higher when the fine particle content is higher in the soil, while it decreases when the fine particle content is lower. The experiment showed that there is an important connection between soil cracking and fine particle content; Mitchell et al. (1976) studied the influence of different mineral compositions and contents on the expansion of soil fractures, and set up 5 sets of crack expansion tests for different clay minerals. By analyzing crack indicators, it was found that the higher the content of clay minerals, the higher the degree of crack expansion. This indicates that the composition and content of clay minerals have an important impact on the dry shrinkage cracking properties of soil; Prat et al. (2006) conducted systematic experiments on crack development under different thicknesses of silty clay, and the results showed that as the soil thickness increased, the surface crack rate increased, but the total crack length significantly decreased. When the soil thickness was small, the total crack length increased, but the average crack width was smaller; Susanga et al. (2013) showed that when there are obvious defects and pores in the soil, the probability of crack propagation is higher. When the soil is more uniform, the soil will mainly shrink. From this, it can be concluded that there is currently little research on the crack propagation law under the effect of dry wet cycles, and dry wet cycles are one of the indispensable factors in crack research.

In response to the current problems, this article designed an indoor crack propagation test device and conducted crack propagation experiments on expansive soil under different dry wet cycling effects under natural conditions. The system measured the water evaporation process and the dynamic development process of apparent cracks over time, and obtained the crack propagation law of expansive soil under dry wet cycling effects.

45.2 Crack Development Test Method

Expansive soil test samples are taken from the channel slope of the Middle Route Project of South to North Water Transfer. The natural water content is 18.7%, the natural density is 1.94 g/cm^3, the specific gravity is 2.62, the Free expansion rate is 54%, the liquid limit is 47.6%, and the plastic limit is 17.9%.

The electronic scale is used to test the overall quality change of the sample, in order to obtain the overall moisture content change of the sample; The caliper around the sample is used to observe the changes in the height direction of the sample during the test process; The light source adopts a fixed lighting system; A high-precision area array camera is used to test the variation process of apparent cracks in experiments.

Crush and air dry the retrieved expansive soil sample, pass it through a 2 mm sieve, thoroughly mix it to make a mud sample, and use a vibration device to drive out the bubbles in the mud. The initial moisture content of the mud sample is 55%. Place the mud sample into a test box with dimensions of 10 cm in length and 15 cm in width. This experiment investigated the crack propagation after different dry wet cycles.

The main fracture parameters studied in this article include total fracture length, area fracture rate, and average fracture width. Among them, the total length of cracks is the result of the superposition of crack skeleton lengths at different positions on the surface of the soil sample; The area crack rate is used to characterize the degree of surface crack development in soil samples, which refers to the percentage of the area of crack expansion on the surface of the soil sample compared to the initial total area. The crack expansion area does not include the area of soil sample shrinkage.

45.3 Rack Development Patterns Under Different Dry Wet Cycles

45.3.1 *Water Evaporation Process of Small-Sized Samples*

In the crack propagation test, the overall water content changes over time under different dry wet cycles as shown in Fig. 45.1. From the figure, it can be seen that the water content continuously decreases and tends to stabilize over time. The more dry wet cycles, the greater the slope of the curve, and the faster the water evaporation rate.

The variation of evaporation rate over time under different dry wet cycles is shown in Fig. 45.2. According to the diagram, the evaporation of water in small-scale experiments can be divided into three stages: stable evaporation stage, deceleration rate evaporation stage, and residual evaporation stage.

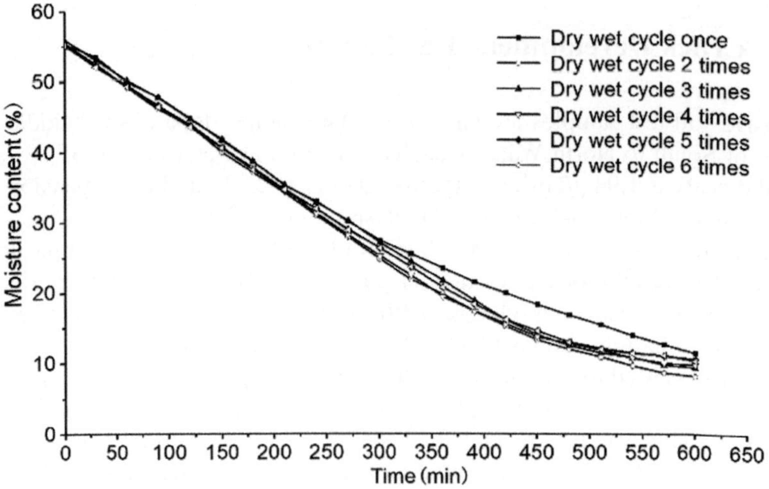

Fig. 45.1 Variation curve of overall moisture content with time under different drying and wetting cycles of small-sized samples

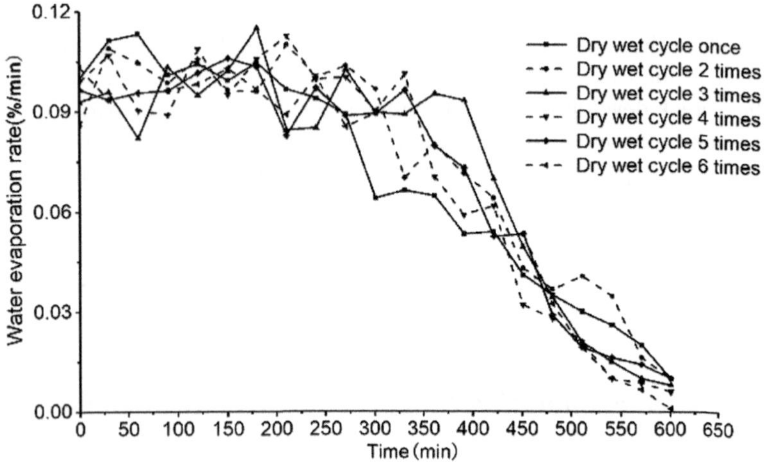

Fig. 45.2 Evaporation rate versus time curves of small-sized samples under different drying and wetting cycles

According to the experimental results, it was found that when the overall moisture content of the small-scale sample evaporates, the moisture content of the sample approaches the shrinkage limit of the soil; In the model experiment, due to the difference in moisture content at different locations, the residual moisture content is very high, and it takes a long time for evaporation to reach the limit level.

45.3.2 *Crack Development Process of Small-Sized Specimens*

As the moisture continues to evaporate, the water content of small-sized samples decreases, and the soil sample shrinks along the plane and depth directions, with the surface of the soil sample being the main evaporation surface. When the water evaporates to a certain extent, small cracks first appear at the weak boundary, which is the stage of crack germination. Micro cracks continue to develop, and different initial cracks develop and intersect, gradually forming skeleton cracks. Skeleton fractures divide soil samples into different sized areas, becoming the most important component of fractures, which is the stage of rapid development of fractures. As time increases, secondary fractures gradually develop along the skeleton fractures in each region. As the water continues to evaporate, the basic morphology of the cracks changes slightly, and the width and depth of the cracks continue to increase until they reach a stable state. After the crack reaches a stable state, the first evaporation process ends.

After the crack reaches a stable state, it enters the wetting process by adding water. Due to the water absorption and expansion characteristics of expansive soil, as the water content increases, the volume of each area of the crack blocks expands, and the small cracks heal, gradually reducing the width of the skeleton crack. At the same time, the strength of the soil sample humidified by water continuously weakens, and some soil particles on the crack surface gradually detach. The soil particles enter the crack and continuously fill the gaps. Therefore, the healing of some cracks is influenced by two factors: on the one hand, the expansion of the soil causes an increase in volume and continuously compresses the crack space, and on the other hand, the soil particles detach from the pathogen and continuously fill the crack area, but the expansion effect dominates. As the expansion and filling effects continue to balance, different segmented areas gradually show a raised state.

Entering the second evaporation process, the initial crack is generated at the weak point of the first skeleton crack. As the water evaporates, the skeleton cracks tend to become apparent. However, the position of the skeleton cracks has undergone slight changes compared to the first time. As the skeleton cracks form, the area where the cracks are divided gradually becomes apparent, and secondary cracks are gradually generated by continuous shrinkage in each area. Subsequently, the width of the cracks gradually increases until it reaches a stable state. During the humidification process, due to the different initial states at the beginning of each cycle, each region exhibits different uplift states.

As the number of wet and dry cycles continues to increase, the skeleton cracks continuously adjust to a stable state, while the number of secondary cracks generated increases. In contrast, as the number of wet and dry cycles increases, the width of the crack gradually decreases, while the total length of the crack gradually increases.

45.3.3 Development Pattern of Cracks in Small-Sized Specimens

Figure 45.3 shows the variation curve of the apparent area crack rate of soil with time after different dry wet cycles. It can be seen from the figure that the area crack rate exhibits an "S" curve relationship with time. Figure 45.4 shows the relationship curve between the crack rate of soil area and the number of dry wet cycles after evaporation stabilization. Under the action of dry wet cycles, the surface area crack rate of the soil gradually increases and tends to stabilize.

Figure 45.5 shows the variation of the total length of cracks over time under different dry wet cycles. As time goes by, the total length of cracks shows a trend of starting to grow, rapidly growing, and gradually stabilizing. As the number of wet and dry cycles increases, the cracking time of the soil sample decreases, and when it reaches a stable state, the total length of the cracks gradually increases and tends to stabilize. Figure 45.6 shows the relationship curve between the total length of cracks and the number of dry wet cycles after evaporation stabilization. As the number of wet and dry cycles increases, the total length of cracks gradually increases.

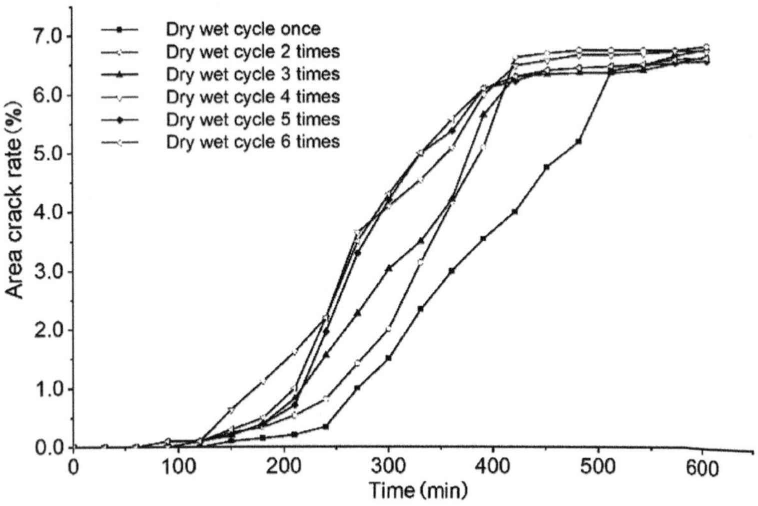

Fig. 45.3 Curve of area porosity of small-sized specimens with time

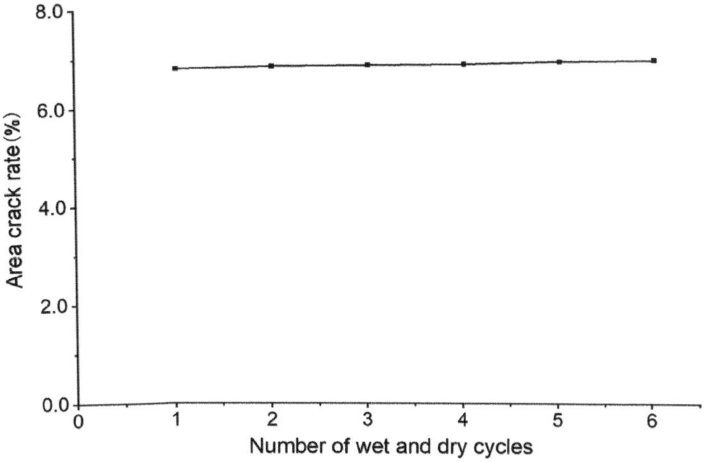

Fig. 45.4 Curve of area fissure ratio and number of drying and wetting cycles for small-sized samples after evaporation and stabilization

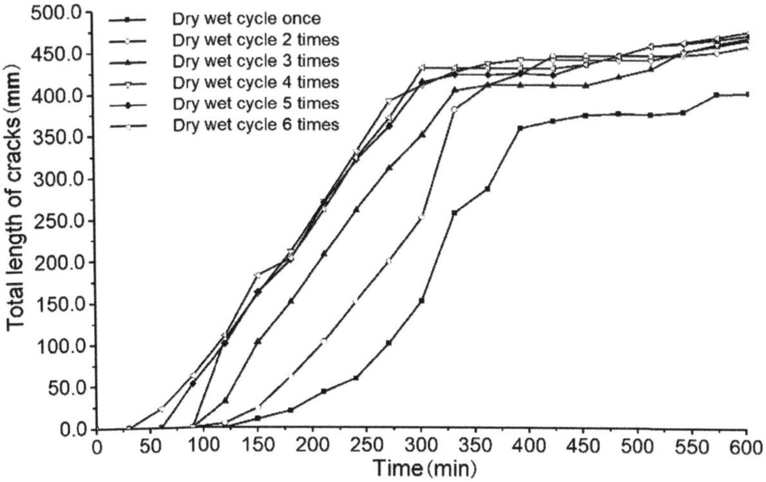

Fig. 45.5 Curve of the total length of the fissures of the small-size specimen as a function of time

Figure 45.7 shows the curve of the average crack width over time for small-sized specimens under different dry wet cycles. Over time, under different dry and wet cycles, the average width of cracks shows a trend of starting to increase, rapidly increasing, and gradually stabilizing. Figure 45.8 shows the variation pattern of the average crack width after experiencing different dry wet cycles and stabilizing them. The average crack width decreases and gradually stabilizes with the increase of dry wet cycles.

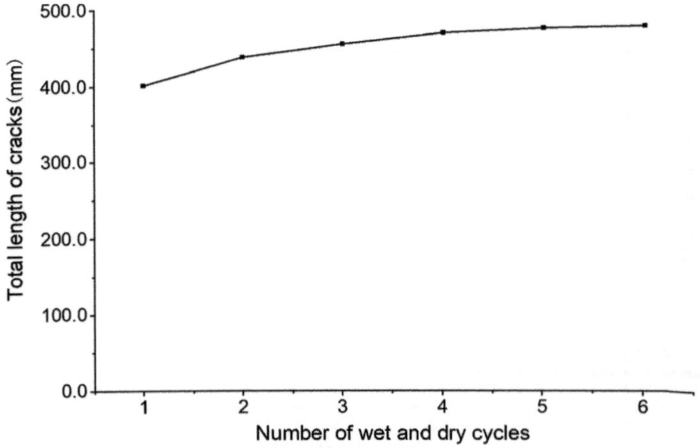

Fig. 45.6 Curve of total fissure length and number of drying and wetting cycles after the evaporation of small-sized samples is stable

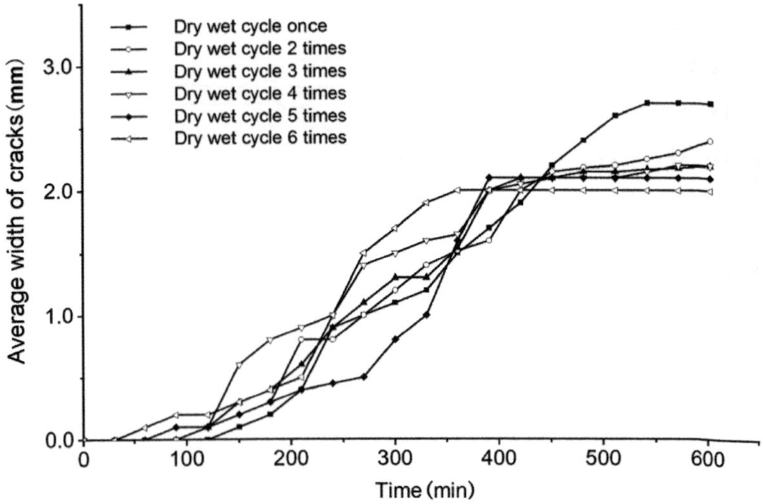

Fig. 45.7 Variation curve of the average fissure width of small size specimens with time

45.4 Discussion

The development of cracks in expansive soil is a mode of soil failure under the action of internal forces. The formation of cracks is a specific manifestation of shrinkage in expansive soil.

Under multiple cycles of drying and wetting, the humidity stress field is dynamically changing with the evaporation and wetting of water. The particle arrangement

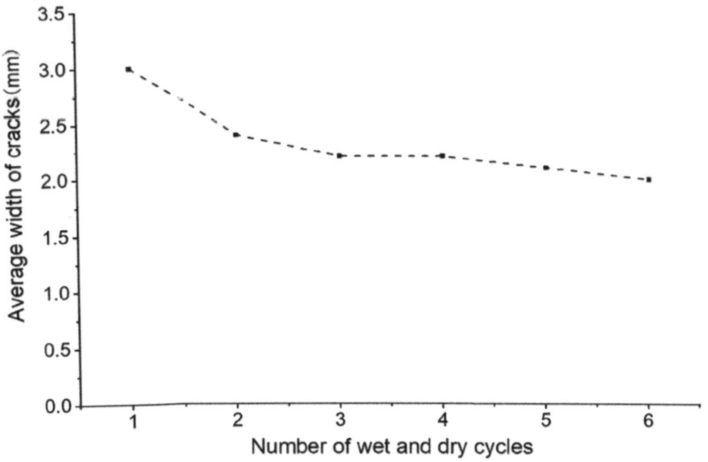

Fig. 45.8 Curve of the average fissure width and the number of drying and wetting cycles after the evaporation of small-sized samples is stabilized

always develops towards a stable direction, and the distribution of humidity stress also develops towards a more advantageous state. As the number of wet and dry cycles increases, the angle between cracks gradually changes to a more stable state, and the distribution of cracks gradually stabilizes after multiple cycles. According to the model test results, the shrinkage volume and crack volume remain stable after different dry wet cycles, indicating that cracks are also a manifestation of soil shrinkage to a certain extent. The development of cracks is due to the continuous balance of humidity and stress. The change in humidity stress is precisely due to the uneven evaporation of water in the soil, resulting in an uneven distribution of water content in the entire soil. So, the formation and development of cracks are caused by the uneven humidity field.

In short, during the evaporation process, the different evaporation rates in space lead to uneven soil shrinkage, which easily leads to the formation of cracks. As the body shrinks further, the width of cracks gradually increases, leading to the development of cracks.

45.5 Conclusion

(1) The evaporation rate over time under different dry and wet cycles can be divided into three stages: stable evaporation stage, deceleration rate evaporation stage, and residual evaporation stage.
(2) As the number of wet and dry cycles continues to increase, the skeleton cracks continuously adjust to a stable state, while the number of secondary cracks generated increases. In contrast, as the number of wet and dry cycles increases,

the width of the crack gradually decreases, while the total length of the crack gradually increases.

(3) Under the action of dry wet cycles, the surface area crack rate of the soil gradually increases and tends to stabilize. As the number of wet and dry cycles increases, the total length of cracks gradually increases. The average crack width decreases and gradually stabilizes with the increase of dry wet cycles.

(4) Under multiple cycles of drying and wetting, the humidity stress field is dynamically changing with the evaporation and wetting of water. The particle arrangement always develops towards a stable direction, and the distribution of humidity stress also develops towards a more advantageous state. As the number of wet and dry cycles increases, the angle between cracks gradually changes to a more stable state, and the distribution of cracks gradually stabilizes after multiple cycles.

Acknowledgements This research was funded by Changjiang Design Group Co., Ltd. Independent Innovation Project, grant number CX2021Z01.

References

An N, Tang C, Xu S et al (2018) Effects of soil characteristics on moisture evaporation. Eng Geol 239:126–135

Hewitt P, Philip L (1999) Problems of clay desiccation incomposite lining systems. Eng Geol 53(2):107–113

Kodikara J, Costa S (2013) Desiccation cracking in clayey soils: mechanisms and modelling. Springer, Berlin, pp 21–32

Krisdani H, Rahardjo H, Leong E (2008) Effects of different drying rates on shrinkage characteristics of a residual soil and soil mixtures. Eng Geol 102(1–2):31–37

Lozada C, Caicedo B, Thorel L (2015) Effects of cracks and desiccation on the bearing capacity of soil deposits. Geotech Lett 5(3):112–117

Mitchell J, Soga K, Mitchell J et al (1976) Fundamentals of soil behavior. Wiley, New York

Nelson J, Miller D (1997) Expansive soils: problems and practice in foundation and pavement engineering. John Wiley, New York

Prat P, Ledesma A, Lakshmikantha M (2006) Size effect in the cracking of drying soil. Fract Nano Eng Mat Struct, 1373–1374

Rayhani M (2007) Desiccation-induced cracking and its effect on the hydraulic conductive. Can Geotech J 44(3):276–283

Susanga C, Jayantha K, Benjamin S (2013) Salient factors controlling desiccation cracking of clay in laboratory experiments. Géotechnique 63(1):18–29

Wang L, Tang C, Shi B et al (2018) Nucleation and propagation mechanisms of soil desiccation cracks. Eng Geol 238:27–35

Yesiller N, Miller C, Inci G et al (2000) Desiccation and cracking behavior of three compacted landfill liner soils. Eng Geol 57(1–2):105–121

Chapter 46
Characterization and Simulation of Swelling Pressure Kinetics for Compacted Bentonite Hydrated by Alkaline Solutions

Zhangrong Liu, Weimin Ye, Linyong Cui, Qiong Wang, and Yonggui Chen

Abstract The swelling pressure (SP) of bentonite material is of great interest for geological disposal of high-level radioactive waste (HLW). Once emplaced as engineering barrier in a HLW repository, the bentonite material would be subjected to infiltration of aggressive alkaline groundwater, resulting in distinctive SP kinetics dramatically different from those hydrated by deionized water. In this study, fifteen SP kinetics of compacted bentonite were collected from literature and divided into three types. For bentonite hydrated by deionized water, the SP ascends continuously with time to a final steady value (type I), or undergoes a mid-stage collapse before reaching a final stable value (type II). While for bentonite hydrated by alkaline solutions, the SP develops following a trend similar to type II and then turn to long-term degradation (type III). It was identified that the SP kinetics was co-governed by the temporal evolutions of accumulated and dissipated wedging pressures (AWP and DWP) within the sample. The AWP resulted from the hydration of montmorillonite and increased continuously with infiltration time. The DWP comprised two components, namely the RDWP attributed to the rearrangement of fabric and the CDWP related to the chemical effects of alkaline solutions. Over time, the RDWP first increased then decreased while the CDWP increased continuously. Accordingly,

Z. Liu (✉) · W. Ye · L. Cui · Q. Wang · Y. Chen
Department of Geotechnical Engineering, College of Civil Engineering, Tongji University, Shanghai, China
e-mail: liuzr@tongji.edu.cn

W. Ye
e-mail: ye_tju@tongji.edu.cn

L. Cui
e-mail: 1510271@tongji.edu.cn

Q. Wang
e-mail: qiong.wang@tongji.edu.cn

Y. Chen
e-mail: cyg@tongji.edu.cn

S. Wang et al. (eds.), *Engineering Geology for a Habitable Earth: IAEG XIV Congress 2023 Proceedings, Chengdu, China*, Environmental Science and Engineering,
https://doi.org/10.1007/978-981-99-9057-3_46

three equations in forms of exponential, Weibull and logistic functions of time were proposed for describing the AWP, RDWP and CDWP, respectively. A simple and practical SP kinetics model was then established as superposition of the three equations and validated against experimental data in literature. Results indicated that the proposed model was capable of satisfactorily reproducing all the three types of SP kinetics ($R^2 > 0.96$) and could be a useful tool for estimating the temporal evolution of swelling pressure during the long-term operation of a HLW repository.

Keywords Bentonite · Swelling pressure kinetics · Alkaline solutions · Wedging pressure · Model

46.1 Introduction

Compacted bentonite is world widely considered as one of the candidate engineered barrier materials for geological disposal of high-level radioactive waste (HLW), thanks to its strong adsorption capacity, low permeability and good swelling ability (Lloret and Villar 2007; Ye et al. 2010). In China, for example, the Gaomiaozi (GMZ) bentonite extracted from the Inner Mongolia Autonomous Region is selected as a buffer/backfill material for China's HLW disposal program. The swelling ability of compacted bentonite is usually characterized by swelling pressure (SP), which is defined as the pressure required to avoid any volume change for soils upon hydration (Sridharan et al. 1986). It is a critical parameter of bentonite as suggested by many HLW concepts such as SKB-3H (Oy 2018). The swelling pressure indicates the sealing performance of bentonite to retard water infiltration and nuclide leakage. The swelling pressure can also indicate the magnitude of hydraulic conductivity to some extent, because they are negatively correlated, i.e., the larger the swelling pressure, the lower the hydraulic conductivity (Cui 2017). Therefore, understanding the SP of compacted bentonite is important for designing and constructing a HLW repository.

Once emplaced in a HLW repository, the initial unsaturated bentonite would be subjected to infiltration of ground water and hydrated under almost constant volume condition, accompanied by progressive development of swelling pressure (Liu et al. 2022a). However, overtime this swelling pressure could decrease sharply to a low or even zero value, due to the montmorillonite in the bentonite could be solved or altered gradually by highly alkaline solution released from concrete degradation (Karnland et al. 2007; Fernández et al. 2009; Liu et al. 2018; Chen et al. 2019). As a result, the temporal evolution of swelling pressure, or the swelling pressure kinetics, of compacted bentonite treated by alkaline solutions were observed to be greatly different from those hydrated by deionized water, as will be detailed in Sect. 46.2.

In fact, due to its low permeability, the saturation of compacted bentonite would be a long-term process. This means that during a long period of time, the bentonite is unsaturated and thus the swelling pressure is incompletely developed. This developing swelling pressure could be a critical factor for safety assessment of early operation stage of the HLW repository. Therefore, except for the final swelling pressure at saturation, the swelling pressure kinetics during hydration are also of great interest.

Extensive models have been established for estimating the final swelling pressure, such as empirical models that simply relate the SP to the dry density or effective montmorillonite dry density of bentonite (Ye et al. 2010; Cui et al. 2012; Wang et al. 2012), physic-chemical models that represent SP as a result of crystalline and osmotic swelling of parallelly stacked montmorillonite platelets (Tripathy et al. 2004; Schanz and Tripathy 2009; Liu 2013), thermodynamic models that take account of the relationship between SP and hydraulic state parameter (e.g., suction) of the specimen (Erzin and Erol 2007; Agus and Schanz 2008) and hydro-mechanical models, including the modified Cam-Clay model (Roscoe and Burland 1968), the Barcelona Basic Model (Alonso et al. 1990), the Barcelona Expansive Model (Alonso et al. 1999) and the elasto-plastic model (Sun et al. 2017), etc. However, to the best of our knowledge, none of these models are capable of predicting the SP kinetics, except for the hydro-mechanical models that solving a complicated coupled process of SP development, structural evolution and water transfer. Recently, an empirical SP kinetics model on this respect was established by Liu et al. (2022a, b). In this model, the SP kinetics was considered as a result of competition between the accumulation and dissipation of wedging pressures (AWP and DWP) at micro scale. Where, the wedging pressure is the stress between montmorillonite platelets or stacks. It was assumed the temporal evolutions of AWP and DWP follow an exponential function and a Weibull distribution, respectively (Liu et al. 2022a, b). The SP was simply estimated by summing up the AWP and DWP. Although this model were verified against various SP kinetics of bentonite materials hydrated by deionized water and saline solutions, unfortunately, it failed to the unique SP kinetics of bentonite hydrated by alkaline solutions, due to a lack of consideration on the second decline of SP induced by the chemical interactions between the bentonite and the alkaline solutions. Consequently, the Liu et al. (2022a, b) model can hardly predicting the long-term evolution of SP for bentonite in a real HLW repository.

The object of this paper is to establish a new SP kinetics model for compacted bentonite hydrated by alkaline solutions. Firstly, fifteen SP kinetics were collected from literature and divided into three types according to their shapes. Then, the Liu et al. (2022a, b) model was introduced with consideration of the competition between the accumulation and dissipation of wedging pressures (AWP and DWP). Afterwards, a new SP kinetics model was developed with consideration of the chemical effects of alkaline solutions. Finally, the new model was validated against experimental data borrowed from literature. Meanwhile, the variation of model parameters with properties of specimen (dry density and initial water content) and infiltration liquid (solution concertation) were analysed.

46.2 Characterization of Three Types of SP Kinetics

Experimental SP kinetics of different bentonite are collected from literature. The specifications of each specimen are listed in Table 46.1. It appears that the SP kinetics not only depend on the properties of the specimen like dry density and initial water content, but also influenced by properties of infiltration liquid. Nevertheless, according to their shapes, the SP kinetics can be divided into three types, i.e., type I, II and III, respectively (Figs. 46.1, 46.2, and 46.3).

For type I, during hydration process the SP increases continuously with time to a final steady value (Fig. 46.1). This type of SP kinetics is commonly observed on

Table 46.1 Specifications of specimens with different types of SP kinetics

Specimen no.	Materials	Dry density (Mg/m^3)	Initial water content (%)	Solution	References
I-1	GMZ-Na-B[a]	1.70	18.4	DIW[b]	Liu (2016)
I-2	Ca-Mg-B	1.50	8.1	DIW	Jadda and Bag (2020)
I-3	Na-B	2.04	6.4	DIW	Komine and Ogata (1994)
I-4	Na-B	1.89	17.2	DIW	Komine and Ogata (1994)
I-5	Calcigel-B	1.58	20.0	DIW	Lang et al. (2019)
II-1	Korean-B	1.70	N/A[c]	DIW	Lee et al. (2012)
II-2	Korean-B	1.60	N/A	DIW	Lee et al. (2012)
II-3	Korean-B	1.50	N/A	DIW	Lee et al. (2012)
II-4	GMZ-Na-B	1.70	10.6	DIW	Ye et al. (2013)
II-5	GMZ-Na-B	1.65	10.2	DIW	Ye et al. (2020)
III-1	GMZ-Na-B	1.70	9.37	YCW[d]	Sun et al. (2018)
III-2	GMZ-Na-B	1.70	11.92	0.1 M NaOH	Liu et al. (2018)
III-3	GMZ-Na-B	1.70	11.92	0.3 M NaOH	Liu et al. (2018)
III-4	GMZ-Na-B	1.70	11.92	0.6 M NaOH	Liu et al. (2018)
III-5	GMZ-Na-B	1.70	11.92	1.0 M NaOH	Liu et al. (2018)

Note [a] bentonite; [b] deionized water; [c] not available; [d] young cement water

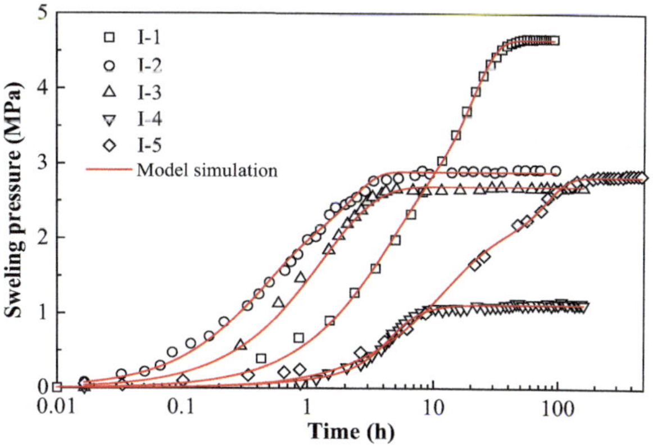

Fig. 46.1 Comparisons of model simulated and experiment measured SP kinetics (type I)

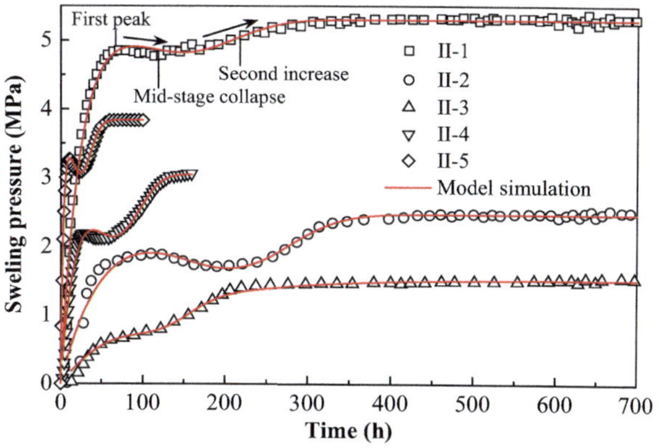

Fig. 46.2 Comparisons of model simulated and experiment measured SP kinetics (type II)

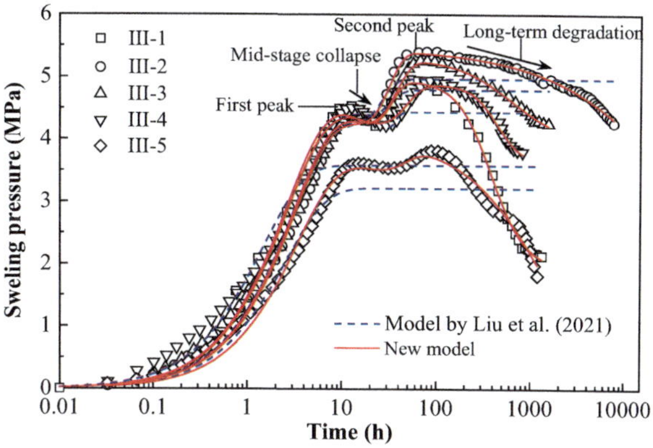

Fig. 46.3 Comparisons of model simulated and experiment measured SP kinetics (type III)

compacted bentonite with low density or high initial water content (Table 46.1). This is because that low-density bentonite contains large amount of inter-aggregate voids for accommodating swelling and exfoliation of aggregates, resulting in a slow development of SP and absence of mid-stage collapse of SP. With high initial water content, the montmorillonite in compacted bentonite was pre-hydrated and the aggregates could be partially exfoliated before the swelling pressure test. This also leads to a slow increase of SP and a lower final SP than those with low initial water content (Fig. 46.1).

For type II, during hydration process the SP firstly increases quickly to a peak value, followed by slight decrease or increase before increasing again. The final SP is larger than the first peak value (Fig. 46.2). This type of SP kinetics is commonly observed on highly compacted bentonite with relatively low initial water content (Table 46.1). With high density, the aggregates in bentonite are densely arranged and swell to compress each other once subjected to water, leading to rapid increase of SP once subjected to hydration (Fig. 46.2). With low initial water content, the montmorillonite could adsorbed up to four layers of tightly arranged water molecules (Saiyouri et al. 2004), causing exfoliation of aggregates and rearrangement of micro-structure of the bentonite and thus resulting in a mid-stage collapse or plateau. Afterwards, the SP increase again due to the contribution of osmotic swelling.

For type III, the SP curve is characterized by two distinct peaks, i.e., there is a mid-stage collapse SP between the two peaks. Moreover, after the second peak the SP decreases continuously to a very low or even zero value (Fig. 46.3). This type of SP kinetics is commonly observed on compacted bentonite hydrated by alkaline solutions (Table 46.1). The development process, including a mid-stage collapse, of SP before its second peak is much like to that of the type II, while the long-term degradation of SP after the second peak may be attributed to the alteration of mineral compositions and micro-structure of the bentonite by the alkaline solutions (Liu et al., 2018; Sun et al. 2018; Chen et al. 2019).

46.3 Model Development

Recently, an empirical model was established by Liu et al. (2022a, b) for modelling the SP kinetics of bentonite and bentonite-based materials. In this model, it was considered that the SP kinetics was resulted from competition between the accumulation and dissipation of wedging pressures (AWP and DWP) within the water-wetting specimen. Wedging pressure means the pressure pushing away the initial tightly-layered montmorillonite during adsorption of water into the inter-layer voids or the DDL domains (Liu et al. 2022a, b). The AWP resulted from the hydration of montmorillonite and increased continuously with infiltration time. The DWP was attributed to the rearrangement of fabric during hydration, which kept growing before a critical water content and afterwards disappeared gradually. Accordingly, it was assumed the temporal evolution of AWP follows an exponential function while that of the DWP obeys to the Weibull distribution (Liu et al. 2022a, b). Thus, the SP was

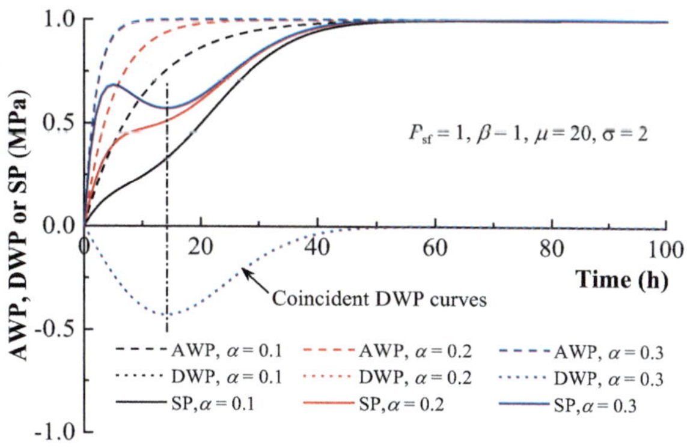

Fig. 46.4 Conceptual diagram of Liu et al. (2022a) model

predicted by summing up the AWP and the DWP:

$$P_s = P_{sf}[1 - \exp(-\alpha t)] - \beta\left(\frac{t}{\mu}\right)^{\sigma - 1}\exp\left[-\left(\frac{t}{\mu}\right)^{\sigma}\right] \tag{46.1}$$

where, P_s is predicted SP. The two terms on the left side of Eq. (46.1) stand for AWP and DWP, respectively. There are five parameters in total. P_{sf} is the final swelling pressure. α is related to the increasing rate of AWP. β, μ and σ are parameters that control the shape and scale of the DWP curve. According to definition, σ is always larger than 1 (Fig. 46.4). The negative sign '−' before β indicates that the contribution of DWP to the SP is negative.

As shown in Fig. 46.4, at a given time, the SP keeps increasing when the AWP dominate against the DWP, otherwise it decreases with time due to the dominated DWP. Moreover, the time when the DWP reaches its peak value just corresponds to the point where the SP reaches a minimum value before increasing again. This model was proved to be able to reproduce the SP kinetics of types I and II satisfactorily with $R^2 > 0.95$ (Liu et al. 2022a, b). However, for the SP kinetics of type III, significant discrepancies between the simulations and measurements were observed (Fig. 46.3). With further insights into the Liu et al. (2022a, b) model (Eq. 46.1), it seems that such discrepancies mostly originate from an inherent defect of the model. As plotted in Fig. 46.4, the Liu et al. (2022a, b) model (Eq. 46.1) can only describe SP kinetics approaching a final steady value, with either one or none mid-stage collapse. This is not a surprise, because in Eq. (46.1), there are always $\exp(-\alpha t) > 0$ and DWP > 0, leading to a predicted P_s not greater than final steady SP (P_{sf}). While for SP kinetics of type III, there is no final steady SP (P_{sf}) and the SP keeps decreasing after the second peak (Fig. 46.3). Therefore, the Liu et al. (2022a, b) model (Eq. 46.1) is incapable of modelling the SP kinetics of type III.

In order to improve the performance of the Liu et al. (2022a, b) model (Eq. 46.1) for modelling the SP kinetics of type III, the long-term degradation of SP should be taken into consideration. It is noted that after the second peak, the SP decrease continuously to a lower value (Fig. 46.3). This decreasing trend, can simply described using a logistic function which is commonly used for characterizing monotonic increasing or decreasing curves. The new model is written as follows:

$$P_s = P_{sp}[1 - \exp(-\alpha t)] - \beta\left(\frac{t}{\mu}\right)^{\sigma-1} \exp\left[-\left(\frac{t}{\mu}\right)^{\sigma}\right] - P_d\left(\frac{1}{1+(t/m)^n} - 1\right) \tag{46.2}$$

where, P_{sp} is the SP at the second peak, P_d is the expected long-term degraded SP after the second peak, m and n are parameters related to the rate of long-term degradation of SP. The smaller the m and the larger the n, the faster the degradation of SP. The third term on the left side of Eq. (46.2) is the newly added logistic function for describing the long-term degradation of SP after the second peak. To prevent confusion, in the new model, the first, second and third terms on the left side of Eq. (46.2) are denoted as AWP (accumulated wedging pressure), RDWP (dissipated wedging pressure induced by fabric-rearrangement) and CDWP (dissipated wedging pressure induced by chemical effects), respectively.

Totally, eight parameters are involved in the new model and all of them have definite physical meanings. A conceptual diagram of the new model is plotted in Fig. 46.5. The contribution of each parameter to the shape or scale of SP, AWP, RDWP and CDWP are marked in the figure. As a modification to the Liu et al. (2022a, b) model, the CDWP represents the long-term degradation of SP due to interactions between the bentonite and the alkaline solutions. From the perspective of morphology, it appears that the newly proposed model can well represent a SP kinetics of type III.

Fig. 46.5 Conceptual diagram of proposed model

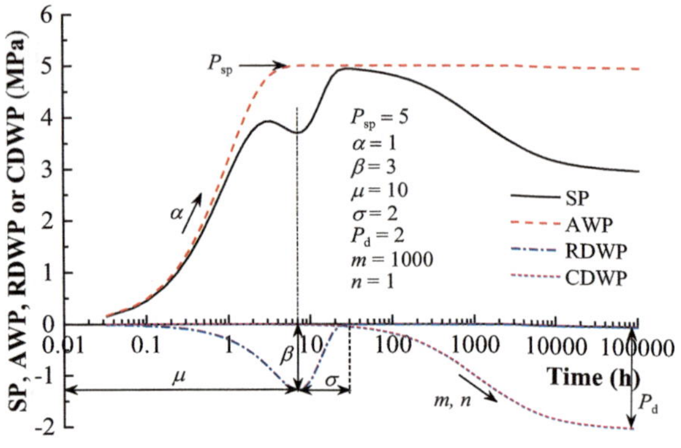

46.4 Model Verification

Fifteen SP kinetics of type I, II and III were collected from literature for verification of the proposed model (Table 46.1). Comparisons of model simulated and experiment measured SP kinetics are shown in Figs. 46.1, 46.2, and 46.3. It appears that the proposed model can reproduce all the three types of SP kinetics with high consistency ($R^2 > 0.96$). Particularly, for SP kinetics of type III, the predicted results of the new model are in good consistent with the experimental measured ones ($R^2 > 0.98$).

The model parameters for reproducing different SP kinetics are listed in Table 46.2. The values of each parameters are highly dependent on specifications of specimens. For a given bentonite, the higher the dry density, the larger the P_{sp} and α. The value of β, however, depends on both dry density and initial water content. The value of μ increases with the increase of the specimen height. Generally, for a given bentonite, the value of σ varies slightly within a narrow range. Considering III-2, III-3 and III-4, it seems that with the increase of solution concertration, the values of P_d and m decrease while the value of n increase. However, opposite trends were observed on parameters of III-5. Such contradiction could be attributed to measurement errors during experiments. Still, the value of m roughly indicates the half-period of the long-term degradation. Therefore, the proposed model can also used for predicting the final SP after long-term degradation.

Table 46.2 Model parameters for reproducing different SP kinetics

Specimen no.	P_{sp} (MPa)	α (h^{-1})	β (MPa)	μ (h)	σ	P_d (MPa)	m (h)	n	R^2
I-1	4.65	0.18	2.43	16.85	1.94	–	–	–	0.99
I-2	2.89	1.89	1.15	1.86	1.88	–	–	–	0.99
I-3	2.67	1.84	1.48	1.71	1.57	–	–	–	0.99
I-4	1.12	0.29	0.64	3.51	2.02	–	–	–	0.99
I-5	2.83	0.07	1.33	62.02	1.92	–	–	–	0.99
II-1	5.30	0.04	1.20	172.84	3.07	–	–	–	0.98
II-2	2.47	0.02	1.93	230.10	3.55	–	–	–	0.96
II-3	1.53	0.01	0.69	134.24	3.80	–	–	–	0.99
II-4	3.05	0.05	2.10	78.76	3.09	–	–	–	0.98
II-5	3.84	0.28	1.94	29.82	2.54	–	–	–	0.99
III-1	4.95	0.36	1.65	25.17	2.38	3.00	381.21	2.39	0.99
III-2	5.46	0.26	2.58	23.39	2.10	5.14	60,757.50	0.62	0.99
III-3	5.29	0.28	2.29	27.72	1.99	1.35	620.97	1.49	0.99
III-4	4.88	0.36	1.40	35.58	1.91	1.27	439.81	3.19	0.98
III-5	3.98	0.28	0.88	35.38	1.88	4.68	1607.23	0.98	0.98

46.5 Conclusions

In this study, fifteen swelling pressure (SP) kinetics of different bentonite are collected from literature and analysed. Based on the Liu et al. (2022a, b) model, a new SP kinetics model was proposed and validated against the fifteen collected SP kinetics. Based on the results obtained in this study, the following conclusions can be drawn.

The SP kinetics of compacted bentonite can be divided into three types. Type I, the SP keeps ascending with time and finally reaches a steady value. Type II, the SP firstly raises rapidly to a peak value, followed by slight decrease or increase before increasing again to a final stable value. Type III, the SP develops following a trend similar to type II and then turn to long-term degradation. Type I and II belongs to bentonite hydrated by deionized water, while type III is observed on bentonite hydrated by alkaline solutions.

The SP kinetics is co-governed by the temporal evolutions of accumulated and dissipated wedging pressures (AWP and DWP) within the sample. The AWP results from the hydration of montmorillonite, while the DWP is contributed by the rearrangement of fabric (RDWP) and the chemical effects of alkaline solutions (CDWP).

A new SP kinetics model was established by superposing the AWP, RDWP and DWP, which were expressed using an exponential, Weibull and logistic functions, respectively. Verification against the fifteen SP kinetics borrowed from literature indicted that the proposed model was able to reproduce all the three types of SP kinetics satisfactorily ($R^2 > 0.96$). Therefore, the proposed model could be a useful tool for estimating the temporal evolution of swelling pressure during the long-term operation of a HLW repository.

Acknowledgements The National Nature Science Foundation of China (42002291 and 42030714) and the China Postdoctoral Science Foundation (2020M671217) are greatly acknowledged.

References

Agus SS, Schanz T (2008) A method for predicting swelling pressure of compacted bentonites. Acta Geotech 3:125–137

Alonso EE, Gens A, Josa A (1990) A constitutive model for partially saturated soils. Géotechnique 40(3):1751–7656

Alonso EE, Vaunat J, Gens A (1999) Modelling the mechanical behaviour of expansive clays. Eng Geol 54(1–2):173–183

Cui YJ (2017) On the hydro-mechanical behaviour of MX80 bentonite-based materials. J Rock Mech Geotech 9(3):565–574

Cui SL, Zhang HY, Zhang M (2012) Swelling characteristics of compacted GMZ bentonite–sand mixtures as a buffer/backfill material in China. Eng Geol 141–142:65–73

Chen YG, Sun Z, Cui YJ, Ye WM, Liu QH (2019) Effect of cement solutions on the swelling pressure of compacted GMZ bentonite at different temperatures. Constr Build Mater 229:116872

Erzin Y, Erol O (2007) Swell pressure prediction by suction methods. Eng Geol 92:133–145

Fernández R, Mäder UK, Rodríguez M, de la Villa VR, Cuevas J (2009) Alteration of compacted bentonite by diffusion of highly alkaline solutions. Eur J Mineral 21:725–735

Jadda K, Bag R (2020) Effect of initial compaction pressure and elevated temperature on swelling pressure of two Indian bentonites. Environ Earth Sci 79:197

Karnland O, Olsson S, Nilsson U, Sellin P (2007) Experimentally determined swelling pressures and geochemical interactions of compacted Wyoming bentonite with highly alkaline solutions. Phys Chem Earth 32:275–286

Komine H, Ogata N (1994) Experimental study on swelling characteristics of compacted bentonite. Can Geotech J 31(4):478–490

Lang LZ, Tripathy S, Baille W, Schanz T, Sridharan A (2019) Linkage between swelling pressure, total suction of saturated bentonites and suction of saturating aqueous solutions. Appl Clay Sci 171:82–91

Lee JO, Lim JG, Kang IM, Kwon S (2012) Swelling pressures of compacted Ca-bentonite. Eng Geol 129–130:20–26

Liu LC (2013) Prediction of swelling pressure of different types of bentonite in dilute solutions. Colloid Surf A-Physicochem Eng Asp 434:303–318

Liu Y (2016) Investigation on the swelling properties and microstructure mechanism of compacted gaomiaozi bentonite. J Eng Geol 24(3):451–458

Liu LN, Chen YG, Ye WM, Cui YJ, Wu DB (2018) Effects of hyperalkaline solutions on the swelling pressure of compacted Gaomiaozi (GMZ) bentonite from the viewpoint of Na+ cations and OH– anions. Appl Clay Sci 161:334–342

Liu ZR, Ye WM, Cui YJ, Zhu HH, Wang Q, Chen YG (2022a) Development of swelling pressure for pellet mixture and compacted block of GMZ bentonite. Constr Build Mater 301:124080

Liu ZR, Ye WM, Cui YJ, Zhu HH, Wang Q, Chen YG (2022b) An empirical swelling pressure kinetics model for bentonite and bentonite-based materials hydrated under constant volume condition. Int J Geomech 22(4):04022024

Lloret A, Villar MV (2007) Advances on the knowledge of the thermo-hydro-mechanical behaviour of heavily compacted "'FEBEX"' bentonite. Phys Chem Earth 32:701–715

Oy P (2018) KBS-3H—design, production and initial state of buffer and filling components. Posiva, Olkiluoto

Roscoe KH, Burland JB (1968) On the generalized stress-strain behavior of 'wet' clay. In: Heyman J (ed) Engineering plasticity. Cambridge University Press, Cambridge, Leckie FA, pp 535–609

Saiyouri N, Tessier D, Hicher PY (2004) Experimental study of swelling in unsaturated compacted clays. Clay Min 39(4):469–479

Schanz T, Tripathy S (2009) Swelling pressure of a divalent-rich bentonite: diffuse double-layer theory revisited. Water Resour Res 45(2):W00C12

Sridharan A, Rao AS, Sivapullaiah PV (1986) Swelling pressure of clays. Geotech Test J 9(1):24–33

Sun WJ, Zong FY, Sun DA, Wei ZF, Schanz T, Behzad F (2017) Swelling prediction of bentonite-sand mixtures in the full range of sand content. Eng Geol 222(18):146–155

Sun Z, Chen YG, Cui YJ, Xu HD, Ye WM, Wu DB (2018) Effect of synthetic water and cement solutions on the swelling pressure of compacted Gaomiaozi (GMZ) bentonite: the Beishan site case, Gansu, China. Eng Geol 244:66–74

Tripathy S, Sridharan A, Schanz T (2004) Swelling pressures of compacted bentonites from diffuse double layer theory. Can Geotech J 41:437–450

Wang Q, Tang AM, Cui YJ, Delage P, Gatmiri B (2012) Experimental study on the swelling behaviour of bentonite/claystone mixture. Eng Geol 124:59–66

Ye WM, Chen YG, Chen B, Wang Q, Wang J (2010) Advances on the knowledge of the buffer/backfill properties of heavily-compacted GMZ bentonite. Eng Geol 116(1–2):12–20

Ye WM, Wan M, Chen B, Chen YG, Cui YJ, Wang J (2013) Temperature effects on the swelling pressure and saturated hydraulic conductivity of the compacted GMZ01 bentonite. Environ Earth Sci 68:281–288

Ye WM, Liu ZR, Cui YJ, Zhang Z, Wang Q, Chen YG (2020) Features and modelling of time-evolution curves of swelling pressure of bentonite. Chin J Geotech Eng 42(1):29–36

Chapter 47
Influence of Initial Moisture Content on the Slaking Features of Expansive Red Clay

Guodong Zhang, Sixiang Ling, and Xiyong Wu

Abstract This paper is aimed to characterize the slaking features of the expansive red clay under the different initial moisture content condition. The slaking experiments of weak expansive red clay were carried out in laboratory. The slaking difference was explained by video photographs, optical microscope photograph, and scanning electron microscope images. The slaking and fractal features of expansive red clay were analyzed under the different initial moisture content of 3%, 8%, 13% and 18%, respectively. The results show that the slaking time is earlier with its lower initial moisture content. The slaking rate is also increased with decreasing initial moisture content. The fissure development degree of expansive red clay in slaking test has presented rapid increase, slow increase, and stable trends characteristic with the increasing soaking time. Under the condition of minimum moisture content in 3%, the maximum fissure development degree can reach 4.6. The index value of maximum fissure width increased in the first half of immersion time, and oscillatory decreased in the second half of immersion time. Macroscopically, the slaking of expansive red clay is more likely to occur under the condition of low moisture content, which is mainly due to the scattered agglomeration units and high percentage of micropores in the expansive red clay. The fractal dimension is positively correlated as a quadratic function with the initial moisture content, which resulted in the slaking more complete and slaking time shorter with lower fractal dimension.

Keywords Expansive red clay · Slaking and deterioration · Initial moisture content · Basic unit body · Fractal dimension

G. Zhang · S. Ling · X. Wu (✉)
Faculty of Geosciences and Engineering, Southwest Jiaotong University, Chengdu 611756, China
e-mail: wuxiyong@126.com

S. Ling
e-mail: lingsx@swjtu.edu.cn

S. Wang et al. (eds.), *Engineering Geology for a Habitable Earth: IAEG XIV Congress 2023 Proceedings, Chengdu, China*, Environmental Science and Engineering,
https://doi.org/10.1007/978-981-99-9057-3_47

47.1 Introduction

Expansive soil is the special soil which is mainly composed of montmorillonite, illite, chlorite and other hydrophilic clay minerals and mixed layer. It is widely distributed around the world and has many characteristics including disintegration, softening, and expansion when the soil absorbs water, whereas shrinkage and cracking when the soil loses water (Ito and Azma 2009; Chen et al. 2014; Gao et al. 2015; Leng et al. 2018). The expansive soil has become an unavoidable soil type for geotechnical engineering construction in expansive soil region. The safety of buildings, roads, railways and other kinds of engineering structures and their geotechnical auxiliary structures, underground pipelines and channel slopes are easy to be destroyed by expansive soil. For example, at the beginning of 21th century, China government planned to build 21,000 km of expressways, nearly 16% of which involved expansive soil hazards. In the middle route of the South-to-North Water Diversion Project, a large number of geological problems such as channel collapse have occurred at expansive soil regions (Wang et al. 2008; Zhang et al. 2015). Meanwhile, the deformation and destruction of expansive soil has caused non-negligible economic losses, even threat to the human life, such as the expansion in the high railway corridor (George et al. 2004; Fredlund et al. 2012; Chen et al. 2014).

At present, many scholars have conducted a large number of experimental studies on the disintegration and microscopic characteristics of expansive soil. (1) Regarding the macroscopic characterization and mechanical properties of soil disintegration, the influencing factors of soil disintegration include moisture content, pore pressure, temperature conditions, etc. (Cernuda et al. 1954; Yamaguchi et al. 1990). For example, Zhou et al. (2019) designed a disintegration test device for red-bed weathered soil and studied the disintegration characteristics under different moisture content and compression conditions. The results show that the rate of soil disintegration will become faster as the initial moisture content and density of soil become lower. Li et al. (2020) simulated the unsaturated effective stress during the disintegration of granite residual soil using the Pore Algorithm (PM) method and found that the disintegration curve of residual soil has obvious two-stage characteristics. The initial disintegration curve is in the shape of "S". Meanwhile, the microstructure is easy to be destroyed when the internal tensile stress of the soil is greater than the inter-particle strength. Tong et al. (2018) analyzed the disintegration amount and disintegration time of sandstone granular soil materials are more sensitive to the changes of water moisture, while mudstone granular soil materials are more sensitive to the relative compactness. (2) Research on the influence of the microstructure of expansive soil on the macroscopic physical and mechanical properties (Smart 1966; Morgenstern and Tchalenko 1967). For example, Tan and Kong (2001) proposed a soil gel cementation structure model and explained its rationality from the perspective of microstructure aiming at the high moisture content, high void ratio, high plasticity, high bearing, and low compressibility. Qu et al. (2002) found that expansive soils have more significant expansion when they have more montmorillonite and illite/montmorillonite mixed layers. It can be seen that the characteristics of expansive soil under the conditions of moisture content, dry density, and compression were studied in previous studies, and the analysis of factors affecting disintegration and the macro-mechanical effects of soil disintegration were considered. However, the research on the disintegration difference of different initial moisture content still remains to be further explored.

With the development of highway and railway construction, engineering problems caused by expansive soil are also one of the important issues faced in the design and construction of railway engineering. Researches tend to pay more attention to engineering disasters caused by medium-strong expansive soil. Among them, the collapse of retaining wall in a section of Nanning-Kunming Railway was caused by expansive soil, which seriously affected the safety of railway operations (Su et al. 2020). In recent years, the construction of high-speed railways in China has been increasingly enhanced, and the high-speed railway ballastless track has extremely strict control of subgrade deformation. Nowadays, high-speed ballastless railways have often undergone a lot of research and improvement to cross the medium-strong expansive soil area, and to a certain extent avoid the arching disease of the roadbed. However, when the high-speed ballastless railway passes through the weakly expansive soil (or even non-expandable) area, the deterioration and disintegration of the expansive soil at the bottom of the high-speed railway subgrade or tunnel seriously affect the safety of high-speed trains (Xiao et al. 2016; Wang et al. 2020). Therefore, controlling the stability of the subgrade has become one of the main problems in high-speed railway construction. At present, the understanding of weak expansive soil is still lacking. Research on the disintegration of weak expansive soil of high-speed railway ballastless track subgrade can provide a scientific basis for the safety of high-speed railway construction and operation. This paper takes the undisturbed weakly expansive red clay in a newly built high-speed railway tunnel in Qingyang, Gansu Province as the research object. The soil samples were carries out the natural disintegration test under different initial moisture content, which can explores the changes of disintegration deterioration parameters over time. The influence of microstructure on its physical and mechanical properties was analyzed. The research results of this paper can be extended to other areas with weakly expansive soils for high-speed railways, providing a reference for the design and construction of high-speed railway subgrades, especially for the construction of weakly expansive geotechnical conditions in arid areas of northern China.

47.2 Experiment

47.2.1 Basic Properties of Soil Samples

The experimental soil sample was taken from a tunnel of a high-speed railway in Qingyang, Gansu province. The soil sample has a light-red color, hard plastic, and compression structures, which has no obvious joints. Through X-ray diffraction, the illite accounted for 14.01% and the illite/montmorillonite mixed layer accounted

Table 47.1 Basic physical properties of red clay samples

Parameters	Mean value
Natural moisture content/(%)	13.0
Dry density/(g/cm^3)	1.91
Liquid limit/(%)	37.2
Plastic limit/(%)	18.8
Void ratio	0.52
Cohesion/(kPa)	38.9
Internal friction angle/(°)	21.2
CEC (NH^{4+})/(mmol/kg)	174
Free expansion rate/(%)	51
Montmorillonite content/(%)	14

for 3.99%. Some experimental data of red clay obtained by geotechnical tests are shown in Table 47.1. According to 'Code for Soil Test of Railway Engineering' (TB 10,038-2012) (2012), the soil samples has follow index: 40 ≤ free expansion rate Fs (%) < 60, 7 < montmorillonite content M (%) < 17, 170 < cation-exchange capacity CEC (NH^{4+}) (mmol/kg) < 260. Therefore, this red clay can be considered as weakly expansive soil. The particle distribution curve is shown in Fig. 47.1. The particles less than 2 mm are accounted for 98.03%. Meanwhile, particles less than 0.075 mm are accounted for 82.8%, which included powder particles with 20.43%, clay particles with 7.89%, and colloidal particles with 54.48%, respectively.

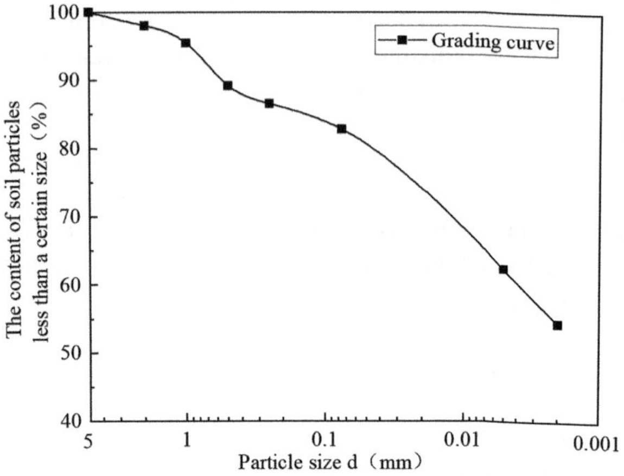

Fig. 47.1 Soil particle distribution curve

47.2.2 Sample Preparation and Initial Moisture Content Settings

The original red clay was sealed and transported back to the laboratory, and then made into a standard ring knife sample with a diameter of 50 mm and a height of 20 mm. In order to explore the influence of different initial moisture content on the disintegration of red clay, a 5% moisture content gradient was set up in this experiment, and the initial moisture content of standard soil samples were 3, 8, 13%, and 18%. In the process of setting different initial moisture content, it was found that the permeability of the red clay sample was poor, and it was difficult to ensure the uniform distribution of moisture inside the soil sample. Hence, in order to ensure the uniformity of the initial moisture content and the soil structure not to be damaged, the soil sample was wrapped in filter paper and buried in quartz sand in this test, and the internal moisture content of the soil sample was controlled by controlling the moisture content of the quartz sand. After being placed in the container, a film is used to cover and seal to prevent water loss. Based on the above, the soil samples were placed in quartz sand with different moisture content, and 4 moisture meters were placed at the center of the soil sample diameter plane, the midpoint of the center and the circumference, the circumference (soil surface) and the interior of the quartz sand respectively. The moisture content was monitored by these 4 moisture meters, and it was found that after the soil sample was placed in the hydrated quartz sand for about 65–80 h, the error of the monitored moisture content was less than 5%. Therefore, in the upcoming test, the soil sample was designed to be placed in quartz sand with different moisture content for 72 h (3 days), and it is believed that the initial moisture content can reach a uniform distribution state in the soil sample.

47.2.3 Experimental Method

The indoor static disintegration method is used for expansive red clay. First, soil samples with different initial moisture content will be put into an open container, and deionized water is added to completely immerse the sample. Secondly, the electron microscope was used to monitor the disintegration test process in real time. The time interval for taking pictures was set to 10 s. After adding distilled water to immerse the rock sample, the disintegrating soil was tracked by video-monitored and photographed microscope. Finally, a typical photograph is selected and analyzed by Image-Pro Plus (IPP) software to obtain the development of cracks in the soil sample during the disintegration of the sample.

47.3 Results and Analysis

47.3.1 Influence of Initial Moisture Content on Disintegration Characteristics

The disintegration test was conducted in the different moisture content, which can explore the natural disintegration pattern of the undisturbed red clay in immersion water. Here, the disintegration photographs under the 3% initial moisture content as an example is shown in Fig. 47.2. When the sample is immersed in deionized water, bubbles appear on the surface of the sample with small particles peeling off instantly, and then the first crack appears in the sample, which rapidly extends and expands to form an empty surface. The crack divides the sample into loose fragments. At the 40 min, the sample was completely degraded, which keep the disperse state with basically unchanged.

When the initial moisture content of red clay increased as 8%, the red clay disintegrated slowly after being immersed in water. After 3 min, the side wall of the sample appeared to fall off, and the first through crack appeared on the surface in 5 min. Subsequently, the sample slowly cracked on both sides of the crack, and completely disintegrated to pieces in 74 min. The sample with 13% moisture content has a slow disintegration rate after immersion in water, and the first crack appears at 240 min. There is no rock collapse and muddy during the disintegration process. The sample is relatively complete, and the crack no longer develops at 360 min. After the sample with 18% moisture content is immersed in water, except for the tiny bubbles on the surface, no particles will peel off and no cracks will be generated, and the final state will remain intact. It can be seen from the above test that the lower the initial moisture content of the sample, the earlier the start of its disintegration. The disintegration

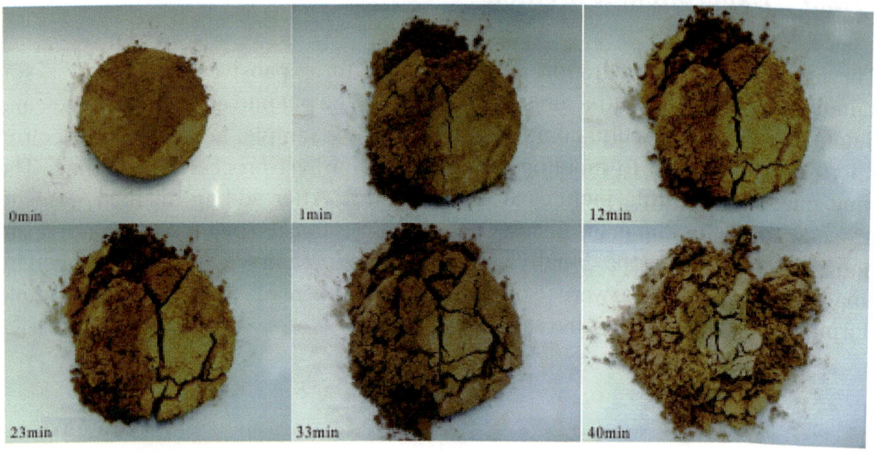

Fig. 47.2 Sample disintegration process under the 3% initial moisture content

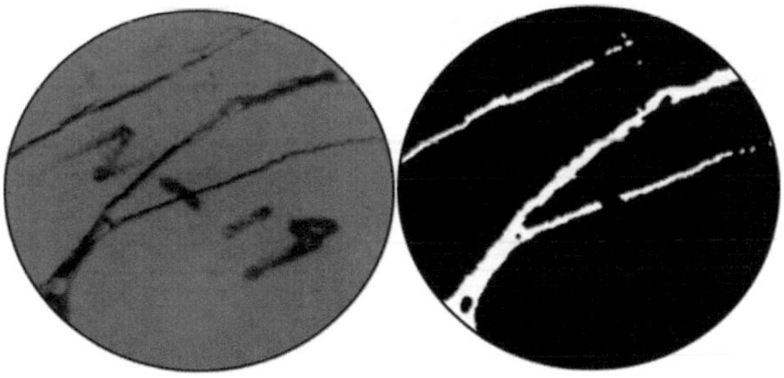

Fig. 47.3 Binary image of cracks

speed of the sample will become faster, and the final disintegration state will be more complete.

The disintegration images in different immersion time is selected to calculate the fissure development through the Image Pro Plus software, as shown in Fig. 47.3. Taking the fissure degree (in this paper, the fissure degree is defined as the ratio of the total fissure area to the original sample area before the sample is completely deteriorated) and the maximum fissure width as the quantitative indicators, the curve of the relationship between the fissure degree, the maximum fissure width and the disintegration time is obtained as shown in Fig. 47.4.

It can be seen from Fig. 47.4 that the fissure degrees of the samples at 3%, 8%, and 13% initial moisture content are 4.6%, 3.72%, and 0.5%, respectively. When the sample does not collapse and can basically maintain the original shape, the relationship curve between the degree of fissure and time shows an exponential change pattern, which displays a rapid increase in the early stage and a slow and stable law in the later stage. This rule is due to the rapid dissolution of soluble matter in the soil at the initial stage, and the anisotropy of the soil material with water entering the pores makes it unable to expand uniformly. Therefore, a large number of cracks are generated in the sample in a short period of time, and the slope of the curve is larger. In the later stage, as the pore space without water is gradually reduced, the slope of the curve becomes smaller. The maximum crack width at 3% moisture content and 8% moisture content gradually increases and then gradually decreases, as the test time progresses. The maximum crack reached 2.77 mm and 2.06 mm at 20 min and 40 min respectively. This article believes that the generation of later cracks has a certain squeezing effect on the initial cracks, which will cause the maximum crack width to gradually compress to a stable state, so the maximum crack width first increases and then decreases with the immersion time. However, the 13% moisture content sample did not undergo further disintegration and deterioration, so it maintained the maximum crack width 0.38 mm until the collapse was completed. Generally, the higher the initial moisture content of the sample, the smaller the crack degree and maximum crack width. When the moisture content reaches 18%, the fissure degree tends to zero.

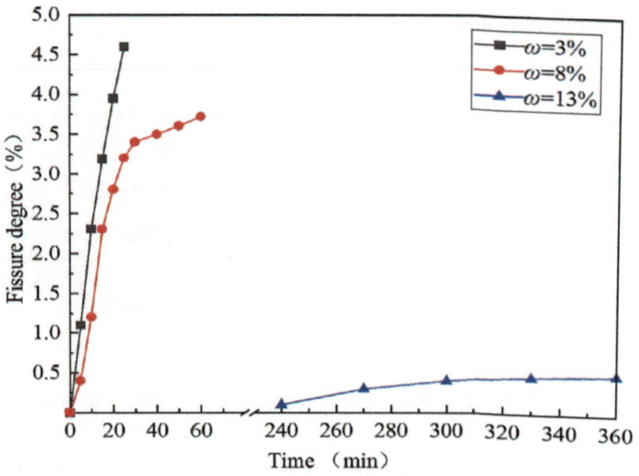

(a) The curve of fissure degree and time

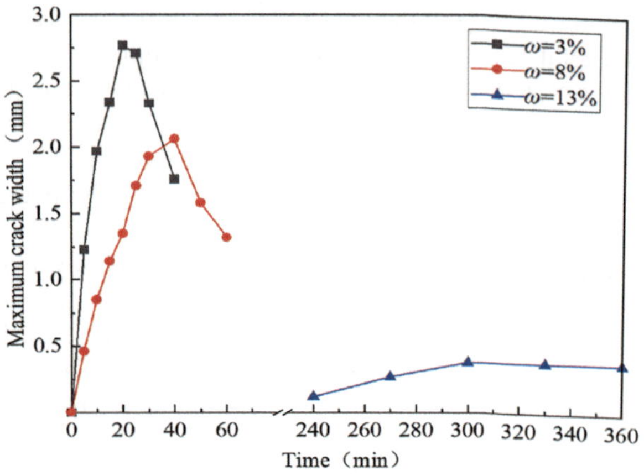

(b) The curve of maximum crack width and time

Fig. 47.4 The relationship between fracture index and time

47.3.2 Unit Characteristics Under Different Moisture Content

Generally, the changes in the macroscopic properties of soil are often controlled by its microstructure. The scanning electron microscope (SEM) was used to observe the surface microstructure of soil under different moisture content. Then, the pore structure and unit characteristics of the red clay can be analyzed from the microscopic images, which try to explain the disintegration difference under the different initial moisture content. Based on the recommendation of Tang et al. (2008), the scanning electron microscope magnification is commonly used as 1500 ± 300. Combining with the morphological features of the experimental soil sample, the 1700 magnification factor was finally used to analyze in IPP software. The microstructure of red clay in different initial moisture content is listed in Fig. 47.5. The microscopic

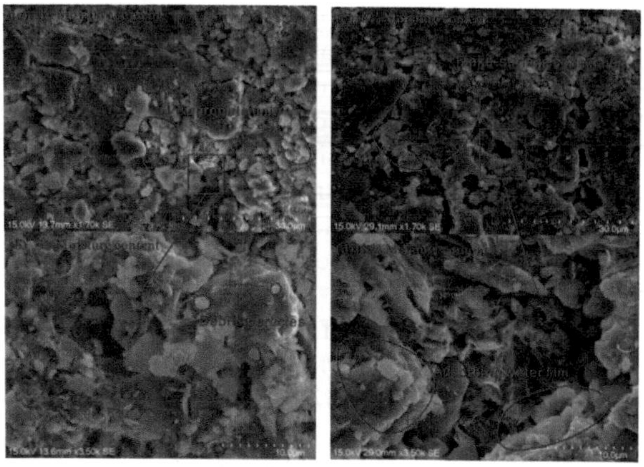

Fig. 47.5 Image of soil microstructure under different moisture content

image shows that the red clay is mainly connected by point-point, edge-edge, surface-surface contact and cemented connection. The connection strength of the basic unit body with the different connected way is lower than the internal unit body, which results in the basic unit is firstly fractured at the connection part (Tan et al. 1994). Therefore, the red clay in this paper has low strength and is prone to deformation. The surface agglomerate units of the sample with low moisture content are more dispersed and there are more debris particles, which makes it easier for water molecules to enter the structure to complete the collapse of the structure. The debris particles decrease with increasing the moisture content of soil sample. Because the fine particles formed on the surface of the rock sample adsorb the water film to wrap the micropores and debris particles, the integral sheet-plane structure has a blocking effect on water molecules on the one hand. Besides, the gravitational action between the water films also makes it is difficult for water molecules to destroy the intergranular structure, and the rock sample finally assumes a relatively stable state.

47.3.3 Fractal Characteristics of Pores Under Different Moisture Content

Fractal theory is mainly employed to state the self-similarity of rock. When the soft rock is immersed and deteriorated, the distribution of disintegrating particles has statistical self-similarity. Therefore, the fractal dimension (*D*) can describe the disintegration process of the red clay (Su et al. 2005). The *D* can be calculated as follow:

$$D = -\lim_{\varepsilon \to 0} \frac{\ln N(\varepsilon)}{\ln \varepsilon} \tag{47.1}$$

where ε is the quantity or quality scale and $N(\varepsilon)$ is the measured value under this scale. The fractal dimension represents the complexity of the pore. The high value of the fractal dimension indicates that the pore structure will be more complex. Nevertheless, the soil mass does not have the same particle distribution characteristics as the rock mass after disintegration, and it is extremely difficult to quantify the disintegration particle distribution. Hence, this paper proposes to investigate the change of pore fractal dimension under different moisture content by counting the pore size distribution before the sample disintegrates based on the quantitative fractal theory.

The scanned images of samples with different moisture content were quantitatively analyzed using IPP software, and the number of different pore diameters extracted is shown in Fig. 47.6. It can be seen from the statistical data that the number of pores is negatively correlated with the size of the pore size, that is, the number of corresponding pores shows a downward trend as the pore size increases. However, as the moisture content increases, some of the fine particles formed on the surface of the rock sample adsorb the water film to wrap the micropores, and then form an integral sheet-plane structure. The existence of this structure also makes part of the pores with smaller pores transformed into pores with larger pores. Therefore, the relationship between the number of pores and the pores in the rock samples with higher moisture content also shows certain irregularities. In order to further explore the influence of pore structure on the disintegration characteristics of red clay under different moisture content, this paper attempts to describe the complexity of pores by using fractal dimension. Figure 47.6 shows that the number of small pores with a pore diameter of less than 10 μm decreases with increasing moisture content. In combination with the previous article, the number of small pores is larger when the moisture content is relatively small. From the perspective of microstructure, the water molecules will be more fully combined with clay minerals after entering the soil structure through the micropores. Then, the expansion potential energy will instantaneously release by the expansive minerals, resulting in the faster disintegration rate in red clay. When the number of large pores is large, water enters the soil along the large pores and cannot be completely combined with clay minerals, which causes the soil to disintegrate slowly. In terms of mechanics, the expansibility of the soil has a certain range. The expansibility of the soil will release part of it at high moisture content, which leads to a lower disintegration of the soil. The shrinkage of the soil results in a larger potential expansion force at low moisture content. Finally, the degree of disintegration will be more severe.

Assume that ε is a certain pore size, and $C(\varepsilon)$ is the number of pores smaller than the pore size. The functional relationship between $C(\varepsilon)$ and ε-D can be obtained by counting the $C(\varepsilon)$ of different ranges of ε, and the value of the fractal dimension D can be calculated based on formula (47.1). Perform a linear fitting under the double logarithmic curve for $C(\varepsilon)$ and ε. The slope value of the fitted curve is the fractal dimension D. The fitting result is shown in Fig. 47.7. The $\ln C(\varepsilon)$ and $\ln\varepsilon$ have a good linear relationship. The slope value of the curve becomes larger with the increase of moisture content, and this increasing interval has an obvious expanding trend. The

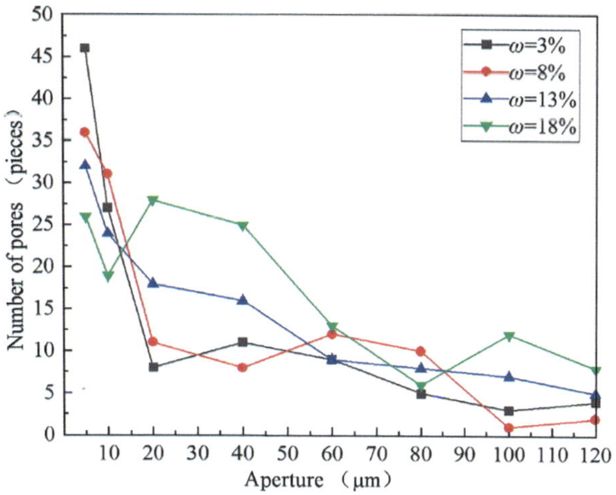

Fig. 47.6 The relationship between pore number and pore size

value of the fractal dimension at 18% moisture content reaches 0.5, which is twice the value of the fractal dimension at 3% moisture content.

Based on the data in Fig. 47.7, the correlation between fractal dimension and moisture content is established as shown in Fig. 47.8. There is an obvious quadratic polynomial positive correlation between the fractal dimension and the moisture content, and the relationship can be expressed as:

$$D = 0.06*10^{-2}\omega^2 + 0.38*10^{-2}\omega + 0.24 \quad (47.2)$$

where $R^2 = 0.980$. $0.06*10^{-2}$, $0.38*10^{-2}$, 0.24 are fitting polynomial coefficients, and their value depends on the change of moisture content. This paper concludes that the fractal dimension of pores tends to become lower when the initial moisture

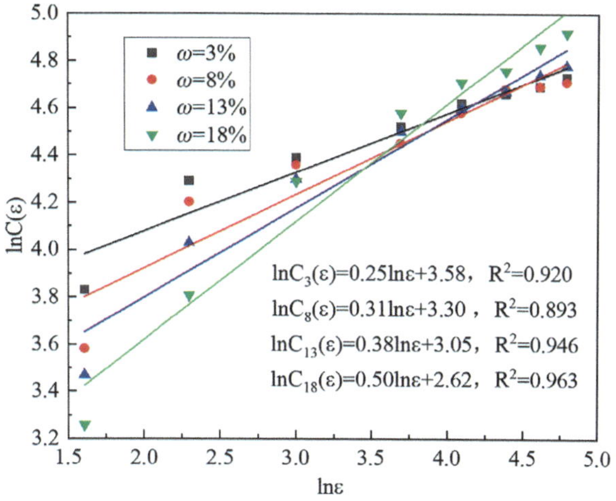

Fig. 47.7 Curve of pore characteristics under different moisture content

Fig. 47.8 Curve of fractal dimension varying with moisture content

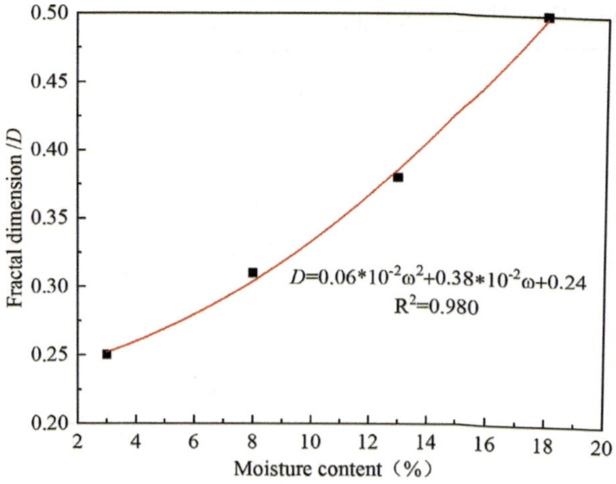

content of the sample is low. Combined with the disintegration characteristics of the sample under low moisture content, there is a positive correlation between the fractal dimension and the degree of disintegration of the sample: the decrease of the pore fractal dimension will cause the expansive soil to deteriorate more severely after encountering water. Therefore, the pore fractal dimension can be used as an evaluation factor for the disintegration susceptibility of expansive red clay when exposed to water. According to the microscopic pore distribution of the undisturbed soil of the subgrade where the project is located, the qualitative evaluation of the intensity of its disintegration has certain guiding significance in actual construction.

47.4 Conclusions

The disintegration features and microstructure of red clay was described through the disintegration experiment, SEM observation, and fractal dimension calculation. The conclusion can be drawn as follow:

(1) The time for the red clay to completely disintegrate and deteriorate is positively correlated with the moisture content, and the degree of cracks and the maximum crack width are also positively correlated with the moisture content. The fissure degrees of the red clay at 3% initial moisture content are 4.6%, and the maximum crack reaches 2.77 mm.

(2) The red clay is mainly connected by point-point, edge-edge, surface-surface contact and cemented connection. At low moisture content, the red clay units are characterized by relatively dispersed debris and aggregate units. As the moisture content increases, more sheet-to-face integral structures wrapped by the adsorbed water film appear inside, which increases the structural stability.

(3) The red clay has a relatively large pore size under low moisture content, and the pore fractal dimension D has a quadratic polynomial positive correlation with the moisture content. The degree of disintegration and degradation of red clay is positively correlated with the fractal dimension, and the disintegration of the soil is weaker when the fractal dimension is higher.

References

Cernuda CF, Smith RM, Vicente-Chandler J (1954) Influence of initial soil moisture condition on resistance of macroaggregates to slaking and water-drop impact. Soil Sci 77(1):19–28

Chen ZW, Jiang GL, Yuan ZH et al (2014) Field soaking tests on expansive soil foundation under low embankment. Chin J Rock Mech Eng 33(08):1609–1618

Fredlund DG, Rahardjo H, Fredlund M (2012) Unsaturated soils mechanics in engineering practice. Wiley, New York

Gao Y, Sun DA, Lv HB (2015) Deformation characteristics of weakly expansive soil and its prediction. Rock Soil Mech 36(03):755–761

George X, Panagiotis K, Ananias T (2004) Road construction onswelling soils: the case of Strymi soils, Rhodope, Thrace, Northern Greece. Bull Eng Geol Env 63:93–101

Ito M, Azma S (2009) Engineering characteristics of a glaciolacustrine clay deposit in a semi-arid climate. Bull Eng Geol Environ 68(4):551–557

Leng T, Tang CS, Xu D et al (2018) Advance on the engineering geological characteristics of expansive soil. J Eng Geol 26(01):112–128

Li CS, An R, Shu RJ et al (2020) Initial-disintegration analysis of granite residual soil and approximate simulation of mathematical morphology. Chin J Rock Mech Eng 39(04):845–854

Morgenstern NR, Tchalenko JS (1967) Microscopic structures in kaolin subjected to direct shear. Geotechnique 17(4):309–328

Qu YX, Zhang YS, Feng YY et al (2002) Quantitative study on the clay mineral composition of expansive soils in china. J Eng Geol 10(S):416–422

Smart P (1966) Optical microscopy and soil structure. Nature 210:1400–1406

Su YH, Zhao MH, Liu XM (2005) Research of fractal mechanism for swelling & collapse of soft rock. Rock Soil Mech 2005(05):728–732

Su PH, Chen SK, Zhao XY et al (2020) Model test on reinforcement mechanism of geogrid-steel tube expansive soil retaining wall for slope stability of Nankun railway. J Eng Geol 28(03):658–666

Tan LR, Kong LW (2001) Fundamental property and microstructure model of red clay. Chinese J Geotech Eng 2001(04):458–462

Tan LR, Zhang MY, Shao WM et al (1994) Microstructure characteristics and engineering properties of expansive soil. Chinese J Geotech Eng 1994(02):48–57

Tang CS, Shi B, Wang BJ (2008) Factors affecting analysis of soil microstructure using SEM. Chinese J Geotech Eng 2008(04):560–565

The Professional Standards Compilation Group of the People's Republic of China (2012) Code for special geotechnical investigation of railway engineering(TB 10038-2012). China Railway Press, Beijing

Tong DK, Feng D, Huang S et al (2018) Experimental investigation of compacted soil disintegration property. Sci Tech Eng 18(15):129–136

Wang NX, Zhang WM, Gu XW et al (2008) Lateral swelling pressure of expansive soil acting on retaining wall due to inundation. J Hydraul Eng 39(5):580–587

Wang R, Cheng JJ, Qi YL et al (2020) Experimental study on micro-expansion mechanism of roadbed filler for high-speed railway ballastless track. J Eng Geol. https://doi.org/10.13544/j.cnki.Jeg2019-467

Xiao XW, Wang LC, Yang JS et al (2016) Cause analysis and treatment scheme for bottom heave of ballsastless track tunnel in nearly horizontally interbedded rock mass with high geostress. China Railway Sci 37(1):78–84

Yamaguchi H, Yoshida K, Kuroshima I et al (1990) Triaxial shear proper-ties of tertiary mudstone crushed by slaking. Soil Mech Found Eng 38:59–66

Zhang WM, Wang NX (2015) Expansive soil foundation of highway. China Architecture and Building Press, Beijing

Zhou CY, Jing XD, Liu Z et al (2019) Disintegration characteristics and modification of weathered soil in red beds in southern china. J Eng Geol 27(06):1253–1261

Chapter 48
Research on the Experiment for Dilatancy of Unsaturated Silty Clay in Different Matrix Suctions

Guo Li-Na, Luo Min-Min, Chen Yun, and Shen Dan-Yi

Abstract Dilatancy is one of the most important mechanical properties of soil, and it makes the deformation characteristics of the soil becoming more complex. Most soil is unsaturated in engineering practice, so it is completely necessary to research the dilatancy of the unsaturated soil considering the matrix suction. This paper studies the direct shear test of the unsaturated silty clay about dilatancy by GDS unsaturated back pressure shear apparatus (UBPS). We have summarized the unique regularity about the dilatancy of unsaturated silty clay with the changing of the matrix suction and got that the matrix suction can resist the vertical load in unsaturated silty clay. When the matrix suction is greater than the vertical load, the cohesive force among soil particles can resist the effect of vertical pressure, so the unsaturated silty clay would appear the phenomenon of dilatancy. When the matrix suction is less than the vertical load, unsaturated silty clay would appear the phenomenon of negative dilatancy (negative dilatancy is shear contraction). The value of dilatancy is increasing with the increase of the matrix suction of unsaturated silty clay under the same vertical load. The matrix suction can directly affect the surface tension of the contractile membrane, thereby affect the bonding strength between the soil particles.

Keywords Matrix suction · Unsaturated silty clay · Dilatancy · The direct shear test

G. Li-Na (✉) · L. Min-Min · C. Yun · S. Dan-Yi
The Architectural Design and Research Institute of Zhejiang University, Hangzhou, China
e-mail: 549244857@qq.com

G. Li-Na · L. Min-Min · C. Yun
Center for Balance Architecture, Zhejiang University, Hangzhou, China

S. Wang et al. (eds.), *Engineering Geology for a Habitable Earth: IAEG XIV Congress 2023 Proceedings, Chengdu, China*, Environmental Science and Engineering,
https://doi.org/10.1007/978-981-99-9057-3_48

48.1 Introduction

Gotechnical material often can produce plastic volume deformation because of the changed order of the particles when sheared, and this kind of characteristic called dilatancy (Zhenghan and Nan 2019). Shear construction can be considered as negative dilatancy (Zhenghan 2014). Dilatancy is one of the most important mechanical properties of soil, and also it is an essential characteristics different from others (Xiao et al. 2018, 2020; Ying-Xiao et al. 2011). Due to the existence of soil dilatancy, it makes the characteristic of soil deformation different from general elastoplastic material and become more complex. Since the soil dilatancy has been acknowledged, it caused general interest among the domestic and foreign researchers, thus made the soil dilatancy research more deeply, and a lot of research results appeared. When summarized the previous research, it can be found that the research methods on the soil dilatancy can be divided into two categories: one is the soil dilatancy be analyzed by the theoretical analysis (Zhang and Salgado 2010; Azizi et al. 2017), and the other is through test methods to get the related law of the soil dilatancy (Zhuang et al. 2017; Milatz et al. 2018; You et al. 2019; Ying-Xiao et al. 2010). Guang-Xin and Rui-Ping (2000) pointed out that generally soil grain changes from low potential energy state to high potential energy state in the process of dilatancy. Guo-Hui and Jing-Jing (2011) directed at the soil comprised by inerratic oval particles, circular and track shape particles as well as irregular particles built a mesoscopic dilatancy model which decided by the soil itself inherent material properties. Majid and Nour (2000) studied the influence in the slope stability analysis caused by the soil dilatancy. At present, soil matrix suction, that is soil saturation, has not been considered mostly in the study on soil dilatancy, and the indoor test has not considered the influence on the soil dilatancy caused by soil saturation as well. However, most soil in the engineering practice is in unsaturated state, therefore, it is necessary to study the dilatancy of the unsaturated soil with considering the matrix suction. But because of the complexity of the properties of unsaturated soil itself and to the requirements of the test instruments make this kind of research is rare. At present, I have seen only Liang-Tong and Charles (2007) studied the potential impact on unsaturated expansive soil caused by the suction. Based on this, the dilatancy of unsaturated silty clay under different matrix suction has been studies with the GDS unsaturated back pressure shear apparatus, and the special law of the dilatancy of unsaturated silty clay changing with matrix suction has been analyzed and summarized, thus beneficial basis has been provided for studying properties of unsaturated soil.

48.2 The Basics of Unsaturated Soil

Unsaturated soil distribution is very extensive, and the soil in the earth's surface that engineering practice closely links, almost all of whom is unsaturated soil. In drought and half dry areas the amount of evaporation is greater than precipitation, which

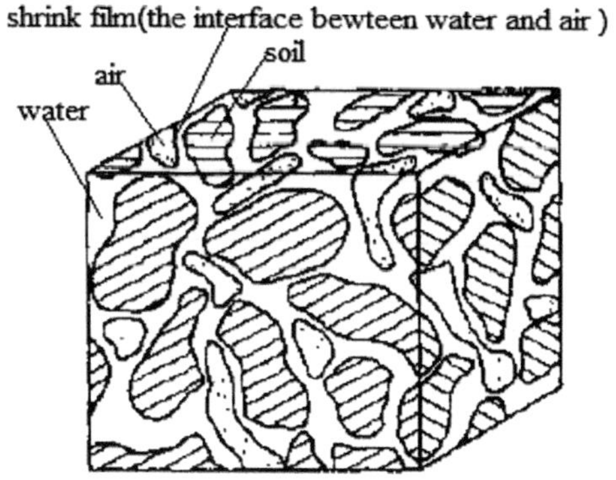

Fig. 48.1 Unsaturated soil units linked together with the gas phase

makes the soil in the regions is strict sense of unsaturated soil; Highway, railway, earth dam of compacted fill are in unsaturated state, that is, unsaturated soil. It can be seen that unsaturated soil is the frequently encountered in engineering practice.

In recent years, along with the understanding of the unsaturated soil gradually thorough, unsaturated soil is regarded as four phase system (as shown in Fig. 48.1). More and more attention of the researchers is attracted by the important role of the fourth phase-water and air interface (i.e., shrink film).

The matrix suction is one of the most important state parameters of unsaturated soil. The general thought of matrix suction of unsaturated soil results from the surface tension of the moisture separating. In the unsaturated soil, surface tension gives the character of elastic thin film character to the interface of the pore water and the pore air, which is similar with the balloon character. In order to maintain balance the interface bands and produces tension in the film when receiving the adsorption of soil particle or pore air pressure and water pressure engender pressure differential in the interface on both sides. Thus pore gas pressure can be greater than pore water pressure in unsaturated soil, which causes the matrix suction, namely:

$$S=u_a-u_w \tag{48.1}$$

where u_a is pore air pressure, u_w is pore water pressure, and S is the matrix suction of unsaturated soil.

The matrix suction influences soil shear strength through the surface tension of the interface of water and air. Considering the force balance of boundary surface we can get:

$$2T_s \sin\beta=2\Delta \mathrm{u} R_s \sin\beta \tag{48.2}$$

where T_s is surface tension, R_s is radius of curvature of film, $2R_s \sin\beta$ is the length of the film that projects in horizontal surface, and Δu is the pressure that exerts on the 2 d thin film (as shown in Figs. 48.2 and 48.3).

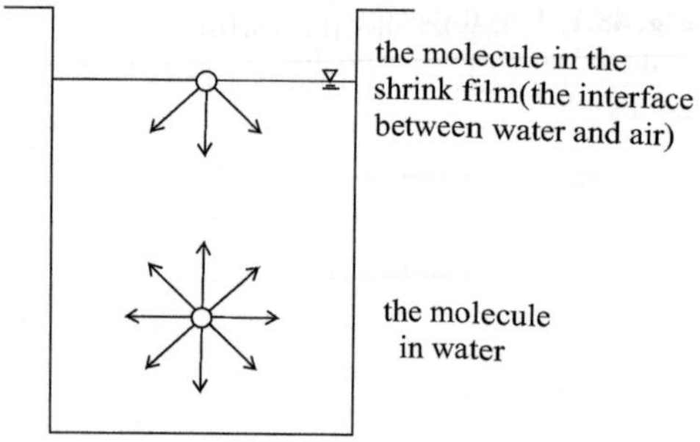

Fig. 48.2 Intermolecular forces on the shrink film and the water

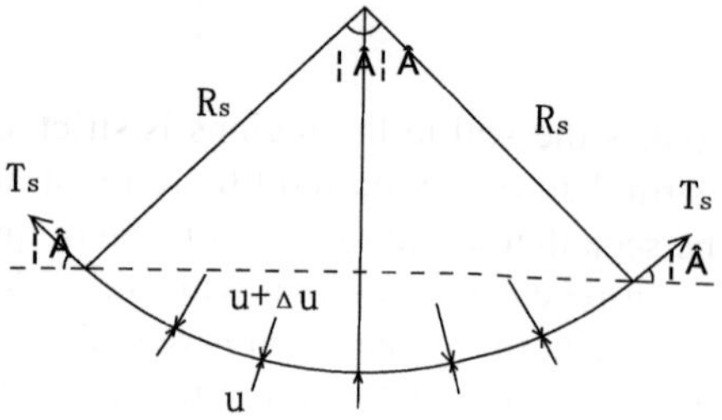

Fig. 48.3 Pressure and surface tension acted on 2D surface

For the saddle-backed warping film (3 d thin film) just as Fig. 48.4, similarly there are:

$$\Delta u = T_s\left(\frac{1}{R_1} + \frac{1}{R_2}\right) \quad (48.3)$$

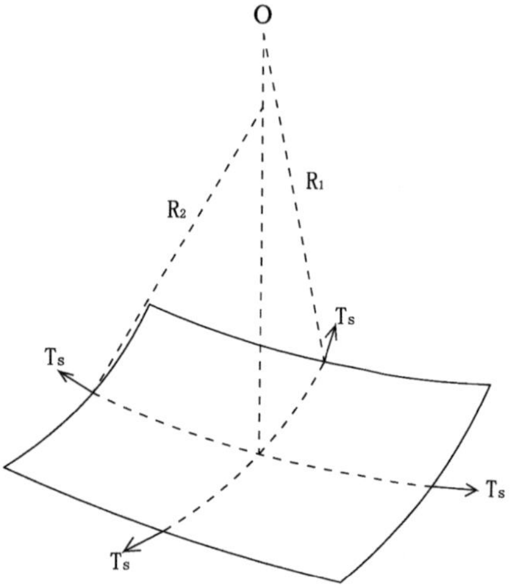

Fig. 48.4 Surface tension of the warping film

where R_1 and R_2 are the radius of curvature for orthogonal plane of warping film. When the values of radius of curvature are same, the formula (48.3) turns to be $\Delta u = \frac{2T_s}{R_s}$.

In the unsaturated soil, the pressure difference for the boundary surface of water and air is from pore air pressure and pore water pressure, namely matrix suction. Thus by the former type we can get:

$$S = (u_a - u_w) = \frac{2T_s}{R_s} \tag{48.4}$$

The formula (48.4) is called Kelvin capillary model equation. The expression of the intergranular force produced by the matrix suction and the surface tension as bellow:

$$H = \pi \mathrm{b}^2 T_\mathrm{s}\left(\frac{1}{R_\mathrm{s}} - \frac{1}{b}\right) + 2\pi b T_s \tag{48.5}$$

There

$$\mathrm{b} = R(\tan\alpha + 1 - \sec\alpha)$$

$$\mathrm{R_s} = R(\sec\alpha - 1)$$

So it created the relationship between the matrix suction and the surface tension. thus it can be seen the matrix suction can directly affect the surface tension of the contractile membrane, thereby affect the bonding strength between the soil particles.

48.3 Soil Samples of the Test

The soil of I landslide of Majiagou is used for testing. The landslide is mainly composed of silty clay clamping macadam which is formed by diluvium, and the composition of the macadam is fine sandstone and mudstone; silty clay comes from diluvium, part of which is artificial accumulation. The size of macadam is uneven, presents angular or subangular, and stacks mixed with clay. The structure of the landslide material is loose, and the permeability is strong. The corresponding earthiness index can be seen in the table below.

48.4 The Introduction of the Equipment

This test was operated in the GDS unsaturated back pressure shear apparatus (UBPS). The normative GDS UBPS tester uses 75 mm × 75 mm square specimen (also can choose other size based on user's needs). Specimen is placed in sample storehouse of shear box, and connected to the driver and shear force sensor. Closing the top of pressure chamber, then exert back pressure. Pressure can be pushed through ceramic plate in the bottom of shear box with the GDS pressure controller, the change of pore water's volume can also be controlled by pressure controller. GDS pressure controller is used in exerting pore air pressure, and the change of air volume is measured by the new GDS high precision measuring device (HAVCD) (Fig. 48.5 for the profile of GDS UBPS). Then, consolidate simples through manual pensile clog or optional feedback control effecter. The specimens enter to the shear stage when simples have been consolidated or matrix suction has met the requirements. All of the above system and tests are carried on under software GDS LAB control.

The matrix suction must be below the pressure of value ceramic plate (10 bar), and imposes on the soil samples by keeping the air pressure and hydraulic pressure exerted in the pressure controller.

It is very important to simulate part of saturated soil by measuring and controlling the matrix suction in the shear process. Therefore, UPBS provides the ideal equipment for many geotechnical engineering problems of today's world, such as the stability problems of soil slopes under the situation of subsaturation.

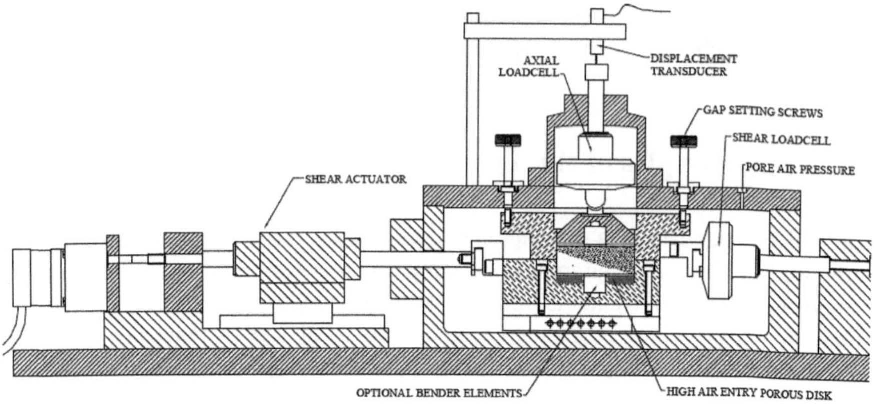

Fig. 48.5 Profile map of GDS UBPS

48.5 Experimental Programs

48.5.1 Aim of the Experiment

To study the dilatancy characteristic of unsaturated soil with different matrix suction in the vertical load supplying 50, 100, 200 and 400 kPa, analyze and summarize the corresponding rule, make groundwork for the further research of the unsaturated soil.

48.5.2 Preparation of the Test Samples

The samples used for unsaturated soil direct shear test are square, with the size of 75 mm × 75 mm × 30 mm. The initial moisture content of Soil samples should be 20.97% and the maximum dry density be1.55 g/cm^3. Firstly, the samples are crushed by a rubber hammer and screened by a 2 mm sieve. Secondly, put the screened soil samples in the oven, weigh the weight when the soil completely dried. Then weigh the corresponding quality distilled water according the moisture content of 20.79%, spray water as well as stir with the sprayer, seal with plastic wrap and lay aside for 1–2 days, prepare samples when soil and water mixed evenly. Finally make saturated samples, get corresponding matrix suction by controlling the hydraulic pressure and air pressure in the GDS tester, and conserve the saturate samples by suction gas.

48.5.3 Method of the Test

The test uses GDS unsaturated back pressure shear apparatus, and takes the method of the consolidated draining shear. Before the experiment, firstly make the ceramic plate saturated by pushing 20 kPa water pressure.

The unsaturated soil direct shear test (also means controlling suction consolidated draining shear) includes three stages as balance of suction, consolidation of equal suction and shear of equal suction. Change the original suction of samples through controlling the air pressure and the water pressure when the samples installed, and make suction evenly distributed in the whole sample, and the stage is called balance of suction. In this stage, firstly exert 25 kPa vertical stress on the sample to ensure the sample and the ceramic plate contacting well.

It is very important to judge whether the suction balances or not, for it affects the reliability of the follow-up test results directly. This experiment adopted judgment standard of suction balance is: the sample's displacement or water absorbing capacity is less than 0.02% of the volume of the sample in 24 h. Specifically for this test, the displacement of the sample is no more than 33.75 mm within 24 h. The time that suction balanced stage needed depends on the difference of targeted suction and initial suction. Consolidate the sample under vertical pressure, generally the consolidated time is no less than 24 h, the vertical deformation is no more than 0.005 mm/h, thus to make the consolidation achieve stability, control the pore air pressure and the pore water pressure for a fixed value when consolidating. Finally, carry on the shear test by ensuring the condition that the suction is not changed. For this test, the shear rate used 0.003 mm/min, and the maximum shear displacement is 12 mm.

48.6 The Results and Analysis of the Test

The unsaturated soil used for the test was divided into four groups: the first group was that the vertical load was 50 kPa and the matrix suctions were 50, 100, 200 kPa; the second group was that the vertical load was 100 kPa and the matrix suctions were 50, 100, 200 kPa; the third group was that the vertical load was 200 kPa and the matrix suctions were 50, 100, 200 kPa; the fourth group was that the vertical load was 400 kPa and the matrix suctions were 50 kPa, 100, 200, 400 kPa. The situation of the test can be seen in Table 48.2, Figs. 48.6 and 48.7.

The test went into shear stage when achieved the precondition that the suction balanced and the requirement of consolidation was satisfied. In the shear process, the curves of the shear dilatancy along with the shear displacement changing made for samples of different matrix suction in the vertical loads as P = 50 kPa, P = 100 kPa, P = 200 kPa, P = 400 kPa are seen as below (Figs. 48.8, 48.9, 48.10, and 48.11):

On the condition that the vertical load is 50 kPa, the characteristics of samples under different matrix suctions can be presented in graph 8: samples which matrix suction is 50 kPa show up the phenomenon of negative dilatancy with the increase of the shear displacement. When the shear displacement reaches to 4 mm the volume of dilatancy is essentially unchanged as the shear displacement increases; samples which matrix suction is 100 kPa at the beginning show negative dilatancy then dilatancy with the increase of the shear displacement, and the inflection point of the the shear displacement is about 2 mm. The volume of dilatancy increases along with the increase of the shear displacement; samples which matrix suction is 200 kPa behave shear dilatancy clearly as the shear displacement increasing, when the shear displacement reaches to 2 mm, the volume of shear dilatancy grows up distinctly.

When the vertical load is 100 kPa, as graph 9 shows, samples which matrix suctions are 50 and 100 kPa appear the phenomenon of negative dilatancy and the volume tends to constant with the increase of the shear displacement; samples which

Table 48.1 Physical properties of the tested soil

Natural moisture content (%)	Natural weight-specific density (kN/m^3)	Dry density (g/cm^3)	Void ratio	Saturation(%)	Liquid limit (%)	Plastic limit (%)	Compression modulus (MPa)
20.79	21.14	1.55	0.680	87.7	31.46	19.23	4.96

Table 48.2 Log sheet of the direct shear tests

Samples		The vertical load (kPa)	The vertical pressure (kN)	The matrix suction (kPa)	Duration of the tests(d)	
Type	Number				Suction balanced + concretion	The stage of shear
The first group	1–1	50	0.281	50	4	3
	1–2			100	4	3
	1–3			200	4	3
The second group	2–1	100	0.563	50	4	3
	2–2			100	4	3
	2–3			200	4	3
The third group	3–1	200	1.125	50	4	3
	3–2			100	4	3
	3–3			200	4	3
The fourth group	4–1	400	2.25	50	4	3
	4–2			100	4	3
	4–3			200	4	3
	4–4			400	4	3

Fig. 48.6 Saturated ceramic plate

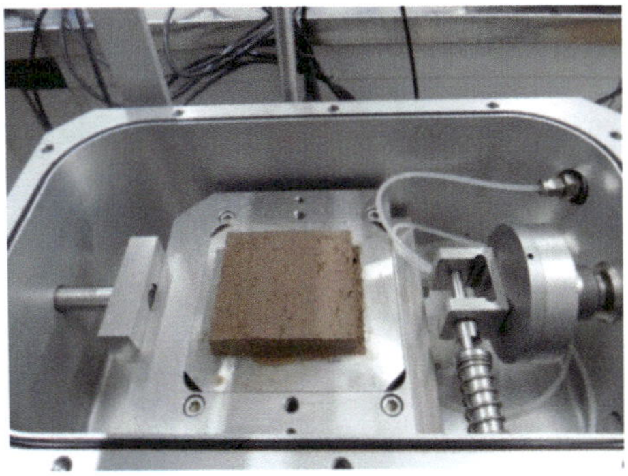

Fig. 48.7 The sample after shear

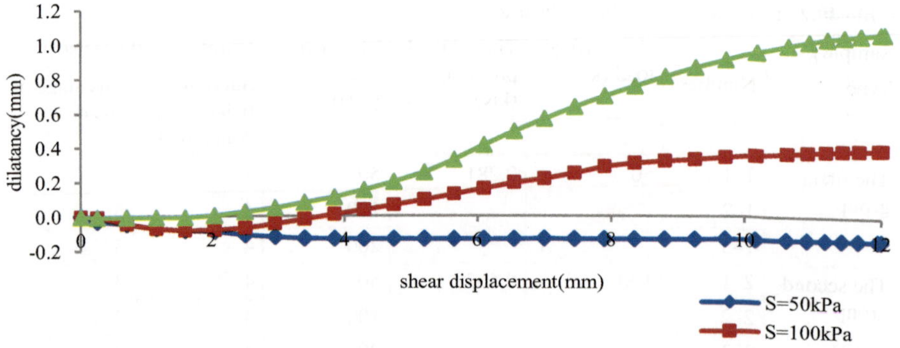

Fig. 48.8 Curve of dilatancy and shear displacement in different matrix suctions when P = 50 kPa

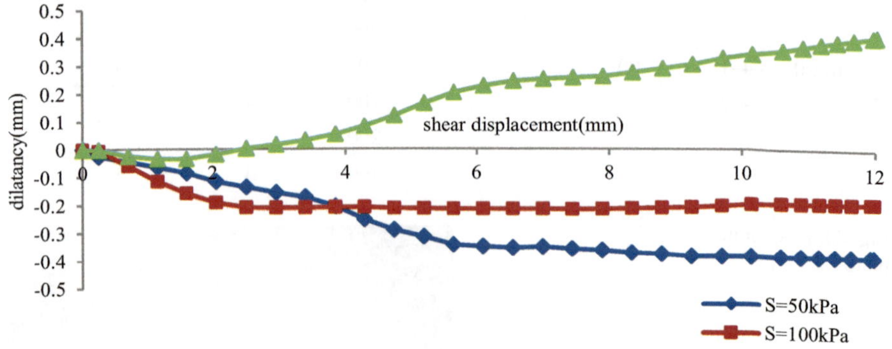

Fig. 48.9 Curve of dilatancy and shear displacement in different matrix suctions when P = 100 kPa

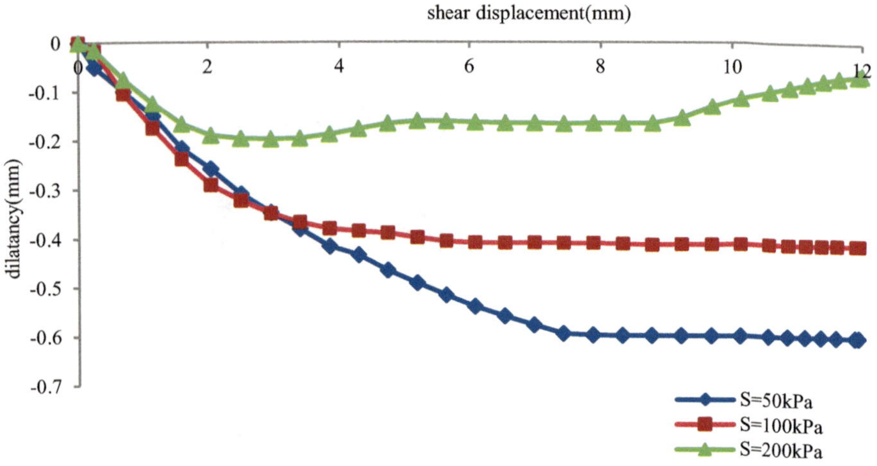

Fig. 48.10 Curve of dilatancy and shear displacement in different matrix suctions when P = 200 kPa

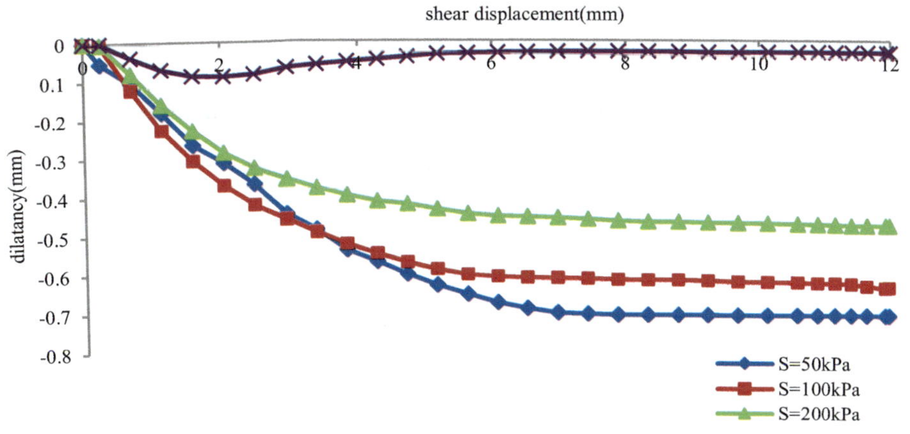

Fig. 48.11 Curve of dilatancy and shear displacement in different matrix suctions when P = 400 kPa

matrix suction is 200 kPa show up obvious dilatancy with the increase of the shear displacement, and when the shear displacement reaches to 2 mm, the volume of dilatancy increases gradually along with the shear displacement increasing.

When the vertical load is 200 kPa, samples which matrix suctions are 50 and 100 kPa appear the phenomenon of obvious negative dilatancy and the volume tends to constant as the shear displacement increases; samples which matrix suction is 200 kPa at the beginning show negative dilatancy then slight dilatancy with the increase of the shear displacement, and when the shear displacement reaches to about 2 mm the behavior of the samples from negative dilatancy turn to dilatancy. All the conditions above are presented in graph 10.

Graph 11 shows that when the vertical load is 400 kPa, samples which matrix suctions are 50, 100 and 200 kPa appear the phenomenon of obvious negative dilatancy with the increase of the shear displacement, and the volume of dilatancy grows as the matrix suction enlarged. Samples which matrix suctions is 400 kPa at the beginning show negative dilatancy then slight dilatancy with the increase of the shear displacement, and when the shear displacement reaches to about 2 mm the behavior of the samples from negative dilatancy turn to dilatancy.

To sum up, when satisfied the requirements of suction balance and consolidation, as the matrix suction is lower than the vertical load, unsaturated silty clay behaves negative dilatancy, the vertical load overrules soil particles' flipping, making the soil particle fill in the pore, thus shows the phenomenon of negative dilatancy; When the matrix suction is greater than the vertical load, unsaturated silty clay shows up dilatancy. That means the cohesive force among soil particles can resist the effect of vertical pressure, which is caused by the matrix suction of unsaturated soil. Thus makes the soil particle can flip and present dilatancy; When the matrix suction is equal to the vertical load, unsaturated silty clay shows either negative dilatancy or first negative dilatancy then dilatancy, that maybe relates to the properties of the samples. Through the contrast analysis of graph 6 to graph 9 we can conclude that in the same matrix suction, for unsaturated soil, the volume of dilatancy decreases with the vertical load increasing.

Analyze the Fig. 48.4, formulas (48.4) and (48.5), we can theoretically know that the larger the matrix suction, the smaller R_s (the radius of curvature of the warping film) is. Thus lead to the cohesive force among soil particles become bigger, while the pore turn to be smaller. Consequently, the shear dilatants of the soil turns to be more obvious for soil particles moving and rolling over mutually rather than filling in the intergranular pore in the shear process. Thus it can be seen: The principle of dilatation is that the ball stress (matrix suction) makes the soil medium compacted and hardened, and the strength increases. The deviant stress (surface tension) causes the volume expansion and strength reduction of the dense soil medium, while the volume reduction and strength increase of the loose soil medium.

48.7 Conclusion

When the requirements of suction balance and consolidation are satisfied, according to the direct shear test of unsaturated silty clay under different matrix suction we can draw the conclusions as followed:

(1) The cohesive force among soil particles can resist the effect of vertical pressure, which is caused by the matrix suction of unsaturated soil. When the matrix suction is greater than vertical load, unsaturated silty clay shows up dilatancy, and the volume of dilatancy increases with the shear displacement increasing; When the matrix suction is lower than vertical load, unsaturated silty clay behaves shows up negative dilatancy, and the volume of dilatancy tends to fixed value with the increase of the shear displacement; When the matrix suction is equal to vertical load, unsaturated silty clay shows either negative dilatancy or first negative dilatancy then dilatancy.

(2) The inflection point of shear displacement for unsaturated silty clay from negative dilatancy to dilatancy is 2 mm.

(3) When under the same vertical loads, the volume of dilatancy for unsaturated silty clay grows with the matrix suction increasing; And when under the same matrix suction (namely saturation), the volume of dilatancy for unsaturated silty clay decreases with the matrix suction increasing.

(4) The larger the matrix suction, the smaller the radius of curvature of the warping film. Thus lead the pore turn to be smaller, while the cohesive force among soil particles become bigger. Consequently, the shear dilatants of the soil turns to be more obvious for soil particles rolling over mutually rather than filling in the intergranular pore in the shear process.

(5) The matrix suction makes the soil medium compacted and hardened, and the strength increases. The surface tension causes the volume expansion and strength reduction of the dense soil medium, while the volume reduction and strength increase of the loose soil medium.

Acknowledgements The authors gratefully acknowledge the funds from the Construction Research Project of Zhejiang Province (Nos: 2021K103).

References

Azizi A, Jommi C, Musso G (2017) A water retention model accounting for the hysteresis induced by hydraulic and mechanical wetting-drying cycles. Comput Geotech 87:86–98

Guang-Xin L, Rui-Ping G (2000) Volume-contraction in unloading of shear tests and reversible dilatation of soils. Chinese J Geotech Eng 22(2):159–161

Guo-Hui L, Jing-Jing C (2011) Microscopic understanding of dilatancy in cohesionless soils. Chinese J Geotech Eng 33(9):1334–1340

Liang-Tong Z, Charles WWN (2007) Effect of suction on shear strength and dilatancy of a kind of unsaturated expansive clay. Chinese J Geotech Eng 29(1):82–87

Majid T, Nour MA (2000) Significance of soil dilatancy in slope analysis. J Geotech Geoenviron Eng, ASCE 26(1):75–80

Milatz M, Törzs T, Nikooee E, Hassanizadeh SM, Grabe J (2018) Theoretical and experimental investigations on the role of transient effects in the water retention behaviour of unsaturated granular soils. Geomech Energy Environ 15:54–64

Xiao X, Guo-Qing C, Jian L et al (2018) The strength and dilatancy characteristics of unsaturated soil at low and tensile stresses. Chin J Rock Mech Eng 37(8):1933–1942

Xiao X, Guo-Qing C, Jian L et al (2020) Peak strength characteristics and failure envelope model of unsaturated soils. Chinese J Geotech Eng 42(07):1211–1219

Ying-Xiao Y, Xiang-Rong Z, Xing-Ping J et al (2010) Triaxial testing study of dilatant characteristics of Qiantangjiang alluvial silts. Rock Soil Mech 31(Supp. 2):48–52

Ying-Xiao Y, Xiao-Nan G, Chuan F et al (2011) Triaxial testing study of dilatant characteristics of Qiantangjiang alluvial unsaturated silts. Rock Soil Mech 32(supp.1):38–43

You G, De-An S, Jun-Ran Z (2019) Soil-water characteristics of unsaturated soils considering initial void ratio and hydraulic path. Chinese J Geotech Eng 41(12):2191–2196

Zhang J, Salgado R (2010) Stress-dilatancy relation for Mohr-Coulomb soils following a non-associated flow rule. Geotech 60(3):223–226

Zhenghan C (2014) On basic theories of unsaturated soils and special soils. Chinese J Geotech Eng 36(02):201–272

Zhenghan C, Nan G (2019) New developments of mechanics and application for unsaturated soils and special soils. Rock Soil Mech 40(01):1–54

Zhuang L, Hassanizadeh SM, Qin C, de Waal A (2017) Experimental investigation of hysteretic dynamic capillarity effect in unsaturated flow. Water Resour Res 53(11):9078–9088

Chapter 49
Effects of Wetting-Drying Cycles on the Strength and Microscopic Characteristics of Ili Loess

Ye Liu, Zhijun Zhou, Hongli Xie, Yubo Ren, and Zhihao Guo

Abstract To investigate the strength degradation and microstructure characteristics of Ili loess under wetting-drying cycles, soil water characteristic curve measurement, triaxial shear test, and scanning electron microscope testing were used to study the changes in water holding characteristics of Ili loess, the effects of different dry densities, dry wet cycling amplitudes and times on the shear strength of Ili loess, and the changes in microstructure of Ili loess before and after dry wet cycling under different dry densities. The results show that the water holding capacity of Ili loess gradually decreases with the increase of wetting-drying cycles; The cohesion of Ili loess increases with the increase of dry density under the same number of wetting-drying cycles, and the trend of cohesion with the increase of dry density is consistent under different number of wetting-drying cycles. However, the relationship between internal friction angle and dry density and number of wetting-drying cycles is not obvious, indicating that wetting-drying cycles have little effect on the relationship between dry density and internal friction angle. In addition, after the wetting-drying cycles, the content of large and medium particles and pores in loess decreases, while the content of small particles and pores increases. Low dry density loess develops towards embedded compaction, while high dry density loess develops towards aerial arrangement.

Keywords Geotechnics · Soil mechanics · Wetting-drying cycle · Ili loess · Strength deterioration

Y. Liu (✉) · Z. Zhou · H. Xie · Y. Ren · Z. Guo
Highway College, Chang'an University, Xi'an 710064, China
e-mail: liuye19980915@163.com

Z. Zhou
e-mail: zhouzhijun@chd.edu.cn

S. Wang et al. (eds.), *Engineering Geology for a Habitable Earth: IAEG XIV Congress 2023 Proceedings, Chengdu, China*, Environmental Science and Engineering,
https://doi.org/10.1007/978-981-99-9057-3_49

49.1 Introduction

Ili loess has a late formation age, special genesis, and different characteristics from common loess. In the Ili region, there is more precipitation in winter and summer, and after multiple cycles of wetting-drying, the strength of shallow loess decreases, and the anti-sliding force is insufficient, leading to deformation and shallow landslides, and other phenomena. Therefore, studying the strength deterioration of Ili loess under wetting-drying cycle action has important practical significance.

Soil water characteristic curve (SWCC) is an important tool to describe unsaturated Soil mechanics, which can be used to obtain strength parameters, permeability coefficient and other important mechanical parameters (Liu et al. 2011). A large number of scholars have conducted research on the impact of different dry wet cycling paths on the soil water characteristic curve, with a focus on exploring the hysteresis characteristics of the curve after dry wet cycling (Likos et al. 2014) and curve morphology (Zhou and Yu 2005), and summarizing the influence of dry density on the soil water characteristic curve of loess (Tonglu et al. 2019). However, due to the narrow applicability of most theoretical models and their reliance on original soil water parameters as inputs, it is difficult to ensure the accuracy of fitting and predicting soil samples after repeated dry wet cycles; On the other hand, some models have problems with too many parameters and are difficult to determine, making them difficult to use.

There have been many studies on the mechanism of strength degradation of loess under different dry wet cycling paths. A large amount of research has found that the strength of soil is concentrated in the shear strength, and related indicators are influenced by factors such as water content (Jian et al. 2017), stress magnitude, stress path (Tang et al. 2016), loading time and rate (Li et al. 2013). In addition, the formulas for shear strength proposed by many experts and scholars (Li et al. 2013) and models (Chen et al. 2022) can also be used to derive the shear strength of unsaturated soil. But most formulas are cumbersome and difficult to apply; The models are mostly limited to the effects of different dry densities and dry wet cycles, and the number of samples studied is relatively small, making it impossible to comprehensively and accurately evaluate the shear strength after wetting-drying cycles. Further research is still needed.

The microstructure of soil directly affects the macroscopic engineering characteristics, and it is necessary to comprehensively study the strength and deformation characteristics of soil from a combination of micro and macro perspectives (Liu et al. 2022; Su 1984). Research has shown that the microstructure of loess can affect the collapsibility of undisturbed loess (He et al. 2003), and the changes in microstructure are closely related to the shear failure of the soil (Ni et al. 2005). Therefore, analyzing the microstructure and failure performance of the sample (Lei et al. 2004) is very important.

Based on a thorough investigation of the field conditions and obtaining the main engineering characteristics of Ili loess, this paper carries out laboratory experiments

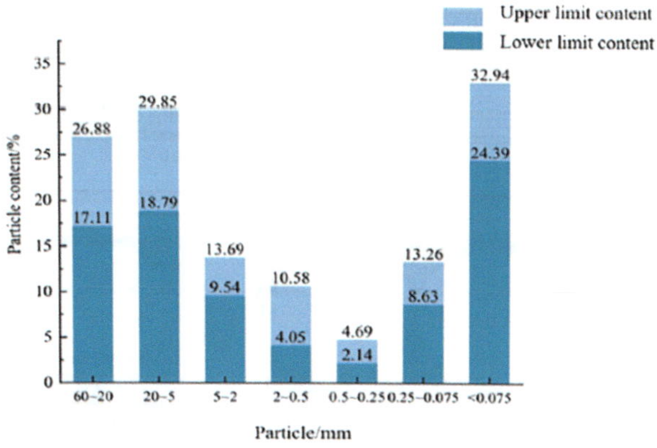

Fig. 49.1 Particle distribution of Ili loess

on macro and micro mechanical properties and microstructure characteristics, considering the influence of dry density, wetting-drying cycle amplitude, and frequency. The degradation law of the strength of the intact and remolded loess under wetting-drying conditions is explored.

49.2 Experimental Plan

49.2.1 Test Soil

The original soil samples and remolded soil samples used in this study were taken from the Naba Highway in the Xinjiang Ili Kazakh Autonomous Prefecture. According to indoor soil tests, the dry density of the soil sample is 1.31–1.72 g/cm^3, the porosity is 37–51%, the liquid limit is 20–45, the plastic limit is 11–24%, and the plasticity index is 9–13. It belongs to silty clay; The coefficient of collapse is between 0.0061 and 0.0146; The content of loess particles is shown in Fig. 49.1, with particles with particle sizes ranging from 60 to 20 mm accounting for 17.11–26.88%, particles with particle sizes ranging from 20 to 5 mm accounting for 18.79–29.85%, and particles with particle sizes less than 0.075 mm accounting for 24.39–32.94%; The content of particles with particle sizes ranging from 2 to 0.5 mm and from 0.5 to 0.25 mm is the lowest, accounting for only 2.14–4.69%.

49.2.2 Soil-Water Characteristic Curve Test

This article uses the filter paper method to measure the SWCC of Ili loess, and determines the matrix suction by using the calibration curve relationship of filter paper at different moisture contents. Firstly, prepare the soil sample with the required

Table 49.1 Sample preparation for soil-water characteristic curve test

Variables	Parameter
Dry density/g/cm^3	1.45/1.55/1.65
Wetting-drying cycle times/time	0/1/3
Water content/%	3/5/7/9/11/14/17/20/25/30

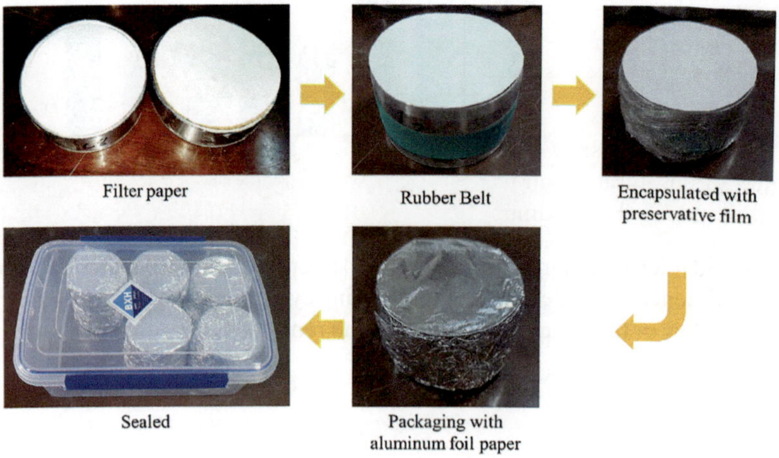

Fig. 49.2 Sample packaging and preservation process

moisture content, let it stand for a period of time, and then seal it in a glass jar for 24 h. Calculate the required amount of soil based on the dry density of the sample, place it in a jack and compact it with a mold. Seal the prepared ring knife sample with cling film and store it for 48 h to evenly distribute the moisture of the sample. The working condition settings of the sample are shown in Table 49.1.

Before the experiment, put Whatman No.42 filter paper into an oven and bake for 24 h to ensure that the filter paper is dry. After cooling, store it in a dryer. Two soil samples with the same moisture content are stacked together as shown in Fig. 49.2. Then, a layer of gauze and a piece of dry filter paper are placed on the top and bottom surfaces of the soil sample, with one filter paper on the top of the container not in contact with the sample. According to ASTM D5298-10 standard (2016), it is recommended to determine the suction in the soil using the calibration curve of the filter paper to obtain the matrix suction of the sample.

49.2.3 *Triaxial Test*

In order to simulate the wetting-drying process under natural conditions, five sets of wetting-drying paths were set during the wetting process of the triaxial test samples,

Table 49.2 Wetting-drying cycle scheme

Test number	Initial water content /%	Dry density/g/cm^3	Cycle path	Cycle times
A	14	1.45/1.55/1.65	14% → 5% → 14%	0, 1, 3, 5
B	17		17% → 5% → 17%	
C	20		20% → 5% → 20%	
D	17		17% → 5% → Saturated → 5%	
E	17	Undisturbed soil	17% → 5% → Saturated → 5%	

with group numbers shown in Table 49.2. Related studies have shown that the pore structure in unsaturated soil has stabilized after three cycles of wetting and drying. Therefore, taking the maximum number of wetting and drying cycles as 5 can meet the requirements for soil stability.

The triaxial shear test was conducted using a GDS stress-strain controlled triaxial apparatus. The test was performed under the conditions of undrained and unconsolidated state. The shear rate was set at 0.8 mm/min, and the confining pressure was controlled at 100 kPa, 200 kPa, and 300 kPa, respectively. The test was terminated when the axial strain $\varepsilon_a = 20\%$. The peak strength of the stress-strain curve was taken as the triaxial shear strength. If no peak was found, the deviator stress corresponding to an axial strain of 15% was taken as the triaxial shear strength.

49.3 Test Result Analysis

49.3.1 Soil-Water Characteristics of Ili Loess Under Wetting-Drying Cycles

Before each cycle, the volumetric water content of the ring knife sample was tested. After multiple wetting-drying cycles, the saturated volumetric water content of the samples is shown in Fig. 49.3. For samples with dry densities of 1.65 g/cm^3, 1.55 g/cm^3, and 1.45 g/cm^3, the saturated volumetric water content after 0, 1, and 3 wetting-drying cycles were 43.92% → 42.24% → 42.63%, 49.05% → 44.78% → 44.60%, and 51.50% → 46.33% → 46.29%, respectively. The experiment found that the first wetting-drying cycle had the greatest impact on the saturated water content of the loess.

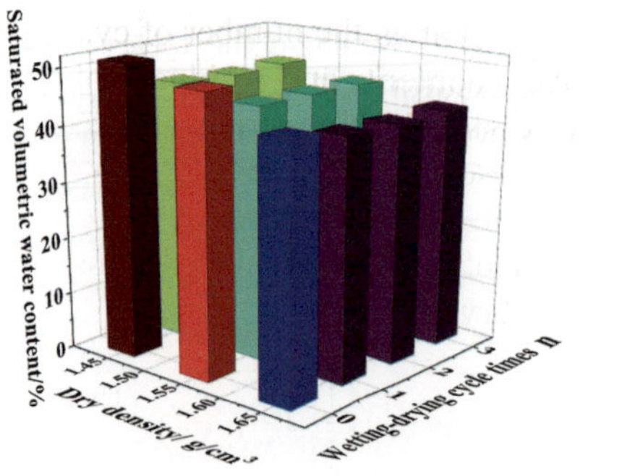

Fig. 49.3 The change of saturated volumetric water content after wetting-drying cycle

49.3.2 SWCC Curve

The classic fitting equations proposed by Van Genuchten and Fredlund were used in the experimental study using the filter paper method in this paper. The measured soil-water characteristic curves are shown in Figs. 49.4 and 49.5.

Observing the fitting results of the graph, both the FX model and VG model can fit the soil-water characteristic curve well. However, the regularity of the parameters in the FX model is not good, and the VG model has higher accuracy, with most of the relevant parameters R2 being greater than 0.99 and having better regularity. Analysis

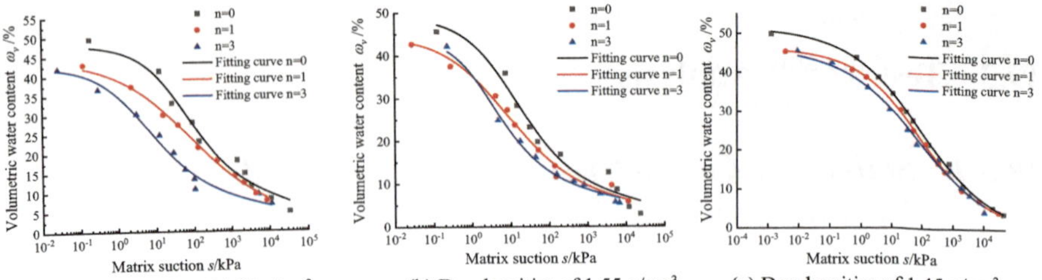

(a) Dry densities of 1.65 g/cm³ (b) Dry densities of 1.55 g/cm³ (c) Dry densities of 1.45 g/cm³

Fig. 49.4 FX model

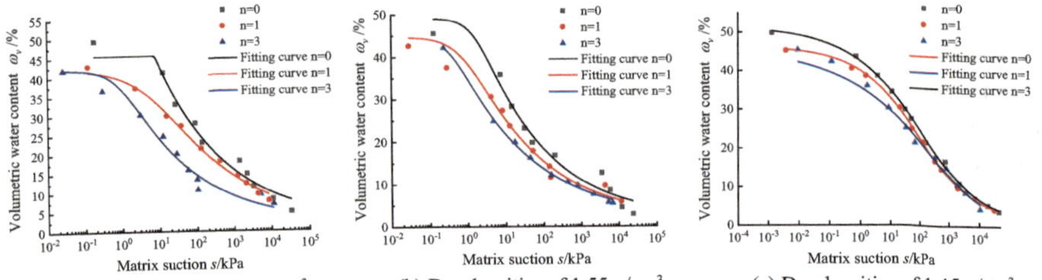

(a) Dry densities of 1.65 g/cm³ (b) Dry densities of 1.55 g/cm³ (c) Dry densities of 1.45 g/cm³

Fig. 49.5 VG model

shows that as the number of cycle increases, the parameters a, saturated volumetric water content, and residual volumetric water content all gradually decrease, but the residual volumetric water content changes little with wetting-drying cycles and increases with dry density. The parameter n related to the soil pore size distribution has the greatest influence on the shape of the SWCC curve, and the value of n is inversely proportional to the absolute value of the slope of the transition section of the curve.

49.3.2.1 Deterioration Analysis of Soil Water Characteristics

Overall, the saturated volume moisture content of the loess sample gradually decreases after the dry wet cycle. From the analysis of Figs. 49.4 and 49.5, it can be seen that the changes in the curves of samples with different dry densities after dry wet cycles are similar, and the soil water characteristic curve after cycling shows an overall downward trend. The dry density has little effect on water retention characteristics. As the dry density increases, the pores become smaller and the residual water content significantly increases; Adjusting the dry density has little effect on the matric suction of Ili loess entering the residual stage, and experiments have shown that the same suction level is maintained to enter the residual stage.

As the dry wet cycle continues, the water holding capacity of the Ili loess gradually decreases. In the saturated section with low matrix suction, the soil water characteristic curve shows a height difference after dry wet cycling, and the height difference is maximum when $n = 1$. After $n = 3$, the water holding capacity tends to stabilize, indicating that the ability of soil particles to adsorb water and the changes in macropores gradually stabilize. As the dry wet cycle progresses, the slope of the soil water characteristic curve first increases and then decreases in the rapid descent section; Afterwards, the position of the curve entering the transition phase gradually shifts to the left, and the decrease in saturated volume water content gradually decreases.

The reason for the above phenomenon is that after the wetting-drying cycles, when the dry density is small, the pores between soil particles are relatively large, and the water holding capacity is poor. After the first wetting-drying cycle, the skeleton structure is rearranged, and the water holding capacity is strengthened. As the wetting-drying cycles continues, the water holding capacity continues to deteriorate, leading to a decrease in the slope of the soil water characteristic curve in the rapid descent section. After multiple wetting-drying cycles, the soil skeleton structure shrinks, but it also promotes the development of small pores in the soil and gradually connects to form large pores, resulting in a decrease in volumetric water content.

49.3.3 *Triaxial Shear Test of Ili Loess Under Wetting-Drying Cycles*

To further investigate the deteriorating effect of wetting-drying cycles on the strength of the Ili loess, the following section analyzes the impact of different initial water contents, dry densities, and wetting-drying cycle paths on the strength of the Ili loess.

49.3.3.1 Shear Strength

Figure 49.6 analyzes the relationship between triaxial shear strength and the number of wetting-drying cycles, showing that the shear strength of the Ili loess decreases with an increase in the number of wetting-drying cycles. The wetting-drying cycle path significantly affects the triaxial shear strength, which shows a decreasing rate of shear strength with a gradual decrease in the number of wetting-drying cycles in all paths. With an increase in the amplitude of wetting-drying cycles, the decay of triaxial shear strength also increases gradually, with path c showing a significantly larger decay value than paths a and b. The slope of the curve of path d shows an increasing trend followed by a decreasing trend when the dry density is greater than 1.45 g/cm^3.

Observing the final degradation degree of triaxial shear strength of Ili loess under different wetting-drying cycle paths, it is found that with an increase in the initial water content, the decrease in shear strength of paths a, b, and c gradually increases. Path d, which is based on path b, shows a larger decrease in shear strength under saturated conditions of wetting-drying cycle tests. The shear strength of path d and path e finally tends to be consistent when the initial water content is 17%, indicating that the undisturbed soil is close to compacted loess after five wetting-drying cycles.

49.3.3.2 Shear Strength Parameters

According to Figs. 49.7 and 49.8, the cohesion of the Ili loess decreases with the increase of the number of wetting-drying cycle times; the degradation of the internal friction angle by wetting-drying cycle is not significant, and the internal friction angle is basically fluctuating within 0 °–4°. Under the same number of cycle times, the cohesion of the Ili loess increases with the increase of dry density, and the trend of cohesion with dry density is consistent under different numbers of cycle times. However, the relationship between the internal friction angle and dry density is not significant, indicating that the effect of wetting-drying cycle on the relationship between dry density and internal friction angle is not significant.

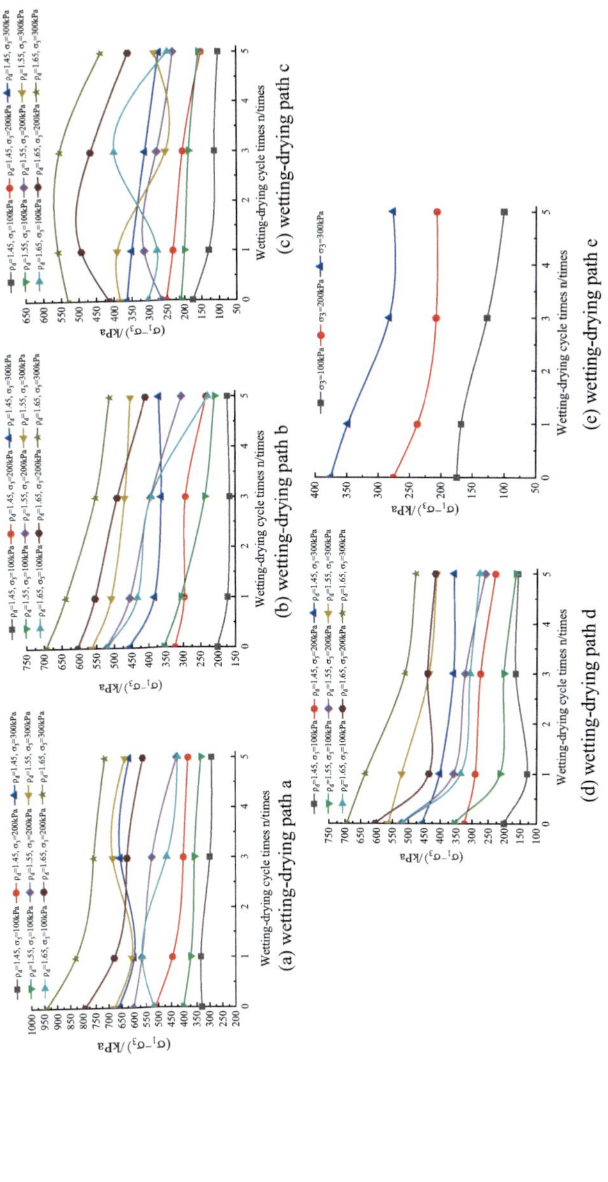

Fig. 49.6 Changes in shear strength of Ili loess

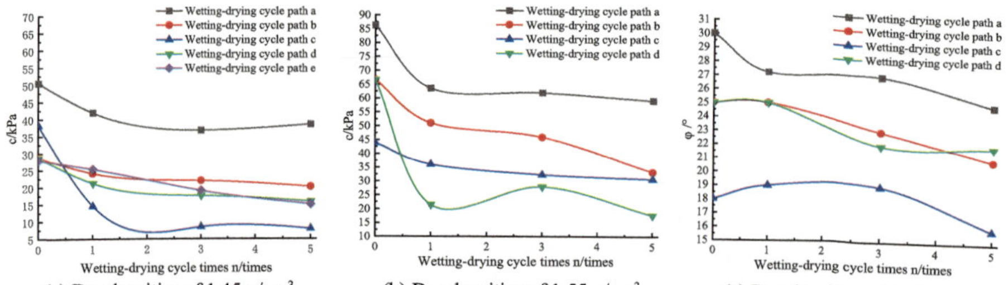

(a) Dry densities of 1.45 g/cm³ (b) Dry densities of 1.55 g/cm³ (c) Dry densities of 1.65 g/cm³

Fig. 49.7 Relationship between c and n

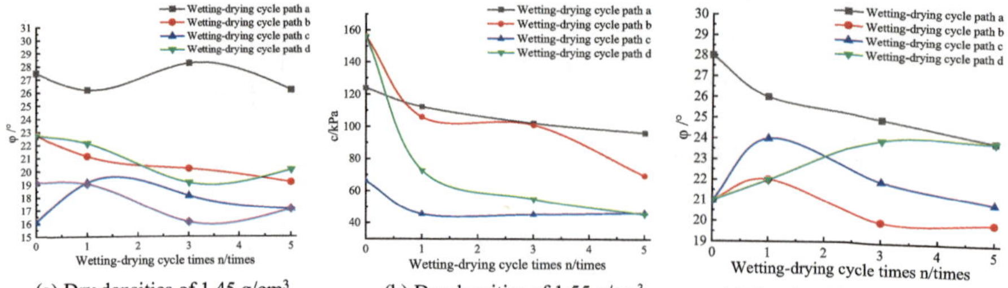

(a) Dry densities of 1.45 g/cm³ (b) Dry densities of 1.55 g/cm³ (c) Dry densities of 1.65 g/cm³

Fig. 49.8 Relationship between φ and n

49.3.3.3 Shear Strength Degradation Analysis

Figure 49.9 analyzes the Degree of Deterioration and development rate of cohesive strength at different dry densities. The results show that under different wet-dry cycle paths and different levels of dry density, the deterioration of cohesive strength shows a trend of rapid increase followed by slow increase. The deterioration rate of cohesive strength of the reshaped loess is fastest when $n = 1$, and the deterioration rate of cohesive strength of the intact soil is fastest when $n = 3$. The reshaped loess with an initial moisture content of 20% at low reshaping degree has the greatest degree of cohesive strength deterioration.

Under medium and high reshaping degrees, the loess under the wetting-drying path d, which undergoes saturation and dehydration, deteriorates the most significantly, and the higher the wetting-drying cycle amplitude A, the higher the degree of cohesive strength deterioration.

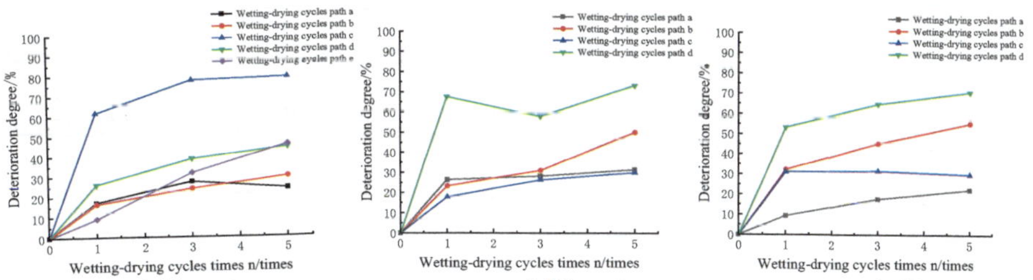

$\rho_d = 1.45$ g/cm³ (b) $\rho_d = 1.55$ g/cm³ (c) $\rho_d = 1.65$ g/cm³

Fig. 49.9 Curves of cohesion degradation degree for different dry densities

49.3.4 Microscopic Changes of Ili Loess After Wetting-Drying Cycles

The particle size, shape, contact mode, and microscopic structure of the pores of Ili loess at various dry density groups were observed at 1000 times magnification. The influence of wetting-drying cycles on the microscopic structure of loess was analyzed, as shown in Figs. 49.10, 49.11, 49.12, and 49.13.

Wetting-drying cycles have changed the pore morphology and quantity of Ili loess. For n = 0 Ili loess, the structure is dense with many small and medium-sized pores and few large pores, fewer microscopic cracks, distinct edges, and small particles and debris scattered around larger particles. For n = 5 Ili loess, there are fewer small and medium-sized pores, many large pores, and multiple microscopic cracks inside. wetting-drying cycles caused the original loose and porous structure of the loess to be lost, with pores and microscopic cracks continually developing and accumulating.

Loess particles became rounder, and there was a higher proportion of small particles and debris, which became embedded and aggregated around the larger particles. This is due to repeated infiltration and dissipation of pore water, gradually weakening the bonding material. The larger particles broke into smaller particles and

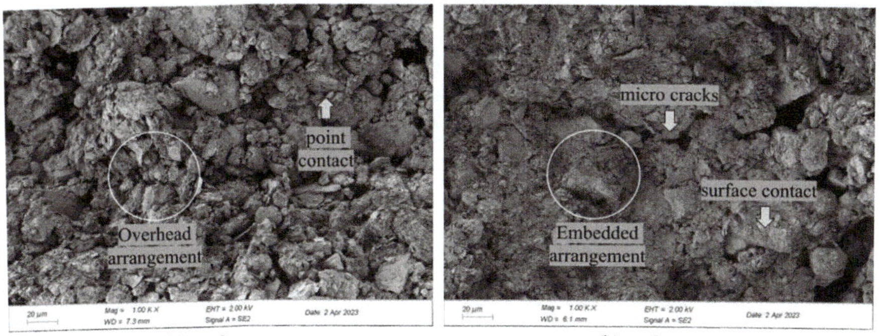

(a) n=0 (b) n=5

Fig. 49.10 Electron microscope images of $\rho_d = 1.45$ g/cm^3

(a) n=0 (b) n=5

Fig. 49.11 Electron microscope images of $\rho_d = 1.55$ g/cm^3

(a) n=0 (b) n=5

Fig. 49.12 Electron microscope images of $\rho_\mathrm{d} = 1.65$ g/cm^3

(a) n=0 (b) n=5

Fig. 49.13 Microscopic images of undisturbed soil

debris, which were adsorbed around the larger particles in an aggregated form after wetting-drying cycles. This phenomenon is particularly evident at lower dry densities.

The wetting-drying cycle has an impact on the distribution and contact mode of loess particles. In the natural state (n = 0), the soil sample has many pores and a relatively loose structure, forming an overhead structure. When the dry density is low, the contact mode is mainly overhead arrangement and point contact, with more pores in the middle; As the dry density increases, the gaps between loess particles become smaller, and the size and connectivity of pores also decrease. The contact mode is mainly embedded arrangement and surface contact. After 5 wetting-drying cycles, when the dry density is low, small particles and debris particles adsorb around large particles, and the contact mode of particles changes from point-to-point or point-to-point to face-to-face contact; When the dry density is high, obvious particle fragmentation can be observed, and some embedded soil skeletons transform into aerial arrangements. Small particles are dispersed between pores, and small and medium-sized pores form large pores.

The reason for the above phenomenon is insufficient compaction of remolded loess at low density. After 5 cycles, the soil skeleton is reconstructed, particles are rearranged, some cementitious substances dissolve and migrate, adsorption capacity is reduced, large particles are broken, small particles are redistributed, and micro cracks appear. The phenomenon of soil shrinkage was indeed observed in the macroscopic wetting-drying cycle test, which can support the above viewpoint.

49.4 Conclusion

Through wetting-drying cycle tests, triaxial shear tests, and microscopic electron microscopy tests, strength degradation analysis was conducted on compacted and undisturbed loess under different dry densities, water contents, and wetting-drying cycle amplitudes. The following conclusions were drawn:

(1) The saturated moisture content of Ili loess decreases most significantly after the first wetting-drying cycle, and then remains stable; The wetting-drying cycle significantly weakens the water holding capacity of the Ili loess, especially in the low suction section, and the position of the first turning point of the soil water characteristic curve gradually shifts to the left.

(2) The greater the amplitude and number of wetting-drying cycles, the faster and greater the attenuation rate of loess shear strength; The triaxial Shear strength of undisturbed soil after five wetting-drying cycles is close to that of remolded loess; The deterioration effect of the wetting-drying cycle mainly targets the cohesion, and the change in internal friction angle is not significant. Moreover, the c value of remolded loess deteriorates the most after the first wetting-drying cycle, while the c value of undisturbed loess deteriorates the most after the third wetting-drying cycle.

(3) During the dry wet cycle, the cementitious substances between soil particles dissolve, the soil particles break, the content of small particles increases, and the pores are filled with micro and small particles. At the same time, the small pores intersect to form larger pores, causing structural damage and continuous accumulation, ultimately leading to a decrease in soil strength; The response of different dry densities to the wetting-drying cycle is different, with low dry density loess developing towards embedded compaction, and high dry density loess developing towards aerial arrangement.

References

ASTM D5298-16 (2016) Standard test method for measurement of soil potential (suction) using filter paper. ASTM international, USA

Chen R, Zhang X, Hao RY(2022) Study on the mechanism and model of strength degradation of geopolymer solidified loess under dry wet cycle. Soil Mech 43(5):1164–1174. https://doi.org/10.16285/j.rsm.2021.1355

He KS et al (2003) Shen the pearl river research on micro deformation and mechanism of structural soil. J Hohai Univ (Natural Science Edition) 2:161–165

Jian WB, Hu HR, Luo YH et al (2017) Experimental study on deterioration of granitic residual soil strength in wetting-drying cycles. J Eng 25(3):592–597. https://doi.org/10.13544/j.cnki.jeg.2017.03.003

Lei SY, Tang WD, Wang XM, Dai JL (2004) CT scanning analysis of the damage and failure process of undisturbed loess. J Lanzhou Univ Tech 4:110–114

Li RJ, Liu JD, Zheng W et al (2013) A bilinear strength formula for structured loess based on tensile strength and shear strength and its application. Chinese J Geotech Eng 35(12):247–252

Likos WJ, Lu N, Godt JW (2014) Hysteresis and uncertainty in soil water-retention curve parameters. J Geotech Geoenviron Eng 140(4):04013050. https://doi.org/10.1061/(ASCE)GT.1943-5606.0001071

Liu FY, Zhang Z, Zhou D et al (2011) The influence of density and dry wet cycling on the soil water characteristic curve of loess. Soil Mech 32(S2):132–136. https://doi.org/10.16285/j.rsm.2011.s2.086

Liu YY, Wang G, Lai HP et al (2022) Study on the relationship between macro and micro parameters of undisturbed loess under dry wet cycling. J Water Res 53(04):421–432. https://doi.org/10.13243/j.cnki.slxb.20210905

Ni WK, Yang HQ, Wang CY (2005) CT detection and analysis of the microstructure damage pattern of undisturbed loess in roadbed. Highway Transp Tech S1:81–83

Su MA (1984) Calculation of foundation and foundation on collapsible loess. Water Resources and Electric Power Publishing House.

Tang CS, Wang DY, Shi B et al (2016) Effect of wetting–drying cycles on profile mechanical behavior of soils with different initial conditions. Catena, 139:105–116. https://doi.org/10.1016/j.catena.2015.12.015

Tonglu L, Fan JW, Xi Y et al (2019) Analysis for effect of microstructure on SWCC of compacted loess. J Eng Geol 27(5):1019–1026. https://doi.org/10.13544/j.cnki.jeg.2019045

Zhou J, Yu JL (2005) Influences affecting the soil-water characteristic curve. J Zhejiang Univ-Sci A 6:797–804. https://doi.org/10.1631/jzus.2005.A0797

Chapter 50
Numerical Investigation of Triaxial Compressive Rock Tests Under CNL and CNS Boundary Conditions Using PFC2D

Peng Sha, Danlian Yang, Yijun Zhang, and Faquan Wu

Abstract Compared to the boundary condition of constant normal load (CNL), constant normal stiffness (CNS) is conventionally considered to represent the mechanical behaviours of rock mass in underground excavations and bolted joints. Previous studies mainly focus on the differences between direct shear strength parameters under CNL and CNS boundary conditions. Comparison study between triaxial compressive test results under CNL and CNS condition has been rarely reported. Concerning the typical rocks, such as Jinping marble and Lac du Bonnet (LdB) granite, this work prepares numerical triaxial compressive tests using PFC2D to investigate the strength and mechanical behaviours under CNL and CNS boundary conditions. The numerical results reveal that the mechanical characteristics vary significantly between two boundary conditions in the triaxial compressive test, especially when the variation of CNS is up to 5 GPa/m. The enhancement effect of confining pressure (EECP) with the CNS modulating from 5 to 15 GPa/m restricts the dilation of rock specimens. Accordingly, compared to the CNL condition, the peak, and the residual strength present growth linearly while the brittle index decreases. Also, the impact of EECP relies on rock type and initial confining pressure. The results indicate that rock mass with low strength under high in-situ stress may suffer more impact of the EECP.

Keywords Triaxial compressive test · Constant normal load · Constant normal stiffness · PFC2D

S. Peng (✉) · Y. Danlian · Z. Yijun · W. Faquan
College of Civil Engineering, Shaoxing University, Shaoxing, Zhejiang, China
e-mail: shapeng@usx.edu.cn

W. Faquan
e-mail: wufaquan@mail.iggcas.ac.cn

S. Peng · W. Faquan
Key Laboratory of Rock Mechanics and Geological Disaster, Zhejiang, China

S. Wang et al. (eds.), *Engineering Geology for a Habitable Earth: IAEG XIV Congress 2023 Proceedings, Chengdu, China*, Environmental Science and Engineering,
https://doi.org/10.1007/978-981-99-9057-3_50

50.1 Introduction

The rock mass excavation is a complex mechanical process, that disturbs the initial stress state and promotes stress redistribution. As a result, different mechanical behaviours are exhibited in rock mass based on two boundary conditions. In slopes and other near-surface gravity-driven environments (Case 1 in Fig. 50.1a), the failure behaviours are produced under unconstrained boundary conditions (constant normal load (CNL)) (Haberfield and Szymakowski 2003). In contrast, rock mass in deep underground excavations and support/bolted engineering (Case 2 and 3 in Fig. 50.1a) cannot deform freely due to the constraint of surrounding rock and support. The normal stress in rock mass varies with the increase of deformation. The stiffness dominates the control of the constraint (see Fig. 50.1b). Therefore, the rock mass behaviours should be produced under constrained boundary conditions (constant normal stiffness (CNS)) (Lee et al. 2014; Shrivastava and Rao 2018; Li et al. 2020).

CNL boundary condition was commonly applied in direct shear testing on rock joints due to relatively simple setup and equipment requirements (Goodman 1976). In 1987, Johnston introduced the principle of CNS boundary condition (Johnston et al. 1987). It was proved to apply to explain the change of normal stress due to rock dilation in the shearing process. Afterwards, the researchers started to investigate the effect of the load boundary conditions on rock discontinuities using the direct shear laboratory test. Jiang et al. (2004) presented a servo-controlled loading system to realize the CNS direct shear test. The normal load Pn applied in the test is defined as a function of normal displacement $\Delta\delta n$ and constant stiffness kn. The test results indicated that the CNS boundary exhibited a significant influence on the shear strength parameters of rock joints. With the improvement of a normal loading system, rock joints can be sheared under three boundary conditions, such as constant normal displacement (CND), CNL and CNS (Brady and Brown 2005; Gutiérrez-Ch et al. 2018; Xia et al. 2020). It is possible to complete laboratory direct shear tests that more accurately resemble site-specific conditions. Mokhtarian et al. (2020) presented the CNS condition exhibited higher values of peak shear strength, friction angle, and

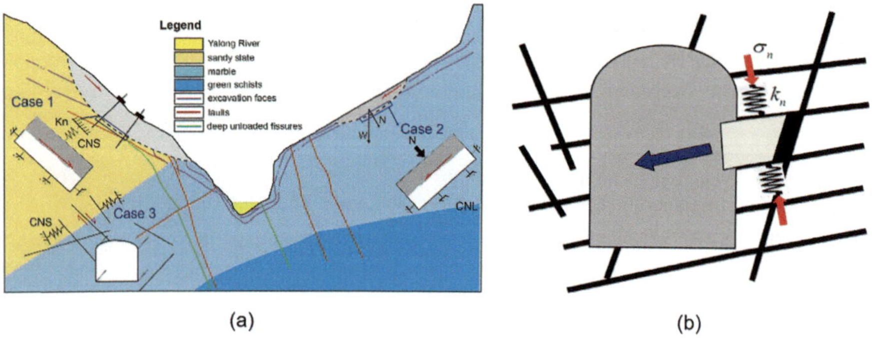

Fig. 50.1 Sketch diagram of the CNL and the CNS boundary conditions in engineering excavation

cohesive strength than the CNL condition, which should lead to underestimated shear strengths. The previous studies concluded that in low and moderate normal stress, the shear behaviours, the strength parameters, and the shear displacement under the CNS boundary are larger than those under the CNL boundary in direct shear tests (Feng et al. 2010; Cui et al. 2020).

Rockmass is composed of two components: intact rock and discontinuities. For intact rock, the mechanical parameters can be determined in the triaxial compression test, which is intended to measure the strength of cylindrical rock specimens as a function of confining pressure (Franklin and Hoek 1970). The apparatus is mainly a composite of a loading device to apply axial load and a triaxial cell to apply confined pressure (ISRM 1978). A stiff loading device is used to apply and measure the axial loading and the hydraulic pump in the triaxial cell is preferably capable of offering and controlling confined pressure. Conventionally, confined pressure must be kept constant under axial loading during tests, which should be considered a CNL boundary condition. However, the CNS boundary condition is considered to better reflect behaviours of rock in deep underground excavations (as shown in Fig. 50.1b) and rock-support systems (Indraratna et al. 1999), which has to date been rarely reported.

Due to the limitation of experimental conditions, the understanding of conventional triaxial compression tests under CNS boundary conditions is relatively few. It is noted that the numerical methods are feasible and economic to quantitatively describe the grain-scale microstructure of rocks and have succeeded in past decades in simulating the process of rock tests (Lan et al. 2010; Li et al. 2017; Qi et al. 2020). Numerical simulation research can not only reduce the cost of materials and machines in laboratory tests but also conduct simulation tests repeatedly on samples in different environments, which is more convenient than laboratory tests. Similarly, recently Shrivastava studied the shear behaviours of filled rock joints under different boundary conditions, analyzed the numerical use of rock shear behaviours under different boundary conditions, and concluded that the strength predicted by the numerical model is very consistent with the test results under low initial normal stress and CNL boundary conditions (Shrivastava and Rao 2022). Therefore, the objective of this study is to explore the effect of CNS boundary conditions on the strength and mechanical behaviours of intact rocks during triaxial compression using Particle Flow Code 2D (PFC2D). The comparison between the CNL and the CNS condition in triaxial compressive rock tests is discussed based on the numerical results.

50.2 Establishment and Calibration of PFC Testing Model

50.2.1 Physical Parameters of Different Rocks

To consider representative rocks in engineering excavations in the numerical model, we take three rocks with different strengths as examples, which correspond to the approximate soft, hard, and very hard rocks, respectively. Figure 50.2 presents the numerical models of rocks, including sandstone (Huang and Yang 2014), Jinping marble (Huang and Li 2017), and Lac du Bonnet (LdB) granite (Cai et al. 2004). It is indicated from Table 50.1 that the mineral composition of the three rocks is quite different. It is taken as those of the LdB granite, i.e., 40% of K-feldspar, 30% of quartz, 20% of plagioclase, and 10% of biotite. On this basis, a series of synthetic crystalline rock specimens are established by PFC2D to conduct triaxial compression tests.

According to the mineral composition of different rocks, three numerical models with the size of 50 mm × 100 mm are established through PFC2D. The grain size of each synthetic specimen is fixed as a similar distribution listed in Table 50.1, different minerals are represented by colours, as shown in Fig. 50.2. Parallel bond model is adopted to simulate the contact behaviours between particles in numerical

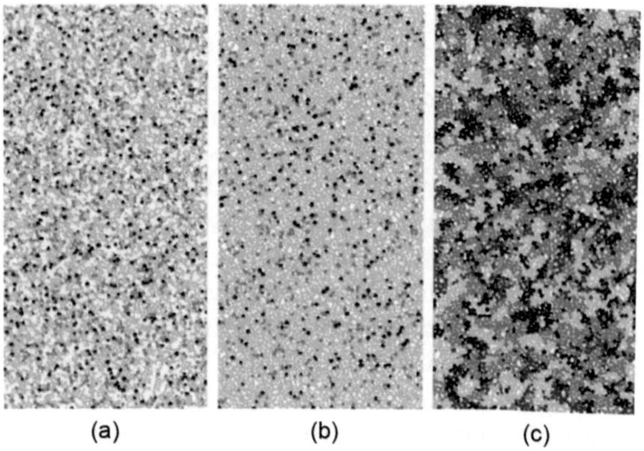

Fig. 50.2 Numerical models of **a** sandstone; **b** marble and **c** granite

Table 50.1 The mineral composition of three rocks

Rock sample	Mineral type	Ratio (%)	Rock sample	Mineral type	Ratio (%)	Rock sample	Mineral type	Ratio (%)
Sandstone	Quartz	60	Marble	Calcite	91	Granite	K-feldspar	40
	Feldspar	20		Dolomite	8		Quartz	30
	Clay cement	15		Mica	1		Plagioclase	20
	Mica	5					Biotite	10

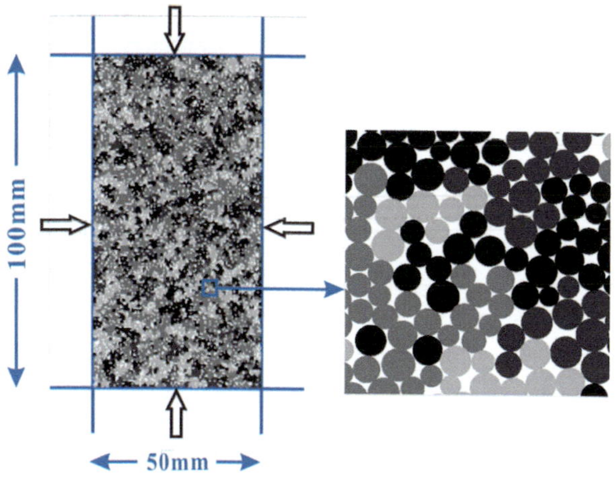

Fig. 50.3 PFC2D model for the triaxial compression test, i.e., LdB granite

tests (Potyondy 2007). For simplicity, the selection of the mechanical properties of different minerals is based on the relative magnitude of parameters among them. Take LdB granite as an example, we initially set the strength parameters of K-feldspar as the counterpart, and the parameters of Quartz, Plagioclase, and Biotite are determined as 1.2, 0.9 and 0.4 times those of K-feldspar (Fig. 50.3).

50.2.2 Calibrations of PFC Model Under Uniaxial Compression

After initially determined, the strength parameters of the mineral grains and the deformation and strength parameters of the grain contacts should be calibrated according to the laboratory data. The selected rock samples in this study have been popularly adopted as standard rock types for rock mechanics research, especially Jinping marble and LdB granite, of which mechanical properties have been extensively investigated by many research institutions in the past decades. Table 50.2 lists the calibrated properties of the synthetic rock specimens by trial and error. Figure 50.4 shows the comparisons between the experimental stress–strain relations and PFC2D results, the simulation curves are consistent with the experimental ones. Table 50.3 shows all the simulation results, such as uniaxial compression strength, elastic modulus and Poisson ratio, which agree well with those in previous studies.

Table 50.2 Calibrated parameters of the numerical model

Meso-parameters	Sandstone	Marble	Granite
Density/kg·m^{-3}	2168	2600	2630
Contact modulus/GPa	10	66	89
Stiffness of grain contact k_n/k_s	3.1	3.9	1.7
Friction coefficient	0.5	0.5	0.5
Parallel bonding modulus/GPa	10	66	89
Parallel bonding stiffness ratio	3.1	3.9	1.7
Normal bonding strength/MPa	29	115	245
Shear bonding strength/MPa	29	115	245

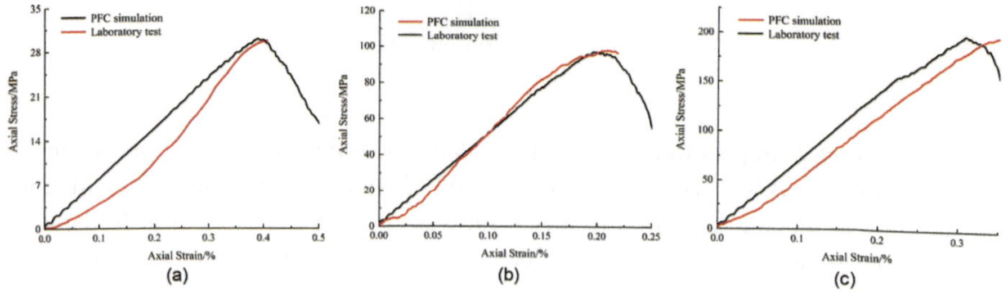

Fig. 50.4 Stress–strain response of **a** sandstone, **b** marble, and **c** granite under triaxial compression and the numerical results by PFC2D

Table 50.3 Comparisons among experimental results and numerical results of three rocks

Rock sample		Mechanical properties		
		Uniaxial compression strength/MPa	Elastic modulus/GPa	Poisson ratio
Sandstone	Experimental data	30.10	8.40	0.290
	Numerical result	31.00	8.79	0.285
	Error	2.99%	4.64%	1.72%
Marble	Experimental data	97.70	51.00	0.200
	Numerical result	94.90	49.10	0.195
	Error	2.86%	1.76%	2.50%
Granite	Experimental data	200.00	69.00	0.260
	Numerical result	204.00	72.10	0.262
	Error	2.00%	4.49%	0.77%

50.2.3 Design of CNS Boundary in Triaxial Compression Test

According to the principle of CNS boundary condition, the normal stress applied to the rock specimen should constantly change over the normal displacement. As

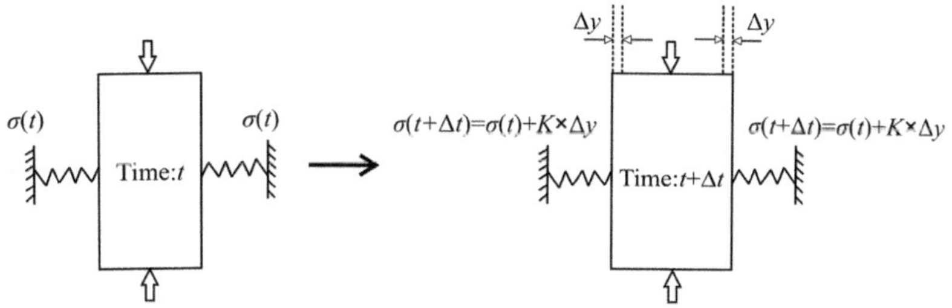

Fig. 50.5 Schematic diagram of CNS boundary condition in triaxial compression test

shown in Fig. 50.5, the normal stress can be calculated as follows,

$$\sigma(t + \Delta t) = \sigma(t) + K \times \Delta y \tag{50.1}$$

where $\sigma(t + \Delta t)$ is the normal stress on the side of the numerical specimen at $(t + \Delta t)$ time, which is determined by Eq. (50.1), $\sigma(t)$ is the normal stress on the side of the specimen at t time, Δt is the time increment, K is the constant stiffness of the side loading boundary (wall unit in this study) and Δy is the increment of the lateral displacement of the specimen. Six variations of constant normal stiffness (0 GPa/m (CNL boundary condition), 0.5 GPa/m, 1.0, 5.0, 10 and 15 GPa/m) are considered in the testing program.

50.3 Numerical Results

50.3.1 Stress–Strain Responses

Triaxial compression simulations were conducted to investigate the stress–strain relations and strength parameters with varying boundary conditions and rock types. Figure 50.6 presents the stress–strain response of synthetic sandstone specimens under varied boundary conditions, including initial confined pressure and stiffness. The curves corresponding to varied normal stiffness overlap with each other when the tests are in the linear elastic stage. For a specific curve of K = 5 GPa/m, the variation trend is almost similar to that of the CNL condition when K is less than 5 GPa/m. then alters with the increase of constant normal stiffness. As K is up to 5 GPa/m, these stress–strain curves are gradually scattered as the axial (lateral) strain increases further, especially when the post-peaking phase transfers from an elastic–plastic state to quasi quasi-plastic state.

Also, it is indicated that the increment of the confined pressure varies significantly with the increase of boundary stiffness in the triaxial compressive test when K is up to 5 GPa/m. According to Eq. (50.1), the confined pressure should be increased with the boundary modulating from CNL to CNS. However, the numerical results suggest

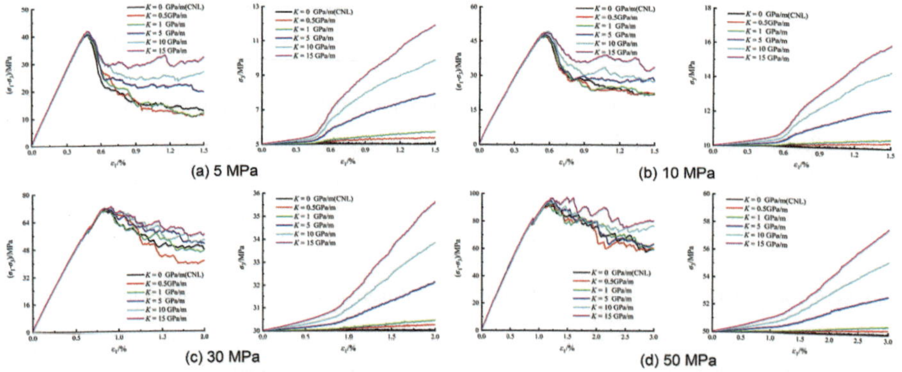

Fig. 50.6 Deviatoric stress–strain curves of sandstone under CNS boundary condition with different confined pressures (5, 10, 30, and 50 MPa), (σ_1–σ_3) is deviatoric stress, and ε_1 is the axial stain obtained in the tests

that the stress–strain response is insensitive to the CNS condition in the triaxial compressive test when K is no more than 5 GPa/m. There exists an enhancement effect of confined pressure (EECP) when K is larger than 5 GPa/m. Furthermore, as the initial confined pressure increased from 5 to 50 MPa, as shown in Fig. 50.6a–d, the difference of the post-peaking phase became smaller, which is close to quasi plastic state. Therefore, the confined pressure can counteract the effect of CNS boundary conditions, which could weaken the EECP.

50.3.2 Strength Parameters

Figure 50.7 shows the increments of the peak stress of three rock types with variations of normal stiffness and confined pressure. For the specimens of sandstone, the peak stress increment generally exhibits linear growth with the increase of the normal stiffness and the confined pressure (Fig. 50.7a). With the increase of boundary stiffness, the EECP restricts the cracks emerge and the dilation of the specimens after the unstable crack growth stage, which consequently promotes the increment of the peak stress. In contrast, the peak stress increment of the other two cases shows almost no relevance to the normal stiffness and the confined pressure (Fig. 50.7b, c). Overall, with a transition from hard rock to soft rock, the relative stiffness between the rock specimen and side load boundary becomes larger. There is a gradual change from irrelevant to linear positive correlation.

In the post-peaking stage, a large number of cracks increase, and more dilation arises in the rock specimen. Meanwhile, the boundary constraint increases with the increase of the normal stiffness. So, the EECP restricts the failure state of rock specimens after peak stress as shown in Fig. 50.6, which can increase the residual stress. Figure 50.8 shows variations of the residual stress increment with different

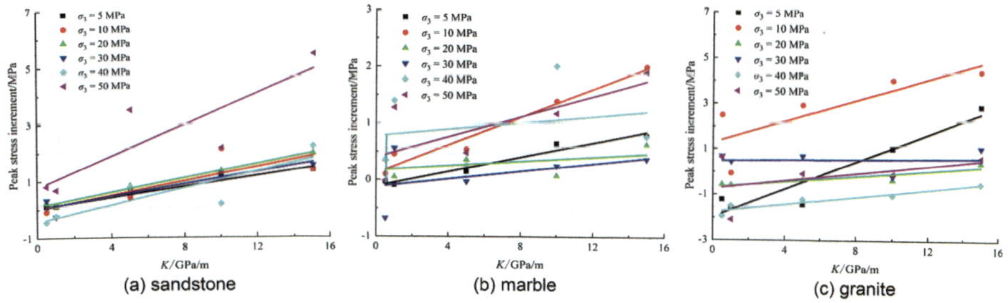

Fig. 50.7 Variations of the peak stress increment with different stiffness and confined pressure

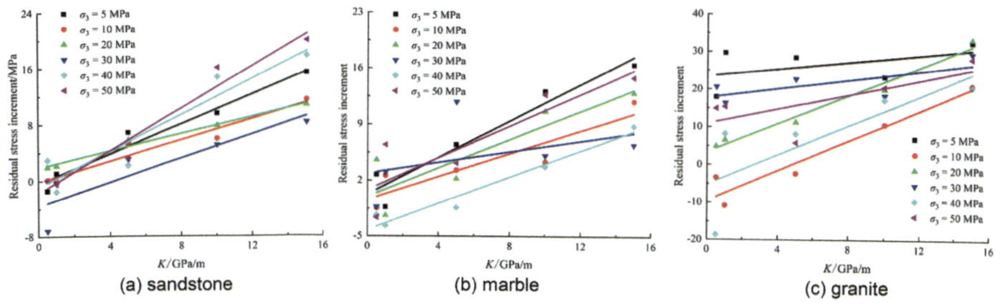

Fig. 50.8 Variations of the residual stress increment with different stiffness and confined pressure

stiffness and confined pressure. Unlike the peak stress increment, the residual stress increment shows a linear positive correlation with the normal stiffness and confined pressure, which is independent of rock type.

50.3.3 Rock Brittleness Index B

Since the EECP can change the post-peaking state from elastic–plastic to quasi-plastic, the rock brittleness index after peak stress should be weakening with the increase of the normal stiffness, which shows the same characteristics as the effect of confined pressure. In this study, we select the definition of rock brittleness index B proposed by Chen et al. (2018), which is as follows,

$$B = \left(\frac{\sigma_p - \sigma_r}{\sigma_p}\right) / \left(\frac{\varepsilon_r - \varepsilon_p}{\varepsilon_p}\right) \tag{50.2}$$

where σ_p, σ_r, ε_r, and ε_p represent peak stress, residual stress, peak strain, and residual strain of the triaxial compression test, respectively.

As shown in Fig. 50.9, it is indicated that the rock brittleness index *B* decreases with the increase of confined pressure, especially when the confined pressure is larger than 30 MPa, which is consistent with that of experimental results. However,

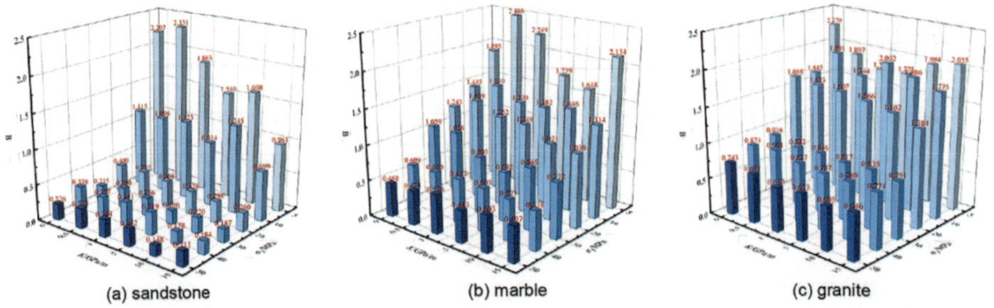

Fig. 50.9 Variations of the rock brittleness index with different stiffness and confined pressure

the effect of normal stiffness increment on B is not significant when the confined pressure increases to 20 MPa. This suggests that the rock brittleness index is more sensitive to confined pressure than the CNS boundary, especially under high confined pressure.

50.4 Discussions

Large deformation, such as excessive overbreak and squeezing deformation is currently becoming one of the most remarkable geotechnical problems during tunnel excavations in Western China. The main causes of the problem are high in-situ stresses and soft rocks (Meng et al. 2013). The conventional method for tunnel design in the soft rock mass is to strengthen the stiffness of the primary support, such as increasing the layers or decreasing the spacing of the steel arch. The numerical results in this study show that rock mass with lower strength under higher confined pressure suffers a more significant impact of the EECP due to the CNS boundary condition. As a result, the rock pressure to the highly rigid support system will increase in the soft surrounding rock under high in-situ stress.

Take Liangshui Tunnel in China as an example, the stratum in the area is mainly phyllite with a uniaxial compression strength of 10–15 MPa. In-situ stress has been measured near the tunnel and the maximal horizontal stress $\sigma_H = 27.16$ MPa. Therefore, the support system of the tunnel was designed with a safer consideration compared to the standard (Liu et al. 2021). The double steel arch was adopted in the section, which significantly enhanced the rigidity of the primary support. Figure 50.10 displays the monitoring results of section DK 358 + 803 of the Liangshui Tunnel (Sha et al. 2015). It is noted that the rock pressure of the primary lining continued to increase for 40 days after the secondary lining was completed. The maximum pressure is up to 1.26 MPa, which is much higher than the upper limit predicted by the Code for Design of Railway Tunnel (2017). This indicates that the EECP has not been involved in the rock pressure calculation.

When tunnelling in squeezing rock, the EECP should be considered due to the CNS boundary condition. The EECP becomes more significant to the stability of a tunnel

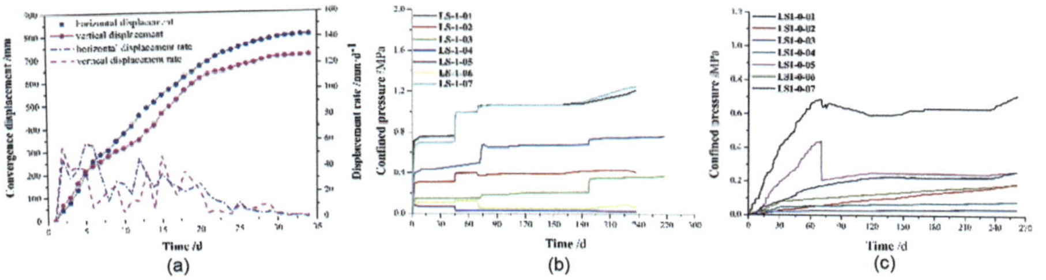

Fig. 50.10 Monitoring results of the section DK 358 + 803 in Liangshui Tunnel

with the increase of relative stiffness between surrounding rock and primary support. Attempts to prevent large deformation using stiff support often result in enormous failure of the shotcrete liner, therefore, the use of a yielding support system has been recommended (Tian et al. 2016). Nevertheless, determining the amount of free deformation of surrounding rock and the initial yielding load remains a challenge. The numerical results in this study may provide a theoretical basis, more attention should be paid to the triaxial compression rock test under the CNS boundary conditions.

50.5 Conclusions

The objective of this study is to analyse the comparison of triaxial compressive rock tests under CNL and CNS boundary conditions using PFC2D. The main conclusions are as follows,

(1) The numerical results reveal that the mechanical characteristics vary significantly between two boundary conditions in the triaxial compressive test, especially when the variation of the CNS becomes larger than 5 GPa/m. With the increase of boundary stiffness, the stress–strain response in post–peaking phase changed from elastic–plastic to quasi-plastic. There also exists the enhancement effect of confined pressure (EECP) when the normal stiffness is larger than 5 GPa/m.
(2) The EECP can restrict the dilation of rock specimens. Compared to the CNL condition, the peak and the residual strength present linear growth while the brittle index decreases. The rock type, the normal stiffness, and the confined pressure can influence the strength parameter in different ways.
(3) The impact of EECP relies on rock strength and initial confining pressure. The results indicate that rock mass with low strength under high in-situ stress may suffer a more significant impact of the EECP. Consequently, more attention should be paid to the triaxial compression rock test under the CNS boundary conditions.

Acknowledgements The authors are grateful for financial support from the China Natural Science Foundation under Grant No. 42007254.

References

Brady BH, Brown ET (2005) Rock mechanics for underground mining. Springer Science and business media
Cai M, Kaiser PK, Tasaka Y et al (2004) Generalized crack initiation and crack damage stress thresholds of brittle rock masses near underground excavations. Int J Rock Mech Min Sci 41(5):833–847
Chen GQ, Zhao C, Wei T et al (2018) Evaluation method of brittle characteristics of rock based on full stress-strain curve and crack initiation stress. Chin J Rock Mech Eng 37(1):51–59
Code for Design of Railway Tunnel. Industry Standard of the People's Republic of China (2017) TB 10003-2016, issued on August 10
Cui GJ, Zhang CQ, Chen JL et al (2020) Effect of bolt inclination angle on shear behaviour of bolted joints under CNL and CNS conditions. J Central South Univ 27(03):937–950
Feng D K, Hou W J, Zhang J M et al (2010) Experimental study of 3D cyclic behaviour of interface under different normal boundary conditions. Rock Soil Mech 31(8):2419–2424+2478
Franklin J, Hoek E (1970) Developments in triaxial testing technique. Rock Mech, 223–228
Goodman RE (1976) Methods of geological engineering in discontinuous rocks. West Publishing Company, West St. Paul
Gutiérrez-Ch JG, Senent S, Melentijevic S et al (2018) Distinct element method simulations of rock-concrete interfaces under different boundary conditions. Eng Geol 240:123–139
Haberfield CM, Szymakowski J (2003) Applications of large scale direct shear testing. Australian Geomech, 29–39
Huang YH, Yang SQ (2014) Particle flow simulation of macro-and meso-mechanical behaviour of red sandstone containing two pre-existing non-coplanar fissures. Chin J Rock Mech Eng 33(8):1644–1653
Huang D, Li XQ (2017) Numerical simulation research on characteristic strength of marble based on the development of microcrack. Rock Soil Mech 1:253–262
Indraratna B, Haque A, Aziz N (1999) Shear behaviour of idealised infilled joints under constant normal stiffness. Geotech 49(3):331–355
ISRM 1978 (1978) Suggested methods for determining the strength of rock materials in triaxial compression. Int J Rock Mech Mining Sci Geomech Abst 15(2):47–51
Jiang Y, Xiao J, Tanabashi Y, Mizokami T (2004) Development of an automated servo-controlled direct shear apparatus applying a constant normal stiffness condition. Int J Rock Mech Min Sci 41(2):275–286
Johnston IW, Lam TSK, Williams AF (1987) Constant normal stiffness direct shear testing for socketed pile design in weak rock. Geotechnique 37(1):83–89
Lan H, Martin CD, Hu B (2010) Effect of heterogeneity of brittle rock on micromechanical extensile behaviour during compression loading. J Geophys Res: Solid Earth 115(B1). https://doi.org/10.1029/2009JB006496
Lee YK, Park JW, Song JJ (2014) Model for the shear behaviour of rock joints under CNL and CNS conditions. Int J Rock Mech Min 70(Complete):252–263
Li J, Konietzky H, Frühwirt T (2017) Voronoi-based DEM simulation approach for sandstone considering grain structure and pore size. Rock Mech Rock Eng 50(10):2749–2761
Liu W, Chen J, Luo Y et al (2021) Deformation behaviors and mechanical mechanisms of double primary linings for large-span tunnels in squeezing rock: a case study. Rock Mech Rock Eng 54:2291–2310

Meng L, Li T, Yun J et al (2013) Characteristics and mechanisms of large deformation in the Zhegu mountain tunnel on the Sichuan-Tibet highway. Tunn Undergr Space Technol 37:157–164

Mokhtarian H, Moomivand H, Moomivand H (2020) Effect of infill material of discontinuities on the failure criterion of rock under triaxial compressive stresses. Theoret Appl Fract Mech 108:102652. https://doi.org/10.1016/j.tafmec.2020.102652

Potyondy DO (2007) Simulating stress corrosion with a bonded-particle model for rock. Int J Rock Mech Min Sci 44(5):677–691

Qi S, Lan H, Martin D et al (2020) Factors controlling the difference in Brazilian and direct tensile strengths of the Lac du Bonnet granite. Rock Mech Rock Eng 53(3):1005–1019

Sha P, Faquan W, Xiang L et al (2015) Squeezing deformation in layered surrounding rock and force characteristics of the support system of a tunnel under high in-situ stress. Rock Soil Mech 36(5):1407–1414

Shrivastava AK, Rao KS (2018) Physical model for the shear behaviour of rock joint under CNL and CNS conditions. Rock Mech Rock Eng 51(1):101–118

Shrivastava AK, Rao KS (2022) Shear behaviour of infilled rock joints under different boundary conditions. In: Proceedings of the Institution of Civil Engineers Geotechnical Engineering, 1–40. https://doi.org/10.1680/jgeen.21.00060

Tian HM, Chen WZ, Yang DS, Wu GJ, Tan XJ (2016) Numerical analysis on the interaction of shotcrete liner with rock for yielding supports. Tunn Undergr Space Technol 54:20–28

Xia CC, Yu QF, Qian X et al (2020) Experimental study of shear-seepage behaviour of rock joints under constant normal stiffness. Rock Soil Mech 41(1):57–66+77 [illegible] Numerical [illegible] [illegible] [illegible]

Meng [illegible] (2021) [illegible] mechanisms of [illegible] formation in [illegible] [illegible]

[illegible] (2020) [illegible] on the failure [illegible] of rock under [illegible] stresses. [illegible] Mech [illegible]

[illegible] (2019) [illegible] model [illegible] Rock Mech [illegible]

[illegible] (2020) Factors controlling the difference in [illegible] strengths of the [illegible] granite. Rock Mech [illegible]

[illegible] (2015) [illegible] deformation of [illegible] surrounding rock and [illegible] characteristics of the support system [illegible] under high [illegible] stress. [illegible] Sci [illegible] Mech [illegible]

Srinivasan [illegible] Rao KS (2018) Physical model for the shear behaviour of rock joint under CNL and CNS conditions. Rock Mech Rock Eng 51(1):[illegible]

Srinivasan [illegible] Rao KS (2020) Shear behaviour of infilled rock joints under different boundary conditions. In: Proceedings of the Institution of Civil Engineers - Geotechnical Engineering, [illegible] https://doi.org/10.1680/jgeen.21.00050

[illegible] (2019) [illegible]

[illegible] (2020) [illegible] study of [illegible] Int J Rock Mech Min Sci [illegible]

GPSR Compliance

The European Union's (EU) General Product Safety Regulation (GPSR) is a set of rules that requires consumer products to be safe and our obligations to ensure this.

If you have any concerns about our products, you can contact us on ProductSafety@springernature.com

In case Publisher is established outside the EU, the EU authorized representative is:

Springer Nature Customer Service Center GmbH
Europaplatz 3
69115 Heidelberg, Germany

Batch number: 10732064

Printed by Printforce, the Netherlands